CRC
Handbook of

Enthalpy Data *of*
Polymer-Solvent
Systems

Christian Wohlfarth

CRC Press
Taylor & Francis Group
Boca Raton London New York

CRC Press is an imprint of the
Taylor & Francis Group, an **informa** business
A TAYLOR & FRANCIS BOOK

Published in 2006 by CRC Press
Taylor & Francis Group
6000 Broken Sound Parkway NW, Suite 300
Boca Raton, FL 33487-2742

© 2006 by Taylor & Francis Group, LLC
CRC Press is an imprint of Taylor & Francis Group, an Informa business

First issued in paperback 2019

No claim to original U.S. Government works

ISBN-13: 978-0-367-45383-1 (pbk)
ISBN-13: 978-0-8493-9361-7 (hbk)

Library of Congress Card Number 2005056860

Library of Congress Cataloging-in-Publication Data

Wohlfarth, C.
    CRC handbook of enthalpy data of polymer-solvent systems / Christian Wohlfarth.
      p. cm.
    Includes bibliographical references and index.
    ISBN 0-8493-9361-2 (alk.)
    1. Polymer solutions--Thermal properties--Handbooks, manuals, etc. 2. Copolymers--Thermal properties--Handbooks, manuals, etc. 3. Nonaqueous solvents--Thermal properties--Handbooks, manuals, etc. I. Title: Enthalpy data of polymer-solvent systems. II. Title.

QD381.9.S65W632 2006
547'.70456--dc22   2005056860

Visit the Taylor & Francis Web site at
http://www.taylorandfrancis.com

and the CRC Press Web site at
http://www.crcpress.com

PREFACE

Knowledge of thermodynamic data of polymer solutions is a necessity for industrial and laboratory processes. Furthermore, such data serve as essential tools for understanding the physical behavior of polymer solutions, for studying intermolecular interactions, and for gaining insights into the molecular nature of mixtures. They also provide the necessary basis for any developments of theoretical thermodynamic models. Scientists and engineers in academic and industrial research need such data and will benefit from a careful collection of existing data. However, no data books or databases detailing enthalpy changes in polymer solutions presently exist. Thus, the *CRC Handbook of Enthalpy Data of Polymer-Solvent Systems* builds a completely new and reliable collection of enthalpy data for polymer solutions from the original literature. It will be a very useful completion to the *CRC Handbook of Thermodynamic Data of Copolymer Solutions*, the *CRC Handbook of Thermodynamic Data of Aqueous Polymer Solutions*, and the *CRC Handbook of Thermodynamic Data of Polymer Solutions at Elevated Pressures* as these contain only a small amount of enthalpic data in comparison to all available data provided in this book.

The book is divided into six chapters: (1) Introduction, (2) Enthalpies of mixing or intermediary enthalpies of dilution, (3) Polymer partial enthalpies of mixing at infinite dilution or polymer first integral enthalpies of solution, (4) Solvent partial enthalpies of mixing measured by calorimetry, (5) Partial molar enthalpies of mixing or solution of solvents in molten polymers from inverse gas-liquid chromatography (IGC), and (6) Table of additional information on enthalpy effects. Finally, appendices quickly route the user to the desired data sets. Thus, the book covers all the necessary areas for researchers and engineers who work in this field.

In comparison with low-molecular systems, the amount of data for polymer solutions is still rather small. About 800 literature sources were perused for the purpose of this book, including some dissertations and diploma papers. About 1770 data sets, i.e., 620 data sets on enthalpies of mixing or intermediary enthalpies of dilution, 970 data sets for polymer partial enthalpies of mixing or solution at infinite dilution, and 180 IGC tables are reported.

Additionally, tables of systems are provided here to lead the reader to further literature sources. Data are included only if numerical values were published or authors provided their numerical results by personal communication (and I wish to thank all those who did so). No digitized data have been included in this data collection. The book is the first complete overview about this subject in the world's literature. The closing day for the data collection was June, 30, 2005. The user who is in need of new additional data sets is kindly invited to ask for new information beyond this book via e-mail at *wohlfarth@chemie.uni-halle.de*. Additionally, the author will be grateful to users who call his attention to mistakes and make suggestions for improvements.

The *CRC Handbook of Enthalpy Data of Polymer-Solvent Systems* will be useful to researchers, specialists, and engineers working in the fields of polymer science, physical chemistry, chemical engineering, material science, biological science and technology, and those developing computerized predictive packages. The book should also be of use as a data source to Ph.D. students and faculty in Chemistry, Physics, Chemical Engineering, Biotechnology, and Materials Science Departments at universities.

Christian Wohlfarth
Merseburg, August 2005

About the Author

Christian Wohlfarth is Associate Professor for Physical Chemistry at Martin Luther University Halle-Wittenberg, Germany. He earned his degree in chemistry in 1974 and wrote his Ph.D. thesis on investigations on the second dielectric virial coefficient and the intermolecular pair potential in 1977, both at Carl Schorlemmer Technical University Merseburg. In 1985, he wrote his habilitation thesis, *Phase Equilibria in Systems with Polymers and Copolymers*, at Technical University Merseburg.

Since then, Dr. Wohlfarth's main research has been related to polymer systems. Currently, his research topics are molecular thermodynamics, continuous thermodynamics, phase equilibria in polymer mixtures and solutions, polymers in supercritical fluids, PVT behavior and equations of state, and sorption properties of polymers, about which he has published approximately 100 original papers. He has written the books: *Vapor-Liquid Equilibria of Binary Polymer Solutions*, *CRC Handbook of Thermodynamic Data of Copolymer Solutions*, *CRC Handbook of Thermodynamic Data of Aqueous Polymer Solutions*, and *CRC Handbook of Thermodynamic Data of Polymer Solutions at Elevated Pressures*.

He is working on the evaluation, correlation, and calculation of thermophysical properties of pure compounds and binary mixtures resulting in six volumes of the *Landolt-Börnstein New Series*. He is a respected contributor to the *CRC Handbook of Chemistry and Physics*.

CONTENTS

6. TABLE OF SYSTEMS FOR ADDITIONAL INFORMATION ON ENTHALPY EFFECTS IN POLYMER SOLUTIONS

APPENDICES

1. INTRODUCTION

1.1. Objectives of the handbook

Knowledge of thermodynamic data of polymer solutions is a necessity for industrial and laboratory processes. Furthermore, such data serve as essential tools for understanding the physical behavior of polymer solutions, for studying intermolecular interactions, and for gaining insights into the molecular nature of mixtures. They also provide the necessary basis for any developments of theoretical thermodynamic models. Scientists and engineers in academic and industrial research need such data and will benefit from a careful collection of existing data. Enthalpies of mixing and solution specially enter into energy balance and phase equilibrium calculations and are related to the structure and the energy of interaction of the molecules. They are important also from the theoretical point of view. However, the database for polymer solutions is still modest in comparison with the enormous amount of data for low-molecular mixtures, and the specialized database for enthalpies of polymer solutions is even smaller.

Basic information on polymers can be found in the *Polymer Handbook* (1999BRA), and among the solution properties, there are also short tables about systems and literature on enthalpy changes and a small table for enthalpies of solution (1999OR1, 1999OR2). The three *CRC Handbooks of Thermodynamic Data of Polymer Solutions* (2001WOH, 2004WOH, and 2005WOH) include chapters on enthalpies, i.e., for copolymer solutions, for aqueous polymer solutions, and for polymer solutions at elevated pressures. However, they provide only a minor amount of all available enthalpy data for polymer-solvent systems. No other data books or databases dedicated specially to enthalpy changes in polymer solutions presently exist. Thus, the intention of the handbook is to fill this gap and to provide scientists and engineers with an up-to-date compilation from the literature of the available enthalpy data of polymer solutions. The handbook does not present theories and models for polymer solution thermodynamics. Other publications (1990BAR, 1990FUJ, 1990KAM, 1999KLE, 1999PRA, and 2001KON) can serve as starting points for investigating those issues.

The data within this book are divided into five chapters following this introduction:

- Enthalpies of mixing or intermediary enthalpies of dilution
- Polymer partial enthalpies of mixing at infinite dilution or polymer first integral enthalpies of solution
- Solvent partial enthalpies of mixing measured by calorimetry
- Partial molar enthalpies of mixing at infinite dilution of solvents and enthalpies of solution of gases/vapors of solvents in molten polymers from inverse gas-liquid chromatography (IGC)
- Table of additional information on enthalpy effects

Data from investigations applying to more than one chapter are divided and appear in the relevant chapters. Data are included only if numerical values were published or authors provided their results by personal communication (and I wish to thank all those who did so). No digitized data have been included in this data collection. Finally, Chapter 6 covers a large number of systems in a table in which additional information on enthalpy effects in polymer solutions can be found.

1.2. Measurement of enthalpy changes in polymer solutions

Experiments on enthalpy changes in binary polymer solutions can be made within common microcalorimeters by applying one of the following three methods:

1. Measurement of the enthalpy change caused by solving a given amount of the solute polymer in an (increasing) amount of solvent, i.e., the solution experiment
2. Measurement of the enthalpy change caused by mixing a given amount of a concentrated polymer solution with an amount of pure solvent, i.e., the dilution experiment
3. Measurement of the enthalpy change caused by mixing a given amount of a liquid/molten polymer with an amount of pure solvent, i.e., the mixing experiment

Care must be taken for polymer solutions with respect to the resolution of the instrument, which has to be higher than for common solutions with larger enthalpic effects. Usually employed calorimeters for such purposes are the Calvet-type calorimeters based on heat-flux principle. Details can be found in 1984HEM and 1994MAR.

In particular, one has to distinguish between the following effects for polymer solutions. The enthalpy of mixing or the enthalpy of solution of a binary system is the amount of heat that must be supplied when n_A mole of pure solvent A and n_B mole of pure polymer B are combined to form a homogeneous mixture/solution in order to keep the total system at constant temperature and pressure.

$$\Delta_M H = n_A H_A + n_B H_B - (n_A H_{0A} + n_B H_{0B}) \tag{1}$$
$$\Delta_{sol} H = n_A H_A + n_B H_B - (n_A H_{0A} + n_B H_{0B}) \tag{2}$$

where:

$\Delta_M H, \Delta_{sol} H$	(extensive) enthalpy of mixing or solution
H_A, H_B	partial molar enthalpy of solvent A or polymer B
H_{0A}, H_{0B}	molar enthalpy of pure solvent A or pure polymer B
n_A, n_B	amount of substance of solvent A or polymer B.

From thermodynamic reasons follows that the change $\Delta_M H$ of the molar (or specific or segment molar) enthalpy in an isothermal-isobaric mixing process is also the molar (or specific or segment molar) excess enthalpy, H^E, of the mixture. The dependence of H^E upon temperature, T, and pressure, P, permits the correlation of such data with excess heat capacities, $C_p{}^E$, and excess volumes, V^E.

$$\left(\partial H^E / \partial T \right)_P = C_P^E \tag{3}$$
$$\left(\partial H^E / \partial P \right)_T = V^E - T \left(\partial V^E / \partial T \right)_P \tag{4}$$

where:

$C_p{}^E$	excess heat capacity
H^E	excess enthalpy $= \Delta_M H =$ enthalpy of mixing
P	pressure
T	(measuring) temperature
V^E	excess volume at temperature T.

The enthalpy effect might be positive (endothermic solution/mixture) or negative (exothermic solution/mixture) depending on the ratio n_A/n_B, i.e., the concentration of the total system. Unfortunately, in some of the older literature, the definition of the sign of the so-called *(integral) heat of solution* is reversed, compared with the enthalpy, occasionally causing some confusion. The value of the enthalpy of solution is dependent on the degree of crystallinity for semicrystalline polymers and, usually to a lesser extent, on the thermal history of glassy polymers. The enthalpy of mixing is independent of any crystalline or glassy aspects of the polymer. Thus, the enthalpy of mixing can be obtained without difficulties only for liquid/molten polymers mixed with a solvent.

The melting enthalpy of the crystallites and/or the glass enthalpy have to be determined additionally by independent measurements. As such a procedure is rather difficult and might cause substantial errors, it is common to measure the intermediary enthalpy of dilution, i.e., the enthalpy effect obtained if solvent A is added to an existing homogeneous polymer solution. The extensive intermediary enthalpy of dilution is the difference between two values of the enthalpy of the polymer solution corresponding to the concentrations of the polymer solution at the beginning and at the end of the dilution process:

$$\Delta_{dil}H^{12} = H^{(2)} - H^{(1)} \tag{5}$$

with

$$H^{(1)} = n_A^{(1)}H_A^{(1)} + n_BH_B^{(1)} \tag{6}$$
$$H^{(2)} = n_A^{(2)}H_A^{(2)} + n_BH_B^{(2)} \tag{7}$$

and

$$n_A^{(2)} = n_A^{(1)} + \Delta n_A \tag{8}$$

where:

$\Delta_{dil}H^{12}$	(extensive) intermediary enthalpy of dilution
$H^{(1)}, H^{(2)}$	enthalpies of the polymer solution before and after the dilution step
$H_A^{(1)}, H_A^{(2)}$	partial molar enthalpies of solvent A before and after the dilution step
$H_B^{(1)}, H_B^{(2)}$	partial molar enthalpies of polymer B before and after the dilution step
$n_A^{(1)}$	amount of solvent in the solution before the dilution step
$n_A^{(2)}$	amount of solvent in the solution after the dilution step
Δn_A	amount of solvent added to solution (1)
n_B	amount of polymer in all solutions.

$\Delta_{dil}H^{12}$ is not directly related to Δ_MH but to $(\partial\Delta_M H / \partial n_A)_{P,T,n_j}$ by:

$$\Delta_{dil}H^{12} = \int_{n_A^{(1)}}^{n_A^{(2)}} (\partial\Delta_M H / \partial n_A)_{P,T,n_j} dn_A \tag{9}$$

The term *integral* is often added to these enthalpy changes to describe changes where finite amounts of substances are mixed. Especially, the *integral enthalpy of solution/mixing for a polymer B* is given in a number of literature sources by applying the following two definitions:

• *per mole* polymer B:

$$^{int}\Delta_{sol}H_B = \Delta_{sol}H/n_B \tag{10}$$
$$^{int}\Delta_MH_B = \Delta_MH/n_B \tag{11}$$

• *per gram* polymer B (where the intensive ΔHs are the specific ones):

$$^{int}\Delta_{sol}H_B = \Delta_{sol}H/m_B \tag{12}$$
$$^{int}\Delta_M H_B = \Delta_M H/m_B \tag{13}$$

where:
$^{int}\Delta_{sol}H_B$ integral enthalpy of solution of polymer B
$^{int}\Delta_M H_B$ integral enthalpy of mixing of polymer B
m_B mass of polymer B
n_B amount of substance of polymer B.

As stated above, the difference between $^{int}\Delta_{sol}H_B$ and $^{int}\Delta_M H_B$ is determined by any enthalpic effects caused from solid-liquid phase transition of the crystallites and/or from glass transition and is zero for liquid/molten polymers.

The term *differential* is sometimes added to enthalpy changes where infinitesimal (i.e., very small) amounts were added to a very large amount of either solution or pure component. These enthalpy changes are usually called *partial* (molar or specific) enthalpies of solution/mixing:

$$\Delta_{sol}H_B = (\partial \Delta_{sol}H / \partial n_B)_{P,T,n_j} = H_B - H_{0B} \tag{14}$$
$$\Delta_M H_B = (\partial \Delta_M H / \partial n_B)_{P,T,n_j} = H_B - H_{0B} \tag{15}$$

with a unit of J/mol. However, for polymer solutions, $\Delta_{sol}H_B$ or $\Delta_M H_B$ is often expressed as the enthalpy change per unit mass of polymer added which can be obtained from the following derivative:

$$\Delta_{sol}H_B = (\partial \Delta_{sol}H / \partial m_B)_{P,T,m_j} \tag{16}$$
$$\Delta_M H_B = (\partial \Delta_M H / \partial m_B)_{P,T,m_j} \tag{17}$$

where:
$\Delta_{sol}H_B$ partial molar (or specific) enthalpy of solution of polymer B
$\Delta_M H_B$ partial molar (or specific) enthalpy of mixing of polymer B

with a unit of J/g. Similar to these definitions one can find results related to one mole of monomers (or base units). The derivative is then made by applying the base mole fraction of the polymer. The partial (molar or specific) enthalpy of solution of the polymer B is equal to the so-called differential enthalpy of solution at finite concentrations which is, for finite concentrations, different from the $^{int}\Delta_{sol}H_B$ or $^{int}\Delta_M H_B$ data as defined above. For example, in the case of a binary mixture, one obtains the relation:

$$\Delta_M H_B = \Delta_M H + (1 - x_B)(\partial \Delta_M H/\partial x_B) \tag{18}$$

which results in different values to $^{int}\Delta_M H_B$.

In the case of adding an infinitesimal amount of polymer to the pure solvent, the partial (molar or specific) enthalpy of solution of the polymer B is properly identified as the partial enthalpy of solution of the polymer at infinite dilution, $\Delta_{sol}H_B^{\infty}$, or the partial enthalpy of mixing of the polymer at infinite dilution, $\Delta_M H_B^{\infty}$. Its value at infinite dilution of the polymer is equal to the so-called *first* integral enthalpy of solution (unfortunately, sometimes referred to more simply as the enthalpy of solution of the polymer, but, as discussed above, identical values can only be obtained for infinite dilution). In practice, the partial (molar or specific) enthalpy of solution of the polymer B is measured by mixing isothermally a large excess of pure solvent and a certain amount of the polymer to form a homogeneous solution.

The state of the polymer before dissolution can significantly affect the enthalpy of solution. An amorphous polymer below its glass transition temperature T_g often dissolves with the release of heat. The enthalpy of solution of a glassy polymer is usually dependent on temperature and, to some extent, on the thermal history of the glass-forming process. An amorphous polymer above T_g can show endothermic or exothermic dissolution behavior depending on the nature of the solvent and the interaction energies involved as is the case for any enthalpy of mixing. The dissolving of a semicrystalline polymer requires an additional amount of heat associated with the disordering of crystalline regions. Consequently, its enthalpy of solution is usually positive and depends on the degree of crystallinity of the polymer sample.

The mathematical definition for the partial molar enthalpies of solution/mixing is given for the solvent A by:

$$\Delta_{sol}H_A = (\partial\Delta_{sol}H / \partial n_A)_{P,T,n_j} = H_A - H_{0A} \tag{19}$$

$$\Delta_M H_A = (\partial\Delta_M H / \partial n_A)_{P,T,n_j} = H_A - H_{0A} \tag{20}$$

where:

$\Delta_{sol}H_A$	partial molar enthalpy of solution of solvent A
$\Delta_M H_A$	partial molar enthalpy of mixing of solvent A (= differential enthalpy of dilution)
n_A	amount of substance of solvent A

again with a unit of J/mol. The partial molar enthalpy of solution/mixing is equal to the so-called differential enthalpy of dilution as a consequence of adding an infinitesimal amount of solvent to the solution/mixture. The *integral enthalpy of dilution for the solvent A* is equivalent to the integral molar enthalpy of mixing for the solvent A as defined by:

$$^{int}\Delta_M H_A = \Delta_M h / n_A \tag{21}$$

and, in the case of adding a very small amount of solvent to the pure polymer, the partial molar enthalpy of solution at infinite dilution of the solvent is obtained. Partial molar enthalpies of mixing (or dilution) of the solvent are included in this data collection only for cases where they were obtained from calorimetric experiments.

Generally, it is known that partial molar enthalpies of mixing (or dilution) of the solvent can also be determined from the temperature dependence of the activity of the solvent, a_A:

$$\Delta_M H_A = R\left[\partial\ln a_A / \partial(1/T)\right]_P \tag{22}$$

where:

a_A	activity of solvent A
P	pressure
T	(measuring) temperature.

Enthalpy data from light scattering, osmometry, vapor pressure or vapor sorption measurements, and demixing experiments can be found in the literature. However, agreement between enthalpy changes measured by calorimetry and results determined from the temperature dependence of solvent activity data is often of limited quality. Therefore, such data are not included here, but, Chapter 6 provides a number of systems in which such additional information on enthalpy effects in polymer solutions can be found.

From engineering and also from scientific aspects, the partial molar enthalpy of mixing at infinite dilution of the solvent in the liquid/molten polymer $\Delta_M H_A^\infty$ is of value. Therefore, data for $\Delta_M H_A^\infty$ determined by *inverse gas-liquid chromatography (IGC)* have been included here.

$$\Delta_M H_A^\infty = R\,[\partial \ln \Omega_A^\infty / \partial(1/T)]_P \tag{23}$$

where:
$\Delta_M H_A^\infty$ partial molar enthalpy of mixing at infinite dilution of solvent A
Ω_A^∞ mass fraction-based activity coefficient of solvent A in the liquid phase with
                            $a_A = w_A \Omega_A$ at infinite dilution.

Additionally, the enthalpies of solution at infinite dilution $\Delta_{sol}H_{A(vap)}^\infty$ of solvent vapor in molten polymers determined by IGC have been included.

$$\Delta_{sol}H_{A(vap)}^\infty = -R\,[\partial \ln V_g^0 / \partial(1/T)]_P \tag{24}$$

where:
$\Delta_{sol}H_{A(vap)}^\infty$ first integral enthalpy of solution of the vapor of solvent A
                            (with $\Delta_{sol}H_{A(vap)}^\infty = \Delta_M H_A^\infty - \Delta_{LV}H_{0A}$)
V_g^0 specific retention volume corrected to $0\,^\circ C$.

The equipment for IGC does not differ in principle very much from that used in analytical GLC. For operating at infinite dilution, the carrier gas flows directly to the column that is inserted into a thermostatted oil bath (to get a more precise temperature control than in a conventional GLC oven). The output of the column is measured with a flame ionization detector or alternately with a thermal conductivity detector. Helium is used today as the carrier gas (nitrogen was used in earlier work). From the difference between the retention time of the injected solvent sample and the retention time of a non-interacting gas (marker gas), thermodynamic equilibrium data can be obtained. Most experiments were done up to now with packed columns, but capillary columns were used too. The experimental conditions must be chosen so that real thermodynamic data can be obtained, i.e., equilibrium bulk absorption conditions. Errors caused by unsuitable gas flow rates, unsuitable polymer loading percentages on the solid support material and support surface effects as well as any interactions between the injected sample and the solid support in packed columns, unsuitable sample size of the injected probes, carrier gas effects, and imprecise knowledge of the real amount of polymer in the column, can be sources of problems, whether data are nominally measured under real thermodynamic equilibrium conditions or not, and have to be eliminated. The sizeable pressure drop through the column must be measured and accounted for. An additional condition for obtaining real thermodynamic equilibrium data is caused by the nature of the polymer sample. Thermodynamic equilibrium data require the polymer to be in a molten state. This means that IGC measurements have to be performed for this purpose well above the glass transition temperature of the amorphous polymer or even above the melting temperature of the crystalline parts of a polymer sample. As a rule, the experimental temperature must exceed the glass transition or the melting temperature by about 50 K.

The *data reduction for infinite dilution IGC* starts with the usually obtained parameters of retention volume or net retention volume which have to be calculated from the measured retention times and the flow rate of the carrier gas at column conditions.

$$V_{net} = V_r - V_{dead}$$ (25)

where:

V_{net}	net retention volume
V_r	retention volume
V_{dead}	retention volume of the inert marker gas, dead retention, gas holdup.

These net retention volumes are reduced to specific retention volumes, V_g^0, by division of equation (1) with the mass of the liquid (here the liquid is the molten polymer). They are corrected for the pressure difference between column inlet and outlet pressure, and reduced to a temperature $T_0 = 273.15$ K.

$$V_g^0 = \left(\frac{V_{net}}{m_B}\right)\left(\frac{T_0}{T}\right)\frac{3(P_{in}/P_{out})^2 - 1}{2(P_{in}/P_{out})^3 - 1}$$ (26)

where:

V_g^0	specific retention volume corrected to 0°C = 273.15 K
m_B	mass of the polymer in the liquid phase within the column
P_{in}	column inlet pressure
P_{out}	column outlet pressure
T	measuring temperature
T_0	reference temperature = 273.15 K.

The activity coefficient at infinite dilution reads, if we neglect interactions to and between carrier gas molecules (which are normally helium):

$$\Omega_A^{\infty} = \left(\frac{RT_0}{V_g^0 M_A P_A^s}\right)\exp\left[\frac{(B_{AA} - V_A^L)(P - P_A^s)}{RT}\right]$$ (27)

where:

B_{AA}	second virial coefficient of pure solvent A at temperature T
M_A	molar mass of solvent A
P_A^s	saturation vapor pressure of pure liquid solvent A at temperature T
R	gas constant
T_0	reference temperature = 273.15 K
V_A^L	molar volume of pure liquid solvent A at temperature T.

More detailed information on the application of IGC to polymer solutions and the corresponding estimation of enthalpic data can be found in a number of books (e.g., 1976NES, 1988NES, 1989LLO, 1989VIL, 1991MUN).

1.3. Guide to the data tables

Characterization of the polymers

Polymers vary by a number of characterization variables. The molar mass and their distribution function are the most important variables. However, tacticity, sequence distribution, branching, and end groups determine their thermodynamic behavior in solution too. Unfortunately, much less information is provided with respect to the polymers that were applied in most of the thermodynamic investigations in the original literature. For copolymers, the chemical distribution and the average chemical composition are also to be given. In many cases, the samples are characterized only by one or two molar mass averages and some additional information (e.g., T_g, ρ, or how and where it was synthesized). Sometimes even this information is missing.

The molar mass averages are defined as follows:

number average M_n

$$M_n = \frac{\sum_i n_{B_i} M_{B_i}}{\sum_i n_{B_i}} = \frac{\sum_i w_{B_i}}{\sum_i w_{B_i} / M_{B_i}} \tag{28}$$

mass average M_w

$$M_w = \frac{\sum_i n_{B_i} M_{B_i}^2}{\sum_i n_{B_i} M_{B_i}} = \frac{\sum_i w_{B_i} M_{B_i}}{\sum_i w_{B_i}} \tag{29}$$

z-average M_z

$$M_z = \frac{\sum_i n_{B_i} M_{B_i}^3}{\sum_i n_{B_i} M_{B_i}^2} = \frac{\sum_i w_{B_i} M_{B_i}^2}{\sum_i w_{B_i} M_{B_i}} \tag{30}$$

viscosity average M_η

$$M_\eta = \left(\frac{\sum_i w_{B_i} M_{B_i}^a}{\sum_i w_{B_i}} \right)^{1/a} \tag{31}$$

where:
a	exponent in the viscosity-molar mass relationship
M_{Bi}	molar mass of polymer species B_i
n_{Bi}	amount of substance of polymer species B_i
w_{Bi}	mass fraction of polymer species B_i.

Measures for the polymer concentration

The following concentration measures are used in the tables of this handbook (where B always denotes the polymer, A denotes the solvent, and in ternary systems C denotes the third component):

mass/volume concentration:

$$c_A = m_A/V \qquad\qquad c_B = m_B/V \tag{32}$$

mass fraction:

$$w_A = m_A/\Sigma\, m_i \qquad w_B = m_B/\Sigma\, m_i \tag{33}$$

mole fraction:

$$x_A = n_A/\Sigma\, n_i \qquad x_B = n_B/\Sigma\, n_i \text{ with } n_i = m_i/M_i \text{ and } M_B = M_n \tag{34}$$

volume fraction:

$$\varphi_A = (m_A/\rho_A)/\Sigma\,(m_i/\rho_i) \qquad \varphi_B = (m_B/\rho_B)/\Sigma\,(m_i/\rho_i) \tag{35}$$

segment fraction:

$$\psi_A = x_A r_A/\Sigma\, x_i r_i \quad \psi_B = x_B r_B/\Sigma\, x_i r_i \text{ usually with } r_A = 1 \tag{36}$$

base mole fraction:

$$z_A = x_A r_A/\Sigma\, x_i r_i \quad z_B = x_B r_B/\Sigma\, x_i r_i \text{ with } r_B = M_B/M_0 \text{ and } r_A = 1 \tag{37}$$

where:

c_A	(mass/volume) concentration of solvent A
c_B	(mass/volume) concentration of polymer B
m_A	mass of solvent A
m_B	mass of polymer B
M_A	molar mass of solvent A
M_B	molar mass of polymer B
M_{Bi}	molar mass of polymer species B_i
M_n	number-average relative molar mass
M_0	molar mass of a basic unit of polymer B
n_A	amount of substance of solvent A
n_B	amount of substance of polymer B
n_{Bi}	amount of substance of polymer species B_i with molar mass M_{Bi}
r_A	segment number of solvent A, usually $r_A = 1$
r_B	segment number of polymer B
V	volume of the liquid solution at temperature T
w_A	mass fraction of solvent A
w_B	mass fraction of polymer B
w_{Bi}	mass fraction of polymer species B_i with molar mass M_{Bi}

x_A	mole fraction of solvent A
x_B	mole fraction of polymer B
z_A	base mole fraction of solvent A
z_B	base mole fraction of polymer B
φ_A	volume fraction of solvent A
φ_B	volume fraction of polymer B
ρ_A	density of solvent A
ρ_B	density of polymer B
ψ_A	segment fraction of solvent A
ψ_B	segment fraction of polymer B.

For high-molecular polymers, a mole fraction is not an appropriate unit to characterize composition. However, for oligomeric products with rather low molar masses, mole fractions were sometimes used. In the common case of a distribution function for the molar mass, $M_B = M_n$ is to be chosen. Mass fraction and volume fraction can be considered as special cases of segment fractions depending on the way by which the segment size is actually determined: $r_i/r_A = M_i/M_A$ or $r_i/r_A = V_i/V_A = (M_i/\rho_i)/(M_A/\rho_A)$, respectively. Classical segment fractions are calculated by applying $r_i/r_A = V_i^{vdW}/V_A^{vdW}$ ratios where hard-core van der Waals volumes, V_i^{vdW}, are taken into account. Their special values depend on the chosen equation of state (or simply some group contribution schemes, e.g., 1968BON, 1990KRE) and have to be specified.

Volume fractions imply a temperature dependence and, as they are defined in equation (35), neglect excess volumes of mixing and, very often, the densities of the polymer in the state of the solution are not known correctly. However, volume fractions can be calculated without the exact knowledge of the polymer molar mass (or its averages).

Base mole fractions are sometimes applied for polymer systems in earlier literature. The value for M_0 is the molar mass of a basic unit of the polymer. Sometimes it is chosen arbitrarily, however, and has to be specified.

Experimental data tables

The data tables in each chapter are provided in order of the names of the polymers. In this data book, mostly source-based polymer names are applied. These names are more common in use, and they are usually given in the original sources, too. Structure-based names, details about their nomenclature can be found in the *Polymer Handbook* (1999BRA), are chosen in some single cases only. CAS index names for polymers are not applied here. Finally, a list of the polymers in the appendix utilizes the names as given in the chapters of this book.

Within types of polymers, the individual samples are ordered by their increasing average molar mass, and, when necessary, systems are ordered by increasing temperature. Within data sets for equal polymers, systems are ordered by the names of the solvents. In ternary systems, ordering is additionally made subsequently according to the name of the third component in the system.

Each data set begins with the lines for the solution components, e.g., in binary systems:

Polymer (B):   **lignin**   **1995BOG**

Characterization: M_n/g.mol^{-1} = 7300, M_w/g.mol^{-1} = 30900, M_z/g.mol^{-1} = 41400,
milled wood lignin from fir (Bjoerkman's lignin),
Arkhangelsk Forest Engineering Institute, Russia

Solvent (A):   **dimethylsulfoxide**   **C_2H_6OS**   **67-68-5**

where the polymer sample is given in the first line together with the reference. The second line provides then the characterization available for the polymer sample. The following line gives the solvent's chemical name, molecular formula, and CAS registry number.

In ternary systems, the following lines are either for a second solvent or a second polymer or another chemical compound, e.g., in ternary systems with two solvents:

Polymer (B):   **poly(γ-benzyl-L-glutamate)**   **1968KA1**

Characterization: M_n/g.mol^{-1} = 290000

Solvent (A):   **dichloroacetic acid**   **$C_2H_2Cl_2O_2$**   **79-43-6**

Solvent (C):   **trichloromethane**   **$CHCl_3$**   **67-66-3**

or, e.g., in ternary systems with a second polymer:

Polymer (B):   **poly(ethylene glycol)**   **2003CO2**

Characterization: M_n/g.mol^{-1} = 275, M_w/g.mol^{-1} = 305,
PEG 300, Fluka AG, Buchs, Switzerland

Solvent (A):   **anisole**   **C_7H_8O**   **100-66-3**

Polymer (C):   **poly(ethylene glycol)**   **2003CO2**

Characterization: M_n/g.mol^{-1} = 365, M_w/g.mol^{-1} = 401,
PEG 400, Fluka AG, Buchs, Switzerland

The originally measured data for each single system are then listed together with some comment lines if necessary. The data are usually given as published, but temperatures are always given in K. Pressures are usually recalculated into kPa or MPa and enthalpy data are always recalculated into J or kJ, if necessary.

Final day for including data into this book was June, 30, 2005.

1.4. List of symbols

a	exponent in the viscosity-molar mass relationship
a_A	activity of solvent A
B_{AA}	second virial coefficient of pure solvent A at temperature T
c_A	(mass/volume) concentration of solvent A
c_B	(mass/volume) concentration of polymer B
C_p^E	excess heat capacity
H^E	excess enthalpy = $\Delta_M H$ = enthalpy of mixing
H_A	partial molar enthalpy of solvent A
H_B	partial molar (or specific) enthalpy of polymer B
H_{0A}	molar (or specific) enthalpy of pure solvent A
H_{0B}	molar (or specific) enthalpy of pure polymer B
$\Delta_{dil}H^{12}$	intermediary enthalpy of dilution (= $H^{(2)} - H^{(1)}$)
$\Delta_M H$	enthalpy of mixing
$\Delta_{sol}H$	enthalpy of solution
$^{int}\Delta_M H_A$	integral enthalpy of mixing of solvent A (= integral enthalpy of dilution)
$\Delta_M H_A$	partial molar enthalpy of mixing of solvent A (= differential enthalpy of dilution)
$\Delta_M H_A^{\infty}$	partial molar enthalpy of mixing at infinite dilution of solvent A
$^{int}\Delta_{sol}H_A$	integral enthalpy of solution of solvent A
$\Delta_{sol}H_A$	partial molar enthalpy of solution of solvent A
$\Delta_{sol}H_A^{\infty}$	first integral enthalpy of solution of solvent A (= $\Delta_M H_A^{\infty}$ in the case of liquid/molten polymers and a liquid solvent, i.e., it is different from the values for solutions of solvent vapors or gases in a liquid/molten polymer $\Delta_{sol}H_{A(vap)}^{\infty}$)
$\Delta_{sol}H_{A(vap)}^{\infty}$	first integral enthalpy of solution of the vapor of solvent A (with $\Delta_{sol}H_{A(vap)}^{\infty} = \Delta_M H_A^{\infty} - \Delta_{LV}H_{0A}$)
$\Delta_{LV}H_{0A}$	molar enthalpy of vaporization of pure solvent A at temperature T
$^{int}\Delta_M H_B$	integral enthalpy of mixing of polymer B
$\Delta_M H_B$	partial molar (or specific) enthalpy of mixing of polymer B
$\Delta_M H_B^{\infty}$	partial molar (or specific) enthalpy of mixing at infinite dilution of polymer B
$^{int}\Delta_{sol}H_B$	integral enthalpy of solution of polymer B
$\Delta_{sol}H_B$	partial molar (or specific) enthalpy of solution of polymer B
$\Delta_{sol}H_B^{\infty}$	first integral enthalpy of solution of polymer B ($\Delta_M H_B^{\infty}$ in the case of liquid/molten B)
m_A	mass of solvent A
m_B	mass of polymer B
M	relative molar mass
M_A	molar mass of solvent A
M_B	molar mass of polymer B
M_n	number-average relative molar mass
M_w	mass-average relative molar mass
M_η	viscosity-average relative molar mass
M_z	z-average relative molar mass
M_0	molar mass of a basic unit of polymer B
MI	melting index
n_A	amount of substance of solvent A
n_B	amount of substance of polymer B

P	pressure
$P_A{}^s$	saturation vapor pressure of pure liquid solvent A at temperature T
P_{in}	column inlet pressure in IGC
P_{out}	column outlet pressure in IGC
R	gas constant
r_A	segment number of solvent A, usually $r_A = 1$
r_B	segment number of polymer B
T	(measuring) temperature
T_g	glass transition temperature
T_m	equilibrium melting temperature
T_0	reference temperature (= 273.15 K)
V, V_{spez}	volume or specific volume at temperature T
$V_A{}^L$	molar volume of pure liquid solvent A at temperature T
V^E	excess volume at temperature T
V_r, V_{net}	retention volume and net retention volume in IGC
V_{dead}	retention volume of the (inert) marker gas, dead retention, gas holdup in IGC
$V_g{}^0$	specific retention volume corrected to 0°C in IGC
w_A	mass fraction of solvent A
w_B	mass fraction of polymer B
x_A	mole fraction of solvent A
x_B	mole fraction of polymer B
z_A	base mole fraction of solvent A
z_B	base mole fraction of polymer B
γ_A	activity coefficient of solvent A in the liquid phase with activity $a_A = x_A \gamma_A$
φ_A	volume fraction of solvent A
φ_B	volume fraction of polymer B
ρ_A	density of solvent A
ρ_B	density of polymer B
ψ_A	segment fraction of solvent A
ψ_B	segment fraction of polymer B
π	osmotic pressure
Ω_A	mass fraction-based activity coefficient of solvent A in the liquid phase with activity $a_A = w_A \Omega_A$
$\Omega_A{}^\infty$	mass fraction-based activity coefficient of solvent A at infinite dilution

1.5. References

1968BON   Bondi, A., *Physical Properties of Molecular Crystals, Liquids and Glasses*, J. Wiley & Sons, New York, 1968.

1976NES   Nesterov, A.E. and Lipatov, Yu.S., *Obrashchennaya Gasovaya Khromatografiya v Termodinamike Polimerov*, Naukova Dumka, Kiev, 1976.

1984HEM   Hemminger, W. and Höhne, G., *Calorimetry: Fundamentals and Practice*, Verlag Chemie, Weinheim, 1984.

1988NES   Nesterov, A.E., *Obrashchennaya Gasovaya Khromatografiya Polimerov*, Naukova Dumka, Kiev, 1988.

1989LLO   Lloyd, D.R., Ward, T.C., Schreiber, H.P., and Pizana, C.C., Eds., *Inverse Gas Chromatography*, ACS Symposium Series 391, American Chemical Society, Washington, 1989.

1989VIL   Vilcu, R. and Leca, M., *Polymer Thermodynamics by Gas Chromatography*, Elsevier, Amsterdam, 1989.

1990BAR   Barton, A.F.M., *CRC Handbook of Polymer-Liquid Interaction Parameters and Solubility Parameters*, CRC Press, Boca Raton, 1990.

1990FUJ   Fujita, H., *Polymer Solutions*, Elsevier, Amsterdam, 1990.

1990KAM   Kamide, K., *Thermodynamics of Polymer Solutions*, Elsevier, Amsterdam, 1990.

1990KRE   [Van] Krevelen, D.W., *Properties of Polymers*, 3rd ed., Elsevier, Amsterdam, 1990.

1991MUN   Munk, P., Polymer characterization using inverse gas chromatography, in *Modern Methods of Polymer Characterization*, Barth, H.G. and Mays, J.W., Eds., J. Wiley & Sons, New York, 1991, 151.

1994MAR   Marsh, K.N., Ed., *Experimental Thermodynamics, Volume 4, Solution Calorimetry*, Blackwell Science, Oxford, 1994.

1994WOH   Wohlfarth, C., *Vapour-Liquid Equilibrium Data of Binary Polymer Solutions: Physical Science Data*, 44, Elsevier, Amsterdam, 1994.

1999BRA   Brandrup, J., Immergut, E.H., and Grulke, E.A., Eds., *Polymer Handbook*, 4th ed., J. Wiley & Sons, New York, 1999.

1999KLE   Klenin, V.J., *Thermodynamics of Systems Containing Flexible-Chain Polymers*, Elsevier, Amsterdam, 1999.

1999OR1   Orwoll, R.A., Heat, entropy and volume changes for polymer-liquid mixtures, VII/649, in *Polymer Handbook*, 4th ed., J. Wiley & Sons, New York, 1999.

1999OR2   Orwoll, R.A., Heats of solution of some common polymers, VII/671, in *Polymer Handbook*, 4th ed., J. Wiley & Sons, New York, 1999.

1999PRA   Prausnitz, J.M., Lichtenthaler, R.N., and de Azevedo, E.G., *Molecular Thermodynamics of Fluid Phase Equilibria*, 3rd ed., Prentice Hall, Upper Saddle River, NJ, 1999.

2001KON   Koningsveld, R., Stockmayer, W.H., and Nies, E., *Polymer Phase Diagrams*, Oxford University Press, Oxford, 2001.

2001WOH   Wohlfarth, C., *CRC Handbook of Thermodynamic Data of Copolymer Solutions*, CRC Press, Boca Raton, 2001.

2004WOH   Wohlfarth, C., *CRC Handbook of Thermodynamic Data of Aqueous Polymer Solutions*, CRC Press, Boca Raton, 2004.

2005WOH   Wohlfarth, C., *CRC Handbook of Thermodynamic Data of Polymer Solutions at Elevated Pressures*, Taylor & Francis, CRC Press, Boca Raton, 2005.

2. ENTHALPIES OF MIXING OR INTERMEDIARY ENTHALPIES OF DILUTION

2.1. Experimental data

Polymer (B):	**cellulose tricarbanilate**						**1975TAG**
Characterization:	$M_w/\text{g.mol}^{-1} = 150000$						
Solvent (A):	**cyclohexanol**	$C_6H_{12}O$					**108-93-0**

$T/\text{K} = 341.15$

w_B	0.005	0.012	0.017	0.020	0.030	0.043	0.060
$\Delta_M H/(\text{J/g})$	−0.2	−0.4	−0.5	−0.6	−0.9	−1.3	−1.9

$T/\text{K} = 354.15$

w_B	0.005	0.012	0.017	0.020	0.030	0.043	0.060
$\Delta_M H/(\text{J/g})$	−0.2	−0.3	−0.4	−0.5	−0.8	−1.1	−1.5

Comments: Data were derived from measurements of intermediary enthalpy of dilution. Complete $\Delta_M H$ curves are given in a figure in the original source.

Polymer (B):	**cellulose tricarbanilate**			**1975TAG**
Characterization:	$M_w/\text{g.mol}^{-1} = 150000$			
Solvent (A):	**5-nonanone**	$C_9H_{18}O$		**502-56-7**

$T/\text{K} = 319.15$

w_B	0.005	0.020	0.030	0.043
$\Delta_M H/(\text{J/g})$	−0.4	−1.5	−2.3	−3.3

Comments: Data were derived from measurements of intermediary enthalpy of dilution. Complete $\Delta_M H$ curves are given in a figure in the original source.

Polymer (B):	**decamethyltetrasiloxane**		**1975TAN**
Characterization:	$M/\text{g.mol}^{-1} = 310.7$, Dow Corning Silicones		
Solvent (A):	**2,2-dimethylbutane**	C_6H_{14}	**75-83-2**

$T/\text{K} = 298.15$

$$\Delta_M H/(\text{J/mol}) = x_B(1 - x_B)[229 - 74(1 - 2x_B) + 37(1 - 2x_B)^2]$$

Comments: Experimental data are given only in a figure in the original source.

Polymer (B):   **decamethyltetrasiloxane**   **1975TAN**
Characterization: M/g.mol^{-1} = 310.7, Dow Corning Silicones
Solvent (A):   **n-dodecane**   **C$_{12}$H$_{26}$**   **112-40-3**

T/K = 298.15

$\Delta_M H$/(J/mol) = $x_B(1 - x_B)[2236 - 247(1 - 2x_B) + 152(1 - 2x_B)^2]$

Comments: Experimental data are given only in a figure in the original source.

Polymer (B):   **decamethyltetrasiloxane**   **1975TAN**
Characterization: M/g.mol^{-1} = 310.7, Dow Corning Silicones
Solvent (A):   **2,2,4,4,6,8,8-heptamethylnonane**   **C$_{16}$H$_{34}$**   **4390-04-9**

T/K = 298.15

$\Delta_M H$/(J/mol) = $x_B(1 - x_B)[1415 - 118(1 - 2x_B) + 120(1 - 2x_B)^2]$

Comments: Experimental data are given only in a figure in the original source.

Polymer (B):   **decamethyltetrasiloxane**   **1975TAN**
Characterization: M/g.mol^{-1} = 310.7, Dow Corning Silicones
Solvent (A):   **n-hexadecane**   **C$_{16}$H$_{34}$**   **544-76-3**

T/K = 298.15

$\Delta_M H$/(J/mol) = $x_B(1 - x_B)[3237 - 113(1 - 2x_B) + 196(1 - 2x_B)^2]$

Comments: Experimental data are given only in a figure in the original source.

Polymer (B):   **decamethyltetrasiloxane**   **1975TAN**
Characterization: M/g.mol^{-1} = 310.7, Dow Corning Silicones
Solvent (A):   **n-hexane**   **C$_6$H$_{14}$**   **110-54-3**

T/K = 298.15

$\Delta_M H$/(J/mol) = $x_B(1 - x_B)[538 - 368(1 - 2x_B) - 20(1 - 2x_B)^2]$

Comments: Experimental data are given only in a figure in the original source.

Polymer (B):   **decamethyltetrasiloxane**   **1975TAN**
Characterization: M/g.mol^{-1} = 310.7, Dow Corning Silicones
Solvent (A):   **n-octane**   **C$_8$H$_{18}$**   **111-65-9**

T/K = 298.15

$\Delta_M H$/(J/mol) = $x_B(1 - x_B)[1163 - 243(1 - 2x_B) + 195(1 - 2x_B)^2]$

Comments: Experimental data are given only in a figure in the original source.

Polymer (B):   **decamethyltetrasiloxane**   **1975TAN**
Characterization: M/g.mol^{-1} = 310.7, Dow Corning Silicones
Solvent (A):   **2,2,4,6,6-pentamethylheptane**   **$C_{12}H_{26}$**   **13475-82-6**

T/K = 298.15

$\Delta_M H$/(J/mol) = $x_B(1 - x_B)[1069 - 224(1 - 2x_B) + 105(1 - 2x_B)^2]$

Comments: Experimental data are given only in a figure in the original source.

Polymer (B):   **decamethyltetrasiloxane**   **1975TAN**
Characterization: M/g.mol^{-1} = 310.7, Dow Corning Silicones
Solvent (A):   **2,2,4-trimethylpentane C_8H_{18}**   **540-84-1**

T/K = 298.15

$\Delta_M H$/(J/mol) = $x_B(1 - x_B)[767 - 195(1 - 2x_B) + 178(1 - 2x_B)^2]$

Comments: Experimental data are given only in a figure in the original source.

Polymer (B):   **dextran**   **1979BAS, 1980BAS**
Characterization: M_n/g.mol^{-1} = 8200, M_w/g.mol^{-1} = 10400,
                               fractionated in the laboratory
Solvent (A):   **dimethylsulfoxide**   **C_2H_6OS**   **67-68-5**

T/K = 298.15 $\varphi_B^{(1)}$ = 0.25 $\varphi_B^{(2)}$ = 0.01 $\Delta_{dil} H^{12}$/(J/g polymer) = −5.3

Comments: $\varphi_B^{(1)}$ and $\varphi_B^{(2)}$ denote the volume fractions of the polymer in the solution before and after
              the dilution process. Additional data are given in a figure in the original source.

Polymer (B):   **dextran**   **1979BAS, 1980BAS**
Characterization: M_n/g.mol^{-1} = 75900, M_w/g.mol^{-1} = 101000,
                               fractionated in the laboratory
Solvent (A):   **dimethylsulfoxide**   **C_2H_6OS**   **67-68-5**

T/K = 298.15 $\varphi_B^{(1)}$ = 0.12 $\varphi_B^{(2)}$ = 0.04 $\Delta_{dil} H^{12}$/(J/g polymer) = −0.58

Comments: $\varphi_B^{(1)}$ and $\varphi_B^{(2)}$ denote the volume fractions of the polymer in the solution before and after
              the dilution process. Additional data are given in a figure in the original source.

Polymer (B):   **dextran**   **1979BAS, 1980BAS**
Characterization: M_n/g.mol^{-1} = 8200, M_w/g.mol^{-1} = 10400,
                               fractionated in the laboratory
Solvent (A):   **water**   **H_2O**   **7732-18-5**

T/K = 298.15 $\varphi_B^{(1)}$ = 0.25 $\varphi_B^{(2)}$ = 0.01 $\Delta_{dil} H^{12}$/(J/g polymer) = −0.68

Comments: $\varphi_B^{(1)}$ and $\varphi_B^{(2)}$ denote the volume fractions of the polymer in the solution before and after
              the dilution process. Additional data are given in a figure in the original source.

Polymer (B):	dextran		**1979BAS, 1980BAS**
Characterization:	M_n/g.mol^{-1} = 75900, M_w/g.mol^{-1} = 101000, fractionated in the laboratory		
Solvent (A):	water	H$_2$O	**7732-18-5**

T/K = 298.15 $w_B^{(1)}$ = 0.12 $w_B^{(2)}$ = 0.04 $\Delta_{dil}H^{12}$/(J/g polymer) = –0.07

Comments: $w_B^{(1)}$ and $w_B^{(2)}$ denote the mass fractions of the polymer in the solution before and after the dilution process. Additional data are given in a figure in the original source.

Polymer (B):	dextran		**1995GRO, 1995TIN**
Characterization:	M_n/g.mol^{-1} = 179347, M_w/g.mol^{-1} = 507000 T-500, Pharmacia Fine Chemicals, Uppsala, Sweden		
Solvent (A):	water	H$_2$O	**7732-18-5**

T/K = 298.15

m_A'/g	39.54	30.30	39.54	39.41	39.33	39.45	39.55	29.91	39.19
m_B'/g	0.0	12.98	0.0	0.0	0.0	0.0	0.0	12.81	0.0
m_A''/g	6.820	8.588	4.706	4.930	5.126	4.561	4.267	8.730	6.849
m_B''/g	2.293	0.0	3.136	3.287	5.124	5.147	5.216	0.0	2.935
$\Delta_{dil}H^{12}$/J	–1.6	–1.8	–3.3	–4.0	–9.4	–10.4	–13.6	–1.7	–2.0

m_A'/g	39.47	39.39	39.36	39.18	39.05
m_B'/g	0.0	0.0	0.0	0.0	0.0
m_A''/g	6.050	5.369	5.128	4.831	4.510
m_B''/g	4.032	3.579	5.126	5.452	5.515
$\Delta_{dil}H^{12}$/J	–3.6	–5.1	–9.9	–10.5	–15.2

T/K = 333.15

m_A'/g	38.82	39.03	38.50	38.81
m_B'/g	0.0	0.0	0.0	0.0
m_A''/g	5.271	4.884	5.874	4.648
m_B''/g	3.568	4.884	3.976	4.647
$\Delta_{dil}H^{12}$/J	–4.8	–13.4	–5.4	–14.8

Comments: $\Delta_{dil}H^{12}$ is the extensive enthalpy change in the result of the mixing process obtained for the given masses in the table, where the superscripts ' and " designate the two solutions placed in the two parts of a mixing cell, respectively.

Polymer (B):	di(ethylene glycol) dibutyl ether		**1985ALK**
Characterization:	M/g.mol^{-1} = 218.34		
Solvent (A):	n-dodecane	C$_{12}$H$_{26}$	**112-40-3**

T/K = 304.15

m_A/g	0.102	0.199	0.276	0.338	0.431	0.490	0.579	0.657	0.733
m_B/g	0.884	0.864	0.729	0.594	0.525	0.402	0.313	0.217	0.125
$\Delta_M H$/J	1.51	2.59	3.27	3.60	3.87	3.60	3.28	2.65	1.84

Comments: $\Delta_M H$ is the extensive quantity measured for the given masses of polymer and solvent.

Polymer (B):   **di(ethylene glycol) dibutyl ether**   **1985ALK**
Characterization: $M/\text{g.mol}^{-1} = 218.34$
Solvent (A):   **n-hexadecane** $C_{16}H_{34}$ **544-76-3**

$T/K = 304.15$

m_A/g	0.098	0.178	0.253	0.332	0.412	0.479	0.561	0.637	0.714
m_B/g	0.824	0.737	0.650	0.561	0.473	0.380	0.296	0.205	0.116
$\Delta_M H/\text{J}$	1.57	2.25	3.20	3.60	3.86	3.65	3.47	2.93	1.90

Comments: $\Delta_M H$ is the extensive quantity measured for the given masses of polymer and solvent.

Polymer (B):   **di(ethylene glycol) dibutyl ether**   **1985ALK**
Characterization: $M/\text{g.mol}^{-1} = 218.34$
Solvent (A):   **bis(2-methoxyethyl) ether** $C_6H_{14}O_3$ **111-96-6**

$T/K = 304.15$

m_A/g	0.866	0.775	0.683	0.584	0.489	0.395	0.301	0.215	0.111
m_B/g	0.092	0.181	0.268	0.359	0.444	0.525	0.634	0.712	0.794
$\Delta_M H/\text{J}$	0.83	1.47	1.92	2.25	2.35	2.27	2.04	1.68	0.98

Comments: $\Delta_M H$ is the extensive quantity measured for the given masses of polymer and solvent.

Polymer (B):   **di(ethylene glycol) dibutyl ether**   **1985ALK**
Characterization: $M/\text{g.mol}^{-1} = 218.34$
Solvent (A):   **n-tetradecane** $C_{14}H_{30}$ **629-59-4**

$T/K = 304.15$

m_A/g	0.097	0.175	0.251	0.323	0.402	0.476	0.554	0.631	0.706
m_B/g	0.826	0.737	0.646	0.550	0.472	0.381	0.295	0.204	0.155
$\Delta_M H/\text{J}$	1.52	2.42	3.15	3.45	3.76	3.65	3.32	2.73	1.77

Comments: $\Delta_M H$ is the extensive quantity measured for the given masses of polymer and solvent.

Polymer (B):   **di(ethylene glycol) dihexyl ether**   **1985ALK**
Characterization: $M/\text{g.mol}^{-1} = 274.44$
Solvent (A):   **bis(2-methoxyethyl) ether** $C_6H_{14}O_3$ **111-96-6**

$T/K = 304.15$

m_A/g	0.850	0.645	0.468	0.288	0.104
m_B/g	0.106	0.291	0.463	9,633	0.890
$\Delta_M H/\text{J}$	1.56	3.17	3.78	3.20	1.50

Comments: $\Delta_M H$ is the extensive quantity measured for the given masses of polymer and solvent.

Polymer (B):   **hexamethyldisiloxane**   **1974DIC**
Characterization: $M/\text{g.mol}^{-1} = 162.4$
Solvent (A):   **n-hexane** C_6H_{14} **110-54-3**

$T/\text{K} = 298.15$

x_A	0.0339	0.1115	0.1753	0.2583	0.3429	0.4482	0.5811	0.6588
$\Delta_M H/(\text{J/mol})$	16	50	65	88	105	124	129	118

x_A	0.7674	0.8278	0.8391	0.8745
$\Delta_M H/(\text{J/mol})$	94	79	72	60

Polymer (B):   **hexamethyldisiloxane**   **1997MCL**
Characterization: $M/\text{g.mol}^{-1} = 162.4$
Solvent (A):   **tetradecafluorohexane** C_6F_{14} **355-42-0**

$T/\text{K} = 298.15$

x_A	0.035	0.092	0.141	0.188	0.241	0.293	0.339	0.378
$\Delta_M H/(\text{J/mol})$	362.6	878.9	1234.7	1488.6	1703.9	1891.1	1987.5	2047.5

x_A	0.424	0.466	0.509	0.536	0.503	0.527	0.563	0.608
$\Delta_M H/(\text{J/mol})$	2095.0	2126.7	2133.1	2151.3	2154.8	2148.9	2131.5	2095.6

x_A	0.648	0.698	0.754	0.805	0.850	0.906	0.959
$\Delta_M H/(\text{J/mol})$	2042.1	1944.8	1732.2	1497.7	1233.1	958.6	473.0

Polymer (B):   **lignin**   **1995BOG**
Characterization: $M_n/\text{g.mol}^{-1} = 1700$, $M_w/\text{g.mol}^{-1} = 8800$, $M_z/\text{g.mol}^{-1} = 24900$,
                          milled wood lignin from fir (Bjoerkman's lignin),
                          Arkhangelsk Forest Engineering Institute, Russia
Solvent (A):   **dimethylsulfoxide** C_2H_6OS **67-68-5**

$T/\text{K} = 298.15$

m_A/g	100	100	100	100	100
m_B/g	0.834	3.057	3.741	4.541	8.299
$\Delta_{dil}H^{12}/(\text{J/g polymer})$	−0.196	−0.378	−0.301	−0.542	−0.703

Comments: The final state is at infinite dilution.

Polymer (B):   **lignin**   **1995BOG**
Characterization: $M_n/\text{g.mol}^{-1} = 7300$, $M_w/\text{g.mol}^{-1} = 30900$, $M_z/\text{g.mol}^{-1} = 41400$,
                          milled wood lignin from fir (Bjoerkman's lignin),
                          Arkhangelsk Forest Engineering Institute, Russia
Solvent (A):   **dimethylsulfoxide** C_2H_6OS **67-68-5**

$T/\text{K} = 298.15$

m_A/g	100	100	100	100	100
m_B/g	1.459	2.811	4.352	4.653	5.291
$\Delta_{dil}H^{12}/(\text{J/g polymer})$	−0.730	−1.150	−1.577	−1.724	−1.907

Comments: The final state is at infinite dilution.

Polymer (B):   **maltotriose**   **2002COO**
Characterization: $M/\text{g.mol}^{-1} = 504.4$
Solvent (A):   **water**   **H_2O**   **7732-18-5**

$T/\text{K} = 318.15$

x_B	0.00128	0.00195	0.00303	0.00345	0.00482	0.00541	0.01259
$\Delta_M H/(\text{J/mol})$	−25	−39	−61	−69	−96	−109	−249

Polymer (B):   **natural rubber**   **1959MAR**
Characterization: −
Solvent (A):   **benzene**   **C_6H_6**   **71-43-2**

$T/\text{K} = 298.15$

$$\Delta_M H/(\text{J/mol}) = 8.314\, x_A \varphi_B (74.54 + 44.72 \varphi_B - 11.93 \varphi_B{}^2 + 71.56 \varphi_B{}^3)$$

Polymer (B):   **nitrocellulose**   **1998KSI**
Characterization: 13.2 % nitrogen, degree of subsitution = 2.65
Solvent (A):   **1,3-diethyl-1,3-diphenylurea**   **$C_{17}H_{20}N_2O$**   **85-98-3**

$T/\text{K} = 348$

w_A	0.9920	0.9841	0.9682	0.9363	0.8730	0.7976	0.6010	0.4943
$\Delta_M H/(\text{J/base mol})$	−650	−750	−1230	−1440	−1850	−2550	−4010	−4420

w_A	0.4577	0.3143	0.2150	0.1155	0.0578	0.0289
$\Delta_M H/(\text{J/base mol})$	−3910	−2470	−1340	−1150	−690	−210

Comments: Molar mass of one subunit of nitrocellulose is equal to 281.56 g/mol.

Polymer (B):   **nitrocellulose**   **2004WOL**
Characterization: $M_n/\text{g.mol}^{-1} = 54000$, 13.2 % nitrogen, degree of subsitution = 2.65,
                        ZTS "Pronit" factory, Pionki, Poland
Solvent (A):   **2,4-dinitrotoluene**   **$C_7H_6N_2O_4$**   **121-14-2**

$T/\text{K} = 343.7$

w_A	0.878	0.750	0.601	0.501	0.401	0.320	0.248	0.125
$\Delta_M H/(\text{J/base mol})$	−264	−376	−628	−1379	−1332	−968	−957	−785

Comments: Molar mass of one subunit of nitrocellulose is equal to 281.56 g/mol.

Polymer (B):   **nitrocellulose**   **2002KSI**
Characterization: $M_n/\text{g.mol}^{-1} = 54000$, 13.2 % nitrogen, degree of subsitution = 2.65,
                        ZTS "Pronit" factory, Pionki, Poland
Solvent (A):   **2,6-dinitrotoluene**   **$C_7H_6N_2O_4$**   **606-20-2**

$T/\text{K} = 341.0$

continued

continued

w_A	0.9510	0.8983	0.7012	0.5992	0.5000	0.4274	0.3015	0.2164
$\Delta_M H$/(J/base mol)	−390	−560	−2000	−1970	−3410	−2810	−1530	−360

w_A	0.1171	0.0483
$\Delta_M H$/(J/base mol)	−170	−280

Comments: Molar mass of one subunit of nitrocellulose is equal to 281.56 g/mol.

Polymer (B):	**nitrocellulose**	**1941TAG**
Characterization:	M_η/g.mol^{-1} = 23000, 11.83 % nitrogen	
Solvent (A):	**2-propanone** **C$_3$H$_6$O**	**67-64-1**

T/K = room temperature

$w_B^{(1)}$ = 0.1436 $w_B^{(2)}$ = 0.0025 $\Delta_{dil}H^{12}$/(J/g polymer) = −17.6

Comments: $w_B^{(1)}$ and $w_B^{(2)}$ denote the mass fractions of the polymer in the solution before and after the dilution process.

Polymer (B):	**nitrocellulose**	**1985RAB**
Characterization:	M_η/g.mol^{-1} = 73000, 11.9 % nitrogen	
Solvent (A):	**triacetin** **C$_9$H$_{14}$O$_6$**	**102-76-1**

T/K = 305.5

z_A	0.05	0.15	0.25	0.35	0.45	0.55	0.65	0.75
$\Delta_M H$/(kJ/base mol)	−3.33	−6.00	−8.19	−9.12	−8.90	−7.39	−5.77	−4.13

z_A	0.85
$\Delta_M H$/(kJ/base mol)	−2.45

Polymer (B):	**nylon 6**	**1956MIK**
Characterization:	Capron fibre	
Solvent (A):	**formic acid (95.01%) CH$_2$O$_2$**	**64-18-6**

T/K = room temperature

$w_B^{(1)}$		0.15	0.15	0.15	0.15	0.15	0.15
$w_B^{(2)}$		0.0246	0.0238	0.0204	0.0194	0.0196	0.0420
$\Delta_{dil}H^{12}$/(J/g polymer)		5.06	5.44	5.10	5.31	6.23	5.98

Comments: $w_B^{(1)}$ and $w_B^{(2)}$ denote the mass fractions of the polymer in the solution before and after the dilution process.

Polymer (B):	**octamethylcyclotetrasiloxane**	**1970MAR**
Characterization:	M/g.mol^{-1} = 296.6	
Solvent (A):	**cyclopentane** **C$_5$H$_{10}$**	**287-92-3**

T/K = 291.15

continued

continued

x_B	0.00257	0.00743	0.01529	0.03063	0.04551	0.05993	0.07393	0.08752
$\Delta_M H/(\text{J/mol})$	3.4	9.8	19.9	38.7	55.8	71.5	85.8	98.9

x_B	0.10071	0.11980	0.13809	0.15564	0.17249	0.19394	0.21431	0.23172
$\Delta_M H/(\text{J/mol})$	110.9	126.9	141.0	153.3	164.1	176.3	186.4	195.2

x_B	0.23367	0.25188	0.27653	0.30613	0.33285	0.36469	0.40325	0.45094
$\Delta_M H/(\text{J/mol})$	194.8	202.8	210.7	217.9	222.5	225.8	226.9	224.4

x_B	0.48953	0.53534	0.59602	0.65862	0.74432	0.85566	0.92484	0.97198
$\Delta_M H/(\text{J/mol})$	219.3	210.4	195.8	172.9	137.4	82.1	44.1	16.7

$T/K = 298.15$

x_B	0.00506	0.01240	0.02705	0.04127	0.05509	0.06853	0.08154	0.09422
$\Delta_M H/(\text{J/mol})$	6.5	15.8	33.6	50.0	64.9	78.8	91.4	103.1

x_B	0.10655	0.11857	0.13024	0.15271	0.17405	0.19433	0.21365	0.25288
$\Delta_M H/(\text{J/mol})$	113.8	123.6	132.6	148.5	161.9	173.3	182.8	198.9

x_B	0.27599	0.30375	0.33771	0.38023	0.40528	0.43505	0.46881	0.50821
$\Delta_M H/(\text{J/mol})$	206.0	212.6	218.2	221.5	221.6	220.2	216.8	210.6

x_B	0.55393	0.61090	0.67879	0.76511	0.80594	0.87734	0.90103	0.02717
$\Delta_M H/(\text{J/mol})$	200.5	184.3	160.3	123.8	104.6	68.5	55.5	41.4

x_B	0.94842
$\Delta_M H/(\text{J/mol})$	29.3

$T/K = 308.15$

x_B	0.00527	0.01278	0.02728	0.04166	0.05536	0.06904	0.08201	0.09741
$\Delta_M H/(\text{J/mol})$	6.6	15.7	32.6	48.4	62.6	76.1	88.1	98.9

x_B	0.10973	0.12479	0.15265	0.17444	0.19406	0.21293	0.25275	0.27579
$\Delta_M H/(\text{J/mol})$	198.9	126.7	142.2	155.3	165.8	174.8	189.4	196.2

x_B	0.30352	0.33684	0.37910	0.43441	0.46802	0.50819	0.55485	0.61093
$\Delta_M H/(\text{J/mol})$	202.5	207.6	210.7	209.4	206.1	199.9	190.0	174.6

x_B	0.67963	0.76574	0.87683	0.94541
$\Delta_M H/(\text{J/mol})$	151.4	116.4	64.7	29.6

Polymer (B):   **octamethylcyclotetrasiloxane**   **1975TAN**

Characterization: $M/\text{g.mol}^{-1} = 296.6$, Dow Corning Silicones

Solvent (A):   **2,2-dimethylbutane C_6H_{14}**   **75-83-2**

$T/K = 298.15$

$\Delta_M H/(\text{J/mol}) = x_B(1 - x_B)[711 - 231(1 - 2x_B) + 62(1 - 2x_B)^2]$

Comments: Experimental data are given only in a figure in the original source.

Polymer (B):   **octamethylcyclotetrasiloxane**   **1975TAN**
Characterization: $M/\text{g.mol}^{-1} = 296.6$, Dow Corning Silicones
Solvent (A):   **n-dodecane** $C_{12}H_{26}$ **112-40-3**

$T/K = 298.15$

$\Delta_M H/(\text{J/mol}) = x_B(1 - x_B)[2928 - 277(1 - 2x_B) + 222(1 - 2x_B)^2]$

Comments: Experimental data are given only in a figure in the original source.

Polymer (B):   **octamethylcyclotetrasiloxane**   **1975TAN**
Characterization: $M/\text{g.mol}^{-1} = 296.6$, Dow Corning Silicones
Solvent (A):   **2,2,4,4,6,8,8-heptamethylnonane** $C_{16}H_{34}$ **4390-04-9**

$T/K = 298.15$

$\Delta_M H/(\text{J/mol}) = x_B(1 - x_B)[2216 - 126(1 - 2x_B) + 141(1 - 2x_B)^2]$

Comments: Experimental data are given only in a figure in the original source.

Polymer (B):   **octamethylcyclotetrasiloxane**   **1975TAN**
Characterization: $M/\text{g.mol}^{-1} = 296.6$, Dow Corning Silicones
Solvent (A):   **n-hexadecane** $C_{16}H_{34}$ **544-76-3**

$T/K = 298.15$

$\Delta_M H/(\text{J/mol}) = x_B(1 - x_B)[3968 - 56(1 - 2x_B) + 353(1 - 2x_B)^2]$

Comments: Experimental data are given only in a figure in the original source.

Polymer (B):   **octamethylcyclotetrasiloxane**   **1975TAN**
Characterization: $M/\text{g.mol}^{-1} = 296.6$, Dow Corning Silicones
Solvent (A):   **n-hexane** C_6H_{14} **110-54-3**

$T/K = 298.15$

$\Delta_M H/(\text{J/mol}) = x_B(1 - x_B)[1188 - 370(1 - 2x_B) + 223(1 - 2x_B)^2]$

Comments: Experimental data are given only in a figure in the original source.

Polymer (B):   **octamethylcyclotetrasiloxane**   **1975TAN**
Characterization: $M/\text{g.mol}^{-1} = 296.6$, Dow Corning Silicones
Solvent (A):   **n-octane** C_8H_{18} **111-65-9**

$T/K = 298.15$

$\Delta_M H/(\text{J/mol}) = x_B(1 - x_B)[1829 - 453(1 - 2x_B) + 46(1 - 2x_B)^2]$

Comments: Experimental data are given only in a figure in the original source.

Polymer (B):   **octamethylcyclotetrasiloxane**   **1975TAN**
Characterization: M/g.mol^{-1} = 296.6, Dow Corning Silicones
Solvent (A):   **2,2,4,6,6-pentamethylheptane**   **C$_{12}$H$_{26}$**   **13475-82-6**

T/K = 298.15

$\Delta_M H$/(J/mol) = $x_B(1 - x_B)[1801 - 279(1 - 2x_B) + 57(1 - 2x_B)^2]$

Comments: Experimental data are given only in a figure in the original source.

Polymer (B):   **octamethylcyclotetrasiloxane**   **1975TAN**
Characterization: M/g.mol^{-1} = 296.6, Dow Corning Silicones
Solvent (A):   **2,2,4-trimethylpentane**   **C$_8$H$_{18}$**   **540-84-1**

T/K = 298.15

$\Delta_M H$/(J/mol) = $x_B(1 - x_B)[1439 - 414(1 - 2x_B) + 74(1 - 2x_B)^2]$

Comments: Experimental data are given only in a figure in the original source.

Polymer (B):   **penta(ethylene glycol) dimethyl ether**   **1999HER**
Characterization: M/g.mol^{-1} = 268, > 99.5 %, Clariant
Solvent (A):   **methanol**   **CH$_4$O**   **67-56-1**

T/K = 303.15

x_A	0.0804	0.1711	0.2083	0.3039	0.3835	0.4567	0.4983	0.5639
$\Delta_M H$/(J/mol)	119.8	269.2	299.4	395.6	496.8	529.7	535.8	527.0

x_A	0.6347	0.7031	0.7997	0.9017
$\Delta_M H$/(J/mol)	514.4	468.1	323.8	166.6

Polymer (B):   **poly(acrylic acid)**   **1988AZU**
Characterization: M_w/g.mol^{-1} = 250000, Scientific Polymer Products, Inc., Ontario, NY
Solvent (A):   **ethanol**   **C$_2$H$_6$O**   **64-17-5**

T/K = 298.15

$\varphi_B^{(1)}$	0.0253	0.0253	0.0253	0.0253	0.0253	0.0253	0.0253
$\varphi_B^{(2)}$	0.0186	0.0160	0.0143	0.0127	0.0101	0.0076	0.0059
$\Delta_{dil} H^{12}$/(J/mol solvent)	−0.516	−0.350	−0.282	−0.321	−0.274	−0.106	−0.096

$\varphi_B^{(1)}$	0.0253	0.0321	0.0321	0.0321	0.0321	0.0321	0.0321
$\varphi_B^{(2)}$	0.0042	0.0182	0.0161	0.0128	0.0096	0.0075	0.0054
$\Delta_{dil} H^{12}$/(J/mol solvent)	−0.078	−0.580	−0.485	−0.445	−0.410	−0.301	−0.125

$\varphi_B^{(1)}$	0.0321	0.0343	0.0343	0.0343	0.0343	0.0343	0.0343
$\varphi_B^{(2)}$	0.0021	0.0194	0.0172	0.0137	0.0103	0.0080	0.0057
$\Delta_{dil} H^{12}$/(J/mol solvent)	−0.073	−0.738	−0.863	−0.674	−0.479	−0.379	−0.296

$\varphi_B^{(1)}$	0.0397	0.0397	0.0397	0.0397
$\varphi_B^{(2)}$	0.0291	0.0251	0.0225	0.0199
$\Delta_{dil} H^{12}$/(J/mol solvent)	−1.780	−1.530	−1.620	−1.540

continued

continued

Comments: The table provides the ratio of $\Delta_{dil}H^{12}/\Delta n_A$, i.e., the enthalpy change caused by diluting the primary solution by 1 mol solvent, where $\varphi_B^{(1)}$ denotes the volume fraction of the polymer in the starting solution and $\varphi_B^{(2)}$ denotes the volume fraction after the dilution process.

Polymer (B):	**poly(acrylic acid)**						**1988AZU**
Characterization:	M_w/g.mol^{-1} = 2000, Scientific Polymer Products, Inc., Ontario, NY						
Solvent (A):	**water**		**H$_2$O**				**7732-18-5**

T/K = 298.15

$\varphi_B^{(1)}$	0.0273	0.0273	0.0273	0.0273	0.0559	0.0559	0.0559
$\varphi_B^{(2)}$	0.0191	0.0164	0.0136	0.0109	0.0336	0.0280	0.0224
$\Delta_{dil}H^{12}$/(J/mol solvent)	0.227	0.199	0.514	0.129	0.671	0.568	0.468
$\varphi_B^{(1)}$	0.0559	0.0559	0.0856	0.0856	0.0856	0.0856	0.0856
$\varphi_B^{(2)}$	0.0168	0.0112	0.0559	0.0514	0.0428	0.0342	0.0257
$\Delta_{dil}H^{12}$/(J/mol solvent)	0.361	0.250	1.450	1.270	1.100	0.913	0.729

Comments: The table provides the ratio of $\Delta_{dil}H^{12}/\Delta n_A$, i.e., the enthalpy change caused by diluting the primary solution by 1 mol solvent, where $\varphi_B^{(1)}$ denotes the volume fraction of the polymer in the starting solution and $\varphi_B^{(2)}$ denotes the volume fraction after the dilution process.

Polymer (B):	**poly(acrylic acid)**						**1988AZU**
Characterization:	M_w/g.mol^{-1} = 5000, Scientific Polymer Products, Inc., Ontario, NY						
Solvent (A):	**water**		**H$_2$O**				**7732-18-5**

T/K = 298.15

$\varphi_B^{(1)}$	0.0346	0.0346	0.0346	0.0346	0.0346	0.0346	0.0739
$\varphi_B^{(2)}$	0.0277	0.0242	0.0173	0.0138	0.0104	0.0069	0.0665
$\Delta_{dil}H^{12}$/(J/mol solvent)	0.324	0.258	0.205	0.164	0.131	0.085	1.480
$\varphi_B^{(1)}$	0.0739	0.0739	0.0739	0.0739	0.0739	0.0739	0.107
$\varphi_B^{(2)}$	0.0591	0.0444	0.0370	0.0296	0.0222	0.0148	0.0748
$\Delta_{dil}H^{12}$/(J/mol solvent)	1.270	0.942	0.803	0.658	0.519	0.367	2.190
$\varphi_B^{(1)}$	0.107	0.107					
$\varphi_B^{(2)}$	0.0534	0.320					
$\Delta_{dil}H^{12}$/(J/mol solvent)	1.590	1.040					

Comments: The table provides the ratio of $\Delta_{dil}H^{12}/\Delta n_A$, i.e., the enthalpy change caused by diluting the primary solution by 1 mol solvent, where $\varphi_B^{(1)}$ denotes the volume fraction of the polymer in the starting solution and $\varphi_B^{(2)}$ denotes the volume fraction after the dilution process.

Polymer (B):	**poly(acrylic acid)**						**1988AZU**
Characterization:	M_w/g.mol^{-1} = 170000, Scientific Polymer Products, Inc., Ontario, NY						
Solvent (A):	**water**		**H$_2$O**				**7732-18-5**

T/K = 298.15

continued

continued

$\varphi_B^{(1)}$	0.0334	0.0334	0.0334	0.0334	0.0334	0.0609	0.0609
$\varphi_B^{(2)}$	0.0234	0.0200	0.0167	0.0134	0.0100	0.0426	0.0366
$\Delta_{dil}H^{12}$/(J/mol solvent)	0.234	0.200	0.170	0.140	0.109	0.761	0.642

$\varphi_B^{(1)}$	0.0609	0.0609	0.0886	0.0886	0.0886	0.0886
$\varphi_B^{(2)}$	0.0305	0.0244	0.0532	0.0443	0.0355	0.0266
$\Delta_{dil}H^{12}$/(J/mol solvent)	0.542	0.443	1.330	1.090	0.882	0.660

Comments: The table provides the ratio of $\Delta_{dil}H^{12}/\Delta n_A$, i.e., the enthalpy change caused by diluting the primary solution by 1 mol solvent, where $\varphi_B^{(1)}$ denotes the volume fraction of the polymer in the starting solution and $\varphi_B^{(2)}$ denotes the volume fraction after the dilution process.

Polymer (B):	**poly(acrylic acid)**						**1988AZU**
Characterization:	M_w/g.mol^{-1} = 250000, Scientific Polymer Products, Inc., Ontario, NY						
Solvent (A):	**water**		**H$_2$O**				**7732-18-5**

T/K = 298.15

$\varphi_B^{(1)}$	0.0210	0.0210	0.0210	0.0281	0.0281	0.0281	0.0281
$\varphi_B^{(2)}$	0.0168	0.0084	0.0063	0.0225	0.0197	0.0169	0.0141
$\Delta_{dil}H^{12}$/(J/mol solvent)	0.068	0.034	0.023	0.128	0.104	0.081	0.064

$\varphi_B^{(1)}$	0.0281	0.0281	0.0353	0.0353	0.0353	0.0353	0.0353
$\varphi_B^{(2)}$	0.0112	0.0084	0.0282	0.0247	0.0247	0.0212	0.0177
$\Delta_{dil}H^{12}$/(J/mol solvent)	0.052	0.037	0.264	0.198	0.198	0.148	0.127

$\varphi_B^{(1)}$	0.0353	0.0353	0.0353	0.0426	0.0426	0.0426	0.0647
$\varphi_B^{(2)}$	0.0141	0.0106	0.0071	0.0383	0.170	0.085	0.453
$\Delta_{dil}H^{12}$/(J/mol solvent)	0.089	0.070	0.043	0.434	0.127	0.062	0.977

$\varphi_B^{(1)}$	0.0647	0.0647	0.0647	0.0647	0.0647	0.0647	0.0721
$\varphi_B^{(2)}$	0.388	0.324	0.259	0.194	0.129	0.065	0.0361
$\Delta_{dil}H^{12}$/(J/mol solvent)	0.770	0.577	0.447	0.310	0.206	0.094	0.843

$\varphi_B^{(1)}$	0.0721	0.0721	0.0721	0.0721
$\varphi_B^{(2)}$	0.0288	0.0216	0.0144	0.0072
$\Delta_{dil}H^{12}$/(J/mol solvent)	0.626	0.430	0.279	0.129

Comments: The table provides the ratio of $\Delta_{dil}H^{12}/\Delta n_A$, i.e., the enthalpy change caused by diluting the primary solution by 1 mol solvent, where $\varphi_B^{(1)}$ denotes the volume fraction of the polymer in the starting solution and $\varphi_B^{(2)}$ denotes the volume fraction after the dilution process.

Polymer (B):	**poly(acrylic acid)**						**1971CAR**
Characterization:	M_η/g.mol^{-1} = 300000						
Solvent (A):	**water**		**H$_2$O**				**7732-18-5**

T/K = 298.15

$\varphi_B^{(1)}$	0.0888	0.0609	0.0433	0.0301	0.0216	0.0150	0.0108	0.0076
$\varphi_B^{(2)}$	0.0664	0.0460	0.0325	0.0227	0.0161	0.0113	0.0081	0.0057
$\Delta_{dil}H^{12}$/(J/g polymer)	0.3808	0.2323	0.1919	0.1159	0.0901	0.0622	0.0306	0.0333

continued

continued

$T/K = 298.15$

The following data were measured with 0.2 N HCl.

$\varphi_B^{(1)}$		0.125	0.0515	0.0251	0.0123	0.0062
$\varphi_B^{(2)}$			0.0897	0.0381	0.0188	0.0093	0.0045
$\Delta_{dil}H^{12}$/(J/g polymer)	0.344	0.123	0.075	0.034	0.028

Comments: $\varphi_B^{(1)}$ and $\varphi_B^{(2)}$ denote the volume fractions of the polymer in the solution before and after the dilution process.

Polymer (B):	**poly(γ-benzyl-L-glutamate)**		**1968KA1**
Characterization:	M_n/g.mol^{-1} = 290000		
Solvent (A):	**dichloroacetic acid**	**C$_2$H$_2$Cl$_2$O$_2$**	**79-43-6**

$T/K = 298.15$

$V^{(1)}$/ml	50	60	70	20	25	10	15
$\varphi_B^{(1)}$	0.042	0.032	0.030	0.040	0.032	0.045	0.035
$V^{(2)}$/ml	55	70	80	40	45	20	20
$\varphi_B^{(2)}$	0.038	0.030	0.026	0.020	0.018	0.023	0.028
$\Delta_{dil}H^{12}$/J	0.021	0.021	0.017	0.096	0.038	0.033	0.013

Comments: $\Delta_{dil}H^{12}$ is the extensive quantity obtained for a given total volume change from $V^{(1)}$ to $V^{(2)}$, where $\varphi_B^{(1)}$ and $\varphi_B^{(2)}$ denote the volume fractions of the polymer in the solution before and after the dilution process.

Polymer (B):	**poly(γ-benzyl-L-glutamate)**		**1968KA1**
Characterization:	M_n/g.mol^{-1} = 290000		
Solvent (A):	**dichloroacetic acid**	**C$_2$H$_2$Cl$_2$O$_2$**	**79-43-6**
Solvent (C):	**trichloromethane**	**CHCl$_3$**	**67-66-3**

$T/K = 298.15$

$V^{(1)}$/ml	10.0	11.0	12.0	13.0	14.0	15.0	16.0
$\varphi_A^{(1)}$	0.980	0.899	0.816	0.753	0.700	0.653	0.612
$\varphi_B^{(1)}$	0.0200	0.0181	0.0166	0.0153	0.0142	0.0133	0.0125
$\varphi_C^{(1)}$	0.000	0.090	0.166	0.230	0.286	0.333	0.375
$V^{(2)}$/ml	11.0	12.0	13.0	14.0	15.0	16.0	17.0
$\varphi_A^{(2)}$	0.899	0.816	0.753	0.700	0.653	0.612	0.576
$\varphi_B^{(2)}$	0.0181	0.0166	0.0153	0.0142	0.0133	0.0125	0.0117
$\varphi_C^{(2)}$	0.090	0.166	0.230	0.286	0.333	0.375	0.411
$\Delta_{dil}H^{12}$/J	−5.35	−4.44	−1.95	−2.13	−2.89	−2.83	−2.63

Comments: $\Delta_{dil}H^{12}$ is the extensive quantity obtained by the successive addition of 1.0 ml trichloromethane to the solution, where $\varphi_B^{(1)}$ and $\varphi_B^{(2)}$ denote the volume fractions of the polymer in the solution before and after the dilution step. The mass of the polymer in the initial binary solution in dichloroacetic acid was 0.2856 g.

Polymer (B):   **poly(γ-benzyl-L-glutamate)**   1972RAI
Characterization: M_w/g.mol^{-1} = 310000
Solvent (A):   **N,N-dimethylformamide** C_3H_7NO 68-12-2

$T/K = 313$

φ_B	0.02	0.05	0.05	0.08	0.093	0.12	0.20	0.30
$\Delta_M H$/(J/base mol)	21	42	46	75	113	151	259	247

Polymer (B):   **poly(γ-benzyl-L-glutamate)**   1968KA1
Characterization: M_n/g.mol^{-1} = 290000
Solvent (A):   **trichloromethane** $CHCl_3$ 67-66-3

$T/K = 298.15$

$V^{(1)}$/ml	20	25	30	50	55	60	60
$\varphi_B^{(1)}$	0.050	0.040	0.033	0.050	0.045	0.041	0.040
$V^{(2)}$/ml	25	30	35	55	60	70	80
$\varphi_B^{(2)}$	0.040	0.033	0.028	0.045	0.041	0.035	0.030
$\Delta_{dil} H^{12}$/J	0.046	0.026	0.013	0.050	0.033	0.054	0.113

Comments: $\Delta_{dil} H^{12}$ is the extensive quantity obtained for a given total volume change from $V^{(1)}$ to $V^{(2)}$, where $\varphi_B^{(1)}$ and $\varphi_B^{(2)}$ denote the volume fractions of the polymer in the solution before and after the dilution process.

Polymer (B):   **polybutadiene**   1959JES
Characterization: –
Solvent (A):   **benzene** C_6H_6 71-43-2

$T/K = 300.05$

φ_B	0.5096	0.5854	0.7244	0.8332	0.8912
$\Delta_M H$/(J/ml)	3.096	2.890	2.407	1.633	0.986

Polymer (B):   **1,4-*cis*-polybutadiene**   1966KA2
Characterization: M_n/g.mol^{-1} = 200000, Bridgestone Tire Co., Ltd.
Solvent (A):   **benzene** C_6H_6 71-43-2

$T/K = 298.15$

$V^{(1)}$/cm^3	20.0	30.0	40.0	20.0	20.0	10.0	20.0	10.0	10.0
$\varphi_B^{(1)}$	0.242	0.161	0.121	0.242	0.242	0.265	0.133	0.265	0.361
$V^{(2)}$/cm^3	30.0	40.0	50.0	40.0	50.0	20.0	30.0	30.0	20.0
$\varphi_B^{(2)}$	0.161	0.121	0.097	0.121	0.097	0.133	0.088	0.088	0.181
$\Delta_{dil} H^{12}$/J	0.410	0.226	0.138	0.640	0.778	0.427	0.146	0.573	0.669

$V^{(1)}$/cm^3	20.0	30.0	10.0	10.0	20.0	40.0	20.0	20.0	15.0
$\varphi_B^{(1)}$	0.181	0.121	0.361	0.361	0.115	0.056	0.115	0.318	0.318
$V^{(2)}$/cm^3	30.0	40.0	30.0	40.0	40.0	60.0	60.0	50.0	25.0
$\varphi_B^{(2)}$	0.121	0.090	0.121	0.090	0.056	0.038	0.038	0.127	0.190
$\Delta_{dil} H^{12}$/J	0.255	0.121	0.929	1.046	0.130	0.042	0.176	0.448	0.619

continued

continued

Comments: $\Delta_{dil}H^{12}$ is the extensive quantity obtained for a given total volume change from $V^{(1)}$ to $V^{(2)}$, where $\varphi_B^{(1)}$ and $\varphi_B^{(2)}$ denote the volume fractions of the polymer in the solution before and after the dilution process.

Polymer (B):	**1,4-*cis*-polybutadiene**							**1966KA2**
Characterization:	$M_n/\text{g.mol}^{-1} = 200000$, Bridgestone Tire Co., Ltd.							
Solvent (A):	**toluene**			**C₇H₈**				**108-88-3**

$T/K = 299.15$

$V^{(1)}/\text{cm}^3$	20.0	30.0	40.0	20.0	20.0	16.0	26.0	16.0	50.0
$\varphi_B^{(1)}$	0.318	0.212	0.159	0.318	0.318	0.367	0.225	0.367	0.127
$V^{(2)}/\text{cm}^3$	30.0	40.0	50.0	40.0	50.0	26.0	46.0	46.0	70.0
$\varphi_B^{(2)}$	0.212	0.159	0.127	0.159	0.127	0.225	0.127	0.127	0.091
$\Delta_{dil}H^{12}/\text{J}$	−0.644	−0.259	−0.167	−1.029	−1.197	−0.749	−0.293	−1.025	−0.205

$V^{(1)}/\text{cm}^3$	20.0	15.0	15.0	25.0	15.0	55.0	15.0	25.0
$\varphi_B^{(1)}$	0.318	0.318	0.235	0.141	0.235	0.064	0.235	0.134
$V^{(2)}/\text{cm}^3$	70.0	25.0	25.0	55.0	55.0	85.0	85.0	35.0
$\varphi_B^{(2)}$	0.091	0.190	0.144	0.064	0.064	0.041	0.041	0.096
$\Delta_{dil}H^{12}/\text{J}$	−1.402	−0.335	−0.188	−0.155	−0.339	−0.042	−0.385	−0.088

Comments: $\Delta_{dil}H^{12}$ is the extensive quantity obtained for a given total volume change from $V^{(1)}$ to $V^{(2)}$, where $\varphi_B^{(1)}$ and $\varphi_B^{(2)}$ denote the volume fractions of the polymer in the solution before and after the dilution process.

Polymer (B):	**poly(butadiene-*co*-styrene)**							**1955TA2**
Characterization:	10.0 wt% styrene, synthesized in the laboratory							
Solvent (A):	**benzene**			**C₆H₆**				**71-43-2**

$T/K = 293.65$

w_B	0.1	0.2	0.3	0.4	0.5	0.7	0.8	0.99
$\Delta_M H/(\text{J/g})$	0.33	0.63	0.92	1.26	1.59	2.43	2.55	0.42

Polymer (B):	**poly(butadiene-*co*-styrene)**							**1955TA2**
Characterization:	30.0 wt% styrene, synthesized in the laboratory							
Solvent (A):	**benzene**			**C₆H₆**				**71-43-2**

$T/K = 293.65$

w_B	0.1	0.2	0.3	0.4	0.5	0.685	0.943	0.960
$\Delta_M H/(\text{J/g})$	0.17	0.38	0.54	0.71	0.88	1.21	0.92	0.67

Polymer (B):	**poly(butadiene-*co*-styrene)**							**1955TA2**
Characterization:	60.0 wt% styrene, synthesized in the laboratory							
Solvent (A):	**benzene**			**C₆H₆**				**71-43-2**

$T/K = 293.65$

continued

continued

w_B	0.1	0.2	0.3	0.4	0.5	0.6	0.7	0.9
$\Delta_M H/(J/g)$	0.0	0.0	0.0	0.0	0.0	0.0	0.0	0.0

Polymer (B): **poly(butadiene-*co*-styrene)** **1955TA2**
Characterization: 90.0 wt% styrene, synthesized in the laboratory
Solvent (A): **benzene** **C_6H_6** **71-43-2**

$T/K = 293.65$

w_B	0.10	0.20	0.30	0.44	0.55	0.65	0.79	0.81
$\Delta_M H/(J/g)$	−0.50	−1.00	−1.46	−2.13	−2.72	−3.18	−3.89	−4.02

w_B	0.86	0.93	0.95	0.95	0.96
$\Delta_M H/(J/g)$	−4.22	−3.39	−2.80	−2.59	−1.88

Polymer (B): **poly(butyl methacrylate)** **1987KYO**
Characterization: $M_w/\text{g.mol}^{-1} = 200000$, Scientific Polymer Products, Inc., Ontario, NY
Solvent (A): **2-butanone** **C_4H_8O** **78-93-3**

$T/K = 298.15$

$\varphi_B^{(1)}$	0.1192	0.1192	0.1192	0.1192	0.1192	0.1192	0.1192
$\varphi_B^{(2)}$	0.0119	0.0238	0.0357	0.0477	0.0596	0.0739	0.0834
$\Delta_{dil} H^{12}/(J/\text{mol solvent})$	0.368	0.736	1.068	1.460	1.885	2.279	2.647

$\varphi_B^{(1)}$	0.1192	0.1192
$\varphi_B^{(2)}$	0.0953	0.1072
$\Delta_{dil} H^{12}/(J/\text{mol solvent})$	3.032	3.485

Comments: The table provides the ratio of $\Delta_{dil} H^{12}/\Delta n_A$, i.e., the enthalpy change caused by diluting the primary solution by 1 mol solvent, where $\varphi_B^{(1)}$ denotes the volume fraction of the polymer in the starting solution and $\varphi_B^{(2)}$ denotes the volume fraction after the dilution process.

Polymer (B): **poly(butyl methacrylate)** **1972MAS**
Characterization: $M_n/\text{g.mol}^{-1} = 358$, synthesized in the laboratory
Solvent (A): **trichloromethane** **$CHCl_3$** **67-66-3**

$T/K = 298.15$

$$\Delta_M H/(J/\text{cm}^3) = 4.184 \varphi_B(1 - \varphi_B)[-66.787 + 30.597(1 - 2\varphi_B) + 3.657(1 - 2\varphi_B)^2 - 7.978(1 - 2\varphi_B)^3]$$

Polymer (B): **poly(butyl methacrylate)** **1972MAS**
Characterization: $M_n/\text{g.mol}^{-1} = 500$, synthesized in the laboratory
Solvent (A): **trichloromethane** **$CHCl_3$** **67-66-3**

$T/K = 298.15$

$$\Delta_M H/(J/\text{cm}^3) = 4.184 \varphi_B(1 - \varphi_B)[-71.806 + 27.409(1 - 2\varphi_B) + 9.612(1 - 2\varphi_B)^2]$$

Polymer (B):   **poly(butyl methacrylate)**   **1972MAS**

Characterization: $M_n/\text{g.mol}^{-1} = 642$, synthesized in the laboratory

Solvent (A):   **trichloromethane**   **CHCl$_3$**   **67-66-3**

$T/\text{K} = 298.15$

$$\Delta_M H/(\text{J/cm}^3) = 4.184\,\varphi_B(1 - \varphi_B)[-72.059 + 31.434(1 - 2\varphi_B) + 14.493(1 - 2\varphi_B)^2]$$

Polymer (B):   **polycarbonate bisphenol-A**   **1968BRU**

Characterization: DP = 1.9, synthesized in the laboratory

Solvent (A):   **1,4-dioxane**   **C$_4$H$_8$O$_2$**   **123-91-1**

$T/\text{K} = 310.15$

m_B/g	1.900	1.425	0.950	0.7125
$c_B^{(1)}/(\text{g/cm}^3)$	0.0982	0.0736	0.0491	0.0368
$\Delta_{dil}H^{12}/\text{J}$	−0.3004	−0.1556	−0.0715	−0.0427
$\Delta_{dil}H^{12}/[m_B(c_B^{(1)} - c_B^{(2)})]$	−3.22	−2.97	−3.05	−3.26

Comments: $c_B^{(1)}$ and $c_B^{(2)}$ denote the concentrations of the polymer in the solution before and after the dilution process.

Polymer (B):   **polycarbonate bisphenol-A**   **1968BRU**

Characterization: DP = 3.6, synthesized in the laboratory

Solvent (A):   **1,4-dioxane**   **C$_4$H$_8$O$_2$**   **123-91-1**

$T/\text{K} = 310.15$

m_B/g	1.900	1.425	0.950	0.7125
$c_B^{(1)}/(\text{g/cm}^3)$	0.0982	0.0736	0.0491	0.0368
$\Delta_{dil}H^{12}/\text{J}$	−0.3904	−0.4024	−0.1004	−0.0527
$\Delta_{dil}H^{12}/[m_B(c_B^{(1)} - c_B^{(2)})]$	−4.18	−4.02	−4.31	−4.02

Comments: $c_B^{(1)}$ and $c_B^{(2)}$ denote the concentrations of the polymer in the solution before and after the dilution process.

Polymer (B):   **polycarbonate bisphenol-A**   **1968BRU**

Characterization: DP = 9.9, synthesized in the laboratory

Solvent (A):   **1,4-dioxane**   **C$_4$H$_8$O$_2$**   **123-91-1**

$T/\text{K} = 310.15$

m_B/g	1.900	1.425	0.950	0.7125
$c_B^{(1)}/(\text{g/cm}^3)$	0.0982	0.0736	0.0491	0.0368
$\Delta_{dil}H^{12}/\text{J}$	−0.5309	−0.2891	−0.1305	−0.0690
$\Delta_{dil}H^{12}/[m_B(c_B^{(1)} - c_B^{(2)})]$	−5.69	−5.52	−5.61	−5.27

Comments: $c_B^{(1)}$ and $c_B^{(2)}$ denote the concentrations of the polymer in the solution before and after the dilution process.

Polymer (B):   **polycarbonate bisphenol-A**   **1968BRU**
Characterization: DP = 20.9, synthesized in the laboratory
Solvent (A):   **1,4-dioxane**   **$C_4H_8O_2$**   **123-91-1**

$T/K = 310.15$

m_B/g	1.900	1.425	0.950	0.7125
$c_B^{(1)}$/(g/cm^3)	0.0982	0.0736	0.0491	0.0368
$\Delta_{dil}H^{12}$/J	−0.5230	−0.3046	−0.1268	−0.0732
$\Delta_{dil}H^{12}/[m_B(c_B^{(1)} − c_B^{(2)})]$	−5.61	−5.82	−5.44	−5.56

Comments: $c_B^{(1)}$ and $c_B^{(2)}$ denote the concentrations of the polymer in the solution before and after the dilution process.

Polymer (B):   **poly(dimethylsiloxane)**   **1975TAN**
Characterization: M_n/g.mol^{-1} = 19000, Dow Corning Silicones
Solvent (A):   **2,2-dimethylbutane**   **C_6H_{14}**   **75-83-2**

$T/K = 298.15$

$\Delta_M H/(\text{J/cm}^3) = \varphi_B(1 − \varphi_B)[−0.24 + 0.24(1 − 2\varphi_B)]$

Comments: Experimental data are given only in a figure in the original source.

Polymer (B):   **poly(dimethylsiloxane)**   **1975TAN**
Characterization: M_n/g.mol^{-1} = 19000, Dow Corning Silicones
Solvent (A):   **n-dodecane**   **$C_{12}H_{26}$**   **112-40-3**

$T/K = 298.15$

$\Delta_M H/(\text{J/cm}^3) = \varphi_B(1 − \varphi_B)[8.23 + 2.21(1 − 2\varphi_B) − 0.38(1 − 2\varphi_B)^2]$

Comments: Experimental data are given only in a figure in the original source.

Polymer (B):   **poly(dimethylsiloxane)**   **1975TAN**
Characterization: M_n/g.mol^{-1} = 19000, Dow Corning Silicones
Solvent (A):   **2,2,4,4,6,8,8-heptamethylnonane**   **$C_{16}H_{34}$**   **4390-04-9**

$T/K = 298.15$

$\Delta_M H/(\text{J/cm}^3) = \varphi_B(1 − \varphi_B)[5.23 + 0.37(1 − 2\varphi_B) − 0.79(1 − 2\varphi_B)^2]$

Comments: Experimental data are given only in a figure in the original source.

Polymer (B):   **poly(dimethylsiloxane)**   **1975TAN**
Characterization: M_n/g.mol^{-1} = 19000, Dow Corning Silicones
Solvent (A):   **n-hexadecane**   **$C_{16}H_{34}$**   **544-76-3**

$T/K = 298.15$

$\Delta_M H/(\text{J/cm}^3) = \varphi_B(1 − \varphi_B)[9.50 + 2.26(1 − 2\varphi_B) + 0.88(1 − 2\varphi_B)^2]$

Comments: Experimental data are given only in a figure in the original source.

Polymer (B):   **poly(dimethylsiloxane)**   **1975TAN**
Characterization: M_n/g.mol^{-1} = 19000, Dow Corning Silicones
Solvent (A):   **n-hexane**   **C$_6$H$_{14}$**   **110-54-3**

T/K = 298.15

$\Delta_M H/(\text{J/cm}^3) = \varphi_B(1 - \varphi_B)[1.82 + 0.89(1 - 2\varphi_B) + 0.09(1 - 2\varphi_B)^2]$

Comments: Experimental data are given only in a figure in the original source.

Polymer (B):   **poly(dimethylsiloxane)**   **1954OST**
Characterization: M_η/g.mol^{-1} = 10100
Solvent (A):   **octamethylcyclotetrasiloxane**   **C$_8$H$_{24}$O$_4$Si$_4$**   **556-67-2**

T/K = 298.15 $\varphi_B = 0.50$ $\Delta_M H/(\text{J/cm}^3) = 0.00 \pm 0.04$

Polymer (B):   **poly(dimethylsiloxane)**   **1975TAN**
Characterization: M_n/g.mol^{-1} = 19000, Dow Corning Silicones
Solvent (A):   **n-octane**   **C$_8$H$_{18}$**   **111-65-9**

T/K = 298.15

$\Delta_M H/(\text{J/cm}^3) = \varphi_B(1 - \varphi_B)[4.91 + 1.55(1 - 2\varphi_B) - 0.54(1 - 2\varphi_B)^2]$

Comments: Experimental data are given only in a figure in the original source.

Polymer (B):   **poly(dimethylsiloxane)**   **1975TAN**
Characterization: M_n/g.mol^{-1} = 19000, Dow Corning Silicones
Solvent (A):   **2,2,4,6,6-pentamethylheptane**   **C$_{12}$H$_{26}$**   **13475-82-6**

T/K = 298.15

$\Delta_M H/(\text{J/cm}^3) = \varphi_B(1 - \varphi_B)[4.24 + 0.56(1 - 2\varphi_B) - 0.33(1 - 2\varphi_B)^2]$

Comments: Experimental data are given only in a figure in the original source.

Polymer (B):   **poly(dimethylsiloxane)**   **1975TAN**
Characterization: M_n/g.mol^{-1} = 19000, Dow Corning Silicones
Solvent (A):   **2,2,4-trimethylpentane C$_8$H$_{18}$**   **540-84-1**

T/K = 298.15

$\Delta_M H/(\text{J/cm}^3) = \varphi_B(1 - \varphi_B)[2.87 + 0.69(1 - 2\varphi_B) - 0.02(1 - 2\varphi_B)^2]$

Comments: Experimental data are given only in a figure in the original source.

Polymer (B):   **poly(ethylene glycol)**   **2003CO2**
Characterization: M_n/g.mol^{-1} = 184, M_w/g.mol^{-1} = 204,
                           PEG 200, Fluka AG, Buchs, Switzerland
Solvent (A):   **anisole**   **C$_7$H$_8$O**   **100-66-3**

continued

continued

$T/K = 308.15$

x_B	0.0260	0.0506	0.0740	0.0963	0.1379	0.1758	0.2423	0.2989
$\Delta_M H/(\text{J/mol})$	116.5	193.6	243.1	273.5	298.5	298.9	275.4	253.6

x_B	0.3901	0.4898	0.5613	0.6574	0.7190	0.7933	0.8365	0.8848
$\Delta_M H/(\text{J/mol})$	222.5	179.8	151.6	97.7	66.0	29.1	15.4	8.1

x_B	0.9389
$\Delta_M H/(\text{J/mol})$	4.3

Polymer (B):	**poly(ethylene glycol)**		**2003CO2**
Characterization:	$M_n/\text{g.mol}^{-1} = 275$, $M_w/\text{g.mol}^{-1} = 305$,		
	PEG 300, Fluka AG, Buchs, Switzerland		
Solvent (A):	**anisole**	$\mathbf{C_7H_8O}$	**100-66-3**

$T/K = 308.15$

x_B	0.0140	0.0183	0.0360	0.0531	0.0695	0.1008	0.1301	0.1832
$\Delta_M H/(\text{J/mol})$	40.6	51.1	87.3	111.1	128.8	141.6	141.6	120.5

x_B	0.2301	0.3097	0.4023	0.4729	0.5737	0.6421	0.7291	0.7821
$\Delta_M H/(\text{J/mol})$	97.6	59.3	32.7	11.1	−32.6	−63.0	−84.9	−91.1

x_B	0.8433	0.9150	0.9349
$\Delta_M H/(\text{J/mol})$	−82.8	−44.9	−29.4

Polymer (B):	**poly(ethylene glycol)**		**2003CO2**
Characterization:	$M_n/\text{g.mol}^{-1} = 365$, $M_w/\text{g.mol}^{-1} = 401$,		
	PEG 400, Fluka AG, Buchs, Switzerland		
Solvent (A):	**anisole**	$\mathbf{C_7H_8O}$	**100-66-3**

$T/K = 308.15$

x_B	0.0104	0.0138	0.0273	0.0404	0.0532	0.0777	0.1009	0.1442
$\Delta_M H/(\text{J/mol})$	18.7	23.9	40.0	49.9	53.1	52.8	40.4	4.5

x_B	0.1833	0.2520	0.3358	0.4025	0.5027	0.5740	0.6691	0.7294
$\Delta_M H/(\text{J/mol})$	−34.2	−95.0	−149.3	−174.7	−200.0	−215.1	−223.9	−216.1

x_B	0.8017	0.8899	0.9151
$\Delta_M H/(\text{J/mol})$	−185.7	−107.2	−79.5

Polymer (B):	**poly(ethylene glycol)**		**2003CO2**
Characterization:	$M_n/\text{g.mol}^{-1} = 555$, $M_w/\text{g.mol}^{-1} = 588$,		
	PEG 600, Fluka AG, Buchs, Switzerland		
Solvent (A):	**anisole**	$\mathbf{C_7H_8O}$	**100-66-3**

$T/K = 308.15$

continued

continued

x_B	0.0182	0.0270	0.0357	0.0526	0.0689	0.0999	0.1288	0.1816
$\Delta_M H$/(J/mol)	−33.9	−49.1	−63.2	−88.3	−112.0	−148.3	−180.2	−230.3

x_B	0.2498	0.3074	0.3997	0.4703	0.5711	0.6397	0.7270	0.8419
$\Delta_M H$/(J/mol)	−272.4	−304.2	−357.4	−385.7	−386.8	−369.5	−311.6	−191.2

x_B	0.9244
$\Delta_M H$/(J/mol)	−86.0

Polymer (B):   **poly(ethylene glycol)**   2003CO2
Characterization: M_n/g.mol^{-1} = 275, M_w/g.mol^{-1} = 305,
                         PEG 300, Fluka AG, Buchs, Switzerland
Solvent (A):   **anisole**   **C$_7$H$_8$O**   100-66-3
Polymer (C):   **poly(ethylene glycol)**   2003CO2
Characterization: M_n/g.mol^{-1} = 365, M_w/g.mol^{-1} = 401,
                         PEG 400, Fluka AG, Buchs, Switzerland

Comments: The weight fraction ratio of polymers is kept constant to get a mixture with
              M_w/g.mol^{-1} = 340 and M_n/g.mol^{-1} = 296.

T/K = 308.15

x_{B+C}	0.0452	0.0865	0.1244	0.1593	0.2213	0.2748	0.3624	0.4311
$\Delta_M H$/(J/mol)	50.8	84.8	107.5	121.5	127.1	120.2	86.6	49.8

x_{B+C}	0.5320	0.6305	0.6946	0.7733	0.8198	0.8722	0.9009	0.9317
$\Delta_M H$/(J/mol)	−2.5	−36.0	−50.1	−75.6	−97.1	−117.0	−116.0	−93.9

x_{B+C}	0.9647	0.9733
$\Delta_M H$/(J/mol)	−42.0	−26.1

Polymer (B):   **poly(ethylene glycol)**   2003CO2
Characterization: M_n/g.mol^{-1} = 184, M_w/g.mol^{-1} = 204,
                         PEG 200, Fluka AG, Buchs, Switzerland
Solvent (A):   **anisole**   **C$_7$H$_8$O**   100-66-3
Polymer (C):   **poly(ethylene glycol)**   2003CO2
Characterization: M_n/g.mol^{-1} = 365, M_w/g.mol^{-1} = 401,
                         PEG 400, Fluka AG, Buchs, Switzerland

Comments: The weight fraction ratio of polymers is kept constant to get a mixture with
              M_w/g.mol^{-1} = 361 and M_n/g.mol^{-1} = 296.

T/K = 308.15

x_{B+C}	0.0452	0.0865	0.1244	0.1573	0.2213	0.2748	0.3678	0.4311
$\Delta_M H$/(J/mol)	48.9	81.4	103.2	115.5	122.6	114.2	80.0	46.1

x_{B+C}	0.5320	0.6305	0.6945	0.7733	0.8197	0.8722	0.9009	0.9317
$\Delta_M H$/(J/mol)	−5.8	−39.2	−51.9	−75.3	−96.0	−115.7	−116.3	−99.6

x_{B+C}	0.9647	0.9732
$\Delta_M H$/(J/mol)	−51.4	−36.3

Polymer (B):   **poly(ethylene glycol)**   **2003CO2**

Characterization: $M_n/\text{g.mol}^{-1} = 184$, $M_w/\text{g.mol}^{-1} = 204$,

      PEG 200, Fluka AG, Buchs, Switzerland

Solvent (A):   **anisole**   **C$_7$H$_8$O**   **100-66-3**

Polymer (C):   **poly(ethylene glycol)**   **2003CO2**

Characterization: $M_n/\text{g.mol}^{-1} = 555$, $M_w/\text{g.mol}^{-1} = 588$,

      PEG 600, Fluka AG, Buchs, Switzerland

Comments: The weight fraction ratio of polymers is kept constant to get a mixture with $M_w/\text{g.mol}^{-1} = 437$ and $M_n/\text{g.mol}^{-1} = 289$.

$T/\text{K} = 308.15$

x_{B+C}	0.0452	0.0865	0.1244	0.1593	0.2213	0.2748	0.3624	0.4310
$\Delta_M H/(\text{J/mol})$	44.6	73.9	92.6	102.0	108.1	100.2	69.0	37.9

x_{B+C}	0.5320	0.6304	0.6945	0.7733	0.8197	0.8722	0.9009	0.9317
$\Delta_M H/(\text{J/mol})$	−8.5	−41.9	−58.2	−87.2	−109.8	−128.3	−126.0	−105.0

x_{B+C}	0.9646	0.9732
$\Delta_M H/(\text{J/mol})$	−54.0	−37.0

Polymer (B):   **poly(ethylene glycol)**   **1974BA2**

Characterization: $M_n/\text{g.mol}^{-1} = 200$, Sanyo Kasei Co. Ltd., Japan

Solvent (A):   **benzene**   **C$_6$H$_6$**   **71-43-2**

$T/\text{K} = 298.15$

$V^{(1)}/\text{ml}$	5.0	6.0	10.0	15.0	17.0	18.0	23.0	25.0
$\varphi_B^{(1)}$	1.000	1.000	1.000	0.667	0.588	0.556	0.435	0.400
$V^{(2)}/\text{ml}$	11.0	13.0	15.0	22.0	25.0	23.0	29.0	30.0
$\varphi_B^{(2)}$	0.455	0.462	0.667	0.455	0.400	0.435	0.345	0.333
$\Delta_{dil} H^{12}/\text{J}$	22.38	26.79	23.39	27.00	24.64	19.64	16.96	13.22

$V^{(1)}/\text{ml}$	29.0	30.0	30.0
$\varphi_B^{(1)}$	0.345	0.333	0.333
$V^{(2)}/\text{ml}$	33.0	35.0	36.0
$\varphi_B^{(2)}$	0.303	0.286	0.278
$\Delta_{dil} H^{12}/\text{J}$	10.30	10.22	10.22

Comments: $\Delta_{dil} H^{12}$ is the extensive quantity obtained for a given total volume change from $V^{(1)}$ to $V^{(2)}$, where $\varphi_B^{(1)}$ and $\varphi_B^{(2)}$ denote the volume fractions of the polymer in the solution before and after the dilution process.

Polymer (B):   **poly(ethylene glycol)**   **1976LA1**

Characterization: $M_n/\text{g.mol}^{-1} = 200$,

      fractionated samples supplied by Union Carbide Corp.

Solvent (A):   **benzene**   **C$_6$H$_6$**   **71-43-2**

$T/\text{K} = 321.35$

continued

continued

w_B	0.05	0.1	0.2	0.3	0.4	0.5	0.6	0.7
$\Delta_M H$/(J/mol)	355.6	464.4	556.5	598.3	610.9	564.8	493.7	435.1

w_B	0.8	0.9
$\Delta_M H$/(J/mol)	330.5	117.8

Polymer (B):	**poly(ethylene glycol)**							**1974BA2**
Characterization:	M_n/g.mol^{-1} = 300, Sanyo Kasei Co. Ltd., Japan							
Solvent (A):	**benzene**			$\mathbf{C_6H_6}$				**71-43-2**

T/K = 298.15

$V^{(1)}$/ml	10.0	10.0	15.0	16.0	16.0	17.0	22.0	23.0
$\varphi_B^{(1)}$	1.000	1.000	0.667	0.625	0.625	0.588	0.455	0.435
$V^{(2)}$/ml	15.0	16.0	22.0	24.0	27.0	26.0	31.0	35.0
$\varphi_B^{(2)}$	0.667	0.625	0.455	0.417	0.370	0.385	0.323	0.286
$\Delta_{dil} H^{12}$/J	20.29	20.19	19.79	17.52	18.76	19.55	19.63	19.54

$V^{(1)}$/ml	26.0	27.0	27.0	31.0
$\varphi_B^{(1)}$	0.385	0.370	0.370	0.323
$V^{(2)}$/ml	34.0	36.0	36.0	42.0
$\varphi_B^{(2)}$	0.294	0.278	0.278	0.238
$\Delta_{dil} H^{12}$/J	10.74	9.121	12.48	13.22

Comments: $\Delta_{dil} H^{12}$ is the extensive quantity obtained for a given total volume change from $V^{(1)}$ to $V^{(2)}$, where $\varphi_B^{(1)}$ and $\varphi_B^{(2)}$ denote the volume fractions of the polymer in the solution before and after the dilution process.

Polymer (B):	**poly(ethylene glycol)**							**1966LA1**
Characterization:	M_n/g.mol^{-1} = 335,							
	fractionated samples supplied by Union Carbide Corp.							
Solvent (A):	**benzene**			$\mathbf{C_6H_6}$				**71-43-2**

T/K = 300.05

w_B	0.03236	0.0731	0.1249	0.2207	0.2613	0.4152	0.7309	0.8471
$\Delta_M H$/(J/mol)	39.3	73.6	107.5	143.5	125.5	106.7	−31.0	−58.6

w_B	0.9183	0.9610
$\Delta_M H$/(J/mol)	−79.5	−108.8

Polymer (B):	**poly(ethylene glycol)**		**1979KOL, 1981KOL**
Characterization:	M_n/g.mol^{-1} = 384 ± 10, Hoechst AG, Germany		
Solvent (A):	**benzene**	$\mathbf{C_6H_6}$	**71-43-2**

T/K = 303.15

$c_B^{(1)}$/(g/cm^3) = 0.10 $c_B^{(2)}$/(g/cm^3) = 0.02 $\Delta_{dil} H^{12}$/(J/g polymer) = 13.3

Comments: $c_B^{(1)}$ and $c_B^{(2)}$ denote the concentrations of the polymer in the solution before and after the dilution process.

Polymer (B): poly(ethylene glycol)   1976LA1
Characterization: M_n/g.mol^{-1} = 400,
   fractionated samples supplied by Union Carbide Corp.
Solvent (A): benzene C_6H_6 71-43-2

T/K = 321.35

w_B	0.05	0.1	0.2	0.3	0.4	0.5	0.6	0.7
$\Delta_M H$/(J/mol)	236.4	324.3	372.4	397.5	397.5	370.3	324.3	261.5

w_B	0.8	0.9
$\Delta_M H$/(J/mol)	188.3	104.6

Polymer (B): poly(ethylene glycol)   1979KOL, 1981KOL
Characterization: M_n/g.mol^{-1} = 560 ± 12, Hoechst AG, Germany
Solvent (A): benzene C_6H_6 71-43-2

T/K = 303.15

$c_B^{(1)}$/(g/cm^3)	0.10	0.011	0.014	0.020	0.0205	0.031	0.040	0.045
$c_B^{(2)}$/(g/cm^3)	0.02	0.005	0.005	0.005	0.005	0.005	0.005	0.005
$\Delta_{dil}H^{12}$/(J/g polymer)	4.8	1.8	3.2	-0.85	1.96	2.7	3.0	2.85

$c_B^{(1)}$/(g/cm^3)	0.050	0.070	0.090	0.100	0.120	0.151
$c_B^{(2)}$/(g/cm^3)	0.005	0.005	0.005	0.005	0.005	0.005
$\Delta_{dil}H^{12}$/(J/g polymer)	2.8	5.7	7.1	7.8	8.1	10.0

Comments: $c_B^{(1)}$ and $c_B^{(2)}$ denote the concentrations of the polymer in the solution before and after the dilution process.

Polymer (B): poly(ethylene glycol)   1974BA2
Characterization: M_n/g.mol^{-1} = 600, Sanyo Kasei Co. Ltd., Japan
Solvent (A): benzene C_6H_6 71-43-2

T/K = 298.15

$V^{(1)}$/ml	10.0	17.0	17.0	20.0	22.0	24.0	25.0	26.0
$\varphi_B^{(1)}$	0.400	0.588	0.588	0.200	0.455	0.167	0.400	0.385
$V^{(2)}$/ml	20.0	25.0	26.0	28.0	30.0	34.0	34.0	34.0
$\varphi_B^{(2)}$	0.200	0.400	0.385	0.143	0.333	0.118	0.294	0.294
$\Delta_{dil}H^{12}$/J	8.096	20.51	22.10	1.561	12.25	2.389	10.22	6.996

$V^{(1)}$/ml	27.0	28.0
$\varphi_B^{(1)}$	0.370	0.143
$V^{(2)}$/ml	32.0	33.0
$\varphi_B^{(2)}$	0.313	0.121
$\Delta_{dil}H^{12}$/J	4.619	0.828

Comments: $\Delta_{dil}H^{12}$ is the extensive quantity obtained for a given total volume change from $V^{(1)}$ to $V^{(2)}$, where $\varphi_B^{(1)}$ and $\varphi_B^{(2)}$ denote the volume fractions of the polymer in the solution before and after the dilution process.

Polymer (B):   **poly(ethylene glycol)**   **1966LA1**
Characterization: $M_n/\text{g.mol}^{-1} = 654$,
                           fractionated samples supplied by Union Carbide Corp.
Solvent (A):   **benzene**   **C₆H₆**   **71-43-2**

$T/\text{K} = 300.05$

w_B	0.0268	0.0695	0.1232	0.2191	0.2588	0.4109	0.7179	0.8338
$\Delta_M H/(\text{J/mol})$	55.2	48.9	43.5	7.1	−18.0	−102.5	−489.5	−389.1

w_B	0.9150	0.9528
$\Delta_M H/(\text{J/mol})$	−305.4	−238.5

Polymer (B):   **poly(ethylene glycol)**   **1966LA1**
Characterization: $M_n/\text{g.mol}^{-1} = 920$,
                           fractionated samples supplied by Union Carbide Corp.
Solvent (A):   **benzene**   **C₆H₆**   **71-43-2**

$T/\text{K} = 300.05$

w_B	0.0254	0.0533	0.0694	0.0877	0.1101	0.1234
$\Delta_M H/(\text{J/mol})$	187.0	368.6	525.9	632.2	754.8	828.4

Comments: These data include the enthalpy change caused by melting the crystalline parts of the polymer when mixing the polymer and the solvent to form the solution.

Polymer (B):   **poly(ethylene glycol)**   **1976LA1**
Characterization: $M_n/\text{g.mol}^{-1} = 990$,
                           fractionated samples supplied by Union Carbide Corp.
Solvent (A):   **benzene**   **C₆H₆**   **71-43-2**

$T/\text{K} = 321.35$

w_B	0.05	0.1	0.2	0.3	0.4	0.5	0.6	0.7
$\Delta_M H/(\text{J/mol})$	244.8	324.3	456.1	535.6	569.0	581.6	577.4	552.3

w_B	0.8	0.9
$\Delta_M H/(\text{J/mol})$	460.2	297.1

Polymer (B):   **poly(ethylene glycol)**   **1974BA2**
Characterization: $M_n/\text{g.mol}^{-1} = 1000$, Sanyo Kasei Co. Ltd., Japan
Solvent (A):   **benzene**   **C₆H₆**   **71-43-2**

$T/\text{K} = 298.15$

$V^{(1)}/\text{ml}$	10.0	10.0	18.0	19.0	33.0	33.0	51.0
$\varphi_B^{(1)}$	0.300	0.303	0.128	0.159	0.303	0.303	0.196
$V^{(2)}/\text{ml}$	13.0	16.0	27.0	27.0	39.0	40.0	60.0
$\varphi_B^{(2)}$	0.231	0.189	0.047	0.059	0.256	0.250	0.166
$\Delta_{dil} H^{12}/\text{J}$	2.222	5.029	0.452	0.720	5.678	5.581	4.004

continued

continued

Comments: $\Delta_{dil}H^{12}$ is the extensive quantity obtained for a given total volume change from $V^{(1)}$ to $V^{(2)}$, where $\varphi_B^{(1)}$ and $\varphi_B^{(2)}$ denote the volume fractions of the polymer in the solution before and after the dilution process.

Polymer (B):   **poly(ethylene glycol)**   **1979KOL, 1981KOL**
Characterization: $M_n/\text{g.mol}^{-1} = 1050 \pm 60$, Hoechst AG, Germany
Solvent (A):   **benzene**   **C_6H_6**   **71-43-2**

$T/\text{K} = 303.15$

$c_B^{(1)}/(\text{g/cm}^3)$ 0.10
$c_B^{(2)}/(\text{g/cm}^3)$ 0.02
$\Delta_{dil}H^{12}/(\text{J/g polymer})$ 1.7

Comments: $c_B^{(1)}$ and $c_B^{(2)}$ denote the concentrations of the polymer in the solution before and after the dilution process.

Polymer (B):   **poly(ethylene glycol)**   **1966LA1**
Characterization: $M_n/\text{g.mol}^{-1} = 1350$,
    fractionated samples supplied by Union Carbide Corp.
Solvent (A):   **benzene**   **C_6H_6**   **71-43-2**

$T/\text{K} = 300.05$

w_B	0.0135	0.0256	0.0523	0.0710	0.0857	0.1096
$\Delta_M H/(\text{J/mol})$	186.6	312.1	622.6	826.3	1039	1401

Comments: These data include the enthalpy change caused by melting the crystalline parts of the polymer when mixing the polymer and the solvent to form the solution.

Polymer (B):   **poly(ethylene glycol)**   **1976LA1**
Characterization: $M_n/\text{g.mol}^{-1} = 1460$,
    fractionated samples supplied by Union Carbide Corp.
Solvent (A):   **benzene**   **C_6H_6**   **71-43-2**

$T/\text{K} = 321.35$

w_B	0.05	0.1	0.2	0.3	0.4	0.5	0.6	0.7
$\Delta_M H/(\text{J/mol})$	198.7	297.1	334.7	334.7	261.5	125.5	−31.4	−182.0

w_B	0.8	0.9
$\Delta_M H/(\text{J/mol})$	−307.5	−355.6

Polymer (B):   **poly(ethylene glycol)**   **1979KOL, 1981KOL**
Characterization: $M_n/\text{g.mol}^{-1} = 1610 \pm 90$, Hoechst AG, Germany
Solvent (A):   **benzene**   **C_6H_6**   **71-43-2**

$T/\text{K} = 303.15$

$c_B^{(1)}/(\text{g/cm}^3) = 0.10$ $c_B^{(2)}/(\text{g/cm}^3) = 0.02$ $\Delta_{dil}H^{12}/(\text{J/g polymer}) = -0.35$

Polymer (B):   **poly(ethylene glycol)**   **1979KOL, 1981KOL**

Characterization: $M_n/\text{g.mol}^{-1} = 1940 \pm 150$, Hoechst AG, Germany

Solvent (A):   **benzene**   **C_6H_6**   **71-43-2**

$T/\text{K} = 303.15$

$c_B^{(1)}/(\text{g/cm}^3) = 0.10$ $c_B^{(2)}/(\text{g/cm}^3) = 0.02$ $\Delta_{dil}H^{12}/(\text{J/g polymer}) = -0.90$

Comments: $c_B^{(1)}$ and $c_B^{(2)}$ denote the concentrations of the polymer in the solution before and after the dilution process.

Polymer (B):   **poly(ethylene glycol)**   **1966LA1**

Characterization: $M_n/\text{g.mol}^{-1} = 2585$,

                     fractionated samples supplied by Union Carbide Corp.

Solvent (A):   **benzene**   **C_6H_6**   **71-43-2**

$T/\text{K} = 300.05$

w_B	0.0135	0.0252	0.0484	0.0649	0.0903	0.1081
$\Delta_M H/(\text{J/mol})$	233.0	362.7	694.5	988.3	1407.5	1572

Comments: These data include the enthalpy change caused by melting the crystalline parts of the polymer when mixing the polymer and the solvent to form the solution.

Polymer (B):   **poly(ethylene glycol)**   **1976LA1**

Characterization: $M_n/\text{g.mol}^{-1} = 4150$,

                     fractionated samples supplied by Union Carbide Corp.

Solvent (A):   **benzene**   **C_6H_6**   **71-43-2**

$T/\text{K} = 321.35$

w_B	0.05	0.1	0.2	0.3	0.4	0.5	0.6	0.7
$\Delta_M H/(\text{J/mol})$	878.6	1674	3849	6192	9121	13723	20836	31296

Comments: These data include the enthalpy change caused by melting the crystalline parts of the polymer when mixing the polymer and the solvent to form the solution.

Polymer (B):   **poly(ethylene glycol)**   **1979KOL, 1981KOL**

Characterization: $M_n/\text{g.mol}^{-1} = 4330 \pm 300$, Hoechst AG, Germany

Solvent (A):   **benzene**   **C_6H_6**   **71-43-2**

$T/\text{K} = 303.15$

$c_B^{(1)}/(\text{g/cm}^3) = 0.10$ $c_B^{(2)}/(\text{g/cm}^3) = 0.02$ $\Delta_{dil}H^{12}/(\text{J/g polymer}) = -2.3$

Comments: $c_B^{(1)}$ and $c_B^{(2)}$ denote the concentrations of the polymer in the solution before and after the dilution process.

Polymer (B): **poly(ethylene glycol)** 1979KOL, 1981KOL
Characterization: $M_n/\text{g.mol}^{-1} = 5850 \pm 250$, Hoechst AG, Germany
Solvent (A): **benzene** C_6H_6 71-43-2

$T/\text{K} = 303.15$

$c_B^{(1)}/(\text{g/cm}^3) = 0.10$ $c_B^{(2)}/(\text{g/cm}^3) = 0.02$ $\Delta_{dil}H^{12}/(\text{J/g polymer}) = -2.3$

Comments: $c_B^{(1)}$ and $c_B^{(2)}$ denote the concentrations of the polymer in the solution before and after the dilution process.

Polymer (B): **poly(ethylene glycol)** 1979KOL, 1981KOL
Characterization: $M_n/\text{g.mol}^{-1} = 9950 \pm 700$, Hoechst AG, Germany
Solvent (A): **benzene** C_6H_6 71-43-2

$T/\text{K} = 303.15$

$c_B^{(1)}/(\text{g/cm}^3) = 0.10$ $c_B^{(2)}/(\text{g/cm}^3) = 0.02$ $\Delta_{dil}H^{12}/(\text{J/g polymer}) = -2.0$

Comments: $c_B^{(1)}$ and $c_B^{(2)}$ denote the concentrations of the polymer in the solution before and after the dilution process.

Polymer (B): **poly(ethylene glycol)** 1979KOL, 1981KOL
Characterization: $M_\eta/\text{g.mol}^{-1} = 43400$, Hoechst AG, Germany
Solvent (A): **benzene** C_6H_6 71-43-2

$T/\text{K} = 303.15$

$c_B^{(1)}/(\text{g/cm}^3)$	0.10	0.02	0.04	0.06	0.08	0.10
$c_B^{(2)}/(\text{g/cm}^3)$	0.02	0.005	0.005	0.005	0.005	0.005
$\Delta_{dil}H^{12}/(\text{J/g polymer})$	−2.3	−0.65	−1.43	−2.15	−2.6	−2.9

Comments: $c_B^{(1)}$ and $c_B^{(2)}$ denote the concentrations of the polymer in the solution before and after the dilution process.

Polymer (B): **poly(ethylene glycol)** 2004FRA
Characterization: $M_n/\text{g.mol}^{-1} = 192$, $M_w/\text{g.mol}^{-1} = 223$, $\rho = 1.11300 \text{ g/cm}^3$ (308 K),
PEG 200, Fluka AG, Buchs, Switzerland
Solvent (A): **benzyl alcohol** C_7H_8O 100-51-6

$T/\text{K} = 308.15$

x_B	0.0232	0.0454	0.0666	0.0869	0.1249	0.1598	0.2220	0.2756
$\Delta_M H/(\text{J/mol})$	−157.8	−260.5	−353.0	−417.0	−507.9	−572.7	−627.1	−648.6

x_B	0.3635	0.4613	0.5330	0.6313	0.6954	0.7740	0.8203	0.8726
$\Delta_M H/(\text{J/mol})$	−678.0	−701.3	−705.9	−666.2	−617.0	−482.6	−397.4	−281.7

x_B	0.9320
$\Delta_M H/(\text{J/mol})$	−137.7

Polymer (B):   **poly(ethylene glycol)**   **2004FRA**

Characterization: M_n/g.mol^{-1} = 274, M_w/g.mol^{-1} = 304, ρ = 1.11391 g/cm^3 (308 K),
                          PEG 300, Fluka AG, Buchs, Switzerland

Solvent (A):   **benzyl alcohol**   **C$_7$H$_8$O**   **100-51-6**

T/K = 308.15

x_B	0.0157	0.0309	0.0456	0.0599	0.0872	0.1130	0.1604	0.2029
$\Delta_M H$/(J/mol)	−132.1	−253.6	−350.1	−439.7	−567.2	−667.4	−800.2	−863.3

x_B	0.2764	0.3644	0.4332	0.5341	0.6045	0.6963	0.7535	0.8210
$\Delta_M H$/(J/mol)	−940.4	−974.4	−977.2	−964.3	−905.2	−752.5	−629.9	−464.7

x_B	0.9017
$\Delta_M H$/(J/mol)	−209.7

Polymer (B):   **poly(ethylene glycol)**   **2004FRA**

Characterization: M_n/g.mol^{-1} = 365, M_w/g.mol^{-1} = 402, ρ = 1.11413 g/cm^3 (308 K),
                          PEG 400, Fluka AG, Buchs, Switzerland

Solvent (A):   **benzyl alcohol**   **C$_7$H$_8$O**   **100-51-6**

T/K = 308.15

x_B	0.0120	0.0236	0.0351	0.0462	0.0677	0.0883	0.1269	0.1623
$\Delta_M H$/(J/mol)	−112.6	−213.5	−304.3	−386.6	−527.6	−633.4	−807.9	−918.5

x_B	0.2252	0.3037	0.3676	0.4658	0.5376	0.6356	0.6993	0.7772
$\Delta_M H$/(J/mol)	−1050.3	−1149.8	−1167.2	−1114.0	−1049.7	−915.0	−818.3	−625.3

x_B	0.8746
$\Delta_M H$/(J/mol)	−357.7

Polymer (B):   **poly(ethylene glycol)**   **2004FRA**

Characterization: M_n/g.mol^{-1} = 544, M_w/g.mol^{-1} = 587, ρ = 1.11440 g/cm^3 (308 K),
                          PEG 600, Fluka AG, Buchs, Switzerland

Solvent (A):   **benzyl alcohol**   **C$_7$H$_8$O**   **100-51-6**

T/K = 308.15

x_B	0.0082	0.0163	0.0242	0.0320	0.0472	0.0620	0.0902	0.1167
$\Delta_M H$/(J/mol)	−108.1	−205.2	−293.8	−373.0	−507.9	−618.4	−789.5	−886.5

x_B	0.1655	0.2293	0.2839	0.3730	0.4423	0.5433	0.6133	0.7041
$\Delta_M H$/(J/mol)	−1029.2	−1134.1	−1195.8	−1256.5	−1245.8	−1142.3	−994.8	−745.8

x_B	0.8264
$\Delta_M H$/(J/mol)	−409.3

Polymer (B):   **poly(ethylene glycol)**   **1992WOE**

Characterization: $M_n/\text{g.mol}^{-1} = 200$

Solvent (A):   **1-butanol** $C_4H_{10}O$ **71-36-3**

$T/K = 303.15$

x_B	0.10063	0.17904	0.28938	0.29783	0.39140	0.39489	0.50730	0.59035
$\Delta_M H/(\text{J/mol})$	620.2	906.7	1132.1	1143.0	1218.5	1208.3	1186.3	1095.0

x_B	0.70599	0.79662	0.89785
$\Delta_M H/(\text{J/mol})$	856.3	640.7	349.3

Polymer (B):   **poly(ethylene glycol)**   **1971KAG**

Characterization: $M_n/\text{g.mol}^{-1} = 300$

Solvent (A):   **1-butanol** $C_4H_{10}O$ **71-36-3**

$T/K = 303.15$

$V^{(1)}/\text{cm}^3$	10.0	10.0	19.0	16.0	31.0	30.0	10.0	10.0	25.0
$\varphi_B^{(1)}$	1.000	1.000	0.526	0.625	0.323	0.333	0.900	0.500	0.200
$V^{(2)}/\text{cm}^3$	19.0	16.0	30.0	31.0	42.0	49.0	20.0	25.0	30.0
$\varphi_B^{(2)}$	0.526	0.625	0.333	0.323	0.238	0.204	0.450	0.200	0.167
$\Delta_{dil} H^{12}/\text{J}$	113.81	89.826	81.617	111.26	39.99	64.358	126.02	67.856	6.971

Comments: $\Delta_{dil} H^{12}$ is the extensive quantity obtained for a given total volume change from $V^{(1)}$ to $V^{(2)}$, where $\varphi_B^{(1)}$ and $\varphi_B^{(2)}$ denote the volume fractions of the polymer in the solution before and after the dilution process.

Polymer (B):   **poly(ethylene glycol)**   **1992WOE**

Characterization: $M_n/\text{g.mol}^{-1} = 300$

Solvent (A):   **1-butanol** $C_4H_{10}O$ **71-36-3**

$T/K = 303.15$

x_B	0.09658	0.09928	0.19246	0.27880	0.33830	0.34416	0.39516	0.51524
$\Delta_M H/(\text{J/mol})$	863.6	879.1	1304.9	1491.0	1558.7	1573.9	1578.2	1448.3

x_B	0.54690	0.61617	0.64700	0.71248	0.72347	0.79813	0.80826	0.90078
$\Delta_M H/(\text{J/mol})$	1402.2	1251.1	1187.1	1031.4	978.1	743.1	713.6	368.5

x_B	0.90270
$\Delta_M H/(\text{J/mol})$	351.2

Polymer (B):   **poly(ethylene glycol)**   **1981GON**

Characterization: $M_n/\text{g.mol}^{-1} = 400$

Solvent (A):   **1-butanol** $C_4H_{10}O$ **71-36-3**

$T/K = 298.15$

x_B	0.0147	0.0161	0.6118
$\Delta_M H/(\text{J/mol})$	225.6	247.0	1428.1

continued

continued

$T/K = 313.15$

x_B	0.0151	0.0289	0.5716
$\Delta_M H/$(J/mol)	225.4	243.6	1592.7

Polymer (B):	**poly(ethylene glycol)**		**1971KAG**
Characterization:	$M_n/$g.mol^{-1} = 400		
Solvent (A):	**1-butanol**	**C$_4$H$_{10}$O**	**71-36-3**

$T/K = 303.15$

$V^{(1)}/$cm^3	10.0	18.0	15.0	23.0	30.0	42.0	10.0	20.0	30.0
$\varphi_B^{(1)}$	1.000	0.556	0.667	0.435	0.333	0.238	1.000	0.500	0.333
$V^{(2)}/$cm^3	18.0	23.0	30.0	30.0	38.0	50.0	20.0	35.0	42.0
$\varphi_B^{(2)}$	0.556	0.435	0.333	0.333	0.263	0.200	0.500	0.286	0.238
$\Delta_{dil} H^{12}/$J	80.098	41.350	98.546	36.196	28.326	15.489	102.07	94.035	36.438

Comments: $\Delta_{dil}H^{12}$ is the extensive quantity obtained for a given total volume change from $V^{(1)}$ to $V^{(2)}$, where $\varphi_B^{(1)}$ and $\varphi_B^{(2)}$ denote the volume fractions of the polymer in the solution before and after the dilution process.

Polymer (B):	**poly(ethylene glycol)**		**1971KAG**
Characterization:	$M_n/$g.mol^{-1} = 600		
Solvent (A):	**1-butanol**	**C$_4$H$_{10}$O**	**71-36-3**

$T/K = 303.15$

$V^{(1)}/$cm^3	10.0	10.0	16.0	26.0	20.0	31.0	30.0
$\varphi_B^{(1)}$	1.000	1.000	0.625	0.385	0.500	0.323	0.333
$V^{(2)}/$cm^3	19.0	16.0	20.0	31.0	30.0	50.0	40.0
$\varphi_B^{(2)}$	0.526	0.625	0.500	0.323	0.333	0.200	0.250
$\Delta_{dil} H^{12}/$J	93.613	30.237	28.012	26.121	62.216	56.417	32.782

Comments: $\Delta_{dil}H^{12}$ is the extensive quantity obtained for a given total volume change from $V^{(1)}$ to $V^{(2)}$, where $\varphi_B^{(1)}$ and $\varphi_B^{(2)}$ denote the volume fractions of the polymer in the solution before and after the dilution process.

Polymer (B):	**poly(ethylene glycol)**		**1992WOE**
Characterization:	$M_n/$g.mol^{-1} = 600		
Solvent (A):	**1-butanol**	**C$_4$H$_{10}$O**	**71-36-3**

$T/K = 303.15$

x_B	0.10188	0.10489	0.20575	0.21290	0.29542	0.39931	0.41010	0.42822
$\Delta_M H/$(J/mol)	1556.9	1608.5	2101.7	2129.1	2197.2	2121.5	2142.4	2083.5

x_B	0.55613	0.60835	0.61017	0.70890	0.71122	0.80532	0.81269
$\Delta_M H/$(J/mol)	1748.1	1567.5	1572.6	1262.4	1210.2	818.5	823.4

Polymer (B):   **poly(ethylene glycol)**   **2004COM**

Characterization: M_n/g.mol^{-1} = 184, M_w/g.mol^{-1} = 204, ρ = 1.11300 g/cm^3 (308 K),

                        PEG 200, Fluka AG, Buchs, Switzerland

Solvent (A):   **diethyl carbonate**   **C$_5$H$_{10}$O$_3$**   **105-58-8**

T/K = 308.15

w_B	0.0462	0.0883	0.1285	0.1622	0.2251	0.2792	0.3675	0.4364
$\Delta_M H$/(J/g)	2.2500	3.7604	4.7847	5.3326	6.0421	6.2517	6.2221	5.9893
w_B	0.5374	0.6355	0.6991	0.7771	0.8229	0.8746	0.9029	0.9331
$\Delta_M H$/(J/g)	5.5006	4.8860	4.5453	3.9002	3.4827	2.7554	2.1865	1.6892
w_B	0.9654							
$\Delta_M H$/(J/g)	0.8982							

Polymer (B):   **poly(ethylene glycol)**   **2004COM**

Characterization: M_n/g.mol^{-1} = 365, M_w/g.mol^{-1} = 401, ρ = 1.11437 g/cm^3 (308 K),

                        PEG 400, Fluka AG, Buchs, Switzerland

Solvent (A):   **diethyl carbonate**   **C$_5$H$_{10}$O$_3$**   **105-58-8**

T/K = 308.15

w_B	0.0462	0.0883	0.1269	0.1624	0.2253	0.2794	0.3677	0.4366
$\Delta_M H$/(J/g)	1.2924	2.2705	2.8784	3.2769	3.7787	3.9915	4.0903	4.0021
w_B	0.5377	0.6357	0.6994	0.7773	0.8231	0.8747	0.9030	0.9332
$\Delta_M H$/(J/g)	3.7017	3.3699	3.0450	2.6312	2.3508	1.7339	1.4958	1.0972
w_B	0.9654							
$\Delta_M H$/(J/g)	0.6151							

Polymer (B):   **poly(ethylene glycol)**   **2004COM**

Characterization: M_n/g.mol^{-1} = 544, M_w/g.mol^{-1} = 587, ρ = 1.11440 g/cm^3 (308 K),

                        PEG 600, Fluka AG, Buchs, Switzerland

Solvent (A):   **diethyl carbonate**   **C$_5$H$_{10}$O$_3$**   **105-58-8**

T/K = 308.15

w_B	0.0462	0.0884	0.1270	0.1624	0.2253	0.2794	0.3678	0.4367
$\Delta_M H$/(J/g)	0.9125	1.5337	1.9289	2.1967	2.5003	2.6669	2.7635	2.6968
w_B	0.5378	0.6358	0.6994	0.7773	0.8231	0.8747	0.9030	0.9332
$\Delta_M H$/(J/g)	2.5771	2.3642	2.1708	1.8881	1.6332	1.2845	1.0263	0.7187
w_B	0.9614							
$\Delta_M H$/(J/g)	0.4500							

Polymer (B):   **poly(ethylene glycol)**   **2002COM**
Characterization: M_n/g.mol^{-1} = 192, M_w/g.mol^{-1} = 224,
                          Fluka AG, Buchs, Switzerland
Solvent (A):   **1,2-dimethoxyethane   C$_4$H$_{10}$O$_2$**   **110-71-4**

T/K = 298.15

x_B	0.0248	0.0484	0.0710	0.0924	0.1325	0.1692	0.2340	0.2894
$\Delta_M H$/(J/mol)	67.0	126.5	176.0	220.8	290.3	339.8	403.7	429.0

x_B	0.3793	0.4782	0.5500	0.6470	0.7097	0.7857	0.8302	0.8800
$\Delta_M H$/(J/mol)	431.7	393.3	351.3	288.8	248.7	201.1	160.2	130.0

x_B	0.9362
$\Delta_M H$/(J/mol)	79.1

Polymer (B):   **poly(ethylene glycol)**   **2002COM**
Characterization: M_n/g.mol^{-1} = 408, M_w/g.mol^{-1} = 447,
                          Fluka AG, Buchs, Switzerland
Solvent (A):   **1,2-dimethoxyethane   C$_4$H$_{10}$O$_2$**   **110-71-4**

T/K = 298.15

x_B	0.0119	0.0234	0.0347	0.0458	0.0672	0.0876	0.1259	0.1610
$\Delta_M H$/(J/mol)	40.3	75.2	106.9	134.6	180.6	218.4	262.4	297.1

x_B	0.2236	0.3016	0.3654	0.4635	0.5353	0.6334	0.6973	0.7756
$\Delta_M H$/(J/mol)	317.4	330.7	316.8	297.3	270.3	227.6	199.1	145.5

x_B	0.8736	0.9120
$\Delta_M H$/(J/mol)	73.0	45.1

Polymer (B):   **poly(ethylene glycol)**   **2002COM**
Characterization: M_n/g.mol^{-1} = 192, M_w/g.mol^{-1} = 224,
                          Fluka AG, Buchs, Switzerland
Solvent (A):   **dimethoxymethane   C$_3$H$_8$O$_2$**   **109-87-5**

T/K = 298.15

x_B	0.0212	0.0416	0.0611	0.0798	0.1151	0.1478	0.2065	0.2575
$\Delta_M H$/(J/mol)	140.0	256.0	348.5	426.4	534.9	612.9	700.2	733.9

x_B	0.3423	0.4384	0.5100	0.6095	0.6755	0.7574	0.8063	0.8620
$\Delta_M H$/(J/mol)	737.8	706.8	650.7	555.3	466.9	339.2	247.7	159.3

x_B	0.9259
$\Delta_M H$/(J/mol)	61.0

Polymer (B):   **poly(ethylene glycol)**   **2002COM**

Characterization: M_n/g.mol^{-1} = 408, M_w/g.mol^{-1} = 447,

                          Fluka AG, Buchs, Switzerland

Solvent (A):   **dimethoxymethane**   **C$_3$H$_8$O$_2$**   **109-87-5**

T/K = 298.15

x_B	0.0101	0.0200	0.0297	0.0393	0.0601	0.0756	0.1092	0.1405
$\Delta_M H$/(J/mol)	69.2	130.1	185.0	235.0	332.2	384.4	479.9	534.2

x_B	0.1969	0.2689	0.3290	0.4238	0.4951	0.5954	0.6623	0.7464
$\Delta_M H$/(J/mol)	581.9	575.3	559.5	518.2	490.9	435.4	381.3	297.3

x_B	0.8548	0.8983
$\Delta_M H$/(J/mol)	147.1	87.0

Polymer (B):   **poly(ethylene glycol)**   **2003CO1**

Characterization: M_n/g.mol^{-1} = 192, M_w/g.mol^{-1} = 223,

                          PEG 200, Fluka AG, Buchs, Switzerland

Solvent (A):   **dimethylsulfoxide**   **C$_2$H$_6$OS**   **67-68-5**

T/K = 308.15

x_B	0.0171	0.0336	0.0496	0.0650	0.0944	0.1221	0.1726	0.2175
$\Delta_M H$/(J/mol)	−104.0	−197.9	−277.9	−346.7	−463.0	−556.8	−678.8	−762.2

x_B	0.2944	0.3850	0.4548	0.5559	0.6253	0.7146	0.7694	0.8335
$\Delta_M H$/(J/mol)	−844.5	−886.4	−900.6	−875.5	−822.6	−729.8	−638.4	−504.6

x_B	0.9092
$\Delta_M H$/(J/mol)	−301.1

Polymer (B):   **poly(ethylene glycol)**   **2003CO1**

Characterization: M_n/g.mol^{-1} = 274, M_w/g.mol^{-1} = 304,

                          PEG 300, Fluka AG, Buchs, Switzerland

Solvent (A):   **dimethylsulfoxide**   **C$_2$H$_6$OS**   **67-68-5**

T/K = 308.15

x_B	0.0120	0.0238	0.0353	0.0465	0.0681	0.0888	0.1276	0.1631
$\Delta_M H$/(J/mol)	−93.3	−181.7	−248.2	−316.9	−420.1	−496.9	−606.3	−676.7

x_B	0.2263	0.3051	0.3691	0.4674	0.5392	0.6371	0.7006	0.7783
$\Delta_M H$/(J/mol)	−763.3	−852.1	−906.8	−971.5	−981.6	−925.8	−836.8	−666.2

x_B	0.8753	0.9035
$\Delta_M H$/(J/mol)	−390.7	−310.7

Polymer (B):   **poly(ethylene glycol)**   **2003CO1**
Characterization: M_n/g.mol^{-1} = 365, M_w/g.mol^{-1} = 402,
                          PEG 400, Fluka AG, Buchs, Switzerland
Solvent (A):   **dimethylsulfoxide   C$_2$H$_6$OS**   **67-68-5**

T/K = 308.15

x_B	0.0091	0.0180	0.0267	0.0353	0.0521	0.0682	0.0990	0.1277
$\Delta_M H$/(J/mol)	−70.0	−149.2	−219.9	−274.8	−354.6	−441.7	−554.4	−625.3

x_B	0.1801	0.2479	0.2923	0.3973	0.4677	0.5687	0.6374	0.7250
$\Delta_M H$/(J/mol)	−717.6	−792.5	−849.6	−949.5	−1010.3	−1035.6	−991.6	−867.6

x_B	0.8406	0.8755
$\Delta_M H$/(J/mol)	−565.9	−460.0

Polymer (B):   **poly(ethylene glycol)**   **2003CO1**
Characterization: M_n/g.mol^{-1} = 544, M_w/g.mol^{-1} = 587,
                          PEG 600, Fluka AG, Buchs, Switzerland
Solvent (A):   **dimethylsulfoxide   C$_2$H$_6$OS**   **67-68-5**

T/K = 308.15

x_B	0.0060	0.0119	0.0178	0.0236	0.0349	0.0460	0.0675	0.0880
$\Delta_M H$/(J/mol)	−50.1	−111.7	−167.9	−204.5	−283.5	−346.9	−442.6	−517.2

x_B	0.1264	0.1784	0.2245	0.3027	0.3667	0.4648	0.5366	0.6346
$\Delta_M H$/(J/mol)	−607.2	−683.2	−745.8	−861.9	−971.8	−1071.9	−1081.9	−999.2

x_B	0.7765	0.8224
$\Delta_M H$/(J/mol)	−685.9	−580.0

Polymer (B):   **poly(ethylene glycol)**   **2003OTT**
Characterization: M_n/g.mol^{-1} = 192, M_w/g.mol^{-1} = 224, ρ = 1.12098 g/cm^3 (298 K),
                          PEG 200, Fluka AG, Buchs, Switzerland
Solvent (A):   **1,4-dioxane   C$_4$H$_8$O$_2$**   **123-91-1**

T/K = 288.15

x_B	0.0203	0.0399	0.0587	0.0767	0.1108	0.1425	0.1995	0.2493
$\Delta_M H$/(J/mol)	58.3	106.4	147.4	182.7	236.6	273.2	319.2	336.4

x_B	0.3326	0.4279	0.4992	0.5993	0.6660	0.7494	0.7995	0.8568
$\Delta_M H$/(J/mol)	369.5	360.1	353.8	321.6	284.6	223.8	180.9	131.8

x_B	0.9229
$\Delta_M H$/(J/mol)	65.1

T/K = 298.15

x_B	0.0204	0.0400	0.0589	0.0770	0.1112	0.1429	0.2001	0.2501
$\Delta_M H$/(J/mol)	82.6	150.4	200.1	249.6	318.9	366.7	408.5	431.9

continued

continued

x_B	0.3336	0.4289	0.5002	0.6002	0.6668	0.7502	0.8001	0.8573
$\Delta_M H/(J/mol)$	438.7	426.3	417.5	379.2	342.1	284.4	224.6	170.4

x_B	0.9231
$\Delta_M H/(J/mol)$	87.6

$T/K = 313.15$

x_B	0.0205	0.0402	0.0592	0.0774	0.1118	0.1436	0.2010	0.2511
$\Delta_M H/(J/mol)$	90.2	170.1	235.3	295.2	369.0	438.3	509.2	543.0

x_B	0.3348	0.4303	0.5016	0.6016	0.6681	0.7512	0.8010	0.8580
$\Delta_M H/(J/mol)$	557.7	535.5	512.0	463.1	423.1	360.4	312.6	245.2

x_B	0.9235
$\Delta_M H/(J/mol)$	147.4

Polymer (B):	**poly(ethylene glycol)**						**1965LAK**	
Characterization:	$M_n/g.mol^{-1} = 335$,							
	fractionated samples supplied by Union Carbide Corp.							
Solvent (A):	**1,4-dioxane**		$C_4H_8O_2$				**123-91-1**	

$T/K = 300.05$

w_B	0.02728	0.08105	0.1245	0.1652	0.2027	0.6809	0.7219	0.7759
$\Delta_M H/(J/mol)$	17.6	17.6	23.4	26.8	18.8	175.7	175.3	163.2

w_B	0.8453	0.9164
$\Delta_M H/(J/mol)$	30.5	−88.7

Polymer (B):	**poly(ethylene glycol)**						**1965LAK**	
Characterization:	$M_n/g.mol^{-1} = 654$,							
	fractionated samples supplied by Union Carbide Corp.							
Solvent (A):	**1,4-dioxane**		$C_4H_8O_2$				**123-91-1**	

$T/K = 300.05$

w_B	0.02751	0.05294	0.07815	0.1228	0.1640	0.7823	0.8388	0.9160
$\Delta_M H/(J/mol)$	21.3	27.6	74.5	28.5	19.2	36.4	−46.0	−50.2

Polymer (B):	**poly(ethylene glycol)**					**1965LAK**	
Characterization:	$M_n/g.mol^{-1} = 920$,						
	fractionated samples supplied by Union Carbide Corp.						
Solvent (A):	**1,4-dioxane**		$C_4H_8O_2$			**123-91-1**	

$T/K = 300.05$

w_B	0.03485	0.05360	0.07803	0.1280	0.1671	0.1973
$\Delta_M H/(J/mol)$	109.2	355.2	655.2	1038	1409	1837

Comments: These data include the enthalpy change caused by melting the crystalline parts of the polymer when mixing the polymer and the solvent to form the solution.

Polymer (B):	**poly(ethylene glycol)**					**1965LAK**

Characterization: $M_n/\text{g.mol}^{-1} = 1350$,
fractionated samples supplied by Union Carbide Corp.

Solvent (A):   **1,4-dioxane** $C_4H_8O_2$ **123-91-1**

$T/K = 300.05$

w_B	0.02912	0.05910	0.08141	0.1358	0.1535	0.1977
$\Delta_M H/(\text{J/mol})$	361.1	765.7	1016	1676	2122	2621

Comments: These data include the enthalpy change caused by melting the crystalline parts of the polymer when mixing the polymer and the solvent to form the solution.

Polymer (B):	**poly(ethylene glycol)**					**1965LAK**

Characterization: $M_n/\text{g.mol}^{-1} = 2825$,
fractionated samples supplied by Union Carbide Corp.

Solvent (A):   **1,4-dioxane** $C_4H_8O_2$ **123-91-1**

$T/K = 300.05$

w_B	0.01870	0.02818	0.04050	0.05287	0.06606	0.07806	0.1228	0.1642
$\Delta_M H/(\text{J/mol})$	21.7	412.1	493.7	719.6	962.3	1280	2163	2983

w_B	0.1923
$\Delta_M H/(\text{J/mol})$	3347

Comments: These data include the enthalpy change caused by melting the crystalline parts of the polymer when mixing the polymer and the solvent to form the solution.

Polymer (B):	**poly(ethylene glycol)**					**2003OTT**

Characterization: $M_n/\text{g.mol}^{-1} = 192$, $M_w/\text{g.mol}^{-1} = 224$, $\rho = 1.12098$ g/cm^3 (298 K),
PEG 200, Fluka AG, Buchs, Switzerland

Solvent (A):   **1,3-dioxolane** $C_3H_6O_2$ **646-06-0**

$T/K = 288.15$

x_B	0.0167	0.0328	0.0484	0.0635	0.0923	0.1194	0.1690	0.2132
$\Delta_M H/(\text{J/mol})$	45.4	86.0	128.1	162.0	213.5	263.7	334.5	362.0

x_B	0.2891	0.3790	0.4485	0.5495	0.6193	0.7093	0.7649	0.8299
$\Delta_M H/(\text{J/mol})$	393.0	382.0	356.6	303.3	264.3	211.8	173.0	128.1

x_B	0.9071
$\Delta_M H/(\text{J/mol})$	75.3

$T/K = 298.15$

x_B	0.0167	0.0329	0.0486	0.0637	0.0927	0.1198	0.1696	0.2140
$\Delta_M H/(\text{J/mol})$	77.0	143.1	198.9	244.0	314.9	368.9	417.4	436.5

x_B	0.2901	0.3800	0.4496	0.5507	0.6203	0.7103	0.7657	0.8306
$\Delta_M H/(\text{J/mol})$	442.5	412.3	390.1	342.3	308.4	245.8	197.6	138.3

x_B	0.9075
$\Delta_M H/(\text{J/mol})$	59.4

continued

continued

$T/K = 313.15$

x_B	0.0168	0.0331	0.0489	0.0641	0.0932	0.1206	0.1706	0.2151
$\Delta_M H/$(J/mol)	89.0	164.2	228.9	283.3	368.3	423.2	499.7	532.5

x_B	0.2914	0.3817	0.4513	0.5524	0.6220	0.7117	0.7669	0.8316
$\Delta_M H/$(J/mol)	545.5	523.8	488.1	422.3	374.7	285.8	232.1	157.3

x_B	0.9080
$\Delta_M H/$(J/mol)	71.0

Polymer (B):	**poly(ethylene glycol)**	**2005BIG**
Characterization:	$M_n/$g.mol^{-1} = 192, $M_w/$g.mol^{-1} = 224, ρ = 1.11284 g/cm^3 (308 K), PEG 200, Fluka AG, Buchs, Switzerland	
Solvent (A):	**ethanol** $\quad$ **C$_2$H$_6$O**	**64-17-5**

$T/K = 308.15$

x_B	0.0133	0.0263	0.0389	0.0512	0.0749	0.0974	0.1393	0.1775
$\Delta_M H/$(J/mol)	70.2	131.9	191.5	230.5	316.0	363.4	441.7	486.0

x_B	0.2446	0.3270	0.3930	0.4927	0.5643	0.6602	0.7214	0.7953
$\Delta_M H/$(J/mol)	521.5	545.7	569.3	567.0	556.6	500.3	439.4	343.3

x_B	0.8860	0.9120
$\Delta_M H/$(J/mol)	176.3	126.3

Polymer (B):	**poly(ethylene glycol)**	**2005BIG**
Characterization:	$M_n/$g.mol^{-1} = 192, $M_w/$g.mol^{-1} = 224, ρ = 1.11284 g/cm^3 (308 K), PEG 200, Fluka AG, Buchs, Switzerland	
Solvent (A):	**ethanol** $\quad$ **C$_2$H$_6$O**	**64-17-5**
Polymer (C):	**poly(ethylene glycol)**	
Characterization:	$M_n/$g.mol^{-1} = 365, $M_w/$g.mol^{-1} = 402, ρ = 1.11413 g/cm^3 (308 K), PEG 400, Fluka AG, Buchs, Switzerland	

Comments: The weight fraction ratio of polymers w_B/w_C = 300/700 is kept constant to get a mixture with $M_w/$g.mol^{-1} = 360 and $M_n/$g.mol^{-1} = 295.

$T/K = 308.15$

x_{B+C}	0.0141	0.0319	0.0471	0.0618	0.0899	0.1163	0.1649	0.2084
$\Delta_M H/$(J/mol)	120.2	255.5	360.0	443.5	583.6	701.5	824.1	899.8

x_{B+C}	0.2832	0.3722	0.4414	0.5424	0.6124	0.7033	0.7599	0.8258
$\Delta_M H/$(J/mol)	961.9	1029.1	1033.8	1011.1	952.9	828.6	700.4	538.0

x_{B+C}	0.9046
$\Delta_M H/$(J/mol)	288.1

Polymer (B):   **poly(ethylene glycol)**   **2005BIG**
Characterization: M_n/g.mol^{-1} = 192, M_w/g.mol^{-1} = 224, ρ = 1.11284 g/cm^3 (308 K),
                         PEG 200, Fluka AG, Buchs, Switzerland
Solvent (A):   **ethanol**   **C$_2$H$_6$O**   **64-17-5**
Polymer (C):   **poly(ethylene glycol)**
Characterization: M_n/g.mol^{-1} = 554, M_w/g.mol^{-1} = 587, ρ = 1.11396 g/cm^3 (308 K),
                         PEG 600, Fluka AG, Buchs, Switzerland

Comments: The weight fraction ratio of polymers w_B/w_C = 550/550 is kept constant to get a mixture
with M_w/g.mol^{-1} = 444 and M_n/g.mol^{-1} = 290.

T/K = 308.15

x_{B+C}	0.0164	0.0323	0.0477	0.0625	0.0910	0.1177	0.1668	0.2106
$\Delta_M H$/(J/mol)	139.6	256.7	362.4	459.4	577.0	672.7	789.2	861.3
x_{B+C}	0.2859	0.3753	0.4447	0.5457	0.6156	0.7061	0.7624	0.8277
$\Delta_M H$/(J/mol)	905.1	958.4	974.4	971.5	931.8	806.3	684.3	509.4
x_{B+C}	0.9057							
$\Delta_M H$/(J/mol)	258.6							

Polymer (B):   **poly(ethylene glycol)**   **1976LA1**
Characterization: M_n/g.mol^{-1} = 200,
                         fractionated samples supplied by Union Carbide Corp.
Solvent (A):   **ethanol**   **C$_2$H$_6$O**   **64-17-5**

T/K = 321.35

w_B	0.05	0.1	0.2	0.3	0.4	0.5	0.6	0.7
$\Delta_M H$/(J/mol)	46.0	75.3	159	226	276	276	339	377
w_B	0.8	0.9						
$\Delta_M H$/(J/mol)	364	251						

Polymer (B):   **poly(ethylene glycol)**   **2005BIG**
Characterization: M_n/g.mol^{-1} = 274, M_w/g.mol^{-1} = 304, ρ = 1.11358 g/cm^3 (308 K),
                         PEG 300, Fluka AG, Buchs, Switzerland
Solvent (A):   **ethanol**   **C$_2$H$_6$O**   **64-17-5**

T/K = 308.15

x_B	0.0089	0.0177	0.0264	0.0348	0.0514	0.0674	0.0978	0.1262
$\Delta_M H$/(J/mol)	70.3	132.1	188.1	238.8	334.4	393.5	495.5	562.6
x_B	0.1781	0.2454	0.3024	0.3940	0.4643	0.5653	0.6342	0.7223
$\Delta_M H$/(J/mol)	651.1	685.7	736.6	745.3	728.6	688.6	599.8	486.5
x_B	0.8387	0.9611						
$\Delta_M H$/(J/mol)	285.4	82.4						

Polymer (B):   **poly(ethylene glycol)**   **2005BIG**

Characterization: $M_n/\text{g.mol}^{-1} = 274$, $M_w/\text{g.mol}^{-1} = 304$, $\rho = 1.11358$ g/cm^3 (308 K),

                    PEG 300, Fluka AG, Buchs, Switzerland

Solvent (A):   **ethanol**   **C$_2$H$_6$O**   **64-17-5**

Polymer (C):   **poly(ethylene glycol)**

Characterization: $M_n/\text{g.mol}^{-1} = 365$, $M_w/\text{g.mol}^{-1} = 402$, $\rho = 1.11413$ g/cm^3 (308 K),

                    PEG 400, Fluka AG, Buchs, Switzerland

Comments: The weight fraction ratio of polymers $w_B/w_C = 600/270$ is kept constant to get a mixture with $M_w/\text{g.mol}^{-1} = 340$ and $M_n/\text{g.mol}^{-1} = 296$.

$T/\text{K} = 308.15$

x_{B+C}	0.0161	0.0318	0.0469	0.0616	0.0896	0.1160	0.1645	0.2078
$\Delta_M H/(\text{J/mol})$	137.2	270.4	357.5	431.6	580.7	675.0	802.7	878.6

x_{B+C}	0.2825	0.3714	0.4405	0.5415	0.6116	0.7026	0.7593	0.8253
$\Delta_M H/(\text{J/mol})$	950.2	999.4	1002.3	982.8	935.6	806.1	691.7	522.2

x_{B+C}	0.9043
$\Delta_M H/(\text{J/mol})$	266.0

Polymer (B):   **poly(ethylene glycol)**   **1971KAG**

Characterization: $M_n/\text{g.mol}^{-1} = 300$

Solvent (A):   **ethanol**   **C$_2$H$_6$O**   **64-17-5**

$T/\text{K} = 303.15$

$V^{(1)}/\text{cm}^3$	5.0	10.0	10.0	10.0	15.0	14.0	29.0	25.0	65.0
$\varphi_B^{(1)}$	1.000	1.000	1.000	1.000	0.667	0.714	0.345	0.400	0.154
$V^{(2)}/\text{cm}^3$	10.0	15.0	17.0	14.0	20.0	29.0	34.0	31.0	80.0
$\varphi_B^{(2)}$	0.500	0.667	0.588	0.714	0.500	0.345	0.294	0.323	0.125
$\Delta_{dil} H^{12}/\text{J}$	50.38	60.62	80.40	56.69	43.26	110.12	21.33	27.64	13.06

Comments: $\Delta_{dil} H^{12}$ is the extensive quantity obtained for a given total volume change from $V^{(1)}$ to $V^{(2)}$, where $\varphi_B^{(1)}$ and $\varphi_B^{(2)}$ denote the volume fractions of the polymer in the solution before and after the dilution process.

Polymer (B):   **poly(ethylene glycol)**   **1966LA1**

Characterization: $M_n/\text{g.mol}^{-1} = 335$,

                    fractionated samples supplied by Union Carbide Corp.

Solvent (A):   **ethanol**   **C$_2$H$_6$O**   **64-17-5**

$T/\text{K} = 300.05$

w_B	0.0582	0.1171	0.2142	0.2795	0.3450	0.5053	0.7344	0.8453
$\Delta_M H/(\text{J/mol})$	53.1	−32.2	−25.5	38.9	25.1	124	409	448

w_B	0.9073	0.9624
$\Delta_M H/(\text{J/mol})$	337	201

Polymer (B):   **poly(ethylene glycol)**   **2005BIG**
Characterization: $M_n/\text{g.mol}^{-1} = 365$, $M_w/\text{g.mol}^{-1} = 402$, $\rho = 1.11413$ g/cm^3 (308 K),
                          PEG 400, Fluka AG, Buchs, Switzerland
Solvent (A):   **ethanol**   **C$_2$H$_6$O**   **64-17-5**

$T/\text{K} = 308.15$

x_B	0.0068	0.0135	0.0202	0.0267	0.0396	0.0521	0.0761	0.0990
$\Delta_M H/(\text{J/mol})$	75.1	145.3	208.7	265.6	374.3	437.6	556.5	635.4

x_B	0.1415	0.1983	0.2480	0.3309	0.3974	0.4973	0.5688	0.6643
$\Delta_M H/(\text{J/mol})$	727.1	793.0	852.4	901.3	932.0	922.2	851.7	688.2

x_B	0.7983	0.8408	0.9600
$\Delta_M H/(\text{J/mol})$	394.7	335.1	143.1

Polymer (B):   **poly(ethylene glycol)**   **1971KAG**
Characterization: $M_n/\text{g.mol}^{-1} = 400$
Solvent (A):   **ethanol**   **C$_2$H$_6$O**   **64-17-5**

$T/\text{K} = 303.15$

$V^{(1)}/\text{cm}^3$	10.0	20.0	15.0	40.0	55.0	55.0	5.0	10.0	10.0
$\varphi_B^{(1)}$	1.000	0.500	0.667	0.250	0.182	0.182	1.000	1.000	1.000
$V^{(2)}/\text{cm}^3$	20.0	25.0	30.0	55.0	65.0	75.0	9.0	17.0	20.0
$\varphi_B^{(2)}$	0.500	0.400	0.333	0.182	0.154	0.133	0.556	0.588	0.500
$\Delta_{dil}H^{12}/\text{J}$	113.34	29.52	89.83	25.18	9.740	21.55	50.23	77.86	106.02

Comments: $\Delta_{dil}H^{12}$ is the extensive quantity obtained for a given total volume change from $V^{(1)}$ to $V^{(2)}$, where $\varphi_B^{(1)}$ and $\varphi_B^{(2)}$ denote the volume fractions of the polymer in the solution before and after the dilution process.

Polymer (B):   **poly(ethylene glycol)**   **1976LA1**
Characterization: $M_n/\text{g.mol}^{-1} = 400$,
                          fractionated samples supplied by Union Carbide Corp.
Solvent (A):   **ethanol**   **C$_2$H$_6$O**   **64-17-5**

$T/\text{K} = 321.35$

w_B	0.05	0.1	0.2	0.3	0.4	0.5	0.6	0.7
$\Delta_M H/(\text{J/mol})$	33.5	54.4	100	201	314	448	602	753

w_B	0.8	0.9
$\Delta_M H/(\text{J/mol})$	820	715

Polymer (B):   **poly(ethylene glycol)**   **1981GON**
Characterization: $M_n/\text{g.mol}^{-1} = 400$
Solvent (A):   **ethanol**   **C$_2$H$_6$O**   **64-17-5**

continued

continued

$T/K = 298.15$

x_B	0.0086	0.0093	0.0104	0.0191	0.2546		
$\Delta_M H/$(J/mol)	82.73	85.89	89.91	90.18	534.9		

$T/K = 313.15$

x_B	0.0096	0.0104	0.0196	0.0197	0.0201	0.1432	0.2572
$\Delta_M H/$(J/mol)	64.56	71.77	77.12	79.81	81.53	273.2	613.2

Polymer (B):   **poly(ethylene glycol)**   **2005BIG**

Characterization: $M_n/\text{g.mol}^{-1} = 554$, $M_w/\text{g.mol}^{-1} = 587$, $\rho = 1.11396$ g/cm^3 (308 K),
PEG 600, Fluka AG, Buchs, Switzerland

Solvent (A):   **ethanol**   **C$_2$H$_6$O**   **64-17-5**

$T/K = 308.15$

x_B	0.0047	0.0093	0.0138	0.0184	0.0277	0.0361	0.0532	0.0697
$\Delta_M H/$(J/mol)	70.0	135.1	196.8	252.8	347.6	438.9	576.6	677.7

x_B	0.1011	0.1444	0.1836	0.2523	0.3103	0.4029	0.4736	0.5744
$\Delta_M H/$(J/mol)	811.9	923.7	985.0	1094.6	1194.1	1266.6	1263.6	1091.0

x_B	0.6514	0.7297	0.7826	0.9100				
$\Delta_M H/$(J/mol)	849.1	581.5	431.2	239.0				

Polymer (B):   **poly(ethylene glycol)**   **1971KAG**

Characterization: $M_n/\text{g.mol}^{-1} = 600$

Solvent (A):   **ethanol**   **C$_2$H$_6$O**   **64-17-5**

$T/K = 303.15$

$V^{(1)}/\text{cm}^3$	10.0	10.0	15.0	19.0	25.0
$\varphi_B^{(1)}$	1.000	1.000	0.667	0.526	0.400
$V^{(2)}/\text{cm}^3$	16.0	17.0	20.0	25.0	40.0
$\varphi_B^{(2)}$	0.667	0.588	0.500	0.400	0.250
$\Delta_{dil} H^{12}/$J	53.01	76.90	40.97	27.50	40.77

Comments: $\Delta_{dil} H^{12}$ is the extensive quantity obtained for a given total volume change from $V^{(1)}$ to $V^{(2)}$, where $\varphi_B^{(1)}$ and $\varphi_B^{(2)}$ denote the volume fractions of the polymer in the solution before and after the dilution process.

Polymer (B):   **poly(ethylene glycol)**   **1966LA1**

Characterization: $M_n/\text{g.mol}^{-1} = 654$,
fractionated samples supplied by Union Carbide Corp.

Solvent (A):   **ethanol**   **C$_2$H$_6$O**   **64-17-5**

$T/K = 300.05$

w_B	0.0523	0.1153	0.2085	0.2816	0.3308	0.5024	0.7972	0.8445
$\Delta_M H/$(J/mol)	16.3	18.4	41.8	84.1	56.9	171	690	674

continued

continued

w_B	0.9134	0.9629
$\Delta_M H$/(J/mol)	707	448

Polymer (B):	**poly(ethylene glycol)**					**1966LA1**

Characterization: M_n/g.mol^{-1} = 920,
                        fractionated samples supplied by Union Carbide Corp.

Solvent (A):	**ethanol**		**C₂H₆O**			**64-17-5**

T/K = 300.05

w_B	0.0288	0.0623	0.0979	0.1379	0.1809	0.2094
$\Delta_M H$/(J/mol)	79.1	184	372	525.5	686	887

Comments: These data include the enthalpy change caused by melting the crystalline parts of the polymer when mixing the polymer and the solvent to form the solution.

Polymer (B):	**poly(ethylene glycol)**							**1976LA1**

Characterization: M_n/g.mol^{-1} = 990,
                        fractionated samples supplied by Union Carbide Corp.

Solvent (A):	**ethanol**			**C₂H₆O**				**64-17-5**

T/K = 321.35

w_B	0.05	0.1	0.2	0.3	0.4	0.5	0.6	0.7
$\Delta_M H$/(J/mol)	71.1	75.3	163	276	402	515	674	795

w_B	0.8	0.9
$\Delta_M H$/(J/mol)	996	987

Polymer (B):	**poly(ethylene glycol)**					**1966LA1**

Characterization: M_n/g.mol^{-1} = 1350,
                        fractionated samples supplied by Union Carbide Corp.

Solvent (A):	**ethanol**		**C₂H₆O**			**64-17-5**

T/K = 300.05

w_B	0.0298	0.0585	0.1042	0.1375	0.1710	0.2055
$\Delta_M H$/(J/mol)	164	438.5	761.5	1092	1461	1730

Comments: These data include the enthalpy change caused by melting the crystalline parts of the polymer when mixing the polymer and the solvent to form the solution.

Polymer (B):	**poly(ethylene glycol)**					**1976LA1**

Characterization: M_n/g.mol^{-1} = 1460,
                        fractionated samples supplied by Union Carbide Corp.

Solvent (A):	**ethanol**		**C₂H₆O**			**64-17-5**

continued

continued

$T/K = 321.35$

w_B	0.05	0.1	0.2	0.3	0.4	0.5	0.6	0.7
$\Delta_M H/(\text{J/mol})$	46.0	75.3	163	276	410	577	761.5	962

w_B	0.8	0.9
$\Delta_M H/(\text{J/mol})$	1255	1372

Polymer (B):	**poly(ethylene glycol)**		**1966LA1**
Characterization:	$M_n/\text{g.mol}^{-1} = 2585$,		
	fractionated samples supplied by Union Carbide Corp.		
Solvent (A):	**ethanol**	**C_2H_6O**	**64-17-5**

$T/K = 300.05$

w_B	0.0307	0.0643	0.1063	0.1369	0.1784	0.2038
$\Delta_M H/(\text{J/mol})$	225	631	987	1281	1857	2133

Comments: These data include the enthalpy change caused by melting the crystalline parts of the polymer when mixing the polymer and the solvent to form the solution.

Polymer (B):	**poly(ethylene glycol)**		**1981GON**
Characterization:	$M_n/\text{g.mol}^{-1} = 4000$		
Solvent (A):	**ethanol**	**C_2H_6O**	**64-17-5**

$T/K = 313.15$

x_B	0.00012	0.00040
$\Delta_M H/(\text{J/mol})$	69.4	125.5

Polymer (B):	**poly(ethylene glycol)**		**1976LA1**
Characterization:	$M_n/\text{g.mol}^{-1} = 4150$,		
	fractionated samples supplied by Union Carbide Corp.		
Solvent (A):	**ethanol**	**C_2H_6O**	**64-17-5**

$T/K = 321.35$

w_B	0.05	0.1	0.2	0.3	0.4	0.5	0.6
$\Delta_M H/(\text{J/mol})$	586	1046	2155	3577	5460	7761	11276

Comments: These data include the enthalpy change caused by melting the crystalline parts of the polymer when mixing the polymer and the solvent to form the solution.

Polymer (B):	**poly(ethylene glycol)**		**1976LA1**
Characterization:	$M_n/\text{g.mol}^{-1} = 200$,		
	fractionated samples supplied by Union Carbide Corp.		
Solvent (A):	**methanol**	**CH_4O**	**67-56-1**

Continued

continued

$T/K = 321.35$

w_B	0.05	0.1	0.2	0.3	0.4	0.5	0.6	0.7
$\Delta_M H/(J/mol)$	27.2	20.9	27.2	46.0	66.9	98.3	140.2	133.9

w_B	0.8	0.9
$\Delta_M H/(J/mol)$	104.6	64.9

Polymer (B):	**poly(ethylene glycol)**		**1992WOE**
Characterization:	$M_n/\text{g.mol}^{-1} = 200$		
Solvent (A):	**methanol**	**CH$_4$O**	**67-56-1**

$T/K = 303.15$

x_B	0.19808	0.29874	0.40898	0.42390	0.51963	0.61864	0.70020
$\Delta_M H/(J/mol)$	37.06	40.80	39.92	39.71	36.78	28.95	22.79

Polymer (B):	**poly(ethylene glycol)**		**1992WOE**
Characterization:	$M_n/\text{g.mol}^{-1} = 300$		
Solvent (A):	**methanol**	**CH$_4$O**	**67-56-1**

$T/K = 303.15$

x_B	0.19940	0.31699	0.46320	0.48869	0.50663	0.60645	0.71720	0.71728
$\Delta_M H/(J/mol)$	87.80	96.67	90.48	86.04	84.08	68.76	46.49	45.55

Polymer (B):	**poly(ethylene glycol)**		**1971KAG**
Characterization:	$M_n/\text{g.mol}^{-1} = 300$		
Solvent (A):	**methanol**	**CH$_4$O**	**67-56-1**

$T/K = 303.15$

$V^{(1)}/\text{cm}^3$	5.0	5.0	5.0	5.0	10.0	15.0
$\varphi_B^{(1)}$	1.000	1.000	1.000	1.000	1.000	1.000
$V^{(2)}/\text{cm}^3$	10.0	15.0	20.0	25.0	15.0	20.0
$\varphi_B^{(2)}$	0.500	0.333	0.250	0.200	0.667	0.750
$\Delta_{dil}H^{12}/\text{J}$	6.653	14.31	16.44	29.33	31.84	42.97

Comments: $\Delta_{dil}H^{12}$ is the extensive quantity obtained for a given total volume change from $V^{(1)}$ to $V^{(2)}$, where $\varphi_B^{(1)}$ and $\varphi_B^{(2)}$ denote the volume fractions of the polymer in the solution before and after the dilution process.

Polymer (B):	**poly(ethylene glycol)**		**1966LA1**
Characterization:	$M_n/\text{g.mol}^{-1} = 335,$		
	fractionated samples supplied by Union Carbide Corp.		
Solvent (A):	**methanol**	**CH$_4$O**	**67-56-1**

continued

continued

$T/K = 300.05$

w_B	0.0519	0.0954	0.1717	0.2368	0.2770	0.4364	0.7172	0.8344
$\Delta_M H/(\text{J/mol})$	−2.09	−5.44	+3.35	46.44	73.22	116.3	58.99	76.99

w_B	0.9122	0.9633
$\Delta_M H/(\text{J/mol})$	87.03	18.83

Polymer (B):	**poly(ethylene glycol)**	**1971KAG**
Characterization:	$M_n/\text{g.mol}^{-1} = 400$	
Solvent (A):	**methanol**　　　　　CH_4O	**67-56-1**

$T/K = 303.15$

$V^{(1)}/\text{cm}^3$	5.0	5.0	5.0	5.0
$\varphi_B^{(1)}$	1.000	1.000	1.000	1.000
$V^{(2)}/\text{cm}^3$	15.0	20.0	25.0	10.0
$\varphi_B^{(2)}$	0.333	0.250	0.200	0.500
$\Delta_{dil} H^{12}/\text{J}$	12.72	16.40	35.69	9.121

Comments:　$\Delta_{dil} H^{12}$ is the extensive quantity obtained for a given total volume change from $V^{(1)}$ to $V^{(2)}$, where $\varphi_B^{(1)}$ and $\varphi_B^{(2)}$ denote the volume fractions of the polymer in the solution before and after the dilution process.

Polymer (B):	**poly(ethylene glycol)**	**1976LA1**
Characterization:	$M_n/\text{g.mol}^{-1} = 400$,	
	fractionated samples supplied by Union Carbide Corp.	
Solvent (A):	**methanol**　　　　　CH_4O	**67-56-1**

$T/K = 321.35$

w_B	0.05	0.1	0.2	0.3	0.4	0.5	0.6	0.7
$\Delta_M H/(\text{J/mol})$	25.10	29.29	33.47	60.67	92.05	127.6	163.2	192.5

w_B	0.8	0.9
$\Delta_M H/(\text{J/mol})$	209.2	198.7

Polymer (B):	**poly(ethylene glycol)**	**1971KAG**
Characterization:	$M_n/\text{g.mol}^{-1} = 600$	
Solvent (A):	**methanol**　　　　　CH_4O	**67-56-1**

$T/K = 303.15$

$V^{(1)}/\text{cm}^3$	5.0	20.0	5.0	10.0	10.0	10.0	10.0
$\varphi_B^{(1)}$	1.000	1.000	1.000	1.000	1.000	1.000	1.000
$V^{(2)}/\text{cm}^3$	10.0	22.0	25.0	30.0	12.0	15.0	17.0
$\varphi_B^{(2)}$	0.500	0.910	0.200	0.429	0.833	0.668	0.590
$\Delta_{dil} H^{12}/\text{J}$	25.02	26.11	53.35	87.36	17.82	38.16	34.27

Comments:　$\Delta_{dil} H^{12}$ is the extensive quantity obtained for a given total volume change from $V^{(1)}$ to $V^{(2)}$, where $\varphi_B^{(1)}$ and $\varphi_B^{(2)}$ denote the volume fractions of the polymer in the solution before and after the dilution process.

Polymer (B):   **poly(ethylene glycol)**   **1992WOE**
Characterization: M_n/g.mol^{-1} = 600
Solvent (A):   **methanol**   **CH$_4$O**   **67-56-1**

T/K = 303.15

x_B	0.00940	0.03638	0.05159	0.08504	0.09699	0.19767	0.21196	0.31561
$\Delta_M H$/(J/mol)	18.79	75.45	113.55	163.55	178.72	232.69	237.13	217.07

x_B	0.34095	0.36325	0.52077
$\Delta_M H$/(J/mol)	200.85	190.84	122.77

Polymer (B):   **poly(ethylene glycol)**   **1966LA1**
Characterization: M_n/g.mol^{-1} = 654,
     fractionated samples supplied by Union Carbide Corp.
Solvent (A):   **methanol**   **CH$_4$O**   **67-56-1**

T/K = 300.05

w_B	0.0525	0.0939	0.1720	0.2366	0.2826	0.4367	0.7277	0.8420
$\Delta_M H$/(J/mol)	−0.84	74.06	65.69	66.94	74.48	133.9	194.1	264.4

w_B	0.9057	0.9578
$\Delta_M H$/(J/mol)	267.8	188.3

Polymer (B):   **poly(ethylene glycol)**   **1966LA1**
Characterization: M_n/g.mol^{-1} = 920,
     fractionated samples supplied by Union Carbide Corp.
Solvent (A):   **methanol**   **CH$_4$O**   **67-56-1**

T/K = 300.05

w_B	0.0308	0.0541	0.0969	0.1362	0.1729	0.2072
$\Delta_M H$/(J/mol)	79.91	133.9	318.4	446.9	509.2	630.9

Comments: These data include the enthalpy change caused by melting the crystalline parts of the polymer when mixing the polymer and the solvent to form the solution.

Polymer (B):   **poly(ethylene glycol)**   **1976LA1**
Characterization: M_n/g.mol^{-1} = 990,
     fractionated samples supplied by Union Carbide Corp.
Solvent (A):   **methanol**   **CH$_4$O**   **67-56-1**

T/K = 321.35

w_B	0.05	0.1	0.2	0.3	0.4	0.5	0.6	0.7
$\Delta_M H$/(J/mol)	29.29	33.47	46.02	75.31	115.1	161.1	215.5	269.9

w_B	0.8	0.9
$\Delta_M H$/(J/mol)	332.6	347.3

Polymer (B):   **poly(ethylene glycol)**   **1966LA1**
Characterization: M_n/g.mol^{-1} = 1350,
                                fractionated samples supplied by Union Carbide Corp.
Solvent (A):   **methanol**   **CH$_4$O**   **67-56-1**

T/K = 300.05

w_B	0.0292	0.0570	0.0976	0.1380	0.1716	0.2050
$\Delta_M H$/(J/mol)	134.3	284.9	500.4	755.6	1018	1193

Comments: These data include the enthalpy change caused by melting the crystalline parts of the polymer when mixing the polymer and the solvent to form the solution.

Polymer (B):   **poly(ethylene glycol)**   **1976LA1**
Characterization: M_n/g.mol^{-1} = 1460,
                                fractionated samples supplied by Union Carbide Corp.
Solvent (A):   **methanol**   **CH$_4$O**   **67-56-1**

T/K = 321.35

w_B	0.05	0.1	0.2	0.3	0.4	0.5	0.6	0.7
$\Delta_M H$/(J/mol)	29.29	32.56	58.58	92.05	133.9	184.1	240.6	303.3

w_B	0.8	0.9
$\Delta_M H$/(J/mol)	364.0	366.1

Polymer (B):   **poly(ethylene glycol)**   **1966LA1**
Characterization: M_n/g.mol^{-1} = 2585,
                                fractionated samples supplied by Union Carbide Corp.
Solvent (A):   **methanol**   **CH$_4$O**   **67-56-1**

T/K = 300.05

w_B	0.0309	0.0538	0.0995	0.1728	0.2352	0.2824
$\Delta_M H$/(J/mol)	195.8	492.9	711.7	1556	2113	2343

Comments: These data include the enthalpy change caused by melting the crystalline parts of the polymer when mixing the polymer and the solvent to form the solution.

Polymer (B):   **poly(ethylene glycol)**   **1976LA1**
Characterization: M_n/g.mol^{-1} = 4150,
                                fractionated samples supplied by Union Carbide Corp.
Solvent (A):   **methanol**   **CH$_4$O**   **67-56-1**

T/K = 321.35

w_B	0.05	0.1	0.2	0.3	0.4	0.5	0.6
$\Delta_M H$/(J/mol)	502.1	836.8	1757	3054	4728	7950	10125

Comments: These data include the enthalpy change caused by melting the crystalline parts of the polymer when mixing the polymer and the solvent to form the solution.

Polymer (B):   **poly(ethylene glycol)**   **2004CA1**
Characterization: M_n/g.mol^{-1} = 192, M_w/g.mol^{-1} = 223, ρ = 1.11300 g/cm^3 (308 K)
                          PEG 200, Fluka AG, Buchs, Switzerland
Solvent (A):   **2-phenylethanol**   **C$_8$H$_{10}$O**   **60-12-8**

T/K = 308.15

x_B	0.0268	0.0522	0.0763	0.0992	0.1418	0.1805	0.2483	0.3058
$\Delta_M H$/(J/mol)	−92.0	−161.3	−212.8	−251.8	−302.7	−322.1	−352.1	−369.6

x_B	0.3979	0.4978	0.5693	0.6647	0.7255	0.7986	0.8409	0.8880
$\Delta_M H$/(J/mol)	−381.8	−395.9	−399.9	−369.8	−340.9	−275.8	−222.6	−158.8

x_B	0.9407
$\Delta_M H$/(J/mol)	−82.0

Polymer (B):   **poly(ethylene glycol)**   **2004CA1**
Characterization: M_n/g.mol^{-1} = 274, M_w/g.mol^{-1} = 304, ρ = 1.11391 g/cm^3 (308 K),
                          PEG 300, Fluka AG, Buchs, Switzerland
Solvent (A):   **2-phenylethanol**   **C$_8$H$_{10}$O**   **60-12-8**

T/K = 308.15

x_B	0.0181	0.0355	0.0524	0.0687	0.0996	0.1285	0.1811	0.2277
$\Delta_M H$/(J/mol)	−87.0	−155.9	−212.7	−266.9	−311.8	−348.0	−389.9	−411.2

x_B	0.3066	0.3989	0.4694	0.5703	0.6389	0.7264	0.7797	0.8415
$\Delta_M H$/(J/mol)	−446.1	−477.9	−480.8	−460.9	−421.2	−344.9	−283.8	−217.1

x_B	0.9139
$\Delta_M H$/(J/mol)	−130.1

Polymer (B):   **poly(ethylene glycol)**   **2004CA1**
Characterization: M_n/g.mol^{-1} = 365, M_w/g.mol^{-1} = 402, ρ = 1.11413 g/cm^3 (308 K),
                          PEG 400, Fluka AG, Buchs, Switzerland
Solvent (A):   **2-phenylethanol**   **C$_8$H$_{10}$O**   **60-12-8**

T/K = 308.15

x_B	0.0138	0.0273	0.0404	0.0531	0.0720	0.1009	0.1440	0.1832
$\Delta_M H$/(J/mol)	−81.0	−149.8	−205.2	−248.1	−307.6	−365.9	−422.6	−453.9

x_B	0.2517	0.3355	0.4023	0.5024	0.5738	0.6688	0.7292	0.8015
$\Delta_M H$/(J/mol)	−490.7	−529.5	−555.3	−530.9	−493.0	−385.5	−307.7	−214.4

x_B	0.8898
$\Delta_M H$/(J/mol)	−110.1

Polymer (B):   **poly(ethylene glycol)**   **2004CA1**
Characterization: M_n/g.mol^{-1} = 554, M_w/g.mol^{-1} = 587, ρ = 1.11440 g/cm^3 (308 K),
                          PEG 600, Fluka AG, Buchs, Switzerland
Solvent (A):   **2-phenylethanol**   **C$_8$H$_{10}$O**   **60-12-8**

continued

continued

T/K = 308.15

x_B	0.0095	0.0188	0.0279	0.0368	0.0543	0.0711	0.1030	0.1327
$\Delta_M H$/(J/mol)	−71.3	−133.3	−196.9	−234.4	−314.1	−362.9	−437.8	−481.5

x_B	0.1866	0.2561	0.3146	0.4078	0.4787	0.5794	0.6474	0.7337
$\Delta_M H$/(J/mol)	−526.9	−562.3	−588.0	−608.1	−586.9	−496.4	−396.9	−248.6

x_B	0.8464
$\Delta_M H$/(J/mol)	−105.1

Polymer (B):	**poly(ethylene glycol)**	**2004CA3**
Characterization:	M_n/g.mol^{-1} = 192, M_w/g.mol^{-1} = 223, ρ = 1.11300 g/cm^3 (308 K)	
	PEG 200, Fluka AG, Buchs, Switzerland	
Solvent (A):	**3-phenyl-1-propanol** **C$_9$H$_{12}$O**	**122-97-4**

T/K = 308.15

x_B	0.0303	0.0588	0.0857	0.1111	0.1579	0.2000	0.2727	0.3334
$\Delta_M H$/(J/mol)	−63.1	−100.5	−124.6	−136.9	−150.0	−152.9	−153.8	−154.8

x_B	0.4287	0.5295	0.6000	0.6923	0.7500	0.8182	0.8571	0.9000
$\Delta_M H$/(J/mol)	−157.3	−159.4	−153.5	−138.8	−120.9	−99.7	−88.8	−69.9

x_B	0.9474
$\Delta_M H$/(J/mol)	−45.1

Polymer (B):	**poly(ethylene glycol)**	**2004CA3**
Characterization:	M_n/g.mol^{-1} = 274, M_w/g.mol^{-1} = 304, ρ = 1.11391 g/cm^3 (308 K),	
	PEG 300, Fluka AG, Buchs, Switzerland	
Solvent (A):	**3-phenyl-1-propanol** **C$_9$H$_{12}$O**	**122-97-4**

T/K = 308.15

x_B	0.0205	0.0402	0.0591	0.0772	0.1115	0.1434	0.2007	0.2507
$\Delta_M H$/(J/mol)	−51.7	−89.3	−114.0	−131.6	−157.4	−169.3	−171.6	−173.8

x_B	0.3342	0.4297	0.5010	0.6010	0.6676	0.7508	0.8007	0.8577
$\Delta_M H$/(J/mol)	−17l.1	−171.4	−166.3	−147.1	−125.1	−95.3	−81.1	−60.4

x_B	0.9234
$\Delta_M H$/(J/mol)	−43.4

Polymer (B):	**poly(ethylene glycol)**	**2004CA3**
Characterization:	M_n/g.mol^{-1} = 365, M_w/g.mol^{-1} = 402, ρ = 1.11413 g/cm^3 (308 K),	
	PEG 400, Fluka AG, Buchs, Switzerland	
Solvent (A):	**3-phenyl-1-propanol** **C$_9$H$_{12}$O**	**122-97-4**

T/K = 308.15

x_B	0.0157	0.0308	0.0456	0.0598	0.0872	0.1129	0.1604	0.2029
$\Delta_M H$/(J/mol)	−40.5	−72.4	−98.5	−118.1	−147.0	−165.4	−180.1	−185.1

continued

continued

x_B	0.2764	0.3643	0.4331	0.5340	0.6044	0.6962	0.7534	0.8209
$\Delta_M H/(J/mol)$	−190.0	−189.0	−183.1	−163.3	−145.5	−106.1	−81.0	−58.6

x_B	0.9017
$\Delta_M H/(J/mol)$	−37.3

Polymer (B):	**poly(ethylene glycol)**	**2004CA3**
Characterization:	$M_n/\text{g.mol}^{-1} = 554$, $M_w/\text{g.mol}^{-1} = 587$, $\rho = 1.11440 \text{ g/cm}^3$ (308 K), PEG 600, Fluka AG, Buchs, Switzerland	
Solvent (A):	**3-phenyl-1-propanol $C_9H_{12}O$**	**122-97-4**

$T/K = 308.15$

x_B	0.0107	0.0213	0.0315	0.0432	0.0612	0.0799	0.1153	0.1479
$\Delta_M H/(J/mol)$	−33.1	−62.6	−87.4	−110.4	−137.1	−157.6	−179.3	−192.5

x_B	0.2067	0.2811	0.3426	0.4387	0.5103	0.6099	0.6758	0.7577
$\Delta_M H/(J/mol)$	−197.0	−200.6	−202.1	−200.3	−183.2	−140.9	−101.3	−60.9

x_B	0.8621	0.9450
$\Delta_M H/(J/mol)$	−38.0	−32.13

Polymer (B):	**poly(ethylene glycol)**	**1981GON**
Characterization:	$M_n/\text{g.mol}^{-1} = 400$	
Solvent (A):	**1-propanol C_3H_8O**	**71-23-8**

$T/K = 298.15$

x_B	0.0119	0.0123	0.0239	0.0245	0.1699
$\Delta_M H/(J/mol)$	65.32	71.77	105.3	116.8	135.4

$T/K = 313.15$

x_B	0.0098	0.0101	0.0117	0.0222	0.1726	0.2927
$\Delta_M H/(J/mol)$	97.2	101.8	117.6	125.0	300.1	690.4

Polymer (B):	**poly(ethylene glycol)**	**2004CA4**
Characterization:	$M_n/\text{g.mol}^{-1} = 192$, $M_w/\text{g.mol}^{-1} = 223$, $\rho = 1.11284 \text{ g/cm}^3$ (308 K) PEG 200, Fluka AG, Buchs, Switzerland	
Solvent (A):	**propylene carbonate $C_4H_6O_3$**	**108-32-7**

$T/K = 308.15$

x_B	0.0203	0.0398	0.0585	0.0765	0.1105	0.1422	0.1991	0.2489
$\Delta_M H/(J/mol)$	158.4	274.7	361.6	423.2	490.3	520.2	534.8	520.3

x_B	0.3320	0.4272	0.4986	0.5986	0.6654	0.7490	0.7991	0.8565
$\Delta_M H/(J/mol)$	490.5	474.4	464.2	419.6	367.0	288.4	227.2	161.7

x_B	0.9227
$\Delta_M H/(J/mol)$	93.2

Polymer (B):   **poly(ethylene glycol)**   **2004CA4**
Characterization: M_n/g.mol^{-1} = 274, M_w/g.mol^{-1} = 304, ρ = 1.11358 g/cm^3 (308 K),
                       PEG 300, Fluka AG, Buchs, Switzerland
Solvent (A):   **propylene carbonate   C$_4$H$_6$O$_3$**   **108-32-7**

T/K = 308.15

x_B	0.0143	0.0282	0.0418	0.0549	0.0802	0.1041	0.1484	0.1885
$\Delta_M H$/(J/mol)	65.0	116.6	155.6	183.0	220.4	241.4	243.3	232.0
x_B	0.2584	0.3434	0.4108	0.5112	0.5824	0.6766	0.7361	0.8071
$\Delta_M H$/(J/mol)	199.6	167.9	148.2	111.9	74.9	12.1	−21.7	−36.0
x_B	0.8932	0.9177						
$\Delta_M H$/(J/mol)	−20.6	−9.6						

Polymer (B):   **poly(ethylene glycol)**   **2004CA4**
Characterization: M_n/g.mol^{-1} = 365, M_w/g.mol^{-1} = 402, ρ = 1.11489 g/cm^3 (308 K),
                       PEG 400, Fluka AG, Buchs, Switzerland
Solvent (A):   **propylene carbonate   C$_4$H$_6$O$_3$**   **108-32-7**

T/K = 308.15

x_B	0.0079	0.0157	0.0317	0.0418	0.0614	0.0802	0.1158	0.14S5
$\Delta_M H$/(J/mol)	18.0	34.0	59.3	70.4	81.5	85.0	79.1	63.5
x_B	0.2074	0.2820	0.3437	0.4399	0.5115	0.6111	0.6768	0.7586
$\Delta_M H$/(J/mol)	35.6	−2.3	−17.5	−39.4	−58.0	−94.5	−121.8	−133.7
x_B	0.8627	0.8934						
$\Delta_M H$/(J/mol)	−91.0	−70.0						

Polymer (B):   **poly(ethylene glycol)**   **2004CA4**
Characterization: M_n/g.mol^{-1} = 554, M_w/g.mol^{-1} = 587, ρ = 1.11396 g/cm^3 (308 K),
                       PEG 600, Fluka AG, Buchs, Switzerland
Solvent (A):   **propylene carbonate   C$_4$H$_6$O$_3$**   **108-32-7**

T/K = 308.15

x_B	0.0139	0.0208	0.0275	0.0407	0.0536	0.0782	0.1016	0.1471
$\Delta_M H$/(J/mol)	−25.5	−36.2	−47.7	−62.6	−86.9	−120.8	−151.0	−198.6
x_B	0.2055	0.2565	0.3410	0.4082	0.5086	0.5798	0.6743	0.8054
$\Delta_M H$/(J/mol)	−249.8	−290.0	−346.6	−382.3	−417.9	−416.1	−385.7	−267.1
x_B	0.8466	0.9501						
$\Delta_M H$/(J/mol)	−220.0	−68.0						

Polymer (B):   **poly(ethylene glycol)**   **1976LA1**
Characterization: M_n/g.mol^{-1} = 200,
                       fractionated samples supplied by Union Carbide Corp.
Solvent (A):   **tetrachloromethane   CCl$_4$**   **56-23-5**

continued

continued

$T/K = 321.35$

w_B	0.05	0.1	0.2	0.3	0.4	0.5	0.6	0.7
$\Delta_M H/(\text{J/mol})$	251.0	246.9	221.8	215.5	198.7	140.2	43.93	−37.66

w_B	0.8	0.9
$\Delta_M H/(\text{J/mol})$	−66.94	−60.67

Polymer (B):	**poly(ethylene glycol)**	**1965LAK**
Characterization:	$M_n/\text{g.mol}^{-1} = 335$,	
	fractionated samples supplied by Union Carbide Corp.	
Solvent (A):	**tetrachloromethane CCl$_4$**	**56-23-5**

$T/K = 300.05$

w_B	0.01258	0.05925	0.1146	0.1606	0.2090	0.7280	0.7842	0.8361
$\Delta_M H/(\text{J/mol})$	9.20	88.70	6.28	−69.45	−38.49	−372.4	−460.2	−355.6

w_B	0.9058
$\Delta_M H/(\text{J/mol})$	−309.6

Polymer (B):	**poly(ethylene glycol)**	**1985COR, 1995KIL**
Characterization:	$M_n/\text{g.mol}^{-1} = 395$, $M_w/\text{g.mol}^{-1} = 430$,	
	$\rho = 1.1184$ g/cm^3 (303 K), Hoechst AG, Germany	
Solvent (A):	**tetrachloromethane CCl$_4$**	**56-23-5**

$T/K = 303.15$

w_B	0.00028	0.00029	0.00068	0.00069	0.0107	0.0108	0.0146	0.0147	0.0185
$\Delta_M H/(\text{J/g})$	0.065	0.065	0.134	0.137	0.196	0.199	0.249	0.253	0.295

w_B	0.0185	0.0261	0.0262	0.0336	0.0337	0.0410	0.0411	0.0475	0.0484
$\Delta_M H/(\text{J/g})$	0.300	0.368	0.374	0.423	0.428	0.463	0.468	0.488	0.497

w_B	0.0551	0.0555	0.0556	0.0626	0.0626	0.0695	0.0696	0.0765	0.0773
$\Delta_M H/(\text{J/g})$	0.513	0.512	0.518	0.532	0.530	0.535	0.541	0.545	0.547

w_B	0.0810	0.0899	0.0916	0.0953	0.1054	0.1091	0.1188	0.1225	0.1291
$\Delta_M H/(\text{J/g})$	0.548	0.544	0.547	0.541	0.536	0.524	0.519	0.501	0.487

w_B	0.1318	0.1327	0.1356	0.1444	0.1465	0.1482	0.1566	0.1599	0.1605
$\Delta_M H/(\text{J/g})$	0.495	0.491	0.473	0.468	0.461	0.442	0.438	0.426	0.408

w_B	0.1686	0.1724	0.1728	0.1854	0.1951	0.2080	0.2205	0.2326	0.2443
$\Delta_M H/(\text{J/g})$	0.406	0.372	0.391	0.354	0.318	0.270	0.223	0.175	0.128

w_B	0.2515	0.2557	0.2667	0.2702	0.2879	0.3180	0.3559	0.3987	0.4472
$\Delta_M H/(\text{J/g})$	0.093	0.083	0.037	0.013	−0.067	−0.142	−0.309	−0.511	−0.725

w_B	0.5024	0.5633	0.5654	0.5741	0.6285	0.6477	0.6930	0.7135	0.7963
$\Delta_M H/(\text{J/g})$	−0.940	−1.108	−1.125	−1.192	−1.236	−1.287	−1.325	−1.296	−1.267

w_B	0.8092	0.8743	0.9208	0.9439	0.9732	0.9901
$\Delta_M H/(\text{J/g})$	−1.133	−0.947	−0.610	−0.473	−0.240	−0.087

Polymer (B):   **poly(ethylene glycol)**   **1993ZEL**

Characterization: $M_n/\text{g.mol}^{-1} = 400$, $M_w/\text{g.mol}^{-1} = 420$,

                          $\rho = 1.1182 \text{ g/cm}^3$ (303K), Hoechst AG, Germany

Solvent (A):   **tetrachloromethane**   **CCl$_4$**   **56-23-5**

$T/\text{K} = 303.15$

$$\Delta_M H/(\text{J/g}) = w_B(1 - w_B)[-12.57 + 2.99(1 - 2w_B)] - 6.33w_B \ln(w_B)$$

$T/\text{K} = 318.15$

$$\Delta_M H/(\text{J/g}) = w_B(1 - w_B)[-6.08 + 5.92(1 - 2w_B) + 2.20(1 - 2w_B)^2 + 1.14(1 - 2w_B)^3] - 4.42w_B \ln(w_B)$$

$T/\text{K} = 333.15$

$$\Delta_M H/(\text{J/g}) = w_B(1 - w_B)[5.03 + 3.31(1 - 2w_B)] + 2.42w_B \ln(w_B)$$

Polymer (B):   **poly(ethylene glycol)**   **1976LA1**

Characterization: $M_n/\text{g.mol}^{-1} = 400$,

                          fractionated samples supplied by Union Carbide Corp.

Solvent (A):   **tetrachloromethane**   **CCl$_4$**   **56-23-5**

$T/\text{K} = 321.35$

w_B	0.05	0.1	0.2	0.3	0.4	0.5	0.6	0.7
$\Delta_M H/(\text{J/mol})$	184.1	255.2	286.6	267.8	156.9	+4.18	−156.9	−261.5

w_B	0.8	0.9
$\Delta_M H/(\text{J/mol})$	−259.4	−188.3

Polymer (B):   **poly(ethylene glycol)**   **1985COR, 1995KIL**

Characterization: $M_n/\text{g.mol}^{-1} = 560$, $M_w/\text{g.mol}^{-1} = 590$,

                          $\rho = 1.1183 \text{ g/cm}^3$ (303 K), Hoechst AG, Germany

Solvent (A):   **tetrachloromethane**   **CCl$_4$**   **56-23-5**

$T/\text{K} = 303.15$

w_B	0.0011	0.0028	0.0035	0.0050	0.0051	0.0059	0.0073	0.0082	0.0094
$\Delta_M H/(\text{J/g})$	0.017	0.035	0.053	0.073	0.062	0.083	0.086	0.107	0.119
w_B	0.0096	0.0106	0.0118	0.0129	0.0168	0.0180	0.0206	0.0244	0.0282
$\Delta_M H/(\text{J/g})$	0.109	0.129	0.129	0.148	0.176	0.186	0.200	0.220	0.237
w_B	0.0307	0.0357	0.0430	0.0472	0.0503	0.0552	0.0574	0.0631	0.0708
$\Delta_M H/(\text{J/g})$	0.249	0.263	0.279	0.288	0.289	0.295	0.294	0.296	0.295
w_B	0.0714	0.0785	0.0933	0.1077	0.1217	0.1288	0.1392	0.1461	0.1561
$\Delta_M H/(\text{J/g})$	0.292	0.288	0.263	0.229	0.187	0.163	0.127	0.103	0.064
w_B	0.1659	0.1755	0.1848	0.1940	0.2604	0.2724	0.2889	0.3024	0.3212
$\Delta_M H/(\text{J/g})$	0.024	−0.015	−0.055	−0.094	−0.361	−0.416	−0.492	−0.555	−0.643
w_B	0.3402	0.3503	0.3816	0.4216	0.4664	0.5033	0.5098	0.5361	0.5630
$\Delta_M H/(\text{J/g})$	−0.733	−0.781	−0.917	−1.079	−1.203	−1.287	−1.332	−1.358	−1.428

continued

continued

w_B	0.5637	0.5946	0.6308	0.7141	0.8021	0.8567	0.9038	0.9398	0.9542
$\Delta_M H/(\text{J/g})$	−1.428	−1.459	−1.472	−1.432	−1.223	−1.050	−0.777	−0.531	−0.462

w_B	0.9765	0.9838	0.9913
$\Delta_M H/(\text{J/g})$	−0.232	−0.160	−0.089

Polymer (B):	**poly(ethylene glycol)**	**1993ZEL**

Characterization: $M_n/\text{g.mol}^{-1} = 590$, $M_w/\text{g.mol}^{-1} = 615$,
$\rho = 1.1183$ g/cm^3 (303K), Hoechst AG, Germany

Solvent (A): **tetrachloromethane   CCl$_4$** **56-23-5**

$T/K = 303.15$

$$\Delta_M H/(\text{J/g}) = w_B(1 - w_B)[-11.06 + 2.23(1 - 2w_B) + 0.41(1 - 2w_B)^2 + 1.68(1 - 2w_B)^3] - 4.26 w_B \ln(w_B)$$

$T/K = 318.15$

$$\Delta_M H/(\text{J/g}) = w_B(1 - w_B)[-7.57 + 3.19(1 - 2w_B) + 0.90(1 - 2w_B)^2] - 3.78 w_B \ln(w_B)$$

Polymer (B):	**poly(ethylene glycol)**	**1965LAK**

Characterization: $M_n/\text{g.mol}^{-1} = 654$,
fractionated samples supplied by Union Carbide Corp.

Solvent (A): **tetrachloromethane   CCl$_4$** **56-23-5**

$T/K = 300.05$

w_B	0.01694	0.06155	0.1127	0.1671	0.2050	0.7158	0.7817	0.8454
$\Delta_M H/(\text{J/mol})$	+7.95	+3.35	−85.35	−129.3	−155.2	−799.1	−778.2	−1343

w_B	0.9148
$\Delta_M H/(\text{J/mol})$	−987.4

Polymer (B):	**poly(ethylene glycol)**	**1965LAK**

Characterization: $M_n/\text{g.mol}^{-1} = 920$,
fractionated samples supplied by Union Carbide Corp.

Solvent (A): **tetrachloromethane   CCl$_4$** **56-23-5**

$T/K = 300.05$

w_B	0.00635	0.01246	0.01842	0.03409
$\Delta_M H/(\text{J/mol})$	57.32	226.4	272.0	312.1

Comments: These data include the enthalpy change caused by melting the crystalline parts of the polymer when mixing the polymer and the solvent to form the solution.

Polymer (B):   **poly(ethylene glycol)**   **1976LA1**
Characterization: $M_n/\text{g.mol}^{-1} = 990$,
     fractionated samples supplied by Union Carbide Corp.
Solvent (A):   **tetrachloromethane**   **CCl₄**   **56-23-5**

$T/\text{K} = 321.35$

w_B	0.05	0.1	0.2	0.3	0.4	0.5	0.6	0.7
$\Delta_M H/(\text{J/mol})$	138.1	184.1	280.3	370.3	464.4	397.5	156.9	−8.4

w_B	0.8	0.9
$\Delta_M H/(\text{J/mol})$	−131.8	−238.5

Polymer (B):   **poly(ethylene glycol)**   **1985COR, 1995KIL**
Characterization: $M_n/\text{g.mol}^{-1} = 1000$, Hoechst AG, Germany
Solvent (A):   **tetrachloromethane**   **CCl₄**   **56-23-5**

$T/\text{K} = 303.15$

$w_B^{(1)}$	0.4058	0.3686	0.3334	0.3015	0.2697	0.2444	0.2236	0.2174
$w_B^{(2)}$	0.3686	0.3334	0.3015	0.2697	0.2444	0.2236	0.2070	0.1967
$\Delta_{dil} H^{12}/(\text{J/g polymer})$	0.016	0.052	0.096	0.150	0.166	0.164	0.156	0.178

$w_B^{(1)}$	0.1967	0.1772	0.1597	0.1439	0.1195	0.1085	0.0980	0.0885
$w_B^{(2)}$	0.1772	0.1597	0.1439	0.1298	0.1085	0.0980	0.0885	0.0799
$\Delta_{dil} H^{12}/(\text{J/g polymer})$	0.190	0.227	0.242	0.235	0.238	0.229	0.217	0.217

$w_B^{(1)}$	0.0799	0.4850	0.4456	0.3969	0.3541	0.3163	0.2830	0.2297
$w_B^{(2)}$	0.0719	0.4456	0.3969	0.3541	0.3163	0.2830	0.2530	0.2078
$\Delta_{dil} H^{12}/(\text{J/g polymer})$	0.207	−0.045	−0.039	0.030	0.096	0.140	0.175	0.186

$w_B^{(1)}$	0.2078	0.1871	0.1686	0.1519	0.1370	0.0778	0.0706	0.0638
$w_B^{(2)}$	0.1871	0.1686	0.1519	0.1370	0.1236	0.0706	0.0638	0.0577
$\Delta_{dil} H^{12}/(\text{J/g polymer})$	0.201	0.236	0.247	0.222	0.237	0.182	0.168	0.176

$w_B^{(1)}$	0.0804	0.0740	0.0614
$w_B^{(2)}$	0.0740	0.0670	0.0559
$\Delta_{dil} H^{12}/(\text{J/g polymer})$	0.152	0.141	0.128

Comments: $w_B^{(1)}$ and $w_B^{(2)}$ denote the mass fractions of the polymer in the solution before and after the dilution process.

Polymer (B):   **poly(ethylene glycol)**   **1965LAK**
Characterization: $M_n/\text{g.mol}^{-1} = 1350$,
     fractionated samples supplied by Union Carbide Corp.
Solvent (A):   **tetrachloromethane**   **CCl₄**   **56-23-5**

$T/\text{K} = 300.05$

w_B	0.00605	0.00840	0.01200	0.001917
$\Delta_M H/(\text{J/mol})$	145.2	161.1	233.0	327.2

Comments: These data include the enthalpy change caused by melting the crystalline parts of the polymer when mixing the polymer and the solvent to form the solution.

Polymer (B):   **poly(ethylene glycol)**   **1976LA1**
Characterization: $M_n/\text{g.mol}^{-1} = 1460$,
                          fractionated samples supplied by Union Carbide Corp.
Solvent (A):   **tetrachloromethane   CCl₄**   **56-23-5**

$T/\text{K} = 321.35$

w_B	0.05	0.1	0.2	0.3	0.4	0.5	0.6	0.7
$\Delta_M H/(\text{J/mol})$	272.0	288.7	351.5	422.6	456.1	439.3	156.9	−96.2

w_B	0.8	0.9
$\Delta_M H/(\text{J/mol})$	−272.0	−410.0

Polymer (B):   **poly(ethylene glycol)**   **1965LAK**
Characterization: $M_n/\text{g.mol}^{-1} = 2585$,
                          fractionated samples supplied by Union Carbide Corp.
Solvent (A):   **tetrachloromethane   CCl₄**   **56-23-5**

$T/\text{K} = 300.05$

w_B	0.00310	0.00639	0.00889	0.01270
$\Delta_M H/(\text{J/mol})$	269.4	332.6	384.1	523.0

Comments: These data include the enthalpy change caused by melting the crystalline parts of the polymer when mixing the polymer and the solvent to form the solution.

Polymer (B):   **poly(ethylene glycol)**   **1976LA1**
Characterization: $M_n/\text{g.mol}^{-1} = 4150$,
                          fractionated samples supplied by Union Carbide Corp.
Solvent (A):   **tetrachloromethane   CCl₄**   **56-23-5**

$T/\text{K} = 321.35$

w_B	0.05	0.1	0.2	0.3	0.4	0.5	0.6
$\Delta_M H/(\text{J/mol})$	1755	3515	6945	11630	18240	26025	11300

Comments: These data include the enthalpy change caused by melting the crystalline parts of the polymer when mixing the polymer and the solvent to form the solution.

Polymer (B):   **poly(ethylene glycol)**   **1979KOL, 1981KOL**
Characterization: $M_\eta/\text{g.mol}^{-1} = 43400$, Hoechst AG, Germany
Solvent (A):   **tetrachloromethane   CCl₄**   **56-23-5**

$T/\text{K} = 303.15$

$c_B^{(1)}/(\text{g/cm}^3)$ 0.10
$c_B^{(2)}/(\text{g/cm}^3)$ 0.005
$\Delta_{dil} H^{12}/(\text{J/g polymer})$ −2.8

Comments: $c_B^{(1)}$ and $c_B^{(2)}$ denote the concentrations of the polymer in the solution before and after the dilution process.

Polymer (B):   **poly(ethylene glycol)**   **2003OTT**

Characterization: $M_n/\text{g.mol}^{-1} = 192$, $M_w/\text{g.mol}^{-1} = 224$, $\rho = 1.12098$ g/cm^3 (298 K),
PEG 200, Fluka AG, Buchs, Switzerland

Solvent (A):   **tetrahydrofuran**   **C_4H_8O**   **109-99-9**

$T/\text{K} = 288.15$

x_B	0.0194	0.0381	0.0560	0.0733	0.1061	0.1367	0.1919	0.2404
$\Delta_M H/(\text{J/mol})$	92.0	169.9	232.3	282.8	358.0	418.7	479.3	506.9

x_B	0.3220	0.4161	0.4871	0.5876	0.6551	0.7403	0.7916	0.8507
$\Delta_M H/(\text{J/mol})$	511.6	499.9	475.2	428.2	373.5	295.0	234.6	158.0

x_B	0.9194
$\Delta_M H/(\text{J/mol})$	71.1

$T/\text{K} = 298.15$

x_B	0.0195	0.0383	0.0563	0.0737	0.1066	0.1373	0.1927	0.2414
$\Delta_M H/(\text{J/mol})$	104.2	190.2	260.9	318.3	406.4	466.6	526.8	546.4

x_B	0.3232	0.4174	0.4884	0.5888	0.6563	0.7103	0.7925	0.8514
$\Delta_M H/(\text{J/mol})$	560.0	544.6	520.7	480.0	440.5	394.8	307.6	224.0

x_B	0.9197
$\Delta_M H/(\text{J/mol})$	118.3

$T/\text{K} = 313.15$

x_B	0.0197	0.0386	0.0567	0.0742	0.1074	0.1382	0.1940	0.2428
$\Delta_M H/(\text{J/mol})$	104.2	192.0	265.2	327.2	423.4	492.8	582.2	624.6

x_B	0.3249	0.4194	0.4905	0.5908	0.6581	0.7428	0.7938	0.8524
$\Delta_M H/(\text{J/mol})$	658.4	653.0	625.3	560.1	499.9	394.9	325.2	231.1

x_B	0.9203
$\Delta_M H/(\text{J/mol})$	117.2

Polymer (B):   **poly(ethylene glycol)**   **2003OTT**

Characterization: $M_n/\text{g.mol}^{-1} = 192$, $M_w/\text{g.mol}^{-1} = 224$, $\rho = 1.12098$ g/cm^3 (298 K),
PEG 200, Fluka AG, Buchs, Switzerland

Solvent (A):   **tetrahydropyran**   **$C_5H_{10}O$**   **142-68-7**

$T/\text{K} = 288.15$

x_B	0.0232	0.0430	0.0665	0.0867	0.1246	0.1595	0.2216	0.2751
$\Delta_M H/(\text{J/mol})$	141.7	240.6	335.9	400.8	482.7	524.1	573.8	582.9

x_B	0.3629	0.4608	0.5325	0.6308	0.6949	0.7736	0.8200	0.8725
$\Delta_M H/(\text{J/mol})$	571.6	553.7	545.1	530.9	485.5	402.4	335.6	240.4

x_B	0.9318
$\Delta_M H/(\text{J/mol})$	124.0

continued

continued

$T/K = 298.15$

x_B	0.0232	0.0455	0.0667	0.0870	0.1250	0.1601	0.2223	0.2760
$\Delta_M H/(J/mol)$	148.7	264.7	354.9	424.5	524.3	580.4	623.9	630.8

x_B	0.3639	0.4615	0.5335	0.6317	0.6958	0.7743	0.8206	0.8728
$\Delta_M H/(J/mol)$	622.7	608.5	583.4	558.4	529.0	441.7	366.5	267.7

x_B	0.9321
$\Delta_M H/(J/mol)$	141.2

$T/K = 313.15$

x_B	0.0234	0.0458	0.0671	0.0875	0.1258	0.1610	0.2235	0.2773
$\Delta_M H/(J/mol)$	160.0	289.1	395.3	479.0	613.0	678.2	760.0	785.8

x_B	0.3653	0.4635	0.5352	0.6333	0.6972	0.7755	0.8216	0.8736
$\Delta_M H/(J/mol)$	787.1	757.7	726.7	666.9	603.7	522.9	436.9	330.0

x_B	0.9325
$\Delta_M H/(J/mol)$	189.0

Polymer (B):   **poly(ethylene glycol)**   **1993ZEL**
Characterization: $M_n/\text{g.mol}^{-1} = 400$, $M_w/\text{g.mol}^{-1} = 420$,
                      $\rho = 1.1182$ g/cm^3 (303K), Hoechst AG, Germany
Solvent (A):   **trichloromethane**   **CHCl₃**   **67-66-3**

$T/K = 303.15$

$\Delta_M H/(J/g) = w_B(1 - w_B)[-92.66 - 29.09(1 - 2w_B) - 4.34(1 - 2w_B)^2 + 21.95(1 - 2w_B)^3 + 24.81(1 - 2w_B)^4]$

Polymer (B):   **poly(ethylene glycol)**   **1993ZEL**
Characterization: $M_n/\text{g.mol}^{-1} = 590$, $M_w/\text{g.mol}^{-1} = 615$,
                      $\rho = 1.1183$ g/cm^3 (303K), Hoechst AG, Germany
Solvent (A):   **trichloromethane**   **CHCl₃**   **67-66-3**

$T/K = 303.15$

$\Delta_M H/(J/g) = w_B(1 - w_B)[-91.85 - 54.82(1 - 2w_B) + 10.28(1 - 2w_B)^2 + 48.49(1 - 2w_B)^3]$

Polymer (B):   **poly(ethylene glycol)**   **1967KA1**
Characterization: $M_n/\text{g.mol}^{-1} = 200$
Solvent (A):   **water**   **H₂O**   **7732-18-5**

$T/K = 298.15$

$V^{(1)}/\text{cm}^3$	14.05	19.38	6.85	11.12	15.39	35.66	45.72	6.25	6.25
$\varphi_B^{(1)}$	1.000	1.000	1.000	0.616	0.445	0.191	0.148	1.000	1.000
$V^{(2)}/\text{cm}^3$	19.07	28.89	11.12	15.39	25.50	45.72	55.91	16.38	26.38
$\varphi_B^{(2)}$	0.713	0.670	0.616	0.445	0.268	0.148	0.121	0.381	0.236
$\Delta_{dil}H^{12}/\text{J}$	−463.1	−781.3	−321.4	−135.1	−133.1	−33.1	−21.9	−456.6	−549.8

continued

continued

$V^{(1)}/cm^3$	3.10	3.10
$\varphi_B^{(1)}$	1.000	1.000
$V^{(2)}/cm^3$	23.25	53.25
$\varphi_B^{(2)}$	0.133	0.058
$\Delta_{dil} H^{12}/J$	−156.3	−68.95

Comments: $\Delta_{dil} H^{12}$ is the extensive quantity obtained for a given total volume change from $V^{(1)}$ to $V^{(2)}$, where $\varphi_B^{(1)}$ and $\varphi_B^{(2)}$ denote the volume fractions of the polymer in the solution before and after the dilution process.

Polymer (B):	**poly(ethylene glycol)**	**1983LAK**
Characterization:	$M_n/g.mol^{-1} = 200$,	
	fractionated samples supplied by Union Carbide Corp.	
Solvent (A):	**water** **H$_2$O**	**7732-18-5**

$T/K = 321.35$

w_B	0.00356	0.0108	0.0227	0.0489	0.0733	0.0932	0.193	0.285
$\Delta_M H/(J/mol)$	−5.48	−26.6	−53.6	−116	−172	−255	−527	−862

w_B	0.356	0.412	0.529	0.601	0.686	0.786	0.888	0.979
$\Delta_M H/(J/mol)$	−1090	−1260	−1840	−2060	−2290	−2430	−2120	−849

Polymer (B):	**poly(ethylene glycol)**	**1957MAL**
Characterization:	$M_n/g.mol^{-1} = 300$, Oxirane Ltd., Manchester	
Solvent (A):	**water** **H$_2$O**	**7732-18-5**

$T/K = 353.45$

w_B	0.287	0.475	0.630	0.719	0.820	0.897	0.948
$\Delta_M H/(J/g)$	−26.32	−33.68	−31.13	−27.57	−22.01	−12.97	−5.94

Polymer (B):	**poly(ethylene glycol)**	**1961CUN**
Characterization:	$M_n/g.mol^{-1} = 300$, Oxirane Ltd., Manchester	
Solvent (A):	**water** **H$_2$O**	**7732-18-5**

$T/K = 300.05$

w_B	0.130	0.209	0.296	0.323	0.392	0.411	0.478	0.520
$\Delta_M H/(J/g)$	−18.83	−30.79	−40.42	−41.63	−49.37	−49.79	−53.14	−53.97

w_B	0.636	0.722	0.816	0.886
$\Delta_M H/(J/g)$	−53.56	−43.93	−32.01	−23.22

Polymer (B):	**poly(ethylene glycol)**							**1966LA2**
Characterization:	M_n/g.mol^{-1} = 335,							
	fractionated samples supplied by Union Carbide Corp.							
Solvent (A):	**water**		**H$_2$O**					**7732-18-5**

T/K = 300.05

w_B	0.02147	0.03852	0.07416	0.1190	0.1670	0.2310	0.3337	0.4022
$\Delta_M H$/(J/mol)	−57.90	−110.5	−204.0	−335.6	−434.2	−693.3	−1078	−1311

Polymer (B):	**poly(ethylene glycol)**							**1983LAK**
Characterization:	M_n/g.mol^{-1} = 400,							
	fractionated samples supplied by Union Carbide Corp.							
Solvent (A):	**water**		**H$_2$O**					**7732-18-5**

T/K = 321.35

w_B	0.00505	0.00948	0.0139	0.0232	0.0476	0.0723	0.0997	0.143
$\Delta_M H$/(J/mol)	−12.8	−22.1	−41.7	−58.6	−134	−228	−287	−477

w_B	0.221	0.295	0.362	0.424	0.476	0.556	0.662	0.807
$\Delta_M H$/(J/mol)	−770	−1080	−1360	−1640	−1790	−2150	−2590	−2900

w_B	0.846	0.877	0.976	0.981
$\Delta_M H$/(J/mol)	−3000	−2710	−1590	−1270

Polymer (B):	**poly(ethylene glycol)**						**1995GRO, 1995TIN**	
Characterization:	M_n/g.mol^{-1} = 414.5, Polymer Standard Services, Mainz, Germany							
Solvent (A):	**water**		**H$_2$O**				**7732-18-5**	

T/K = 298.15

$m_A{}'$/g	28.17	3.371	22.17	28.25	20.74	6.353	9.333	28.04	12.42
$m_B{}'$/g	0.0	27.34	7.390	0.0	8.901	25.24	21.90	0.0	18.24
$m_A{}''$/g	16.62	0.0	19.63	14.75	20.19	0.0	0.0	11.07	0.0
$m_B{}''$/g	4.158	22.34	0.0	6.302	0.0	22.65	22.52	10.95	22.63
$\Delta_{dil} H^{12}$/J	−42.2	−59.8	−71.8	−112	−115	−136	−277	−416	−528

$m_A{}'$/g	28.52	15.28	28.14	27.92	28.86	18.05	39.63	28.02	0.0
$m_B{}'$/g	0.0	15.28	0.0	0.0	0.0	12.03	0.0	0.0	44.43
$m_A{}''$/g	8.913	0.0	6.671	6.713	6.757	0.0	0.0	4.557	8.458
$m_B{}''$/g	13.37	22.92	15.63	15.69	15.77	22.63	9.667	18.25	0.0
$\Delta_{dil} H^{12}$/J	−705	−865	−1081	−1091	−1106	−1240	−1472	−1585	−1646

$m_A{}'$/g	0.0	0.0	23.31	28.23	25.83	28.01	0.0	27.92	0.0
$m_B{}'$/g	44.43	44.27	5.810	0.0	2.867	0.0	31.90	0.0	31.70
$m_A{}''$/g	8.458	8.423	0.0	2.247	0.0	0.0	19.94	0.0	20.03
$m_B{}''$/g	0.0	0.0	22.44	20.47	22.71	22.70	0.0	22.60	0.0
$\Delta_{dil} H^{12}$/J	−1664	−1698	−2065	−2137	−2441	−2745	−2763	−2720	−2793

continued

continued

m_A'/g	28.23	3.496	22.03	20.68	6.366	9.387	27.92	12.48	27.92
m_B'/g	0.0	28.36	7.342	8.874	25.29	21.90	0.0	18.71	0.0
m_A''/g	16.95	0.0	19.96	20.12	0.0	0.0	11.23	0.0	8.912
m_B''/g	4.239	22.36	0.0	0.0	22.72	22.60	11.12	22.46	13.37
$\Delta_{dil}H^{12}/J$	−42.7	−61.6	−72.5	−114	−134	−277	−413	−518	−700

m_A'/g	15.22	28.25	28.61	28.16	18.16	39.54	28.11	0.0	20.75
m_B'/g	15.22	0.0	0.0	0.0	12.10	0.0	0.0	44.27	8.900
m_A''/g	0.0	6.619	6.718	6.626	0.0	0.0	4.585	8.423	0.0
m_B''/g	22.69	15.50	15.68	15.52	22.20	9.574	18.36	0.0	22.56
$\Delta_{dil}H^{12}/J$	−859	−1050	−1085	−1094	−1229	−1380	−1585	−1638	−1649

m_A'/g	20.58	23.30	28.31	25.68
m_B'/g	8.827	5.817	0.0	2.851
m_A''/g	0.0	0.0	2.240	0.0
m_B''/g	22.66	22.50	20.41	22.73
$\Delta_{dil}H^{12}/J$	−1668	−2058	−2119	−2423

$T/K = 333.15$

m_A'/g	39.03	13.05	38.93	0.0	39.02	12.87	39.05	0.0
m_B'/g	0.0	30.44	0.0	43.24	0.0	30.01	0.0	43.54
m_A''/g	2.735	8.085	0.0	8.150	2.835	8.245	0.0	8.342
m_B''/g	6.376	0.0	8.993	0.0	6.608	0.0	8.862	0.0
$\Delta_{dil}H^{12}/J$	−447	−588	−1096	−1273	−468	−608	−1101	−1281

Comments: $\Delta_{dil}H^{12}$ is the extensive enthalpy change in the result of the mixing process obtained for the given masses in the table, where the superscripts ' and " designate the two solutions placed in the two parts of a mixing cell, respectively.

Polymer (B):	**poly(ethylene glycol)**		**1979KOL, 1981KOL**
Characterization:	$M_n/g.mol^{-1} = 560 \pm 12$, Hoechst AG, Germany		
Solvent (A):	**water**	H_2O	**7732-18-5**

$T/K = 303.15$

$c_B^{(1)}/(g/cm^3)$	0.1	0.02	0.010	0.020	0.031	0.051	0.101	0.200
$c_B^{(2)}/(g/cm^3)$	0.005	0.005	0.005	0.005	0.005	0.005	0.005	0.005
$\Delta_{dil}H^{12}/(J/g\ polymer)$	−7.3	−1.16	−0.5	−1.1	−2.1	−3.0	−7.3	−14.7

$c_B^{(1)}/(g/cm^3)$	0.404	0.450	0.564	1.12
$c_B^{(2)}/(g/cm^3)$	0.005	0.005	0.005	0.005
$\Delta_{dil}H^{12}/(J/g\ polymer)$	−31.6	−32.3	−43.0	−83.3

Comments: $c_B^{(1)}$ and $c_B^{(2)}$ denote the concentrations of the polymer in the solution before and after the dilution process.

Polymer (B):	**poly(ethylene glycol)**		**1967KA1**
Characterization:	$M_n/g.mol^{-1} = 600$		
Solvent (A):	**water**	H_2O	**7732-18-5**

continued

continued

$T/K = 298.15$

$V^{(1)}/cm^3$	5.0	10.0	25.0	40.0	5.0	5.0	5.0	25.0	30.0
$\varphi_B^{(1)}$	1.000	0.500	0.200	0.125	1.000	1.000	1.000	0.200	0.167
$V^{(2)}/cm^3$	10.0	15.0	30.0	45.0	25.0	35.0	45.0	30.0	35.0
$\varphi_B^{(2)}$	0.500	0.333	0.167	0.111	0.200	0.143	0.111	0.167	0.143
$\Delta_{dil}H^{12}/J$	−235.4	−70.75	−14.39	−5.90	−358.6	−383.8	−399.0	−14.39	−9.96

Comments: $\Delta_{dil}H^{12}$ is the extensive quantity obtained for a given total volume change from $V^{(1)}$ to $V^{(2)}$, where $\varphi_B^{(1)}$ and $\varphi_B^{(2)}$ denote the volume fractions of the polymer in the solution before and after the dilution process.

Polymer (B):	**poly(ethylene glycol)**		**1966LA2**
Characterization:	$M_n/g.mol^{-1} = 654$,		
	fractionated samples supplied by Union Carbide Corp.		
Solvent (A):	**water**	**H_2O**	**7732-18-5**

$T/K = 300.05$

w_B	0.01907	0.03857	0.07766	0.1075	0.1668	0.2310	0.3395	0.4131
$\Delta_M H/(J/mol)$	−60.71	−119.7	−240.6	−333.9	−530.08	−775.7	−1184	−1444

Polymer (B):	**poly(ethylene glycol)**		**1966LA2**
Characterization:	$M_n/g.mol^{-1} = 920$,		
	fractionated samples supplied by Union Carbide Corp.		
Solvent (A):	**water**	**H_2O**	**7732-18-5**

$T/K = 300.05$

w_B	0.01320	0.02658	0.05175	0.07574	0.1205	0.1399	0.2543	0.3236
$\Delta_M H/(J/mol)$	−28.87	−68.03	−194.3	−189.5	−302.4	−372.2	−616.7	−737.6

Comments: These data include the enthalpy change caused by melting the crystalline parts of the polymer when mixing the polymer and the solvent to form the solution.

Polymer (B):	**poly(ethylene glycol)**		**1995GRO, 1995TIN**
Characterization:	$M_n/g.mol^{-1} = 943.1$, Polymer Standard Services, Mainz, Germany		
Solvent (A):	**water**	**H_2O**	**7732-18-5**

$T/K = 298.15$

m_A'/g	28.27	28.19	12.29	28.19	23.35	23.34	27.76	12.42	28.22
m_B'/g	0.0	0.0	18.43	0.0	5.838	5.836	0.0	18.62	0.0
m_A''/g	16.91	8.760	19.81	4.515	19.90	20.40	8.793	19.88	4.615
m_B''/g	4.226	13.13	0.0	17.88	0.0	0.0	13.18	0.0	18.28
$\Delta_{dil}H^{12}/J$	−37.9	−715	−769	−1686	−37.0	−38.9	−732	−783	−1755

continued

continued

$T/K = 333.15$

m_A'/g	38.85	12.92	0.0	38.98	38.84	12.82	0.0	38.88
m_B'/g	0.0	30.11	43.60	0.0	0.0	29.905	43.12	0.0
m_A''/g	2.826	8.056	8.191	0.0	2.846	8.052	8.167	0.0
m_B''/g	6.584	0.0	0.0	9.169	6.630	0.0	0.0	9.421
$\Delta_{dil} H^{12}/J$	−491	−647	−1099	−1157	−503	−667	−1104	−1161

Comments: $\Delta_{dil} H^{12}$ is the extensive enthalpy change in the result of the mixing process obtained for the given masses in the table, where the superscripts ' and " designate the two solutions placed in the two parts of a mixing cell, respectively.

Polymer (B):	**poly(ethylene glycol)**	**1983LAK**
Characterization:	$M_n/\text{g.mol}^{-1} = 990$,	
	fractionated samples supplied by Union Carbide Corp.	
Solvent (A):	**water** **H₂O**	**7732-18-5**

$T/K = 321.35$

w_B	0.00364	0.0140	0.0243	0.0357	0.0481	0.0618	0.0817	0.135
$\Delta_M H/(\text{J/mol})$	−6.44	−26.4	−51.9	−105	−126	−165	−227	−416

w_B	0.221	0.300	0.371	0.486	0.576	0.694	0.878	0.972
$\Delta_M H/(\text{J/mol})$	−703	−992	−1240	−1820	−2260	−2670	−3080	−2260

Polymer (B):	**poly(ethylene glycol)**	**1967KA1**
Characterization:	$M_n/\text{g.mol}^{-1} = 1000$	
Solvent (A):	**water** **H₂O**	**7732-18-5**

$T/K = 298.15$

$V^{(1)}/\text{cm}^3$	26.6	46.6	15.3	25.3	50.3	10.5	30.3	30.3	36.6
$\varphi_B^{(1)}$	0.230	0.131	0.341	0.206	0.103	0.096	0.172	0.172	0.167
$V^{(2)}/\text{cm}^3$	36.6	66.6	25.3	30.3	60.3	20.5	50.3	60.3	66.6
$\varphi_B^{(2)}$	0.167	0.092	0.206	0.172	0.086	0.046	0.103	0.086	0.092
$\Delta_{dil} H^{12}/J$	−26.13	−15.43	−49.73	−11.53	−5.548	−2.853	−22.66	−28.21	−30.34

$V^{(1)}/\text{cm}^3$	36.6	10.5
$\varphi_B^{(1)}$	0.167	0.050
$V^{(2)}/\text{cm}^3$	76.6	20.2
$\varphi_B^{(2)}$	0.080	0.026
$\Delta_{dil} H^{12}/J$	−35.65	−0.812

Comments: $\Delta_{dil} H^{12}$ is the extensive quantity obtained for a given total volume change from $V^{(1)}$ to $V^{(2)}$, where $\varphi_B^{(1)}$ and $\varphi_B^{(2)}$ denote the volume fractions of the polymer in the solution before and after the dilution process.

Polymer (B):   **poly(ethylene glycol)**   **1979KOL, 1981KOL**
Characterization: M_n/g.mol^{-1} = 1050 ± 60, Hoechst AG, Germany
Solvent (A):   **water**   **H$_2$O**   **7732-18-5**

T/K = 303.15

$c_B^{(1)}$/(g/cm^3)	0.1	0.02
$c_B^{(2)}$/(g/cm^3)	0.005	0.005
$\Delta_{dil}H^{12}$/(J/g polymer)	−6.1	−0.98

Comments: $c_B^{(1)}$ and $c_B^{(2)}$ denote the concentrations of the polymer in the solution before and after the dilution process.

Polymer (B):   **poly(ethylene glycol)**   **1966LA2**
Characterization: M_n/g.mol^{-1} = 1350,
          fractionated samples supplied by Union Carbide Corp.
Solvent (A):   **water**   **H$_2$O**   **7732-18-5**

T/K = 300.05

w_B	0.03192	0.05559	0.1054	0.1501	0.2275	0.3062	0.4120	0.5130
$\Delta_M H$/(J/mol)	−23.43	−45.44	−79.29	−115.9	−187.0	−200.4	−322.5	−263.3

Comments: These data include the enthalpy change caused by melting the crystalline parts of the polymer when mixing the polymer and the solvent to form the solution.

Polymer (B):   **poly(ethylene glycol)**   **1983LAK**
Characterization: M_n/g.mol^{-1} = 1460,
          fractionated samples supplied by Union Carbide Corp.
Solvent (A):   **water**   **H$_2$O**   **7732-18-5**

T/K = 321.35

w_B	0.00232	0.00678	0.0145	0.0283	0.0376	0.0600	0.0647	0.0826
$\Delta_M H$/(J/mol)	−4.94	−14.7	−36.9	−76.6	−107	−175	−187	−223

w_B	0.120	0.143	0.184	0.272	0.345	0.418	0.557	0.661
$\Delta_M H$/(J/mol)	−366	−435	−548	−866	−1120	−1460	−2330	−2780

w_B	0.771	0.917	0.976
$\Delta_M H$/(J/mol)	−3060	−2960	−2080

Polymer (B):   **poly(ethylene glycol)**   **1979KOL, 1981KOL**
Characterization: M_n/g.mol^{-1} = 1610 ± 90, Hoechst AG, Germany
Solvent (A):   **water**   **H$_2$O**   **7732-18-5**

T/K = 303.15

$c_B^{(1)}$/(g/cm^3)	0.1	0.02
$c_B^{(2)}$/(g/cm^3)	0.005	0.005
$\Delta_{dil}H^{12}$/(J/g polymer)	−5.9	−0.90

Polymer (B):	**poly(ethylene glycol)**					**1979MON**
Characterization:	M_n/g.mol^{-1} = 2000, Merck-Schuchardt, Germany					
Solvent (A):	**water**		**H$_2$O**			**7732-18-5**

T/K = 298.31

$\varphi_B^{(1)}$	0.0747	0.0721	0.0675	0.0634	0.0617	0.567	0.535
$\varphi_B^{(2)}$	0.0721	0.0675	0.0634	0.0617	0.0567	0.0535	0.0497
$\Delta_{dil}H^{12}$/(J/g polymer)	−0.219	−0.461	−0.394	−0.015	−0.472	−0.297	−0.355

Comments: $\varphi_B^{(1)}$ and $\varphi_B^{(2)}$ denote the volume fractions of the polymer in the solution before and after the dilution process.

Polymer (B):	**poly(ethylene glycol)**					**1980SCH**
Characterization:	M_n/g.mol^{-1} = 2000, Merck-Schuchardt, Germany					
Solvent (A):	**water**		**H$_2$O**			**7732-18-5**

T/K = 298.31

$w_B^{(1)}$	0.1088	0.1052	0.0987	0.0929	0.0905	0.0834
$m^{(1)}$/g	29.03	30.03	32.02	34.01	34.90	37.89
$w_B^{(2)}$	0.1052	0.0987	0.0929	0.0905	0.0834	0.0788
$m^{(2)}$/g	30.03	32.02	34.01	34.90	37.89	40.08
$\Delta_{dil}H^{12}$/J	−0.693	−1.391	−1.245	−0.465	−1.493	−0.940

Comments: $\Delta_{dil}H^{12}$ is here an extensive quantity obtained for a given total mass change from $m^{(1)}$ to $m^{(2)}$. $w_B^{(1)}$ and $w_B^{(2)}$ are the mass fractions of the polymer in the solution before and after dilution.

Polymer (B):	**poly(ethylene glycol)**						**1979MON**
Characterization:	M_η/g.mol^{-1} = 2300, Serva, Heidelberg, Germany						
Solvent (A):	**water**		**H$_2$O**				**7732-18-5**

T/K = 298.31

$\varphi_B^{(1)}$	0.0606	0.0584	0.0408	0.0390	0.0367	0.0346	0.0269	0.0458
$\varphi_B^{(2)}$	0.0584	0.0563	0.0394	0.0367	0.0346	0.0327	0.0261	0.0442
$\Delta_{dil}H^{12}$/(J/g polymer)	−0.192	−0.175	−0.129	−0.192	−0.157	−0.163	−0.060	−0.146
$\varphi_B^{(1)}$	0.1529	0.1482	0.1438	0.1397	0.1266	0.1234	0.1028	0.0958
$\varphi_B^{(2)}$	0.1482	0.1438	0.1397	0.1358	0.1234	0.1203	0.0922	0.0917
$\Delta_{dil}H^{12}$/(J/g polymer)	−0.412	−0.405	−0.391	−0.368	−0.306	−0.296	−0.351	−0.420
$\varphi_B^{(1)}$	0.0526	0.0473	0.0364	0.0338	0.0473	0.0313	0.1214	0.1214
$\varphi_B^{(2)}$	0.0473	0.0430	0.0338	0.0313	0.0430	0.0289	0.0529	0.0531
$\Delta_{dil}H^{12}$/(J/g polymer)	−0.402	−0.336	−0.212	−0.173	−0.334	−0.167	−6.49	−6.57

T/K = 304.80

$\varphi_B^{(1)}$	0.1353	0.1221	0.1188	0.1127	0.1098	0.0964	0.0943	0.0923
$\varphi_B^{(2)}$	0.1333	0.1188	0.1156	0.1098	0.1032	0.0943	0.0923	0.0904
$\Delta_{dil}H^{12}$/(J/g polymer)	−0.153	−0.246	−0.250	−0.222	−0.578	−0.188	−0.161	−0.165

continued

continued

$\varphi_B^{(1)}$	0.0904	0.0877	0.0812	0.0783	0.0683	0.0638	0.0590	0.0453
$\varphi_B^{(2)}$	0.0877	0.0860	0.0797	0.0769	0.0638	0.0590	0.0549	0.0423
$\Delta_{dil}H^{12}$/(J/g polymer)	−0.250	−0.153	−0.122	−0.126	−0.424	−0.431	−0.384	−0.252

$T/K = 330.48$

$\varphi_B^{(1)}$	0.0718	0.0694	0.0374	0.0345	0.1629	0.1609	0.1588	0.1380
$\varphi_B^{(2)}$	0.0636	0.0670	0.0345	0.0321	0.1609	0.1588	0.1569	0.1366
$\Delta_{dil}H^{12}$/(J/g polymer)	−0.266	−0.197	−0.219	−0.193	−0.211	−0.208	−0.203	−0.153

$\varphi_B^{(1)}$	0.1049	0.1032	0.1015
$\varphi_B^{(2)}$	0.1032	0.1015	0.0999
$\Delta_{dil}H^{12}$/(J/g polymer)	−0.160	−0.164	−0.153

Comments: $\varphi_B^{(1)}$ and $\varphi_B^{(2)}$ denote the volume fractions of the polymer in the solution before and after the dilution process.

Polymer (B):	**poly(ethylene glycol)**		**1980SCH**
Characterization:	M_η/g.mol^{-1} = 2300, Serva, Heidelberg, Germany		
Solvent (A):	**water**	**H$_2$O**	**7732-18-5**

$T/K = 298.31$

$w_B^{(1)}$	0.0897	0.0858	0.0605	0.0579	0.0545	0.0514	0.0402	0.0678	0.0654
$m^{(1)}$/g	54.29	56.28	45.14	47.13	50.12	53.11	64.88	54.69	56.69
$w_B^{(2)}$	0.0859	0.0829	0.0585	0.0546	0.0515	0.0487	0.0390	0.0655	0.0611
$m^{(2)}$/g	56.70	58.26	46.68	50.01	53.05	56.04	66.86	56.64	60.67
$\Delta_{dil}H^{12}$/J	−0.935	−0.847	−0.352	−0.525	−0.430	−0.446	−0.156	−0.541	−0.973

$w_B^{(1)}$	0.2145	0.2085	0.2027	0.1973	0.1799	0.1756	0.1478	0.1383	0.1273
$m^{(1)}$/g	34.22	35.22	36.22	37.21	40.80	41.80	43.45	46.44	50.43
$w_B^{(2)}$	0.2086	0.2028	0.1974	0.1925	0.1758	0.1716	0.1429	0.1326	0.1192
$m^{(2)}$/g	35.19	36.21	37.19	38.19	41.76	42.76	44.92	48.42	53.87
$\Delta_{dil}H^{12}$/J	−3.028	−2.977	−2.870	−2.703	−2.244	−2.169	−2.255	−2.701	−3.700

$w_B^{(1)}$	0.0775	0.0699	0.0541	0.0503	0.0699	0.0466	0.0775	0.0637	0.0637
$m^{(1)}$/g	36.88	40.86	52.83	56.82	40.87	61.31	36.88	44.85	44.86
$w_B^{(2)}$	0.0699	0.0637	0.0503	0.0467	0.0637	0.0431	0.0699	0.0585	0.0585
$m^{(2)}$/g	40.86	44.82	56.80	61.25	44.83	66.23	40.86	48.84	48.85
$\Delta_{dil}H^{12}$/J	−1.150	−0.959	−0.605	−0.493	−0.955	−0.477	−1.190	−0.819	−0.850

$T/K = 304.80$

$w_B^{(1)}$	0.1915	0.1739	0.1695	0.1612	0.1574	0.1391	0.1362	0.1334	0.1308
$m^{(1)}$/g	34.47	37.96	38.96	40.95	41.95	47.47	48.47	49.46	50.46
$w_B^{(2)}$	0.1889	0.1696	0.1653	0.1574	0.1484	0.1362	0.1334	0.1308	0.1271
$m^{(2)}$/g	34.94	38.93	39.96	41.93	44.48	48.47	49.46	50.46	51.93
$\Delta_{dil}H^{12}$/J	−1.011	−1.327	−1.653	−1.468	−3.814	−1.239	−1.060	−1.091	−1.652

$w_B^{(1)}$	0.1271	0.1180	0.1139	0.1035	0.0936	0.0867	0.0670
$m^{(1)}$/g	51.96	55.95	57.94	29.39	31.39	33.88	43.85
$w_B^{(2)}$	0.1247	0.1160	0.1121	0.0966	0.0867	0.0809	0.0627
$m^{(2)}$/g	52.94	56.93	58.90	31.41	33.87	36.32	46.85
$\Delta_{dil}H^{12}$/J	−1.009	−0.803	−0.834	−1.246	−1.265	−1.127	−0.741

continued

continued

$T/K = 330.48$

$w_B^{(1)}$	0.1014	0.0555	0.0514	0.2276	0.2249	0.2223	0.1951	0.1506	0.1486
$m^{(1)}/g$	57.47	36.82	39.81	41.95	42.45	42.95	48.93	63.39	64.38
$w_B^{(2)}$	0.0981	0.0514	0.0478	0.2251	0.2224	0.2199	0.1933	0.1484	0.1461
$m^{(2)}/g$	59.40	39.80	42.70	42.42	42.93	43.42	49.38	64.31	65.34
$\Delta_{dil} H^{12}/J$	−1.148	−0.448	−0.395	−2.011	−1.986	−1.938	−1.457	−1.530	−1.564

$w_B^{(1)}$	0.1460
$m^{(1)}/g$	65.38
$w_B^{(2)}$	0.1439
$m^{(2)}/g$	66.32
$\Delta_{dil} H^{12}/J$	−1.464

Comments: $\Delta_{dil} H^{12}$ is here an extensive quantity obtained for a given total mass change from $m^{(1)}$ to $m^{(2)}$. $w_B^{(1)}$ and $w_B^{(2)}$ denote the mass fractions of the polymer in the solution before and after the dilution process.

Polymer (B):	**poly(ethylene glycol)**		**1966LA2**
Characterization:	$M_n/g.mol^{-1} = 2825$,		
	fractionated samples supplied by Union Carbide Corp.		
Solvent (A):	**water**	H_2O	**7732-18-5**

$T/K = 300.05$

w_B	0.01965	0.03855	0.1075	0.1669	0.2384	0.2929	0.3854	0.4142
$\Delta_M H/(J/mol)$	1.63	5.40	28.24	60.0	130.2	204.3	325.0	518.0

Comments: These data include the enthalpy change caused by melting the crystalline parts of the polymer when mixing the polymer and the solvent to form the solution.

Polymer (B):	**poly(ethylene glycol)**		**1957MAL**
Characterization:	$M_n/g.mol^{-1} = 3000$, Oxirane Ltd., Manchester		
Solvent (A):	**water**	H_2O	**7732-18-5**

$T/K = 353.45$

w_B	0.102	0.303	0.517	0.719	0.780	0.847	0.904
$\Delta_M H/(J/g)$	−9.71	−27.57	−34.06	−24.77	−19.50	−12.01	−5.56

Comments: These data include the enthalpy change caused by melting the crystalline parts of the polymer when mixing the polymer and the solvent to form the solution.

Polymer (B):	**poly(ethylene glycol)**		**1979KOL, 1981KOL**
Characterization:	$M_n/g.mol^{-1} = 3200$, Hoechst AG, Germany		
Solvent (A):	**water**	H_2O	**7732-18-5**

$T/K = 303.15$

$c_B^{(1)}/(g/cm^3)$	0.1	0.02
$c_B^{(2)}/(g/cm^3)$	0.005	0.005
$\Delta_{dil} H^{12}/(J/g \text{ polymer})$	−5.1	−0.71

Comments: $c_B^{(1)}$ and $c_B^{(2)}$ denote the concentrations of the polymer in the solution before and after the dilution process.

Polymer (B):	**poly(ethylene glycol)**						**1967KA1**	
Characterization:	$M_n/\text{g.mol}^{-1} = 4000$							
Solvent (A):	**water**		H_2O				**7732-18-5**	

$T/\text{K} = 298.15$

$V^{(1)}/\text{cm}^3$	30.4	35.5	51.9	20.1	20.1	20.1	20.1	61.9	60.5
$\varphi_B^{(1)}$	0.177	0.152	0.244	0.270	0.270	0.270	0.270	0.205	0.060
$V^{(2)}/\text{cm}^3$	35.5	40.5	61.9	30.3	35.4	40.5	45.6	71.9	70.5
$\varphi_B^{(2)}$	0.152	0.133	0.205	0.177	0.152	0.133	0.166	0.174	0.051
$\Delta_{dil}H^{12}/\text{J}$	−4.703	−3.757	−20.10	−18.95	−23.66	−27.41	−30.13	−12.54	−1.079

Comments: $\Delta_{dil}H^{12}$ is the extensive quantity obtained for a given total volume change from $V^{(1)}$ to $V^{(2)}$, where $\varphi_B^{(1)}$ and $\varphi_B^{(2)}$ denote the volume fractions of the polymer in the solution before and after the dilution process.

Polymer (B):	**poly(ethylene glycol)**						**1983LAK**	
Characterization:	$M_n/\text{g.mol}^{-1} = 4150$,							
	fractionated samples supplied by Union Carbide Corp.							
Solvent (A):	**water**		H_2O				**7732-18-5**	

$T/\text{K} = 321.35$

w_B	0.00333	0.0108	0.0229	0.0504	0.0768	0.112	0.123	0.221
$\Delta_M H/\text{(J/mol)}$	4.52	12.1	21.8	94.1	64.9	109	84.9	169

w_B	0.296	0.360	0.412	0.458
$\Delta_M H/\text{(J/mol)}$	257	481	690	921

Comments: These data include the enthalpy change caused by melting the crystalline parts of the polymer when mixing the polymer and the solvent to form the solution.

Polymer (B):	**poly(ethylene glycol)**						**1957MAL**	
Characterization:	$M_n/\text{g.mol}^{-1} = 5000$, Oxirane Ltd., Manchester							
Solvent (A):	**water**		H_2O				**7732-18-5**	

$T/\text{K} = 353.45$

w_B	0.206	0.336	0.480	0.548	0.675	0.755	0.825	0.887	0.953
$\Delta_M H/\text{(J/g)}$	−20.04	−29.41	−34.31	−33.10	−25.86	−20.46	13.60	−5.82	−1.26

Polymer (B):	**poly(ethylene glycol)**						**1979KOL, 1981KOL**	
Characterization:	$M_n/\text{g.mol}^{-1} = 5850 \pm 250$, Hoechst AG, Germany							
Solvent (A):	**water**		H_2O				**7732-18-5**	

$T/\text{K} = 303.15$

$c_B^{(1)}/\text{(g/cm}^3)$		0.1	0.01	0.02	0.03	0.101	0.200	0.302	0.404
$c_B^{(2)}/\text{(g/cm}^3)$			0.005	0.005	0.005	0.005	0.005	0.005	0.005
$\Delta_{dil}H^{12}/\text{(J/g polymer)}$		−4.6	−0.40	−0.60	−0.94	−4.55	−10.0	−17.7	−23.1

Comments: $c_B^{(1)}$ and $c_B^{(2)}$ denote the concentrations of the polymer in the solution before and after the dilution process.

Polymer (B):   **poly(ethylene glycol)**   **1995GRO, 1995TIN**

Characterization: $M_n/\text{g.mol}^{-1} = 6230$, $M_w/\text{g.mol}^{-1} = 6480$

Lot 664762, Hoechst AG, Germany

Solvent (A):   **water**   **H$_2$O**   **7732-18-5**

$T/\text{K} = 298.15$

m_A'/g	26.99	39.61	36.37	28.01	25.59	24.34	39.70	32.97	32.81
m_B'/g	1.438	0.0	4.040	0.0	2.843	4.297	0.0	8.241	8.202
m_A''/g	20.35	7.743	8.319	18.39	19.74	20.21	6.963	8.477	8.690
m_B''/g	0.0	0.861	0.0	2.043	0.0	0.0	1.741	0.0	0.0
$\Delta_{dil}H^{12}/\text{J}$	−1.1	−2.6	−3.8	−5.3	−5.9	−16.1	−17.1	−21.4	−22.0
m_A'/g	23.36	28.09	23.81	22.98	39.67	21.92	29.45	29.02	20.46
m_B'/g	5.436	0.0	5.970	6.269	0.0	7.274	12.62	12.43	8.312
m_A''/g	20.29	16.78	19.83	20.36	6.186	20.02	8.323	8.693	20.05
m_B''/g	0.0	4.195	0.0	0.0	2.652	0.0	0.0	0.0	0.0
$\Delta_{dil}H^{12}/\text{J}$	−29.2	−32.4	−33.5	−41.1	−50.5	−60.2	−63.8	−67.3	−88.6
m_A'/g	25.42	27.96	20.94	20.49	39.78	25.96	25.56	19.67	18.30
m_B'/g	16.95	0.0	8.988	8.794	0.0	17.30	17.04	10.63	12.00
m_A''/g	3.755	14.96	19.84	20.26	5.376	8.388	8.405	20.20	19.91
m_B''/g	5.633	6.411	0.0	0.0	3.586	0.0	0.0	0.0	0.0
$\Delta_{dil}H^{12}/\text{J}$	−92.8	−95.5	−100	−100	−112	−151	−154	−156	−205
m_A'/g	28.00	39.60	39.66	18.50	18.20	16.52	21.74	21.94	17.13
m_B'/g	0.0	0.0	0.0	12.22	12.13	28.12	21.73	21.93	13.94
m_A''/g	13.07	4.666	4.610	20.54	20.09	6.569	8.292	9.597	20.18
m_B''/g	8.708	4.663	4.607	0.0	0.0	2.814	0.0	0.0	0.0
$\Delta_{dil}H^{12}/\text{J}$	−217	−219	−222	−229	−231	−268	−309	−316	−323
m_A'/g	16.63	39.65	19.98	28.26	15.41	15.28	15.64	14.55	27.80
m_B'/g	13.67	0.0	24.42	0.0	15.37	15.11	15.64	16.00	0.0
m_A''/g	20.56	3.700	8.377	11.00	19.58	20.15	19.63	20.08	10.16
m_B''/g	0.0	5.554	0.0	10.98	0.0	0.0	0.0	0.0	12.24
$\Delta_{dil}H^{12}/\text{J}$	−328	−384	−423	−425	−441	−446	−456	−521	−557
m_A'/g	17.45	14.21	13.90	16.28	28.99	28.17	12.56	28.07	11.68
m_B'/g	26.17	17.18	16.98	27.72	0.0	0.0	18.83	0.0	19.88
m_A''/g	8.550	19.99	19.99	8.498	8.954	9.033	19.95	8.362	20.11
m_B''/g	0.0	0.0	0.0	0.0	13.42	13.56	0.0	14.24	0.0
$\Delta_{dil}H^{12}/\text{J}$	−570	−606	−620	−683	−746	−766	−839	−899	−988
m_A'/g	26.93	39.60	36.08	25.17	24.28	39.96	32.91	32.82	23.55
m_B'/g	1.435	0.0	4.008	2.797	4.286	0.0	8.226	8.205	5.480
m_A''/g	20.05	7.785	8.642	20.45	20.06	6.888	8.419	8.507	19.96
m_B''/g	0.0	0.865	0.0	0.0	0.0	1.722	0.0	0.0	0.0
$\Delta_{dil}H^{12}/\text{J}$	−1.4	−3.1	−3.8	−5.9	−16.0	−16.9	−21.1	−21.7	−28.7
m_A'/g	28.19	23.71	22.58	39.80	22.38	29.42	29.22	20.63	25.44
m_B'/g	0.0	5.946	6.159	0.0	7.426	12.60	12.53	8.379	16.96
m_A''/g	16.67	19.99	20.27	6.069	19.98	8.216	8.537	20.06	3.738
m_B''/g	4.167	0.0	0.0	2.602	0.0	0.0	0.0	0.0	5.608
$\Delta_{dil}H^{12}/\text{J}$	−31.9	−33.4	−40.0	−49.3	−59.8	−63.8	−65.3	−88.0	−90.3

continued

continued

m_A'/g	28.14	20.48	20.84	39.77	26.00	25.53	19.57	18.00	28.22
m_B'/g	0.0	.903	8.947	0.0	17.33	17.02	10.57	11.80	0.0
m_A''/g	14.77	19.94	19.98	5.362	8.491	8.481	20.57	19.84	13.10
m_B''/g	6.332	0.0	0.0	3.576	0.0	0.0	0.0	0.0	8.729
$\Delta_{dil}H^{12}/J$	−94.6	−97.4	−100	−109	−151	−153	−155	−203	−217

m_A'/g	24.12	18.22	23.95	18.15	16.60	22.05	21.65	16.55	17.13
m_B'/g	19.74	12.03	19.59	12.10	28.26	22.04	21.64	13.61	13.94
m_A''/g	8.716	19.89	8.687	20.05	6.543	8.359	8.312	19.97	20.03
m_B''/g	0.0	0.0	0.0	0.0	2.803	0.0	0.0	0.0	0.0
$\Delta_{dil}H^{12}/J$	−219	−221	−223	−231	−265	−308	−311	−322	−327

m_A'/g	39.71	28.17	19.97	39.74	40.01	15.57	15.58	14.55	27.97
m_B'/g	0.0	0.0	24.44	0.0	0.0	15.56	15.59	16.00	0.0
m_A''/g	3.754	10.80	8.561	3.462	3.407	19.88	20.03	20.08	10.02
m_B''/g	5.634	10.79	0.0	5.898	5.805	0.0	0.0	0.0	12.07
$\Delta_{dil}H^{12}/J$	−382	−414	−425	−440	−443	−454	−458	−525	−560

m_A'/g	17.44	14.14	16.34	12.96	28.20	21.60	28.19	11.56
m_B'/g	26.16	17.10	27.82	17.90	0.0	18.89	0.0	19.68
m_A''/g	8.506	19.97	8.366	20.07	8.845	19.93	8.230	20.13
m_B''/g	0.0	0.0	0.0	0.0	13.28	0.0	14.02	0.0
$\Delta_{dil}H^{12}/J$	−576	−611	−637	−727	−746	−831	−881	−975

$T/K = 333.15$

m_A'/g	25.00	38.80	19.01	39.22	39.06	12.80	39.22	39.11	8.812
m_B'/g	16.66	0.0	23.23	0.0	0.0	29.84	0.0	0.0	35.19
m_A''/g	8.200	3.995	6.762	3.804	1.848	8.202	1.834	0.882	8.284
m_B''/g	0.0	4.876	0.0	5.682	7.447	0.0	7.389	7.938	0.0
$\Delta_{dil}H^{12}/J$	−120	−236	−288	−324	−693	−708	−727	−945	−963

m_A'/g	8.630	2.128	25.01	39.03	39.07	18.87	39.40	12.84	38.72
m_B'/g	34.47	40.44	16.67	0.0	0.0	23.06	0.0	29.92	0.0
m_A''/g	8.268	8.443	8.325	4.037	3.778	8.314	1.814	8.510	0.907
m_B''/g	0.0	0.0	0.0	4.928	5.643	0.0	7.237	0.0	8.163
$\Delta_{dil}H^{12}/J$	−980	−1130	−120	−240	−319	−339	−701	−718	−934

m_A'/g	39.14	38.74	2.151
m_B'/g	0.0	0.0	40.87
m_A''/g	0.916	0.918	8.462
m_B''/g	8.231	8.248	0.0
$\Delta_{dil}H^{12}/J$	−948	−968	−1045

Comments: $\Delta_{dil}H^{12}$ is the extensive enthalpy change in the result of the mixing process obtained for the given masses in the table, where the superscripts ' and " designate the two solutions placed in the two parts of a mixing cell, respectively.

Polymer (B):   **poly(ethylene glycol)**   **1979KOL, 1981KOL**
Characterization: $M_n/\text{g.mol}^{-1} = 9950 \pm 700$, Hoechst AG, Germany
Solvent (A):   **water**   **H₂O**   **7732-18-5**

continued

continued

$T/K = 303.15$

$c_B^{(1)}/(g/cm^3)$	0.1	0.02
$c_B^{(2)}/(g/cm^3)$	0.005	0.005
$\Delta_{dil}H^{12}/(J/g$ polymer)	−4.0	−0.54

Comments: $c_B^{(1)}$ and $c_B^{(2)}$ denote the concentrations of the polymer in the solution before and after the dilution process.

Polymer (B):   **poly(ethylene glycol)**   **1967KA1**

Characterization: $M_n/g.mol^{-1} = 20000$

Solvent (A):   **water**   **H_2O**   **7732-18-5**

$T/K = 298.15$

$V^{(1)}/cm^3$	25.6	35.6	45.6	55.6	15.6	15.6	25.3	10.0	10.0
$\varphi_B^{(1)}$	0.211	0.152	0.119	0.096	0.346	0.346	0.190	0.301	0.325
$V^{(2)}/cm^3$	35.6	45.6	55.6	65.6	35.6	45.6	35.3	25.0	25.0
$\varphi_B^{(2)}$	0.152	0.119	0.096	0.081	0.152	0.119	0.136	0.120	0.130
$\Delta_{dil}H^{12}/J$	−4.368	−2.954	−1.397	−1.058	−19.29	−22.25	−4.222	−8.309	−10.05

Comments: $\Delta_{dil}H^{12}$ is the extensive quantity obtained for a given total volume change from $V^{(1)}$ to $V^{(2)}$, where $\varphi_B^{(1)}$ and $\varphi_B^{(2)}$ denote the volume fractions of the polymer in the solution before and after the dilution process.

Polymer (B):   **poly(ethylene glycol)**   **1979KOL, 1981KOL**

Characterization: $M_n/g.mol^{-1} = 20300$, Hoechst AG, Germany

Solvent (A):   **water**   **H_2O**   **7732-18-5**

$T/K = 303.15$

$c_B^{(1)}/(g/cm^3)$	0.1	0.02
$c_B^{(2)}/(g/cm^3)$	0.005	0.005
$\Delta_{dil}H^{12}/(J/g$ polymer)	−3.5	−0.44

Comments: $c_B^{(1)}$ and $c_B^{(2)}$ denote the concentrations of the polymer in the solution before and after the dilution process.

Polymer (B):   **poly(ethylene glycol)**   **1979MON**

Characterization: $M_n/g.mol^{-1} = 30000$

Solvent (A):   **water**   **H_2O**   **7732-18-5**

$T/K = 298.31$

$\varphi_B^{(1)}$	0.0661	0.0579	0.0507	0.0457	0.0411
$\varphi_B^{(2)}$	0.0623	0.0507	0.0457	0.0411	0.0373
$\Delta_{dil}H^{12}/(J/g$ polymer)	−0.133	−0.246	−0.219	−0.183	−0.016

Comments: $\varphi_B^{(1)}$ and $\varphi_B^{(2)}$ denote the volume fractions of the polymer in the solution before and after the dilution process.

Polymer (B):	**poly(ethylene glycol)**						**1979KOL, 1981KOL**	
Characterization:	M_η/g.mol^{-1} = 34500, Hoechst AG, Germany							
Solvent (A):	**water**			**H$_2$O**			**7732-18-5**	

T/K = 303.15

$c_B^{(1)}$/(g/cm^3)	0.1	0.02
$c_B^{(2)}$/(g/cm^3)	0.005	0.005
$\Delta_{dil}H^{12}$/(J/g polymer)	−3.4	−0.44

Comments: $c_B^{(1)}$ and $c_B^{(2)}$ denote the concentrations of the polymer in the solution before and after the dilution process.

Polymer (B):	**poly(ethylene glycol)**						**1995GRO, 1995TIN**	
Characterization:	M_n/g.mol^{-1} = 39000, Hoechst AG, Germany							
Solvent (A):	**water**			**H$_2$O**			**7732-18-5**	

T/K = 298.15

m_A'/g	26.70	39.68	36.38	28.31	26.21	39.71	92.91	28.02	23.00
m_B'/g	1.471	0.0	4.042	0.0	2.975	0.0	8.227	0.0	5.835
m_A''/g	20.00	7.799	8.732	18.43	20.06	6.892	8.405	16.78	20.29
m_B''/g	0.0	0.867	0.0	2.048	0.0	1.721	0.0	4.200	0.0
$\Delta_{dil}H^{12}$/J	−0.7	−2.3	−3.4	−5.3	−5.9	−16.5	−20.9	−31.9	−33.7
m_A'/g	39.92	21.85	29.40	29.29	28.21	20.53	39.67	19.23	25.43
m_B'/g	0.0	6.971	12.60	12.55	0.0	8.601	0.0	10.17	16.95
m_A''/g	6.186	20.09	8.285	4.550	14.94	20.55	5.395	20.16	8.436
m_B''/g	2.652	0.0	0.0	5.130	6.397	0.0	3.597	0.0	0.0
$\Delta_{dil}H^{12}$/J	−48.6	−59.3	−62.5	−92.5	−94.0	−94.6	−109	−146	−150
m_A'/g	25.64	18.44	27.97	39.46	39.68	44.35	16.70	21.94	16.84
m_B'/g	17.09	11.04	0.0	0.0	0.0	0.0	13.33	21.94	14.33
m_A''/g	8.619	20.68	13.04	4.725	4.342	4.384	20.22	8.198	20.20
m_B''/g	0.0	0.0	8.683	4.729	4.894	4.943	0.0	0.0	0.0
$\Delta_{dil}H^{12}$/J	−155	−185	−215	−220	−256	−274	−303	−305	−336
m_A'/g	20.58	16.60	26.72	39.77	36.38	28.13	25.60	39.74	32.67
m_B'/g	23.21	15.19	1.473	0.0	4.043	0.0	2.907	0.0	8.166
m_A''/g	8.634	19.85	20.22	7.693	8.390	18.49	19.89	6.894	8.525
m_B''/g	0.0	0.0	0.0	0.855	0.0	2.054	0.0	1.721	0.0
$\Delta_{dil}H^{12}$/J	−380.	−399.	−1.1	−2.8	−3.9	−5.7	−6.3	−16.6	−21.1
m_A'/g	28.17	23.06	39.66	22.24	29.16	27.99	20.51	29.36	39.38
m_B'/g	0.0	5.851	0.0	7.094	12.50	0.0	8.594	12.58	0.0
m_A''/g	16.68	20.04	6.262	20.78	8.570	14.84	20.47	4.369	5.617
m_B''/g	4.174	0.0	2.684	0.0	0.0	6.356	0.0	4.927	3.746
$\Delta_{dil}H^{12}$/J	−32.0	−34.3	−49.3	−60.5	−64.2	−93.4	−94.1	−96.5	−114
m_A'/g	19.18	28.21	18.36	28.19	39.65	44.52	39.64	16.92	21.75
m_B'/g	10.15	0.0	10.99	0.0	0.0	0.0	0.0	13.29	21.74
m_A''/g	20.20	10.97	21.07	12.96	4.645	4.485	4.341	20.48	8.240
m_B''/g	0.0	10.97	0.0	8.636	4.649	5.057	5.893	0.0	0.0
$\Delta_{dil}H^{12}$/J	−146	−411	−184	−210	−217	−253	−260	−300	−303

continued

continued

m_A'/g	16.74	20.48	15.98
m_B'/g	13.97	23.09	14.62
m_A''/g	19.99	8.324	20.14
m_B''/g	0.0	0.0	0.0
$\Delta_{dil} H^{12}$/J	−334	−379	−390

T/K = 333.15

m_A'/g	18.90	18.67	38.84	39.28	38.45	12.10	12.81	8.554	18.56
m_B'/g	23.09	22.82	0.0	0.0	0.0	28.22	29.88	34.22	22.67
m_A''/g	8.688	8.177	2.165	3.089	1.436	8.990	8.268	8.382	8.748
m_B''/g	0.0	0.0	5.050	7.159	5.744	0.0	0.0	0.0	0.0
$\Delta_{dil} H^{12}$/J	−352	−363	−426	−535	−609	−656	−699	−1064	−359

m_A'/g	39.03	39.06	39.40	38.99	38.43	8.239
m_B'/g	0.0	0.0	0.0	0.0	0.0	32.96
m_A''/g	2.412	2.326	1.604	1.576	1.629	8.699
m_B''/g	5.627	5.391	6.406	6.305	6.507	0.0
$\Delta_{dil} H^{12}$/J	−402	−428	−562	−614	−670	−852

Comments: $\Delta_{dil} H^{12}$ is the extensive enthalpy change in the result of the mixing process obtained for the given masses in the table, where the superscripts ' and " designate the two solutions placed in the two parts of a mixing cell, respectively.

Polymer (B):	**poly(ethylene glycol)**		**1979KOL, 1981KOL**
Characterization:	M_η/g.mol^{-1} = 43400, Hoechst AG, Germany		
Solvent (A):	**water**	**H₂O**	**7732-18-5**

T/K = 303.15

$c_B^{(1)}$/(g/cm³)	0.10	0.010	0.030	0.051	0.101	0.198	0.300	0.416
$c_B^{(2)}$/(g/cm³)	0.005	0.005	0.005	0.005	0.005	0.005	0.005	0.005
$\Delta_{dil} H^{12}$/(J/g polymer)	−3.9	−0.22	−0.90	−1.7	−3.9	−10.0	−18.1	−21.4

Comments: $c_B^{(1)}$ and $c_B^{(2)}$ denote the concentrations of the polymer in the solution before and after the dilution process.

Polymer (B):	**poly(ethylene glycol)**		**1967KA1**
Characterization:	M_n/g.mol^{-1} = 100000		
Solvent (A):	**water**	**H₂O**	**7732-18-5**

T/K = 298.15

$V^{(1)}$/cm³	15.0	35.0	45.0	50.0	70.0	30.0	20.0	50.0	30.0
$\varphi_B^{(1)}$	0.127	0.222	0.172	0.154	0.110	0.156	0.186	0.154	0.154
$V^{(2)}$/cm³	25.0	40.0	50.0	60.0	80.0	50.0	30.0	80.0	70.0
$\varphi_B^{(2)}$	0.076	0.194	0.154	0.130	0.097	0.093	0.124	0.097	0.066
$\Delta_{dil} H^{12}$/J	−0.397	−1.331	−0.891	−1.515	−0.661	−2.477	−1.874	−3.318	−3.322

continued

continued

$V^{(1)}/cm^3$	20.0	60.0	10.0	40.0
$\varphi_B^{(1)}$	0.186	0.130	0.101	0.194
$V^{(2)}/cm^3$	50.0	70.0	30.0	45.0
$\varphi_B^{(2)}$	0.074	0.110	0.050	0.172
$\Delta_{dil}H^{12}/J$	−3.138	−1.142	−0.791	−1.167

Comments: $\Delta_{dil}H^{12}$ is the extensive quantity obtained for a given total volume change from $V^{(1)}$ to $V^{(2)}$, where $\varphi_B^{(1)}$ and $\varphi_B^{(2)}$ denote the volume fractions of the polymer in the solution before and after the dilution process.

Polymer (B):	**poly(ethylene glycol)**							**1967KA1**
Characterization:	$M_n/g.mol^{-1} = 600000$							
Solvent (A):	**water**			**H₂O**				**7732-18-5**

$T/K = 298.15$

$V^{(1)}/cm^3$	30.5	40.5	50.0	60.0	50.1	55.1	60.1	65.1	30.3
$\varphi_B^{(1)}$	0.085	0.064	0.051	0.043	0.065	0.059	0.054	0.050	0.070
$V^{(2)}/cm^3$	40.5	50.5	60.5	70.5	55.1	60.1	65.1	70.1	40.3
$\varphi_B^{(2)}$	0.064	0.051	0.043	0.037	0.059	0.054	0.050	0.046	0.055
$\Delta_{dil}H^{12}/J$	−0.372	−0.213	−0.134	−0.096	−0.134	−0.100	−0.092	−0.079	−0.264

$V^{(1)}/cm^3$	40.3	50.3	30.5	30.5	30.5	50.1	50.1	50.1	30.3
$\varphi_B^{(1)}$	0.053	0.040	0.085	0.085	0.085	0.065	0.065	0.065	0.070
$V^{(2)}/cm^3$	50.3	60.3	50.5	60.5	70.5	60.1	65.1	70.1	50.3
$\varphi_B^{(2)}$	0.040	0.033	0.051	0.043	0.037	0.054	0.050	0.046	0.040
$\Delta_{dil}H^{12}/J$	−0.130	−0.084	−0.590	−0.724	−0.820	−0.230	−0.322	−0.289	−0.393

$V^{(1)}/cm^3$	30.3	40.2	50.2
$\varphi_B^{(1)}$	0.070	0.050	0.040
$V^{(2)}/cm^3$	60.3	50.2	60.2
$\varphi_B^{(2)}$	0.033	0.040	0.033
$\Delta_{dil}H^{12}/J$	−0.477	−0.163	−0.092

Comments: $\Delta_{dil}H^{12}$ is the extensive quantity obtained for a given total volume change from $V^{(1)}$ to $V^{(2)}$, where $\varphi_B^{(1)}$ and $\varphi_B^{(2)}$ denote the volume fractions of the polymer in the solution before and after the dilution process.

Polymer (B):	**poly(ethylene glycol) dimethyl ether**							**1994EST**
Characterization:	$M_n/g.mol^{-1} = 283$, PEGDME 250, a mixture of oligomers of n = 3 to 9, Fluka AG, Buchs, Switzerland							
Solvent (A):	**methanol**			**CH₄O**				**67-56-1**

$T/K = 303.15$

x_A	0.0403	0.0857	0.1252	0.1538	0.1814	0.2003	0.2506	0.2954
$\Delta_M H/(J/mol)$	67.0	147.3	206.7	249.4	287.6	315.1	378.8	433.0

x_A	0.3528	0.4078	0.4404	0.4882	0.5498	0.5675	0.6242	0.6539
$\Delta_M H/(J/mol)$	490.6	525.7	545.8	572.1	587.0	586.6	568.8	563.8

x_A	0.6924	0.7544	0.8037	0.8477	0.8913	0.9413	0.9529
$\Delta_M H/(J/mol)$	534.2	486.7	411.5	352.9	265.1	136.2	112.7

Polymer (B):   **poly(ethylene glycol) dimethyl ether**   **1997LOP**
Characterization: M_n/g.mol^{-1} = 280, PEGDME 250, a mixture of oligomers
                          of n = 3 to 9, Aldrich Chem. Co., Inc., Milwaukee, WI
Solvent (A):   **methanol**   **CH$_4$O**   **67-56-1**

T/K = 298.15

x_A	0.0925	0.1208	0.1751	0.2318	0.3046	0.3235	0.3727	0.4518
$\Delta_M H$/(J/mol)	137.2	171.8	257.9	331.4	437.5	450.4	511.8	581.1
x_A	0.4782	0.5240	0.5952	0.6006	0.6409	0.7236	0.7436	0.7894
$\Delta_M H$/(J/mol)	604.9	621.1	616.7	622.2	613.7	548.5	533.8	479.4
x_A	0.8072	0.8603	0.8821	0.9285	0.9427	0.9471	0.9599	0.9655
$\Delta_M H$/(J/mol)	442.5	333.5	291.5	179.3	142.6	132.2	98.1	78.0
x_A	0.9666	0.9829						
$\Delta_M H$/(J/mol)	72.0	35.7						

Polymer (B):   **poly(ethylene glycol) dimethyl ether**   **1999LOP**
Characterization: M_n/g.mol^{-1} = 280, PEGDME 250, a mixture of oligomers
                          of n = 3 to 9, Aldrich Chem. Co., Inc., Milwaukee, WI
Solvent (A):   **methanol**   **CH$_4$O**   **67-56-1**

T/K = 323.15 P/MPa = 8.0

x_A	0.1464	0.2180	0.2992	0.3981	0.4999	0.5002	0.5951	0.7016
$\Delta_M H$/(J/mol)	115.8	221.8	378.4	515.4	591.7	604.7	612.3	571.3
x_A	0.7972	0.8559	0.8987					
$\Delta_M H$/(J/mol)	470.5	363.3	267.7					

T/K = 373.15 P/MPa = 8.0

x_A	0.0829	0.1202	0.1549	0.1875	0.2181	0.2605	0.2994	0.3572
$\Delta_M H$/(J/mol)	214.6	296.2	367.4	419.8	474.4	547.3	603.6	684.8
x_A	0.3983	0.4264	0.5000	0.5415	0.5950	0.6537	0.7016	0.7590
$\Delta_M H$/(J/mol)	732.2	765.6	823.4	836.5	840.7	788.1	754.8	674.5
x_A	0.7972	0.8558	0.8987	0.9315				
$\Delta_M H$/(J/mol)	608.6	476.8	349.0	236.6				

T/K = 423.15 P/MPa = 8.0

x_A	0.0829	0.1202	0.1550	0.1875	0.2181	0.2606	0.2994	0.3573
$\Delta_M H$/(J/mol)	184.7	271.1	349.5	423.8	482.0	563.2	630.3	716.7
x_A	0.3984	0.4526	0.4999	0.5415	0.5950	0.6400	0.7016	0.7590
$\Delta_M H$/(J/mol)	775.6	864.9	905.4	922.7	895.8	882.7	828.0	745.7
x_A	0.7972	0.8558	0.8987	0.9314				
$\Delta_M H$/(J/mol)	672.3	516.3	366.2	245.8				

Polymer (B):	poly(ethylene glycol) dimethyl ether	1993ZEL
Characterization:	M_n/g.mol^{-1} = 250	
Solvent (A):	tetrachloromethane CCl$_4$	56-23-5

T/K = 303.15

$\Delta_M H/(\text{J/g}) = w_B(1 - w_B)[-11.13 - 0.50(1 - 2w_B) - 1.31(1 - 2w_B)^2 + 1.21(1 - 2w_B)^3]$

T/K = 318.15

$\Delta_M H/(\text{J/g}) = w_B(1 - w_B)[-9.09 - 0.54(1 - 2w_B) + 0.84(1 - 2w_B)^2 + 1.15(1 - 2w_B)^3]$

Polymer (B):	poly(ethylene glycol) dimethyl ether	1969MAL
Characterization:	M_n/g.mol^{-1} = 350 ±25, ρ = 1.080 g/cm^3 (278.68 K), prepared from Carbowax 350 in the laboratory	
Solvent (A):	tetrachloromethane CCl$_4$	56-23-5

T/K = 278.68

m_A/g	0.947	0.563	0.385	0.509	0.506	0.254	0.297	0.145
m_B/g	0.1095	0.1513	0.1979	0.1178	0.6068	0.4432	0.8150	0.8708
φ_B	0.1480	0.2876	0.4358	0.5522	0.6431	0.7239	0.8104	0.9002
$\Delta_M H/(\text{J/cm}^3)$	−2.7	−4.5	−5.2	−5.5	−5.4	−4.9	−3.9	−2.6

Polymer (B):	poly(ethylene glycol) dimethyl ether	1993ZEL
Characterization:	M_n/g.mol^{-1} = 398	
Solvent (A):	tetrachloromethane CCl$_4$	56-23-5

T/K = 303.15

$\Delta_M H/(\text{J/g}) = w_B(1 - w_B)[-13.28 + 0.35(1 - 2w_B) + 0.62(1 - 2w_B)^2]$

Polymer (B):	poly(ethylene glycol) dimethyl ether	1985COR, 1995KIL
Characterization:	M_n/g.mol^{-1} = 520, M_w/g.mol^{-1} = 550, ρ = 1.0712 g/cm^3 (298 K)	
Solvent (A):	tetrachloromethane CCl$_4$	56-23-5

T/K = 303.15

w_B	0.0764	0.0898	0.1055	0.1239	0.1455	0.1708	0.2114	0.2547
$\Delta_M H/(\text{J/g})$	−0.885	−1.028	−1.191	−1.377	−1.580	−1.809	−2.151	−2.464

w_B	0.3076	0.3724	0.4526	0.5528	0.6797	0.7519	0.8581	0.9595
$\Delta_M H/(\text{J/g})$	−2.780	−3.074	−3.269	−3.287	−2.922	−2.484	−1.595	−0.507

Polymer (B):	poly(ethylene glycol) dimethyl ether	1989MOE
Characterization:	M_n/g.mol^{-1} = 520, M_w/g.mol^{-1} = 550, ρ = 1.0712 g/cm^3 (298 K)	
Solvent (A):	tetrachloromethane CCl$_4$	56-23-5

continued

continued

$T/K = 303.15$

w_B	0.0116	0.0332	0.0736	0.1108	0.1451	0.1769	0.2173	0.2539	0.2928
$\Delta_M H/(J/g)$	−0.134	−0.375	−0.814	−1.186	−1.482	−1.745	−2.049	−2.299	−2.561

w_B	0.3027	0.3342	0.3456	0.3892	0.4294	0.4788	0.5412	0.5785	0.6214
$\Delta_M H/(J/g)$	−2.580	−2.782	−2.813	−3.010	−3.136	−3.230	−3.265	−3.232	−3.134

w_B	0.6712	0.7297	0.7993	0.8590	0.9285	0.9676	0.9884
$\Delta_M H/(J/g)$	−2.954	−2.648	−2.183	−1.660	−0.920	−0.428	−0.156

Polymer (B):	**poly(ethylene glycol) dimethyl ether**	**1993ZEL**
Characterization:	$M_n/\text{g.mol}^{-1} = 520$, $M_w/\text{g.mol}^{-1} = 550$, $\rho = 1.0712$ g/cm^3 (298 K)	
Solvent (A):	**tetrachloromethane CCl$_4$**	**56-23-5**

$T/K = 303.15$

$$\Delta_M H/(J/g) = w_B(1 - w_B)[-12.98 + 1.53(1 - 2w_B) + 0.39(1 - 2w_B)^2 - 0.62(1 - 2w_B)^3]$$

Polymer (B):	**poly(ethylene glycol) dimethyl ether**	**1993ZEL**
Characterization:	$M_n/\text{g.mol}^{-1} = 250$	
Solvent (A):	**trichloromethane CHCl$_3$**	**67-66-3**

$T/K = 303.15$

$$\Delta_M H/(J/g) = w_B(1 - w_B)[-125.03 - 54.22(1 - 2w_B) - 4.74(1 - 2w_B)^2]$$

Polymer (B):	**poly(ethylene glycol) dimethyl ether**	**1969MAL**
Characterization:	$M_n/\text{g.mol}^{-1} = 350 \pm 25$, $\rho = 1.080$ g/cm^3 (278.68 K),	
	prepared from Carbowax 350 in the laboratory	
Solvent (A):	**trichloromethane CHCl$_3$**	**67-66-3**

$T/K = 278.68$

m_A/g	0.994	0.895	0.740	0.588	0.536	0.603	0.526	0.415
m_B/g	0.1175	0.3164	0.3245	0.3613	0.3534	0.5434	0.5953	0.6388
φ_B	0.1424	0.3319	0.3812	0.4633	0.4809	0.5586	0.6139	0.6838
$\Delta_M H/(J/cm^3)$	−23.5	−47.0	−45.3	−49.6	−49.8	−48.3	−43.6	−37.1

m_A/g	0.269	0.132
m_B/g	0.6831	0.8446
φ_B	0.7810	0.8999
$\Delta_M H/(J/cm^3)$	−27.4	−13.3

Polymer (B):	**poly(ethylene glycol) dimethyl ether**	**1993ZEL**
Characterization:	$M_n/\text{g.mol}^{-1} = 520$, $M_w/\text{g.mol}^{-1} = 550$, $\rho = 1.0712$ g/cm^3 (298 K)	
Solvent (A):	**trichloromethane CHCl$_3$**	**67-66-3**

$T/K = 303.15$

$$\Delta_M H/(J/g) = w_B(1 - w_B)[-125.5 - 49.31(1 - 2w_B) + 7.64(1 - 2w_B)^2 + 32.25(1 - 2w_B)^3]$$

Polymer (B):	poly(ethylene glycol) dimethyl ether						1997LOP
Characterization:	M_n/g.mol^{-1} = 280, PEGDME 250, a mixture of oligomers of n = 3 to 9, Aldrich Chem. Co., Inc., Milwaukee, WI						
Solvent (A):	2,2,2-trifluoroethanol C$_2$H$_3$F$_3$O						75-89-8

T/K = 298.15

x_A	0.0290	0.1213	0.1689	0.2325	0.2725	0.3441	0.3596	0.4018
$\Delta_M H$/(J/mol)	−246	−984	−1371	−1857	−2115	−2675	−2858	−3054

x_A	0.4074	0.4614	0.5307	0.5971	0.6579	0.6965	0.7777	0.8004
$\Delta_M H$/(J/mol)	−3118	−3438	−3818	−4098	−4184	−4267	−4017	−3982

x_A	0.8556	0.9014	0.9475	0.9660
$\Delta_M H$/(J/mol)	−3499	−2814	−1898	−1355

Polymer (B):	poly(ethylene glycol) monododecyl ether						1980KUR
Characterization:	M_n/g.mol^{-1} = 595, surfactant						
Solvent (A):	water		H$_2$O				7732-18-5

T/K	303.15	308.15	313.15	318.15	323.15	328.15	333.15	338.15
φ_B	0.0522	0.0521	0.0519	0.0517	0.0513	0.0509	0.0503	0.0495
$\Delta_M H$/(J/mol solvent)	−74.94	−73.97	−72.26	−68.49	−63.14	−58.58	−53.51	−51.46

T/K	343.15	348.15	353.15
φ_B	0.0484	0.0476	0.0469
$\Delta_M H$/(J/mol solvent)	−48.28	−47.66	−46.15

Polymer (B):	poly(ethylene glycol) monomethyl ether						2002RUI
Characterization:	M_n/g.mol^{-1} = 340, ρ = 1.085 g/cm^3 (298 K)						
Solvent (A):	1-butanol		C$_4$H$_{10}$O				71-36-3

T/K = 298.15

x_B	0.0336	0.0403	0.0882	0.1399	0.1730	0.2648	0.3315	0.4171
$\Delta_M H$/(J/mol)	450	523	950	1301	1484	1797	1859	1840

x_B	0.4931	0.5899	0.7284	0.8052	0.9049	0.9544
$\Delta_M H$/(J/mol)	1733	1526	1086	816	417	190

Polymer (B):	poly(ethylene glycol) monomethyl ether						2002RUI
Characterization:	M_n/g.mol^{-1} = 340, ρ = 1.085 g/cm^3 (298 K)						
Solvent (A):	1-pentanol		C$_5$H$_{12}$O				71-41-0

T/K = 298.15

x_B	0.0184	0.0350	0.0520	0.1025	0.1643	0.2174	0.3221	0.3892
$\Delta_M H$/(J/mol)	276	494	683	1166	1555	1795	2044	2083

x_B	0.4608	0.5135	0.6054	0.6709	0.7993	0.8781	0.9356
$\Delta_M H$/(J/mol)	2010	1938	1731	1500	960	588	359

Polymer (B):   **poly(ethylene glycol) monomethyl ether**   **2002RUI**
Characterization: M_n/g.mol^{-1} = 340, ρ = 1.085 g/cm^3 (298 K)
Solvent (A):   **1-propanol**   **C$_3$H$_8$O**   **71-23-8**

T/K = 298.15

x_B	0.0130	0.0257	0.0436	0.0696	0.0805	0.1308	0.1498	0.1921
$\Delta_M H$/(J/mol)	164	315	501	724	811	1107	1198	1348
x_B	0.2426	0.2781	0.3551	0.4071	0.4231	0.4959	0.5802	0.5937
$\Delta_M H$/(J/mol)	1483	1546	1547	1545	1551	1449	1272	1249
x_B	0.6511	0.7569	0.7858	0.8580	0.9084			
$\Delta_M H$/(J/mol)	1097	800	718	491	314			

Polymer (B):   **poly(ethylene glycol) monomethyl ether**   **1989MOE, 1995KIL**
Characterization: M_n/g.mol^{-1} = 353, M_w/g.mol^{-1} = 377, ρ = 1.0845 g/cm^3 (298 K)
Solvent (A):   **tetrachloromethane**   **CCl$_4$**   **56-23-5**

T/K = 303.15

w_B	0.0031	0.0089	0.0185	0.0351	0.0530	0.0703	0.1029	0.1438	0.1811
$\Delta_M H$/(J/g)	0.024	0.068	0.121	0.163	0.156	0.117	−0.009	−0.233	−0.466
w_B	0.2154	0.2406	0.2468	0.2728	0.2759	0.3032	0.3413	0.3770	0.4211
$\Delta_M H$/(J/g)	−0.687	−0.803	−0.890	−1.015	−1.073	−1.209	−1.433	−1.633	−1.849
w_B	0.4769	0.5285	0.5927	0.6747	0.7811	0.8480	0.9274	0.9730	0.9925
$\Delta_M H$/(J/g)	−2.066	−2.203	−2.288	−2.249	−1.887	−1.464	−0.787	−0.301	−0.073

Polymer (B):   **poly(ethylene glycol) monomethyl ether**   **1993ZEL**
Characterization: M_n/g.mol^{-1} = 353, M_w/g.mol^{-1} = 377, ρ = 1.0845 g/cm^3 (298 K)
Solvent (A):   **tetrachloromethane**   **CCl$_4$**   **56-23-5**

T/K = 303.15

$$\Delta_M H/(J/g) = w_B(1 - w_B)[-15.03 + 2.78(1 - 2w_B) + 1.10(1 - 2w_B)^2] - 4.68w_B \ln(w_B)$$

T/K = 318.15

$$\Delta_M H/(J/g) = w_B(1 - w_B)[-4.23 + 3.47(1 - 2w_B) + 0.25(1 - 2w_B)^2] - 0.64w_B \ln(w_B)$$

Polymer (B):   **poly(ethylene glycol) monomethyl ether**   **1985COR, 1995KIL**
Characterization: M_n/g.mol^{-1} = 550, M_w/g.mol^{-1} = 580, ρ = 1.0991 g/cm^3 (298 K)
Solvent (A):   **tetrachloromethane**   **CCl$_4$**   **56-23-5**

T/K = 303.15

w_B	0.0150	0.0265	0.0579	0.0922	0.1256	0.1577	0.2015	0.2578	0.3117
$\Delta_M H$/(J/g)	0.002	−0.009	−0.096	−0.248	−0.428	−0.619	−0.880	−1.246	−1.572
w_B	0.3616	0.3635	0.4131	0.4193	0.4429	0.5384	0.6407	0.7105	0.7716
$\Delta_M H$/(J/g)	−1.800	−1.851	−2.080	−2.052	−2.134	−2.335	−2.308	−2.121	−1.854
w_B	0.8286	0.8825	0.9509	0.9992					
$\Delta_M H$/(J/g)	−1.527	−1.129	−0.509	−0.009					

Polymer (B): **poly(ethylene glycol) monomethyl ether** **1993ZEL**
Characterization: M_n/g.mol^{-1} = 550, M_w/g.mol^{-1} = 580, ρ = 1.0991 g/cm^3 (298 K)
Solvent (A): **tetrachloromethane** **CCl$_4$** **56-23-5**

T/K = 303.15

$$\Delta_M H/(J/g) = w_B(1 - w_B)[-10.86 + 2.91(1 - 2w_B) + 2.47(1 - 2w_B)^2 + 1.22(1 - 2w_B)^3] - 1.21 w_B \ln(w_B)$$

T/K = 318.15

$$\Delta_M H/(J/g) = w_B(1 - w_B)[-7.31 + 1.92(1 - 2w_B) + 0.88(1 - 2w_B)^2] - 2.35 w_B \ln(w_B)$$

Polymer (B): **poly(ethylene glycol) monomethyl ether** **1993ZEL**
Characterization: M_n/g.mol^{-1} = 353, M_w/g.mol^{-1} = 377, ρ = 1.0845 g/cm^3 (298 K)
Solvent (A): **trichloromethane** **CHCl$_3$** **67-66-3**

T/K = 303.15

$$\Delta_M H/(J/g) = w_B(1 - w_B)[-115.44 - 53.69(1 - 2w_B) + 10.51(1 - 2w_B)^2 + 33.53(1 - 2w_B)^3]$$

Polymer (B): **poly(ethylene glycol) monomethyl ether** **1993ZEL**
Characterization: M_n/g.mol^{-1} = 550, M_w/g.mol^{-1} = 580, ρ = 1.0991 g/cm^3 (298 K)
Solvent (A): **trichloromethane** **CHCl$_3$** **67-66-3**

T/K = 303.15

$$\Delta_M H/(J/g) = w_B(1 - w_B)[-114.09 - 40.83(1 - 2w_B) + 4.26(1 - 2w_B)^2 + 33.44(1 - 2w_B)^3]$$

Polymer (B): **poly(ethylene oxide)** **1978KOL**
Characterization: M_n/g.mol^{-1} = 4000, M_w/M_n < 1.07, Merck-Schuchardt, Germany
Solvent (A): **benzene** **C$_6$H$_6$** **71-43-2**

T/K = 298.15

$c_B^{(1)}$/(mol/m^3)	15.0	15.0	15.0	15.0	15.0
$c_B^{(2)}$/(mol/m^3)	6.0	9.0	2.8	3.3	4.6
$\Delta_{dil} H^{12}$/(kJ/mol polymer)	−26.2	−10.4	−21.1	−20.8	−16.0

Comments: $c_B^{(1)}$ and $c_B^{(2)}$ denote the concentrations of the polymer in the solution before and after the dilution process.

Polymer (B): **poly(ethylene oxide)** **1978KOL**
Characterization: M_n/g.mol^{-1} = 6000, M_w/M_n < 1.07, Merck-Schuchardt, Germany
Solvent (A): **benzene** **C$_6$H$_6$** **71-43-2**

T/K = 298.15

$c_B^{(1)}$/(mol/m^3)	8.8	8.8	8.8	8.8	8.8
$c_B^{(2)}$/(mol/m^3)	6.2	5.7	3.8	1.9	0.5
$\Delta_{dil} H^{12}$/(kJ/mol polymer)	−8.4	−6.9	−15.7	−17.3	−20.9

Comments: $c_B^{(1)}$ and $c_B^{(2)}$ denote the concentrations of the polymer in the solution before and after the dilution process.

Polymer (B): **poly(ethylene oxide)** **1978KOL**
Characterization: $M_n/\text{g.mol}^{-1} = 15000$, $M_w/M_n < 1.07$, Merck-Schuchardt, Germany
Solvent (A): **benzene** **C_6H_6** **71-43-2**

$T/\text{K} = 298.15$

$c_B^{(1)}/(\text{mol/m}^3)$	3.4	3.4	3.4	3.4	3.4	3.4
$c_B^{(2)}/(\text{mol/m}^3)$	2.3	1.8	0.9	0.6	0.4	0.2
$\Delta_{dil} H^{12}/(\text{kJ/mol polymer})$	-11.7	-27.2	-62.3	-64.8	-138.3	-92.0

Comments: $c_B^{(1)}$ and $c_B^{(2)}$ denote the concentrations of the polymer in the solution before and after the dilution process.

Polymer (B): **poly(ethylene oxide)** **1978KOL**
Characterization: $M_n/\text{g.mol}^{-1} = 20000$, $M_w/M_n < 1.07$, Merck-Schuchardt, Germany
Solvent (A): **benzene** **C_6H_6** **71-43-2**

$T/\text{K} = 298.15$

$c_B^{(1)}/(\text{mol/m}^3)$	1.7	1.7	1.7	1.7	1.7	1.7
$c_B^{(2)}/(\text{mol/m}^3)$	1.0	1.2	0.5	0.3	0.2	0.1
$\Delta_{dil} H^{12}/(\text{kJ/mol polymer})$	-34.0	-38.9	-40.0	-207.0	-168.2	-55.0

Comments: $c_B^{(1)}$ and $c_B^{(2)}$ denote the concentrations of the polymer in the solution before and after the dilution process.

Polymer (B): **poly(ethylene oxide)** `1978KOL`
Characterization: $M_n/\text{g.mol}^{-1} = 40000$, $M_w/M_n < 1.07$, Merck-Schuchardt, Germany
Solvent (A): **benzene** **C_6H_6** **71-43-2**

$T/\text{K} = 298.15$

$c_B^{(1)}/(\text{mol/m}^3)$	1.0	1.0	1.0	1.0	1.0
$c_B^{(2)}/(\text{mol/m}^3)$	0.6	0.5	0.3	0.2	0.1
$\Delta_{dil} H^{12}/(\text{kJ/mol polymer})$	-48.5	-110.3	-188.7	-246.3	-514.6

Comments: $c_B^{(1)}$ and $c_B^{(2)}$ denote the concentrations of the polymer in the solution before and after the dilution process.

Polymer (B): **poly(ethylene oxide)** **1979MOR**
Characterization: $M_n/\text{g.mol}^{-1} = 7400$
Solvent (A): **benzene** **C_6H_6** **71-43-2**
Solvent (C): **n-heptane** **C_7H_{16}** **142-82-5**

$T/\text{K} = 303.15$

x_C	0.0290	0.0518	0.0904	0.1040	0.1880	0.2143	0.2311	0.2787
$\Delta_M H/(\text{J/mol})$	+7.95	+4.90	+13.8	+2.43	+5.23	+33.3	-5.31	-727.7

x_C	0.3266	0.3465	0.4510	0.4673	0.4726	0.5621	0.6391	0.6488
$\Delta_M H/(\text{J/mol})$	-683.0	-650.3	-584.7	-514.9	-546.6	-433.1	-348.4	-342.5

x_C	0.7369	0.8382	0.8847
$\Delta_M H/(\text{J/mol})$	-267.6	-158.9	-122.3

Comments: The enthalpy of mixing is given for a benzene solution of PEO of constant concentration of 7.5502 wt% (13.339 mol%) PEO with a varying amount of n-heptane (mole fraction).

Polymer (B):	**poly(ethylene oxide)**				**1972MA2**
Characterization:	M_n/g.mol^{-1} = 6000, Carbowax 6000				
Solvent (A):	**dichloromethane**		**CH₂Cl₂**		**75-09-2**

T/K = 303.15

$\varphi_B^{(1)}$	0.0639	0.1065	0.1616	0.1939	0.2424
$\varphi_B^{(2)}$	0.0160	0.0266	0.0404	0.0277	0.0606
$\Delta_{dil}H^{12}$/(J/g polymer)	−2.15	−4.23	−7.66	−10.55	−12.68

Comments: $\varphi_B^{(1)}$ and $\varphi_B^{(2)}$ denote the volume fractions of the polymer in the solution before and after the dilution process.

Polymer (B):	**poly(ethylene oxide)**				**1972MA2**
Characterization:	M_n/g.mol^{-1} = 6000, Carbowax 6000				
Solvent (A):	**trichloromethane**		**CHCl₃**		**67-66-3**

T/K = 303.15

$\varphi_B^{(1)}$	0.0390	0.0636	0.1244	0.1901	0.2851
$\varphi_B^{(2)}$	0.0097	0.0159	0.0311	0.0271	0.0407
$\Delta_{dil}H^{12}$/(J/g polymer)	−1.99	−3.37	−7.03	−13.94	−25.16

Comments: $\varphi_B^{(1)}$ and $\varphi_B^{(2)}$ denote the volume fractions of the polymer in the solution before and after the dilution process.

Polymer (B):	**poly(ethylene oxide)**							**1972MA2**
Characterization:	M_n/g.mol^{-1} = 6000, Carbowax 6000							
Solvent (A):	**water**		**H₂O**					**7732-18-5**

T/K = 303.15

$\varphi_B^{(1)}$	0.0179	0.0346	0.0474	0.0474	0.0474	0.0619	0.0995	0.1133
$\varphi_B^{(2)}$	0.0149	0.0289	0.0090	0.0406	0.0068	0.0530	0.0911	0.0283
$\Delta_{dil}H^{12}$/(J/g polymer)	−0.13	−0.29	−1.95	−0.32	−2.16	−0.42	−0.46	−4.94

Comments: $\varphi_B^{(1)}$ and $\varphi_B^{(2)}$ denote the volume fractions of the polymer in the solution before and after the dilution process.

Polymer (B):	**poly(ethylene oxide)-b-poly(propylene oxide) diblock copolymer**							**1989MOE, 1990KIL**
Characterization:	M_n/g.mol^{-1} = 1700, 58.0 wt% propylene oxide, 16.8 units propylene oxide, 16.1 units ethylene oxide, ρ = 1.0335 g/cm^3 (298 K)							
Solvent (A):	**tetrachloromethane**		**CCl₄**					**56-23-5**

T/K = 303.15

w_B	0.0180	0.0408	0.0628	0.0838	0.1235	0.1602	0.1942	0.2370	0.2402
$\Delta_M H$/(J/g)	−0.285	−0.618	−0.925	−1.205	−1.685	−2.096	−2.437	−2.856	−2.897

continued

continued

w_B	0.2763	0.2936	0.3121	0.3449	0.3804	0.4241	0.4792	0.5306	0.5942
$\Delta_M H/$(J/g)	−3.178	−3.287	−3.436	−3.578	−3.728	−3.844	−3.902	−3.860	−3.690

w_B	0.6751	0.7246	0.7818	0.8487	0.9283	0.9739	0.9935		
$\Delta_M H/$(J/g)	−3.324	−3.010	−2.564	−1.917	−0.950	−0.298	−0.078		

Polymer (B):   **poly(ethylene oxide)-b-poly(propylene**   **1985COR, 1995KIL**
                          oxide)-b-poly(ethylene oxide) triblock copolymer

Characterization: M_n/g.mol^{-1} = 1900, 67 wt% propylene oxide, 21.7 units propylene
                          oxide, 14.1 units ethylene oxide, ρ = 1.0111 g/cm^3 (298 K)
                          PE 4300, BASF AG, Germany

Solvent (A):   **tetrachloromethane   CCl$_4$**   **56-23-5**

T/K = 303.15

w_B	0.0076	0.0147	0.0251	0.0353	0.0453	0.0551	0.0658	0.0763	0.0866
$\Delta_M H/$(J/g)	−0.067	−0.134	−0.236	−0.340	−0.443	−0.546	−0.663	−0.776	−0.889

w_B	0.0966	0.1064	0.1160	0.1265	0.1368	0.1564	0.1687	0.1806	0.1935
$\Delta_M H/$(J/g)	−0.998	−1.104	−1.209	−1.321	−1.429	−1.612	−1.718	−1.822	−1.932

w_B	0.2059	0.2180	0.2297	0.2411	0.2567	0.2852	0.3131	0.3506	0.3587
$\Delta_M H/$(J/g)	−2.037	−2.139	−2.235	−2.327	−2.425	−2.653	−2.818	−2.990	−3.021

w_B	0.3750	0.3865	0.3912	0.4101	0.4215	0.4524	0.4923	0.5305	0.5581
$\Delta_M H/$(J/g)	−3.092	−3.153	−3.162	−3.226	−3.245	−3.329	−3.352	−3.331	−3.296

w_B	0.5930	0.6263	0.6578	0.7062	0.8022	0.8708	0.9475	0.9821	
$\Delta_M H/$(J/g)	−3.194	−3.043	−2.871	−2.592	−1.982	−1.381	−0.584	−0.198	

Polymer (B):   **poly(ethylene oxide)-b-poly(propylene**   **1985COR, 1995KIL**
                          oxide)-b-poly(ethylene oxide) triblock copolymer

Characterization: M_n/g.mol^{-1} = 1100, 87 wt% propylene oxide, 14.0 units propylene
                          oxide,2.8 units ethylene oxide, ρ = 1.0108 g/cm^3 (298 K)
                          PE 3100, BASF AG, Germany

Solvent (A):   **tetrachloromethane   CCl$_4$**   **56-23-5**

T/K = 303.15

w_B	0.0040	0.0110	0.0215	0.0316	0.0449	0.0577	0.0850	0.1107	0.1339
$\Delta_M H/$(J/g)	−0.021	−0.065	−0.136	−0.209	−0.309	−0.407	−0.624	−0.837	−0.967

w_B	0.1573	0.1936	0.2284	0.2729	0.3432	0.4134	0.4892	0.5528	0.6162
$\Delta_M H/$(J/g)	−1.123	−1.331	−1.503	−1.694	−1.924	−2.094	−2.138	−2.158	−2.045

w_B	0.7140	0.8178	0.9076	0.9389	0.9723				
$\Delta_M H/$(J/g)	−1.817	−1.455	−0.829	−0.566	−0.254				

Polymer (B):   **poly(ethylene oxide)-b-poly(propylene**   **1985COR, 1995KIL**
     oxide)-b-poly(ethylene oxide) triblock copolymer

Characterization: $M_n/\text{g.mol}^{-1} = 1800$, 93 wt% propylene oxide, 28.6 units propylene
     oxide, 2.8 units ethylene oxide, $\rho = 1.0378$ g/cm^3 (298 K)
     PE 6100, BASF AG, Germany

Solvent (A):   **tetrachloromethane**   **CCl₄**   **56-23-5**

$T/\text{K} = 303.15$

w_B	0.0027	0.0043	0.0389	0.0613	0.0896	0.1207	0.1498	0.1770	0.2272
$\Delta_M H/(\text{J/g})$	−0.024	−0.038	−0.361	−0.564	−0.834	−1.130	−1.394	−1.617	−2.048

w_B	0.2726	0.3285	0.3974	0.4479	0.5449	0.6677	0.7389	0.8270	0.9053
$\Delta_M H/(\text{J/g})$	−2.367	−2.668	−2.907	−2.999	−2.945	−2.535	−2.130	−1.496	−0.837

Polymer (B):   **polyisobutylene**   **1958KAB**

Characterization: $M_n/\text{g.mol}^{-1} = 34000$, fractionated in the laboratory from
     Enjay Vistanex LM-MH-225

Solvent (A):   **benzene**   **C₆H₆**   **71-43-2**

$T/\text{K} = 298.15$

$n_B r_B$	0.001727	0.001889	0.005881	0.005969	0.006417	0.010463	0.005168
$\varphi_B^{(1)}$	0.0836	0.0836	0.1591	0.1591	0.1591	0.2807	0.2807
$\varphi_B^{(2)}$	0.0687	0.0693	0.1210	0.1214	0.1227	0.2137	0.2198
$\Delta_{dil} H^{12}/\text{J}$	0.025	0.025	0.356	0.397	0.385	0.891	0.381

$n_B r_B$	0.012192	0.015087	0.004799	0.014831	0.009801
$\varphi_B^{(1)}$	0.2807	0.3432	0.4378	0.4378	0.4378
$\varphi_B^{(2)}$	0.2201	0.2717	0.3149	0.3264	0.3759
$\Delta_{dil} H^{12}/\text{J}$	1.017	2.280	0.833	2.477	1.243

Comments: $\Delta_{dil} H^{12}$ is the extensive quantity obtained for the dilution process, where $n_B r_B$ is the product
     of the polymer amount of substance with the segment number of the polymer. $\varphi_B^{(1)}$ and
     $\varphi_B^{(2)}$ are the volume fractions of the polymer in the solution before and after the dilution
     process.

Polymer (B):   **polyisobutylene**   **1960WAT**

Characterization: $M_n/\text{g.mol}^{-1} = 63000$, fractionated in the laboratory from
     commercial Vistanex MH

Solvent (A):   **benzene**   **C₆H₆**   **71-43-2**

$T/\text{K} = 298.15$

φ_B	0.00714	0.0139	0.0169	0.0191	0.0230	0.0249	0.0279	0.0286
$\Delta_M H/(\text{J/cm}^3)$	0.129	0.254	0.308	0.347	0.410	0.443	0.509	0.506

φ_B	0.0347	0.0367	0.0377	0.0593
$\Delta_M H/(\text{J/cm}^3)$	0.610	0.643	0.679	1.041

Polymer (B):   **polyisobutylene**   **1958KAB**

Characterization: $M_\eta/\text{g.mol}^{-1} = 82000$, fractionated in the laboratory from
Enjay Vistanex LM-MH-225

Solvent (A):   **benzene**   **C_6H_6**   **71-43-2**

$T/\text{K} = 298.15$

$n_B r_B$	0.003580	0.015913	0.005018	0.006469
$\varphi_B^{(1)}$	0.2015	0.2015	0.2768	0.3635
$\varphi_B^{(2)}$	0.1572	0.1597	0.1992	0.2879
$\Delta_{dil} H^{12}/\text{J}$	0.163	0.556	0.393	0.561

Comments: $\Delta_{dil} H^{12}$ is the extensive quantity obtained for the dilution process, where $n_B r_B$ is the product of the polymer amount of substance with the segment number of the polymer. $\varphi_B^{(1)}$ and $\varphi_B^{(2)}$ are the volume fractions of the polymer in the solution before and after the dilution process.

Polymer (B):   **polyisobutylene**   **1958KAB**

Characterization: $M_\eta/\text{g.mol}^{-1} = 252000$, fractionated in the laboratory from
Enjay Vistanex LM-100

Solvent (A):   **benzene**   **C_6H_6**   **71-43-2**

$T/\text{K} = 298.15$

$n_B r_B$	0.004805	0.010010	0.003583	0.003633	0.003641	0.003744
$\varphi_B^{(1)}$	0.1270	0.2656	0.7251	0.8217	0.8945	0.9357
$\varphi_B^{(2)}$	0.0978	0.2051	0.3872	0.4358	0.4726	0.4936
$\Delta_{dil} H^{12}/\text{J}$	0.213	0.866	1.954	3.017	3.791	5.309

Comments: $\Delta_{dil} H^{12}$ is the extensive quantity obtained for the dilution process, where $n_B r_B$ is the product of the polymer amount of substance with the segment number of the polymer. $\varphi_B^{(1)}$ and $\varphi_B^{(2)}$ are the volume fractions of the polymer in the solution before and after the dilution process.

Polymer (B):   **polyisobutylene**   **1958KAB**

Characterization: $M_\eta/\text{g.mol}^{-1} = 392000$, fractionated in the laboratory from
Enjay Vistanex LM-100

Solvent (A):   **benzene**   **C_6H_6**   **71-43-2**

$T/\text{K} = 298.15$

$n_B r_B$	0.003777	0.003777	0.005994	0.008485	0.001842	0.001153	0.001282
$\varphi_B^{(1)}$	0.1420	0.1420	0.1890	0.1890	0.6180	0.6420	0.6510
$\varphi_B^{(2)}$	0.0902	0.0934	0.1282	0.1328	0.3219	0.3295	0.3345
$\Delta_{dil} H^{12}/\text{J}$	0.247	0.297	0.418	0.749	0.778	0.653	0.778

$n_B r_B$	0.001251	0.001159	0.002190	0.001783
$\varphi_B^{(1)}$	0.6620	0.6880	0.7970	0.8220
$\varphi_B^{(2)}$	0.3422	0.3523	0.4137	0.4236
$\Delta_{dil} H^{12}/\text{J}$	0.791	0.661	1.791	1.389

Comments: $\Delta_{dil} H^{12}$ is the extensive quantity obtained for the dilution process, where $n_B r_B$ is the product of the polymer amount of substance with the segment number of the polymer. $\varphi_B^{(1)}$ and $\varphi_B^{(2)}$ are the volume fractions of the polymer in the solution before and after the dilution process.

Polymer (B):   **polyisobutylene**   **1960WAT**
Characterization: M_η/g.mol^{-1} = 63000, fractionated in the laboratory from
          commercial Vistanex MH
Solvent (A):   **chlorobenzene**   **C₆H₅Cl**   **108-90-7**

T/K = 298.15

φ_B	0.00559	0.00914	0.0108	0.0121	0.0151	0.0161	0.0178	0.0218
$\Delta_M H$/(J/cm^3)	0.0644	0.1075	0.1280	0.1456	0.1757	0.1941	0.2146	0.2623

φ_B	0.0226	0.0278	0.0286	0.0379
$\Delta_M H$/(J/cm^3)	0.2590	0.3301	0.3464	0.4477

Polymer (B):   **polyisobutylene**   **1960WAT**
Characterization: M_η/g.mol^{-1} = 252000, fractionated in the laboratory from
          commercial Vistanex L-100
Solvent (A):   **chlorobenzene**   **C₆H₅Cl**   **108-90-7**

T/K = 298.15

φ_B	0.00420	0.00594	0.00671	0.00749	0.00869	0.00917	0.0115	0.0137
$\Delta_M H$/(J/cm^3)	0.0485	0.0736	0.0824	0.0874	0.1042	0.1088	0.1452	0.1632

φ_B	0.0152	0.0155	0.0210	0.0210
$\Delta_M H$/(J/cm^3)	0.1833	0.1854	0.2594	0.2628

Polymer (B):   **polyisobutylene**   **1960WAT**
Characterization: M_η/g.mol^{-1} = 63000, fractionated in the laboratory from
          commercial Vistanex MH
Solvent (A):   **cyclohexane**   **C₆H₁₂**   **110-82-7**

T/K = 298.15

φ_B	0.0321	0.0324
$\Delta_M H$/(J/cm^3)	−0.0172	−0.0276

Polymer (B):   **polyisobutylene**   **1960WAT**
Characterization: M_η/g.mol^{-1} = 252000, fractionated in the laboratory from
          commercial Vistanex L-100
Solvent (A):   **cyclohexane**   **C₆H₁₂**   **110-82-7**

T/K = 298.15

φ_B	0.00688	0.00807	0.0103	0.0121	0.0142	0.0146	0.0147	0.0156
$\Delta_M H$/(J/cm^3)	0.0565	0.0653	0.0812	0.1004	0.1134	0.1146	0.1155	0.1326

φ_B	0.0177	0.0180	0.0207	0.0330
$\Delta_M H$/(J/cm^3)	0.1414	0.1397	0.1690	0.2414

Polymer (B):	**polyisobutylene**		**1960WAT**

Characterization: M_η/g.mol^{-1} = 252000, fractionated in the laboratory from commercial Vistanex L-100

Solvent (A):	**n-heptane**	**C$_7$H$_{16}$**	**142-82-5**

T/K = 298.15

φ_B	0.0104	0.0152	0.0428
$\Delta_M H$/(J/cm^3)	−0.0192	−0.0251	−0.0648

Polymer (B):	**poly(L-lysine)**						**1979BAB**

Characterization: M_n/g.mol^{-1} = 70000, Sigma Chemical Co., Inc., St. Louis, MO

Solvent (A):	**methanol**	**CH$_4$O**	**67-56-1**
Solvent (C):	**water**	**H$_2$O**	**7732-18-5**

T/K = 298.15

| $c_B^{(1)}$/(base mol/l) | 0.20 | was kept constant |

φ_A	0.669	0.752	0.802	0.835	0.859	0.876	0.890	0.901
$\Delta_{dil} H^{12}$/J	−7.646	−4.890	−3.894	−3.171	−2.552	−2.191	−1.724	−1.500

φ_A	0.910	0.918	0.924
$\Delta_{dil} H^{12}$/J	−1.309	−1.095	−0.885

Comments: $\Delta_{dil} H^{12}$ is here an extensive quantity obtained for a given amount of 1 cm^3 pure methanol added successively to a definite volume of solution in which the composition of poly(L-lysine), methanol and water is known.

Polymer (B):	**poly(methacrylic acid)**		**1976DA2**

Characterization: M_η/g.mol^{-1} = 5000, Rohm & Haas, Philadelphia, PA

Solvent (A):	**water**	**H$_2$O**	**7732-18-5**

T/K = 298.15

$n_A^{(1)}$/mol	1.1176	1.2251	1.2785	1.3101	1.2085	1.3028	1.3276	1.3226
$n_B^{(1)}$/base mol	0.0344	0.0278	0.0218	0.0170	0.0145	0.0129	0.0668	0.0033
$n_A^{(2)}$/mol	1.5167	1.6288	1.6842	1.7167	1.6120	1.6906	1.7335	1.7199
$\Delta_{dil} H^{12}$/J	−0.624	−0.298	−0.200	−0.112	−0.093	−0.071	−0.024	−0.005

Comments: $\Delta_{dil} H^{12}$ is here an extensive quantity obtained for a given total amount of solvent added (difference $n_A^{(2)} - n_A^{(1)}$).

Polymer (B):	**poly(methacrylic acid)**		**1976DA2**

Characterization: M_η/g.mol^{-1} = 140000, synthesized in the laboratory

Solvent (A):	**water**	**H$_2$O**	**7732-18-5**

T/K = 298.15

Continued

continued

$n_A^{(1)}$/mol	0.9557	1.1000	1.2108	1.2477	1.2568	1.2555	1.2705	1.3216
$n_B^{(1)}$/base mol	0.0278	0.0282	0.0291	0.0225	0.0172	0.0131	0.0100	0.0034
$n_A^{(2)}$/mol	1.3146	1.4928	1.6082	1.6476	1.6500	1.6591	1.6576	1.7180
$\Delta_{dil}H^{12}$/J	−0.522	−0.415	−0.383	−0.179	−0.066	−0.043	−0.025	0.000

Comments: $\Delta_{dil}H^{12}$ is here an extensive quantity obtained for a given total amount of solvent added
(difference $n_A^{(2)} - n_A^{(1)}$).

Polymer (B):	**poly(methacrylic acid) rubidium salt**							**1976DA2**
Characterization:	M_η/g.mol^{-1} = 5000, salt prepared in the laboratory							
Solvent (A):	**water**			**H$_2$O**				**7732-18-5**

T/K = 298.15

$n_A^{(1)}$/mol	1.1495	1.2285	1.2738	1.2628	1.2576	1.3181	1.3368	1.3058
$n_B^{(1)}$/base mol	0.0314	0.0180	0.0139	0.0105	0.0079	0.0076	0.0054	0.0045
$n_A^{(2)}$/mol	1.5619	1.6401	1.6704	1.6653	1.6656	1.7084	1.7380	1.7101
$\Delta_{dil}H^{12}$/J	0.000	−0.257	−0.212	−0.181	−0.135	−0.091	−0.056	−0.041

$n_A^{(1)}$/mol	1.2557	1.2503	1.2266	1.2800
$n_B^{(1)}$/base mol	0.0033	0.0025	0.0018	0.0007
$n_A^{(2)}$/mol	1.6469	1.6439	1.6263	1.6858
$\Delta_{dil}H^{12}$/J	−0.029	−0.018	−0.011	0.000

Comments: $\Delta_{dil}H^{12}$ is here an extensive quantity obtained for a given total amount of solvent added
(difference $n_A^{(2)} - n_A^{(1)}$).

Polymer (B):	**poly(methacrylic acid) rubidium salt**							**1976DA2**
Characterization:	M_η/g.mol^{-1} = 140000, salt prepared in the laboratory							
Solvent (A):	**water**			**H$_2$O**				**7732-18-5**

T/K = 298.15

$n_A^{(1)}$/mol	1.1541	1.1245	1.0958	1.1812	1.1351	1.2844	1.2240	1.2134
$n_B^{(1)}$/base mol	0.0289	0.0272	0.0205	0.0209	0.0156	0.0130	0.0095	0.0082
$n_A^{(2)}$/mol	1.5464	1.6357	1.4917	1.5840	1.5349	1.6752	1.6179	1.6229
$\Delta_{dil}H^{12}$/J	−0.217	−0.284	−0.267	−0.264	−0.240	−0.218	−0.168	−0.104

$n_A^{(1)}$/mol	1.3047	1.3178	1.3424	1.2581	1.3169	1.4124
$n_B^{(1)}$/base mol	0.0074	0.0057	0.0034	0.0024	0.0019	0.0016
$n_A^{(2)}$/mol	1.7040	1.7181	1.7102	1.6614	1.7225	1.8143
$\Delta_{dil}H^{12}$/J	−0.053	−0.043	−0.025	−0.015	−0.010	−0.005

Comments: $\Delta_{dil}H^{12}$ is here an extensive quantity obtained for a given total amount of solvent added
(difference $n_A^{(2)} - n_A^{(1)}$).

Polymer (B):	**poly(methacrylic acid) sodium salt**							**1976DA2**
Characterization:	M_η/g.mol^{-1} = 5000, salt prepared in the laboratory							
Solvent (A):	**water**			**H$_2$O**				**7732-18-5**

continued

continued

$T/K = 298.15$

$n_A^{(1)}$/mol	1.1764	1.2249	1.2483	1.2598	1.2843	1.2290	1.2527	1.2512
$n_B^{(1)}$/base mol	0.0289	0.0225	0.0171	0.0130	0.0114	0.0096	0.0085	0.0064
$n_A^{(2)}$/mol	1.5772	1.6402	1.6526	1.6649	1.6807	1.6222	1.6629	1.6514
$\Delta_{dil} H^{12}$/J	1.234	0.545	0.210	0.034	0.015	−0.002	−0.005	−0.011

$n_A^{(1)}$/mol	1.3020	1.2816	1.2404	1.2769
$n_B^{(1)}$/base mol	0.0050	0.0031	0.0014	0.0009
$n_A^{(2)}$/mol	1.7136	1.6875	1.6607	1.6756
$\Delta_{dil} H^{12}$/J	−0.015	−0.019	−0.005	0.000

Comments: $\Delta_{dil} H^{12}$ is here an extensive quantity obtained for a given total amount of solvent added (difference $n_A^{(2)} - n_A^{(1)}$).

Polymer (B):	**poly(methacrylic acid) sodium salt**						**1976DA2**	
Characterization:	M_η/g.mol^{-1} = 140000, salt prepared in the laboratory							
Solvent (A):	**water**		**H$_2$O**				**7732-18-5**	

$T/K = 298.15$

$n_A^{(1)}$/mol	1.1504	1.2055	1.2412	1.2558	1.2177	1.2510	1.2541	1.2716
$n_B^{(1)}$/base mol	0.0291	0.0224	0.0172	0.0132	0.0097	0.0083	0.0063	0.0046
$n_A^{(2)}$/mol	1.5619	1.6123	1.6389	1.6577	1.6049	1.6573	1.6536	1.6648
$\Delta_{dil} H^{12}$/J	2.659	1.244	0.590	0.272	0.164	0.063	0.010	0.005

$n_A^{(1)}$/mol	1.2543	1.2985	1.2134
$n_B^{(1)}$/base mol	0.0036	0.0022	0.0014
$n_A^{(2)}$/mol	1.6560	1.6972	1.6162
$\Delta_{dil} H^{12}$/J	0.002	0.000	−0.002

Comments: $\Delta_{dil} H^{12}$ is here an extensive quantity obtained for a given total amount of solvent added (difference $n_A^{(2)} - n_A^{(1)}$).

Polymer (B):	**poly(methyl methacrylate)**						**1969TAK**	
Characterization:	M_η/g.mol^{-1} = 120000, 100% isotactic sample							
Solvent (A):	**benzene**		**C$_6$H$_6$**				**71-43-2**	

$T/K = 298.15$

$\varphi_B^{(1)}$	0.0913	0.0870	0.0375	0.0442	0.0294	0.0294	0.0203	0.0166
$\varphi_B^{(2)}$	0.0375	0.0442	0.0247	0.0294	0.0203	0.0166	0.0148	0.0118
$\Delta_{dil} H^{12}$/(J/g solution)	−0.0475	−0.0545	−0.0109	−0.0116	0.0134	0.0158	0.00084	0.00208

$\varphi_B^{(1)}$	0.0118	0.0635	0.0783	0.0325
$\varphi_B^{(2)}$	0.0057	0.0144	0.0034	0.0013
$\Delta_{dil} H^{12}$/(J/g solution)	0.00184	0.0000	0.0106	0.00807

Comments: $\varphi_B^{(1)}$ and $\varphi_B^{(2)}$ denote the volume fractions of the polymer in the solution before and after the dilution process.

Polymer (B):   **poly(methyl methacrylate)**   **1968DAO**
Characterization: M_η/g.mol^{-1} = 182000, fractionated in the laboratory
Solvent (A):   **benzene**   **C$_6$H$_6$**   **71-43-2**

T/K = 298.15

$\varphi_B^{(1)}$	0.1395	0.1395	0.0953	0.0658	0.0658	0.0457	0.0457	0.0318
$\varphi_B^{(2)}$	0.1157	0.1106	0.0744	0.0526	0.0528	0.0366	0.0366	0.0254
$\Delta_{dil}H^{12}$/(J/g polymer)	0.0233	0.0351	0.0414	0.0540	0.0565	0.0715	0.0341	0.0502

$\varphi_B^{(1)}$	0.0318	0.0200	0.0200
$\varphi_B^{(2)}$	0.0252	0.0174	0.0171
$\Delta_{dil}H^{12}$/(J/g polymer)	0.0607	0.0469	0.0400

Comments: $\varphi_B^{(1)}$ and $\varphi_B^{(2)}$ denote the volume fractions of the polymer in the solution before and after the dilution process.

Polymer (B):   **poly(methyl methacrylate)**   **1968DAO**
Characterization: M_η/g.mol^{-1} = 182000, fractionated in the laboratory
Solvent (A):   **chlorobenzene**   **C$_6$H$_5$Cl**   **108-90-7**

T/K = 298.15

$\varphi_B^{(1)}$	0.1477	0.1005	0.1005	0.0696	0.0696	0.0482	0.0337	0.0337
$\varphi_B^{(2)}$	0.1039	0.0791	0.0782	0.0541	0.0542	0.0380	0.0263	0.0262
$\Delta_{dil}H^{12}$/(J/g polymer)	0.0061	0.0072	0.0048	0.0186	0.0123	0.0010	0.0169	0.0073

$\varphi_B^{(1)}$	0.0234	0.0234
$\varphi_B^{(2)}$	0.0184	0.0180
$\Delta_{dil}H^{12}$/(J/g polymer)	0.0145	0.0234

Comments: $\varphi_B^{(1)}$ and $\varphi_B^{(2)}$ denote the volume fractions of the polymer in the solution before and after the dilution process.

Polymer (B):   **poly(methyl methacrylate)**   **1968DAO**
Characterization: M_η/g.mol^{-1} = 182000, fractionated in the laboratory
Solvent (A):   **1,2-dichlorobenzene**   **C$_6$H$_4$Cl$_2$**   **95-50-1**

T/K = 298.15

$\varphi_B^{(1)}$	0.1539	0.1539	0.1057	0.1057	0.0734	0.0507	0.0507	0.0236
$\varphi_B^{(2)}$	0.1198	0.1210	0.0818	0.0830	0.0573	0.0400	0.0393	0.0185
$\Delta_{dil}H^{12}$/(J/g polymer)	0.0379	0.0452	0.0408	0.0418	0.0372	0.0386	0.0251	0.0180

$\varphi_B^{(1)}$	0.0152	0.0152
$\varphi_B^{(2)}$	0.0118	0.0119
$\Delta_{dil}H^{12}$/(J/g polymer)	0.0084	−0.0020

Comments: $\varphi_B^{(1)}$ and $\varphi_B^{(2)}$ denote the volume fractions of the polymer in the solution before and after the dilution process.

Polymer (B):	poly(methyl methacrylate)							1967KA2

Characterization: $M_n/\text{g.mol}^{-1} = 300000$, atactic,
Mitsubishi Rayon Co. Ltd., Japan

Solvent (A):	1,4-dioxane		$C_4H_8O_2$					123-91-1

$T/\text{K} = 298.15$

$V^{(1)}/\text{cm}^3$	10.0	20.0	20.0	10.0	10.0	20.0	30.0	10.0	20.0
$\varphi_B^{(1)}$	0.188	0.094	0.063	0.126	0.210	0.105	0.070	0.163	0.082
$V^{(2)}/\text{cm}^3$	20.0	30.0	40.0	20.0	20.0	30.0	40.0	20.0	30.0
$\varphi_B^{(2)}$	0.094	0.063	0.047	0.063	0.105	0.070	0.053	0.082	0.054
$\Delta_{dil}H^{12}/\text{J}$	0.406	0.151	0.084	0.192	0.540	0.188	0.100	0.368	0.121
$V^{(1)}/\text{cm}^3$	30.0	10.0	10.0	10.0	10.0	10.0	10.0	20.0	20.0
$\varphi_B^{(1)}$	0.188	0.188	0.188	0.210	0.210	0.163	0.163	0.076	0.076
$V^{(2)}/\text{cm}^3$	40.0	30.0	40.0	30.0	40.0	30.0	40.0	40.0	40.0
$\varphi_B^{(2)}$	0.041	0.063	0.047	0.070	0.053	0.054	0.041	0.038	0.038
$\Delta_{dil}H^{12}/\text{J}$	0.054	0.552	0.632	0.728	0.828	0.494	0.548	0.146	0.159

Comments: $\Delta_{dil}H^{12}$ is the extensive quantity obtained for a given total volume change from $V^{(1)}$ to $V^{(2)}$, where $\varphi_B^{(1)}$ and $\varphi_B^{(2)}$ denote the volume fractions of the polymer in the solution before and after the dilution process.

Polymer (B):	poly(methyl methacrylate)							1967KA2

Characterization: $M_n/\text{g.mol}^{-1} = 780000$, atactic,
Mitsubishi Rayon Co. Ltd., Japan

Solvent (A):	1,4-dioxane		$C_4H_8O_2$					123-91-1

$T/\text{K} = 298.15$

$V^{(1)}/\text{cm}^3$	10.0	20.0	10.0	10.0	20.0	30.0	10.0	10.0	10.0
$\varphi_B^{(1)}$	0.188	0.093	0.186	0.292	0.146	0.097	0.292	0.292	0.156
$V^{(2)}/\text{cm}^3$	20.0	30.0	30.0	20.0	30.0	40.0	30.0	40.0	20.0
$\varphi_B^{(2)}$	0.093	0.062	0.062	0.146	0.097	0.073	0.097	0.073	0.078
$\Delta_{dil}H^{12}/\text{J}$	0.372	0.138	0.510	0.883	0.322	0.146	1.201	1.331	0.297
$V^{(1)}/\text{cm}^3$	20.0	10.0	20.0	10.0	10.0	20.0	10.0	20.0	30.0
$\varphi_B^{(1)}$	0.078	0.150	0.075	0.150	0.200	0.100	0.200	0.070	0.056
$V^{(2)}/\text{cm}^3$	30.0	20.0	30.0	30.0	20.0	30.0	30.0	40.0	50.0
$\varphi_B^{(2)}$	0.052	0.075	0.050	0.050	0.100	0.067	0.067	0.035	0.033
$\Delta_{dil}H^{12}/\text{J}$	0.096	0.289	0.092	0.381	0.490	0.172	0.661	0.126	0.100
$V^{(1)}/\text{cm}^3$	20.0	20.0	30.0	50.0					
$\varphi_B^{(1)}$	0.063	0.060	0.052	0.042					
$V^{(2)}/\text{cm}^3$	40.0	40.0	50.0	70.0					
$\varphi_B^{(2)}$	0.032	0.030	0.031	0.030					
$\Delta_{dil}H^{12}/\text{J}$	0.100	0.105	0.075	0.067					

Comments: $\Delta_{dil}H^{12}$ is the extensive quantity obtained for a given total volume change from $V^{(1)}$ to $V^{(2)}$, where $\varphi_B^{(1)}$ and $\varphi_B^{(2)}$ denote the volume fractions of the polymer in the solution before and after the dilution process.

Polymer (B):   **poly(methyl methacrylate)**   **1980GRA, 1984KIL**
Characterization: M_n/g.mol^{-1} = 28900, M_w/g.mol^{-1} = 35900, atactic,
                     Roehm GmbH, Darmstadt, Germany
Solvent (A):   **4-methyl-2-pentanone   C$_6$H$_{12}$O**   **108-10-1**

T/K = 303.15

$c_B^{(1)}$/(g/100 cm^3)	10.0	10.0	10.0	10.0	10.0	10.0
$c_B^{(2)}$/(g/100 cm^3)	0.0378	0.0777	0.189	0.45	2.69	4.90
$\Delta_{dil}H^{12}$/(J/g polymer)	0.34	0.35	0.31	0.25	0.18	0.12

Comments: $c_B^{(1)}$ and $c_B^{(2)}$ denote the concentrations of the polymer in the solution before and after the
               dilution process.

Polymer (B):   **poly(methyl methacrylate)**   **1980GRA, 1984KIL**
Characterization: M_n/g.mol^{-1} = 137000, M_w/g.mol^{-1} = 215000, atactic,
                     Roehm GmbH, Darmstadt, Germany
Solvent (A):   **4-methyl-2-pentanone   C$_6$H$_{12}$O**   **108-10-1**

T/K = 303.15

$c_B^{(1)}$/(g/100 cm^3)	10.0	10.0	10.0	10.0	10.0	10.0	10.0	10.0
$c_B^{(2)}$/(g/100 cm^3)	0.0378	0.0378	0.0378	0.0756	0.0756	0.15	0.15	0.189
$\Delta_{dil}H^{12}$/(J/g polymer)	0.361	0.364	0.342	0.317	0.315	0.294	0.297	0.285

$c_B^{(1)}$/(g/100 cm^3)	10.0	10.0	10.0
$c_B^{(2)}$/(g/100 cm^3)	0.189	1.29	2.80
$\Delta_{dil}H^{12}$/(J/g polymer)	0.280	0.200	0.175

Comments: $c_B^{(1)}$ and $c_B^{(2)}$ denote the concentrations of the polymer in the solution before and after the
               dilution process.

Polymer (B):   **poly(methyl methacrylate)**   **1968BRU**
Characterization: M_n/g.mol^{-1} = 700, DP = 7, synthesized in the laboratory
Solvent (A):   **2-propanone**   **C$_3$H$_6$O**   **67-64-1**

T/K = 298.15

m_B/g	1.95
$c_B^{(1)}$/(g/cm^3)	0.100
$\Delta_{dil}H^{12}$/J	0.0314
$\Delta_{dil}H^{12}$/[$m_B(c_B^{(1)} - c_B^{(2)})$]	0.322

Comments: $c_B^{(1)}$ and $c_B^{(2)}$ denote the concentrations of the polymer in the solution before and after the
               dilution process.

Polymer (B):   **poly(methyl methacrylate)**   **1968BRU**
Characterization: M_n/g.mol^{-1} = 1100, DP = 11, synthesized in the laboratory
Solvent (A):   **2-propanone**   **C$_3$H$_6$O**   **67-64-1**

continued

continued

$T/K = 298.15$

m_B/g	1.95
$c_B^{(1)}/(g/cm^3)$	0.100
$\Delta_{dil}H^{12}/J$	0.0946
$\Delta_{dil}H^{12}/[m_B(c_B^{(1)} - c_B^{(2)})]$	0.971

Comments: $c_B^{(1)}$ and $c_B^{(2)}$ denote the concentrations of the polymer in the solution before and after the dilution process.

Polymer (B):	poly(methyl methacrylate)			1968BRU
Characterization:	$M_n/g.mol^{-1} = 1700$, DP = 17, synthesized in the laboratory			
Solvent (A):	**2-propanone**	**C_3H_6O**		**67-64-1**

$T/K = 298.15$

m_B/g	0.975	1.465	1.950	2.440
$c_B^{(1)}/(g/cm^3)$	0.050	0.075	0.100	0.125
$\Delta_{dil}H^{12}/J$	0.0254	0.0561	0.103	0.160
$\Delta_{dil}H^{12}/[m_B(c_B^{(1)} - c_B^{(2)})]$	1.042	1.038	1.059	1.046

Comments: $c_B^{(1)}$ and $c_B^{(2)}$ denote the concentrations of the polymer in the solution before and after the dilution process.

Polymer (B):	poly(methyl methacrylate)			1968BRU
Characterization:	$M_n/g.mol^{-1} = 5000$, DP = 50, synthesized in the laboratory			
Solvent (A):	**2-propanone**	**C_3H_6O**		**67-64-1**

$T/K = 298.15$

m_B/g	1.950
$c_B^{(1)}/(g/cm^3)$	0.100
$\Delta_{dil}H^{12}/J$	0.101
$\Delta_{dil}H^{12}/[m_B(c_B^{(1)} - c_B^{(2)})]$	1.038

Comments: $c_B^{(1)}$ and $c_B^{(2)}$ denote the concentrations of the polymer in the solution before and after the dilution process.

Polymer (B):	poly(methyl methacrylate)			1967KA2
Characterization:	$M_n/g.mol^{-1} = 300000$, atactic,			
	Mitsubishi Rayon Co. Ltd., Japan			
Solvent (A):	**2-propanone**	**C_3H_6O**		**67-64-1**

$T/K = 298.15$

$V^{(1)}/cm^3$	7.9	12.9	17.9	22.9	7.9	7.9	7.9	11.3	21.3
$\varphi_B^{(1)}$	0.290	0.178	0.128	0.100	0.290	0.290	0.290	0.115	0.061
$V^{(2)}/cm^3$	12.9	17.9	22.9	32.9	17.9	22.9	32.9	21.3	31.3
$\varphi_B^{(2)}$	0.178	0.128	0.100	0.069	0.128	0.100	0.069	0.061	0.042
$\Delta_{dil}H^{12}/J$	0.456	0.213	0.121	0.121	0.669	0.782	0.925	0.134	0.100

continued

continued

$V^{(1)}/cm^3$	11.3	20.5	40.5	20.5	11.8	21.5	11.8
$\varphi_B^{(1)}$	0.115	0.156	0.079	0.156	0.107	0.058	0.107
$V^{(2)}/cm^3$	31.3	40.5	50.5	50.5	21.8	31.5	31.5
$\varphi_B^{(2)}$	0.042	0.079	0.063	0.063	0.058	0.040	0.040
$\Delta_{dil}H^{12}/J$	0.192	0.498	0.084	0.552	0.113	0.046	0.151

Comments: $\Delta_{dil}H^{12}$ is the extensive quantity obtained for a given total volume change from $V^{(1)}$ to $V^{(2)}$, where $\varphi_B^{(1)}$ and $\varphi_B^{(2)}$ denote the volume fractions of the polymer in the solution before and after the dilution process.

Polymer (B):	**poly(methyl methacrylate)**						**1967KA2**
Characterization:	$M_n/g.mol^{-1} = 780000$, atactic,						
	Mitsubishi Rayon Co. Ltd., Japan						
Solvent (A):	**2-propanone**		C_3H_6O				**67-64-1**

$T/K = 298.15$

$V^{(1)}/cm^3$	20.9	40.9	20.9	12.8	22.8	32.8	12.8	12.8	7.5
$\varphi_B^{(1)}$	0.126	0.064	0.126	0.191	0.107	0.074	0.191	0.191	0.237
$V^{(2)}/cm^3$	40.9	50.9	50.9	22.8	32.8	42.8	32.8	42.8	12.5
$\varphi_B^{(2)}$	0.064	0.052	0.052	0.107	0.074	0.057	0.075	0.057	0.141
$\Delta_{dil}H^{12}/J$	0.238	0.054	0.285	0.293	0.121	0.067	0.414	0.431	0.243

$V^{(1)}/cm^3$	12.5	17.5	22.5	7.5	7.5	7.5	22.0	32.0	22.0
$\varphi_B^{(1)}$	0.142	0.101	0.097	0.237	0.237	0.237	0.094	0.094	0.094
$V^{(2)}/cm^3$	17.5	22.5	32.5	17.5	22.5	32.5	32.0	42.0	42.0
$\varphi_B^{(2)}$	0.101	0.097	0.055	0.101	0.097	0.055	0.064	0.049	0.049
$\Delta_{dil}H^{12}/J$	0.117	0.075	0.079	0.360	0.435	0.515	0.079	0.046	0.121

Comments: $\Delta_{dil}H^{12}$ is the extensive quantity obtained for a given total volume change from $V^{(1)}$ to $V^{(2)}$, where $\varphi_B^{(1)}$ and $\varphi_B^{(2)}$ denote the volume fractions of the polymer in the solution before and after the dilution process.

Polymer (B):	**poly(methyl methacrylate)**					**1980GRA, 1984KIL**
Characterization:	$M_w/g.mol^{-1} = 30000$, 91% isotactic,					
	Roehm GmbH, Darmstadt, Germany					
Solvent (A):	**tetrachloromethane**	CCl_4				**56-23-5**

$T/K = 303.15$

$c_B^{(1)}/(g/100\ cm^3)$	5.48	5.48	5.48	5.48	5.48	0.548	0.548	0.548
$c_B^{(2)}/(g/100\ cm^3)$	0.01	0.066	0.066	0.10	0.20	0.001	0.0021	0.003
$\Delta_{dil}H^{12}/(J/g\ polymer)$	1.14	1.07	1.10	1.05	0.97	−106.9	−33.9	−8.2

$c_B^{(1)}/(g/100\ cm^3)$	0.548	0.548	0.548	0.548	0.548
$c_B^{(2)}/(g/100\ cm^3)$	0.004	0.005	0.0065	0.0084	0.010
$\Delta_{dil}H^{12}/(J/g\ polymer)$	−2.4	3.8	3.4	3.5	2.9

Comments: $c_B^{(1)}$ and $c_B^{(2)}$ denote the concentrations of the polymer in the solution before and after the dilution process.

Polymer (B):   **poly(methyl methacrylate)**   **1980GRA, 1984KIL**
Characterization: M_n/g.mol^{-1} = 689000, M_w/g.mol^{-1} = 782000, atactic,
                          Roehm GmbH, Darmstadt, Germany
Solvent (A):   **tetrachloromethane**   **CCl$_4$**   **56-23-5**

T/K = 303.15

$c_B^{(1)}$/(g/100 cm^3)	5.42	5.42	5.42	5.42	5.42	5.42	0.548	0.548
$c_B^{(2)}$/(g/100 cm^3)	0.05	0.05	0.10	0.10	0.10	0.2	0.001	0.010
$\Delta_{dil}H^{12}$/(J/g polymer)	2.35	2.36	2.42	2.12	2.27	1.93	10.23	2.92

Comments: $c_B^{(1)}$ and $c_B^{(2)}$ denote the concentrations of the polymer in the solution before and after the dilution process.

Polymer (B):   **poly(methyl methacrylate)**   **1971LEW**
Characterization: M_η/g.mol^{-1} = 26400, M_w/M_n = 1.38 by GPC,
                          synthesized in the laboratory
Solvent (A):   **toluene**   **C$_7$H$_8$**   **108-88-3**

T/K = 303.15

$\varphi_B^{(1)}$	0.3228	0.3228	0.3228	0.1658	0.1658	0.0804	0.0451
$\varphi_B^{(2)}$	0.0935	0.0184	0.2316	0.0797	0.0381	0.0409	0.0224
$\Delta_{dil}H^{12}$/(J/g polymer)	0.1459	0.0342	0.1473	0.1275	0.0640	0.0524	0.0244

Comments: $\varphi_B^{(1)}$ and $\varphi_B^{(2)}$ denote the volume fractions of the polymer in the solution before and after the dilution process.

Polymer (B):   **poly(methyl methacrylate)**   **1971LEW**
Characterization: M_η/g.mol^{-1} = 41500, M_w/M_n = 1.82 by GPC,
                          synthesized in the laboratory
Solvent (A):   **toluene**   **C$_7$H$_8$**   **108-88-3**

T/K = 303.15

$\varphi_B^{(1)}$	0.0321	0.0638	0.1160	0.2043	0.2374	0.2374	0.2374	0.2374
$\varphi_B^{(2)}$	0.0158	0.0326	0.0596	0.0878	0.1468	0.0670	0.0417	0.0212
$\Delta_{dil}H^{12}$/(J/g polymer)	0.0243	0.0513	0.1181	0.0733	0.1071	0.0832	0.0060	0.0314

$\varphi_B^{(1)}$	0.2374
$\varphi_B^{(2)}$	0.0910
$\Delta_{dil}H^{12}$/(J/g polymer)	0.0957

Polymer (B):   **poly(methyl methacrylate)**   **1980GRA, 1984KIL**
Characterization: M_n/g.mol^{-1} = 93940, M_w/g.mol^{-1} = 101000, atactic,
                          Roehm GmbH, Darmstadt, Germany
Solvent (A):   **toluene**   **C$_7$H$_8$**   **108-88-3**

continued

continued

$T/K = 303.15$

$c_B^{(1)}/(g/100\ cm^3)$	5.13	5.13	10.0	10.0	10.0	10.0	10.0	10.0
$c_B^{(2)}/(g/100\ cm^3)$	0.094	0.094	0.090	0.186	0.37	0.67	0.89	1.39
$\Delta_{dil}H^{12}/(J/g\ polymer)$	1.050	1.030	1.350	1.080	0.985	0.899	0.764	0.616

$c_B^{(1)}/(g/100\ cm^3)$	10.0	10.0	15.1	15.1
$c_B^{(2)}/(g/100\ cm^3)$	3.55	4.98	0.28	0.28
$\Delta_{dil}H^{12}/(J/g\ polymer)$	0.514	0.376	0.963	0.976

Comments: $c_B^{(1)}$ and $c_B^{(2)}$ denote the concentrations of the polymer in the solution before and after the dilution process.

Polymer (B):   **poly(methyl methacrylate)**   **1967KA2**
Characterization: $M_n/g.mol^{-1} = 300000$, atactic,
                          Mitsubishi Rayon Co. Ltd., Japan
Solvent (A):   **toluene**   **C7H8**   **108-88-3**

$T/K = 303.15$

$V^{(1)}/cm^3$	10.0	15.0	10.0	10.0	10.0	20.0	30.0	10.0	10.0
$\varphi_B^{(1)}$	0.141	0.094	0.125	0.176	0.216	0.108	0.072	0.141	0.216
$V^{(2)}/cm^3$	15.0	20.0	20.0	20.0	20.0	30.0	40.0	20.0	30.0
$\varphi_B^{(2)}$	0.094	0.070	0.063	0.088	0.108	0.072	0.054	0.070	0.072
$\Delta_{dil}H^{12}/J$	0.105	0.050	0.105	0.201	0.322	0.113	0.054	0.151	0.435

$V^{(1)}/cm^3$	10.0	50.0	30.0	20.0	30.0	40.0	50.0	20.0	30.0
$\varphi_B^{(1)}$	0.216	0.056	0.070	0.083	0.069	0.050	0.043	0.080	0.058
$V^{(2)}/cm^3$	40.0	70.0	50.0	40.0	50.0	60.0	70.0	40.0	50.0
$\varphi_B^{(2)}$	0.054	0.040	0.042	0.042	0.042	0.033	0.030	0.040	0.035
$\Delta_{dil}H^{12}/J$	0.490	0.067	0.075	0.109	0.084	0.042	0.096	0.063	0.121

Comments: $\Delta_{dil}H^{12}$ is the extensive quantity obtained for a given total volume change from $V^{(1)}$ to $V^{(2)}$, where $\varphi_B^{(1)}$ and $\varphi_B^{(2)}$ denote the volume fractions of the polymer in the solution before and after the dilution process.

Polymer (B):   **poly(methyl methacrylate)**   **1980GRA, 1984KIL**
Characterization: $M_n/g.mol^{-1} = 689000$, $M_w/g.mol^{-1} = 782000$, atactic,
                          Roehm GmbH, Darmstadt, Germany
Solvent (A):   **toluene**   **C7H8**   **108-88-3**

$T/K = 303.15$

$c_B^{(1)}/(g/100\ cm^3)$	5.02	5.02	10.0	10.0	10.0	10.0	10.0	10.0
$c_B^{(2)}/(g/100\ cm^3)$	0.093	0.70	0.087	0.093	0.19	0.76	0.97	1.26
$\Delta_{dil}H^{12}/(J/g\ polymer)$	1.120	0.990	1.362	1.369	1.133	1.121	1.067	1.000

$c_B^{(1)}/(g/100\ cm^3)$	10.0	10.0	10.0
$c_B^{(2)}/(g/100\ cm^3)$	2.02	3.00	4.90
$\Delta_{dil}H^{12}/(J/g\ polymer)$	0.969	0.3800	0.640

Comments: $c_B^{(1)}$ and $c_B^{(2)}$ denote the concentrations of the polymer in the solution before and after the dilution process.

| **Polymer (B):** | | **poly(methyl methacrylate)** | | | | | | **1967KA2** |

Characterization: $M_n/\text{g.mol}^{-1} = 780000$, atactic,
Mitsubishi Rayon Co. Ltd., Japan

| **Solvent (A):** | | **toluene** | | | **C$_7$H$_8$** | | | **108-88-3** |

$T/\text{K} = 303.15$

$V^{(1)}/\text{cm}^3$	5.0	10.0	10.0	20.0	30.0	10.0	20.0	30.0	5.0
$\varphi_B^{(1)}$	0.253	0.127	0.300	0.150	0.100	0.263	0.132	0.088	0.258
$V^{(2)}/\text{cm}^3$	10.0	15.0	20.0	30.0	40.0	20.0	30.0	40.0	15.0
$\varphi_B^{(2)}$	0.127	0.084	0.150	0.100	0.075	0.132	0.088	0.066	0.084
$\Delta_{dil}H^{12}/\text{J}$	0.188	0.067	0.598	0.209	0.092	0.368	0.163	0.071	0.255

$V^{(1)}/\text{cm}^3$	10.0	10.0	10.0	10.0	30.0	40.0	30.0	30.0	40.0
$\varphi_B^{(1)}$	0.300	0.300	0.263	0.263	0.062	0.047	0.062	0.075	0.056
$V^{(2)}/\text{cm}^3$	30.0	40.0	30.0	40.0	40.0	50.0	50.0	40.0	50.0
$\varphi_B^{(2)}$	0.100	0.075	0.088	0.066	0.047	0.037	0.037	0.056	0.041
$\Delta_{dil}H^{12}/\text{J}$	0.808	0.900	0.531	0.602	0.046	0.025	0.071	0.054	0.033

$V^{(1)}/\text{cm}^3$	30.0	20.0	30.0
$\varphi_B^{(1)}$	0.075	0.073	0.064
$V^{(2)}/\text{cm}^3$	50.0	40.0	50.0
$\varphi_B^{(2)}$	0.041	0.037	0.038
$\Delta_{dil}H^{12}/\text{J}$	0.088	0.075	0.071

Comments: $\Delta_{dil}H^{12}$ is the extensive quantity obtained for a given total volume change from $V^{(1)}$ to $V^{(2)}$, where $\varphi_B^{(1)}$ and $\varphi_B^{(2)}$ denote the volume fractions of the polymer in the solution before and after the dilution process.

| **Polymer (B):** | | **poly(methyl methacrylate)** | | **1972MAS** |

Characterization: $M_\eta/\text{g.mol}^{-1} = 10000$, synthesized in the laboratory

| **Solvent (A):** | | **trichloromethane** | **CHCl$_3$** | **67-66-3** |

$T/\text{K} = 298.15$

$$\Delta_M H/(\text{J/cm}^3) = 4.184\,\varphi_B(1 - \varphi_B)[-76.831 + 29.886(1 - 2\varphi_B)]$$

| **Polymer (B):** | | **poly(methyl methacrylate)** | | **1971LEW** |

Characterization: $M_\eta/\text{g.mol}^{-1} = 26400$, $M_w/M_n = 1.38$ by GPC,
synthesized in the laboratory

| **Solvent (A):** | | **trichloromethane** | **CHCl$_3$** | **67-66-3** |

$T/\text{K} = 303.15$

$\varphi_B^{(1)}$	0.2501	0.1669	0.0728	0.3264	0.0430	0.3272
$\varphi_B^{(2)}$	0.1298	0.0853	0.0375	0.2668	0.0220	0.1144
$\Delta_{dil}H^{12}/(\text{J/g polymer})$	−0.8568	−0.2048	−0.0221	−5.2446	−0.0054	−1.1583

Comments: $\varphi_B^{(1)}$ and $\varphi_B^{(2)}$ denote the volume fractions of the polymer in the solution before and after the dilution process.

Polymer (B): **poly(L-ornithine)** **1979BAB**
Characterization: $M_n/\text{g.mol}^{-1} = 270000$, Sigma Chemical Co., Inc., St. Louis, MO
Solvent (A): **methanol** **CH$_4$O** **67-56-1**
Solvent (C): **water** **H$_2$O** **7732-18-5**

$T/\text{K} = 298.15$

$c_B^{(1)}/(\text{base mol/l})$ 0.20 was kept constant

φ_A	0.660	0.752	0.802	0.859	0.890	0.910	0.924	0.934
$\Delta_{dil}H^{12}/\text{J}$	−7.639	−4.861	−3.833	−5.610	−3.875	−2.809	−2.134	−1.684

φ_A	0.935	0.963	0.974	0.981
$\Delta_{dil}H^{12}/\text{J}$	−0.963	−0.568	−0.738	−0.638

Comments: $\Delta_{dil}H^{12}$ is here an extensive quantity obtained for a given amount of 1 cm^3 pure methanol added successively to a definite volume of solution in which the composition of poly(L-ornithine), methanol and water is known.

Polymer (B): **poly(propylene glycol)** **2005COM**
Characterization: $M_n/\text{g.mol}^{-1} = 502$, $M_w/\text{g.mol}^{-1} = 512$, $\rho = 0.99568$ g/cm^3 (308 K), PPG 400, Fluka AG, Buchs, Switzerland
Solvent (A): **anisole** **C$_7$H$_8$O** **100-66-3**

$T/\text{K} = 308.15$

w_B	0.0406	0.0781	0.1127	0.1448	0.2026	0.2530	0.3369	0.4038
$\Delta_M H/(\text{J/g})$	0.3380	0.5660	0.7212	0.8243	0.9285	0.9576	0.9832	0.9636

w_B	0.5041	0.6040	0.6703	0.7530	0.8026	0.8591	0.8905	0.9242
$\Delta_M H/(\text{J/g})$	0.8792	0.7736	0.6693	0.5335	0.4302	0.3021	0.2417	0.1541

w_B	0.9606
$\Delta_M H/(\text{J/g})$	0.0740

Polymer (B): **poly(propylene glycol)** **2005COM**
Characterization: $M_n/\text{g.mol}^{-1} = 856$, $M_w/\text{g.mol}^{-1} = 884$, $\rho = 0.99353$ g/cm^3 (308 K), PPG 725, Aldrich Chem. Co., Inc., Milwaukee, WI
Solvent (A): **anisole** **C$_7$H$_8$O** **100-66-3**

$T/\text{K} = 308.15$

w_B	0.0405	0.0779	0.1125	0.1446	0.2022	0.2526	0.3364	0.4033
$\Delta_M H/(\text{J/g})$	0.1253	0.1605	0.1496	0.1034	−0.0089	−0.1157	−0.2633	−0.3319

w_B	0.5035	0.6034	0.6698	0.7526	0.8022	0.8589	0.8903	0.9116
$\Delta_M H/(\text{J/g})$	−0.3487	−0.2948	−0.2488	−0.2081	−0.1880	−0.1727	−0.1584	−0.1441

w_B	0.9605
$\Delta_M H/(\text{J/g})$	−0.0856

Polymer (B):   **poly(propylene glycol)**   **2005COM**

Characterization: M_n/g.mol^{-1} = 1334, M_w/g.mol^{-1} = 1374, ρ = 0.99303 g/cm^3 (308 K),
                       PPG 1200, Fluka AG, Buchs, Switzerland

Solvent (A):   **anisole**   **C$_7$H$_8$O**   **100-66-3**

T/K = 308.15

w_B	0.0405	0.0779	0.1124	0.1445	0.2022	0.2525	0.3363	0.4031
$\Delta_M H$/(J/g)	−0.0373	−0.1143	−0.2064	−0.3029	−0.4785	−0.6248	−0.7870	−0.8709

w_B	0.5034	0.6033	0.6696	0.7525	0.8021	0.8588	0.8902	0.9240
$\Delta_M H$/(J/g)	−0.8770	−0.7723	−0.7088	−0.5840	−0.5143	−0.4000	−0.3333	−0.2467

w_B	0.9605
$\Delta_M H$/(J/g)	−0.1500

Polymer (B):   **poly(propylene glycol)**   **2005COM**

Characterization: M_n/g.mol^{-1} = 2004, M_w/g.mol^{-1} = 2063, ρ = 0.99212 g/cm^3 (308 K),
                       PPG 2000, Fluka AG, Buchs, Switzerland

Solvent (A):   **anisole**   **C$_7$H$_8$O**   **100-66-3**

T/K = 308.15

w_B	0.0405	0.0778	0.1124	0.1444	0.2020	0.2523	0.3361	0.4029
$\Delta_M H$/(J/g)	−0.0414	−0.1341	−0.2450	−0.3665	−0.5821	−0.7577	−0.9818	−1.0587

w_B	0.5031	0.6031	0.6694	0.7523	0.8020	0.8587	0.8901	0.9240
$\Delta_M H$/(J/g)	−1.1132	−1.0009	−0.9332	−0.7663	−0.6685	−0.5280	−0.4188	−0.3213

w_B	0.9605
$\Delta_M H$/(J/g)	−0.1776

Polymer (B):   **poly(propylene glycol)**   **1988PAR**

Characterization: M_n/g.mol^{-1} = 150,
                       fractionated samples supplied by Union Carbide Corp.

Solvent (A):   **benzene**   **C$_6$H$_6$**   **71-43-2**

T/K = 321.35

w_B	0.004051	0.01217	0.03221	0.05549	0.08811	0.1265	0.2323	0.3139
$\Delta_M H$/(J/mol)	64.85	153.9	309.6	382.4	504.2	556.9	706.7	794.1

w_B	0.3795	0.4326	0.5539	0.6435	0.7548	0.8899	0.9692
$\Delta_M H$/(J/mol)	855.2	859.8	846.0	824.7	705.8	393.3	107.5

Polymer (B):   **poly(propylene glycol)**   **1988PAR**

Characterization: M_n/g.mol^{-1} = 425,
                       fractionated samples supplied by Union Carbide Corp.

Solvent (A):   **benzene**   **C$_6$H$_6$**   **71-43-2**

continued

continued

T/K = 321.35

w_B	0.005546	0.01413	0.02678	0.07014	0.09204	0.1239	0.1730	0.1955
$\Delta_M H$/(J/mol)	34.00	73.64	93.72	148.5	163.2	207.5	226.8	258.2

w_B	0.2437	0.3467	0.4100	0.5149	0.6018	0.7233	0.8719	0.9662
$\Delta_M H$/(J/mol)	304.2	329.7	364.0	421.7	395.0	341.0	239.3	70.3

w_B	0.9796	0.9917
$\Delta_M H$/(J/mol)	−100.8	−51.88

Polymer (B):	poly(propylene glycol)		1988PAR
Characterization:	M_n/g.mol^{-1} = 2025,		
	fractionated samples supplied by Union Carbide Corp.		
Solvent (A):	benzene	C_6H_6	71-43-2

T/K = 321.35

w_B	0.005587	0.01437	0.03147	0.06114	0.09846	0.1311	0.2273	0.3043
$\Delta_M H$/(J/mol)	19.24	35.56	53.13	56.90	51.04	−29.70	−74.47	−102.9

w_B	0.3716	0.4245	0.5072	0.5968	0.7199	0.8669	0.9684
$\Delta_M H$/(J/mol)	−124.3	−150.6	−228.0	−294.9	−325.5	−351.0	−267.4

Polymer (B):	poly(propylene glycol)		2005COM
Characterization:	M_n/g.mol^{-1} = 502, M_w/g.mol^{-1} = 512, ρ = 0.99568 g/cm^3 (308 K),		
	PPG 400, Fluka AG, Buchs, Switzerland		
Solvent (A):	benzyl alcohol	C_7H_8O	100-51-6

T/K = 308.15

w_B	0.0386	0.0743	0.1075	0.1383	0.1941	0.2431	0.3338	0.4003
$\Delta_M H$/(J/g)	−1.4512	−2.5309	−3.2720	−3.9264	−4.7332	−5.5570	−6.0587	−6.2828

w_B	0.4907	0.5911	0.6583	0.7429	0.7940	0.8525	0.8851	0.9204
$\Delta_M H$/(J/g)	−6.5090	−6.3842	−6.0549	−5.2713	−4.5000	−3.5141	−2.8003	−2.0345

w_B	0.9585
$\Delta_M H$/(J/g)	−1.1108

Polymer (B):	poly(propylene glycol)		2004COM
Characterization:	M_n/g.mol^{-1} = 856, M_w/g.mol^{-1} = 884, ρ = 0.99353 g/cm^3 (308 K),		
	PPG 725, Aldrich Chem. Co., Inc., Milwaukee, WI		
Solvent (A):	benzyl alcohol	C_7H_8O	100-51-6

T/K = 308.15

w_B	0.0501	0.0741	0.1073	0.1381	0.1937	0.2426	0.3246	0.3904
$\Delta_M H$/(J/g)	−1.5703	−2.1438	−2.8024	−3.2949	−4.0821	−4.5440	−5.1556	−5.5623

w_B	0.4901	0.5905	0.6578	0.7425	0.7936	0.8522	0.8849	0.9202
$\Delta_M H$/(J/g)	−5.9504	−5.7229	−5.3709	−4.6936	−3.8561	−2.9500	−2.3624	−1.6612

Polymer (B):   **poly(propylene glycol)**   **2004COM**
Characterization: $M_n/\text{g.mol}^{-1} = 1334$, $M_w/\text{g.mol}^{-1} = 1374$, $\rho = 0.99303$ g/cm^3 (308 K),
     PPG 1200, Fluka AG, Buchs, Switzerland
Solvent (A):   **benzyl alcohol**   **C$_7$H$_8$O**   **100-51-6**

$T/\text{K} = 308.15$

w_B	0.0385	0.0741	0.1072	0.1380	0.1937	0.2425	0.3245	0.3903
$\Delta_M H/(\text{J/g})$	−1.1325	−1.9561	−2.5543	−3.0051	−3.6084	−3.9625	−4.4363	−4.5584

w_B	0.4900	0.5904	0.6577	0.7424	0.7935	0.8522	0.8849	0.9202
$\Delta_M H/(\text{J/g})$	−4.5881	−4.4031	−4.1422	−3.5367	−2.9881	−2.2512	−1.8214	−1.2524

w_B	0.9584
$\Delta_M H/(\text{J/g})$	−0.6567

Polymer (B):   **poly(propylene glycol)**   **2004COM**
Characterization: $M_n/\text{g.mol}^{-1} = 2004$, $M_w/\text{g.mol}^{-1} = 2063$, $\rho = 0.99212$ g/cm^3 (308 K),
     PPG 2000, Fluka AG, Buchs, Switzerland
Solvent (A):   **benzyl alcohol**   **C$_7$H$_8$O**   **100-51-6**

$T/\text{K} = 308.15$

w_B	0.0384	0.0740	0.1071	0.1379	0.1935	0.2424	0.3243	0.3901
$\Delta_M H/(\text{J/g})$	−1.0001	−1.6877	−2.1203	−2.4326	−2.7154	−2.9728	−3.1648	−3.2617

w_B	0.4897	0.5902	0.6575	0.7422	0.7934	0.8521	0.8848	0.9201
$\Delta_M H/(\text{J/g})$	−3.3701	−3.3536	−3.2496	−2.7011	−2.3132	−1.7565	−1.3700	−0.9478

w_B	0.9584
$\Delta_M H/(\text{J/g})$	−0.4746

Polymer (B):   **poly(propylene glycol)**   **2004COM**
Characterization: $M_n/\text{g.mol}^{-1} = 502$, $M_w/\text{g.mol}^{-1} = 512$, $\rho = 0.99580$ g/cm^3 (308 K),
     PPG 400, Fluka AG, Buchs, Switzerland
Solvent (A):   **diethyl carbonate**   **C$_5$H$_{10}$O$_3$**   **105-58-8**

$T/\text{K} = 308.15$

w_B	0.0415	0.0797	0.1150	0.1477	0.2063	0.2573	0.3420	0.4093
$\Delta_M H/(\text{J/g})$	0.6598	1.2068	1.5763	1.8907	2.3938	2.8241	3.1071	3.2146

w_B	0.5097	0.6094	0.6752	0.7572	0.8061	0.8618	0.8927	0.9258
$\Delta_M H/(\text{J/g})$	3.1355	2.9043	2.6971	2.3037	2.0384	1.6397	1.3887	1.0070

w_B	0.9615
$\Delta_M H/(\text{J/g})$	0.5801

Polymer (B):   **poly(propylene glycol)**   **2004COM**
Characterization: $M_n/\text{g.mol}^{-1} = 856$, $M_w/\text{g.mol}^{-1} = 884$, $\rho = 0.99361$ g/cm^3 (308 K),
     PPG 725, Aldrich Chem. Co., Inc., Milwaukee, WI
Solvent (A):   **diethyl carbonate**   **C$_5$H$_{10}$O$_3$**   **105-58-8**

continued

continued

$T/K = 308.15$

w_B	0.0414	0.0795	0.1148	0.1474	0.2059	0.2569	0.3415	0.4087
$\Delta_M H/(\text{J/g})$	0.3214	0.5785	0.7752	0.9317	1.1628	1.3520	1.5007	1.5643

w_B	0.5091	0.6088	0.6747	0.7568	0.8057	0.8615	0.8924	0.9256
$\Delta_M H/(\text{J/g})$	1.6058	1.5630	1.4741	1.2958	1.1336	0.9080	0.7500	0.5268

w_B	0.9614
$\Delta_M H/(\text{J/g})$	0.3099

Polymer (B):	**poly(propylene glycol)**	**2004COM**
Characterization:	$M_n/\text{g.mol}^{-1} = 2004$, $M_w/\text{g.mol}^{-1} = 2063$, $\rho = 0.99215$ g/cm^3 (308 K), PPG 2000, Fluka AG, Buchs, Switzerland	
Solvent (A):	**diethyl carbonate** $C_5H_{10}O_3$	**105-58-8**

$T/K = 308.15$

w_B	0.0414	0.0794	0.1146	0.1472	0.2057	0.2566	0.3412	0.4084
$\Delta_M H/(\text{J/g})$	0.0997	0.1662	0.2261	0.2656	0.3385	0.3855	0.4447	0.4784

w_B	0.5088	0.6085	0.6744	0.7565	0.8056	0.8614	0.8923	0.9255
$\Delta_M H/(\text{J/g})$	0.5166	0.5315	0.5221	0.4829	0.4412	0.3681	0.3142	0.2379

w_B	0.9613
$\Delta_M H/(\text{J/g})$	0.1369

Polymer (B):	**poly(propylene glycol)**	**2004CA2**
Characterization:	$M_n/\text{g.mol}^{-1} = 502$, $M_w/\text{g.mol}^{-1} = 512$, $\rho = 0.99568$ g/cm^3 (298 K), PPG 400, Fluka AG, Buchs, Switzerland	
Solvent (A):	**dimethyl carbonate** $C_3H_6O_3$	**616-38-6**

$T/K = 308.15$

w_B	0.0380	0.0732	0.1060	0.1375	0.1917	0.2402	0.3217	0.3873
$\Delta_M H/(\text{J/g})$	0.72	1.34	1.86	2.30	2.97	3.44	4.21	4.56

w_B	0.4868	0.5873	0.6548	0.7399	0.7914	0.8506	0.8835	0.9192
$\Delta_M H/(\text{J/g})$	4.87	4.73	4.51	3.79	3.28	2.52	2.04	1.47

w_B	0.9599
$\Delta_M H/(\text{J/g})$	0.78

Polymer (B):	**poly(propylene glycol)**	**2004CA2**
Characterization:	$M_n/\text{g.mol}^{-1} = 856$, $M_w/\text{g.mol}^{-1} = 884$, $\rho = 0.99353$ g/cm^3 (298 K), PPG 725, Aldrich Chem. Co., Inc., Milwaukee, WI	
Solvent (A):	**dimethyl carbonate** $C_3H_6O_3$	**616-38-6**

$T/K = 308.15$

continued

continued

w_B	0.0379	0.0731	0.1058	0.1362	0.1913	0.2398	0.3212	0.3868
$\Delta_M H/(J/g)$	0.48	0.86	1.20	1.48	1.87	2.15	2.56	2.76

w_B	0.4862	0.5868	0.6543	0.7395	0.7910	0.8503	0.8833	0.9191
$\Delta_M H/(J/g)$	2.94	2.91	2.77	2.44	2.19	1.78	1.48	1.06

w_B	0.9578
$\Delta_M H/(J/g)$	0.60

Polymer (B):　　　　**poly(propylene glycol)**　　　　　　　　　　　　　**2004CA2**
Characterization:　　　$M_n/\text{g.mol}^{-1} = 1335$, $M_w/\text{g.mol}^{-1} = 1375$, $\rho = 0.99303$ g/cm^3 (298 K),
　　　　　　　　　　　　PPG 1200, Fluka AG, Buchs, Switzerland
Solvent (A):　　　　　**dimethyl carbonate**　　　$C_3H_6O_3$　　　　　　　　**616-38-6**

$T/K = 308.15$

w_B	0.0379	0.0731	0.1057	0.1362	0.1912	0.2397	0.3211	0.3866
$\Delta_M H/(J/g)$	0.31	0.57	0.80	1.04	1.31	1.57	1.89	2.06

w_B	0.4861	0.5867	0.6542	0.7394	0.7910	0.8502	0.8833	0.9190
$\Delta_M H/(J/g)$	2.24	2.11	2.02	1.76	1.53	1.17	0.97	0.70

w_B	0.9578
$\Delta_M H/(J/g)$	0.37

Polymer (B):　　　　**poly(propylene glycol)**　　　　　　　　　　　　　**2004CA2**
Characterization:　　　$M_n/\text{g.mol}^{-1} = 2004$, $M_w/\text{g.mol}^{-1} = 2063$, $\rho = 0.99212$ g/cm^3 (298 K),
　　　　　　　　　　　　PPG 2000, Fluka AG, Buchs, Switzerland
Solvent (A):　　　　　**dimethyl carbonate**　　　$C_3H_6O_3$　　　　　　　　**616-38-6**

$T/K = 308.15$

w_B	0.0379	0.0730	0.1057	0.1361	0.1911	0.2395	0.3209	0.3864
$\Delta_M H/(J/g)$	0.25	0.48	0.68	0.84	1.14	1.32	1.63	1.76

w_B	0.4859	0.5865	0.6540	0.7393	0.7908	0.8501	0.8832	0.9190
$\Delta_M H/(J/g)$	1.91	1.92	1.74	1.57	1.34	1.04	0.86	0.63

w_B	0.9578
$\Delta_M H/(J/g)$	0.33

Polymer (B):　　　　**poly(propylene glycol)**　　　　　　　　　　　　　**1988PAR**
Characterization:　　　$M_n/\text{g.mol}^{-1} = 150$,
　　　　　　　　　　　　fractionated samples supplied by Union Carbide Corp.
Solvent (A):　　　　　**ethanol**　　　　　　　　　C_2H_6O　　　　　　　　　**64-17-5**

$T/K = 321.35$

w_B	0.006579	0.01516	0.03037	0.06095	0.1044	0.1711	0.2827	0.3811
$\Delta_M H/(J/mol)$	13.10	15.56	12.13	29.29	28.45	40.15	21.50	50.20

continued

continued

w_B	0.4509	0.5079	0.5915	0.7137	0.8717	0.9647
$\Delta_M H/(\text{J/mol})$	50.20	37.66	−4.184	−94.56	−123.4	−46.86

Polymer (B):	**poly(propylene glycol)**						**1988PAR**

Characterization: $M_n/\text{g.mol}^{-1} = 425$,
fractionated samples supplied by Union Carbide Corp.

Solvent (A):	**ethanol**		**C$_2$H$_6$O**				**64-17-5**

$T/\text{K} = 321.35$

w_B	0.006561	0.01580	0.03050	0.06366	0.1120	0.1792	0.2979	0.3830
$\Delta_M H/(\text{J/mol})$	19.25	27.15	15.77	3.565	−10.13	−17.15	−9.245	+3.185

w_B	0.4497	0.5101	0.5993	0.7188	0.8643	0.9668
$\Delta_M H/(\text{J/mol})$	9.842	18.79	31.72	69.04	100.8	46.02

Polymer (B):	**poly(propylene glycol)**						**1988PAR**

Characterization: $M_n/\text{g.mol}^{-1} = 2025$,
fractionated samples supplied by Union Carbide Corp.

Solvent (A):	**ethanol**		**C$_2$H$_6$O**				**64-17-5**

$T/\text{K} = 321.35$

w_B	0.00636	0.01618	0.03149	0.06831	0.1124	0.1692	0.2902	0.3794
$\Delta_M H/(\text{J/mol})$	18.83	25.52	16.74	38.49	36.40	54.40	127.2	204.6

w_B	0.4476	0.5117	0.5990	0.7102	0.8548	0.9647
$\Delta_M H/(\text{J/mol})$	275.7	362.3	503.3	757.7	1447	1732

Polymer (B):	**poly(propylene glycol)**						**1960LAK**

Characterization: $M_n/\text{g.mol}^{-1} = 150$, fractionated sample

Solvent (A):	**methanol**		**CH$_4$O**				**67-56-1**

$T/\text{K} = 300.05$

w_B	0.094	0.242	0.359	0.723	0.808	0.850	0.912
$\Delta_M H/(\text{J/mol})$	−27.6	−62.8	−92.0	−184	−188	−175	−138

Polymer (B):	**poly(propylene glycol)**						**1960LAK**

Characterization: $M_n/\text{g.mol}^{-1} = 1120$, fractionated sample

Solvent (A):	**methanol**		**CH$_4$O**				**67-56-1**

$T/\text{K} = 300.05$

w_B	0.143	0.300	0.347	0.432	0.624	0.718	0.793	0.888
$\Delta_M H/(\text{J/mol})$	−22.6	−45.6	−54.4	−62.8	−62.8	−62.8	−58.6	−31.4

Polymer (B):	**poly(propylene glycol)**						**1960LAK**

Characterization: M_n/g.mol^{-1} = 1955, fractionated sample

Solvent (A): **methanol** **CH$_4$O** **67-56-1**

T/K = 300.05

w_B	0.113	0.224	0.316	0.447	0.515	0.599	0.633	0.703
$\Delta_M H$/(J/mol)	−6.28	−8.79	−8.79	+0.42	21.3	43.1	109	138

w_B	0.785	0.908
$\Delta_M H$/(J/mol)	192	272

Polymer (B):	**poly(propylene glycol)**	**1960LAK**

Characterization: M_n/g.mol^{-1} = 3350, fractionated sample

Solvent (A): **methanol** **CH$_4$O** **67-56-1**

T/K = 300.05

w_B	0.177	0.279	0.429	0.485	0.519	0.572	0.666	0.752
$\Delta_M H$/(J/mol)	−7.53	+25.9	117	66.9	184	105	167	259

w_B	0.867
$\Delta_M H$/(J/mol)	485

Polymer (B):	**poly(propylene glycol)**	**2004COM**

Characterization: M_n/g.mol^{-1} = 502, M_w/g.mol^{-1} = 512, ρ = 0.99568 g/cm^3 (308 K), PPG 400, Fluka AG, Buchs, Switzerland

Solvent (A): **3-methylphenol** **C$_7$H$_8$O** **108-39-4**

T/K = 308.15

w_B	0.0390	0.0771	0.1086	0.1397	0.1959	0.2452	0.3276	0.3937
$\Delta_M H$/(J/g)	−6.4001	−11.3675	−14.5166	−16.5652	−19.4365	−21.0420	−22.7067	−23.9142

w_B	0.4935	0.5938	0.6609	0.7451	0.7958	0.8540	0.8863	0.9212
$\Delta_M H$/(J/g)	−24.0542	−23.7458	−22.4902	−19.2771	−15.9040	−12.2612	−9.7311	−6.7412

w_B	0.9690
$\Delta_M H$/(J/g)	−2.7559

Polymer (B):	**poly(propylene glycol)**	**2004COM**

Characterization: M_n/g.mol^{-1} = 856, M_w/g.mol^{-1} = 884, ρ = 0.99353 g/cm^3 (308 K), PPG 725, Aldrich Chem. Co., Inc., Milwaukee, WI

Solvent (A): **3-methylphenol** **C$_7$H$_8$O** **108-39-4**

T/K = 308.15

w_B	0.0399	0.0749	0.1084	0.1394	0.1955	0.2447	0.3271	0.3931
$\Delta_M H$/(J/g)	−6.0144	−10.2115	−13.2212	−15.6891	−18.1268	−19.7253	−21−8070	−22.7730

continued

continued

w_B	0.4929	0.5933	0.6604	0.7447	0.7954	0.8537	0.8861	0.9211
$\Delta_M H/(J/g)$	−23.8469	−23.0024	−21.9621	−18.7092	−16.1036	−12.3613	−9.8615	−6.5422

w_B	0.9589
$\Delta_M H/(J/g)$	−3.6326

Polymer (B):	**poly(propylene glycol)**	**2004COM**

Characterization: $M_n/g.mol^{-1} = 1334$, $M_w/g.mol^{-1} = 1374$, $\rho = 0.99303$ g/cm^3 (308 K), PPG 1200, Fluka AG, Buchs, Switzerland

Solvent (A):	**3-methylphenol**	**C$_7$H$_8$O**	**108-39-4**

$T/K = 308.15$

w_B	0.0389	0.0749	0.1083	0.1394	0.1954	0.2446	0.3270	0.3930
$\Delta_M H/(J/g)$	−5.5241	−9.4224	−12.1858	−14.1083	−16.3566	−17.7879	−19.1706	−20.0227

w_B	0.4928	0.5932	0.6602	0.7446	0.7954	0.8536	0.8860	0.9211
$\Delta_M H/(J/g)$	−20.4230	−19.7431	−18.4124	−15.8178	−13.6144	−10.5000	−8.3600	−6.0502

w_B	0.9589
$\Delta_M H/(J/g)$	−3.0249

Polymer (B):	**poly(propylene glycol)**	**2004COM**

Characterization: $M_n/g.mol^{-1} = 2004$, $M_w/g.mol^{-1} = 2063$, $\rho = 0.99212$ g/cm^3 (308 K), PPG 2000, Fluka AG, Buchs, Switzerland

Solvent (A):	**3-methylphenol**	**C$_7$H$_8$O**	**108-39-4**

$T/K = 308.15$

w_B	0.0389	0.0748	0.1082	0.1392	0.1953	0.2445	0.3268	0.3928
$\Delta_M H/(J/g)$	−4.8533	−8.4446	−10.9589	−12.7009	−15.1078	−16.6011	−18.4001	−19.1242

w_B	0.4926	0.5930	0.6600	0.7444	0.7952	0.8535	0.8859	0.9210
$\Delta_M H/(J/g)$	−19.4638	−18.5314	−17.1122	−14.4000	−12.4110	−9.7425	−7.8140	−5.7925

w_B	0.9590
$\Delta_M H/(J/g)$	−3.0001

Polymer (B):	**poly(propylene glycol)**	**1989MOE, 1995KIL**

Characterization: $M_n/g.mol^{-1} = 396$, $M_w/g.mol^{-1} = 412$, $\rho = 1.0042$ g/cm^3 (298 K)

Solvent (A):	**tetrachloromethane**	**CCl$_4$**	**56-23-5**

$T/K = 303.15$

w_B	0.0084	0.0166	0.0367	0.0744	0.1025	0.1257	0.1716	0.2095
$\Delta_M H/(J/g)$	0.035	0.076	0.151	0.204	0.195	0.169	0.084	−0.042

w_B	0.2441	0.2461	0.2763	0.3126	0.3574	0.4170	0.4741	0.5298
$\Delta_M H/(J/g)$	−0.167	−0.158	−0.283	−0.437	−0.626	−0.857	−1.053	−1.207

w_B	0.5989	0.6696	0.7582	0.8177	0.8855	0.9658	0.9834
$\Delta_M H/(J/g)$	−1.345	−1.352	−1.206	−1.004	−0.709	−0.237	−0.117

Polymer (B): **poly(propylene glycol)**   **1993ZEL**
Characterization: $M_n/\text{g.mol}^{-1} = 396$, $M_w/\text{g.mol}^{-1} = 412$, $\rho = 1.0042$ g/cm^3 (298 K)
Solvent (A): **tetrachloromethane**   **CCl$_4$**   **56-23-5**

$T/\text{K} = 303.15$

$\Delta_M H/(\text{J/g}) = w_B(1 - w_B)[-4.57 + 5.47(1 - 2w_B) + 3.34(1 - 2w_B)^2 + 0.44(1 - 2w_B)^3]$

$T/\text{K} = 318.15$

$\Delta_M H/(\text{J/g}) = w_B(1 - w_B)[-0.16 + 4.85(1 - 2w_B) + 0.48(1 - 2w_B)^2]$

Polymer (B): **poly(propylene glycol)**   **1989MOE, 1995KIL**
Characterization: $M_n/\text{g.mol}^{-1} = 570$, $M_w/\text{g.mol}^{-1} = 590$, $\rho = 1.0026$ g/cm^3 (298 K)
Solvent (A): **tetrachloromethane**   **CCl$_4$**   **56-23-5**

$T/\text{K} = 303.15$

w_B	0.0022	0.0370	0.0696	0.1000	0.1508	0.1962	0.1929	0.2109	0.2326
$\Delta_M H/(\text{J/g})$	−0.010	−0.098	−0.189	−0.291	−0.512	−0.716	−0.674	−0.773	−0.891
w_B	0.2593	0.2756	0.2941	0.3152	0.3396	0.3619	0.3873	0.4165	0.4505
$\Delta_M H/(\text{J/g})$	−1.035	−1.119	−1.212	−1.316	−1.431	−1.528	−1.629	−1.732	−1.832
w_B	0.4817	0.5375	0.6081	0.6800	0.7712	0.9667	0.9239	0.9892	
$\Delta_M H/(\text{J/g})$	−1.902	−1.982	−1.992	−1.860	−1.570	−1.021	−0.620	−0.096	

Polymer (B): **poly(propylene glycol)**   **1989MOE, 1995KIL**
Characterization: $M_n/\text{g.mol}^{-1} = 1900$, $\rho = 1.0003$ g/cm^3 (298 K)
Solvent (A): **tetrachloromethane**   **CCl$_4$**   **56-23-5**

$T/\text{K} = 303.15$

w_B	0.0027	0.0115	0.0289	0.0618	0.0836	0.0985	0.1486	0.2073	0.2713
$\Delta_M H/(\text{J/g})$	−0.020	−0.100	−0.243	−0.591	−0.783	−0.895	−1.359	−1.810	−2.261
w_B	0.2978	0.3403	0.3969	0.4247	0.4568	0.4941	0.5294	0.5701	0.6177
$\Delta_M H/(\text{J/g})$	−2.586	−2.792	−2.978	−3.023	−3.037	−3.041	−2.999	−2.917	−2.792
w_B	0.6397	0.7162	0.7783	0.8225	0.8549	0.9281	0.9696	0.9873	
$\Delta_M H/(\text{J/g})$	−2.707	−2.328	−1.966	−1.622	−1.365	−0.707	−0.306	−0.128	

Polymer (B): **poly(propylene glycol)**   **1993ZEL**
Characterization: $M_n/\text{g.mol}^{-1} = 1900$, $\rho = 1.0003$ g/cm^3 (298 K)
Solvent (A): **tetrachloromethane**   **CCl$_4$**   **56-23-5**

$T/\text{K} = 303.15$

$\Delta_M H/(\text{J/g}) = w_B(1 - w_B)[-12.21 - 1.05(1 - 2w_B) + 2.65(1 - 2w_B)^2 + 2.41(1 - 2w_B)^3]$

$T/\text{K} = 318.15$

$\Delta_M H/(\text{J/g}) = w_B(1 - w_B)[-8.12 - 1.01(1 - 2w_B) - 2.09(1 - 2w_B)^2]$

Polymer (B):   **poly(propylene glycol)**   **1993ZEL**
Characterization: M_n/g.mol^{-1} = 396, M_w/g.mol^{-1} = 412, ρ = 1.0042 g/cm^3 (298 K)
Solvent (A):   **trichloromethane**   **CHCl$_3$**   **67-66-3**

T/K = 303.15

$\Delta_M H/(\text{J/g}) = w_B(1 - w_B)[-85.23 - 17.08(1 - 2w_B) + 13.51(1 - 2w_B)^2 + 19.14(1 - 2w_B)^3]$

Polymer (B):   **poly(propylene glycol)**   **1993ZEL**
Characterization: M_n/g.mol^{-1} = 1900, ρ = 1.0003 g/cm^3 (298 K)
Solvent (A):   **trichloromethane**   **CHCl$_3$**   **67-66-3**

T/K = 303.15

$\Delta_M H/(\text{J/g}) = w_B(1 - w_B)[-86.60 + 3.10(1 - 2w_B) + 2.53(1 - 2w_B)^2]$

Polymer (B):   **poly(propylene glycol)**   **1968LA2**
Characterization: M_n/g.mol^{-1} = 150,
                          fractionated samples supplied by Union Carbide Corp.
Solvent (A):   **water**   **H$_2$O**   **7732-18-5**

T/K = 300.05

w_B	0.01483	0.03418	0.04643	0.08906	0.1425	0.2211	0.4113	0.7867
$\Delta_M H/$(J/mol)	−34.6	−95.8	−129	−242	−395	−615	−1152	−242

Polymer (B):   **poly(propylene glycol)**   **1988PAR**
Characterization: M_n/g.mol^{-1} = 150,
                          fractionated samples supplied by Union Carbide Corp.
Solvent (A):   **water**   **H$_2$O**   **7732-18-5**

T/K = 321.35

w_B	0.00325	0.01104	0.02217	0.05288	0.08175	0.1543	0.2663	0.3511
$\Delta_M H/$(J/mol)	−5.424	−24.12	−61.04	−132.5	−197.1	−377.0	−631.8	−820.5

w_B	0.4784	0.5503	0.6373	0.7558	0.8890	0.9755
$\Delta_M H/$(J/mol)	−1020	−1112	−1205	−1242	−998.7	−0.500

Polymer (B):   **poly(propylene glycol)**   **1961CUN**
Characterization: M_n/g.mol^{-1} = 400, Oxirane Ltd., Manchester
Solvent (A):   **water**   **H$_2$O**   **7732-18-5**

T/K = 300.05

w_B	0.109	0.198	0.314	0.393	0.467	0.477	0.609	0.701
$\Delta_M H/$(J/g)	−16.32	−26.48	−34.31	−36.94	−37.35	−37.66	−33.89	−28.74

w_B	0.791	0.811	0.908
$\Delta_M H/$(J/g)	−22.01	−20.59	−11.13

Polymer (B):   **poly(propylene glycol)**   **1995CAR**
Characterization: $M_n/\text{g.mol}^{-1} = 400$
Solvent (A):   **water**   **H_2O**   **7732-18-5**

$T/\text{K} = 298.15$

w_B	0.02496	0.05003	0.1701	0.2494	0.3801	0.5899	0.7486	0.8949
$\Delta_M H/(\text{J/g})$	−4.12	−8.03	−24.5	−32.6	−38.7	−34.9	−25.8	−12.6

$T/\text{K} = 321.15$

w_B	0.9997	0.1699	0.3791	0.5881	0.7494
$\Delta_M H/(\text{J/g})$	−0.041	−18.9	−26.5	−26.0	−20.2

Polymer (B):   **poly(propylene glycol)**   **1968LA2**
Characterization: $M_n/\text{g.mol}^{-1} = 425$,
                     fractionated samples supplied by Union Carbide Corp.
Solvent (A):   **water**   **H_2O**   **7732-18-5**

$T/\text{K} = 300.05$

w_B	0.00885	0.03336	0.06077	0.1070	0.1352	0.1499	0.2623	0.7372
$\Delta_M H/(\text{J/mol})$	−27.8	−92.9	−172	−297	−351	−415	−692	−1557

w_B	0.7750
$\Delta_M H/(\text{J/mol})$	−1540

Polymer (B):   **poly(propylene glycol)**   **1988PAR**
Characterization: $M_n/\text{g.mol}^{-1} = 425$,
                     fractionated samples supplied by Union Carbide Corp.
Solvent (A):   **water**   **H_2O**   **7732-18-5**

$T/\text{K} = 321.35$

w_B	0.005353	0.01405	0.02872	0.06428	0.1060	0.1369	0.1832	0.2885
$\Delta_M H/(\text{J/mol})$	−9.082	−27.74	−57.74	−150.2	−248.1	−313.4	−430.1	−599.2

w_B	0.3715	0.4399	0.4963	0.5784	0.7016	0.8879	0.9700
$\Delta_M H/(\text{J/mol})$	−709.6	−794.5	−950.6	−1084	−1296	−1217	−755.6

Polymer (B):   **poly(propylene glycol)**   **1968LA2**
Characterization: $M_n/\text{g.mol}^{-1} = 2025$,
                     fractionated samples supplied by Union Carbide Corp.
Solvent (A):   **water**   **H_2O**   **7732-18-5**

$T/\text{K} = 300.05$

w_B	0.04045	0.08389	0.1162	0.1399	0.1868	0.2719	0.3613	0.6545
$\Delta_M H/(\text{J/mol})$	−5.31	−8.79	−6.44	−7.49	−10.2	−7.20	−4.90	+14.3

Comments: These data include the enthalpy change caused by melting the crystalline parts of the polymer when mixing the polymer and the solvent to form the solution.

Polymer (B):　　　　　　**poly(propylene glycol) dimethyl ether**　　　　**1968KER**
Characterization:　　　　$M_n/\text{g.mol}^{-1} = 2050$, synthesized in the laboratory from PPG 2025
Solvent (A):　　　　　　**tetrachloromethane**　　　CCl_4　　　　　　　**56-23-5**

$T/\text{K} = 278.68$

w_B	0.1371	0.2511	0.3835	0.5494	0.7027	0.8458
φ_B	0.2040	0.3511	0.5010	0.6630	0.7923	0.8985
$\Delta_M H/(\text{J/cm}^3)$	−3.10	−4.69	−5.44	−4.69	−3.47	−2.22

Polymer (B):　　　　　　**poly(propylene glycol) dimethyl ether**　　　　**1968KER**
Characterization:　　　　$M_n/\text{g.mol}^{-1} = 2050$, synthesized in the laboratory from PPG 2025
Solvent (A):　　　　　　**trichloromethane**　　　$CHCl_3$　　　　　　　**67-66-3**

$T/\text{K} = 278.68$

w_B	0.1056	0.1989	0.2840	0.3692	0.4202	0.5117	0.5987	0.7243
φ_B	0.1512	0.2725	0.3744	0.4690	0.5223	0.6126	0.6923	0.7985
$\Delta_M H/(\text{J/cm}^3)$	−13.05	−21.88	−29.16	−32.00	−32.80	−32.22	−28.74	−21.92

w_B	0.8638
φ_B	0.9054
$\Delta_M H/(\text{J/cm}^3)$	−10.96

Polymer (B):　　　　　　**polystyrene**　　　　　　　　　　　　　　　**1966KA1**
Characterization:　　　　$M_n/\text{g.mol}^{-1} = 25860$, DP = 244, synthesized in the laboratory
Solvent (A):　　　　　　**anisole**　　　　　　　C_7H_8O　　　　　　　**100-66-3**

$T/\text{K} = 298.15$

$V^{(1)}/\text{cm}^3$	10.0	15.0	20.0	25.0	10.0	15.0	20.0	10.0	10.0
$\varphi_B^{(1)}$	0.175	0.116	0.087	0.069	0.184	0.122	0.091	0.175	0.175
$V^{(2)}/\text{cm}^3$	15.0	20.0	25.0	30.0	15.0	20.0	25.0	20.0	25.0
$\varphi_B^{(2)}$	0.116	0.087	0.069	0.056	0.122	0.091	0.072	0.087	0.069
$\Delta_{dil} H^{12}/\text{J}$	−0.950	−0.640	−0.393	−0.151	−0.950	−0.619	−0.444	−1.590	−1.983

$V^{(1)}/\text{cm}^3$	10.0	10.0	10.0
$\varphi_B^{(1)}$	0.175	0.175	0.184
$V^{(2)}/\text{cm}^3$	30.0	20.0	25.0
$\varphi_B^{(2)}$	0.058	0.091	0.072
$\Delta_{dil} H^{12}/\text{J}$	−2.134	−1.569	−2.013

Comments:　　$\Delta_{dil}H^{12}$ is the extensive quantity obtained for a given total volume change from $V^{(1)}$ to $V^{(2)}$, where $\varphi_B^{(1)}$ and $\varphi_B^{(2)}$ denote the volume fractions of the polymer in the solution before and after the dilution process.

Polymer (B):　　　　　　**polystyrene**　　　　　　　　　　　　　　　**1959SCH**
Characterization:　　　　$M_n/\text{g.mol}^{-1} = 400$
Solvent (A):　　　　　　**benzene**　　　　　　　C_6H_6　　　　　　　**71-43-2**

continued

continued

$T/K = 296.15$

$c_B^{(1)}/(g/cm^3) = 0.100$ $c_B^{(2)}/(g/cm^3) = 0.010$ $\Delta_{dil}H^{12}/(J/g \text{ polymer}) = 0.335$

Comments: $c_B^{(1)}$ and $c_B^{(2)}$ denote the concentrations of the polymer in the solution before and after the
dilution process.

Polymer (B):	**polystyrene**		**1970MO1, 1970MO4**
Characterization:	$M_w/g.mol^{-1} = 600$, $M_w/M_n \le 1.10$,		
	Pressure Chemical Co., Pittsburgh, PA		
Solvent (A):	**benzene**	C_6H_6	**71-43-2**

$T/K = 298.15$

$c_B^{(1)}/(g/cm^3)$	0.503	0.605
$c_B^{(2)}/(g/cm^3)$	0.0017	0.0023
$\Delta_{dil}H^{12}/(J/cm^3)$	0.00833	0.0101

Comments: $c_B^{(1)}$ and $c_B^{(2)}$ denote the concentrations of the polymer in the solution before and after the
dilution process.

Polymer (B):	**polystyrene**		**1970MO1, 1970MO4**
Characterization:	$M_w/g.mol^{-1} = 900$, $M_w/M_n \le 1.10$,		
	Pressure Chemical Co., Pittsburgh, PA		
Solvent (A):	**benzene**	C_6H_6	**71-43-2**

$T/K = 298.15$

$c_B^{(1)}/(g/cm^3)$	0.702
$c_B^{(2)}/(g/cm^3)$	0.0025
$\Delta_{dil}H^{12}/(J/cm^3)$	−0.00322

Comments: $c_B^{(1)}$ and $c_B^{(2)}$ denote the concentrations of the polymer in the solution before and after the
dilution process.

Polymer (B):	**polystyrene**		**1959SCH**
Characterization:	$M_n/g.mol^{-1} = 1800$		
Solvent (A):	**benzene**	C_6H_6	**71-43-2**

$T/K = 296.15$

$c_B^{(1)}/(g/cm^3)$	0.200	0.100	0.100
$c_B^{(2)}/(g/cm^3)$	0.103	0.0514	0.010
$\Delta_{dil}H^{12}/(J/g \text{ polymer})$	−0.134	−0.059	−0.126

Comments: $c_B^{(1)}$ and $c_B^{(2)}$ denote the concentrations of the polymer in the solution before and after the
dilution process.

Polymer (B):   **polystyrene**   **1970MO1, 1970MO4**
Characterization: $M_w/\text{g.mol}^{-1} = 2000$, $M_w/M_n \leq 1.10$,
                          Pressure Chemical Co., Pittsburgh, PA
Solvent (A):   **benzene**   **C_6H_6**   **71-43-2**

$T/\text{K} = 298.15$

$c_B^{(1)}/(\text{g/cm}^3) = 0.501$ $c_B^{(2)}/(\text{g/cm}^3) = 0.0019$ $\Delta_{dil}H^{12}/(\text{J/cm}^3) = -0.0036$

Comments: $c_B^{(1)}$ and $c_B^{(2)}$ denote the concentrations of the polymer in the solution before and after the
              dilution process.

Polymer (B):   **polystyrene**   **1970MO1, 1970MO4**
Characterization: $M_w/\text{g.mol}^{-1} = 4800$, $M_w/M_n \leq 1.10$,
                          Pressure Chemical Co., Pittsburgh, PA
Solvent (A):   **benzene**   **C_6H_6**   **71-43-2**

$T/\text{K} = 298.15$

$c_B^{(1)}/(\text{g/cm}^3) = 0.499$ $c_B^{(2)}/(\text{g/cm}^3) = 0.0019$ $\Delta_{dil}H^{12}/(\text{J/cm}^3) = -0.0034$

Comments: $c_B^{(1)}$ and $c_B^{(2)}$ denote the concentrations of the polymer in the solution before and after the
              dilution process.

Polymer (B):   **polystyrene**   **1959SCH**
Characterization: $M_n/\text{g.mol}^{-1} = 6000$
Solvent (A):   **benzene**   **C_6H_6**   **71-43-2**

$T/\text{K} = 296.15$

$c_B^{(1)}/(\text{g/cm}^3) = 0.100$ $c_B^{(2)}/(\text{g/cm}^3) = 0.010$ $\Delta_{dil}H^{12}/(\text{J/g polymer}) = -0.109$

Comments: $c_B^{(1)}$ and $c_B^{(2)}$ denote the concentrations of the polymer in the solution before and after the
              dilution process.

Polymer (B):   **polystyrene**   **1970MO1, 1970MO4**
Characterization: $M_w/\text{g.mol}^{-1} = 10500$, $M_w/M_n \leq 1.10$,
                          Pressure Chemical Co., Pittsburgh, PA
Solvent (A):   **benzene**   **C_6H_6**   **71-43-2**

$T/\text{K} = 298.15$

$c_B^{(1)}/(\text{g/cm}^3) = 0.497$ $c_B^{(2)}/(\text{g/cm}^3) = 0.0017$ $\Delta_{dil}H^{12}/(\text{J/cm}^3) = -0.0036$

Comments: $c_B^{(1)}$ and $c_B^{(2)}$ denote the concentrations of the polymer in the solution before and after the
              dilution process.

Polymer (B):   **polystyrene**   **1959SCH**
Characterization: $M_n/\text{g.mol}^{-1} = 11000$
Solvent (A):   **benzene**   **C_6H_6**   **71-43-2**

continued

continued

$T/K = 296.15$

$c_B^{(1)}/(g/cm^3)$	0.186	0.099	0.050
$c_B^{(2)}/(g/cm^3)$	0.0956	0.0501	0.0258
$\Delta_{dil}H^{12}/(J/g$ polymer)	−0.0925	−0.0586	−0.0272

Comments: $c_B^{(1)}$ and $c_B^{(2)}$ denote the concentrations of the polymer in the solution before and after the dilution process.

Polymer (B):	**polystyrene**		**1958AMA**
Characterization:	$M_n/g.mol^{-1} = 29150$, DP = 275, synthesized in the laboratory		
Solvent (A):	**benzene**	**C_6H_6**	**71-43-2**

$T/K = 298.15$

$V^{(1)}/cm^3$	4.75	4.85	3.73	4.82	4.79
$\varphi_B^{(1)}$	0.185	0.185	0.185	0.153	0.153
$V^{(2)}/cm^3$	9.75	9.85	8.73	9.82	9.79
$\varphi_B^{(2)}$	0.090	0.091	0.079	0.075	0.075
$\Delta_{dil}H^{12}/J$	0.218	0.255	0.280	0.201	0.142

Comments: $\Delta_{dil}H^{12}$ is the extensive quantity obtained for a given total volume change from $V^{(1)}$ to $V^{(2)}$, where $\varphi_B^{(1)}$ and $\varphi_B^{(2)}$ denote the volume fractions of the polymer in the solution before and after the dilution process.

Polymer (B):	**polystyrene**		**1953TAG**
Characterization:	$M_\eta/g.mol^{-1} = 300000$		
Solvent (A):	**benzene**	**C_6H_6**	**71-43-2**

$T/K = 298.15$

w_B	0.0034	0.0071	0.0141	0.0187	0.0284	0.0368
$\Delta_M H/(J/g)$	−0.130	−0.176	−0.280	−0.397	−0.628	−0.816

Comments: These data include the enthalpy change when mixing the glassy polymer and the solvent to form the mixture.

$T/K = 298.15$

$w_B^{(1)}$	0.973	0.965	0.943	0.925	0.918	0.869	0.826	0.824
$w_B^{(2)}$	0.024	0.019	0.0188	0.0192	0.0184	0.0184	0.0187	0.0187
$\Delta_{dil}H^{12}/(J/g$ polymer)	−14.4	−13.6	−8.79	−9.00	−8.45	−4.85	−2.13	−1.26

$w_B^{(1)}$	0.772	0.709	0.665	0.532	0.102
$w_B^{(2)}$	0.0196	0.0205	0.0222	0.0218	0.00435
$\Delta_{dil}H^{12}/(J/g$ polymer)	0.0	0.0	0.0	0.0	0.0

Comments: $w_B^{(1)}$ and $w_B^{(2)}$ denote the mass fractions of the polymer in the solution before and after the dilution process. Additional data are given in a figure in the original source.

Polymer (B):	**polystyrene**					**1973TAM**	
Characterization:	M_w/g.mol^{-1} = 2100, Pressure Chemical Co., Pittsburgh, PA						
Solvent (A):	**2-butanone**		**C$_4$H$_8$O**			**78-93-3**	

T/K = 298.15

$\varphi_B^{(1)}$	0.2779	0.2704	0.2558	0.2355	0.2111	0.1972	0.1856
$\varphi_B^{(2)}$	0.2704	0.2588	0.2355	0.2145	0.1972	0.1856	0.1729
$\Delta_{dil}H^{12}$/(J/mol solvent)	0.0321	0.0910	0.3344	0.2434	0.1117	0.2419	0.4116

$\varphi_B^{(1)}$	0.1729
$\varphi_B^{(2)}$	0.1618
$\Delta_{dil}H^{12}$/(J/mol solvent)	0.3041

Comments: $\varphi_B^{(1)}$ and $\varphi_B^{(2)}$ denote the volume fractions of the polymer in the solution before and after the dilution process.

Polymer (B):	**polystyrene**					**1973TAM**	
Characterization:	M_w/g.mol^{-1} = 22000, Asahi Dow Chem. Co., Japan						
Solvent (A):	**2-butanone**		**C$_4$H$_8$O**			**78-93-3**	

T/K = 298.15

$\varphi_B^{(1)}$	0.2889	0.2630	0.2419	0.2242	0.2077	0.1965	0.1838
$\varphi_B^{(2)}$	0.2630	0.2419	0.2242	0.2077	0.1965	0.1838	0.1707
$\Delta_{dil}H^{12}$/(J/mol solvent)	−0.1714	0.1715	0.2387	0.2177	0.1442	−0.0516	0.0322

$\varphi_B^{(1)}$	0.1707	0.1593
$\varphi_B^{(2)}$	0.1593	0.1489
$\Delta_{dil}H^{12}$/(J/mol solvent)	−0.0099	0.0011

Comments: $\varphi_B^{(1)}$ and $\varphi_B^{(2)}$ denote the volume fractions of the polymer in the solution before and after the dilution process.

Polymer (B):	**polystyrene**					**1966KA1**	
Characterization:	M_n/g.mol^{-1} = 25860, DP = 244, synthesized in the laboratory						
Solvent (A):	**2-butanone**		**C$_4$H$_8$O**			**78-93-3**	

T/K = 298.15

$V^{(1)}$/cm^3	10.0	15.0	20.0	10.0	15.0	20.0	25.0	10.0
$\varphi_B^{(1)}$	0.222	0.148	0.111	0.222	0.148	0.111	0.088	0.222
$V^{(2)}$/cm^3	20.0	25.0	25.0	15.0	20.0	25.0	30.0	20.0
$\varphi_B^{(2)}$	0.148	0.111	0.088	0.148	0.111	0.088	0.060	0.111
$\Delta_{dil}H^{12}$/J	0.397	0.416	0.130	0.385	0.213	0.117	0.071	0.615

$V^{(1)}$/cm^3	10.0	10.0	10.0	10.0
$\varphi_B^{(1)}$	0.222	0.222	0.222	0.222
$V^{(2)}$/cm^3	25.0	20.0	25.0	35.0
$\varphi_B^{(2)}$	0.088	0.111	0.088	0.060
$\Delta_{dil}H^{12}$/J	0.745	0.598	0.715	0.787

Comments: $\Delta_{dil}H^{12}$ is the extensive quantity obtained for a given total volume change from $V^{(1)}$ to $V^{(2)}$, where $\varphi_B^{(1)}$ and $\varphi_B^{(2)}$ denote the volume fractions of the polymer in the solution before and after the dilution process.

Polymer (B):	polystyrene				1973TAM
Characterization:	M_w/g.mol^{-1} = 57500, Kyoto University, Japan				
Solvent (A):	2-butanone	C$_4$H$_8$O			78-93-3

T/K = 298.15

$\varphi_B^{(1)}$	0.2049	0.1898	0.1748	0.1621	0.1508
$\varphi_B^{(2)}$	0.1898	0.1748	0.1621	0.1508	0.1406
$\Delta_{dil} H^{12}$/(J/mol solvent)	−0.1043	−0.0576	−0.1077	−0.1023	−0.1056

Comments: $\varphi_B^{(1)}$ and $\varphi_B^{(2)}$ denote the volume fractions of the polymer in the solution before and after the dilution process.

Polymer (B):	polystyrene							1956JEN
Characterization:	−							
Solvent (A):	chlorobenzene		C$_6$H$_5$Cl					108-90-7

T/K = 293.15

z_B	0.94	0.87	0.78	0.74	0.73	0.54	0.42	0.35
$\Delta_M H$/(J/base mol)	−1222	−2025	−2510	−2628	−2665	−2067	−1653	−1389

z_B	0.23	0.09
$\Delta_M H$/(J/base mol)	+945	+368

Comments: These data include the enthalpy change when mixing the glassy polymer and the solvent to form the mixture.

z_B	0.94	0.87	0.78	0.74	0.74	0.73	0.54	0.42
$\Delta_M H$/(J/base mol)	368	1071	264	−4.2	−29.3	−71.1	−146	−159

z_B	0.35	0.23	0.09
$\Delta_M H$/(J/base mol)	−146	−126	−46.0

Comments: These data were observed when mixing the equilibrium "liquid" polymer and the solvent to form the mixture, i.e., after substracting the part caused by the glass enthalpy.

Polymer (B):	polystyrene		1959SCH
Characterization:	M_n/g.mol^{-1} = 400		
Solvent (A):	cyclohexane	C$_6$H$_{12}$	110-82-7

T/K = 296.15

$c_B^{(1)}$/(g/cm^3) = 0.080 $c_B^{(2)}$/(g/cm^3) = 0.012 $\Delta_{dil} H^{12}$/(J/g polymer) = 4.06

Comments: $c_B^{(1)}$ and $c_B^{(2)}$ denote the concentrations of the polymer in the solution before and after the dilution process.

Polymer (B):	polystyrene		1959SCH
Characterization:	M_n/g.mol^{-1} = 1100		
Solvent (A):	cyclohexane	C$_6$H$_{12}$	110-82-7

T/K = 296.15

$c_B^{(1)}$/(g/cm^3) = 0.080 $c_B^{(2)}$/(g/cm^3) = 0.012 $\Delta_{dil} H^{12}$/(J/g polymer) = 2.34

Comments: $c_B^{(1)}$ and $c_B^{(2)}$ denote the concentrations of the polymer in the solution before and after the dilution process.

Polymer (B):	**polystyrene**				**1959SCH**
Characterization:	$M_n/\text{g.mol}^{-1} = 1200$				
Solvent (A):	**cyclohexane**		C_6H_{12}		**110-82-7**

$T/\text{K} = 296.15$

$c_B^{(1)}/(\text{g/cm}^3)$	0.100	0.080	0.050	0.025	0.0125
$c_B^{(2)}/(\text{g/cm}^3)$	0.0508	0.012	0.0257	0.0129	0.0065
$\Delta_{dil}H^{12}/(\text{J/g polymer})$	1.34	2.09	0.71	0.42	0.21

Comments: $c_B^{(1)}$ and $c_B^{(2)}$ denote the concentrations of the polymer in the solution before and after the dilution process.

Polymer (B):	**polystyrene**		**1959SCH**
Characterization:	$M_n/\text{g.mol}^{-1} = 2300$		
Solvent (A):	**cyclohexane**	C_6H_{12}	**110-82-7**

$T/\text{K} = 296.15$

$c_B^{(1)}/(\text{g/cm}^3) = 0.080$ $c_B^{(2)}/(\text{g/cm}^3) = 0.012$ $\Delta_{dil}H^{12}/(\text{J/g polymer}) = 1.09$

Comments: $c_B^{(1)}$ and $c_B^{(2)}$ denote the concentrations of the polymer in the solution before and after the dilution process.

Polymer (B):	**polystyrene**		**1959SCH**
Characterization:	$M_n/\text{g.mol}^{-1} = 2800$		
Solvent (A):	**cyclohexane**	C_6H_{12}	**110-82-7**

$T/\text{K} = 296.15$

$c_B^{(1)}/(\text{g/cm}^3) = 0.080$ $c_B^{(2)}/(\text{g/cm}^3) = 0.012$ $\Delta_{dil}H^{12}/(\text{J/g polymer}) = 1.21$

Comments: $c_B^{(1)}$ and $c_B^{(2)}$ denote the concentrations of the polymer in the solution before and after the dilution process.

Polymer (B):	**polystyrene**		**1959SCH**
Characterization:	$M_n/\text{g.mol}^{-1} = 6000$		
Solvent (A):	**cyclohexane**	C_6H_{12}	**110-82-7**

$T/\text{K} = 296.15$

$c_B^{(1)}/(\text{g/cm}^3) = 0.080$ $c_B^{(2)}/(\text{g/cm}^3) = 0.012$ $\Delta_{dil}H^{12}/(\text{J/g polymer}) = 0.63$

Comments: $c_B^{(1)}$ and $c_B^{(2)}$ denote the concentrations of the polymer in the solution before and after the dilution process.

Polymer (B):	**polystyrene**			**1959SCH**
Characterization:	$M_n/\text{g.mol}^{-1} = 11000$			
Solvent (A):	**cyclohexane**		C_6H_{12}	**110-82-7**

$T/\text{K} = 296.15$

$c_B^{(1)}/(\text{g/cm}^3)$	0.100	0.080	0.050	0.025
$c_B^{(2)}/(\text{g/cm}^3)$	0.0502	0.012	0.0239	0.0129
$\Delta_{dil}H^{12}/(\text{J/g polymer})$	0.418	0.753	0.339	0.209

Polymer (B):   **polystyrene**   1959SCH
Characterization: $M_n/\text{g.mol}^{-1} = 14000$
Solvent (A):   **cyclohexane** C_6H_{12} 110-82-7

$T/\text{K} = 296.15$

$c_B^{(1)}/(\text{g/cm}^3)$ 0.080
$c_B^{(2)}/(\text{g/cm}^3)$ 0.012
$\Delta_{dil}H^{12}/(\text{J/g polymer})$ 0.753

Comments: $c_B^{(1)}$ and $c_B^{(2)}$ denote the concentrations of the polymer in the solution before and after the dilution process.

Polymer (B):   **polystyrene**   1959SCH
Characterization: $M_n/\text{g.mol}^{-1} = 17000$
Solvent (A):   **cyclohexane** C_6H_{12} 110-82-7

$T/\text{K} = 296.15$

$c_B^{(1)}/(\text{g/cm}^3)$ 0.080
$c_B^{(2)}/(\text{g/cm}^3)$ 0.012
$\Delta_{dil}H^{12}/(\text{J/g polymer})$ 0.50

Comments: $c_B^{(1)}$ and $c_B^{(2)}$ denote the concentrations of the polymer in the solution before and after the dilution process.

Polymer (B):   **polystyrene**   1958AMA
Characterization: $M_n/\text{g.mol}^{-1} = 29150$, DP = 275, synthesized in the laboratory
Solvent (A):   **cyclohexane** C_6H_{12} 110-82-7

$T/\text{K} = 298.15$

$V^{(1)}/\text{cm}^3$	4.90	9.90	14.90	4.95	9.95	14.95
$\varphi_B^{(1)}$	0.108	0.053	0.036	0.108	0.054	0.036
$V^{(2)}/\text{cm}^3$	9.90	14.90	19.90	9.95	14.95	19.95
$\varphi_B^{(2)}$	0.053	0.036	0.027	0.054	0.036	0.027
$\Delta_{dil}H^{12}/\text{J}$	0.594	0.494	0.414	0.644	0.531	0.372

Comments: $\Delta_{dil}H^{12}$ is the extensive quantity obtained for a given total volume change from $V^{(1)}$ to $V^{(2)}$, where $\varphi_B^{(1)}$ and $\varphi_B^{(2)}$ denote the volume fractions of the polymer in the solution before and after the dilution process.

Polymer (B):   **polystyrene**   1956JEN
Characterization: —
Solvent (A):   **cyclohexane** C_6H_{12} 110-82-7

$T/\text{K} = 293.15$

z_B	0.91	0.88	0.81	0.78	0.70	0.68
$\Delta_M H/(\text{J/base mol})$	−1477	−1289	−1703	−1544	−1197	−1038

Comments: These data include the enthalpy change when mixing the glassy polymer and the solvent to form the mixture.

continued

continued

z_B	0.91	0.88	0.81	0.78	0.70	0.68
$\Delta_M H$/(J/base mol)	858	1841	1175	1230	1293	1381

Comments: These data were observed when mixing the equilibrium "liquid" polymer and the solvent to form the mixture, i.e., after substracting the part caused by the glass enthalpy.

Polymer (B):	**polystyrene**		**1966KA1**
Characterization:	M_n/g.mol^{-1} = 25860, DP = 244, synthesized in the laboratory		
Solvent (A):	**1,4-dioxane**	**C$_4$H$_8$O$_2$**	**123-91-1**

T/K = 298.15

$V^{(1)}$/cm^3	10.0	15.0	20.0	25.0	30.0	10.0	15.0	20.0	25.0
$\varphi_B^{(1)}$	0.216	0.144	0.108	0.086	0.070	0.213	0.142	0.106	0.084
$V^{(2)}$/cm^3	15.0	20.0	25.0	30.0	35.0	15.0	20.0	25.0	30.0
$\varphi_B^{(2)}$	0.144	0.108	0.086	0.070	0.060	0.142	0.106	0.084	0.070
$\Delta_{dil}H^{12}$/J	0.536	0.489	0.226	0.172	0.134	0.435	0.540	0.218	0.192
$V^{(1)}$/cm^3	30.0	10.9	9.7	19.2	24.2	11.1	16.3	10.0	10.0
$\varphi_B^{(1)}$	0.070	0.039	0.053	0.027	0.021	0.054	0.037	0.216	0.216
$V^{(2)}$/cm^3	35.0	21.4	19.2	24.2	29.3	16.3	21.3	20.0	25.0
$\varphi_B^{(2)}$	0.060	0.020	0.027	0.021	0.018	0.037	0.028	0.108	0.086
$\Delta_{dil}H^{12}$/J	0.126	0.029	0.046	0.013	0.008	0.050	0.025	1.025	1.251
$V^{(1)}$/cm^3	10.0	10.0	10.0	10.0	10.0	10.0	9.7	9.7	
$\varphi_B^{(1)}$	0.216	0.216	0.213	0.213	0.213	0.213	0.053	0.053	
$V^{(2)}$/cm^3	30.0	35.0	20.0	25.0	30.0	35.0	24.2	29.3	
$\varphi_B^{(2)}$	0.070	0.060	0.106	0.084	0.070	0.060	0.021	0.018	
$\Delta_{dil}H^{12}$/J	1.423	1.556	0.975	1.192	1.385	1.510	0.059	0.067	

Comments: $\Delta_{dil}H^{12}$ is the extensive quantity obtained for a given total volume change from $V^{(1)}$ to $V^{(2)}$, where $\varphi_B^{(1)}$ and $\varphi_B^{(2)}$ denote the volume fractions of the polymer in the solution before and after the dilution process.

Polymer (B):	**polystyrene**		**1974BA1**
Characterization:	M_η/g.mol^{-1} = 2100, American Commercial Co. Ltd.		
Solvent (A):	**ethyl acetate**	**C$_4$H$_8$O$_2$**	**141-78-6**

T/K = 303.15

$V^{(1)}$/cm^3	1.0	2.2	2.6	3.6	4.2	5.4	6.9
$\varphi_B^{(1)}$	0.314	0.337	0.116	0.084	0.117	0.138	0.107
$V^{(2)}$/cm^3	1.9	3.5	3.6	4.8	5.4	6.9	8.4
$\varphi_B^{(2)}$	0.159	0.210	0.084	0.063	0.138	0.107	0.087
$\Delta_{dil}H^{12}$/(J/mol solvent)	0.159	0.322	0.068	0.068	0.165	0.167	0.097

Comments: $\varphi_B^{(1)}$ and $\varphi_B^{(2)}$ denote the volume fractions of the polymer in the solution before and after the dilution process.

Polymer (B):   **polystyrene**   **1974BA1**
Characterization: M_n/g.mol^{-1} = 4000, American Commercial Co. Ltd.
Solvent (A):   **ethyl acetate**   **C$_4$H$_8$O$_2$**   **141-78-6**

T/K = 303.15

$V^{(1)}$/cm^3	2.0	2.3	2.7	3.6	5.1	5.2	5.7
$\varphi_B^{(1)}$	0.205	0.218	0.218	0.202	0.178	0.140	0.109
$V^{(2)}$/cm^3	6.3	7.6	5.2	6.5	7.0	9.2	10.6
$\varphi_B^{(2)}$	0.080	0.066	0.140	0.111	0.130	0.079	0.059
$\Delta_{dil}H^{12}$/(J/mol solvent)	0.109	0.114	0.180	0.131	0.094	0.157	0.134

$V^{(1)}$/cm^3	7.0	8.6	10.4
$\varphi_B^{(1)}$	0.092	0.059	0.035
$V^{(2)}$/cm^3	8.5	14.9	22.0
$\varphi_B^{(2)}$	0.075	0.035	0.016
$\Delta_{dil}H^{12}$/(J/mol solvent)	0.045	0.028	0.029

Polymer (B):   **polystyrene**   **1974BA1**
Characterization: M_η/g.mol^{-1} = 10000, American Commercial Co. Ltd.
Solvent (A):   **ethyl acetate**   **C$_4$H$_8$O$_2$**   **141-78-6**

T/K = 303.15

$V^{(1)}$/cm^3	1.5	2.5	3.0	3.1	4.5	5.2
$\varphi_B^{(1)}$	0.321	0.241	0.298	0.234	0.135	0.172
$V^{(2)}$/cm^3	4.5	4.5	5.2	5.5	8.8	7.6
$\varphi_B^{(2)}$	0.109	0.135	0.172	0.129	0.070	0.118
$\Delta_{dil}H^{12}$/(J/mol solvent)	−0.272	−0.142	−0.188	−0.164	−0.077	−0.043

$V^{(1)}$/cm^3	5.5	6.3	7.6	7.6	8.8
$\varphi_B^{(1)}$	0.129	0.167	0.118	0.094	0.070
$V^{(2)}$/cm^3	7.6	8.6	9.8	9.9	15.7
$\varphi_B^{(2)}$	0.094	0.123	0.090	0.072	0.039
$\Delta_{dil}H^{12}$/(J/mol solvent)	−0.080	−0.095	−0.078	−0.032	−0.032

Comments: $\varphi_B^{(1)}$ and $\varphi_B^{(2)}$ denote the volume fractions of the polymer in the solution before and after the dilution process.

Polymer (B):   **polystyrene**   **1966KA1**
Characterization: M_n/g.mol^{-1} = 25860, DP = 244, synthesized in the laboratory
Solvent (A):   **ethyl acetate**   **C$_4$H$_8$O$_2$**   **141-78-6**

T/K = 298.15

$V^{(1)}$/cm^3	10.0	15.0	20.0	25.0	10.0	15.0	20.0	25.0	30.
$\varphi_B^{(1)}$	0.186	0.124	0.093	0.073	0.186	0.124	0.083	0.072	0.060
$V^{(2)}$/cm^3	15.0	20.0	25.0	30.0	15.0	20.0	25.0	30.0	35.0
$\varphi_B^{(2)}$	0.124	0.093	0.073	0.060	0.124	0.083	0.072	0.060	0.040
$\Delta_{dil}H^{12}$/J	−0.322	−0.360	−0.088	−0.054	−0.142	−0.264	−0.130	−0.059	−0.059

continued

continued

$V^{(1)}/cm^3$	20.0	40.0	50.0	60.0	20.0	40.0	10.0	10.0	10.0
$\varphi_B^{(1)}$	0.050	0.025	0.020	0.016	0.060	0.030	0.186	0.186	0.186
$V^{(2)}/cm^3$	40.0	50.0	60.0	70.0	40.0	50.0	20.0	25.0	30.0
$\varphi_B^{(2)}$	0.025	0.020	0.016	0.014	0.029	0.024	0.083	0.073	0.060
$\Delta_{dil}H^{12}/J$	−0.046	−0.0054	−0.0017	−0.000	−0.082	−0.0062	−0.590	−0.678	−0.732

$V^{(1)}/cm^3$	10.0	10.0	10.0	10.0	20.0	20.0	20.0
$\varphi_B^{(1)}$	0.186	0.186	0.186	0.186	0.050	0.050	0.059
$V^{(2)}/cm^3$	20.0	25.0	30.0	35.0	50.0	60.0	50.0
$\varphi_B^{(2)}$	0.093	0.072	0.060	0.040	0.020	0.016	0.024
$\Delta_{dil}H^{12}/J$	−0.615	−0.745	−0.803	−0.862	−0.0515	−0.053	−0.089

Comments: $\Delta_{dil}H^{12}$ is the extensive quantity obtained for a given total volume change from $V^{(1)}$ to $V^{(2)}$, where $\varphi_B^{(1)}$ and $\varphi_B^{(2)}$ denote the volume fractions of the polymer in the solution before and after the dilution process.

Polymer (B):	**polystyrene**						**1974BA1**
Characterization:	$M_n/g.mol^{-1} = 200000$, American Commercial Co. Ltd.						
Solvent (A):	**ethyl acetate**		$C_4H_8O_2$				**141-78-6**

$T/K = 303.15$

$V^{(1)}/cm^3$	2.1	2.3	2.3	2.3	4.1	4.3	4.4
$\varphi_B^{(1)}$	0.118	0.127	0.083	0.163	0.059	0.044	0.084
$V^{(2)}/cm^3$	4.1	4.4	4.3	4.4	6.1	6.4	6.5
$\varphi_B^{(2)}$	0.059	0.067	0.044	0.084	0.040	0.030	0.057
$\Delta_{dil}H^{12}/(J/mol\ solvent)$	−0.030	−0.032	−0.016	−0.049	−0.013	−0.007	−0.022

$V^{(1)}/cm^3$	4.4
$\varphi_B^{(1)}$	0.067
$V^{(2)}/cm^3$	6.4
$\varphi_B^{(2)}$	0.046
$\Delta_{dil}H^{12}/(J/mol\ solvent)$	−0.015

Comments: $\varphi_B^{(1)}$ and $\varphi_B^{(2)}$ denote the volume fractions of the polymer in the solution before and after the dilution process.

Polymer (B):	**polystyrene**					**1974BA1**
Characterization:	$M_n/g.mol^{-1} = 670000$, American Commercial Co. Ltd.					
Solvent (A):	**ethyl acetate**		$C_4H_8O_2$			**141-78-6**

$T/K = 303.15$

$V^{(1)}/cm^3$	5.2	9.2	10.2	10.8	15.1	22.0
$\varphi_B^{(1)}$	0.161	0.092	0.167	0.120	0.113	0.070
$V^{(2)}/cm^3$	9.2	13.1	15.1	17.3	19.9	24.5
$\varphi_B^{(2)}$	0.092	0.064	0.113	0.075	0.085	0.063
$\Delta_{dil}H^{12}/(J/mol\ solvent)$	−0.064	−0.043	−0.104	−0.093	−0.094	−0.026

Comments: $\varphi_B^{(1)}$ and $\varphi_B^{(2)}$ denote the volume fractions of the polymer in the solution before and after the dilution process.

Polymer (B):   **polystyrene**   **1959SCH**
Characterization: $M_n/\text{g.mol}^{-1} = 11000$
Solvent (A):   **ethylbenzene**   **C_8H_{10}**   **100-41-4**

$T/K = 296.15$

$c_B^{(1)}/(\text{g/cm}^3) = 0.181$ $c_B^{(2)}/(\text{g/cm}^3) = 0.093$ $\Delta_{dil}H^{12}/(\text{J/g polymer}) = -0.134$

Polymer (B):   **polystyrene**   **1956JEN**
Characterization: –
Solvent (A):   **ethylbenzene**   **C_8H_{10}**   **100-41-4**

$T/K = 293.15$

z_B	0.94	0.88	0.83	0.82	0.76	0.67	0.60	0.48
$\Delta_M H/(\text{J/base mol})$	−2364	−3130	−2950	−2916	−2703	−2381	−2134	−1707

z_B	0.38	0.32	0.17	0.08
$\Delta_M H/(\text{J/base mol})$	−1351	−1138	−602	−368

Comments: These data include the enthalpy change when mixing the glassy polymer and the solvent to form the mixture.

z_B	$0 \le z_B \le 1$
$\Delta_M H/(\text{J/base mol})$	0.0

Comments: Athermal behaviour was observed when mixing the equilibrium "liquid" polymer and the solvent to form the mixture, i.e., after substracting the part caused by the glass enthalpy.

Polymer (B):   **polystyrene**   **1956AM3**
Characterization: –
Solvent (A):   **ethylbenzene**   **C_8H_{10}**   **100-41-4**

$T/K = 298.15$

$V^{(1)}/\text{cm}^3$	4.75	9.75
$\varphi_B^{(1)}$	0.194	0.081
$V^{(2)}/\text{cm}^3$	9.75	14.75
$\varphi_B^{(2)}$	0.081	0.053
$\Delta_{dil}H^{12}/\text{J}$	−0.008	−0.008

Comments: $\Delta_{dil}H^{12}$ was measured for an amount of 5 ml solvent added to the given volume of solution before dilution, $V^{(1)}$, where $\varphi_B^{(1)}$ and $\varphi_B^{(2)}$ denote the volume fractions of the polymer in the solution before and after the dilution process.

Polymer (B):   **polystyrene**   **1972DAV**
Characterization: $M_w/\text{g.mol}^{-1} = 498000$, ArRo Laboratories, Inc., Joliet, IL
Solvent (A):   **tetrachloromethane**   **CCl$_4$**   **56-23-5**

continued

continued

$T/K = 298.15$

$m^{(1)}/g$	20.9585	22.8983	22.8119
$w_B^{(1)}$	0.05608	0.06747	0.07517
$m^{(2)}/g$	36.2988	38.0918	38.0439
$w_B^{(2)}$	0.03238	0.04056	0.04507
$\Delta_{dil}H^{12}/(J/g)$	−0.067	−0.021	−0.105

Comments: $w_B^{(1)}$ and $w_B^{(2)}$ denote the mass fractions of the polymer in the solution before and after the dilution process, $m^{(1)}$ and $m^{(2)}$ are the total masses of both solutions.

Polymer (B):   **polystyrene**   **1970MO1, 1970MO4**
Characterization: $M_w/g.mol^{-1} = 600$, $M_w/M_n \leq 1.10$,
                         Pressure Chemical Co., Pittsburgh, PA
Solvent (A):   **toluene**   **C₇H₈**   **108-88-3**

$T/K = 298.15$

$c_B^{(1)}/(g/cm^3) = 0.0701$ $c_B^{(2)}/(g/cm^3) = 0.0038$ $\Delta_{dil}H^{12}/(J/cm^3) = -0.00519$

Comments: $c_B^{(1)}$ and $c_B^{(2)}$ denote the concentrations of the polymer in the solution before and after the dilution process.

Polymer (B):   **polystyrene**   **1970MO1, 1970MO4**
Characterization: $M_w/g.mol^{-1} = 900$, $M_w/M_n \leq 1.10$,
                         Pressure Chemical Co., Pittsburgh, PA
Solvent (A):   **toluene**   **C₇H₈**   **108-88-3**

$T/K = 298.15$

$c_B^{(1)}/(g/cm^3)$	0.0693	0.0697
$c_B^{(2)}/(g/cm^3)$	0.0026	0.0020
$\Delta_{dil}H^{12}/(J/cm^3)$	−0.00368	−0.00464

Comments: $c_B^{(1)}$ and $c_B^{(2)}$ denote the concentrations of the polymer in the solution before and after the dilution process.

Polymer (B):   **polystyrene**   **1970LEW**
Characterization: $M_n/g.mol^{-1} = 900$, $M_w/g.mol^{-1} = 990$,
                         Pressure Chemical Co., Pittsburgh, PA
Solvent (A):   **toluene**   **C₇H₈**   **108-88-3**

$T/K = 303.15$

$m^{(1)}/g$	0.9408	1.2037	0.2299	0.5667	0.3713	0.4617	0.2428
$\varphi_B^{(1)}$	0.4063	0.2255	0.2255	0.1145	0.1051	0.1051	0.4063
$m^{(2)}/g$	2.2036	2.1004	1.4962	1.9089	1.5835	1.5932	1.5007
$\varphi_B^{(2)}$	0.1662	0.1394	0.0335	0.0693	0.0243	0.0300	0.0618
$\Delta_{dil}H^{12}/J$	−0.3862	−0.1104	−0.0355	−0.0295	−0.0201	−0.0203	−0.1375

Comments: $\Delta_{dil}H^{12}$ is the extensive quantity obtained for a given total mass change from $m^{(1)}$ to $m^{(2)}$, where $\varphi_B^{(1)}$ and $\varphi_B^{(2)}$ denote the volume fractions of the polymer in the solution before and after the dilution process.

Polymer (B):	polystyrene	1970MO1, 1970MO4

Characterization: $M_w/\text{g.mol}^{-1} = 2000$, $M_w/M_n \leq 1.10$,
Pressure Chemical Co., Pittsburgh, PA

Solvent (A): toluene C_7H_8 108-88-3

$T/\text{K} = 298.15$

$c_B^{(1)}/(\text{g/cm}^3)$	0.0711	0.0695
$c_B^{(2)}/(\text{g/cm}^3)$	0.0025	0.0030
$\Delta_{dil} H^{12}/(\text{J/cm}^3)$	−0.00431	−0.00464

Comments: $c_B^{(1)}$ and $c_B^{(2)}$ denote the concentrations of the polymer in the solution before and after the dilution process.

Polymer (B):	polystyrene	1970MO1, 1970MO4

Characterization: $M_w/\text{g.mol}^{-1} = 4800$, $M_w/M_n \leq 1.10$,
Pressure Chemical Co., Pittsburgh, PA

Solvent (A): toluene C_7H_8 108-88-3

$T/\text{K} = 298.15$

$c_B^{(1)}/(\text{g/cm}^3) = 0.0702$ $c_B^{(2)}/(\text{g/cm}^3) = 0.0024$ $\Delta_{dil} H^{12}/(\text{J/cm}^3) = -0.00565$

Comments: $c_B^{(1)}$ and $c_B^{(2)}$ denote the concentrations of the polymer in the solution before and after the dilution process.

Polymer (B):	polystyrene	1970LEW

Characterization: $M_n/\text{g.mol}^{-1} = 4600$, $M_w/\text{g.mol}^{-1} = 5000$,
Waters Associates, Framingham, MA

Solvent (A): toluene C_7H_8 108-88-3

$T/\text{K} = 303.15$

$m^{(1)}/\text{g}$	1.1701	0.1387	0.5579	0.2552	0.6311	0.5779
$\varphi_B^{(1)}$	0.1785	0.1785	0.3837	0.3837	0.3837	0.1048
$m^{(2)}/\text{g}$	1.6565	1.5229	1.7443	1.4970	1.8218	2.0092
$\varphi_B^{(2)}$	0.1250	0.0158	0.1167	0.0616	0.1267	0.0401
$\Delta_{dil} H^{12}/\text{J}$	−0.0368	−0.0126	−0.2360	−0.0842	−0.2362	−0.0159

Comments: $\Delta_{dil} H^{12}$ is the extensive quantity obtained for a given total mass change from $m^{(1)}$ to $m^{(2)}$, where $\varphi_B^{(1)}$ and $\varphi_B^{(2)}$ denote the volume fractions of the polymer in the solution before and after the dilution process.

Polymer (B):	polystyrene	1970MO1, 1970MO4

Characterization: $M_w/\text{g.mol}^{-1} = 10500$, $M_w/M_n \leq 1.10$,
Pressure Chemical Co., Pittsburgh, PA

Solvent (A): toluene C_7H_8 108-88-3

$T/\text{K} = 298.15$

$c_B^{(1)}/(\text{g/cm}^3)$	0.0703	0.0688	0.0698	0.0701	0.0701
$c_B^{(2)}/(\text{g/cm}^3)$	0.0032	0.0029	0.0028	0.0024	0.0024
$\Delta_{dil} H^{12}/(\text{J/cm}^3)$	−0.00498	−0.00385	−0.00381	−0.00628	−0.00565

Polymer (B):	**polystyrene**				**1970LEW**

Characterization: $M_n/\text{g.mol}^{-1} = 10500$, $M_w/\text{g.mol}^{-1} = 10900$,
                        Pressure Chemical Co., Pittsburgh, PA

Solvent (A):	**toluene**		$\mathbf{C_7H_8}$		**108-88-3**

$T/\text{K} = 303.15$

$m^{(1)}/\text{g}$	0.8451	0.4523	1.2802	0.2717	1.0685	0.4727
$\varphi_B^{(1)}$	0.1465	0.1465	0.1465	0.1465	0.0776	0.0775
$m^{(2)}/\text{g}$	1.8607	1.7015	1.7387	1.5687	1.8496	1.7283
$\varphi_B^{(2)}$	0.0652	0.0410	0.1157	0.0267	0.0444	0.0209
$\Delta_{dil}H^{12}/\text{J}$	−0.0216	−0.0202	−0.0193	−0.0107	−0.0084	−0.0054

Comments: $\Delta_{dil}H^{12}$ is the extensive quantity obtained for a given total mass change from $m^{(1)}$ to $m^{(2)}$, where $\varphi_B^{(1)}$ and $\varphi_B^{(2)}$ denote the volume fractions of the polymer in the solution before and after the dilution process.

Polymer (B):	**polystyrene**		**1959SCH**

Characterization: $M_n/\text{g.mol}^{-1} = 11000$

Solvent (A):	**toluene**	$\mathbf{C_7H_8}$	**108-88-3**

$T/\text{K} = 296.15$

$c_B^{(1)}/(\text{g/cm}^3) = 0.100$ $c_B^{(2)}/(\text{g/cm}^3) = 0.050$ $\Delta_{dil}H^{12}/(\text{J/g polymer}) = -0.038$

Polymer (B):	**polystyrene**				**1970LEW**

Characterization: $M_n/\text{g.mol}^{-1} = 19650$, $M_w/\text{g.mol}^{-1} = 19850$,
                        Waters Associates, Framingham, MA

Solvent (A):	**toluene**		$\mathbf{C_7H_8}$		**108-88-3**

$T/\text{K} = 303.15$

$m^{(1)}/\text{g}$	0.9217	0.2213	0.9853	0.8387	0.2873
$\varphi_B^{(1)}$	0.3924	0.3924	0.2590	0.1373	0.2590
$m^{(2)}/\text{g}$	2.1602	1.5413	2.2747	2.2625	1.7005
$\varphi_B^{(2)}$	0.1601	0.027	0.1079	0.0493	0.0811
$\Delta_{dil}H^{12}/\text{J}$	−0.2845	−0.0631	−0.1037	−0.0146	−0.0261

Comments: $\Delta_{dil}H^{12}$ is the extensive quantity obtained for a given total mass change from $m^{(1)}$ to $m^{(2)}$, where $\varphi_B^{(1)}$ and $\varphi_B^{(2)}$ denote the volume fractions of the polymer in the solution before and after the dilution process.

Polymer (B):	**polystyrene**		**1966KA1**

Characterization: $M_n/\text{g.mol}^{-1} = 25860$, DP = 244, synthesized in the laboratory

Solvent (A):	**toluene**	$\mathbf{C_7H_8}$	**108-88-3**

$T/\text{K} = 298.15$

continued

continued

$V^{(1)}$/cm^3	5.0	10.0	15.0	5.0	10.0	15.0	20.0	30.0	40.0
$\varphi_B^{(1)}$	0.176	0.088	0.056	0.166	0.083	0.056	0.052	0.035	0.026
$V^{(2)}$/cm^3	10.0	15.0	20.0	10.0	15.0	20.0	30.0	40.0	50.0
$\varphi_B^{(2)}$	0.088	0.056	0.043	0.083	0.056	0.042	0.035	0.026	0.021
$\Delta_{dil} H^{12}$/J	−0.276	−0.121	−0.038	−0.310	−0.138	−0.059	−0.056	−0.0184	−0.0067

$V^{(1)}$/cm^3	20.0	30.0	40.0	5.0	5.0	5.0	5.0	20.0	20.0
$\varphi_B^{(1)}$	0.048	0.032	0.024	0.176	0.176	0.166	0.166	0.052	0.052
$V^{(2)}$/cm^3	30.0	40.0	50.0	15.0	20.0	15.0	20.0	40.0	50.0
$\varphi_B^{(2)}$	0.032	0.024	0.019	0.058	0.043	0.056	0.042	0.026	0.021
$\Delta_{dil} H^{12}$/J	−0.0364	−0.0146	−0.0	−0.397	−0.435	−0.448	−0.506	−0.0745	−0.0812

$V^{(1)}$/cm^3	20.0
$\varphi_B^{(1)}$	0.048
$V^{(2)}$/cm^3	40.0
$\varphi_B^{(2)}$	0.024
$\Delta_{dil} H^{12}$/J	−0.0515

Comments: $\Delta_{dil} H^{12}$ is the extensive quantity obtained for a given total volume change from $V^{(1)}$ to $V^{(2)}$, where $\varphi_B^{(1)}$ and $\varphi_B^{(2)}$ denote the volume fractions of the polymer in the solution before and after the dilution process.

Polymer (B):	**polystyrene**					**1970LEW**
Characterization:	M_n/g.mol^{-1} = 96200, M_w/g.mol^{-1} = 98200,					
	Waters Associates, Framingham, MA					
Solvent (A):	**toluene**	C_7H_8				**108-88-3**

T/K = 303.15

$m^{(1)}$/g	0.4544	0.8393	1.0481	0.9229	1.0797
$\varphi_B^{(1)}$	0.3284	0.3284	0.2216	0.1217	0.0817
$m^{(2)}$/g	1.7916	1.0969	2.4242	2.2985	2.0562
$\varphi_B^{(2)}$	0.0791	0.2471	0.0929	0.0526	0.0425
$\Delta_{dil} H^{12}$/J	−0.0748	−0.0837	−0.0526	−0.0049	−0.0032

Comments: $\Delta_{dil} H^{12}$ is the extensive quantity obtained for a given total mass change from $m^{(1)}$ to $m^{(2)}$, where $\varphi_B^{(1)}$ and $\varphi_B^{(2)}$ denote the volume fractions of the polymer in the solution before and after the dilution process.

Polymer (B):	**polystyrene**					**1970LEW**
Characterization:	M_n/g.mol^{-1} = 164000					
Solvent (A):	**toluene**	C_7H_8				**108-88-3**

T/K = 303.15

$m^{(1)}$/g	0.8806	0.5336	1.0242	0.2284	0.2371	0.7151
$\varphi_B^{(1)}$	0.3092	0.1165	0.1165	0.3092	0.1165	0.1165
$m^{(2)}$/g	2.0457	1.7338	1.9036	1.3893	1.5107	1.8357
$\varphi_B^{(2)}$	0.1272	0.0351	0.0648	0.0491	0.0178	0.0445
$\Delta_{dil} H^{12}$/J	−0.1544	−0.0085	−0.0221	−0.0161	−0.0136	−0.0362

Polymer (B):	**polystyrene**		**1972DAV**
Characterization:	$M_w/\text{g.mol}^{-1} = 411000$, ArRo Laboratories, Inc., Joliet, IL		
Solvent (A):	**toluene**	**C$_7$H$_8$**	**108-88-3**

$T/\text{K} = 298.15$

$m^{(1)}/\text{g}$	8.7902
$w_B^{(1)}$	0.19695
$m^{(2)}/\text{g}$	21.9636
$w_B^{(2)}$	0.07882
$\Delta_{dil}H^{12}/(\text{J/g})$	−0.037

Comments: $w_B^{(1)}$ and $w_B^{(2)}$ denote the mass fractions of the polymer in the solution before and after the dilution process, $m^{(1)}$ and $m^{(2)}$ are the total masses of both solutions.

Polymer (B):	**polystyrene**		**1956AM3**
Characterization:	−		
Solvent (A):	**toluene**	**C$_7$H$_8$**	**108-88-3**

$T/\text{K} = 298.15$

$V^{(1)}/\text{cm}^3$	4.70	9.70	14.70	4.70	9.70	14.70	9.70	14.70
$\varphi_B^{(1)}$	0.189	0.092	0.060	0.170	0.083	0.055	0.083	0.055
$V^{(2)}/\text{cm}^3$	9.70	14.70	19.70	9.70	14.70	19.70	14.70	19.70
$\varphi_B^{(2)}$	0.092	0.060	0.045	0.083	0.055	0.041	0.055	0.041
$\Delta_{dil}H^{12}/\text{J}$	−0.414	−0.146	−0.105	−0.444	−0.155	−0.059	−0.155	−0.067

Comments: $\Delta_{dil}H^{12}$ was measured for an amount of 5 ml solvent added to the given volume of solution before dilution, $V^{(1)}$, where $\varphi_B^{(1)}$ and $\varphi_B^{(2)}$ denote the volume fractions of the polymer in the solution before and after the dilution process.

Polymer (B):	**polystyrene**		**1956JEN**
Characterization:	−		
Solvent (A):	**toluene**	**C$_7$H$_8$**	**108-88-3**

$T/\text{K} = 293.15$

z_B	0.94	0.90	0.88	0.87	0.73	0.60	0.46	0.34
$\Delta_M H/(\text{J/base mol})$	−2314	−2970	−3130	−3092	−2594	−2134	−1636	−1209

z_B	0.26	0.14	0.07
$\Delta_M H/(\text{J/base mol})$	−883	−498	−251

Comments: These data include the enthalpy change when mixing the glassy polymer and the solvent to form the mixture.

z_B	$0 \leq z_B \leq 1$
$\Delta_M H/(\text{J/base mol})$	0.0

Comments: Athermal behaviour was observed when mixing the equilibrium "liquid" polymer and the solvent to form the mixture, i.e., after substracting the part caused by the glass enthalpy.

Polymer (B): **polystyrene** 1970MO1, 1970MO4
Characterization: $M_w/\text{g.mol}^{-1} = 600$, $M_w/M_n \leq 1.10$,
Pressure Chemical Co., Pittsburgh, PA
Solvent (A): **trichloromethane** **CHCl$_3$** 67-66-3

$T/\text{K} = 298.15$

$c_B^{(1)}/(\text{g/cm}^3) = 0.2003$ $c_B^{(2)}/(\text{g/cm}^3) = 0.0023$ $\Delta_{dil} H^{12}/(\text{J/cm}^3) = -0.124$

Comments: $c_B^{(1)}$ and $c_B^{(2)}$ denote the concentrations of the polymer in the solution before and after the dilution process.

Polymer (B): **polystyrene** 1970MO1, 1970MO4
Characterization: $M_w/\text{g.mol}^{-1} = 900$, $M_w/M_n \leq 1.10$,
Pressure Chemical Co., Pittsburgh, PA
Solvent (A): **trichloromethane** **CHCl$_3$** 67-66-3

$T/\text{K} = 298.15$

$c_B^{(1)}/(\text{g/cm}^3) = 0.1582$ $c_B^{(2)}/(\text{g/cm}^3) = 0.0017$ $\Delta_{dil} H^{12}/(\text{J/cm}^3) = -0.096$

Comments: $c_B^{(1)}$ and $c_B^{(2)}$ denote the concentrations of the polymer in the solution before and after the dilution process.

Polymer (B): **polystyrene** 1970MO1, 1970MO4
Characterization: $M_w/\text{g.mol}^{-1} = 2000$, $M_w/M_n \leq 1.10$,
Pressure Chemical Co., Pittsburgh, PA
Solvent (A): **trichloromethane** **CHCl$_3$** 67-66-3

$T/\text{K} = 298.15$

$c_B^{(1)}/(\text{g/cm}^3) = 0.1513$ $c_B^{(2)}/(\text{g/cm}^3) = 0.0018$ $\Delta_{dil} H^{12}/(\text{J/cm}^3) = -0.127$

Comments: $c_B^{(1)}$ and $c_B^{(2)}$ denote the concentrations of the polymer in the solution before and after the dilution process.

Polymer (B): **polystyrene** 1970MO1, 1970MO4
Characterization: $M_w/\text{g.mol}^{-1} = 4800$, $M_w/M_n \leq 1.10$,
Pressure Chemical Co., Pittsburgh, PA
Solvent (A): **trichloromethane** **CHCl$_3$** 67-66-3

$T/\text{K} = 298.15$

$c_B^{(1)}/(\text{g/cm}^3) = 0.1526$ $c_B^{(2)}/(\text{g/cm}^3) = 0.0019$ $\Delta_{dil} H^{12}/(\text{J/cm}^3) = -0.144$

Comments: $c_B^{(1)}$ and $c_B^{(2)}$ denote the concentrations of the polymer in the solution before and after the dilution process.

Polymer (B): **polystyrene** 1970MO1, 1970MO4
Characterization: $M_w/\text{g.mol}^{-1} = 10500$, $M_w/M_n \leq 1.10$,
Pressure Chemical Co., Pittsburgh, PA
Solvent (A): **trichloromethane** **CHCl$_3$** 67-66-3

$T/\text{K} = 298.15$

$c_B^{(1)}/(\text{g/cm}^3) = 0.1512$ $c_B^{(2)}/(\text{g/cm}^3) = 0.0018$ $\Delta_{dil} H^{12}/(\text{J/cm}^3) = -0.144$

Polymer (B):	polystyrene				**1956AM3**
Characterization:	–				
Solvent (A):	trichloromethane		CHCl$_3$		**67-66-3**

T/K = 298.15

$V^{(1)}$/cm^3	4.75	9.75	14.75	4.70	9.70
$\varphi_B^{(1)}$	0.275	0.131	0.088	0.171	0.083
$V^{(2)}$/cm^3	9.75	14.75	19.75	9.70	14.70
$\varphi_B^{(2)}$	0.131	0.088	0.066	0.083	0.055
$\Delta_{dil}H^{12}$/J	−0.978	−0.510	−0.192	−0.339	−0.117

Comments: $\Delta_{dil}H^{12}$ was measured for an amount of 5 ml solvent added to the given volume of solution before dilution, $V^{(1)}$, where $\varphi_B^{(1)}$ and $\varphi_B^{(2)}$ denote the volume fractions of the polymer in the solution before and after the dilution process.

Polymer (B):	poly(styrene-*co*-butyl methacrylate)					**1987KYO**
Characterization:	M_n/g.mol^{-1} = 185000, M_w/g.mol^{-1} = 311000, 20.6 wt% styrene,					
	Scientific Polymer Products, Inc., Ontario, NY					
Solvent (A):	2-butanone		C$_4$H$_8$O			**78-93-3**

T/K = 298.15

$\varphi_B^{(1)}$	0.0908	0.0908	0.0908	0.0908	0.0908	0.0908
$\varphi_B^{(2)}$	0.0272	0.0454	0.0545	0.0636	0.0724	0.0814
$\Delta_{dil}H^{12}$/(J/mol solvent)	0.564	0.778	0.860	1.056	1.390	1.748

Comments: The table provides the ratio of $\Delta_{dil}H^{12}/\Delta n_A$, i.e., the enthalpy change caused by diluting the primary solution by 1 mol solvent, where $\varphi_B^{(1)}$ denotes the volume fraction of the polymer in the starting solution and $\varphi_B^{(2)}$ denotes the volume fraction after the dilution process.

Polymer (B):	poly(styrene-*co*-butyl methacrylate)					**1987KYO**
Characterization:	M_n/g.mol^{-1} = 121000, M_w/g.mol^{-1} = 395000, 67.7 wt% styrene,					
	Scientific Polymer Products, Inc., Ontario, NY					
Solvent (A):	2-butanone		C$_4$H$_8$O			**78-93-3**

T/K = 298.15

$\varphi_B^{(1)}$	0.0788	0.0788	0.0788	0.0788	0.0788	0.0788	0.0788
$\varphi_B^{(2)}$	0.0221	0.0323	0.0355	0.0497	0.0551	0.0654	0.0607
$\Delta_{dil}H^{12}$/(J/mol solvent)	0.170	0.282	0.261	0.351	0.393	0.399	0.569

Comments: The table provides the ratio of $\Delta_{dil}H^{12}/\Delta n_A$, i.e., the enthalpy change caused by diluting the primary solution by 1 mol solvent, where $\varphi_B^{(1)}$ denotes the volume fraction of the polymer in the starting solution and $\varphi_B^{(2)}$ denotes the volume fraction after the dilution process.

Polymer (B):	poly(styrene-*co*-butyl methacrylate)				**1987KYO**
Characterization:	M_n/g.mol^{-1} = 193000, M_w/g.mol^{-1} = 249000, 80.0 wt% styrene,				
	Scientific Polymer Products, Inc., Ontario, NY				
Solvent (A):	2-butanone		C$_4$H$_8$O		**78-93-3**

continued

continued

$T/\text{K} = 298.15$

$\varphi_B^{(1)}$	0.0465	0.0465	0.0465	0.0465	0.0465	0.0776	0.0776
$\varphi_B^{(2)}$	0.0189	0.0279	0.0326	0.0372	0.0419	0.0380	0.0388
$\Delta_{dil} H^{12}/(\text{J/mol solvent})$	0.104	0.164	0.121	0.228	0.232	0.304	0.326

$\varphi_B^{(1)}$	0.0766	0.0766	0.0766	0.0766
$\varphi_B^{(2)}$	0.0465	0.0543	0.0621	0.0698
$\Delta_{dil} H^{12}/(\text{J/mol solvent})$	0.432	0.555	0.706	0.867

Comments: The table provides the ratio of $\Delta_{dil} H^{12}/\Delta n_A$, i.e., the enthalpy change caused by diluting the primary solution by 1 mol solvent, where $\varphi_B^{(1)}$ denotes the volume fraction of the polymer in the starting solution and $\varphi_B^{(2)}$ denotes the volume fraction after the dilution process.

Polymer (B):   **poly(styrene-*co*-butyl methacrylate)**   **1987KYO**
Characterization: $M_n/\text{g.mol}^{-1} = 176000$, $M_w/\text{g.mol}^{-1} = 308000$, 85.0 wt% styrene, Scientific Polymer Products, Inc., Ontario, NY
Solvent (A):   **2-butanone**   **C₄H₈O**   **78-93-3**

$T/\text{K} = 298.15$

$\varphi_B^{(1)}$	0.0777	0.0777	0.0777	0.0777	0.0777	0.0777
$\varphi_B^{(2)}$	0.0583	0.0505	0.0428	0.0350	0.0272	0.0192
$\Delta_{dil} H^{12}/(\text{J/mol solvent})$	−0.453	−0.328	−0.255	−0.195	−0.169	−0.105

Comments: The table provides the ratio of $\Delta_{dil} H^{12}/\Delta n_A$, i.e., the enthalpy change caused by diluting the primary solution by 1 mol solvent, where $\varphi_B^{(1)}$ denotes the volume fraction of the polymer in the starting solution and $\varphi_B^{(2)}$ denotes the volume fraction after the dilution process.

Polymer (B):   **poly(styrenesulfonic acid)**   **1995PER**
Characterization: –
Solvent (A):   **water**   **H₂O**   **7732-18-5**

$T/\text{K} = 298.15$

$c_B^{(1)}/(\text{base mol/l}) = 0.12$ $c_B^{(2)}/(\text{base mol/l}) = 0.02$ $\Delta_{dil} H^{12}/(\text{J/base mol polymer}) = -369$

Comments: $c_B^{(1)}$ and $c_B^{(2)}$ denote the concentrations of the polymer in the solution before and after the dilution process.

Polymer (B):   **poly(styrenesulfonic acid)**   **1996PER**
Characterization: –
Solvent (A):   **water**   **H₂O**   **7732-18-5**

$T/\text{K} = 298.15$

$c_B^{(1)}/(\text{base mol/l}) = 0.06$ $c_B^{(2)}/(\text{base mol/l}) = 0.02$ $\Delta_{dil} H^{12}/(\text{J/base mol polymer}) = -213$

Comments: $c_B^{(1)}$ and $c_B^{(2)}$ denote the concentrations of the polymer in the solution before and after the dilution process.

Polymer (B):	**poly(styrenesulfonic acid)**						**1976DA1**

Characterization: M_η/g.mol^{-1} = 525000, degree of sulfonation = 1.0

Solvent (A):   **water** H_2O **7732-18-5**

T/K = 298.15

$n_A^{(1)}$/mol	1.0953	1.1212	1.1588	1.1685	1.1655	1.1691	1.2074	1.1978
$n_B^{(1)}$*100/base mol	0.9281	0.7657	0.4662	0.3709	0.2269	0.1775	0.1136	0.0875
$n_A^{(2)}$/mol	1.4857	1.5115	1.5574	1.5637	1.5880	1.5660	1.5989	1.6027
$\Delta_{dil}H^{12}$/J	−1.035	−0.809	−0.516	−0.413	−0.266	−0.215	−0.125	−0.095

$n_A^{(1)}$/mol	1.1564	1.1729	1.1770
$n_B^{(1)}$ *100/base mol	0.0516	0.0413	0.0259
$n_A^{(2)}$/mol	1.5526	1.5704	1.5772
$\Delta_{dil}H^{12}$/J	−0.045	−0.035	0.000

Comments: $\Delta_{dil}H^{12}$ is here an extensive quantity obtained for a given total amount of solvent added (difference $n_A^{(2)} - n_A^{(1)}$).

Polymer (B):	**poly(styrenesulfonic acid)**						**1967SK1**

Characterization: M_η/g.mol^{-1} = 200000, degree of sulfonation = 1.0

Solvent (A):   **water** H_2O **7732-18-5**

T/K = 298.15

$c_B^{(1)}$/(base mol/kg water)	0.381	0.181	0.127	0.0876	0.0611	0.0424	0.0297
$c_B^{(2)}$/(base mol/kg water)	0.181	0.0876	0.0611	0.0424	0.0297	0.0206	0.0144
$\Delta_{dil}H^{12}$/(J/base mol polymer)	−214.6	−182.0	−179.1	−172.0	−168.2	−164.8	−162.3

$c_B^{(1)}$/(base mol/kg water)	0.0100	0.00704	0.00489	0.00344	0.00240	0.00167	0.00117
$c_B^{(2)}$/(base mol/kg water)	0.00489	0.00344	0.00240	0.00167	0.00117	0.000811	0.000571
$\Delta_{dil}H^{12}$/(J/base mol polymer)	−162.8	−160.2	−154.8	−163.2	−198.7	−139.3	−190.0

$c_B^{(1)}$/(base mol/kg water)	0.381	0.181	0.127	0.0876	0.0611	0.0424	0.0297
$c_B^{(2)}$/(base mol/kg water)	0.000398	0.000398	0.000398	0.000398	0.000398	0.000398	0.000398
$\Delta_{dil}H^{12}$/(J/base mol polymer)	−1757	−1544	−1456	−1364	−1276	−1188	−1109

Comments: $c_B^{(1)}$ and $c_B^{(2)}$ denote the concentrations of the polymer in the solution before and after the dilution process.

Polymer (B):	**poly(styrenesulfonic acid) calcium salt**						**1973SKE**

Characterization: M_w/g.mol^{-1} = 100000, degree of sulfonation = 1.0

Solvent (A):   **water** H_2O **7732-18-5**

T/K = 298.15

$c_B^{(1)}$/(base mol/kg water)	0.640	0.310	0.153	0.0756	0.0378	0.0188	0.00936
$c_B^{(2)}$/(base mol/kg water)	0.310	0.153	0.0756	0.0378	0.0188	0.00936	0.00466
$\Delta_{dil}H^{12}$/(J/base mol polymer)	+11.7	−11.3	−17.2	−18.0	−20.1	−22.6	−28.0

$c_B^{(1)}$/(base mol/kg water)	0.640	0.310	0.153	0.0756	0.0378	0.0188	0.00936
$c_B^{(2)}$/(base mol/kg water)	0.00466	0.00466	0.00466	0.00466	0.00466	0.00466	0.00466
$\Delta_{dil}H^{12}$/(J/base mol polymer)	−105.4	−117.2	−105.9	−88.7	−70.7	−50.6	−28.0

Comments: $c_B^{(1)}$ and $c_B^{(2)}$ denote the concentrations of the polymer in the solution before and after the dilution process.

Polymer (B):   **poly(styrenesulfonic acid) cesium salt**   **1970SKE**
Characterization: $M_\eta/\text{g.mol}^{-1} = 200000$, degree of sulfonation = 1.0
Solvent (A):   **water**   **H_2O**   **7732-18-5**

$T/\text{K} = 298.15$

$c_B^{(1)}$/(base mol/kg water)	0.191	0.0935	0.0460	0.0228	0.0113	0.00565	0.00282
$c_B^{(2)}$/(base mol/kg water)	0.0935	0.0460	0.0228	0.0113	0.00565	0.00282	0.00140
$\Delta_{dil}H^{12}$/(J/base mol polymer)	+12.1	−23.0	−41.8	−49.8	−63.2	−78.2	−98.7

$c_B^{(1)}$/(base mol/kg water)	0.191	0.0935	0.0460	0.0228	0.0113	0.00565	0.00282
$c_B^{(2)}$/(base mol/kg water)	0.00070	0.00070	0.00070	0.00070	0.00070	0.00070	0.00070
$\Delta_{dil}H^{12}$/(J/base mol polymer)	−464.4	−477.0	−451.9	−410.0	−359.8	−297.1	−275.6

Comments: $c_B^{(1)}$ and $c_B^{(2)}$ denote the concentrations of the polymer in the solution before and after the dilution process.

Polymer (B):   **poly(styrenesulfonic acid) cesium salt**   **1976DA1**
Characterization: $M_\eta/\text{g.mol}^{-1} = 525000$, degree of sulfonation = 1.0
Solvent (A):   **water**   **H_2O**   **7732-18-5**

$T/\text{K} = 298.15$

$n_A^{(1)}$/mol	1.1190	1.1273	1.2407	1.1636	1.1927	1.1774	1.1881	1.2173
$n_B^{(1)}*100$/base mol	0.8474	0.6150	0.3973	0.2977	0.1854	0.1424	0.0818	0.0704
$n_A^{(2)}$/mol	1.5128	1.5335	1.5482	1.5577	1.5522	1.5797	1.5977	1.6017
$\Delta_{dil}H^{12}$/J	0.429	0.148	0.034	0.000	−0.004	−0.004	−0.005	0.000

Comments: $\Delta_{dil}H^{12}$ is here an extensive quantity obtained for a given total amount of solvent added (difference $n_A^{(2)} - n_A^{(1)}$).

Polymer (B):   **poly(styrenesulfonic acid) cupric salt**   **1995PER**
Characterization: –
Solvent (A):   **water**   **H_2O**   **7732-18-5**

T/K	298.15	308.15	318.15
$c_B^{(1)}$/(base mol/l)	0.12	0.12	0.12
$c_B^{(2)}$/(base mol/l)	0.02	0.02	0.02
$\Delta_{dil}H^{12}$/(J/base mol polymer)	−106	−113	−124

Comments: $c_B^{(1)}$ and $c_B^{(2)}$ denote the concentrations of the polymer in the solution before and after the dilution process.

Polymer (B):   **poly(styrenesulfonic acid) lead salt**   **1995PER**
Characterization: –
Solvent (A):   **water**   **H_2O**   **7732-18-5**

$T/\text{K} = 298.15$

$c_B^{(1)}$/(base mol/l) = 0.08 $c_B^{(2)}$/(base mol/l) = 0.02 $\Delta_{dil}H^{12}$/(J/base mol polymer) = 164

Comments: $c_B^{(1)}$ and $c_B^{(2)}$ denote the concentrations of the polymer in the solution before and after the dilution process.

Polymer (B):	poly(styrenesulfonic acid) iron salt		1996PER
Characterization:	–		
Solvent (A):	water	H$_2$O	7732-18-5

T/K	298.15	308.15	318.15
$c_B^{(1)}$/(base mol/l)	0.06	0.06	0.06
$c_B^{(2)}$/(base mol/l)	0.02	0.02	0.02
$\Delta_{dil}H^{12}$/(J/base mol polymer)	810	591	510

Comments: $c_B^{(1)}$ and $c_B^{(2)}$ denote the concentrations of the polymer in the solution before and after the dilution process.

Polymer (B):	poly(styrenesulfonic acid) lithium salt	1970SKE
Characterization:	M_η/g.mol^{-1} = 200000, degree of sulfonation = 1.0	
Solvent (A):	water H$_2$O	7732-18-5

T/K = 298.15

$c_B^{(1)}$/(base mol/kg water)	0.344	0.165	0.0804	0.0394	0.0194	0.00951	0.00467
$c_B^{(2)}$/(base mol/kg water)	0.165	0.0804	0.0394	0.0194	0.00951	0.00467	0.00231
$\Delta_{dil}H^{12}$/(J/base mol polymer)	−193.3	−160.7	−139.3	−123.0	−144.8	−136.4	143.9

$c_B^{(1)}$/(base mol/kg water)	0.00231	0.00112	0.344	0.165	0.0804	0.0394	0.0194
$c_B^{(2)}$/(base mol/kg water)	0.00112	0.000551	0.000551	0.000551	0.000551	0.000551	0.000551
$\Delta_{dil}H^{12}$/(J/base mol polymer)	−113.4	−119.2	−1276	−1079	−920.5	−782.4	−656.9

$c_B^{(1)}$/(base mol/kg water)	0.00951	0.00467	0.00231	0.00112
$c_B^{(2)}$/(base mol/kg water)	0.000551	0.000551	0.000551	0.000551
$\Delta_{dil}H^{12}$/(J/base mol polymer)	−514.6	−376.6	−234.3	−121.3

Polymer (B):	poly(styrenesulfonic acid) lithium salt	1976DA1
Characterization:	M_η/g.mol^{-1} = 525000, degree of sulfonation = 1.0	
Solvent (A):	water H$_2$O	7732-18-5

T/K = 298.15

$n_A^{(1)}$/mol	1.0340	1.1163	1.1422	1.1605	1.1708	1.1909	1.1764	1.2195
$n_B^{(1)}$*100/base mol	0.9969	0.7764	0.5823	0.4131	0.3092	0.2108	0.1569	0.1123
$n_A^{(2)}$/mol	1.4399	1.5216	1.5631	1.5648	1.5672	1.5933	1.5747	1.6135
$\Delta_{dil}H^{12}$/J	−0.485	−0.369	−0.271	−0.154	−0.098	−0.047	−0.024	−0.014

Comments: $\Delta_{dil}H^{12}$ is here an extensive quantity obtained for a given total amount of solvent added (difference $n_A^{(2)} - n_A^{(1)}$).

Polymer (B):	poly(styrenesulfonic acid) magnesium salt	1973SKE
Characterization:	M_w/g.mol^{-1} = 100000, degree of sulfonation = 1.0	
Solvent (A):	water H$_2$O	7732-18-5

T/K = 298.15

$c_B^{(1)}$/(base mol/kg water)	0.460	0.226	0.111	0.0546	0.0270	0.0134	0.00668
$c_B^{(2)}$/(base mol/kg water)	0.226	0.111	0.0546	0.0270	0.0134	0.00668	0.00334
$\Delta_{dil}H^{12}$/(J/base mol polymer)	−51.5	−44.4	−38.1	−34.7	−36.4	−31.0	−29.3

continued

continued

$c_B^{(1)}$/(base mol/kg water)	0.460	0.226	0.111	0.0546	0.0270	0.0134	0.00668
$c_B^{(2)}$/(base mol/kg water)	0.00334	0.00334	0.00334	0.00334	0.00334	0.00334	0.00334
$\Delta_{dil}H^{12}$/(J/base mol polymer)	−265.3	−213.8	−169.5	−131.4	−96.7	−60.2	−29.3

Comments: $c_B^{(1)}$ and $c_B^{(2)}$ denote the concentrations of the polymer in the solution before and after the dilution process.

Polymer (B):   **poly(styrenesulfonic acid) potassium salt**   **1970SKE**
Characterization: M_η/g.mol^{-1} = 200000, degree of sulfonation = 1.0
Solvent (A):   **water**   **H$_2$O**   **7732-18-5**

T/K = 298.15

$c_B^{(1)}$/(base mol/kg water)	0.350	0.162	0.0801	0.0398	0.0199	0.00995	0.00495
$c_B^{(2)}$/(base mol/kg water)	0.170	0.0801	0.0398	0.0199	0.00995	0.00495	0.00248
$\Delta_{dil}H^{12}$/(J/base mol polymer)	+46.0	−31.0	−49.8	−59.4	−82.8	−110.9	−107.1

$c_B^{(1)}$/(base mol/kg water)	0.00248	0.00124	0.350	0.162	0.0801	0.0398	0.0199
$c_B^{(2)}$/(base mol/kg water)	0.00124	0.00062	0.000620	0.000620	0.000620	0.000620	0.000620
$\Delta_{dil}H^{12}$/(J/base mol polymer)	−117.2	−108.8	−619.2	−665.3	636.0	−585.8	−527.2

$c_B^{(1)}$/(base mol/kg water)	0.00995	0.00495	0.00248	0.00124
$c_B^{(2)}$/(base mol/kg water)	0.000620	0.000620	0.000620	0.000620
$\Delta_{dil}H^{12}$/(J/base mol polymer)	−443.5	−334.7	−225.9	−108.8

Comments: $c_B^{(1)}$ and $c_B^{(2)}$ denote the concentrations of the polymer in the solution before and after the dilution process.

Polymer (B):   **poly(styrenesulfonic acid) potassium salt**   **1976DA1**
Characterization: M_η/g.mol^{-1} = 525000, degree of sulfonation = 1.0
Solvent (A):   **water**   **H$_2$O**   **7732-18-5**

T/K = 298.15

$n_A^{(1)}$/mol	1.1072	1.1515	1.1651	1.1784	1.1887	1.1904	1.2024	1.1947
$n_B^{(1)}$*100/base mol	0.8161	0.6338	0.4456	0.3394	0.2295	0.1848	0.1153	0.0923
$n_A^{(2)}$/mol	1.5060	1.5484	1.5396	1.5764	1.5897	1.5904	1.5949	1.5850
$\Delta_{dil}H^{12}$/J	0.156	0.045	−0.014	−0.023	−0.027	−0.020	−0.007	0.000

Comments: $\Delta_{dil}H^{12}$ is here an extensive quantity obtained for a given total amount of solvent added (difference $n_A^{(2)} - n_A^{(1)}$).

Polymer (B):   **poly(styrenesulfonic acid) rubidium salt**   **1976DA1**
Characterization: M_η/g.mol^{-1} = 525000, degree of sulfonation = 1.0
Solvent (A):   **water**   **H$_2$O**   **7732-18-5**

T/K = 298.15

$n_A^{(1)}$/mol	1.1209	1.1359	1.1585	1.1629	1.1460	1.1838	1.2047	1.2117
$n_B^{(1)}$*100/base mol	0.9076	0.6745	0.4210	0.3180	0.1963	0.1535	0.1014	0.0759
$n_A^{(2)}$/mol	1.5128	1.5335	1.5482	1.5577	1.5522	1.5797	1.5977	1.6017
$\Delta_{dil}H^{12}$/J	0.352	0.199	0.000	−0.011	−0.017	−0.006	−0.003	−0.003

Comments: $\Delta_{dil}H^{12}$ is here an extensive quantity obtained for a given total amount of solvent added (difference $n_A^{(2)} - n_A^{(1)}$).

Polymer (B):   **poly(styrenesulfonic acid) silver salt**   **1995PER**

Characterization: –

Solvent (A):   **water**   **H$_2$O**   **7732-18-5**

$T/K = 298.15$

$c_B^{(1)}$/(base mol/l)   0.13 $c_B^{(2)}$/(base mol/l)   0.02 $\Delta_{dil}H^{12}$/(J/base mol polymer)   −374

Comments: $c_B^{(1)}$ and $c_B^{(2)}$ denote the concentrations of the polymer in the solution before and after the dilution process.

Polymer (B):   **poly(styrenesulfonic acid) sodium salt**   **1967SK1**

Characterization: M_η/g.mol^{-1} = 200000, degree of sulfonation = 1.0

Solvent (A):   **water**   **H$_2$O**   **7732-18-5**

$T/K = 298.15$

$c_B^{(1)}$/(base mol/kg water)	0.430	0.215	0.107	0.0822	0.0537	0.0411	0.0206
$c_B^{(2)}$/(base mol/kg water)	0.215	0.107	0.0537	0.0411	0.0269	0.0206	0.0103
$\Delta_{dil}H^{12}$/(J/base mol polymer)	+33.5	−28.9	−58.2	−64.0	−77.0	−75.7	−118.4

$c_B^{(1)}$/(base mol/kg water)	0.0103	0.00514	0.00257	0.00128	0.000642	0.430	0.215
$c_B^{(2)}$/(base mol/kg water)	0.00514	0.00257	0.00128	0.000642	0.000321	0.000321	0.000321
$\Delta_{dil}H^{12}$/(J/base mol polymer)	−124.3	−144.3	−167.4	−138.1	−121.3	−970.6	−1004

$c_B^{(1)}$/(base mol/kg water)	0.107	0.0822	0.0537	0.0411	0.0206	0.0103	0.00514
$c_B^{(2)}$/(base mol/kg water)	0.000321	0.000321	0.000321	0.000321	0.000321	0.000321	0.000321
$\Delta_{dil}H^{12}$/(J/base mol polymer)	−974.9	−954.0	−916.3	−891.2	−815.9	−694.5	−573.2

Comments: $c_B^{(1)}$ and $c_B^{(2)}$ denote the concentrations of the polymer in the solution before and after the dilution process.

Polymer (B):   **poly(styrenesulfonic acid) sodium salt**   **1976DA1**

Characterization: M_η/g.mol^{-1} = 525000, degree of sulfonation = 1.0

Solvent (A):   **water**   **H$_2$O**   **7732-18-5**

$T/K = 298.15$

$n_A^{(1)}$/mol	1.1078	1.1160	1.1172	1.3623	1.1237	1.1865	1.1454	1.1542
$n_B^{(1)}$*100/base mol	1.0115	0.8416	0.7946	0.6576	0.4359	0.4373	0.3078	0.2401
$n_A^{(2)}$/mol	1.4971	1.5096	1.5014	1.7196	1.5105	1.5297	1.5639	1.5681
$\Delta_{dil}H^{12}$/J	0.142	0.024	0.000	−0.046	−0.050	−0.052	−0.063	−0.053

$n_A^{(1)}$/mol	1.1646	1.1717	1.1972	1.1563
$n_B^{(1)}$*100/base mol	0.2279	0.1295	0.1184	0.0864
$n_A^{(2)}$/mol	1.5504	1.5605	1.5913	1.5553
$\Delta_{dil}H^{12}$/J	−0.044	−0.007	−0.012	0.000

Comments: $\Delta_{dil}H^{12}$ is here an extensive quantity obtained for a given total amount of solvent added (difference $n_A^{(2)} - n_A^{(1)}$).

Polymer (B):   **poly(styrenesulfonic acid) strontium salt**   **1973SKE**
Characterization: M_w/g.mol^{-1} = 100000, degree of sulfonation = 1.0
Solvent (A):   **water**   **H$_2$O**   **7732-18-5**

T/K = 298.15

$c_B^{(1)}$/(base mol/kg water)	0.586	0.284	0.140	0.0690	0.0344	0.0171	0.00852
$c_B^{(2)}$/(base mol/kg water)	0.284	0.140	0.0690	0.0344	0.0171	0.00852	0.00426
$\Delta_{dil} H^{12}$/(J/base mol polymer)	+41.8	+10.0	−3.85	−8.8	−8.8	−10.0	−10.9

$c_B^{(1)}$/(base mol/kg water)	0.586	0.284	0.140	0.0690	0.0344	0.0171	0.00852
$c_B^{(2)}$/(base mol/kg water)	0.00426	0.00426	0.00426	0.00426	0.00426	0.00426	0.00426
$\Delta_{dil} H^{12}$/(J/base mol polymer)	+9.6	−32.2	−42.3	−38.5	−29.7	−20.9	−10.9

Comments: $c_B^{(1)}$ and $c_B^{(2)}$ denote the concentrations of the polymer in the solution before and after the dilution process.

Polymer (B):   **poly(tetramethylene oxide)**   **1982SH2**
Characterization: M_n/g.mol^{-1} = 650, Quaker Oats Corporation
Solvent (A):   **benzene**   **C$_6$H$_6$**   **71-43-2**

T/K = 321.35

$\Delta_M H$/(J/mol) = $4.184\varphi_B(1 - \varphi_B)[520.05 - 740.40(1 - 2\varphi_B) - 677.95(1 - 2\varphi_B)^2 - 198.98(1 - 2\varphi_B)^3]$

Comments: Experimental data are given only in a figure in the original source.

Polymer (B):   **poly(tetramethylene oxide)**   **1985SHA**
Characterization: M_n/g.mol^{-1} = 650, Quaker Oats Corporation
Solvent (A):   **benzene**   **C$_6$H$_6$**   **71-43-2**

T/K = 313.15

$\Delta_M H$/(J/mol) = $\varphi_B(1 - \varphi_B)[-32.91 + 2111.68(1 - 2\varphi_B) - 224.28(1 - 2\varphi_B)^2 + 753.43(1 - 2\varphi_B)^3]$

Comments: Experimental data are given only in a figure in the original source.

Polymer (B):   **poly(tetramethylene oxide)**   **1982SH2**
Characterization: M_n/g.mol^{-1} = 1000, Quaker Oats Corporation
Solvent (A):   **benzene**   **C$_6$H$_6$**   **71-43-2**

T/K = 321.35

$\Delta_M H$/(J/mol) = $4.184\varphi_B(1 - \varphi_B)[321.98 - 405.66(1 - 2\varphi_B) - 765.29(1 - 2\varphi_B)^2 - 422.57(1 - 2\varphi_B)^3]$

Comments: Experimental data are given only in a figure in the original source.

Polymer (B):   **poly(tetramethylene oxide)**   **1985SHA**
Characterization: M_n/g.mol^{-1} = 1000, Quaker Oats Corporation
Solvent (A):   **benzene**   **C$_6$H$_6$**   **71-43-2**

T/K = 313.15

$\Delta_M H$/(J/mol) = $\varphi_B(1 - \varphi_B)[-766.79 + 2191.75(1 - 2\varphi_B) - 879.53(1 - 2\varphi_B)^2 + 1429.86(1 - 2\varphi_B)^3]$

Comments: Experimental data are given only in a figure in the original source.

Polymer (B):   **poly(tetramethylene oxide)**   **1982SH2**
Characterization: M_n/g.mol^{-1} = 2000, Quaker Oats Corporation
Solvent (A):   **benzene**   **C$_6$H$_6$**   **71-43-2**

T/K = 321.35

$\Delta_M H/(\text{J/mol}) = 4.184\,\varphi_B(1 - \varphi_B)[240.74 - 352.94(1 - 2\varphi_B) - 940.57(1 - 2\varphi_B)^2 - 518.00(1 - 2\varphi_B)^3]$

Comments: Experimental data are given only in a figure in the original source.

Polymer (B):   **poly(tetramethylene oxide)**   **1985SHA**
Characterization: M_n/g.mol^{-1} = 2000, Quaker Oats Corporation
Solvent (A):   **benzene**   **C$_6$H$_6$**   **71-43-2**

T/K = 313.15

$\Delta_M H/(\text{J/mol}) = \varphi_B(1 - \varphi_B)[-1466.95 + 2264.37(1 - 2\varphi_B) - 181.74(1 - 2\varphi_B)^2 - 140.05(1 - 2\varphi_B)^3$
$- 2907.76(1 - 2\varphi_B)^4 + 4551.52(1 - 2\varphi_B)^5]$

Comments: Experimental data are given only in a figure in the original source.

Polymer (B):   **poly(tetramethylene oxide)**   **1982SH1**
Characterization: M_n/g.mol^{-1} = 650, Quaker Oats Corporation
Solvent (A):   **cyclohexane**   **C$_6$H$_{12}$**   **110-82-7**

T/K = 321.35

$\Delta_M H/(\text{J/mol}) = 4.184\,\varphi_B(1 - \varphi_B)[1336.99 + 771.45(1 - 2\varphi_B) + 776.56(1 - 2\varphi_B)^2 + 816.72(1 - 2\varphi_B)^3]$

Comments: Experimental data are given only in a figure in the original source.

Polymer (B):   **poly(tetramethylene oxide)**   **1982SH1**
Characterization: M_n/g.mol^{-1} = 1000, Quaker Oats Corporation
Solvent (A):   **cyclohexane**   **C$_6$H$_{12}$**   **110-82-7**

T/K = 321.35

$\Delta_M H/(\text{J/mol}) = 4.184\,\varphi_B(1 - \varphi_B)[1186.42 + 680.46(1 - 2\varphi_B) + 399.59(1 - 2\varphi_B)^2 + 534.02(1 - 2\varphi_B)^3]$

Comments: Experimental data are given only in a figure in the original source.

Polymer (B):   **poly(tetramethylene oxide)**   **1982SH1**
Characterization: M_n/g.mol^{-1} = 2000, Quaker Oats Corporation
Solvent (A):   **cyclohexane**   **C$_6$H$_{12}$**   **110-82-7**

T/K = 321.35

$\Delta_M H/(\text{J/mol}) = 4.184\,\varphi_B(1 - \varphi_B)[1066.95 + 590.15(1 - 2\varphi_B) + 425.05(1 - 2\varphi_B)^2 + 595.91(1 - 2\varphi_B)^3]$

Comments: Experimental data are given only in a figure in the original source.

Polymer (B):   **poly(tetramethylene oxide)**   **1987SHA**
Characterization: M_n/g.mol^{-1} = 650, Quaker Oats Corporation
Solvent (A):   **1,2-dichloroethane**   **C$_2$H$_4$Cl$_2$**   **107-06-2**

T/K = 313.15

$\Delta_M H$/(J/mol) = $\varphi_B(1 - \varphi_B)[-936.18 + 4313.33(1 - 2\varphi_B) - 1972.99(1 - 2\varphi_B)^2 + 598.89(1 - 2\varphi_B)^3]$

Comments: Experimental data are given only in a figure in the original source.

Polymer (B):   **poly(tetramethylene oxide)**   **1987SHA**
Characterization: M_n/g.mol^{-1} = 1000, Quaker Oats Corporation
Solvent (A):   **1,2-dichloroethane**   **C$_2$H$_4$Cl$_2$**   **107-06-2**

T/K = 313.15

$$\Delta_M H/(\text{J/mol}) = \varphi_B(1 - \varphi_B)[-1658.89 + 4503.77(1 - 2\varphi_B) - 5843.82(1 - 2\varphi_B)^2 + 1105.42(1 - 2\varphi_B)^3$$
$$+13454.3(1 - 2\varphi_B)^4 + 3673.52(1 - 2\varphi_B)^5 + 16966.64(1 - 2\varphi_B)^6]$$

Comments: Experimental data are given only in a figure in the original source.

Polymer (B):   **poly(tetramethylene oxide)**   **1987SHA**
Characterization: M_n/g.mol^{-1} = 2000, Quaker Oats Corporation
Solvent (A):   **1,2-dichloroethane**   **C$_2$H$_4$Cl$_2$**   **107-06-2**

T/K = 313.15

$$\Delta_M H/(\text{J/mol}) = \varphi_B(1 - \varphi_B)[-2497.53 + 5321.52(1 - 2\varphi_B) - 5828.24(1 - 2\varphi_B)^2 + 6153.04(1 - 2\varphi_B)^3$$
$$+ 9771.77(1 - 2\varphi_B)^4 - 10527.55(1 - 2\varphi_B)^5 - 19837.00(1 - 2\varphi_B)^6$$
$$+ 18122.79(1 - 2\varphi_B)^7]$$

Comments: Experimental data are given only in a figure in the original source.

Polymer (B):   **poly(tetramethylene oxide)**   **1986SHA**
Characterization: M_n/g.mol^{-1} = 650, Quaker Oats Corporation
Solvent (A):   **1,2-dimethylbenzene**   **C$_8$H$_{10}$**   **95-47-6**

T/K = 313.15

$\Delta_M H$/(J/mol) = $\varphi_B(1 - \varphi_B)[78.38 + 1825.31(1 - 2\varphi_B) - 351.53(1 - 2\varphi_B)^2 + 2266.99(1 - 2\varphi_B)^3]$

Comments: Experimental data are given only in a figure in the original source.

Polymer (B):   **poly(tetramethylene oxide)**   **1986SHA**
Characterization: M_n/g.mol^{-1} = 1000, Quaker Oats Corporation
Solvent (A):   **1,2-dimethylbenzene**   **C$_8$H$_{10}$**   **95-47-6**

T/K = 313.15

$\Delta_M H$/(J/mol) = $\varphi_B(1 - \varphi_B)[-920.32 + 2062.55(1 - 2\varphi_B) - 1335.53(1 - 2\varphi_B)^2 + 2011.21(1 - 2\varphi_B)^3]$

Comments: Experimental data are given only in a figure in the original source.

Polymer (B):   **poly(tetramethylene oxide)**   **1986SHA**
Characterization: $M_n/\text{g.mol}^{-1} = 2000$, Quaker Oats Corporation
Solvent (A):   **1,2-dimethylbenzene**   **C_8H_{10}**   **95-47-6**

$T/\text{K} = 313.15$

$\Delta_M H/(\text{J/mol}) = \varphi_B(1 - \varphi_B)[-1910.44 + 2813.26(1 - 2\varphi_B) - 1170.31(1 - 2\varphi_B)^2 - 310.76(1 - 2\varphi_B)^3$
$- 1504.97(1 - 2\varphi_B)^4 + 3964.53(1 - 2\varphi_B)^5]$

Comments: Experimental data are given only in a figure in the original source.

Polymer (B):   **poly(tetramethylene oxide)**   **1986SHA**
Characterization: $M_n/\text{g.mol}^{-1} = 650$, Quaker Oats Corporation
Solvent (A):   **1,3-dimethylbenzene**   **C_8H_{10}**   **108-38-3**

$T/\text{K} = 313.15$

$\Delta_M H/(\text{J/mol}) = \varphi_B(1 - \varphi_B)[492.55 + 1789.74(1 - 2\varphi_B) - 461.56(1 - 2\varphi_B)^2 + 1362.11(1 - 2\varphi_B)^3$
$+ 953.91(1 - 2\varphi_B)^4]$

Comments: Experimental data are given only in a figure in the original source.

Polymer (B):   **poly(tetramethylene oxide)**   **1986SHA**
Characterization: $M_n/\text{g.mol}^{-1} = 1000$, Quaker Oats Corporation
Solvent (A):   **1,3-dimethylbenzene**   **C_8H_{10}**   **108-38-3**

$T/\text{K} = 313.15$

$\Delta_M H/(\text{J/mol}) = \varphi_B(1 - \varphi_B)[-842.51 + 2404.99(1 - 2\varphi_B) - 706.92(1 - 2\varphi_B)^2 + 2149.32(1 - 2\varphi_B)^3$
$- 782.12(1 - 2\varphi_B)^4 - 1637.12(1 - 2\varphi_B)^5]$

Comments: Experimental data are given only in a figure in the original source.

Polymer (B):   **poly(tetramethylene oxide)**   **1986SHA**
Characterization: $M_n/\text{g.mol}^{-1} = 2000$, Quaker Oats Corporation
Solvent (A):   **1,3-dimethylbenzene**   **C_8H_{10}**   **108-38-3**

$T/\text{K} = 313.15$

$\Delta_M H/(\text{J/mol}) = \varphi_B(1 - \varphi_B)[-1821.46 + 2632.26(1 - 2\varphi_B) - 998.83(1 - 2\varphi_B)^2 + 428.67(1 - 2\varphi_B)^3$
$- 1884.56(1 - 2\varphi_B)^4 + 3259.27(1 - 2\varphi_B)^5]$

Comments: Experimental data are given only in a figure in the original source.

Polymer (B):   **poly(tetramethylene oxide)**   **1986SHA**
Characterization: $M_n/\text{g.mol}^{-1} = 650$, Quaker Oats Corporation
Solvent (A):   **1,4-dimethylbenzene**   **C_8H_{10}**   **106-42-3**

$T/\text{K} = 313.15$

$\Delta_M H/(\text{J/mol}) = \varphi_B(1 - \varphi_B)[-63.59 + 1972.96(1 - 2\varphi_B) - 621.03(1 - 2\varphi_B)^2 + 1452.08(1 - 2\varphi_B)^3$
$+ 45.00(1 - 2\varphi_B)^4]$

Comments: Experimental data are given only in a figure in the original source.

Polymer (B):   **poly(tetramethylene oxide)**   **1986SHA**
Characterization: $M_n/\text{g.mol}^{-1} = 1000$, Quaker Oats Corporation
Solvent (A):   **1,4-dimethylbenzene**   **C$_8$H$_{10}$**   **106-42-3**

$T/\text{K} = 313.15$

$\Delta_M H/(\text{J/mol}) = \varphi_B(1 - \varphi_B)[-970.44 + 2246.90(1 - 2\varphi_B) - 1537.62(1 - 2\varphi_B)^2 + 2097.52(1 - 2\varphi_B)^3]$

Comments: Experimental data are given only in a figure in the original source.

Polymer (B):   **poly(tetramethylene oxide)**   **1986SHA**
Characterization: $M_n/\text{g.mol}^{-1} = 2000$, Quaker Oats Corporation
Solvent (A):   **1,4-dimethylbenzene**   **C$_8$H$_{10}$**   **106-42-3**

$T/\text{K} = 313.15$

$$\Delta_M H/(\text{J/mol}) = \varphi_B(1 - \varphi_B)[-2000.46 + 2263.17(1 - 2\varphi_B) - 1276.72(1 - 2\varphi_B)^2 + 268.20(1 - 2\varphi_B)^3 - 2186.84(1 - 2\varphi_B)^4 + 4278.11(1 - 2\varphi_B)^5]$$

Comments: Experimental data are given only in a figure in the original source.

Polymer (B):   **poly(tetramethylene oxide)**   **1982SH1**
Characterization: $M_n/\text{g.mol}^{-1} = 650$, Quaker Oats Corporation
Solvent (A):   **1,4-dioxane**   **C$_4$H$_8$O$_2$**   **123-91-1**

$T/\text{K} = 321.35$

$\Delta_M H/(\text{J/mol}) = 4.184\varphi_B(1 - \varphi_B)[1112.98 + 792.83(1 - 2\varphi_B) + 360.17(1 - 2\varphi_B)^2 + 58.481(1 - 2\varphi_B)^3]$

Comments: Experimental data are given only in a figure in the original source.

Polymer (B):   **poly(tetramethylene oxide)**   **1983SHA**
Characterization: $M_n/\text{g.mol}^{-1} = 650$, Quaker Oats Corporation
Solvent (A):   **1,4-dioxane**   **C$_4$H$_8$O$_2$**   **123-91-1**

$T/\text{K} = 321.35$

φ_B	0.0737	0.1778	0.2365	0.3064	0.4259	0.5276	0.5592	0.6322
$\Delta_M H/(\text{J/ mol})$	174.5	432.6	577.4	737.2	1006	1174	1241	1305

φ_B	0.7322	0.7790	0.8732	0.8997	0.9517
$\Delta_M H/(\text{J/ mol})$	1295	1233	920.9	780.3	426.8

Polymer (B):   **poly(tetramethylene oxide)**   **1982SH1**
Characterization: $M_n/\text{g.mol}^{-1} = 1000$, Quaker Oats Corporation
Solvent (A):   **1,4-dioxane**   **C$_4$H$_8$O$_2$**   **123-91-1**

$T/\text{K} = 321.35$

$\Delta_M H/(\text{J/mol}) = 4.184\varphi_B(1 - \varphi_B)[870.42 + 736.89(1 - 2\varphi_B) + 139.89(1 - 2\varphi_B)^2 - 275.91(1 - 2\varphi_B)^3]$

Comments: Experimental data are given only in a figure in the original source.

Polymer (B):   **poly(tetramethylene oxide)**   **1983SHA**
Characterization: M_n/g.mol^{-1} = 1000, Quaker Oats Corporation
Solvent (A):   **1,4-dioxane**   **C$_4$H$_8$O$_2$**   **123-91-1**

T/K = 321.35

φ_B	0.0770	0.1361	0.1475	0.2351	0.2946	0.3993	0.4395	0.4982
$\Delta_M H$/(J/ mol)	143.1	236.4	263.6	395.4	520.1	710.9	780.7	878.6

φ_B	0.5232	0.6469	0.7690	0.8123	0.8993	0.9098	0.9546
$\Delta_M H$/(J/ mol)	914.6	1032	964.4	923.4	631.8	489.1	271.1

Polymer (B):   **poly(tetramethylene oxide)**   **1982SH1**
Characterization: M_n/g.mol^{-1} = 2000, Quaker Oats Corporation
Solvent (A):   **1,4-dioxane**   **C$_4$H$_8$O$_2$**   **123-91-1**

T/K = 321.35

$$\Delta_M H/(\text{J/mol}) = 4.184\varphi_B(1 - \varphi_B)[788.18 + 701.02(1 - 2\varphi_B) +143.97(1 - 2\varphi_B)^2 - 242.35(1 - 2\varphi_B)^3]$$

Comments: Experimental data are given only in a figure in the original source.

Polymer (B):   **poly(tetramethylene oxide)**   **1983SHA**
Characterization: M_n/g.mol^{-1} = 2000, Quaker Oats Corporation
Solvent (A):   **1,4-dioxane**   **C$_4$H$_8$O$_2$**   **123-91-1**

T/K = 321.35

φ_B	0.0791	0.1339	0.2157	0.3441	0.4196	0.4662	0.5315	0.6096
$\Delta_M H$/(J/ mol)	124.3	203.7	325.9	535.6	663.6	743.9	846.0	920.1

φ_B	0.6232	0.7049	0.7556	0.8062	0.9385	0.9605
$\Delta_M H$/(J/ mol)	938.9	958.6	912.5	818.8	395.0	221.8

Polymer (B):   **poly(tetramethylene oxide)**   **1985SHA**
Characterization: M_n/g.mol^{-1} = 650, Quaker Oats Corporation
Solvent (A):   **ethylbenzene**   **C$_8$H$_{10}$**   **100-41-4**

T/K = 313.15

$$\Delta_M H/(\text{J/mol}) = \varphi_B(1 - \varphi_B)[584.50 + 1339.91(1 - 2\varphi_B) + 177.49(1 - 2\varphi_B)^2 + 1670.28(1 - 2\varphi_B)^3 + 704.57(1 - 2\varphi_B)^4]$$

Comments: Experimental data are given only in a figure in the original source.

Polymer (B):   **poly(tetramethylene oxide)**   **1985SHA**
Characterization: M_n/g.mol^{-1} = 1000, Quaker Oats Corporation
Solvent (A):   **ethylbenzene**   **C$_8$H$_{10}$**   **100-41-4**

T/K = 313.15

$$\Delta_M H/(\text{J/mol}) = \varphi_B(1 - \varphi_B)[155.57 + 1700.39(1 - 2\varphi_B) - 651.96(1 - 2\varphi_B)^2 + 2238.75(1 - 2\varphi_B)^3]$$

Comments: Experimental data are given only in a figure in the original source.

Polymer (B):   **poly(tetramethylene oxide)**   **1985SHA**
Characterization: $M_n/\text{g.mol}^{-1} = 2000$, Quaker Oats Corporation
Solvent (A):   **ethylbenzene**   **C_8H_{10}**   **100-41-4**

$T/\text{K} = 313.15$

$$\Delta_M H/(\text{J/mol}) = \varphi_B(1 - \varphi_B)[1676.25 + 2384.39(1 - 2\varphi_B) - 6252.19(1 - 2\varphi_B)^2 - 180.35(1 - 2\varphi_B)^3$$
$$- 1825.90(1 - 2\varphi_B)^4 + 4102.20(1 - 2\varphi_B)^5]$$

Comments: Experimental data are given only in a figure in the original source.

Polymer (B):   **poly(tetramethylene oxide)**   **1985SHA**
Characterization: $M_n/\text{g.mol}^{-1} = 650$, Quaker Oats Corporation
Solvent (A):   **propylbenzene**   **C_9H_{12}**   **103-65-1**

$T/\text{K} = 313.15$

$$\Delta_M H/(\text{J/mol}) = \varphi_B(1 - \varphi_B)[1390.50 + 1036.67(1 - 2\varphi_B) + 785.05(1 - 2\varphi_B)^2 + 558.28(1 - 2\varphi_B)^3]$$

Comments: Experimental data are given only in a figure in the original source.

Polymer (B):   **poly(tetramethylene oxide)**   **1985SHA**
Characterization: $M_n/\text{g.mol}^{-1} = 1000$, Quaker Oats Corporation
Solvent (A):   **propylbenzene**   **C_9H_{12}**   **103-65-1**

$T/\text{K} = 313.15$

$$\Delta_M H/(\text{J/mol}) = \varphi_B(1 - \varphi_B)[-121.99 + 1697.23(1 - 2\varphi_B) + 455.10(1 - 2\varphi_B)^2 + 708.99(1 - 2\varphi_B)^3$$
$$- 1436.52(1 - 2\varphi_B)^4]$$

Comments: Experimental data are given only in a figure in the original source.

Polymer (B):   **poly(tetramethylene oxide)**   **1985SHA**
Characterization: $M_n/\text{g.mol}^{-1} = 2000$, Quaker Oats Corporation
Solvent (A):   **propylbenzene**   **C_9H_{12}**   **103-65-1**

$T/\text{K} = 313.15$

$$\Delta_M H/(\text{J/mol}) = \varphi_B(1 - \varphi_B)[-1096.55 + 1919.89(1 - 2\varphi_B) - 569.14(1 - 2\varphi_B)^2 + 275.48(1 - 2\varphi_B)^3$$
$$- 1233.51(1 - 2\varphi_B)^4 + 2521.84(1 - 2\varphi_B)^5]$$

Comments: Experimental data are given only in a figure in the original source.

Polymer (B):   **poly(tetramethylene oxide)**   **1982SH2**
Characterization: $M_n/\text{g.mol}^{-1} = 650$, Quaker Oats Corporation
Solvent (A):   **tetrachloromethane**   **CCl_4**   **56-23-5**

$T/\text{K} = 321.35$

$$\Delta_M H/(\text{J/mol}) = 4.184\varphi_B(1 - \varphi_B)[-421.73 - 718.31(1 - 2\varphi_B) + 282.73(1 - 2\varphi_B)^2 - 979.94(1 - 2\varphi_B)^3]$$

Comments: Experimental data are given only in a figure in the original source.

Polymer (B):   **poly(tetramethylene oxide)**   **1987SHA**
Characterization: M_n/g.mol^{-1} = 650, Quaker Oats Corporation
Solvent (A):   **tetrachloromethane   CCl$_4$**   **56-23-5**

T/K = 313.15

$$\Delta_M H/(\text{J/mol}) = \varphi_B(1 - \varphi_B)[-1824.75 + 4319.83(1 - 2\varphi_B) - 1399.84(1 - 2\varphi_B)^2 - 1318.33(1 - 2\varphi_B)^3$$
$$- 663.00(1 - 2\varphi_B)^4 + 3023.89(1 - 2\varphi_B)^5]$$

Comments: Experimental data are given only in a figure in the original source.

Polymer (B):   **poly(tetramethylene oxide)**   **1982SH2**
Characterization: M_n/g.mol^{-1} = 1000, Quaker Oats Corporation
Solvent (A):   **tetrachloromethane   CCl$_4$**   **56-23-5**

T/K = 321.35

$$\Delta_M H/(\text{J/mol}) = 4.184\varphi_B(1 - \varphi_B)[-523.38 - 716.23(1 - 2\varphi_B) + 58.69(1 - 2\varphi_B)^2 - 1169.25(1 - 2\varphi_B)^3]$$

Comments: Experimental data are given only in a figure in the original source.

Polymer (B):   **poly(tetramethylene oxide)**   **1987SHA**
Characterization: M_n/g.mol^{-1} = 1000, Quaker Oats Corporation
Solvent (A):   **tetrachloromethane   CCl$_4$**   **56-23-5**

T/K = 313.15

$$\Delta_M H/(\text{J/mol}) = \varphi_B(1 - \varphi_B)[-3434.44 + 5741.87(1 - 2\varphi_B) - 297.45(1 - 2\varphi_B)^2 - 4716.03(1 - 2\varphi_B)^3$$
$$- 6423.88(1 - 2\varphi_B)^4 + 1434.70(1 - 2\varphi_B)^5]$$

Comments: Experimental data are given only in a figure in the original source.

Polymer (B):   **poly(tetramethylene oxide)**   **1982SH2**
Characterization: M_n/g.mol^{-1} = 2000, Quaker Oats Corporation
Solvent (A):   **tetrachloromethane   CCl$_4$**   **56-23-5**

T/K = 321.35

$$\Delta_M H/(\text{J/mol}) = 4.184\varphi_B(1 - \varphi_B)[-660.65 - 702.99(1 - 2\varphi_B) - 313.65(1 - 2\varphi_B)^2 - 1570.76(1 - 2\varphi_B)^3]$$

Comments: Experimental data are given only in a figure in the original source.

Polymer (B):   **poly(tetramethylene oxide)**   **1987SHA**
Characterization: M_n/g.mol^{-1} = 2000, Quaker Oats Corporation
Solvent (A):   **tetrachloromethane   CCl$_4$**   **56-23-5**

T/K = 313.15

$$\Delta_M H/(\text{J/mol}) = \varphi_B(1 - \varphi_B)[-2597.89 + 4322.63(1 - 2\varphi_B) - 1505.18(1 - 2\varphi_B)^2 - 2547.31(1 - 2\varphi_B)^3$$
$$- 1726.83(1 - 2\varphi_B)^4 + 2194.94(1 - 2\varphi_B)^5]$$

Comments: Experimental data are given only in a figure in the original source.

Polymer (B):   **poly(tetramethylene oxide)**   **1985SHA**
Characterization: $M_n/\text{g.mol}^{-1} = 650$, Quaker Oats Corporation
Solvent (A):   **toluene**   **C$_7$H$_8$**   **108-88-3**

$T/\text{K} = 313.15$

$\Delta_M H/(\text{J/mol}) = \varphi_B(1 - \varphi_B)[-449.80 + 2388.37(1 - 2\varphi_B) + 77.93(1 - 2\varphi_B)^2 + 748.20(1 - 2\varphi_B)^3]$

Comments: Experimental data are given only in a figure in the original source.

Polymer (B):   **poly(tetramethylene oxide)**   **1985SHA**
Characterization: $M_n/\text{g.mol}^{-1} = 1000$, Quaker Oats Corporation
Solvent (A):   **toluene**   **C$_7$H$_8$**   **108-88-3**

$T/\text{K} = 313.15$

$\Delta_M H/(\text{J/mol}) = \varphi_B(1 - \varphi_B)[-1497.02 + 2648.19(1 - 2\varphi_B) - 737.09(1 - 2\varphi_B)^2 + 1040.69(1 - 2\varphi_B)^3$
$\qquad\qquad\qquad - 925.33(1 - 2\varphi_B)^4 + 1469.70(1 - 2\varphi_B)^5]$

Comments: Experimental data are given only in a figure in the original source.

Polymer (B):   **poly(tetramethylene oxide)**   **1985SHA**
Characterization: $M_n/\text{g.mol}^{-1} = 2000$, Quaker Oats Corporation
Solvent (A):   **toluene**   **C$_7$H$_8$**   **108-88-3**

$T/\text{K} = 313.15$

$\Delta_M H/(\text{J/mol}) = \varphi_B(1 - \varphi_B)[-2316.62 + 3163.38(1 - 2\varphi_B) - 657.66(1 - 2\varphi_B)^2 - 919.49(1 - 2\varphi_B)^3$
$\qquad\qquad\qquad - 3678.74(1 - 2\varphi_B)^4 + 6231.33(1 - 2\varphi_B)^5]$

Comments: Experimental data are given only in a figure in the original source.

Polymer (B):   **poly(tetramethylene oxide)**   **1987SHA**
Characterization: $M_n/\text{g.mol}^{-1} = 650$, Quaker Oats Corporation
Solvent (A):   **1,1,1-trichloroethane**   **C$_2$H$_3$Cl$_3$**   **71-55-6**

$T/\text{K} = 313.15$

$\Delta_M H/(\text{J/mol}) = \varphi_B(1 - \varphi_B)[-1823.95 - 4126.63(1 - 2\varphi_B) - 1690.97(1 - 2\varphi_B)^2 - 3503.60(1 - 2\varphi_B)^3$
$\qquad\qquad\qquad - 2368.90(1 - 2\varphi_B)^4 + 3238.79(1 - 2\varphi_B)^5 + 6651.38(1 - 2\varphi_B)^6]$

Comments: Experimental data are given only in a figure in the original source.

Polymer (B):   **poly(tetramethylene oxide)**   **1987SHA**
Characterization: $M_n/\text{g.mol}^{-1} = 1000$, Quaker Oats Corporation
Solvent (A):   **1,1,1-trichloroethane**   **C$_2$H$_3$Cl$_3$**   **71-55-6**

$T/\text{K} = 313.15$

$\Delta_M H/(\text{J/mol}) = \varphi_B(1 - \varphi_B)[-2626.90 - 5191.45(1 - 2\varphi_B) - 2981.67(1 - 2\varphi_B)^2 + 524.25(1 - 2\varphi_B)^3$
$\qquad\qquad\qquad - 705.79(1 - 2\varphi_B)^4 + 3576.62(1 - 2\varphi_B)^5]$

Comments: Experimental data are given only in a figure in the original source.

Polymer (B):   **poly(tetramethylene oxide)**   **1987SHA**
Characterization: M_n/g.mol^{-1} = 2000, Quaker Oats Corporation
Solvent (A):   **1,1,1-trichloroethane   C$_2$H$_3$Cl$_3$**   **71-55-6**

T/K = 313.15

$\Delta_M H$/(J/mol) = $\varphi_B(1 - \varphi_B)[-3667.38 + 4733.26(1 - 2\varphi_B) + 5293.65(1 - 2\varphi_B)^2 + 5825.13(1 - 2\varphi_B)^3$
$- 5743.19(1 - 2\varphi_B)^4 - 8210.13(1 - 2\varphi_B)^5 + 11223.39(1 - 2\varphi_B)^6$
$- 133310.13(1 - 2\varphi_B)^7]$

Comments: Experimental data are given only in a figure in the original source.

Polymer (B):   **poly(tetramethylene oxide)**   **1986SHA**
Characterization: M_n/g.mol^{-1} = 650, Quaker Oats Corporation
Solvent (A):   **1,3,5-trimethylbenzene C$_9$H$_{12}$**   **108-67-8**

T/K = 313.15

$\Delta_M H$/(J/mol) = $\varphi_B(1 - \varphi_B)[+1534.00 + 1300.96(1 - 2\varphi_B) + 989.91(1 - 2\varphi_B)^2 + 550.09(1 - 2\varphi_B)^3]$

Comments: Experimental data are given only in a figure in the original source.

Polymer (B):   **poly(tetramethylene oxide)**   **1986SHA**
Characterization: M_n/g.mol^{-1} = 1000, Quaker Oats Corporation
Solvent (A):   **1,3,5-trimethylbenzene C$_9$H$_{12}$**   **108-67-8**

T/K = 313.15

$\Delta_M H$/(J/mol) = $\varphi_B(1 - \varphi_B)[+523.17 + 1469.71(1 - 2\varphi_B) - 276.28(1 - 2\varphi_B)^2 + 1000.33(1 - 2\varphi_B)^3]$

Comments: Experimental data are given only in a figure in the original source.

Polymer (B):   **poly(tetramethylene oxide)**   **1986SHA**
Characterization: M_n/g.mol^{-1} = 2000, Quaker Oats Corporation
Solvent (A):   **1,3,5-trimethylbenzene C$_9$H$_{12}$**   **108-67-8**

T/K = 313.15

$\Delta_M H$/(J/mol) = $\varphi_B(1 - \varphi_B)[-691.16 + 1633.04(1 - 2\varphi_B) - 126.51(1 - 2\varphi_B)^2 + 1415.81(1 - 2\varphi_B)^3$
$- 1111.68(1 - 2\varphi_B)^4]$

Comments: Experimental data are given only in a figure in the original source.

Polymer (B):   **poly(vinyl acetate)**   **1968KA2**
Characterization: M_n/g.mol^{-1} = 41300, DP = 480
Solvent (A):   **benzene**   **C$_6$H$_6$**   **71-43-2**

T/K = 298.15

continued

continued

$V^{(1)}$/cm^3	10.0	20.0	30.0	20.0	30.0	20.0	30.0	40.0	50.0
$\varphi_B^{(1)}$	0.265	0.128	0.088	0.212	0.141	0.197	0.131	0.098	0.078
$V^{(2)}$/cm^3	20.0	30.0	40.0	30.0	50.0	30.0	40.0	50.0	60.0
$\varphi_B^{(2)}$	0.128	0.088	0.066	0.141	0.085	0.131	0.098	0.078	0.065
$\Delta_{dil} H^{12}$/J	0.682	0.272	0.142	0.590	0.510	0.577	0.301	0.188	0.105

Comments: $\Delta_{dil} H^{12}$ is the extensive quantity obtained for a given total volume change from $V^{(1)}$ to $V^{(2)}$, where $\varphi_B^{(1)}$ and $\varphi_B^{(2)}$ denote the volume fractions of the polymer in the solution before and after the dilution process.

Polymer (B):	**poly(vinyl acetate)**								**1968KA2**
Characterization:	M_n/g.mol^{-1} = 159000, DP = 1850								
Solvent (A):	**benzene**			**C$_6$H$_6$**					**71-43-2**

T/K = 298.15

$V^{(1)}$/cm^3	10.0	20.0	30.0	30.0	10.0	10.0	20.0	10.0	10.0
$\varphi_B^{(1)}$	0.220	0.110	0.095	0.080	0.196	0.156	0.078	0.220	0.156
$V^{(2)}$/cm^3	20.0	30.0	40.0	50.0	30.0	20.0	30.0	30.0	30.0
$\varphi_B^{(2)}$	0.110	0.077	0.074	0.050	0.065	0.078	0.052	0.077	0.052
$\Delta_{dil} H^{12}$/J	0.364	0.167	0.134	0.142	0.523	0.230	0.071	0.531	0.301

$V^{(1)}$/cm^3	30.0
$\varphi_B^{(1)}$	0.105
$V^{(2)}$/cm^3	50.0
$\varphi_B^{(2)}$	0.063
$\Delta_{dil} H^{12}$/J	0.234

Comments: $\Delta_{dil} H^{12}$ is the extensive quantity obtained for a given total volume change from $V^{(1)}$ to $V^{(2)}$, where $\varphi_B^{(1)}$ and $\varphi_B^{(2)}$ denote the volume fractions of the polymer in the solution before and after the dilution process.

Polymer (B):	**poly(vinyl acetate)**								**1968KA2**
Characterization:	M_n/g.mol^{-1} = 210700, DP = 2450								
Solvent (A):	**benzene**			**C$_6$H$_6$**					**71-43-2**

T/K = 298.15

$V^{(1)}$/cm^3	10.0	20.0	10.0	30.0	10.0	40.0	20.0	30.0	20.0
$\varphi_B^{(1)}$	0.196	0.098	0.092	0.098	0.151	0.062	0.085	0.123	0.176
$V^{(2)}$/cm^3	20.0	30.0	20.0	40.0	20.0	50.0	30.0	50.0	30.0
$\varphi_B^{(2)}$	0.098	0.065	0.046	0.073	0.076	0.050	0.056	0.074	0.117
$\Delta_{dil} H^{12}$/J	0.222	0.084	0.046	0.096	0.151	0.033	0.059	0.247	0.218

$V^{(1)}$/cm^3	30.0
$\varphi_B^{(1)}$	0.117
$V^{(2)}$/cm^3	40.0
$\varphi_B^{(2)}$	0.088
$\Delta_{dil} H^{12}$/J	0.125

Comments: $\Delta_{dil} H^{12}$ is the extensive quantity obtained for a given total volume change from $V^{(1)}$ to $V^{(2)}$, where $\varphi_B^{(1)}$ and $\varphi_B^{(2)}$ denote the volume fractions of the polymer in the solution before and after the dilution process.

Polymer (B):   **poly(vinyl acetate)**   **1968KA2**
Characterization: $M_n/\text{g.mol}^{-1} = 41300$, DP = 480
Solvent (A):   **2-butanone**   **C$_4$H$_8$O**   **78-93-3**

$T/\text{K} = 298.15$

$V^{(1)}/\text{cm}^3$	10.0	20.0	30.0	10.0	20.0	30.0	10.0	20.0	30.0
$\varphi_B^{(1)}$	0.256	0.128	0.085	0.205	0.104	0.086	0.167	0.237	0.158
$V^{(2)}/\text{cm}^3$	20.0	30.0	40.0	20.0	30.0	50.0	30.0	30.0	40.0
$\varphi_B^{(2)}$	0.128	0.085	0.064	0.104	0.068	0.051	0.055	0.158	0.059
$\Delta_{dil}H^{12}/\text{J}$	−0.874	−0.251	−0.134	−0.510	−0.184	−0.188	−0.452	−1.021	−0.222

$V^{(1)}/\text{cm}^3$	40.0	10.0	10.0	10.0	10.0
$\varphi_B^{(1)}$	0.156	0.256	0.256	0.205	0.205
$V^{(2)}/\text{cm}^3$	70.0	30.0	40.0	30.0	50.0
$\varphi_B^{(2)}$	0.089	0.085	0.064	0.068	0.051
$\Delta_{dil}H^{12}/\text{J}$	−1.000	−1.125	−1.259	−0.695	−0.883

Comments: $\Delta_{dil}H^{12}$ is the extensive quantity obtained for a given total volume change from $V^{(1)}$ to $V^{(2)}$, where $\varphi_B^{(1)}$ and $\varphi_B^{(2)}$ denote the volume fractions of the polymer in the solution before and after the dilution process.

Polymer (B):   **poly(vinyl acetate)**   **1968KA2**
Characterization: $M_n/\text{g.mol}^{-1} = 159000$, DP = 1850
Solvent (A):   **2-butanone**   **C$_4$H$_8$O**   **78-93-3**

$T/\text{K} = 298.15$

$V^{(1)}/\text{cm}^3$	10.0	20.0	10.0	20.0	20.0	10.0	20.0	10.0	20.0
$\varphi_B^{(1)}$	0.210	0.105	0.198	0.099	0.125	0.157	0.087	0.231	0.116
$V^{(2)}/\text{cm}^3$	20.0	30.0	20.0	30.0	40.0	20.0	30.0	20.0	30.0
$\varphi_B^{(2)}$	0.105	0.070	0.099	0.066	0.063	0.079	0.058	0.116	0.077
$\Delta_{dil}H^{12}/\text{J}$	−0.481	−0.151	−0.368	−0.117	−0.322	−0.222	−0.088	−0.619	−0.184

$V^{(1)}/\text{cm}^3$	30.0	10.0	10.0	10.0
$\varphi_B^{(1)}$	0.085	0.210	0.198	0.231
$V^{(2)}/\text{cm}^3$	40.0	30.0	30.0	30.0
$\varphi_B^{(2)}$	0.064	0.070	0.066	0.077
$\Delta_{dil}H^{12}/\text{J}$	−0.105	−0.632	−0.485	−0.803

Comments: $\Delta_{dil}H^{12}$ is the extensive quantity obtained for a given total volume change from $V^{(1)}$ to $V^{(2)}$, where $\varphi_B^{(1)}$ and $\varphi_B^{(2)}$ denote the volume fractions of the polymer in the solution before and after the dilution process.

Polymer (B):   **poly(vinyl acetate)**   **1968KA2**
Characterization: $M_n/\text{g.mol}^{-1} = 210700$, DP = 2450
Solvent (A):   **2-butanone**   **C$_4$H$_8$O**   **78-93-3**

$T/\text{K} = 298.15$

continued

continued

$V^{(1)}/cm^3$	10.0	10.0	10.0	20.0	20.0	10.0	30.0	40.0	10.0
$\varphi_B^{(1)}$	0.156	0.115	0.186	0.093	0.075	0.135	0.084	0.069	0.109
$V^{(2)}/cm^3$	20.0	20.0	20.0	30.0	30.0	20.0	40.0	50.0	20.0
$\varphi_B^{(2)}$	0.078	0.058	0.093	0.062	0.050	0.068	0.063	0.055	0.055
$\Delta_{dil}H^{12}/J$	−0.192	−0.100	−0.276	−0.096	−0.059	−0.126	−0.084	−0.050	−0.096

$V^{(1)}/cm^3$	50.0
$\varphi_B^{(1)}$	0.082
$V^{(2)}/cm^3$	60.0
$\varphi_B^{(2)}$	0.068
$\Delta_{dil}H^{12}/J$	−0.084

Comments: $\Delta_{dil}H^{12}$ is the extensive quantity obtained for a given total volume change from $V^{(1)}$ to $V^{(2)}$, where $\varphi_B^{(1)}$ and $\varphi_B^{(2)}$ denote the volume fractions of the polymer in the solution before and after the dilution process.

Polymer (B): **poly(vinyl acetate)** 1951MEA

Characterization: low molecular weight, $\rho = 1.159$ g/ml (293.15 K), Gelva 09, Canadian Resins and Chemicals Ltd.

Solvent (A): **dibutyl adipate** $C_{14}H_{26}O_4$ 105-99-7

$T/K = 293.15$

$V^{(1)}/cm^3$	1.67	1.67	1.67	1.67
$\varphi_B^{(1)}$	0.5988	0.5988	0.5988	0.5988
$V^{(2)}/cm^3$	3.00	4.00	5.00	6.00
$\varphi_B^{(2)}$	0.3333	0.2500	0.2000	0.1667
$\Delta_{dil}H^{12}/(J/cm^3$ polymer)	1.17	1.63	1.90	2.07

Comments: $\varphi_B^{(1)}$ and $\varphi_B^{(2)}$ denote the volume fractions of the polymer in the solution before and after the dilution process.

Polymer (B): **poly(vinyl acetate)** 1951MEA

Characterization: low molecular weight, $\rho = 1.159$ g/ml (293.15 K), Gelva 09, Canadian Resins and Chemicals Ltd.

Solvent (A): **dibutyl malonate** $C_{11}H_{20}O_4$ 1190-39-2

$T/K = 293.15$

$V^{(1)}/cm^3$	1.67	1.67	1.67	1.67
$\varphi_B^{(1)}$	0.5988	0.5988	0.5988	0.5988
$V^{(2)}/cm^3$	3.00	4.00	5.00	6.00
$\varphi_B^{(2)}$	0.3333	0.2500	0.2000	0.1667
$\Delta_{dil}H^{12}/(J/cm^3$ polymer)	0.44	0.59	0.65	0.69

Comments: $\varphi_B^{(1)}$ and $\varphi_B^{(2)}$ denote the volume fractions of the polymer in the solution before and after the dilution process.

Polymer (B):	**poly(vinyl acetate)**					**1951MEA**
Characterization:	low molecular weight, $\rho = 1.159$ g/ml (293.15 K),					
	Gelva 09, Canadian Resins and Chemicals Ltd.					
Solvent (A):	**dibutyl succinate**	**C$_{12}$H$_{22}$O$_4$**				**141-03-7**

$T/K = 293.15$

$V^{(1)}/\text{cm}^3$	1.67	1.67	1.67	1.67	1.67
$\varphi_B^{(1)}$	0.5988	0.5988	0.5988	0.5988	0.5988
$V^{(2)}/\text{cm}^3$	2.00	2.50	3.00	3.50	4.00
$\varphi_B^{(2)}$	0.5000	0.4000	0.3333	0.2857	0.2500
$\Delta_{dil}H^{12}/(\text{J/cm}^3 \text{ polymer})$	0.38	0.67	0.92	1.07	1.21

Comments: $\varphi_B^{(1)}$ and $\varphi_B^{(2)}$ denote the volume fractions of the polymer in the solution before and after the dilution process.

Polymer (B):	**poly(vinyl acetate)**							**1954DAO**
Characterization:	$M_n/\text{g.mol}^{-1} = 85000$, fractionated in the laboratory							
Solvent (A):	**1,2-dichloroethane**	**C$_2$H$_4$Cl$_2$**						**107-06-2**

$T/K = 298.35$

φ_B	0.001	0.002	0.003	0.004	0.006	0.009	0.011	0.013
$\Delta_M H/(\text{J/cm}^3)$	−0.0130	−0.0268	−0.0435	−0.0669	−0.0967	−0.121	−0.142	−0.165

φ_B	0.016
$\Delta_M H/(\text{J/cm}^3)$	−0.156

Polymer (B):	**poly(vinyl acetate)**					**1951MEA**
Characterization:	low molecular weight, $\rho = 1.159$ g/ml (293.15 K),					
	Gelva 09, Canadian Resins and Chemicals Ltd.					
Solvent (A):	**diethyl sebacate**	**C$_{14}$H$_{26}$O$_4$**				**110-40-7**

$T/K = 293.15$

$V^{(1)}/\text{cm}^3$	1.67	1.67	1.67	1.67	1.67
$\varphi_B^{(1)}$	0.5988	0.5988	0.5988	0.5988	0.5988
$V^{(2)}/\text{cm}^3$	2.00	2.50	3.00	3.50	4.00
$\varphi_B^{(2)}$	0.5000	0.4000	0.3333	0.2857	0.2500
$\Delta_{dil}H^{12}/(\text{J/cm}^3 \text{ polymer})$	0.63	1.21	1.63	1.92	2.18

Comments: $\varphi_B^{(1)}$ and $\varphi_B^{(2)}$ denote the volume fractions of the polymer in the solution before and after the dilution process.

Polymer (B):	**poly(vinyl acetate)**					**1951MEA**
Characterization:	low molecular weight, $\rho = 1.159$ g/ml (293.15 K),					
	Gelva 09, Canadian Resins and Chemicals Ltd.					
Solvent (A):	**diethyl oxalate**	**C$_6$H$_{10}$O$_4$**				**95-92-1**

continued

continued

$T/K = 293.15$

$V^{(1)}/cm^3$	1.67	1.67	1.67	1.67
$\varphi_B^{(1)}$	0.5988	0.5988	0.5988	0.5988
$V^{(2)}/cm^3$	3.00	4.00	5.00	6.00
$\varphi_B^{(2)}$	0.3333	0.2500	0.2000	0.1667
$\Delta_{dil}H^{12}/(J/cm^3$ polymer)	−0.80	−1.03	−1.14	−1.19

Comments: $\varphi_B^{(1)}$ and $\varphi_B^{(2)}$ denote the volume fractions of the polymer in the solution before and after the dilution process.

Polymer (B):   **poly(vinyl acetate)**   **1955PAR**
Characterization: $M_\eta/g.mol^{-1} = 29000$, fractionated in the laboratory from a
                         commercial sample of Gelva 15, Canadian Resins and Chemicals Ltd.
Solvent (A):   **methanol**   **CH$_4$O**   **67-56-1**

$T/K = 298.15$

$\varphi_A\varphi_B$	0.00793	0.01198	0.01567	0.01695	0.01809	0.01928	0.01982
$\Delta_M H/(J/cm^3)$	0.237	0.359	0.473	0.515	0.552	0.540	0.573

Comments: Only the product of volume fractions is given in the original source.

Polymer (B):   **poly(vinyl acetate)**   **1954DAO**
Characterization: $M_\eta/g.mol^{-1} = 85000$, fractionated in the laboratory
Solvent (A):   **methanol**   **CH$_4$O**   **67-56-1**

$T/K = 298.35$

φ_B	0.002	0.004	0.009	0.012	0.019	0.021	0.030
$\Delta_M H/(J/cm^3)$	0.071	0.138	0.318	0.418	0.611	0.678	0.920

Polymer (B):   **poly(vinyl acetate)**   **1955PAR**
Characterization: $M_\eta/g.mol^{-1} = 255000$, fractionated in the laboratory from a
                         commercial sample of Gelva 15, Canadian Resins and Chemicals Ltd.
Solvent (A):   **methanol**   **CH$_4$O**   **67-56-1**

$T/K = 298.15$

$\varphi_A\varphi_B$	0.00402	0.00706	0.00799	0.00992	0.01103	0.01198	0.01291	0.0142
$\Delta_M H/(J/cm^3)$	0.124	0.207	0.227	0.296	0.320	0.325	0.359	0.380

$\varphi_A\varphi_B$	0.01570	0.01958
$\Delta_M H/(J/cm^3)$	0.423	0.544

Comments: Only the product of volume fractions is given in the original source.

Polymer (B):   poly(vinyl acetate)   **1975TAG**
Characterization: $M_\eta/\text{g.mol}^{-1} = 320000$
Solvent (A):   **methanol**   **CH₄O**   **67-56-1**

$T/\text{K} = 298.15$

w_B	0.03	0.05	0.08	0.10	0.12
$\Delta_M H/(\text{J/g})$	1.3	2.3	3.6	4.5	5.4

$T/\text{K} = 318.15$

w_B	0.03	0.05	0.08	0.10	0.12
$\Delta_M H/(\text{J/g})$	0.9	1.5	2.5	3.1	3.7

$T/\text{K} = 328.15$

w_B	0.03	0.05	0.08	0.10	0.12
$\Delta_M H/(\text{J/g})$	0.5	0.9	1.4	1.8	2.2

Comments: Data were derived from measurements of intermediary enthalpy of dilution. Complete $\Delta_M H$ curves are given in a figure in the original source.

Polymer (B):   **poly(vinyl acetate)**   **1955PAR**
Characterization: $M_\eta/\text{g.mol}^{-1} = 536000$, fractionated in the laboratory from a commercial sample of Gelva 15, Canadian Resins and Chemicals Ltd.
Solvent (A):   **methanol**   **CH₄O**   **67-56-1**

$T/\text{K} = 298.15$

$\varphi_A \varphi_B$	0.00438	0.00501	0.00644	0.00756	0.00882	0.01198	0.01593	0.01932
$\Delta_M H/(\text{J/cm}^3)$	0.127	0.140	0.195	0.210	0.236	0.310	0.439	0.523

$\varphi_A \varphi_B$	0.0235
$\Delta_M H/(\text{J/cm}^3)$	0.653

Comments: Only the product of volume fractions is given in the original source.

Polymer (B):   **poly(vinyl acetate)**   **1954DAO**
Characterization: $M_\eta/\text{g.mol}^{-1} = 85000$, fractionated in the laboratory
Solvent (A):   **1,1,2,2-tetrachloroethane**   **C₂H₂Cl₄**   **79-34-5**

$T/\text{K} = 298.35$

φ_B	0.0009	0.0013	0.0016	0.0048	0.0065	0.0081
$\Delta_M H/(\text{J/cm}^3)$	−0.0582	−0.0895	−0.161	−0.381	−0.527	−0.611

Polymer (B):   **poly(vinyl acetate)**   **1955PAR**
Characterization: $M_\eta/\text{g.mol}^{-1} = 29000$, fractionated in the laboratory from a commercial sample of Gelva 15, Canadian Resins and Chemicals Ltd.
Solvent (A):   **tetrachloromethane**   **CCl₄**   **56-23-5**

continued

continued

$T/K = 298.15$

$\varphi_A \varphi_B$	0.00249	0.00350	0.00453	0.00504	0.00598	0.00696	0.00996	0.01087
$\Delta_M H/(\text{J/cm}^3)$	−0.220	−0.313	−0.413	−0.452	−0.548	−0.628	−0.895	−0.962

$\varphi_A \varphi_B$	0.01214
$\Delta_M H/(\text{J/cm}^3)$	−1.084

Comments: Only the product of volume fractions is given in the original source.

Polymer (B):   **poly(vinyl acetate)**   **1955PAR**
Characterization: $M_\eta/\text{g.mol}^{-1} = 191000$, fractionated in the laboratory from a
                             commercial sample of Gelva 15, Canadian Resins and Chemicals Ltd.
Solvent (A):   **tetrachloromethane   CCl$_4$**   **56-23-5**

$T/K = 298.15$

$\varphi_A \varphi_B$	0.00102	0.00202	0.00252	0.00300	0.00349	0.00379	0.00401	0.00507
$\Delta_M H/(\text{J/cm}^3)$	−0.0887	−0.169	−0.220	−0.263	−0.299	−0.319	−0.337	−0.431

$\varphi_A \varphi_B$	0.00601
$\Delta_M H/(\text{J/cm}^3)$	−0.510

Comments: Only the product of volume fractions is given in the original source.

Polymer (B):   **poly(vinyl acetate)**   **1955PAR**
Characterization: $M_\eta/\text{g.mol}^{-1} = 255000$, fractionated in the laboratory from a
                             commercial sample of Gelva 15, Canadian Resins and Chemicals Ltd.
Solvent (A):   **tetrachloromethane   CCl$_4$**   **56-23-5**

$T/K = 298.15$

$\varphi_A \varphi_B$	0.00126	0.00201	0.00244	0.00299	0.00363	0.00395	0.00495	0.00504
$\Delta_M H/(\text{J/cm}^3)$	−0.111	−0.175	−0.203	−0.252	−0.302	−0.333	−0.431	−0.427

$\varphi_A \varphi_B$	0.00643
$\Delta_M H/(\text{J/cm}^3)$	−0.548

Comments: Only the product of volume fractions is given in the original source.

Polymer (B):   **poly(vinyl acetate)**   **1955PAR**
Characterization: $M_\eta/\text{g.mol}^{-1} = 536000$, fractionated in the laboratory from a
                             commercial sample of Gelva 15, Canadian Resins and Chemicals Ltd.
Solvent (A):   **tetrachloromethane   CCl$_4$**   **56-23-5**

$T/K = 298.15$

continued

continued

$\varphi_A\varphi_B$	0.00104	0.00153	0.00175	0.00203	0.00213	0.00252	0.00352	0.00469
$\Delta_M H/(\text{J/cm}^3)$	−0.095	−0.126	−0.154	−0.174	−0.182	−0.192	−0.279	−0.382

$\varphi_A\varphi_B$	0.0057	0.00649	0.00684
$\Delta_M H/(\text{J/cm}^3)$	−0.464	−0.536	−0.577

Comments: Only the product of volume fractions is given in the original source.

Polymer (B):	**poly(vinyl acetate)**		**1957SCH**
Characterization:	$M_\eta/\text{g.mol}^{-1} = 390000$, fractionated in the laboratory		
Solvent (A):	**toluene**	**C$_7$H$_8$**	**108-88-3**

$T/\text{K} = 298.15$

$\varphi_A^{(1)}\varphi_B^{(1)}$	0.01576	0.01576	0.01576	0.01576	0.01576	0.01576	0.01576	0.01576
$\varphi_A^{(2)}\varphi_B^{(2)}$	0.008617	0.008484	0.011839	0.011868	0.005256	0.005107	0.003987	0.003935
$\Delta_{dil}H^{12}/(\text{J/cm}^3)$	47.82	50.38	25.10	28.12	72.13	69.71	82.55	81.63

$\varphi_A^{(1)}\varphi_B^{(1)}$	0.008219	0.008219	0.008219	0.008219
$\varphi_A^{(2)}\varphi_B^{(2)}$	0.002048	0.002008	0.006179	0.006226
$\Delta_{dil}H^{12}/(\text{J/cm}^3)$	44.22	40.88	11.76	14.48

Comments: Only the product of volume fractions is given in the original source. $\varphi_B^{(1)}$ and $\varphi_B^{(2)}$ denote the volume fractions of the polymer in the solution before and after the dilution process.

Polymer (B):	**poly(vinyl acetate)**		**1957SCH**
Characterization:	$M_\eta/\text{g.mol}^{-1} = 880000$, fractionated in the laboratory		
Solvent (A):	**toluene**	**C$_7$H$_8$**	**108-88-3**

$T/\text{K} = 298.15$

$\varphi_A^{(1)}\varphi_B^{(1)}$	0.015818	0.015818	0.015818	0.015818	0.015818	0.015818	0.015818	0.015818
$\varphi_A^{(2)}\varphi_B^{(2)}$	0.008682	0.008560	0.012021	0.011992	0.005234	0.005060	0.003983	0.003922
$\Delta_{dil}H^{12}/(\text{J/cm}^3)$	42.59	44.89	19.83	18.49	65.10	68.07	76.90	72.76

$\varphi_A^{(1)}\varphi_B^{(1)}$	0.008161	0.008161	0.008161	0.008161	0.013563	0.013563	0.013563	0.013563
$\varphi_A^{(2)}\varphi_B^{(2)}$	0.002032	0.002007	0.006149	0.006208	0.007441	0.007296	0.009130	0.009324
$\Delta_{dil}H^{12}/(\text{J/cm}^3)$	40.96	36.57	10.13	11.75	38.74	40.17	21.84	25.27

$T/\text{K} = 308.15$

$\varphi_A^{(1)}\varphi_B^{(1)}$	0.015818	0.015818	0.015818	0.015818	0.015818	0.015818
$\varphi_A^{(2)}\varphi_B^{(2)}$	0.008628	0.008007	0.011976	0.011917	0.005284	0.005053
$\Delta_{dil}H^{12}/(\text{J/cm}^3)$	38.03	38.20	18.54	18.33	50.63	55.77

Comments: Only the product of volume fractions is given in the original source. $\varphi_B^{(1)}$ and $\varphi_B^{(2)}$ denote the volume fractions of the polymer in the solution before and after the dilution process.

Polymer (B): $\qquad$ **poly(vinyl acetate-*co*-vinyl alcohol)** $\qquad$ **1956AM2**

Characterization: $\qquad$ M_n/g.mol^{-1} = 53000, 0.4 mol% vinyl acetate, made from
$\qquad$ poly(vinyl alcohol) of DP = 1200, Mitsubishi Rayon Co. Ltd., Japan

Solvent (A): $\qquad$ **water** $\qquad$ **H$_2$O** $\qquad$ **7732-18-5**

T/K = 298.15

$V^{(1)}$/cm^3	9.00	9.51
$\varphi_B^{(1)}$	0.099	0.109
$V^{(2)}$/cm^3	19.00	19.51
$\varphi_B^{(2)}$	0.047	0.053
$\Delta_{dil} H^{12}$/J	-0.607	-0.741

Polymer (B): $\qquad$ **poly(vinyl acetate-*co*-vinyl alcohol)** $\qquad$ **1956AM1**

Characterization: $\qquad$ M_n/g.mol^{-1} = 19800, 1.7 mol% vinyl acetate

Solvent (A): $\qquad$ **water** $\qquad$ **H$_2$O** $\qquad$ **7732-18-5**

T/K = 303.15

$V^{(1)}$/cm^3	15.0	20.0	25.0
$\varphi_B^{(1)}$	0.155	0.115	0.092
$V^{(2)}$/cm^3	20.0	25.0	30.0
$\varphi_B^{(2)}$	0.115	0.092	0.077
$\Delta_{dil} H^{12}$/J	-1.305	-1.054	-0.720

Comments: $\quad$ $\Delta_{dil} H^{12}$ is the extensive quantity obtained for a given total volume change from $V^{(1)}$ to $V^{(2)}$, where $\varphi_B^{(1)}$ and $\varphi_B^{(2)}$ denote the volume fractions of the polymer in the solution before and after the dilution process.

Polymer (B): $\qquad$ **poly(vinyl acetate-*co*-vinyl alcohol)** $\qquad$ **1956AM2**

Characterization: $\qquad$ M_n/g.mol^{-1} = 53000, 7.8 mol% vinyl acetate, made from
$\qquad$ poly(vinyl alcohol) of DP = 1200, Mitsubishi Rayon Co. Ltd., Japan

Solvent (A): $\qquad$ **water** $\qquad$ **H$_2$O** $\qquad$ **7732-18-5**

T/K = 298.15

$V^{(1)}$/cm^3	9.57	8.57
$\varphi_B^{(1)}$	0.077	0.077
$V^{(2)}$/cm^3	19.57	18.57
$\varphi_B^{(2)}$	0.038	0.0355
$\Delta_{dil} H^{12}$/J	-0.699	-0.632

Comments: $\quad$ $\Delta_{dil} H^{12}$ is the extensive quantity obtained for a given total volume change from $V^{(1)}$ to $V^{(2)}$, where $\varphi_B^{(1)}$ and $\varphi_B^{(2)}$ denote the volume fractions of the polymer in the solution before and after the dilution process.

Polymer (B):   **poly(vinyl acetate-*co*-vinyl alcohol)**   **1956AM2**
Characterization: M_n/g.mol^{-1} = 53000, 15.6 mol% vinyl acetate, made from
                         poly(vinyl alcohol) of DP = 1200, Mitsubishi Rayon Co. Ltd., Japan
Solvent (A):   **water**   **H$_2$O**   **7732-18-5**

T/K = 298.15

$V^{(1)}$/cm^3	9.91	9.94
$\varphi_B^{(1)}$	0.052	0.052
$V^{(2)}$/cm^3	19.91	19.94
$\varphi_B^{(2)}$	0.026	0.026
$\Delta_{dil}H^{12}$/J	−0.167	−0.172

Comments: $\Delta_{dil}H^{12}$ is the extensive quantity obtained for a given total volume change from $V^{(1)}$ to $V^{(2)}$, where $\varphi_B^{(1)}$ and $\varphi_B^{(2)}$ denote the volume fractions of the polymer in the solution before and after the dilution process.

Polymer (B):   **poly(vinyl acetate-*co*-vinyl alcohol)**   **1956AM2**
Characterization: M_n/g.mol^{-1} = 53000, 19.4 mol% vinyl acetate, made from
                         poly(vinyl alcohol) of DP = 1200, Mitsubishi Rayon Co. Ltd., Japan
Solvent (A):   **water**   **H$_2$O**   **7732-18-5**

T/K = 298.15

$V^{(1)}$/cm^3	9.86	9.94
$\varphi_B^{(1)}$	0.038	0.038
$V^{(2)}$/cm^3	19.86	19.94
$\varphi_B^{(2)}$	0.019	0.019
$\Delta_{dil}H^{12}$/J	−0.071	−0.063

Comments: $\Delta_{dil}H^{12}$ is the extensive quantity obtained for a given total volume change from $V^{(1)}$ to $V^{(2)}$, where $\varphi_B^{(1)}$ and $\varphi_B^{(2)}$ denote the volume fractions of the polymer in the solution before and after the dilution process.

Polymer (B):   **poly(vinyl alcohol)**   **1956AM1**
Characterization: M_n/g.mol^{-1} = 26400, < 0.3 mol% vinyl acetate
Solvent (A):   **water**   **H$_2$O**   **7732-18-5**

T/K = 303.15

$V^{(1)}$/cm^3	15.0	20.0	25.0	30.0	20.0	20.0	25.0	30.0	40.0
$\varphi_B^{(1)}$	0.0997	0.0740	0.0589	0.0489	0.1429	0.0933	0.0740	0.0613	0.0453
$V^{(2)}$/cm^3	20.0	25.0	30.0	35.0	25.0	25.0	30.0	35.0	45.0
$\varphi_B^{(2)}$	0.0740	0.0589	0.0489	0.0418	0.1129	0.0740	0.0614	0.0453	0.0365
$\Delta_{dil}H^{12}$/J	−0.469	−0.310	−0.192	−0.042	−0.803	−0.289	−0.209	−0.264	−0.184

continued

continued

$V^{(1)}$/cm^3	20.0	25.0	20.0	25.0	30.0	40.0	25.0	30.0	35.0
$\varphi_B^{(1)}$	0.143	0.114	0.094	0.075	0.062	0.046	0.114	0.094	0.081
$V^{(2)}$/cm^3	25.0	30.0	25.0	30.0	40.0	50.0	30.0	35.0	40.0
$\varphi_B^{(2)}$	0.114	0.094	0.075	0.062	0.046	0.037	0.094	0.081	0.071
$\Delta_{dil} H^{12}$/J	−0.192	−0.119	−0.069	−0.050	−0.063	−0.044	−0.120	−0.090	−0.078

$V^{(1)}$/cm^3	40.0	45.0
$\varphi_B^{(1)}$	0.071	0.063
$V^{(2)}$/cm^3	45.0	50.0
$\varphi_B^{(2)}$	0.063	0.056
$\Delta_{dil} H^{12}$/J	−0.046	−0.037

Comments: $\Delta_{dil} H^{12}$ is the extensive quantity obtained for a given total volume change from $V^{(1)}$ to $V^{(2)}$, where $\varphi_B^{(1)}$ and $\varphi_B^{(2)}$ denote the volume fractions of the polymer in the solution before and after the dilution process.

Polymer (B): **poly(vinylbenzyltrimethylammonium) bromide** **1995PER**
Characterization: –
Solvent (A): **water** **H$_2$O** **7732-18-5**

T/K	298.15	308.15	318.15
$c_B^{(1)}$/(base mol/l)	0.17	0.17	0.17
$c_B^{(2)}$/(base mol/l)	0.02	0.02	0.02
$\Delta_{dil} H^{12}$/(J/base mol polymer)	950	640	470

Comments: $c_B^{(1)}$ and $c_B^{(2)}$ denote the concentrations of the polymer in the solution before and after the dilution process.

Polymer (B): **poly(vinylbenzyltrimethylammonium) chloride** **1995PER**
Characterization: –
Solvent (A): **water** **H$_2$O** **7732-18-5**

T/K	298.15	308.15	318.15
$c_B^{(1)}$/(base mol/l)	0.18	0.18	0.18
$c_B^{(2)}$/(base mol/l)	0.02	0.02	0.02
$\Delta_{dil} H^{12}$/(J/base mol polymer)	810	560	390

Comments: $c_B^{(1)}$ and $c_B^{(2)}$ denote the concentrations of the polymer in the solution before and after the dilution process.

Polymer (B): **poly(vinylbenzyltrimethylammonium) hydroxide** **1995PER**
Characterization: –
Solvent (A): **water** **H$_2$O** **7732-18-5**

T/K	298.15	308.15	318.15
$c_B^{(1)}$/(base mol/l)	0.19	0.19	0.19
$c_B^{(2)}$/(base mol/l)	0.02	0.02	0.02
$\Delta_{dil} H^{12}$/(J/base mol polymer)	550	420	320

Comments: $c_B^{(1)}$ and $c_B^{(2)}$ denote the concentrations of the polymer in the solution before and after the dilution process.

Polymer (B):   **poly(vinylbenzyltrimethylammonium) iodide**   **1995PER**
Characterization: –
Solvent (A):   **water**   **H$_2$O**   **7732-18-5**

T/K	298.15	308.15	318.15
$c_B^{(1)}$/(base mol/l)	0.12	0.12	0.12
$c_B^{(2)}$/(base mol/l)	0.02	0.02	0.02
$\Delta_{dil}H^{12}$/(J/base mol polymer)	810	480	390

Comments: $c_B^{(1)}$ and $c_B^{(2)}$ denote the concentrations of the polymer in the solution before and after the dilution process.

Polymer (B):   **poly(1-vinyl-N-butylpyridinium) bromide**   **1954SCH**
Characterization: M_η/g.mol^{-1} = 160000
Solvent (A):   **water**   **H$_2$O**   **7732-18-5**

$T/K = 298.15$

$c_B^{(2)}$/(base mol/l) = 0.00000246   was kept constant for all measurements

$c_B^{(1)}$/(base mol/l)	0.00488	0.02500	0.04928	0.19796	1.07912	2.00435	4.07555
$\Delta_{dil}H^{12}$/(J/base mol polymer)	−1485	−3749	−5268	−5933	−6749	−6293	−6180

Comments: $c_B^{(1)}$ and $c_B^{(2)}$ denote the concentrations of the polymer in the solution before and after the dilution process.

Polymer (B):   **poly(N-vinylcarbazole)**   **1968BRU**
Characterization: M_n/g.mol^{-1} = 425, DP = 2.2, synthesized in the laboratory
Solvent (A):   **benzene**   **C$_6$H$_6$**   **71-43-2**

$T/K = 310.15$

m_B/g	1.900	1.425	0.950	0.475
$c_B^{(1)}$/(g/cm^3)	0.0980	0.0734	0.0490	0.0245
$\Delta_{dil}H^{12}$/J	0.9644	0.5251	0.2460	0.05715
$\Delta_{dil}H^{12}$/[$m_B(c_B^{(1)} - c_B^{(2)})$]	10.38	10.04	10.59	10.04

Comments: $c_B^{(1)}$ and $c_B^{(2)}$ denote the concentrations of the polymer in the solution before and after the dilution process.

Polymer (B):   **poly(N-vinylcarbazole)**   **1968BRU**
Characterization: M_n/g.mol^{-1} = 750, DP = 3.9, synthesized in the laboratory
Solvent (A):   **benzene**   **C$_6$H$_6$**   **71-43-2**

$T/K = 310.15$

m_B/g	1.900	1.425	0.950	0.475
$c_B^{(1)}$/(g/cm^3)	0.0980	0.0734	0.0490	0.0245
$\Delta_{dil}H^{12}$/J	0.7782	0.4657	0.2021	0.04833
$\Delta_{dil}H^{12}$/[$m_B(c_B^{(1)} - c_B^{(2)})$]	8.37	8.91	8.70	8.49

Comments: $c_B^{(1)}$ and $c_B^{(2)}$ denote the concentrations of the polymer in the solution before and after the dilution process.

Polymer (B):   **poly(N-vinylcarbazole)**   **1968BRU**
Characterization: $M_n/\text{g.mol}^{-1} = 1370$, DP = 7.1, synthesized in the laboratory
Solvent (A):   **benzene**   **C$_6$H$_6$**   **71-43-2**

$T/\text{K} = 310.15$

m_B/g	1.900	1.425	0.950	0.475
$c_B^{(1)}/(\text{g/cm}^3)$	0.0980	0.0734	0.0490	0.0245
$\Delta_{dil}H^{12}/\text{J}$	0.6381	0.3389	0.1577	0.0381
$\Delta_{dil}H^{12}/[m_B(c_B^{(1)} - c_B^{(2)})]$	6.86	6.49	6.78	6.65

Comments: $c_B^{(1)}$ and $c_B^{(2)}$ denote the concentrations of the polymer in the solution before and after the dilution process.

Polymer (B):   **poly(N-vinylcarbazole)**   **1968BRU**
Characterization: $M_n/\text{g.mol}^{-1} = 1740$, DP = 9.9, synthesized in the laboratory
Solvent (A):   **benzene**   **C$_6$H$_6$**   **71-43-2**

$T/\text{K} = 310.15$

m_B/g	1.900	1.425	0.950	0.475
$c_B^{(1)}/(\text{g/cm}^3)$	0.0980	0.0734	0.0490	0.0245
$\Delta_{dil}H^{12}/\text{J}$	0.5368	0.3084	0.1402	0.0343
$\Delta_{dil}H^{12}/[m_B(c_B^{(1)} - c_B^{(2)})]$	5.77	5.90	6.02	6.02

Comments: $c_B^{(1)}$ and $c_B^{(2)}$ denote the concentrations of the polymer in the solution before and after the dilution process.

Polymer (B):   **poly(N-vinylcarbazole)**   **1968BRU**
Characterization: $M_n/\text{g.mol}^{-1} = 2200$, DP = 11.4, synthesized in the laboratory
Solvent (A):   **benzene**   **C$_6$H$_6$**   **71-43-2**

$T/\text{K} = 310.15$

m_B/g	1.900	1.425	0.950	0.7125
$c_B^{(1)}/(\text{g/cm}^3)$	0.0980	0.0734	0.0490	0.03675
$\Delta_{dil}H^{12}/\text{J}$	0.4904	0.2669	0.1264	0.0678
$\Delta_{dil}H^{12}/[m_B(c_B^{(1)} - c_B^{(2)})]$	5.27	5.10	5.44	5.19

Polymer (B):   **poly(N-vinylcarbazole)**   **1968BRU**
Characterization: $M_n/\text{g.mol}^{-1} = 3000$, DP = 15.4, synthesized in the laboratory
Solvent (A):   **benzene**   **C$_6$H$_6$**   **71-43-2**

$T/\text{K} = 310.15$

m_B/g	1.900	1.425	0.950	0.7125
$c_B^{(1)}/(\text{g/cm}^3)$	0.0980	0.0734	0.0490	0.03675
$\Delta_{dil}H^{12}/\text{J}$	0.3774	0.2318	0.0933	0.0598
$\Delta_{dil}H^{12}/[m_B(c_B^{(1)} - c_B^{(2)})]$	4.06	4.44	4.02	4.56

Comments: $c_B^{(1)}$ and $c_B^{(2)}$ denote the concentrations of the polymer in the solution before and after the dilution process.

Polymer (B):   **poly(N-vinylcarbazole)**   **1968BRU**
Characterization: M_n/g.mol^{-1} = 5300, DP = 27.5, synthesized in the laboratory
Solvent (A):   **benzene**   **C$_6$H$_6$**   **71-43-2**

T/K = 310.15

m_B/g	1.900	1.425	0.950	0.7125
$c_B^{(1)}$/(g/cm^3)	0.0980	0.0734	0.0490	0.03675
$\Delta_{dil}H^{12}$/J	0.2879	0.1569	0.0661	0.0427
$\Delta_{dil}H^{12}/[m_B(c_B^{(1)} - c_B^{(2)})]$	3.10	3.18	2.85	3.26

Comments: $c_B^{(1)}$ and $c_B^{(2)}$ denote the concentrations of the polymer in the solution before and after the dilution process.

Polymer (B):   **poly(N-vinylcarbazole)**   **1968BRU**
Characterization: M_n/g.mol^{-1} = 46400, DP = 240, synthesized in the laboratory
Solvent (A):   **benzene**   **C$_6$H$_6$**   **71-43-2**

T/K = 310.15

m_B/g	1.900	1.425	0.950	0.7125
$c_B^{(1)}$/(g/cm^3)	0.0980	0.0734	0.0490	0.03675
$\Delta_{dil}H^{12}$/J	0.1870	0.0787	0.0485	0.00644
$\Delta_{dil}H^{12}/[m_B(c_B^{(1)} - c_B^{(2)})]$	2.01	1.51	2.09	1.13

Comments: $c_B^{(1)}$ and $c_B^{(2)}$ denote the concentrations of the polymer in the solution before and after the dilution process.

Polymer (B):   **poly(vinyl chloride)**   **2002SAF**
Characterization: M_η/g.mol^{-1} = 85000
Solvent (A):   **bis(2-ethylhexyl) phthalate**   **C$_{24}$H$_{38}$O$_4$**   **117-81-7**

T/K = 298.15 w_B = 0.72 $\Delta_M H$/(J/g) = −13.5

Comments: The data pair corresponds to the minimum of the enthalpy of mixing curve.

Polymer (B):   **poly(vinyl chloride)**   **1991NOV**
Characterization: M_η/g.mol^{-1} = 130000, T_g/K = 355
Solvent (A):   **bis(2-ethylhexyl) phthalate**   **C$_{24}$H$_{38}$O$_4$**   **117-81-7**

T/K = 300

z_B	0.90	0.80	0.70	0.60	0.50	0.40	0.30	0.20
$\Delta_M H$/(J/base mol)	−1540	−1400	−1230	−1050	−845	−672	−520	−346

z_B	0.10
$\Delta_M H$/(J/base mol)	−170

Comments: These data include the enthalpy change when mixing the glassy polymer and the solvent to form the mixture.

continued

continued

w_B	0.10	0.20	0.33	0.50	0.70	0.81	0.91	0.96
$\Delta_M H/(J/g)$	−11.8	−17.6	−17.2	−14.0	−8.40	−5.42	−2.50	−1.04

Comments: These data include the enthalpy change when mixing the glassy polymer and the solvent to form the mixture.

w_B	0.10	0.20	0.33	0.50	0.70	0.81	0.91	0.96
$\Delta_M H/(J/g)$	+1.62	+2.86	−0.14	−1.27	−0.78	−0.48	−0.19	−0.07

Comments: These data were observed when mixing the equilibrium "liquid" polymer and the solvent to form the mixture, i.e., after substracting the part caused by the glass enthalpy.

Polymer (B):   **poly(vinyl chloride)**   **2002SAF**
Characterization: $M_n/\text{g.mol}^{-1} = 85000$
Solvent (A):   **bis(2,5,5-trimethylhexyl) phthalate** $C_{26}H_{42}O_4$ **53445-26-4**

$T/K = 298.15$ $w_B = 0.72$ $\Delta_M H/(J/g) = -13.5$

Comments: The data pair corresponds to the minimum of the enthalpy of mixing curve.

Polymer (B):   **poly(vinyl chloride)**   **1972MA3**
Characterization: $M_n/\text{g.mol}^{-1} = 23200$, B.F. Goodrich Chemical Company
Solvent (A):   **cyclohexanone** $C_6H_{10}O$ **108-94-1**

$T/K = 303.15$

$\varphi_B^{(1)}$	0.0402	0.0546	0.1002	0.3634	0.4251	0.5228
$\varphi_B^{(2)}$	0.0067	0.0091	0.0167	0.0119	0.0171	0.0113
$\Delta_{dil} H^{12}/(J/g \text{ polymer})$	−0.236	−0.315	−0.492	−3.054	−2.615	−4.146

Comments: $\varphi_B^{(1)}$ and $\varphi_B^{(2)}$ denote the volume fractions of the polymer in the solution before and after the dilution process.

Polymer (B):   **poly(vinyl chloride)**   **1972MA3**
Characterization: $M_n/\text{g.mol}^{-1} = 53500$, B.F. Goodrich Chemical Company
Solvent (A):   **cyclohexanone** $C_6H_{10}O$ **108-94-1**

$T/K = 303.15$

$\varphi_B^{(1)}$	0.0272	0.0375	0.0508
$\varphi_B^{(2)}$	0.0045	0.0062	0.0085
$\Delta_{dil} H^{12}/(J/g \text{ polymer})$	−0.142	−0.198	−0.234

Comments: $\varphi_B^{(1)}$ and $\varphi_B^{(2)}$ denote the volume fractions of the polymer in the solution before and after the dilution process.

Polymer (B):   **poly(vinyl chloride)**   **2002SAF**
Characterization: $M_\eta/\text{g.mol}^{-1} = 85000$
Solvent (A):   **dibutyl phthalate** $C_{16}H_{22}O_4$ **84-74-2**

$T/\text{K} = 298.15$

$w_B = 0.75$
$\Delta_M H/(\text{J/g}) = -14.3$

Comments: The data pair corresponds to the minimum of the enthalpy of mixing curve.

Polymer (B):   **poly(vinyl chloride)**   **1991NOV**
Characterization: $M_\eta/\text{g.mol}^{-1} = 130000$, $T_g/\text{K} = 355$
Solvent (A):   **dibutyl phthalate** $C_{16}H_{22}O_4$ **84-74-2**

$T/\text{K} = 300$

z_B	0.90	0.80	0.70	0.60	0.50	0.40	0.30	0.20
$\Delta_M H/(\text{J/base mol})$	−1360	−1370	−1200	−1030	−855	−680	−510	−340

z_B	0.10
$\Delta_M H/(\text{J/base mol})$	−170

Comments: These data include the enthalpy change when mixing the glassy polymer and the solvent to
form the mixture.

Polymer (B):   **poly(vinyl chloride)**   **2002SAF**
Characterization: $M_\eta/\text{g.mol}^{-1} = 85000$
Solvent (A):   **didodecyl phthalate** $C_{32}H_{54}O_4$ **2432-90-8**

$T/\text{K} = 298.15$ $w_B = 0.70$ $\Delta_M H/(\text{J/g}) = -12.1$

Comments: The data pair corresponds to the minimum of the enthalpy of mixing curve.

Polymer (B):   **poly(vinyl chloride)**   **2002SAF**
Characterization: $M_\eta/\text{g.mol}^{-1} = 85000$
Solvent (A):   **diethyl phthalate** $C_{12}H_{14}O_4$ **84-66-2**

$T/\text{K} = 298.15$ $w_B = 0.78$ $\Delta_M H/(\text{J/g}) = -14.8$

Comments: The data pair corresponds to the minimum of the enthalpy of mixing curve.

Polymer (B):   **poly(vinyl chloride)**   **2002SAF**
Characterization: $M_\eta/\text{g.mol}^{-1} = 85000$
Solvent (A):   **dimethyl phthalate** $C_{10}H_{10}O_4$ **131-11-3**

$T/\text{K} = 298.15$ $w_B = 0.78$ $\Delta_M H/(\text{J/g}) = -13.7$

Comments: The data pair corresponds to the minimum of the enthalpy of mixing curve.

Polymer (B):   **poly(vinyl chloride)**   **1991NOV**

Characterization: $M_\eta/\text{g.mol}^{-1} = 130000$, $T_g/\text{K} = 355$

Solvent (A):   **dimethyl phthalate** $C_{10}H_{10}O_4$ **131-11-3**

$T/\text{K} = 300$

z_B	0.90	0.80	0.70	0.60	0.50	0.40	0.30	0.20
$\Delta_M H/(\text{J/base mol})$	−1240	−1250	−1090	−930	−780	−625	−460	−310

z_B	0.10
$\Delta_M H/(\text{J/base mol})$	−155

Comments: These data include the enthalpy change when mixing the glassy polymer and the solvent to form the mixture.

Polymer (B):   **poly(vinyl chloride)**   **1977KUS**

Characterization: −

Solvent (A):   **dioctyl phthalate** $C_{24}H_{38}O_4$ **117-84-0**

T/K	303.15	323.15	343.15
z_B	0.5	0.5	0.5
$\Delta_M H/(\text{J/g polymer})$	0.25	0.54	0.54

Polymer (B):   **poly(vinyl chloride)**   **2002SAF**

Characterization: $M_\eta/\text{g.mol}^{-1} = 85000$

Solvent (A):   **ditridecyl phthalate** $C_{32}H_{54}O_4$ **119-06-2**

$T/\text{K} = 298.15$ $w_B = 0.70$ $\Delta_M H/(\text{J/g}) = -12.1$

Comments: The data pair corresponds to the minimum of the enthalpy of mixing curve.

Polymer (B):   **poly(vinyl chloride)**   **1977KUS**

Characterization: −

Solvent (A):   **naphthyltolylmethane** $C_{18}H_{16}$ **30306-53-7**

T/K	303.15	323.15	343.15
z_B	0.5	0.5	0.5
$\Delta_M H/(\text{J/g polymer})$	0.17	1.17	5.61

Polymer (B):   **poly(vinyl chloride)**   **1972MA3**

Characterization: $M_n/\text{g.mol}^{-1} = 23200$, B.F. Goodrich Chemical Company

Solvent (A):   **tetrahydrofuran** C_4H_8O **109-99-9**

$T/\text{K} = 303.15$

$\varphi_B^{(1)}$	0.0691
$\varphi_B^{(2)}$	0.0173
$\Delta_{dil} H^{12}/(\text{J/g polymer})$	−0.707

Comments: $\varphi_B^{(1)}$ and $\varphi_B^{(2)}$ denote the volume fractions of the polymer in the solution before and after the dilution process.

Polymer (B):   **poly(vinyl chloride)**   **1972MA3**
Characterization: M_n/g.mol^{-1} = 53500, B.F. Goodrich Chemical Company
Solvent (A):   **tetrahydrofuran**   **C$_4$H$_8$O**   **109-99-9**

T/K = 303.15

$\varphi_B^{(1)}$	0.0351	0.0488	0.0519
$\varphi_B^{(2)}$	0.0088	0.0122	0.0130
$\Delta_{dil}H^{12}$/(J/g polymer)	−0.364	−0.561	−0.703

Comments: $\varphi_B^{(1)}$ and $\varphi_B^{(2)}$ denote the volume fractions of the polymer in the solution before and after the dilution process.

Polymer (B):   **poly(1-vinyl-2-pyrrolidinone)**   **1981OGA**
Characterization: M_w/g.mol^{-1} = 10000, Tokyo Kasei Kogyo Co. Ltd., Japan
Solvent (A):   **1-butanol**   **C$_4$H$_{10}$O**   **71-36-3**

T/K = 298.15

$\varphi_B^{(1)}$	0.0480	0.0480	0.0480	0.0480	0.0712	0.0712	0.0712
$\varphi_B^{(2)}$	0.0240	0.0288	0.0192	0.0096	0.0356	0.0427	0.0570
$\Delta_{dil}H^{12}$/(J/mol solvent)	6.140	7.281	4.908	2.505	12.06	14.15	19.27

$\varphi_B^{(1)}$	0.0712	0.0712
$\varphi_B^{(2)}$	0.0285	0.0142
$\Delta_{dil}H^{12}$/(J/mol solvent)	9.755	5.310

Comments: $\varphi_B^{(1)}$ and $\varphi_B^{(2)}$ denote the volume fractions of the polymer in the solution before and after the dilution process.

Polymer (B):   **poly(1-vinyl-2-pyrrolidinone)**   **1981OGA**
Characterization: M_w/g.mol^{-1} = 40000, Tokyo Kasei Kogyo Co. Ltd., Japan
Solvent (A):   **1-butanol**   **C$_4$H$_{10}$O**   **71-36-3**

T/K = 298.15

$\varphi_B^{(1)}$	0.0566	0.0566	0.0566	0.0566	0.0566	0.0479	0.0479
$\varphi_B^{(2)}$	0.0283	0.0339	0.0226	0.0113	0.0170	0.0240	0.0288
$\Delta_{dil}H^{12}$/(J/mol solvent)	8.787	11.20	7.178	3.931	5.249	6.949	8.433

$\varphi_B^{(1)}$	0.0479	0.0479	0.0479	0.0479	0.0479	0.0361	0.0361
$\varphi_B^{(2)}$	0.0384	0.0192	0.0144	0.0096	0.0336	0.0289	0.0216
$\Delta_{dil}H^{12}$/(J/mol solvent)	11.56	5.491	4.252	3.011	9.939	7.229	5.161

$\varphi_B^{(1)}$	0.0361	0.0361	0.0361	0.0264	0.0264	0.0264	0.0264
$\varphi_B^{(2)}$	0.0144	0.0180	0.0072	0.0132	0.0211	0.0158	0.0106
$\Delta_{dil}H^{12}$/(J/mol solvent)	3.405	4.265	1.759	2.445	4.073	2.972	1.893

$\varphi_B^{(1)}$	0.0264	0.0264	0.0264	0.0264	0.0681	0.0681	0.0681
$\varphi_B^{(2)}$	0.0079	0.0053	0.0185	0.0132	0.0340	0.0545	0.0477
$\Delta_{dil}H^{12}$/(J/mol solvent)	1.446	0.988	3.564	2.448	11.21	17.98	15.21

continued

continued

$\varphi_B^{(1)}$	0.0681	0.0681	0.0681	0.0681	0.0421	0.0421	0.0421
$\varphi_B^{(2)}$	0.0272	0.0204	0.0136	0.0409	0.0211	0.0337	0.0295
$\Delta_{dil}H^{12}$/(J/mol solvent)	8.804	6.850	4.965	12.80	5.993	9.908	8.466

$\varphi_B^{(1)}$	0.0421	0.0421	0.0421	0.0421	0.0421	0.0421
$\varphi_B^{(2)}$	0.0253	0.0168	0.0126	0.0084	0.0211	0.0253
$\Delta_{dil}H^{12}$/(J/mol solvent)	7.189	4.939	3.674	2.483	6.029	7.419

Comments: $\varphi_B^{(1)}$ and $\varphi_B^{(2)}$ denote the volume fractions of the polymer in the solution before and after the dilution process.

Polymer (B):	**poly(1-vinyl-2-pyrrolidinone)**	**1981OGA**
Characterization:	M_w/g.mol^{-1} = 10000, Tokyo Kasei Kogyo Co. Ltd., Japan	
Solvent (A):	**ethanol** **C$_2$H$_6$O**	**64-17-5**

T/K = 298.15

$\varphi_B^{(1)}$	0.0747	0.0747	0.0747	0.0747	0.0370	0.0370	0.0370
$\varphi_B^{(2)}$	0.0149	0.0299	0.0448	0.0597	0.0185	0.0074	0.0148
$\Delta_{dil}H^{12}$/(J/mol solvent)	1.243	2.293	3.419	4.406	0.777	0.334	0.622

$\varphi_B^{(1)}$	0.0370	0.0370	0.0370	0.0370
$\varphi_B^{(2)}$	0.0222	0.0296	0.0111	0.0259
$\Delta_{dil}H^{12}$/(J/mol solvent)	0.944	1.241	0.466	1.086

Comments: $\varphi_B^{(1)}$ and $\varphi_B^{(2)}$ denote the volume fractions of the polymer in the solution before and after the dilution process.

Polymer (B):	**poly(1-vinyl-2-pyrrolidinone)**	**1981OGA**
Characterization:	M_w/g.mol^{-1} = 40000, Tokyo Kasei Kogyo Co. Ltd., Japan	
Solvent (A):	**ethanol** **C$_2$H$_6$O**	**64-17-5**

T/K = 298.15

$\varphi_B^{(1)}$	0.0762	0.0762	0.0762	0.0762	0.0762	0.0387	0.0387
$\varphi_B^{(2)}$	0.0381	0.0152	0.0305	0.0457	0.0610	0.0077	0.0155
$\Delta_{dil}H^{12}$/(J/mol solvent)	3.232	1.447	2.592	3.844	4.643	0.437	0.805

$\varphi_B^{(1)}$	0.0387	0.0387	0.0387	0.0387
$\varphi_B^{(2)}$	0.0232	0.0310	0.0193	0.0116
$\Delta_{dil}H^{12}$/(J/mol solvent)	1.258	1.661	1.019	0.623

Comments: $\varphi_B^{(1)}$ and $\varphi_B^{(2)}$ denote the volume fractions of the polymer in the solution before and after the dilution process.

Polymer (B):   **poly(1-vinyl-2-pyrrolidinone)**   **1981OGA**
Characterization: M_w/g.mol^{-1} = 10000, Tokyo Kasei Kogyo Co. Ltd., Japan
Solvent (A):   **1-propanol**   **C$_3$H$_8$O**   **71-23-8**

T/K = 298.15

$\varphi_B^{(1)}$	0.0655	0.0655	0.0655	0.0655	0.0655	0.0655	0.0655
$\varphi_B^{(2)}$	0.0327	0.0524	0.0393	0.0262	0.0131	0.0131	0.0196
$\Delta_{dil}H^{12}$/(J/mol solvent)	6.967	10.22	8.233	5.664	2.838	2.744	4.034

$\varphi_B^{(1)}$	0.0655	0.0381	0.0381	0.0381	0.0381	0.0381
$\varphi_B^{(2)}$	0.0458	0.0190	0.0228	0.0305	0.0152	0.0076
$\Delta_{dil}H^{12}$/(J/mol solvent)	8.850	2.558	3.068	4.113	2.005	0.944

Comments: $\varphi_B^{(1)}$ and $\varphi_B^{(2)}$ denote the volume fractions of the polymer in the solution before and after the dilution process.

Polymer (B):   **poly(1-vinyl-2-pyrrolidinone)**   **1981OGA**
Characterization: M_w/g.mol^{-1} = 40000, Tokyo Kasei Kogyo Co. Ltd., Japan
Solvent (A):   **1-propanol**   **C$_3$H$_8$O**   **71-23-8**

T/K = 298.15

$\varphi_B^{(1)}$	0.0364	0.0364	0.0364	0.0364	0.0364	0.0364	0.0655
$\varphi_B^{(2)}$	0.0182	0.0219	0.0291	0.0146	0.0073	0.0109	0.0327
$\Delta_{dil}H^{12}$/(J/mol solvent)	2.183	2.579	3.579	1.764	0.991	1.369	6.366

$\varphi_B^{(1)}$	0.0655	0.0655	0.0655	0.0655	0.0655	0.0346	0.346
$\varphi_B^{(2)}$	0.0393	0.0524	0.0262	0.0131	0.0196	0.0173	0.0208
$\Delta_{dil}H^{12}$/(J/mol solvent)	7.744	10.75	4.912	2.669	3.906	1.910	2.320

$\varphi_B^{(1)}$	0.346	0.346	0.346	0.346
$\varphi_B^{(2)}$	0.0277	0.0138	0.0104	0.0069
$\Delta_{dil}H^{12}$/(J/mol solvent)	3.007	1.526	1.150	0.772

Comments: $\varphi_B^{(1)}$ and $\varphi_B^{(2)}$ denote the volume fractions of the polymer in the solution before and after the dilution process.

Polymer (B):   **tetra(ethylene glycol)**   **2003CO2**
Characterization: M/g.mol^{-1} = 194, Aldrich Chem. Co., Inc., Milwaukee, WI
Solvent (A):   **anisole**   **C$_7$H$_8$O**   **100-66-3**

T/K = 308.15

x_B	0.0257	0.0500	0.0732	0.0953	0.1365	0.1740	0.2401	0.2964
$\Delta_M H$/(J/mol)	77.4	127.2	163.3	192.9	222.8	227.8	222.4	207.0

x_B	0.3873	0.4868	0.5583	0.6547	0.7166	0.7914	0.8349	0.8835
$\Delta_M H$/(J/mol)	189.0	153.8	125.8	85.2	55.3	27.2	15.5	4.1

x_B	0.9382
$\Delta_M H$/(J/mol)	2.2

Comments: Enthalpies of mixing for mixtures of anisole + diethylene glycol or + triethylene glycol are given in the original source.

Polymer (B):	tetra(ethylene glycol)						2004FRA
Characterization:	$M/\text{g.mol}^{-1}$ = 194, Aldrich Chem. Co., Inc., Milwaukee, WI						
Solvent (A):	**benzyl alcohol**		C_7H_8O				**100-51-6**

T/K = 308.15

x_B	0.0244	0.0475	0.0697	0.0908	0.1303	0.1665	0.2305	0.2853
$\Delta_M H/(\text{J/mol})$	−157.9	−279.3	−366.0	−437.9	−532.6	−591.5	−640.6	−680.5

x_B	0.3747	0.4734	0.5451	0.6425	0.7056	0.7824	0.8274	0.8778
$\Delta_M H/(\text{J/mol})$	−708.0	−733.9	−727.3	−672.3	−609.4	−476.8	−385.6	−258.8

x_B	0.9349
$\Delta_M H/(\text{J/mol})$	−121.2

Comments: Enthalpies of mixing for benzyl alcohol + 1,2-ethanediol, + diethylene glycol or + triethylene glycol are given in the original source.

Polymer (B):	tetra(ethylene glycol)						2002RUI
Characterization:	$M/\text{g.mol}^{-1}$ = 194, Aldrich Chem. Co., Inc., Milwaukee, WI						
Solvent (A):	**1-butanol**		$C_4H_{10}O$				**71-36-3**

T/K = 298.15

x_B	0.0475	0.0988	0.1531	0.2308	0.2582	0.3040	0.3942	0.4454
$\Delta_M H/(\text{J/mol})$	492	922	1252	1583	1678	1787	1918	1925

x_B	0.5011	0.5590	0.6371	0.7118	0.7995	0.8860	0.9700
$\Delta_M H/(\text{J/mol})$	1860	1736	1570	1344	1011	610	150

Polymer (B):	tetra(ethylene glycol)						2003CO1
Characterization:	$M/\text{g.mol}^{-1}$ = 194, Aldrich Chem. Co., Inc., Milwaukee, WI						
Solvent (A):	**dimethylsulfoxide**		C_2H_6OS				**67-68-5**

T/K = 308.15

x_B	0.0169	0.0332	0.0490	0.0643	0.0934	0.1208	0.1709	0.2155
$\Delta_M H/(\text{J/mol})$	−110.0	−201.9	−289.5	−365.4	−494.1	−602.7	−739.9	−838.7

x_B	0.2919	0.3818	0.4519	0.5529	0.6225	0.7121	0.7673	0.8318
$\Delta_M H/(\text{J/mol})$	−922.0	−950.3	−945.3	−904.2	−846.2	−744.0	−675.5	−555.9

x_B	0.9082
$\Delta_M H/(\text{J/mol})$	−351.1

Comments: Enthalpies of mixing for DMSO + 1,2-ethanediol, + 1,2-propanediol, + diethylene glycol or + triethylene glycol are given in the original source.

Polymer (B):   **tetra(ethylene glycol)**   **2005BIG**
Characterization: $M/\text{g.mol}^{-1} = 194$, Aldrich Chem. Co., Inc., Milwaukee, WI
Solvent (A):   **ethanol**   **C$_2$H$_6$O**   **64-17-5**

$T/\text{K} = 308.15$

x_B	0.0243	0.0475	0.0697	0.0908	0.1303	0.1664	0.2305	0.2853
$\Delta_M H/(\text{J/mol})$	126.9	229.9	315.8	380.8	478.6	544.4	635.7	678.9

x_B	0.3746	0.4734	0.5451	0.6425	0.7056	0.7824	0.8276	0.8779
$\Delta_M H/(\text{J/mol})$	736.1	761.5	747.2	690.0	622.1	500.1	416.0	292.9

x_B	0.9350
$\Delta_M H/(\text{J/mol})$	153.1

Polymer (B):   **tetra(ethylene glycol)**   **2002RUI**
Characterization: $M/\text{g.mol}^{-1} = 194$, Aldrich Chem. Co., Inc., Milwaukee, WI
Solvent (A):   **1-pentanol**   **C$_5$H$_{12}$O**   **71-41-0**

$T/\text{K} = 298.15$

x_B	0.0538	0.1106	0.1649	0.1984	0.2296	0.2667	0.3312	0.3791
$\Delta_M H/(\text{J/mol})$	584	1070	1383	1573	1695	1792	1915	2015

x_B	0.4637	0.5081	0.5348	0.6049	0.6780	0.7401	0.8021	0.8842
$\Delta_M H/(\text{J/mol})$	2012	1983	1958	1811	1584	1371	1079	675

Polymer (B):   **tetra(ethylene glycol)**   **2004CA1**
Characterization: $M/\text{g.mol}^{-1} = 194$, Aldrich Chem. Co., Inc., Milwaukee, WI
Solvent (A):   **2-phenylethanol**   **C$_8$H$_{10}$O**   **60-12-8**

$T/\text{K} = 308.15$

x_B	0.0281	0.0546	0.0797	0.1036	0.1478	0.1878	0.2575	0.3161
$\Delta_M H/(\text{J/mol})$	−121.0	−204.2	−256.7	−287.5	−327.2	−341.7	−348.1	−349.2

x_B	0.4095	0.5099	0.5811	0.6754	0.7351	0.8063	0.8473	0.8928
$\Delta_M H/(\text{J/mol})$	−359.2	−367.2	−353.7	−334.6	−303.0	−253.5	−212.2	−175.9

x_B	0.9433
$\Delta_M H/(\text{J/mol})$	−108.1

Comments: Enthalpies of mixing for 2-phenylethanol + 1,2-ethanediol, + 1,2-propanediol, + diethylene
glycol or + triethylene glycol are given in the original source.

Polymer (B):   **tetra(ethylene glycol)**   **2004CA3**
Characterization: $M/\text{g.mol}^{-1} = 194$, Aldrich Chem. Co., Inc., Milwaukee, WI
Solvent (A):   **3-phenyl-1-propanol**   **C$_9$H$_{12}$O**   **122-97-4**

$T/\text{K} = 308.15$

continued

continued

x_B	0.0318	0.0616	0.0896	0.1160	0.1644	0.2079	0.2824	0.3441
$\Delta_M H/$(J/mol)	−70.2	−114.0	−140.3	−156.9	−171.1	−175.1	−173.7	−174.4

x_B	0.4405	0.5416	0.6116	0.7025	0.7590	0.8253	0.8630	0.9043
$\Delta_M H/$(J/mol)	−175.4	−169.9	−157.6	−138.7	−122.1	−105.6	−88.3	−71.1

x_B	0.9497
$\Delta_M H/$(J/mol)	−48.3

Comments: Enthalpies of mixing for 3-phenyl-1-propanol + 1,2-ethanediol, + diethylene glycol or + triethylene glycol are given in the original source.

Polymer (B): **tetra(ethylene glycol)** **2002RUI**
Characterization: $M/$g.mol^{-1} = 194, Aldrich Chem. Co., Inc., Milwaukee, WI
Solvent (A): **1-propanol** **C$_3$H$_8$O** **71-23-8**

$T/$K = 298.15

x_B	0.0178	0.0377	0.0556	0.0785	0.1009	0.1465	0.1727	0.2468
$\Delta_M H/$(J/mol)	17.6	338	537	705	817	1093	1230	1419

x_B	0.3455	0.4245	0.5718	0.6302	0.7422	0.8328	0.8903	0.9135
$\Delta_M H/$(J/mol)	1610	1647	1506	1392	1071	746	507	405

Polymer (B): **tetra(ethylene glycol)** **2004CA4**
Characterization: $M/$g.mol^{-1} = 194, Aldrich Chem. Co., Inc., Milwaukee, WI
Solvent (A): **propylene carbonate** **C$_4$H$_6$O$_3$** **108-32-7**

$T/$K = 308.15

x_B	0.0201	0.0393	0.0578	0.0757	0.1094	0.1407	0.1972	0.2466
$\Delta_M H/$(J/mol)	120.7	213.7	285.3	342.1	425.4	463.1	502.8	521.4

x_B	0.3293	0.4242	0.4956	0.5957	0.6627	0.7467	0.7971	0.8550
$\Delta_M H/$(J/mol)	511.0	499.3	482.8	417.6	357.6	261.6	212.1	144.2

x_B	0.9218
$\Delta_M H/$(J/mol)	76.9

Comments: Enthalpies of mixing for propylene carbonate + 1,2-ethanediol, + 1,2-propanediol, + diethylene glycol or + triethylene glycol are given in the original source.

Polymer (B): **tetra(ethylene glycol)** **1993ZEL**
Characterization: $M/$g.mol^{-1} = 194.23
Solvent (A): **tetrachloromethane** **CCl$_4$** **56-23-5**

$T/$K = 303.15

$$\Delta_M H/(\text{J/g}) = w_B(1 - w_B)[-1.55 + 3.48(1 - 2w_B) - 1.31(1 - 2w_B)^2] - 1.47w_B \ln(w_B)$$

$T/$K = 318.15

$$\Delta_M H/(\text{J/g}) = w_B(1 - w_B)[-6.32 - 0.02(1 - 2w_B) - 7.87(1 - 2w_B)^2] - 7.19w_B \ln(w_B)$$

continued

continued

$T/K = 333.15$

$\Delta_M H/(J/g) = w_B(1 - w_B)[10.10 + 0.74(1 - 2w_B)] - 10.02 w_B \ln(w_B)$

Polymer (B):	tetra(ethylene glycol)		1993ZEL
Characterization:	$M/g.mol^{-1} = 194.23$		
Solvent (A):	trichloromethane	CHCl$_3$	67-66-3

$T/K = 303.15$

$\Delta_M H/(J/g) = w_B(1 - w_B)[-78.47 - 19.40(1 - 2w_B) + 11.97(1 - 2w_B)^2 + 9.07(1 - 2w_B)^3]$

Polymer (B):	tetra(ethylene glycol)		1987BIR
Characterization:	$M/g.mol^{-1} = 194.23$		
Solvent (A):	water	H$_2$O	7732-18-5

$T/K = 298.15$

w_A	0.0146	0.0348	0.0542	0.0728	0.0994	0.1482	0.1962	0.2403	0.2764
$\Delta_M H/(J/g)$	−3.982	−9.298	−13.74	−17.69	−22.80	−30.72	−37.21	−42.44	−45.60
w_A	0.3084	0.3467	0.3849	0.4185	0.4418	0.4466	0.4683	0.5000	0.5343
$\Delta_M H/(J/g)$	−48.00	−50.00	−51.87	−52.72	−53.19	−53.01	−53.09	−52.63	−51.68
w_A	0.5737	0.6195	0.6705	0.7250	0.7800	0.8429	0.8912	0.9313	0.9758
$\Delta_M H/(J/g)$	−49.91	−47.10	−43.14	−37.91	−31.84	−23.70	−16.79	−10.65	−3.842

Comments: Experimental data for mixtures of water and 1,2-ethanediol, di(ethylene glycol) and tri(ethylene glycol) can be found in the original source.

Polymer (B):	tetra(ethylene glycol) diethyl ether		1972NAK
Characterization:	$M/g.mol^{-1} = 250.33$		
Solvent (A):	water	H$_2$O	7732-18-5

$T/K = 298.15$

x_A	0.0106	0.0346	0.0662	0.0973	0.1456	0.1962	0.2964	0.3708
$\Delta_M H/(J/mol)$	−531.4	−1323	−1708	−1813	−1772	−1662	−1236	−941.4
x_A	0.4383	0.4829	0.5168	0.5954	0.6881	0.7880	0.8865	
$\Delta_M H/(J/mol)$	−682.4	−577.0	−463.2	−242.7	−53.1	+48.5	+67.8	

Polymer (B):	tetra(ethylene glycol) dimethyl ether		1999BUR
Characterization:	$M/g.mol^{-1} = 222$		
Solvent (A):	n-dodecane	C$_{12}$H$_{26}$	112-40-3

$T/K = 298.15$

x_B	0.051	0.103	0.171	0.203	0.242	0.744	0.813	0.857
$\Delta_M H/(J/mol)$	552.8	1030.2	1322.6	1431.2	1561.8	1579.8	1392.1	1201.3

continued

continued

x_B	0.910
$\Delta_M H$/(J/mol)	830.2

T/K = 323.15

x_B	0.108	0.209	0.309	0.395	0.509	0.607	0.706	0.805
$\Delta_M H$/(J/mol)	1162.1	1822.3	2220.3	2376.6	2432.1	2376.2	2116.4	1681.0

x_B	0.885
$\Delta_M H$/(J/mol)	925.5

Comments: Experimental data for mixtures of n-dodecane and the dimethyl ethers of 1,2-ethanediol, di(ethylene glycol) and tri(ethylene glycol) as well as di(ethylene glycol) dibutyl ether can be found in the original source.

Polymer (B):	**tetra(ethylene glycol) dimethyl ether**	**1997LOP**
Characterization:	M/g.mol^{-1} = 222, Aldrich Chem. Co., Inc., Milwaukee, WI	
Solvent (A):	**methanol** **CH$_4$O**	**67-56-1**

T/K = 298.15

x_A	0.0108	0.0666	0.1320	0.1742	0.1899	0.2130	0.2153	0.2384
$\Delta_M H$/(J/mol)	14.8	129.7	234.0	286.0	313.6	338.5	342.7	377.0

x_A	0.2895	0.3151	0.3398	0.3530	0.4362	0.4806	0.4935	0.5422
$\Delta_M H$/(J/mol)	428.4	454.8	477.4	498.2	557.1	579.3	581.1	586.0

x_A	0.5432	0.5790	0.6335	0.6546	0.6750	0.7341	0.7707	0.8279
$\Delta_M H$/(J/mol)	586.3	579.4	566.6	557.7	538.2	483.8	435.1	359.9

x_A	0.8498	0.8926	0.9158	0.9581	0.9650
$\Delta_M H$/(J/mol)	309.7	234.8	175.7	76.2	60.3

Polymer (B):	**tetra(ethylene glycol) dimethyl ether**	**1999LOP**
Characterization:	M/g.mol^{-1} = 222, Aldrich Chem. Co., Inc., Milwaukee, WI	
Solvent (A):	**methanol** **CH$_4$O**	**67-56-1**

T/K = 323.15 P/MPa = 8.0

x_A	0.0686	0.1000	0.1441	0.1849	0.2227	0.2907	0.3500	0.4022
$\Delta_M H$/(J/mol)	134.9	193.3	273.5	336.1	392.5	478.7	538.0	588.1

x_A	0.4023	0.4699	0.4700	0.5091	0.5608	0.6056	0.6567	0.7000
$\Delta_M H$/(J/mol)	577.6	605.6	611.3	616.5	614.0	604.6	586.0	548.7

x_A	0.7538	0.7977	0.8550	0.8988	0.9333
$\Delta_M H$/(J/mol)	481.6	418.8	312.2	220.4	141.3

T/K = 373.15 P/MPa = 8.0

x_A	0.0686	0.1001	0.1442	0.1850	0.2229	0.2909	0.3502	0.4024
$\Delta_M H$/(J/mol)	163.3	242.5	348.6	431.3	499.3	611.0	690.9	734.0

continued

continued

x_A	0.4025	0.4701	0.5093	0.5610	0.6058	0.6569	0.7002	0.7540
$\Delta_M H/(\text{J/mol})$	744.3	779.5	791.8	797.9	787.0	755.7	715.5	643.2

x_A	0.7978	0.8551	0.8989	0.9334
$\Delta_M H/(\text{J/mol})$	564.8	430.5	310.8	205.7

$T/\text{K} = 423.15$ $P/\text{MPa} = 8.0$

x_A	0.0686	0.1001	0.1442	0.1850	0.2229	0.2909	0.3502	0.4024
$\Delta_M H/(\text{J/mol})$	196.2	279.1	382.6	468.9	545.3	668.0	755.4	806.2

x_A	0.4024	0.4700	0.5092	0.5610	0.6057	0.6568	0.6991	0.7001
$\Delta_M H/(\text{J/mol})$	812.3	861.7	577.9	899.4	894.8	871.2	820.0	837.0

x_A	0.7002	0.7539	0.7977	0.8449	0.8989	0.9334
$\Delta_M H/(\text{J/mol})$	811.8	707.0	616.3	485.2	322.7	210.5

Polymer (B):   **tetra(ethylene glycol) dimethyl ether**   **1993ZEL**
Characterization: $M/\text{g.mol}^{-1} = 222.28$
Solvent (A):   **tetrachloromethane   CCl$_4$**   **56-23-5**

$T/\text{K} = 303.15$

$\Delta_M H/(\text{J/g}) = w_B(1 - w_B)[-11.41 + 1.46(1 - 2w_B) - 1.66(1 - 2w_B)^2 - 3.47(1 - 2w_B)^3 + 2.77(1 - 2w_B)^4]$

$T/\text{K} = 318.15$

$\Delta_M H/(\text{J/g}) = w_B(1 - w_B)[-9.06 - 1.34(1 - 2w_B) - 0.03(1 - 2w_B)^2 + 2.01(1 - 2w_B)^3 + 0.30(1 - 2w_B)^4]$

$T/\text{K} = 333.15$

$\Delta_M H/(\text{J/g}) = w_B(1 - w_B)[-8.45]$

Polymer (B):   **tetra(ethylene glycol) dimethyl ether**   **1993ZEL**
Characterization: $M/\text{g.mol}^{-1} = 222.28$
Solvent (A):   **trichloromethane   CHCl$_3$**   **67-66-3**

$T/\text{K} = 303.15$

$\Delta_M H/(\text{J/g}) = w_B(1 - w_B)[-121.23 - 61.37(1 - 2w_B) + 47.47(1 - 2w_B)^2 + 30.97(1 - 2w_B)^3]$

Polymer (B):   **tetra(ethylene glycol) dimethyl ether**   **1987BIR**
Characterization: $M/\text{g.mol}^{-1} = 222$
Solvent (A):   **water   H$_2$O**   **7732-18-5**

$T/\text{K} = 298.15$

w_A	0.0125	0.0247	0.0366	0.0383	0.0596	0.0739	0.0919	0.1225
$\Delta_M H/(\text{J/g})$	+0.2490	−0.0430	−0.787	−0.804	−3.142	−4.828	−7.536	−11.89

continued

continued

w_A	0.1319	0.1663	0.1684	0.2051	0.2059	0.2421	0.2750	0.3108
$\Delta_M H/(J/g)$	−13.80	−19.10	−19.82	−25.93	−25.82	−31.63	−36.75	−42.16

w_A	0.3424	0.3713	0.4014	0.4286	0.4535	0.4764	0.4884	0.4949
$\Delta_M H/(J/g)$	−46.39	−49.84	−53.00	−55.30	−56.97	−58.06	−58.89	−58.57

w_A	0.5128	0.5398	0.5697	0.6032	0.6409	0.6836	0.7324	0.7872
$\Delta_M H/(J/g)$	−59.51	−59.74	−59.38	−58.21	−56.00	−52.47	−47.08	−39.58

w_A	0.8369	0.8914	0.9556
$\Delta_M H/(J/g)$	−31.56	−21.58	−8.925

Comments: Experimental data for mixtures of water and the dimethyl ethers of 1,2-ethanediol, di(ethylene glycol) and tri(ethylene glycol) can be found in the original source.

2.2. References

1941TAG   Tager, A. and Kargin, V., Lösungs- und Quellungsvorgang von Zelluloseestern, *Acta Physicochim. URSS*, 14, 713, 1941.

1951MEA   Meares, P., The energies of dilution of poly(vinyl acetate) solutions (the data were taken from the tables of the 6th ed. of Landolt-Börnstein), *Trans. Faraday Soc.*, 47, 699, 1951.

1953TAG   Tager, A.A. and Dombek, Zh.S., Thermodynamic study of polystyrene solutions (Russ.), *Kolloidn. Zh.*, 15, 69, 1953.

1954DAO   Daoust, H. and Rinfret, M., Microcalorimetric studies of poly(vinyl acetate) solutions, *Can. J. Chem.*, 32, 492, 1954.

1954OST   Osthoff, R.C. and Grubb, W.T., Physical properties of organosilicon compounds. III. Thermodynamic properties of octamethylcyclotetrasiloxane, *J. Amer. Chem. Soc.*, 76, 399, 1954.

1954SCH   Schulze, W., Verdünnungswärmen eines Polyelektrolyten in Wasser bei kleinen Konzentrationen, *Z. Elektrochem.*, 58, 165, 1954.

1955PAR   Parent, M. and Rinfret, M., Microcalorimetric determination of the critical concentration and the molecular dimensions of polyvinyl acetate in solution, *Can. J. Chem.*, 33, 971, 1955.

1955TA1   Tager, A.A. and Kosova, L.K., Thermodynamic studies of copolymer solutions 2. Thermodynamic study of butadiene-acrylonitrile copolymer solutions (Russ.), *Kolloidn. Zh.*, 17, 391, 1955.

1955TA2   Tager, A.A., Kosova, L.K., Karlinskaya, D.Yu., and Yurina, I.A., Thermodynamic studies of copolymer solutions 1. Thermodynamic study of butadiene-styrene copolymer solutions (Russ.), *Kolloidn. Zh.*, 17, 315, 1955.

1956AM1   Amaya, K. and Fujishiro, R., Heats of dilution of polyvinylalcohol solutions I, *Bull. Chem. Soc. Japan*, 29, 361, 1956.

1956AM2   Amaya, K. and Fujishiro, R., Heats of dilution of polyvinylalcohol solutions II, *Bull. Chem. Soc. Japan*, 29, 830, 1956.

1956AM3   Amaya, K. and Fujishiro, R., Heats of dilution of polystyrene solutions I, *Bull. Chem. Soc. Japan*, 29, 270, 1956.

1956JEN   Jenckel, E. and Gorke, K., Zur Kalorimetrie von Hochpolymeren. II. Die integralen Verdünnungswärmen der Lösungen des Polystyrols in Äthylbenzol, Toluol, Chlorbenzol und Cyclohexan, *Z. Elektrochem.*, 60, 579, 1956.

1956MIK   Mikhailov, N.V. and Fainberg, E.Z., Issledovanie struktury sinteticheskikh poliamidnykh volokon 6. Integral'naya teplota rastvoreniya kapronovogo volokna v muravinoi kislote, *Kolloidn. Zh.*, 18, 44, 1956.

1957MAL   Malcolm, G.N. and Rowlinson, J.S., The thermodynamic properties of aqueous solutions of polyethylene glycol, polypropylene glycol, and dioxane, *Trans. Faraday Soc.*, 53, 921, 1957.

1957SCH   Schuurmans, J.L. and Hermans, J.J., Heats of mixing in the system poly(vinyl acetate)-toluene, *J. Phys. Chem.*, 61, 1496, 1957.

1958AMA   Amaya, K. and Fujishiro, R., Heats of dilution of polystyrene solutions. II, *Bull. Chem. Soc. Japan*, 31, 19, 1958.

1958KAB   Kabayama, M.A. and Daoust, H., Heats of dilution of the polyisobutylene-benzene system, *J. Phys. Chem.*, 62, 1127, 1958.

1959JES   Jessup, R.S., Heat of mixing of polybutadiene and benzene, *J. Res. Natnl. Bur. Stand.* 62, 1, 1959.

1959MAR   Maron, S.H. and Nakajima, N., A theory of the thermodynamic behavior of nonelectrolyte solutions. II. Application to the system rubber-benzene, *J. Polym. Sci.*, 40, 59, 1959.

1959SCH   Schulz, G.V. and Horbach, A., Kalorimetrische Messungen zur Thermodynamik von Polystyrollösungen, *Z. Phys. Chem. N. F.*, 22, 377, 1959.

1960LAK   Lakhanpal, M.L. and Conway, B.E., Studies on polyoxypropylene glycols. Part I. Vapor pressures and heats of mixing in the systems polyglycols-methanol, *J. Polym. Sci.*, 46, 75, 1960.

1960WAT   Watters, C., Daoust, H., and Rinfret, M., Heats of mixing of polyisobutylene with some organic solvents, *Can. J. Chem.*, 38, 1087, 1960.

1961CUN   Cunninghame, R.G. and Malcolm, G.N., The heats of mixing of aqueous solutions of polypropylene glycol and polyethylene glycol, *J. Phys. Chem.*, 65, 1454, 1961.

1965LAK   Lakhanpal, M.L., Lal, M., and Sharma, R.K., Studies on polyoxyethylene glycols. Part I - Heats of mixing of polyoxyethylene glycols with dioxane and carbon tetrachloride, *Indian J. Chem.*, 3, 547, 1965.

1966KA1   Kagemoto, A., Murakami, S., and Fujishiro, R., The heats of dilution of atactic polystyrene solutions, *Bull. Chem. Soc. Japan*, 39, 15, 1966.

1966KA2   Kagemoto, A., Murakami, S., and Fujishiro, R., The heats of dilution of *cis*-polybutadiene solutions in benzene and toluene, *Bull. Chem. Soc. Japan*, 39, 1814, 1966.

1966LA1   Lakhanpal, M.L., Taneja, H.L., and Sharma, R.K., Studies on polyoxyethylene glycols. Part II - Heats of mixing of polyoxyethylene glycols with benzene, methanol and ethanol, *Indian J. Chem.*, 4, 12, 1966.

1966LA2   Lakhanpal, M.L., Kapoor, V., Sharma, R.K., and Sharma, S.C., Studies on polyoxyethylene glycols. Part III - Heats of mixing of polyoxyethylene glycols-water systems, *Indian J. Chem.*, 4, 59, 1966.

1967KA1   Kagemoto, A., Murakami, S., and Fujishiro, R., The heats of dilution of the poly(ethylene oxide)/water solutions, *Makromol. Chem.*, 105, 154, 1967.

1967KA2   Kagemoto, A., Murakami, S., and Fujishiro, R., The heats of dilution in atactic poly(methyl methacrylate) solutions, *Bull. Chem. Soc. Japan*, 40, 11, 1967.

1967SK1   Skerjanc, J., Dolar, D., and Leskovsek, D., Heats of dilution of polyelectrolyte solutions. I. Polystyrenesulphonic acid and its sodium salt, *Z. Phys. Chem., N. F.*, 56, 207, 1967.

1967SK2   Skerjanc, J., Dolar, D., and Leskovsek, D., Heats of dilution of polyelectrolyte solutions. II. Zinc polystyrenesulphonate, *Z. Phys. Chem., N. F.*, 56, 218, 1967.

1968DAO   Daoust, H. and Hade, A., Effect of polarity on the heats of dilution of poly(methyl methacrylate) solutions, *Polymer*, 9, 47, 1968.

1968KA1   Kagemoto, A. and Fujishiro, R., The heat of the coil-helix transition of poly(γ-benzyl-L-glutamate) in the solution, *Makromol. Chem.*, 114, 139, 1968.

1968KA2   Kagemoto, A. and Fujishiro, R., The heat of dilution of poly(vinyl acetate) in benzene and methyl ethyl ketone, *Bull. Chem. Soc. Japan*, 41, 2201, 1968.

1968KER   Kershaw, R.W. and Malcolm, G.N., Thermodynamics of solutions of polypropylene oxide in chloroform and in carbon tetrachloride, *Trans. Faraday Soc.*, 64, 323, 1968.

1968LA1   Lakhanpal, M.L., Chhina, K.S., and Sharma, S.C., Thermodynamic properties of aqueous solutions of polyoxethyleneglycols, *Indian J. Chem.*, 6, 505, 1968.

1968LA2   Lakhanpal, M.L., Singh, H.G., Singh, H., and Sharma, S.C., A comparative study of heats of mixing of polyoxypropylene glycol-water and polyoxypropylene glycol-water systems, *Indian J. Chem.*, 6, 95, 1968.

1968BRU   Bruns, W., Mehdorn, F., Mirus, K., and Ueberreiter, K., Beiträge zur Thermodynamik von Polymerlösungen. 9. Mitteilung. Verdünnungswärmen von Oligomerlösungen, *Kolloid Z. Z. Polym.*, 224, 17, 1968.

1969MAL   Malcolm, G.N., Baird, C.E., Bruce, G.R., Cheyne, K.G., Kershaw, R.W., and Pratt, M.C., Thermodynamics of polyether solutions, *J. Polym. Sci.: Part A-2*, 7, 1495, 1969.

1969TAK   Takagi, S. and Fujishiro, R., Heats of dilution of isotactic poly(methyl methacrylate) solutions in benzene, *Rep. Progr. Polym. Phys. Japan*, 12, 39, 1969.

1970LEW   Lewis, G. and Johnson, A.F., Interpretation of the heat of dilution of polymer solutions, *Polymer*, 11, 336, 1970.

1970MAR   Marsh, K.N., Enthalpies of mixing and excess Gibbs free energies of mixtures of octamethylcyclotetrasiloxane + cyclopentane at 291.15, 298.15, and 308.15 K, *J. Chem. Thermodyn.*, 2, 359, 1970.

1970MO1   Morimoto, S., Calorimetric investigations on polymer solutions. IV. Heats of dilution of polystyrene solutions and their molecular weight dependence, *Bull. Res. Inst. Polym. Textil.*, 90(3), 38, 1970.

1970MO4   Morimoto, S., Heats of dilution of polystyrene solutions and their molecular weight dependence (Jap.), *Nippon Kagaku Zasshi*, 91, 117, 1970.

1970SKE   Skerjanc, J., Dolar, D., and Leskovsek, D., Heats of dilution of polyelectrolyte solutions III., *Z. Phys. Chem., N. F.*, 70, 31, 1970.

1971CAR   Cartier, J.-P. and Daoust, H., Chaleur de dilution de solutions aqueuses d'acide poly-acrylique, de son sel de sodium et de l'acide propionique, *Can. J. Chem.*, 49, 3935, 1971.

1971KAG   Kagemoto, A., Itoi, Y., Baba, Y., and Fujishiro, R., The heats of dilution of the oligomeric ethylene oxide/alcohol solutions, *Makromol. Chem.*, 150, 255, 1971.

1971LEW   Lewis, G. and Johnson, A.F., Heat of dilution of polymer solutions. Part II. Atactic poly(methyl methacrylate) solutions, *J. Chem. Soc. (A)*, 3524, 1971.

1972DAV   Davalloo, P., Gainer, J.L., and Hall, K.R., Enthalpies of solution in complex systems. Albumin + KCl(aq), polystyrene + toluene, and polystyrene + carbon tetrachloride, *J. Chem. Thermodyn.*, 4, 691, 1972.

1972MA1   Maron, S.H. and Filisko, F.E., A modified Tian-Calvet microcalorimeter for polymer solution measurements, *J. Macromol. Sci.-Phys. B*, 6, 57, 1972.

1972MA2   Maron, S.H. and Filisko, F.E., Heats of solution and dilution for polyethylene oxide in several solvents, *J. Macromol. Sci.-Phys. B*, 6, 79, 1972.

1972MA3   Maron, S.H. and Filisko, F.E., Heats of solution and dilution for polyvinyl chloride in cyclohexanone and tetrahydrofuran, *J. Macromol. Sci.-Phys. B*, 6, 413, 1972.

1972MAS   Masa, Z., Biros, J., Trekoval, J., and Pouchly, J., Specific interactions in solutions of polymers. IV. A calorimetric study of butyl methacrylate oligomers in chloroform, *J. Polym. Sci.: Part C*, 39, 219, 1972.

1972NAK   Nakayama, H., Thermodynamic properties of an aqueous solution of tetraethylene glycol diethyl ether, *Bull. Chem. Soc. Japan*, 45, 1371, 1972.

1972RAI   Rai, J.H. and Miller, W.G., DSC studies of the system poly(γ-benzyl-L-glutamate)-dimethylformamide, *J. Phys. Chem.*, 76, 1081, 1972.

1973SKE   Skerjanc, J., Hocevar, S., and Dolar, D., Heats of dilution of polyelectrolyte solutions. IV. Alkaline-earth polystyrenesulphonates, *Z. Phys. Chem., N. F.*, 86, 311, 1973.

1973TAM   Tamura, K., Murakami, S., and Fujishiro, R., Estimation of heat of dilution of polymer solution: a trial on a correction of the heat of stirring ascribed to the viscosity difference before and after dilution, *Polymer*, 14, 237, 1973.

1974BA1   Baba, Y., Katayama, H., and Kagemoto, A., Heats of dilution of atactic polystyrene in ethylacetate, *Makromol. Chem.*, 175, 209, 1974.

1974BA2   Baba, Y., Katayama, H., and Kagemoto, A., The heats of dilution of the oligomeric ethylene oxide-benzene system, *Polym. J.*, 6, 230, 1974.

1974DIC   Dickinson, E., McLure, I.A., and Powell, B.H., Thermodynamics of alkane-dimethyl siloxane mixtures 2. Vapor pressures and enthalpies of mixing, *J. Chem. Soc., Faraday Trans. I*, 70, 2321, 1974.

1975TAG   Tager, A.A. and Bessonov, Yu.S., Thermodynamic study of solutions of poly(vinyl acetate) and cellulose tricarbanilate in the precritical region (Russ.), *Vysokomol. Soedin., Ser. A*, 17, 2377, 1975.

1975TAN   Tancrede, P., Patterson, D., and Lam, V.-T., Thermodynamic effects of orientation order in chain-molecule mixtures, *J. Chem. Soc., Faraday Trans. II*, 71, 985, 1975.

1976DA1   Daoust, H. and Hade, A., Effect of cation size on heats of dilution of aqueous solutions of alkaline poly(styrenesulfonates), *Macromolecules*, 9, 608, 1976.

1976DA2   Daoust, H. and Lajoie, A., Chaleurs de dilution de solutions aqueuses d'acide polymethacrylique et de ses sels de sodium et de rubidium, *Can. J. Chem.*, 54, 1853, 1976.

1976LA1   Lakhanpal, M.L., Sharma, S.C., Krishan, B., and Parashar, R.N., Enthalpies of mixing of polyoxyethylene glycols in benzene, carbon tetrachloride, methyl alcohol and ethyl alcohol, *Indian J. Chem.*, 14A, 642, 1976.

1977KUS   Kushch, N.D., Moshchinskaya, N.K., Buryak, I.P., and Kushch, P.P., Calorimetric investigation of the system poly(vinyl chloride)-plasticizer (Russ.), *VINITI Depos. Doc. No.* 1568-77, 1977.

1978KOL   Kolomeer, M.G., Taganov, N.G., Bilgov, N.G., Vainshtein, E.F., and Entelis, S.G., Termokhimicheskoe issledovanie assotsiatsii polioksietilenov raslichnoi molekulyarnoi massy, *Termodin. Organ. Soedin.*, 7, 58, 1978.

1979BAB   Baba, Y., Kagemoto, A., and Fujishiro, R., Calorimetric studies in the helix-coil transition of poly(amino acid)s in methanol-water solvent mixtures, *Makromol. Chem.*, 180, 2221, 1979.

1979BAS   Basedow, A.M. and Ebert, K.H., Production, characterization, and solution properties of dextran fractions of narrow molecular weight distributions, *J. Polym. Sci. Polym. Symp.*, 66, 101, 1979.

1979KOL   Koller, J., Kalorimetrische Untersuchungen der inter- und intramolekularen Wechselwirkungen von Polyethylenglycolen in wässriger und benzolischer Lösung, *Dissertation*, TU München, 1979.

1979MON   Monshausen, F.W., Kalorimetrische Messungen am System Wasser - Polyethylenoxid, *Dissertation*, RWTH Aachen, 1979.

1979MOR   Morimoto, S. and Ohtani, N., Phase separation enthalpies of polymer solids in ternary polymer-solvent-nonsolvent systems (Jap.), *Kenkyu Hokoku Seni Kobunshi Zairyo Kenkyushu*, 119, 35, 1979.

1980BAS   Basedow, A.M., Ebert, K.H., and Feigenbutz, W., Polymer-solvent interactions. Dextrans in water and DMSO, *Makromol. Chem.*, 181, 1071, 1980.

1980GRA   Graun, K., Kalorimetrische Untersuchung der Wechselwirkungen von taktischen und atakti-schen Polymethylmethacrylaten in verdünnter Lösung, *Dissertation*, TU München, 1980.

1980KUR   Kuroiwa, S., Matsuda, H., and Fujimatsu, H., Temperature dependence of solution properties of nonionic surface active agent, *Nippon Kagaku Kaishi*, (3), 362, 1980.

1980SCH   Schönert, H. and Monshausen, F., Calorimetric measurements on dilute solutions of polyethylene oxide in water, *Coll. Polym. Sci.*, 258, 578, 1980.

1981GON   Gontschignorowijn, G., Mischungsenthalpien von Polyethylenglycollösungen in Alkoholen, *Diploma Paper*, TH Leuna-Merseburg, 1981.

1981KOL   Koller, J. and Killmann, E., Lösungs- und Verdünnungswärmen von Polyethylenglykolen in Wasser und Benzol, 1. Experimentelle Ergebnisse, *Makromol. Chem.*, 182, 3579, 1981.

1981OGA   Ogawa, H., Baba, Y., and Kagemoto, A., The heats of dilution of poly(1-vinyl-2-pyrrolidone)/alcohol systems, *Makromol. Chem.*, 182, 2495, 1981.

1982SH1   Sharma, S.C., Mahajan, R., Sharma, V.K., and Lakhanpal, M.L., Enthalpies of mixing of poly(tetramethylene oxides) with dioxane and cyclohexane, *Indian J. Chem.*, 21A, 682, 1982.

1982SH2   Sharma, S.C., Mahajan, R., Sharma, V.K., and Lakhanpal, M.L., Enthalpies of mixing of poly(tetramethylene oxides) with benzene and carbon tetrachloride, *Indian J. Chem.*, 21A, 685, 1982.

1983LAK   Lakhanpal, M.L. and Parashar, R.N., Enthalpies of mixing for aqueous solutions of polyoxyethylene glycols, *Indian J. Chem.*, 22A, 48, 1983.

1983SHA   Sharma, S.C. and Lakhanpal, M.L., Thermodynamics of mixtures of poly(tetramethylene oxide)s and 1,4-dioxane, *J. Polym. Sci.: Polym. Phys. Ed.*, 21, 353, 1983.

1984KIL   Killmann, E. and Graun, K., Microcalorimetric studies on enthalpic interactions of poly(methyl methacrylate) of varying tacticity in solution (Ger.), *Makromol. Chem.*, 185, 1199, 1984.

1985ALK   Al-Kafaji, J.K.H., Ariffin, Z., Cope, J., and Booth, C., Enthalpy and volume changes on mixing diethylene glycol di-n-alkyl ethers with diethylene glycol dimethyl ether or n-alkanes, *J. Chem. Soc., Faraday Trans I*, 81, 223, 1985.

1985COR   Cordt, F., Die Energetik der Wechselwirkungen von Oligomeren des Ethylenoxids und dessen Copolymeren mit Propylenoxid in CCl$_4$-Lösung, *Dissertation*, TU München, 1985.

1985RAB   Rabinovich, I.B., Khlyustova, T.B., and Mochalov, A.N., Physico-chemical analysis of mixtures of cellulose nitrate with triacetin and thermodynamics of their mixing (Russ.), *Vysokomol. Soedin., Ser. A*, 27, 1724, 1985.

1985SHA   Sharma, S.C. and Sharma, V.K., Enthalpies of mixing of poly(tetramethylene oxide) with benzene, methylbenzene, ethylbenzene and propylbenzene at 313.15 K, *Indian J. Chem.*, 24A, 292, 1985.

1986SHA   Sharma, S.C., Bhalla, S., and Sharma, V.K., Enthalpies of mixing of poly(tetramethylene oxide) with o-, m-, and p-xylenes and mesitylene, *Indian J. Chem.*, 25A, 131, 1986.

1987BIR   Biros, J., Pouchly, J., and Zivny, A., A calorimetric investigation of interactions in aqueous solutions of poly(oxyethylene) 1. Heats of mixing of oligomeric models, *Makromol.Chem.*, 188, 379, 1987.

1987KYO   Kyohmen, M., Inoue, K., Baba, Y., Kagemoto, A., and Beatty, Ch.L., Heats of dilution of poly[styrene-*ran*-(butyl methacrylate)] solutions measured with an automatic flow microcalorimeter, *Makromol. Chem.*, 188, 2721, 1987.

1987SHA   Sharma, S.C., Syngal, M., and Sharma, V.K., Enthalpies and excess volumes of mixing of poly(tetramethylene oxide) fractions with tetrachloromethane, 1,2-dichloroethane and 1,1,1-trichloroethane at 313.15 K, *Indian J. Chem.*, 26A, 285, 1987.

1988AZU   Azuma, H., Hanada, K., Yoshikawa, Y., Baba, Y., and Kagemoto, A., Heats of dilution of water-soluble polymer solutions, *Thermochim. Acta*, 123, 271, 1988.

1988PAR   Parashar, R. and Sharma, S.C., Enthalpies of mixing of polyoxypropyleneglycols with benzene, ethanol and water, *Indian J. Chem.*, 27A, 1092, 1988.

1989MOE   Möller, F., Energetik der Wechselwirkungen von Oligomeren des Ethylen- und Propylen-oxids und deren Cooligomeren in Mischung mit CCl$_4$, *Dissertation*, TU München, 1989.

1990KIL   Killmann, E., Cordt, F., and Möller, F., Thermodynamics of mixing ethylene oxide oligomers with different end groups in tetrachloromethane, *Makromol. Chem.*, 191, 2929, 1990.

1991NOV   Novoselova, N.V., Tsvetkova, L.Ya., Lebedev, Yu.A., and Muroshnichenko, E.A., Enthalpy of mixing of esters of phthalic acid and poly(vinyl chloride) (Russ.), *Zh. Obshch. Khim.*, 61, 75, 1991.

1992WOE   Wörfel, S., Mischungsenthalpien von Polyethylenglycol + n-Alkohol-Systemen, *Diploma Paper*, TH Leuna-Merseburg, 1992.

1993ZEL   Zellner, H., Mischungsthermodynamik und Energetik der Wechselwirkungen von Polyethern in Lösungsmitteln unterschiedlicher Polarität, *Dissertation*, TU München, 1993.

1994EST   Esteve, X., Boer, D., Patil, K.R., Chaudhari, S.K., and Coronas, A., Densities, viscosities, and enthalpies of mixing of the binary system methanol + polyethylene glycol 250 dimethyl ether at 303.15 K, *J. Chem. Eng. Data*, 39, 767, 1994.

1995BOG   Bogolitsyn, K.G. and Volkova, N.N., Enthalpies of dilution of the system lignin-dimethyl sulfoxide at 298.15 K, *Macromol. Chem. Phys.*, 196, 369, 1995.

1995CAR   Carlsson, M., Hallen, D., and Linse, P., Mixing enthalpy and phase separation in a poly(propylene oxide)-water system, *J. Chem. Soc., Faraday Trans.*, 91, 2081, 1995.

1995GRO   Grossmann, C., Tintinger, R., Zhu, J., and Maurer, G., Aqueous two-phase systems of poly(ethylene glycol) and dextran – experimental results and modeling of thermodynamic properties, *Fluid Phase Equil.*, 106, 111, 1995.

1995KIL   Killmann, E., Cordt, F., Möller, F., and Zellner, H., Thermodynamics of mixing propylene oxide oligomers with different end groups of statistical and block cooligomers of ethylene oxide and propylene oxide in tetrachloromethane, *Macromol. Chem. Phys.*, 196, 47, 1995.

1995PER   Peregudov, Yu.S., Amelin, A.N., and Perelygin, V.M., Thermodynamic characteristics of interactions between poly(styrenesulfonic acid), poly(vinylbenzyltrimethylammonium) hydroxide, and their salt forms with water (Russ.), *Vysokomol. Soedin., Ser. B*, 37, 302, 1995.

1995TIN   Tintinger, R., Thermodynamische Eigenschaften ausgewählter wässriger Zwei-Phasen Systeme, *Dissertation*, Universität Kaiserslautern, 1995.

1996PER   Peregudov, Yu.S., Amelin, A.N., and Perelygin, V.M., Termodinamicheskie kharakteristiki rastvorov polistirolsul'fokisloty i ee zhelznoi formy, *Izv. Vyssh. Uchebn. Zav., Khim. Khim. Tekhnol.*, 39, 63, 1996.

1997LOP   Lopez, E.R., Garcia, J., Coronas, A., and Fernandez, J., Experimental and predicted excess enthalpies of the working pairs (methanol or trifluoroethanol + polyglycol ethers) for absorption cycles, *Fluid Phase Equil.*, 133, 229, 1997.

1997MCL   McLure, I.A., Edmonds, B., and Mokhtari, A., Thermodynamics of linear dimethylsiloxane-perfluoroalkane mixtures Part 2. Excess volumes at 298.15 and 303.15 K and excess enthalpies at 298.15 K of hexamethyldisiloxane-tetradecafluorohexane, *J. Chem. Soc., Faraday Trans.*, 93, 257, 1997.

1998KSI   Ksiqzczak, A. and Ksiqzczak, T., Thermochemistry of the binary system nitrocellulose + *s*-diethyldiphenylurea, *J. Therm. Anal. Calorim.*, 54, 323, 1998.

1999BUR   Burgdorf, R., Zocholl, A., Arlt, W., and Knapp, H., Thermophysical properties of binary liquid mixtures of polyether and n-alkane at 298.15 and 323.15 K: heat of mixing, heat capacity, viscosity, density and thermal conductivity, *Fluid Phase Equil.*, 164, 225, 1999.

1999HER   Herraiz, J., Shen, S., Fernandez, J., and Coronas, A., Thermodynamic properties of methanol + some polyethylene glycol dimethyl ether by UNIFAC and DISQUAC group-contribution models for absorption heat pumps, *Fluid Phase Equil.*, 155, 327, 1999.

1999LOP   Lopez, E.R., Coxam, J.-Y., Fernandez, J., and Grolier, J.-P.E., Pressure and temperature dependence of excess enthalpies of methanol + tetraethylene glycol dimethyl ether and methanol + polyethylene glycol dimethyl ether 250, *J. Chem. Eng. Data*, 44, 1409, 1999.

2002COM   Comelli, F., Ottani, S., Francesconi, R., and Castellari, C., Densities, viscosities, refractive indices, and excess molar enthalpies of binary mixtures containing poly(ethylene glycol) 200 and 400 + dimethoxymethane and + 1,2-dimethoxyethane at 298.15 K, *J. Chem. Eng. Data*, 47, 1226, 2002.

2002COO   Cooke, S.A., Jonsdottir, S.O., and Westh, P., A thermodynamic study of glucose and related oligomers in aqueous solution: vapor pressures and enthalpies of mixing, *J. Chem. Eng. Data*, 47, 1185, 2002.

2002KSI   Ksiqzczak, A. and Wolszakiewicz, T., Thermochemistry of the binary system nitrocellulose + 2,6-dinitrotoluene, *J. Therm. Anal. Calorim.*, 67, 751, 2002.

2002RUI   Ruiz Holgado, M.E.F. de, Fernandez, J., Paz Andrade, M.I., and Arancibia, E.L., Excess molar enthalpies of mixtures of methyl derivatives of polyethylene glycol with 1-alkanol at 298.15 K and 101.3 kPa, *Can. J. Chem.*, 80, 462, 2002.

2002SAF   Safronov, A.P. and Somova, T.V., Thermodynamics of poly(vinyl chloride) mixing with phthalate plasticizers (Russ.), *Vysokomol. Soedin., Ser. A*, 44, 2014, 2002.

2003CO1   Comelli, F., Ottani, S., Francesconi, R., and Castellari, C., Excess molar enthalpies of binary mixtures containing glycols or polyglycols + dimethyl sulfoxide at 308.15 K, *J. Chem. Eng. Data*, 48, 995, 2003.

2003CO2   Comelli, F., Ottani, S., Vitalini, D., and Francesconi, R., A calorimetric study of binary mixtures containing some glycols and polyglycols + anisole at 308.15 K and atmospheric pressure, *Thermochim. Acta*, 407, 85, 2003.

2003OTT   Ottani, S., Francesconi, R., Comelli, F., and Castellari, C., Excess molar enthalpies of binary mixtures containing poly(ethylene glycol) 200 + four cyclic ethers at (288.15, 298.15 and 313.15) K and at atmospheric pressure, *Thermochim. Acta*, 401, 87, 2003.

2004CA1   Castellari, C., Francesconi, R., and Comelli, F., Excess molar enthalpies and hydrogen bonding in binary mixtures containing glycols or poly(ethylene glycols) and 2-phenylethyl alcohol at 308.15 K and atmospheric pressure, *J. Chem. Eng. Data*, 49, 1032, 2004.

2004CA2   Castellari, C., Vitalini, D., Comelli, F., and Francesconi, R., Effect of excess enthalpies on binary mixtures containing propylene glycols and poly(propylene glycols) + dimethyl carbonate at 308.15 K, *Thermochim. Acta*, 412, 125, 2004.

2004CA3   Castellari, C., Francesconi, R., and Comelli, F., Excess molar enthalpies and hydrogen bonding in binary mixtures containing glycols of poly(ethylene glycols) and 3-phenylpropyl alcohol at 308.15 K and atmospheric pressure, *Thermochim. Acta*, 424, 69, 2004.

2004CA4   Castellari, C., Comelli, F., and Francesconi, R., Excess molar enthalpies of binary mixtures containing glycols or poly(ethylene glycol)s + propylene carbonate at 308.15 K, *Thermochim. Acta*, 413, 249, 2004.

2004COM   Comelli, F. and Ottani, S., Excess enthalpies, densities, viscosities, and refractive indices of binary mixtures involving some poly(glycols) + diethyl carbonate at 308.15 K, *J. Chem. Eng. Data*, 49, 970, 2004.

2004FRA   Francesconi, R., Castellari, C., Comelli, F., and Ottani, S., Thermodynamic study of binary mixtures containing glycols or polyethylene glycols + benzyl alcohol at 308.15 K, *J. Chem. Eng. Data*, 49, 363, 2004.

2004WOL   Wolszakiewicz, T., Ksiqzczak, A., and Ksiqzczak, T., Thermochemistry of the binary system nitrocellulose + 2,4-dinitrotoluene, *J. Therm. Anal. Calorim.*, 77, 353, 2004.

2005BIG   Bigi, A., Comelli, F., Excess molar enthalpies of binary mixtures containing ethylene glycols of poly(ethylene glycols) + ethyl alcohol at 308.15 K and atmospheric pressure (experimental data by A. Bigi), *Thermochim. Acta*, 430, 191, 2005.

2005COM   Comelli, F. and Ottani, S., Excess molar enthalpies of binary mixtures containing poly(propylene glycols) + benzyl alcohol, or + m-cresol, or + anisole at 308.15 K and at atmospheric pressure, *Thermochim. Acta*, 430, 123, 2005.

3. POLYMER PARTIAL ENTHALPIES OF MIXING (AT INFINITE DILUTION) OR POLYMER (FIRST) INTEGRAL ENTHALPIES OF SOLUTION

3.1. Experimental data

Polymer (B):   **benzylcellulose**   **1951GLI, 1956GLI**
Characterization: –
Solvent (A):   **benzene** C_6H_6 **71-43-2**

$T/K = 298.15$ $\Delta_{sol}H_B^{\infty} = -10.5$ J/(g polymer)

Polymer (B):   **benzylcellulose**   **1956STR, 1958SLO**
Characterization: intrinsic viscosity = 1.40
Solvent (A):   **cyclohexanone** $C_6H_{10}O$ **108-94-1**

$T/K = 298.15$ $\Delta_{sol}H_B^{\infty} = -14.52$ J/(g polymer)

Comments: The final concentration is 1 g polymer/100 g solvent.

Polymer (B):   **benzylcellulose**   **1956STR, 1958SLO**
Characterization: intrinsic viscosity = 1.40
Solvent (A):   **trichloromethane** $CHCl_3$ **67-66-3**

$T/K = 298.15$ $\Delta_{sol}H_B^{\infty} = -37.45$ J/(g polymer)

Comments: The final concentration is 1 g polymer/100 g solvent.

Polymer (B):   **cellulose**   **1999NOV**
Characterization: DP = 395
Solvent (A):   **N-methylmorpholine-N-oxide** $C_5H_{11}NO_2$ **7529-22-8**
Solvent (C):   **water** H_2O **7732-18-5**

$w_A/w_C = 0.17/0.83$ was kept constant

T/K	323.15	343.15	353.15	363.15
$\Delta_{sol}H_B$/(J/g polymer)	−60	−80	−82	−100

Comments: The final concentration is 0.2 wt%. $\Delta_{sol}H_B$ values at higher polymer concentrations are given in the original source.

Polymer (B):	**cellulose**		**2001NOV**
Characterization:	DP = 424, 95.8 % α-cellulose		
Solvent (A):	**N-methylmorpholine-N-oxide**		
	monohydrate	**$C_5H_{11}NO_2.H_2O$**	**70187-32-5**

$T/K = 348.15$ $\qquad$ $\Delta_{sol}H_B^{\infty} = -142.8$ J/(g polymer)

Comments: The final concentration is 0.8 wt%.

Polymer (B):	**cellulose**		**2001NOV**
Characterization:	DP = 424, 95.8 % α-cellulose		
Solvent (A):	**N-methylmorpholine-N-oxide**		
	monohydrate	**$C_5H_{11}NO_2.H_2O$**	**70187-32-5**
Solvent (C):	**N,N-dimethylacetamide**	**C_4H_9NO**	**127-19-5**

$T/K = 348.15$

w_A/w_C	924/76	795/205	713/287	517/483
$\Delta_{sol}H_B^{\infty}$/(J/g polymer)	−121.6	−134.7	−138.1	−137.1

Comments: The final concentration is 0.8 to 0.9 wt%.

Polymer (B):	**cellulose**		**2001NOV**
Characterization:	DP = 424, 95.8 % α-cellulose		
Solvent (A):	**N-methylmorpholine-N-oxide**		
	monohydrate	**$C_5H_{11}NO_2.H_2O$**	**70187-32-5**
Solvent (C):	**N,N-dimethylformamide**	**C_3H_7NO**	**68-12-2**

$T/K = 348.15$

w_A/w_C	904/96	810/190	528/472
$\Delta_{sol}H_B^{\infty}$/(J/g polymer)	−151.7	−136.7	−136.4

Comments: The final concentration is 0.8 to 0.9 wt%.

Polymer (B):	**cellulose**		**2001NOV**
Characterization:	DP = 424, 95.8 % α-cellulose		
Solvent (A):	**N-methylmorpholine-N-oxide**		
	monohydrate	**$C_5H_{11}NO_2.H_2O$**	**70187-32-5**
Solvent (C):	**dimethylsulfoxide**	**C_2H_6OS**	**67-68-5**

$T/K = 348.15$

w_A/w_C	892/108	855/145	784/216	585/415	476/524
$\Delta_{sol}H_B^{\infty}$/(J/g polymer)	−155.0	−171.2	−243.9	−159.6	−145.7

Comments: The final concentration is 0.8 to 0.9 wt%.

Polymer (B):   **cellulose acetate**   **1951TA1**
Characterization: 52.2% acetate
Solvent (A):   **formic acid** CH_2O_2 **64-18-6**

$T/K = 298.15$ $\Delta_{sol} H_B^\infty = -30.1$ J/(g polymer)

Comments: The final concentration is 1 g polymer/100 g solvent.

Polymer (B):   **cellulose acetate**   **1951TA1**
Characterization: 55.8% acetate
Solvent (A):   **formic acid** CH_2O_2 **64-18-6**

$T/K = 298.15$ $\Delta_{sol} H_B^\infty = -43.9$ J/(g polymer)

Comments: The final concentration is 1 g polymer/100 g solvent.

Polymer (B):   **cellulose acetate**   **1934LIE**
Characterization: 52.5% acetate
Solvent (A):   **methyl acetate** $C_3H_6O_2$ **79-20-9**

$T/K =$ room temperature $\Delta_{sol} H_B^\infty = -79.5$ J/(g polymer)

Polymer (B):   **cellulose acetate**   **1956STR, 1958SLO**
Characterization: 48 wt% acetate
Solvent (A):   **2-propanone** C_3H_6O **67-64-1**

$T/K = 298.15$ $\Delta_{sol} H_B^\infty = -35.4$ J/(g polymer)

Comments: The final concentration is 1 g polymer/100 g solvent.

Polymer (B):   **cellulose acetate**   **1951TA1**
Characterization: 52.2% acetate
Solvent (A):   **2-propanone** C_3H_6O **67-64-1**

$T/K = 298.15$ $\Delta_{sol} H_B^\infty = -29.7$ J/(g polymer)

Comments: The final concentration is 1 g polymer/100 g solvent.

Polymer (B):   **cellulose acetate**   **1951TA1**
Characterization: 55.8% acetate
Solvent (A):   **2-propanone** C_3H_6O **67-64-1**

$T/K = 298.15$ $\Delta_{sol} H_B^\infty = -26.35$ J/(g polymer)

Comments: The final concentration is 1 g polymer/100 g solvent.

Polymer (B):   **cellulose acetate**   **1941TAG**
Characterization: 56 wt% acetate
Solvent (A):   **2-propanone**   **C$_3$H$_6$O**   **67-64-1**

T/K = 298.15 $\Delta_{sol}H_B$ = −28.87 J/(g polymer)

Comments: The final concentration is 0.29 wt%.

Polymer (B):   **cellulose acetate**   **1956STR, 1958SLO**
Characterization: 56 wt% acetate
Solvent (A):   **2-propanone**   **C$_3$H$_6$O**   **67-64-1**

T/K = 298.15 $\Delta_{sol}H_B^{\infty}$ = −44.56 J/(g polymer)

Comments: The final concentration is 1 g polymer/100 g solvent.

Polymer (B):   **cellulose triacetate**   **1941TAG**
Characterization: −
Solvent (A):   **2-propanone**   **C$_3$H$_6$O**   **67-64-1**

T/K = 298.15 $\Delta_{sol}H_B$ = −29.3 J/(g polymer)

Comments: The final concentration is 2 wt%.

Polymer (B):   **cellulose triacetate**   **1941TAG**
Characterization: −
Solvent (A):   **trichloromethane**   **CHCl$_3$**   **67-66-3**

T/K = 298.15 $\Delta_{sol}H_B$ = −47.3 J/(g polymer)

Comments: The final concentration is 0.68 wt%.

Polymer (B):   **decamethyltetrasiloxane**   **1968MOR**
Characterization: M/g.mol^{-1} = 311, fraction from a commercial silicone,
                                          KF 96, Shinetsu Chemicals Co., fractionated in the laboratory,
                                          η = 1.5 cSt, T_b = 192 °C
Solvent (A):   **benzene**   **C$_6$H$_6$**   **71-43-2**

T/K = 298.15 $\Delta_M H_B^{\infty}$ = 6297 J/mol

Polymer (B):   **decamethyltetrasiloxane**   **1968MOR**
Characterization: M/g.mol^{-1} = 311, fraction from a commercial silicone,
                                          KF 96, Shinetsu Chemicals Co., fractionated in the laboratory,
                                          η = 1.5 cSt, T_b = 192 °C
Solvent (A):   **1,3-dimethylbenzene**   **C$_8$H$_{10}$**   **108-38-3**

T/K = 298.15 $\Delta_M H_B^{\infty}$ = 3213 J/mol

Polymer (B):   **decamethyltetrasiloxane**   **1975TAN**
Characterization: $M/\text{g.mol}^{-1} = 310.7$, Dow Corning Silicones
Solvent (A):   **2,2-dimethylbutane**   **C$_6$H$_{14}$**   **75-83-2**

$T/\text{K} = 298.15$ $\Delta_M H_B^\infty = 192$ J/mol

Polymer (B):   **decamethyltetrasiloxane**   **1975TAN**
Characterization: $M/\text{g.mol}^{-1} = 310.7$, Dow Corning Silicones
Solvent (A):   **n-dodecane**   **C$_{12}$H$_{26}$**   **112-40-3**

$T/\text{K} = 298.15$ $\Delta_M H_B^\infty = 2141$ J/mol

Polymer (B):   **decamethyltetrasiloxane**   **1968MOR**
Characterization: $M/\text{g.mol}^{-1} = 311$, fraction from a commercial silicone,
                       KF 96, Shinetsu Chemicals Co., fractionated in the laboratory,
                       $\eta = 1.5$ cSt, $T_b = 192$ °C
Solvent (A):   **ethylbenzene**   **C$_8$H$_{10}$**   **100-41-4**

$T/\text{K} = 298.15$ $\Delta_M H_B^\infty = 3895$ J/mol

Polymer (B):   **decamethyltetrasiloxane**   **1975TAN**
Characterization: $M/\text{g.mol}^{-1} = 310.7$, Dow Corning Silicones
Solvent (A):   **2,2,4,4,6,8,8-heptamethylnonane**   **C$_{16}$H$_{34}$**   **4390-04-9**

$T/\text{K} = 298.15$ $\Delta_M H_B^\infty = 1417$ J/mol

Polymer (B):   **decamethyltetrasiloxane**   **1968MOR**
Characterization: $M/\text{g.mol}^{-1} = 311$, fraction from a commercial silicone,
                       KF 96, Shinetsu Chemicals Co., fractionated in the laboratory,
                       $\eta = 1.5$ cSt, $T_b = 192$ °C
Solvent (A):   **n-heptane**   **C$_7$H$_{16}$**   **142-82-5**

$T/\text{K} = 298.15$ $\Delta_M H_B^\infty = 1092$ J/mol

Polymer (B):   **decamethyltetrasiloxane**   **1975TAN**
Characterization: $M/\text{g.mol}^{-1} = 310.7$, Dow Corning Silicones
Solvent (A):   **n-hexadecane**   **C$_{16}$H$_{34}$**   **544-76-3**

$T/\text{K} = 298.15$ $\Delta_M H_B^\infty = 3320$ J/mol

Polymer (B):   **decamethyltetrasiloxane**   **1975TAN**
Characterization: $M/\text{g.mol}^{-1} = 310.7$, Dow Corning Silicones
Solvent (A):   **n-hexane**   **C$_6$H$_{14}$**   **110-54-3**

$T/\text{K} = 298.15$ $\Delta_M H_B^\infty = 150$ J/mol

Polymer (B):	**decamethyltetrasiloxane**	**1975TAN**
Characterization:	M/g.mol^{-1} = 310.7, Dow Corning Silicones	
Solvent (A):	**n-octane** **C$_8$H$_{18}$**	**111-65-9**

T/K = 298.15 $\Delta_M H_B^\infty$ = 1115 J/mol

Polymer (B):	**decamethyltetrasiloxane**	**1975TAN**
Characterization:	M/g.mol^{-1} = 310.7, Dow Corning Silicones	
Solvent (A):	**2,2,4,6,6-pentamethylheptane C$_{12}$H$_{26}$**	**13475-82-6**

T/K = 298.15 $\Delta_M H_B^\infty$ = 950 J/mol

Polymer (B):	**decamethyltetrasiloxane**	**1975TAN**
Characterization:	M/g.mol^{-1} = 310.7, Dow Corning Silicones	
Solvent (A):	**2,2,4-trimethylpentane C$_8$H$_{18}$**	**540-84-1**

T/K = 298.15 $\Delta_M H_B^\infty$ = 750 J/mol

Polymer (B):	**dextran**	**1979BAS, 1980BAS**
Characterization:	M_n/g.mol^{-1} = 8200, M_w/g.mol^{-1} = 10400, fractionated in the laboratory	
Solvent (A):	**dimethylsulfoxide C$_2$H$_6$OS**	**67-68-5**

T/K = 298.15 $\Delta_{sol} H_B^\infty$ = −185 J/(g polymer)

Comments: The final concentration is 0.1 g polymer/25 ml solvent.

Polymer (B):	**dextran**	**1979BAS, 1980BAS**
Characterization:	M_n/g.mol^{-1} = 75900, M_w/g.mol^{-1} = 101000, fractionated in the laboratory	
Solvent (A):	**dimethylsulfoxide C$_2$H$_6$OS**	**67-68-5**

T/K = 298.15 $\Delta_{sol} H_B^\infty$ = −187 J/(g polymer)

Comments: The final concentration is 0.1 g polymer/25 ml solvent.

Polymer (B):	**dextran**	**1979BAS, 1980BAS**
Characterization:	M_n/g.mol^{-1} = 75900, M_w/g.mol^{-1} = 101000, fractionated in the laboratory	
Solvent (A):	**1,2-ethanediol C$_2$H$_6$O$_2$**	**107-21-1**

T/K = 298.15 $\Delta_{sol} H_B^\infty$ = −98 J/(g polymer)

Comments: The final concentration is 0.1 g polymer/25 ml solvent.

Polymer (B):   **dextran**   **1979BAS, 1980BAS**
Characterization: M_n/g.mol^{-1} = 75900, M_w/g.mol^{-1} = 101000,
                        fractionated in the laboratory
Solvent (A):   **formamide**   **CH$_3$NO**   **75-12-7**

T/K = 298.15 $\Delta_{sol} H_B^\infty$ = −228 J/(g polymer)

Comments: The final concentration is 0.1 g polymer/25 ml solvent.

Polymer (B):   **dextran**   **1979BAS, 1980BAS**
Characterization: M_n/g.mol^{-1} = 8200, M_w/g.mol^{-1} = 10400,
                        fractionated in the laboratory
Solvent (A):   **water**   **H$_2$O**   **7732-18-5**

T/K = 298.15 $\Delta_{sol} H_B^\infty$ = −140 J/(g polymer)

Polymer (B):   **dextran**   **1979BAS, 1980BAS**
Characterization: M_n/g.mol^{-1} = 75900, M_w/g.mol^{-1} = 101000,
                        fractionated in the laboratory
Solvent (A):   **water**   **H$_2$O**   **7732-18-5**

T/K = 298.15 $\Delta_{sol} H_B^\infty$ = −150 J/(g polymer)

Comments: The final concentration is 0.1 g polymer/25 ml solvent.

Polymer (B):   **dextran**   **1976KIS**
Characterization: ρ = 1.303 g/cm^3, completely amorphous sample
Solvent (A):   **water**   **H$_2$O**   **7732-18-5**

T/K = 298.15 $\Delta_{sol} H_B^\infty$ = −123.4 J/(g polymer)

Polymer (B):   **gelatine**   **1958MEE**
Characterization: −
Solvent (A):   **water**   **H$_2$O**   **7732-18-5**

T/K = 293.15 $\Delta_{sol} H_B^\infty$ = −92.0 J/(g polymer)

T/K = 323.15 $\Delta_{sol} H_B^\infty$ = −62.8 J/(g polymer)

Comments: Additional data were given by a figure in the original source, also for solutions containing
                urea.

Polymer (B):   **guttapercha**   **1956LIP**
Characterization: −
Solvent (A):   **trichloromethane**   **CHCl$_3$**   **67-66-3**

T/K = 303.15 $\Delta_{sol} H_B^\infty$ = 47.3 J/(g polymer)   (non-oriented sample)

T/K = 303.15 $\Delta_{sol} H_B^\infty$ = 46.0 J/(g polymer)   (oriented sample, drawn 200%)

Polymer (B):   **hexamethyldisiloxane**   **1968MOR**
Characterization: M/g.mol^{-1} = 162, fraction from a commercial silicone,
                          KF 96, Shinetsu Chemicals Co., fractionated in the laboratory,
                          η = 0.65 cSt, T_b = 99.5 °C
Solvent (A):   **benzene**   **C$_6$H$_6$**   **71-43-2**

T/K = 298.15 $\Delta_M H_B^\infty$ = 5339 J/mol

Polymer (B):   **hexamethyldisiloxane**   **1968MOR**
Characterization: M/g.mol^{-1} = 162, fraction from a commercial silicone,
                          KF 96, Shinetsu Chemicals Co., fractionated in the laboratory,
                          η = 0.65 cSt, T_b = 99.5 °C
Solvent (A):   **1,3-dimethylbenzene**   **C$_8$H$_{10}$**   **108-38-3**

T/K = 298.15 $\Delta_M H_B^\infty$ = 2234 J/mol

Polymer (B):   **hexamethyldisiloxane**   **1968MOR**
Characterization: M/g.mol^{-1} = 162, fraction from a commercial silicone,
                          KF 96, Shinetsu Chemicals Co., fractionated in the laboratory,
                          η = 0.65 cSt, T_b = 99.5 °C
Solvent (A):   **ethylbenzene**   **C$_8$H$_{10}$**   **100-41-4**

T/K = 298.15 $\Delta_M H_B^\infty$ = 2640 J/mol

Polymer (B):   **hexamethyldisiloxane**   **1968MOR**
Characterization: M/g.mol^{-1} = 162, fraction from a commercial silicone,
                          KF 96, Shinetsu Chemicals Co., fractionated in the laboratory,
                          η = 0.65 cSt, T_b = 99.5 °C
Solvent (A):   **n-heptane**   **C$_7$H$_{16}$**   **142-82-5**

T/K = 298.15 $\Delta_M H_B^\infty$ = 920 J/mol

Polymer (B):   **natural rubber**   **1956GLI**
Characterization: –
Solvent (A):   **benzene**   **C$_6$H$_6$**   **71-43-2**

T/K = 298.15 $\Delta_{sol} H_B^\infty$ = 9.96 J/(g polymer)

Polymer (B):   **natural rubber**   **1956STR, 1958SLO**
Characterization: –
Solvent (A):   **benzene**   **C$_6$H$_6$**   **71-43-2**

T/K = 298.15 $\Delta_{sol} H_B^\infty$ = 11.7 J/(g polymer)

Comments: The final concentration is 1 g polymer/100 g solvent.

Polymer (B):	nitrocellulose		1941TAG
Characterization:	M_η/g.mol^{-1} = 16600		
Solvent (A):	2-butanone	C$_4$H$_8$O	78-93-3

T/K = 298.15 $\Delta_{sol} H_B^\infty$ = −80.3 J/(g polymer)

Polymer (B):	nitrocellulose		1941TAG
Characterization:	M_η/g.mol^{-1} = 23000, 11.83 % nitrogen		
Solvent (A):	2-butanone	C$_4$H$_8$O	78-93-3

T/K = 298.15 $\Delta_{sol} H_B^\infty$ = −80.75 J/(g polymer)

Polymer (B):	nitrocellulose		1941TAG
Characterization:	M_η/g.mol^{-1} = 40000, 11.97 % nitrogen		
Solvent (A):	2-butanone	C$_4$H$_8$O	78-93-3

T/K = 298.15 $\Delta_{sol} H_B^\infty$ = −80.75 J/(g polymer)

Polymer (B):	nitrocellulose		1956MEE
Characterization:	–		
Solvent (A):	butyl acetate	C$_6$H$_{12}$O$_2$	123-86-4

T/K	293.15	298.15	303.15	308.15	313.15	318.15	323.15
$\Delta_{sol} H_B^\infty$/(J/g polymer)	−74.5	−74.5	−73.6	−71.1	−65.3	−58.6	−53.6

T/K	328.15
$\Delta_{sol} H_B^\infty$/(J/g polymer)	−50.2

Polymer (B):	nitrocellulose		1957GAL
Characterization:	12% nitrogen		
Solvent (A):	butyl acetate	C$_6$H$_{12}$O$_2$	123-86-4

T/K	298.15	313.15	333.15	343.15	353.15
$\Delta_{sol} H_B$/(J/g polymer)	−73.2	−66.9	−58.6	−54.8	−47.3

Comments: The final concentration is about 1.5 g/150 ml.

Polymer (B):	nitrocellulose		1957GAL
Characterization:	12% nitrogen		
Solvent (A):	dibutyl phthalate	C$_{16}$H$_{22}$O$_4$	84-74-2

T/K	273.15	298.15	313.15	333.15
$\Delta_{sol} H_B$/(J/g polymer)	−44.8	−45.6	−46.0	−41.8

Comments: The final concentration is about 1.5 g/150 ml.

Polymer (B):	nitrocellulose						**1937PAP**
Characterization:	11.5% nitrogen						
Solvent (A):	**ethanol**		C_2H_6O				**64-17-5**
Solvent (C):	**diethyl ether**		$C_4H_{10}O$				**60-29-7**

T/K = room temperature

Comments: The partial specific enthalpies of solution at infinite dilution were measured for solutions of nitrocellulose in solvent mixtures with given volume-fraction ratios.

φ_A/φ_C	1/0	8/2	5/5	4/6	35/65	25/75	15/85
$\Delta_{sol}H_B^{\infty}$/(J/g polymer)	−45.6	−52.7	−61.1	−61.5	−63.6	−61.9	−60.7

φ_A/φ_C	1/9	0/1
$\Delta_{sol}H_B^{\infty}$/(J/g polymer)	−63.6	−61.5

Polymer (B):	nitrocellulose						**1956MEE**
Characterization:	−						
Solvent (A):	**ethyl acetate**		$C_4H_8O_2$				**141-78-6**

T/K	293.15	298.15	303.15	308.15	313.15	318.15	323.15
$\Delta_{sol}H_B^{\infty}$/(J/g polymer)	−76.1	−75.3	−68.6	−61.1	−54.0	−50.2	−50.2

T/K	328.15
$\Delta_{sol}H_B^{\infty}$/(J/g polymer)	−50.2

Polymer (B):	nitrocellulose						**1956MEE**
Characterization:	−						
Solvent (A):	**methanol**		CH_4O				**67-56-1**

T/K	293.15	298.15	303.15	308.15	313.15	318.15	323.15
$\Delta_{sol}H_B^{\infty}$/(J/g polymer)	−68.6	−56.1	−50.2	−50.2	−50.2	−50.2	−50.2

T/K	328.15
$\Delta_{sol}H_B^{\infty}$/(J/g polymer)	−50.2

Polymer (B):	nitrocellulose	**1941TAG**
Characterization:	M_η/g.mol^{-1} = 23000, 11.83 % nitrogen	
Solvent (A):	**2,4-pentanedione** $\quad$ $C_5H_8O_2$	**123-54-6**

T/K = 298.15 $\qquad$ $\Delta_{sol}H_B^{\infty}$ = −73.6 J/(g polymer)

Polymer (B):	nitrocellulose	**1941TAG**
Characterization:	M_η/g.mol^{-1} = 23000, 11.83 % nitrogen	
Solvent (A):	**2-pentanone** $\quad$ $C_5H_{10}O$	**107-87-9**

T/K = 298.15 $\qquad$ $\Delta_{sol}H_B^{\infty}$ = −63.6 J/(g polymer)

Polymer (B):	nitrocellulose		1935KAR
Characterization:	11.5% nitrogen		
Solvent (A):	**2-propanone**	C_3H_6O	67-64-1

$T/K = 298.15$ $\Delta_{sol} H_B^\infty = -82.8$ J/(g polymer)

Polymer (B):	nitrocellulose		1941TAG
Characterization:	$M_\eta/\text{g.mol}^{-1} = 23000$, 11.83 % nitrogen		
Solvent (A):	**2-propanone**	C_3H_6O	67-64-1

$T/K = 298.15$ $\Delta_{sol} H_B^\infty = -68.2$ J/(g polymer)

Polymer (B):	nitrocellulose		1951GLI, 1956GLI
Characterization:	–		
Solvent (A):	**2-propanone**	C_3H_6O	67-64-1

$T/K = 298.15$ $\Delta_{sol} H_B^\infty = -71.1$ J/(g polymer)

Polymer (B):	nitrocellulose		1956MEE
Characterization:	–		
Solvent (A):	**2-propanone**	C_3H_6O	67-64-1

T/K	293.15	298.15	303.15	308.15	313.15	318.15	323.15
$\Delta_{sol} H_B^\infty/$(J/g polymer)	−75.3	−73.6	−60.2	−51.5	−50.2	−50.2	−50.2

T/K	328.15
$\Delta_{sol} H_B^\infty/$(J/g polymer)	−50.2

Polymer (B):	nitrocellulose		1956STR, 1958SLO
Characterization:	11.9% nitrogen		
Solvent (A):	**2-propanone**	C_3H_6O	67-64-1

$T/K = 298.15$ $\Delta_{sol} H_B^\infty = -78.58$ J/(g polymer)

Comments: The final concentration is 1 g polymer/100 g solvent.

Polymer (B):	nitrocellulose		1957GAL
Characterization:	12% nitrogen		
Solvent (A):	**2-propanone**	C_3H_6O	67-64-1

T/K	273.15	298.15	313.15	333.15
$\Delta_{sol} H_B/$(J/g polymer)	−74.9	−75.3	−64.8	−50.2

Comments: The final concentration is about 1.5 g/150 ml.

Polymer (B):   **nitrocellulose**   **1935KAR**
Characterization: 11.5% nitrogen
Solvent (A):   **pyridine** C_5H_5N **110-86-1**

$T/K = 298.15$ $\Delta_{sol}H_B^{\infty} = -105.6$ J/(g polymer)

Polymer (B):   **nitrocellulose**   **1957GAL**
Characterization: 12% nitrogen
Solvent (A):   **tris(4-methylphenyl) phosphate** $C_{21}H_{21}O_4P$ **78-32-0**

T/K	298.15	313.15	333.15	343.15	353.15
$\Delta_{sol}H_B$/(J/g polymer)	−16.3	−28.0	−41.4	−44.4	−47.3

Comments: The final concentration is about 1.5 g/150 ml.

Polymer (B):   **nylon 6**   **1956MIK**
Characterization: Capron fibre
Solvent (A):   **formic acid (95.01%)** CH_2O_2 **64-18-6**

T/K = room temperature $\Delta_{sol}H_B = -52.7$ J/g polymer   (non-oriented sample)

T/K = room temperature $\Delta_{sol}H_B = -48.1$ J/g polymer   (oriented fibre)

Polymer (B):   **nylon 6**   **1956LIP**
Characterization: –
Solvent (A):   **tricresol** C_7H_8O **1319-77-3**

$T/K = 323.55$ $\Delta_{sol}H_B^{\infty} = -66.5$ J/(g polymer)   (non-oriented sample)

$T/K = 323.55$ $\Delta_{sol}H_B^{\infty} = -68.6$ J/(g polymer)   (oriented sample, drawn 200%)

$T/K = 345.55$ $\Delta_{sol}H_B^{\infty} = -65.7$ J/(g polymer)   (non-oriented sample)

$T/K = 345.55$ $\Delta_{sol}H_B^{\infty} = -62.8$ J/(g polymer)   (oriented sample, drawn 200%)

Polymer (B):   **octaacetylcellobiose**   **1941TAG**
Characterization: –
Solvent (A):   **2-propanone** C_3H_6O **67-64-1**

$T/K = 298.15$ $\Delta_{sol}H_B = 25.1$ J/(g polymer)

Comments: The final concentration is 2 wt%.

Polymer (B):   **octaacetylcellobiose**   **1941TAG**
Characterization: –
Solvent (A):   **trichloromethane** $CHCl_3$ **67-66-3**

$T/K = 298.15$ $\Delta_{sol}H_B = -4.60$ J/(g polymer)

Comments: The final concentration is 0.68 wt%.

Polymer (B):   octamethylcyclotetrasiloxane   **1975TAN**
Characterization: M/g.mol^{-1} = 296.6, Dow Corning Silicones
Solvent (A):   **2,2-dimethylbutane   C$_6$H$_{14}$**   **75-83-2**

T/K = 298.15 $\Delta_M H_B^\infty$ = 542 J/mol

Polymer (B):   octamethylcyclotetrasiloxane   **1975TAN**
Characterization: M/g.mol^{-1} = 296.6, Dow Corning Silicones
Solvent (A):   **n-dodecane   C$_{12}$H$_{26}$**   **112-40-3**

T/K = 298.15 $\Delta_M H_B^\infty$ = 2873 J/mol

Polymer (B):   octamethylcyclotetrasiloxane   **1975TAN**
Characterization: M/g.mol^{-1} = 296.6, Dow Corning Silicones
Solvent (A):   **2,2,4,4,6,8,8-heptamethylnonane   C$_{16}$H$_{34}$**   **4390-04-9**

T/K = 298.15 $\Delta_M H_B^\infty$ = 2231 J/mol

Polymer (B):   octamethylcyclotetrasiloxane   **1975TAN**
Characterization: M/g.mol^{-1} = 296.6, Dow Corning Silicones
Solvent (A):   **n-hexadecane   C$_{16}$H$_{34}$**   **544-76-3**

T/K = 298.15 $\Delta_M H_B^\infty$ = 4265 J/mol

Polymer (B):   octamethylcyclotetrasiloxane   **1975TAN**
Characterization: M/g.mol^{-1} = 296.6, Dow Corning Silicones
Solvent (A):   **n-hexane   C$_6$H$_{14}$**   **110-54-3**

T/K = 298.15 $\Delta_M H_B^\infty$ = 1041 J/mol

Polymer (B):   octamethylcyclotetrasiloxane   **1975TAN**
Characterization: M/g.mol^{-1} = 296.6, Dow Corning Silicones
Solvent (A):   **n-octane   C$_8$H$_{18}$**   **111-65-9**

T/K = 298.15 $\Delta_M H_B^\infty$ = 1422 J/mol

Polymer (B):   octamethylcyclotetrasiloxane   **1975TAN**
Characterization: M/g.mol^{-1} = 296.6, Dow Corning Silicones
Solvent (A):   **2,2,4,6,6-pentamethylheptane   C$_{12}$H$_{26}$**   **13475-82-6**

T/K = 298.15 $\Delta_M H_B^\infty$ = 1579 J/mol

Polymer (B):   octamethylcyclotetrasiloxane   **1975TAN**
Characterization: M/g.mol^{-1} = 296.6, Dow Corning Silicones
Solvent (A):   **2,2,4-trimethylpentane C$_8$H$_{18}$**   **540-84-1**

T/K = 298.15 $\Delta_M H_B^\infty$ = 1099 J/mol

Polymer (B):   **octamethyltrisiloxane**   **1968MOR**
Characterization: M/g.mol^{-1} = 237, fraction from a commercial silicone,
                          KF 96, Shinetsu Chemicals Co., fractionated in the laboratory,
                          η = 1.0 cSt, T_b = 152 °C
Solvent (A):   **benzene**   **C$_6$H$_6$**   **71-43-2**

T/K = 298.15 $\Delta_M H_B^\infty$ = 6205 J/mol

Polymer (B):   **octamethyltrisiloxane**   **1968MOR**
Characterization: M/g.mol^{-1} = 237, fraction from a commercial silicone,
                          KF 96, Shinetsu Chemicals Co., fractionated in the laboratory,
                          η = 1.0 cSt, T_b = 152 °C
Solvent (A):   **1,3-dimethylbenzene**   **C$_8$H$_{10}$**   **108-38-3**

T/K = 298.15 $\Delta_M H_B^\infty$ = 3251 J/mol

Polymer (B):   **octamethyltrisiloxane**   **1968MOR**
Characterization: M/g.mol^{-1} = 237, fraction from a commercial silicone,
                          KF 96, Shinetsu Chemicals Co., fractionated in the laboratory,
                          η = 1.0 cSt, T_b = 152 °C
Solvent (A):   **ethylbenzene**   **C$_8$H$_{10}$**   **100-41-4**

T/K = 298.15 $\Delta_M H_B^\infty$ = 3945 J/mol

Polymer (B):   **octamethyltrisiloxane**   **1968MOR**
Characterization: M/g.mol^{-1} = 237, fraction from a commercial silicone,
                          KF 96, Shinetsu Chemicals Co., fractionated in the laboratory,
                          η = 1.0 cSt, T_b = 152 °C
Solvent (A):   **n-heptane**   **C$_7$H$_{16}$**   **142-82-5**

T/K = 298.15 $\Delta_M H_B^\infty$ = 1105 J/mol

Polymer (B):   **oxyethylcellulose**   **1982MEE**
Characterization: 3 mole ethylene oxide per 1 mole cellulosic units
Solvent (A):   **water**   **H$_2$O**   **7732-18-5**

T/K = 318.15 $\Delta_{sol} H_B^\infty$ = –55.3 J/(g polymer)

Comments: The final concentration is 0.13 polymer/20 g solution.

Polymer (B):   **penta(ethylene glycol)**   **1998OHT**
Characterization: M/g.mol^{-1} = 238.3
Solvent (A):   **water**   **H$_2$O**   **7732-18-5**

T/K = 283.15 $\Delta_M H_B^\infty$ = –41.31 ± 0.041 kJ/mol

continued

continued

$T/K = 288.15$	$\Delta_M H_B^\infty = -40.67 \pm 0.033$ kJ/mol
$T/K = 293.15$	$\Delta_M H_B^\infty = -39.73 \pm 0.030$ kJ/mol
$T/K = 298.15$	$\Delta_M H_B^\infty = -38.77 \pm 0.024$ kJ/mol
$T/K = 303.15$	$\Delta_M H_B^\infty = -37.61 \pm 0.024$ kJ/mol
$T/K = 308.15$	$\Delta_M H_B^\infty = -36.51 \pm 0.016$ kJ/mol
$T/K = 313.15$	$\Delta_M H_B^\infty = -35.32 \pm 0.026$ kJ/mol

Polymer (B): **polyacrylonitrile** **1955TA1**
Characterization: –
Solvent (A): **benzene** C_6H_6 **71-43-2**

$T/K = 298.15$ $\Delta_{sol} H_B^\infty = 0.0$ J/(g polymer)

Comments: The final concentration is about 0.4 g polymer/132 g solvent.

Polymer (B): **polyacrylonitrile** **1955TA1**
Characterization: –
Solvent (A): **N,N-dimethylformamide** C_3H_7NO **68-12-2**

$T/K = 298.15$ $\Delta_{sol} H_B^\infty = -21.3$ J/(g polymer)

Comments: The final concentration is about 1 g polymer/100 g solvent.

Polymer (B): **polyacrylonitrile** **1959ZEL**
Characterization: –
Solvent (A): **N,N-dimethylformamide** C_3H_7NO **68-12-2**

T/K	295.15	295.15	295.15	295.15	308.15	308.15	323.15
$\Delta_{sol} H_B^\infty$/(J/g polymer)	−27.57	−24.81	−20.50	−19.66	−19.25	−14.85	−15.48

T/K	323.15	338.15
$\Delta_{sol} H_B^\infty$/(J/g polymer)	−10.59	−10.04

Comments: The final concentration is about 1 g polymer/165 g solvent.

Polymer (B): **polyacrylonitrile** **1964DIE**
Characterization: –
Solvent (A): **N,N-dimethylformamide** C_3H_7NO **68-12-2**

$T/K = 293.15$ $\Delta_{sol} H_B^\infty = -12.0$ J/(g polymer)

Polymer (B): **polyacrylonitrile** **1964ZVE**
Characterization: –
Solvent (A): **N,N-dimethylformamide** C_3H_7NO **68-12-2**

continued

continued

$T/K = 298.15$ $\Delta_{sol}H_B^{\infty} = -42.7$ J/(g polymer)

Comments: The final concentration is about 1 g polymer/100 g solvent.

Polymer (B):	**polyacrylonitrile**	**1976PET**
Characterization:	–	
Solvent (A):	**N,N-dimethylformamide** C$_3$H$_7$NO	**68-12-2**

$T/K = 323.15$ $\Delta_{sol}H_B^{\infty} = -14.6$ J/(g polymer)

Polymer (B):	**polyacrylonitrile**	**1964ZVE**
Characterization:	–	
Solvent (A):	**dimethylsulfoxide** C$_2$H$_6$OS	**67-68-5**

$T/K = 298.15$ $\Delta_{sol}H_B^{\infty} = -70.3$ J/(g polymer)

Comments: The final concentration is about 1 g polymer/100 g solvent.

Polymer (B):	**poly(acrylonitrile-*co*-butadiene)**	**1955TA1**
Characterization:	18.0 wt% acrylonitrile, Buna 18	
Solvent (A):	**benzene** C$_6$H$_6$	**71-43-2**

$T/K = 298.15$ $\Delta_{sol}H_B^{\infty} = 0.0$ J/(g polymer)

Comments: The final concentration is about 2 g polymer/132 g solvent.

Polymer (B):	**poly(acrylonitrile-*co*-butadiene)**	**1950TAG, 1951TA2**
Characterization:	26.0 wt% acrylonitrile, Buna 26	
Solvent (A):	**benzene** C$_6$H$_6$	**71-43-2**

$T/K = 298.15$ $\Delta_{sol}H_B^{\infty} = 5.82$ J/(g polymer)

Comments: The final concentration is about 0.55 g polymer/100 g solution.

Polymer (B):	**poly(acrylonitrile-*co*-butadiene)**	**1955TA1**
Characterization:	26.0 wt% acrylonitrile, Buna 26	
Solvent (A):	**benzene** C$_6$H$_6$	**71-43-2**

$T/K = 298.15$ $\Delta_{sol}H_B^{\infty} = -1.88$ J/(g polymer)

Comments: The final concentration is about 3 g polymer/132 g solvent.

Polymer (B):	**poly(acrylonitrile-*co*-butadiene)**	**1950TAG, 1951TA2**
Characterization:	40.0 wt% acrylonitrile, Buna 40	
Solvent (A):	**benzene** C$_6$H$_6$	**71-43-2**

$T/K = 298.15$ $\Delta_{sol}H_B^{\infty} = -7.0$ J/(g polymer)

Comments: The final concentration is about 0.85 g polymer/100 g solution.

Polymer (B):   **poly(acrylonitrile-*co*-butadiene)**   **1955TA1**
Characterization: 40.0 wt% acrylonitrile, Buna 40
Solvent (A):   **benzene**   **C$_6$H$_6$**   **71-43-2**

$T/K = 298.15$ $\Delta_{sol} H_B^\infty = -2.93$ J/(g polymer)

Comments: The final concentration is about 3 g polymer/88 g solvent.

Polymer (B):   **poly(acrylonitrile-*co*-isoprene)**   **1976PET**
Characterization: 85.0 mol% acrylonitrile
Solvent (A):   **N,N-dimethylformamide**   **C$_3$H$_7$NO**   **68-12-2**

$T/K = 323.15$ $\Delta_{sol} H_B^\infty = -31.8$ J/(g polymer)

Comments: Additional information is given for varying isoprene contents between 2 and 22 mol% in the polymer in Fig. 1 of the original source, 1976PET.

Polymer (B):   **poly(acrylonitrile-*co*-vinyl chloride)**   **1959ZEL**
Characterization: 13.0 wt% acrylonitrile
Solvent (A):   **N,N-dimethylformamide**   **C$_3$H$_7$NO**   **68-12-2**

T/K	295.15	308.15	308.15	323.15	323.15	338.15	338.15
$\Delta_{sol} H_B^\infty$/(J/g polymer)	−37.95	−22.18	−21.34	−19.25	−16.95	−15.48	−14.64

T/K	353.15	353.15
$\Delta_{sol} H_B^\infty$/(J/g polymer)	−12.97	−11.30

Comments: The final concentration is about 1 g polymer/165 g solvent.

Polymer (B):   **poly(acrylonitrile-*co*-vinyl chloride)**   **1959ZEL**
Characterization: 29.0 wt% acrylonitrile
Solvent (A):   **N,N-dimethylformamide**   **C$_3$H$_7$NO**   **68-12-2**

T/K	295.15	295.15	308.15	308.15	323.15	323.15	338.15
$\Delta_{sol} H_B^\infty$/(J/g polymer)	−43.93	−40.21	−27.61	−26.78	−22.18	−20.08	−20.50

T/K	338.15	353.15	353.15
$\Delta_{sol} H_B^\infty$/(J/g polymer)	−17.99	−16.32	−15.90

Comments: The final concentration is about 1 g polymer/165 g solvent.

Polymer (B):   **poly(acrylonitrile-*co*-vinyl chloride)**   **1959ZEL**
Characterization: 40.0 wt% acrylonitrile
Solvent (A):   **N,N-dimethylformamide**   **C$_3$H$_7$NO**   **68-12-2**

T/K	295.15	295.15	308.15	323.15	323.15	338.15	338.15
$\Delta_{sol} H_B^\infty$/(J/g polymer)	−47.28	−46.65	−30.54	−28.87	−26.36	−20.08	−19.66

T/K	353.15
$\Delta_{sol} H_B^\infty$/(J/g polymer)	−17.15

Comments: The final concentration is about 1 g polymer/165 g solvent.

Polymer (B):	**poly(acrylonitrile-*co*-vinylidene chloride)**		**1955ZAZ**
Characterization:	unspecified contents of acrylonitrile, fibre "Saniv", USSR		
Solvent (A):	**2-propanone**	**C_3H_6O**	**67-64-1**

$T/K = 298.15$ $\Delta_{sol}H_B^\infty = -28.03$ J/(g polymer)

$T/K = 298.15$ $\Delta_{sol}H_B^\infty = -19.25$ J/(g polymer)   (fibre drawn 156%)

$T/K = 298.15$ $\Delta_{sol}H_B^\infty = -15.92$ J/(g polymer)   (fibre drawn 300%)

$T/K = 298.15$ $\Delta_{sol}H_B^\infty = -14.23$ J/(g polymer)   (fibre drawn 400%)

Polymer (B):	**poly(m-benzamide)**		**1987KOJ**
Characterization:	$M_n/\text{g.mol}^{-1} = 441$, crystalline sample, synthesized in the laboratory		
Solvent (A):	**N,N-dimethylacetamide**	**C_4H_9NO**	**127-19-5**

$T/K = 293.15$ $\Delta_{sol}H_B^\infty = -(1.2 \pm 0.2)$ kJ/mol

Polymer (B):	**poly(m-benzamide)**		**1987KOJ**
Characterization:	$M_n/\text{g.mol}^{-1} = 562$, synthesized in the laboratory		
Solvent (A):	**N,N-dimethylacetamide**	**C_4H_9NO**	**127-19-5**

$T/K = 293.15$ $\Delta_{sol}H_B^\infty = -(23.7 \pm 1.1)$ kJ/mol   crystalline sample

$T/K = 293.15$ $\Delta_{sol}H_B^\infty = -(55.2 \pm 0.8)$ kJ/mol   amorphous sample

Polymer (B):	**poly(m-benzamide)**		**1987KOJ**
Characterization:	$M_n/\text{g.mol}^{-1} = 683$, crystalline sample, synthesized in the laboratory		
Solvent (A):	**N,N-dimethylacetamide**	**C_4H_9NO**	**127-19-5**

$T/K = 293.15$ $\Delta_{sol}H_B^\infty = -(16.9 \pm 0.8)$ kJ/mol

Polymer (B):	**poly(m-benzamide)**		**1987KOJ**
Characterization:	$M_n/\text{g.mol}^{-1} = 441$, crystalline sample, synthesized in the laboratory		
Solvent (A):	**tetramethylurea**	**$C_5H_{12}N_2O$**	**632-22-4**

$T/K = 293.15$ $\Delta_{sol}H_B^\infty = -(8.8 \pm 1.0)$ kJ/mol

Polymer (B):	**poly(m-benzamide)**		**1987KOJ**
Characterization:	$M_n/\text{g.mol}^{-1} = 562$, synthesized in the laboratory		
Solvent (A):	**tetramethylurea**	**$C_5H_{12}N_2O$**	**632-22-4**

$T/K = 293.15$ $\Delta_{sol}H_B^\infty = -(33.8 \pm 1.0)$ kJ/mol   crystalline sample

$T/K = 293.15$ $\Delta_{sol}H_B^\infty = -(65.1 \pm 0.7)$ kJ/mol   amorphous sample

---

| **Polymer (B):** | **poly(m-benzamide)** | **1987KOJ** |

Characterization: $M_n/\text{g.mol}^{-1} = 683$, crystalline sample, synthesized in the laboratory

| **Solvent (A):** | **1,1,3,3-tetramethylurea** $\quad$ **C$_5$H$_{12}$N$_2$O** | **632-22-4** |

$T/\text{K} = 293.15$ $\qquad$ $\Delta_{sol}H_B^\infty = -(23.8 \pm 1.5)$ kJ/mol

---

| **Polymer (B):** | **poly(γ-benzyl-L-glutamate)** | **1966GIA** |

Characterization: $M_w/\text{g.mol}^{-1} = 160000$

| **Solvent (A):** | **dichloroacetic acid** $\quad$ **C$_2$H$_2$Cl$_2$O$_2$** | **79-43-6** |

$T/\text{K} = 303.15$ $\qquad$ $\Delta_{sol}H_B^\infty = -8368$ J/(base mol polymer)

Comments: The final concentration is about 0.007 mol monomer/1000 cm^3 solvent.
A graph is given in the original source for $\Delta_{sol}H_B^\infty$ in the binary mixture with 1,2-dichlorethane.

| **Polymer (B):** | **poly(γ-benzyl-L-glutamate)** | **1966GIA** |

Characterization: $M_w/\text{g.mol}^{-1} = 160000$

| **Solvent (A):** | **1,2-dichloroethane** $\quad$ **C$_2$H$_4$Cl$_2$** | **107-06-2** |

$T/\text{K} = 303.15$ $\qquad$ $\Delta_{sol}H_B^\infty = -377$ J/(base mol polymer)

Comments: The final concentration is about 0.007 mol monomer/1000 cm^3 solvent.
A graph is given in the original source for $\Delta_{sol}H_B^\infty$ in the binary mixture with dichloroacetic acid.

| **Polymer (B):** | **poly(γ-benzyl-D-glutamate)** | **1975BAB** |

Characterization: DP = 1320

| **Solvent (A):** | **dichloroacetic acid** $\quad$ **C$_2$H$_2$Cl$_2$O$_2$** | **79-43-6** |

| **Polymer (C):** | **poly(γ-benzyl-L-glutamate)** |

Characterization: DP = 1750

$T/\text{K} = 303.15$ $\qquad$ $\Delta_{sol}H_B^\infty = -5510$ J/(base mol polymer)

Comments: The mixture of both polymers is made with a ratio of D/L = 1/1.

$T/\text{K} = 303.15$ $\qquad$ $\Delta_{sol}H_B^\infty = -6020$ J/(base mol polymer)

Comments: The mixture of both polymers is made with a ratio of D/L = 3/1.

| **Polymer (B):** | **poly(γ-benzyl-D-glutamate)** | **1975BAB** |

Characterization: DP = 1320

| **Solvent (A):** | **1,2-dichloroethane** $\quad$ **C$_2$H$_4$Cl$_2$** | **107-06-2** |

| **Polymer (C):** | **poly(γ-benzyl-L-glutamate)** |

Characterization: DP = 1750

| **Solvent (D):** | **dichloroacetic acid** $\quad$ **C$_2$H$_2$Cl$_2$O$_2$** | **79-43-6** |

$T/\text{K} = 303.15$

φ_A/φ_D	0.9/0.1	0.8/0.2	0.7/0.3	0.5/0.5	0.4/0.6	0.3/0.7	0.2/0.8
$\Delta_{sol}H_{B+C}^\infty$/(J/base mol polymer)	−1890	−1890	−2450	−2390	−2890	−3320	−4310

Comments: The mixture of both polymers is made with a ratio of D/L = 1/1.

continued

continued

$T/K = 303.15$

φ_A / φ_D	0.9/0.1	0.8/0.2	0.7/0.3	0.5/0.5	0.3/0.7	0.1/0.9
$\Delta_{sol}H_{B+C}^\infty$/(J/base mol polymer)	−3180	−3480	−3860	−4100	−4630	−5890

Comments: The mixture of both polymers is made with a ratio of D/L = 3/1.

Polymer (B):   **poly(bisphenol A-isophthaloyl chloride-*co*-**   **1978SOK**
                                  terephthaloyl chloride)
Characterization: 50 mol% terephthaloyl chloride, synthesized in the laboratory
Solvent (A):   **N,N-dimethylacetamide**   **C₄H₉NO**   **127-19-5**

$T/K = 298.15$ $\Delta_{sol}H_B^\infty = -56.5$ J/(g polymer)   (amorphous sample)

Polymer (B):   **poly(bisphenol A-isophthaloyl chloride-*co*-**   **1978SOK**
                                  terephthaloyl chloride)
Characterization: 50 mol% terephthaloyl chloride, synthesized in the laboratory
Solvent (A):   **1,1,2,2-tetrachloroethane**   **C₂H₂Cl₄**   **79-34-5**

$T/K = 298.15$ $\Delta_{sol}H_B^\infty = 72.5$ J/(g polymer)   (amorphous sample)
$T/K = 298.15$ $\Delta_{sol}H_B^\infty = 41.9$ J/(g polymer)   (partially crystalline sample)

Polymer (B):   **polybutadiene**   **1950TAG, 1951TA2**
Characterization: –
Solvent (A):   **benzene**   **C₆H₆**   **71-43-2**

$T/K = 298.15$ $\Delta_{sol}H_B^\infty = 6.1$ J/(g polymer)

Comments: The final concentration is about 0.22 g polymer/100 g solution.

Polymer (B):   **polybutadiene**   **1956STR, 1958SLO**
Characterization: –
Solvent (A):   **benzene**   **C₆H₆**   **71-43-2**

$T/K = 298.15$ $\Delta_{sol}H_B^\infty = 7.1$ J/(g polymer)

Comments: The final concentration is 1 g polymer/100 g solvent.

Polymer (B):   **polybutadiene**   **1958TA3**
Characterization: –
Solvent (A):   **benzene**   **C₆H₆**   **71-43-2**

$T/K = 298.15$ $\Delta_{sol}H_B^\infty = 10.5$ J/(g polymer)

Polymer (B):	**1,4-*cis*-polybutadiene**		**1979PH2**
Characterization:	sample of low molecular mass		
Solvent (A):	**cyclohexane**	C_6H_{12}	**110-82-7**

$T/K = 298.15$ $\Delta_M H_B^\infty = 5.4$ J/(g polymer)

Polymer (B):	**1,4-*cis*-polybutadiene**		**1979PH2**
Characterization:	sample of low molecular mass		
Solvent (A):	**cyclooctane**	C_8H_{16}	**292-64-8**

$T/K = 298.15$ $\Delta_M H_B^\infty = 5.8$ J/(g polymer)

Polymer (B):	**1,4-*cis*-polybutadiene**		**1979PH2**
Characterization:	sample of low molecular mass		
Solvent (A):	**cyclopentane**	C_5H_{10}	**287-92-3**

$T/K = 298.15$ $\Delta_M H_B^\infty = <0.1$ J/(g polymer)

Polymer (B):	**1,4-*cis*-polybutadiene**		**1979PH2**
Characterization:	sample of low molecular mass		
Solvent (A):	***cis*-decahydronaphthalene**	$C_{10}H_{18}$	**493-01-6**

$T/K = 298.15$ $\Delta_M H_B^\infty = 4.2$ J/(g polymer)

Polymer (B):	**1,4-*cis*-polybutadiene**		**1979PH2**
Characterization:	sample of low molecular mass		
Solvent (A):	***trans*-decahydronaphthalene**	$C_{10}H_{18}$	**493-02-7**

$T/K = 298.15$ $\Delta_M H_B^\infty = 2.6$ J/(g polymer)

Polymer (B):	**1,4-*cis*-polybutadiene**		**1979PH2**
Characterization:	sample of low molecular mass		
Solvent (A):	**3,3-diethylpentane**	C_9H_{20}	**4032-86-4**

$T/K = 298.15$ $\Delta_M H_B^\infty = 5.2$ J/(g polymer)

Polymer (B):	**1,4-*cis*-polybutadiene**		**1979PH2**
Characterization:	sample of low molecular mass		
Solvent (A):	**2,2-dimethylpentane**	C_7H_{16}	**590-35-2**

$T/K = 298.15$ $\Delta_M H_B^\infty = 4.1$ J/(g polymer)

Polymer (B):	**1,4-*cis*-polybutadiene**		**1979PH2**
Characterization:	sample of low molecular mass		
Solvent (A):	**2,3-dimethylpentane**	C_7H_{16}	**565-59-3**

$T/K = 298.15$ $\Delta_M H_B^\infty = 4.5$ J/(g polymer)

Polymer (B):   **1,4-*cis*-polybutadiene**   1979PH2
Characterization: sample of low molecular mass
Solvent (A):   **2,4-dimethylpentane** C_7H_{16} 108-08-7

$T/K = 298.15$ $\Delta_M H_B^\infty = 3.2$ J/(g polymer)

Polymer (B):   **1,4-*cis*-polybutadiene**   1979PH2
Characterization: sample of low molecular mass
Solvent (A):   **3,3-dimethylpentane** C_7H_{16} 562-49-2

$T/K = 298.15$ $\Delta_M H_B^\infty = 3.2$ J/(g polymer)

Polymer (B):   **1,4-*cis*-polybutadiene**   1979PH2
Characterization: sample of low molecular mass
Solvent (A):   **n-dodecane** $C_{12}H_{26}$ 112-40-3

$T/K = 298.15$ $\Delta_M H_B^\infty = 4.2$ J/(g polymer)

Polymer (B):   **1,4-*cis*-polybutadiene**   1979PH2
Characterization: sample of low molecular mass
Solvent (A):   **3-ethylpentane** C_7H_{16} 617-78-7

$T/K = 298.15$ $\Delta_M H_B^\infty = 3.7$ J/(g polymer)

Polymer (B):   **1,4-*cis*-polybutadiene**   1979PH2
Characterization: sample of low molecular mass
Solvent (A):   **2,2,4,4,6,8,8-heptamethylnonane** $C_{16}H_{34}$ 4390-04-9

$T/K = 298.15$ $\Delta_M H_B^\infty = 4.8$ J/(g polymer)

Polymer (B):   **1,4-*cis*-polybutadiene**   1979PH2
Characterization: sample of low molecular mass
Solvent (A):   **n-hexadecane** $C_{16}H_{34}$ 544-76-3

$T/K = 298.15$ $\Delta_M H_B^\infty = 4.9$ J/(g polymer)

Polymer (B):   **1,4-*cis*-polybutadiene**   1979PH2
Characterization: sample of low molecular mass
Solvent (A):   **3-methylhexane** C_7H_{16} 589-34-4

$T/K = 298.15$ $\Delta_M H_B^\infty = 3.6$ J/(g polymer)

Polymer (B):   **1,4-*cis*-polybutadiene**   1979PH2
Characterization: sample of low molecular mass
Solvent (A):   **n-octane** C_8H_{18} 111-65-9

$T/K = 298.15$ $\Delta_M H_B^\infty = 4.3$ J/(g polymer)

Polymer (B):	**1,4-*cis*-polybutadiene**	1979PH2
Characterization:	sample of low molecular mass	
Solvent (A):	**2,2,4,6,6-pentamethylheptane** $C_{12}H_{26}$	13475-82-6

$T/K = 298.15$ $\Delta_M H_B^\infty = 5.0$ J/(g polymer)

Polymer (B):	**1,4-*cis*-polybutadiene**	1979PH1, 1979PH2
Characterization:	sample of low molecular mass	
Solvent (A):	**2,2,4,4-tetramethylpentane** C_9H_{20}	1070-87-7

$T/K = 298.15$ $\Delta_M H_B^\infty = 5.8$ J/(g polymer)

Polymer (B):	**1,4-*cis*-polybutadiene**	1979PH2
Characterization:	sample of low molecular mass	
Solvent (A):	**2,3,3,4-tetramethylpentane** C_9H_{20}	16747-38-9

$T/K = 298.15$ $\Delta_M H_B^\infty = 5.1$ J/(g polymer)

Polymer (B):	**polybutadiene**	1958TA2, 1958TA3
Characterization:	–	
Solvent (A):	**2,2,4-trimethylpentane** C_8H_{18}	540-84-1

$T/K = 298.15$ $\Delta_{sol} H_B^\infty = 1.09$ J/(g polymer)

Polymer (B):	**poly(butadiene-co-styrene)**	1955TA2
Characterization:	10.0 wt% styrene, synthesized in the laboratory	
Solvent (A):	**benzene** C_6H_6	71-43-2

$T/K = 293.65$ $\Delta_{sol} H_B^\infty = 4.94$ J/(g polymer)

Comments: The final concentration is about 2.7 g polymer/100 ml solvent.

Polymer (B):	**poly(butadiene-co-styrene)**	1955TA2
Characterization:	30.0 wt% styrene, synthesized in the laboratory	
Solvent (A):	**benzene** C_6H_6	71-43-2

$T/K = 293.65$ $\Delta_{sol} H_B^\infty = 3.01$ J/(g polymer)

Comments: The final concentration is about 2.7 g polymer/100 ml solvent.

Polymer (B):	**poly(butadiene-co-styrene)**	1956STR, 1958SLO
Characterization:	30.0 wt% styrene, intrinsic viscosity = 1.50	
Solvent (A):	**benzene** C_6H_6	71-43-2

$T/K = 298.15$ $\Delta_{sol} H_B^\infty = 3.05$ J/(g polymer)

Comments: The final concentration is 1 g polymer/100 g solvent.

Polymer (B):	**poly(butadiene-co-styrene)**	**1955TA2**
Characterization:	50.0 wt% styrene, synthesized in the laboratory	
Solvent (A):	**benzene** C_6H_6	**71-43-2**

$T/K = 293.65$ $\Delta_{sol}H_B^{\infty} = 1.80$ J/(g polymer)

Comments: The final concentration is about 4 g polymer/100 ml solvent.

Polymer (B):	**poly(butadiene-co-styrene)**	**1955TA2**
Characterization:	60.0 wt% styrene, synthesized in the laboratory	
Solvent (A):	**benzene** C_6H_6	**71-43-2**

$T/K = 293.65$ $\Delta_{sol}H_B^{\infty} = 0.0$ J/(g polymer)

Comments: The final concentration is about 4 g polymer/100 ml solvent.

Polymer (B):	**poly(butadiene-co-styrene)**	**1955TA2**
Characterization:	70.0 wt% styrene, synthesized in the laboratory	
Solvent (A):	**benzene** C_6H_6	**71-43-2**

$T/K = 293.65$ $\Delta_{sol}H_B^{\infty} = 0.0$ J/(g polymer)

Comments: The final concentration is about 4.4 g polymer/100 ml solvent.

Polymer (B):	**poly(butadiene-co-styrene)**	**1950TAG, 1951TA2**
Characterization:	75.0 wt% styrene, synthesized in the laboratory	
Solvent (A):	**benzene** C_6H_6	**71-43-2**

$T/K = 298.15$ $\Delta_{sol}H_B^{\infty} = 1.50$ J/(g polymer)

Comments: The final concentration is about 3 g polymer/100 ml solvent.

Polymer (B):	**poly(butadiene-co-styrene)**	**1955TA2**
Characterization:	80.0 wt% styrene, synthesized in the laboratory	
Solvent (A):	**benzene** C_6H_6	**71-43-2**

$T/K = 293.65$ $\Delta_{sol}H_B^{\infty} = -0.59$ J/(g polymer)

Comments: The final concentration is about 3.5 g polymer/150 ml solvent.

Polymer (B):	**poly(butadiene-co-styrene)**	**1955TA2**
Characterization:	90.0 wt% styrene, synthesized in the laboratory	
Solvent (A):	**benzene** C_6H_6	**71-43-2**

$T/K = 293.65$ $\Delta_{sol}H_B^{\infty} = -4.94$ J/(g polymer)

Comments: The final concentration is about 2 g polymer/150 ml solvent.

Polymer (B):	**poly(butadiene-co-styrene)**	**1958TA2**
Characterization:	10.0 wt% styrene, synthesized in the laboratory	
Solvent (A):	**ethylbenzene** $\quad$ **C_8H_{10}**	**100-41-4**
Solvent (C):	**2,2,4-trimethylpentane** C_8H_{18}	**540-84-1**

$T/K = 298.15$ $\qquad$ $\Delta_{sol} H_B^\infty = 1.46$ J/(g polymer)

Comments: $\qquad$ The solvent mixture concentration is 10 wt% C_8H_{10}/90 wt% C_8H_{18}.

Polymer (B):	**poly(butadiene-co-styrene)**	**1958TA2**
Characterization:	30.0 wt% styrene, synthesized in the laboratory	
Solvent (A):	**ethylbenzene** $\quad$ **C_8H_{10}**	**100-41-4**
Solvent (C):	**2,2,4-trimethylpentane** C_8H_{18}	**540-84-1**

$T/K = 298.15$ $\qquad$ $\Delta_{sol} H_B^\infty = 0.75$ J/(g polymer)

Comments: $\qquad$ The solvent mixture concentration is 30 wt% C_8H_{10}/70 wt% C_8H_{18}.

Polymer (B):	**poly(butadiene-co-styrene)**	**1958TA2**
Characterization:	50.0 wt% styrene, synthesized in the laboratory	
Solvent (A):	**ethylbenzene** $\quad$ **C_8H_{10}**	**100-41-4**
Solvent (C):	**2,2,4-trimethylpentane** C_8H_{18}	**540-84-1**

$T/K = 298.15$ $\qquad$ $\Delta_{sol} H_B^\infty = 0.54$ J/(g polymer)

Comments: $\qquad$ The solvent mixture concentration is 50 wt% C_8H_{10}/50 wt% C_8H_{18}.

Polymer (B):	**poly(butadiene-co-styrene)**	**1958TA2**
Characterization:	60.0 wt% styrene, synthesized in the laboratory	
Solvent (A):	**ethylbenzene** $\quad$ **C_8H_{10}**	**100-41-4**
Solvent (C):	**2,2,4-trimethylpentane** C_8H_{18}	**540-84-1**

$T/K = 298.15$ $\qquad$ $\Delta_{sol} H_B^\infty = 0.42$ J/(g polymer)

Comments: $\qquad$ The solvent mixture concentration is 60 wt% C_8H_{10}/40 wt% C_8H_{18}.

Polymer (B):	**poly(butadiene-co-styrene)**	**1958TA2**
Characterization:	70.0 wt% styrene, synthesized in the laboratory	
Solvent (A):	**ethylbenzene** $\quad$ **C_8H_{10}**	**100-41-4**
Solvent (C):	**2,2,4-trimethylpentane** C_8H_{18}	**540-84-1**

$T/K = 298.15$ $\qquad$ $\Delta_{sol} H_B^\infty = 0.33$ J/(g polymer)

Comments: $\qquad$ The solvent mixture concentration is 70 wt% C_8H_{10}/30 wt% C_8H_{18}.

Polymer (B):	**poly(butadiene-co-styrene)**	**1958TA2**
Characterization:	80.0 wt% styrene, synthesized in the laboratory	
Solvent (A):	**ethylbenzene** C_8H_{10}	**100-41-4**
Solvent (C):	**2,2,4-trimethylpentane** C_8H_{18}	**540-84-1**

$T/K = 298.15$ $\Delta_{sol}H_B^\infty = 0.21$ J/(g polymer)

Comments: The solvent mixture concentration is 80 wt% C_8H_{10}/20 wt% C_8H_{18}.

Polymer (B):	**poly(butadiene-co-styrene)**	**1958TA2**
Characterization:	90.0 wt% styrene, synthesized in the laboratory	
Solvent (A):	**ethylbenzene** C_8H_{10}	**100-41-4**
Solvent (C):	**2,2,4-trimethylpentane** C_8H_{18}	**540-84-1**

$T/K = 298.15$ $\Delta_{sol}H_B^\infty = 0.13$ J/(g polymer)

Comments: The solvent mixture concentration is 90 wt% C_8H_{10}/10 wt% C_8H_{18}.

Polymer (B):	**poly(1-butene)**	**1979PH2**
Characterization:	M_η/g.mol^{-1} = 20000, atactic	
Solvent (A):	**cyclohexane** C_6H_{12}	**110-82-7**

$T/K = 298.15$ $\Delta_M H_B^\infty = 1.0$ J/(g polymer)

Polymer (B):	**poly(1-butene)**	**1979PH2**
Characterization:	M_η/g.mol^{-1} = 20000, atactic	
Solvent (A):	**cyclooctane** C_8H_{16}	**292-64-8**

$T/K = 298.15$ $\Delta_M H_B^\infty = 1.8$ J/(g polymer)

Polymer (B):	**poly(1-butene)**	**1979PH2**
Characterization:	M_η/g.mol^{-1} = 20000, atactic	
Solvent (A):	**cyclopentane** C_5H_{10}	**287-92-3**

$T/K = 298.15$ $\Delta_M H_B^\infty = -2.9$ J/(g polymer)

Polymer (B):	**poly(1-butene)**	**1979PH2**
Characterization:	M_η/g.mol^{-1} = 20000, atactic	
Solvent (A):	***cis*-decahydronaphthalene** $C_{10}H_{18}$	**493-01-6**

$T/K = 298.15$ $\Delta_M H_B^\infty = <0.1$ J/(g polymer)

Polymer (B):	**poly(1-butene)**	**1979PH2**
Characterization:	M_η/g.mol^{-1} = 20000, atactic	
Solvent (A):	***trans*-decahydronaphthalene** $C_{10}H_{18}$	**493-02-7**

$T/K = 298.15$ $\Delta_M H_B^\infty = -2.0$ J/(g polymer)

Polymer (B):	**poly(1-butene)**	**1973DEL**
Characterization:	M_η/g.mol^{-1} = 20000, atactic	
Solvent (A):	**n-decane** $C_{10}H_{22}$	**124-18-5**

T/K = 298.15 $\Delta_M H_B^\infty$ = 1.21 J/(g polymer)

Polymer (B):	**poly(1-butene)**	**1979PH2**
Characterization:	M_η/g.mol^{-1} = 20000, atactic	
Solvent (A):	**3,3-diethylpentane** C_9H_{20}	**4032-86-4**

T/K = 298.15 $\Delta_M H_B^\infty$ = –2.6 J/(g polymer)

Polymer (B):	**poly(1-butene)**	**1979PH2**
Characterization:	M_η/g.mol^{-1} = 20000, atactic	
Solvent (A):	**2,2-dimethylpentane** C_7H_{16}	**590-35-2**

T/K = 298.15 $\Delta_M H_B^\infty$ = –4.0 J/(g polymer)

Polymer (B):	**poly(1-butene)**	**1979PH2**
Characterization:	M_η/g.mol^{-1} = 20000, atactic	
Solvent (A):	**2,3-dimethylpentane** C_7H_{16}	**565-59-3**

T/K = 298.15 $\Delta_M H_B^\infty$ = –2.8 J/(g polymer)

Polymer (B):	**poly(1-butene)**	**1979PH2**
Characterization:	M_η/g.mol^{-1} = 20000, atactic	
Solvent (A):	**2,4-dimethylpentane** C_7H_{16}	**108-08-7**

T/K = 298.15 $\Delta_M H_B^\infty$ = –2.3 J/(g polymer)

Polymer (B):	**poly(1-butene)**	**1979PH2**
Characterization:	M_η/g.mol^{-1} = 20000, atactic	
Solvent (A):	**3,3-dimethylpentane** C_7H_{16}	**562-49-2**

T/K = 298.15 $\Delta_M H_B^\infty$ = –2.2 J/(g polymer)

Polymer (B):	**poly(1-butene)**	**1973DEL**
Characterization:	M_η/g.mol^{-1} = 20000, atactic	
Solvent (A):	**n-dodecane** $C_{12}H_{26}$	**112-40-3**

T/K = 298.15 $\Delta_M H_B^\infty$ = 2.13 J/(g polymer)

Polymer (B):	**poly(1-butene)**	**1979PH2**
Characterization:	M_η/g.mol^{-1} = 20000, atactic	
Solvent (A):	**n-dodecane** $C_{12}H_{26}$	**112-40-3**

T/K = 298.15 $\Delta_M H_B^\infty$ = 2.1 J/(g polymer)

Polymer (B):   **poly(1-butene)**   **1979PH2**
Characterization: M_η/g.mol^{-1} = 20000, atactic
Solvent (A):   **3-ethylpentane**   **C$_7$H$_{16}$**   **617-78-7**

T/K = 298.15 $\Delta_M H_B^\infty$ = -2.8 J/(g polymer)

Polymer (B):   **poly(1-butene)**   **1979PH2**
Characterization: M_η/g.mol^{-1} = 20000, atactic
Solvent (A):   **2,2,4,4,6,8,8-heptamethylnonane**   **C$_{16}$H$_{34}$**   **4390-04-9**

T/K = 298.15 $\Delta_M H_B^\infty$ = 0.1 J/(g polymer)

Polymer (B):   **poly(1-butene)**   **1973DEL**
Characterization: M_η/g.mol^{-1} = 20000, atactic
Solvent (A):   **n-heptane**   **C$_7$H$_{16}$**   **142-82-5**

T/K = 298.15 $\Delta_M H_B^\infty$ = 0.04 J/(g polymer)

Polymer (B):   **poly(1-butene)**   **1979PH1**
Characterization: M_η/g.mol^{-1} = 20000, atactic
Solvent (A):   **n-heptane**   **C$_7$H$_{16}$**   **142-82-5**

T/K = 298.15 $\Delta_M H_B^\infty$ = 0.0 J/(g polymer)

Polymer (B):   **poly(1-butene)**   **1973DEL**
Characterization: M_η/g.mol^{-1} = 20000, atactic
Solvent (A):   **n-hexadecane**   **C$_{16}$H$_{34}$**   **544-76-3**

T/K = 298.15 $\Delta_M H_B^\infty$ = 3.85 J/(g polymer)

Polymer (B):   **poly(1-butene)**   **1979PH1**
Characterization: M_η/g.mol^{-1} = 20000, atactic
Solvent (A):   **n-hexadecane**   **C$_{16}$H$_{34}$**   **544-76-3**

T/K = 298.15 $\Delta_M H_B^\infty$ = 3.9 J/(g polymer)

Polymer (B):   **poly(1-butene)**   **1973DEL**
Characterization: M_η/g.mol^{-1} = 20000, atactic
Solvent (A):   **n-hexane**   **C$_6$H$_{14}$**   **110-54-3**

T/K = 298.15 $\Delta_M H_B^\infty$ = -1.25 J/(g polymer)

Polymer (B):   **poly(1-butene)**   **1979PH2**
Characterization: M_η/g.mol^{-1} = 20000, atactic
Solvent (A):   **3-methylhexane**   **C$_7$H$_{16}$**   **589-34-4**

T/K = 298.15 $\Delta_M H_B^\infty$ = -2.1 J/(g polymer)

Polymer (B):   **poly(1-butene)**   **1973DEL**
Characterization: $M_\eta/\text{g.mol}^{-1} = 20000$, atactic
Solvent (A):   **n-nonane**   **C$_9$H$_{20}$**   **111-84-2**

$T/\text{K} = 298.15$ $\Delta_M H_B^\infty = 0.92$ J/(g polymer)

Polymer (B):   **poly(1-butene)**   **1979PH1**
Characterization: $M_\eta/\text{g.mol}^{-1} = 20000$, atactic
Solvent (A):   **n-nonane**   **C$_9$H$_{20}$**   **111-84-2**

$T/\text{K} = 298.15$ $\Delta_M H_B^\infty = 0.9$ J/(g polymer)

Polymer (B):   **poly(1-butene)**   **1973DEL**
Characterization: $M_\eta/\text{g.mol}^{-1} = 20000$, atactic
Solvent (A):   **n-octane**   **C$_8$H$_{18}$**   **111-65-9**

$T/\text{K} = 298.15$ $\Delta_M H_B^\infty = 0.42$ J/(g polymer)

Polymer (B):   **poly(1-butene)**   **1979PH1**
Characterization: $M_\eta/\text{g.mol}^{-1} = 20000$, atactic
Solvent (A):   **n-octane**   **C$_8$H$_{18}$**   **111-65-9**

$T/\text{K} = 298.15$ $\Delta_M H_B^\infty = 0.4$ J/(g polymer)

Polymer (B):   **poly(1-butene)**   **1979PH1**
Characterization: $M_\eta/\text{g.mol}^{-1} = 20000$, atactic
Solvent (A):   **2,2,4,6,6-pentamethylheptane**   **C$_{12}$H$_{26}$**   **13475-82-6**

$T/\text{K} = 298.15$ $\Delta_M H_B^\infty = 0.6$ J/(g polymer)

Polymer (B):   **poly(1-butene)**   **1973DEL**
Characterization: $M_\eta/\text{g.mol}^{-1} = 20000$, atactic
Solvent (A):   **n-pentane**   **C$_5$H$_{12}$**   **109-66-0**

$T/\text{K} = 298.15$ $\Delta_M H_B^\infty = -2.55$ J/(g polymer)

Polymer (B):   **poly(1-butene)**   **1973DEL**
Characterization: $M_\eta/\text{g.mol}^{-1} = 20000$, atactic
Solvent (A):   **n-tetradecane**   **C$_{14}$H$_{30}$**   **629-59-4**

$T/\text{K} = 298.15$ $\Delta_M H_B^\infty = 2.72$ J/(g polymer)

Polymer (B):   **poly(1-butene)**   **1979PH2**
Characterization: $M_\eta/\text{g.mol}^{-1} = 20000$, atactic
Solvent (A):   **2,2,4,4-tetramethylpentane**   **C$_9$H$_{20}$**   **1070-87-7**

$T/\text{K} = 298.15$ $\Delta_M H_B^\infty = -1.4$ J/(g polymer)

Polymer (B):   **poly(1-butene)**   **1979PH2**
Characterization: M_η/g.mol^{-1} = 20000, atactic
Solvent (A):   **2,3,3,4-tetramethylpentane**   **C$_9$H$_{20}$**   **16747-38-9**

T/K = 298.15 $\Delta_M H_B^\infty$ = −2.2 J/(g polymer)

Polymer (B):   **poly(1-butene)**   **1979PH1**
Characterization: M_η/g.mol^{-1} = 20000, atactic
Solvent (A):   **2,2,4-trimethylpentane C$_8$H$_{18}$**   **540-84-1**

T/K = 298.15 $\Delta_M H_B^\infty$ = −0.5 J/(g polymer)

Polymer (B):   **poly(butyl acrylate)**   **1956STR, 1958SLO**
Characterization: intrinsic viscosity = 1.00
Solvent (A):   **2-propanone**   **C$_3$H$_6$O**   **67-64-1**

T/K = 298.15 $\Delta_{sol} H_B^\infty$ = 0.84 J/(g polymer)

Comments: The final concentration is 1 g polymer/100 g solvent.

Polymer (B):   **poly(butyl methacrylate)**   **1996SHI**
Characterization: M_n/g.mol^{-1} = 91300, M_w/g.mol^{-1} = 210000
Solvent (A):   **cyclohexanone**   **C$_6$H$_{10}$O**   **108-94-1**

T/K = 304.15 $\Delta_{sol} H_B^\infty$ = 7.7 J/(g polymer)   (glassy state polymer)

T/K = 304.15 $\Delta_M H_B^\infty$ = 8.2 J/(g polymer)   (liquid state polymer)

Comments: $\Delta_M H_B^\infty$ is corrected for the enthalpy from the glass transition.

Polymer (B):   **poly(butyl methacrylate)**   **1956STR, 1958SLO**
Characterization: intrinsic viscosity = 5.1
Solvent (A):   **2-propanone**   **C$_3$H$_6$O**   **67-64-1**

T/K = 298.15 $\Delta_{sol} H_B^\infty$ = 19.52 J/(g polymer)

Comments: The final concentration is 1 g polymer/100 g solvent.

Polymer (B):   **poly(butyl methacrylate)**   **1972MAS**
Characterization: M_n/g.mol^{-1} = 358, synthesized in the laboratory
Solvent (A):   **trichloromethane**   **CHCl$_3$**   **67-66-3**

T/K = 298.15 $\Delta_M H_B^\infty$ = −169.5 J/cm^3

Polymer (B):   **poly(butyl methacrylate)**   **1972MAS**
Characterization: M_n/g.mol^{-1} = 500, synthesized in the laboratory
Solvent (A):   **trichloromethane**   **CHCl$_3$**   **67-66-3**

T/K = 298.15 $\Delta_M H_B^\infty$ = −145.5 J/cm^3

Polymer (B):	**poly(butyl methacrylate)**	**1972MAS**
Characterization:	M_n/g.mol^{-1} = 642, synthesized in the laboratory	
Solvent (A):	**trichloromethane** **CHCl$_3$**	**67-66-3**

T/K = 298.15 $\Delta_M H_B^\infty$ = −109.3 J/cm^3

Polymer (B):	**poly(butyl methacrylate-*co*-isobutyl methacrylate)**	**1996SAT**
Characterization:	M_w/g.mol^{-1} = 150000, 50 wt% isobutyl methacrylate, synthesized in the laboratory	
Solvent (A):	**cyclohexanone** **C$_6$H$_{10}$O**	**108-94-1**

T/K = 298.15 $\Delta_{sol} H_B^\infty$ = 5.9 J/(g polymer)   (glassy state polymer)

T/K = 298.15 $\Delta_M H_B^\infty$ = 14.0 J/(g polymer)   (liquid state polymer)

Comments: $\Delta_M H_B^\infty$ is corrected for the enthalpy from the glass transition.

Polymer (B):	**poly(butyl methacrylate-*co*-methyl methacrylate)**	**1996SHI**
Characterization:	M_n/g.mol^{-1} = 109000, M_w/g.mol^{-1} = 250000, 55 wt% methyl methacrylate, synthesized in the laboratory	
Solvent (A):	**cyclohexanone** **C$_6$H$_{10}$O**	**108-94-1**

T/K = 298.15 $\Delta_{sol} H_B^\infty$ = −5.4 J/(g polymer)   (glassy state polymer)

T/K = 298.15 $\Delta_M H_B^\infty$ = +9.1 J/(g polymer)   (liquid state polymer)

Comments: $\Delta_M H_B^\infty$ is corrected for the enthalpy from the glass transition.

Polymer (B):	**poly(ε-carbobenzoxy-L-lysine)**	**1970GIA**
Characterization:	M_w/g.mol^{-1} = 75000-150000	
Solvent (A):	**1,2-dichloroethane** **C$_2$H$_4$Cl$_2$**	**107-06-2**

T/K = 303.15 $\Delta_{sol} H_B^\infty$ = −8800 J/(base mol polymer)

Comments: The final concentration is about 0.1 wt%. A graph is given in the original source for $\Delta_{sol} H_B^\infty$ for the binary mixture of 1,2-dichloroethane with dichloroacetic acid.

Polymer (B):	**poly(ε-carbobenzoxy-L-lysine-*co*-L-phenylalanine)**	**1970GIA**
Characterization:	25 mol% L-phenylalanine	
Solvent (A):	**1,2-dichloroethane** **C$_2$H$_4$Cl$_2$**	**107-06-2**

T/K = 303.15 $\Delta_{sol} H_B^\infty$ = −6900 J/(base mol polymer)

Comments: The final concentration is about 0.1 wt%. A graph is given in the original source for $\Delta_{sol} H_B^\infty$ for the binary mixture of 1,2-dichloroethane with dichloroacetic acid.

Polymer (B):   **poly(ε-carbobenzoxy-L-lysine-*co*-L-phenylalanine)**   **1970GIA**
Characterization: 50 mol% L-phenylalanine
Solvent (A):   **1,2-dichloroethane**   **C₂H₄Cl₂**   **107-06-2**

$T/K = 303.15$ $\Delta_{sol}H_B^{\infty} = -6500$ J/(base mol polymer)

Comments: The final concentration is about 0.1 wt%. A graph is given in the original source for $\Delta_{sol}H_B^{\infty}$ for the binary mixture of 1,2-dichloroethane with dichloroacetic acid.

Polymer (B):   **poly(ε-carbobenzoxy-L-lysine-*co*-L-phenylalanine)**   **1970GIA**
Characterization: 75 mol% L-phenylalanine
Solvent (A):   **1,2-dichloroethane**   **C₂H₄Cl₂**   **107-06-2**

$T/K = 303.15$ $\Delta_{sol}H_B^{\infty} = -5860$ J/(base mol polymer)

Comments: The final concentration is about 0.1 wt%. A graph is given in the original source for $\Delta_{sol}H_B^{\infty}$ for the binary mixture of 1,2-dichloroethane with dichloroacetic acid.

Polymer (B):   **polychloroprene**   **1950TAG, 1951TA2**
Characterization: –
Solvent (A):   **benzene**   **C₆H₆**   **71-43-2**

$T/K = 298.15$ $\Delta_{sol}H_B^{\infty} = 0.50$ J/(g polymer)

Comments: The final concentration is about 1 g polymer/100 g solution.

Polymer (B):   **poly(1,3-cyclohexadiene)**   **1965NAU**
Characterization: –
Solvent (A):   **benzene**   **C₆H₆**   **71-43-2**

$T/K = 293.15$ $\Delta_{sol}H_B = 2.43$ J/(mol.l⁻¹)

Polymer (B):   **poly(1,3-cyclohexadiene)**   **1965NAU**
Characterization: –
Solvent (A):   **toluene**   **C₇H₈**   **108-88-3**

$T/K = 293.15$ $\Delta_{sol}H_B = 2.34$ J/(mol.l⁻¹)

Polymer (B):   **poly(2,6-dimethyl-1,4-phenylene oxide)**   **1974FI1**
Characterization: General Electric Co.
Solvent (A):   **1,2-dichlorobenzene**   **C₆H₄Cl₂**   **95-50-1**

$T/K = 400$ $\Delta_{sol}H_B = -18$ J/(g polymer)   (semicrystalline sample)

$T/K = 400$ $\Delta_{sol}H_B = -29$ J/(g polymer)   (quenched sample)

Comments: Additional data were given in Fig. 1 in the original source, 1974FI1.

Polymer (B):	**poly(2,6-dimethyl-1,4-phenylene oxide)**	**1988AUK**
Characterization:	M_n/g.mol^{-1} = 17000, M_w/g.mol^{-1} = 46400, T_g/K = 492	
Solvent (A):	**1,2-dichlorobenzene** **C$_6$H$_4$Cl$_2$**	**95-50-1**

T/K = 303.05 $\Delta_{sol} H_B$ = 55.2 J/(g polymer)

Polymer (B):	**poly(dimethylsiloxane)**	**1968MOR**
Characterization:	M_n/g.mol^{-1} = 13000, η = 1.0 cSt, KF 96, Shinetsu Chemicals Co.	
Solvent (A):	**benzene** **C$_6$H$_6$**	**71-43-2**

T/K = 298.15 $\Delta_M H_B^\infty$ = 11.21 J/(g polymer)

Polymer (B):	**poly(dimethylsiloxane)**	**1973CHA**
Characterization:	M_η/g.mol^{-1} = 20000, SE-76, General Electric Co.	
Solvent (A):	**benzene** **C$_6$H$_6$**	**71-43-2**

T/K = 298.15 $\Delta_M H_B^\infty$ = 13.48 J/(g polymer)

Polymer (B):	**poly(dimethylsiloxane)**	**1964DEL**
Characterization:	M_η/g.mol^{-1} = 100000, commercial Dow Corning DC 200 silicone	
Solvent (A):	**benzene** **C$_6$H$_6$**	**71-43-2**

T/K = 298.15 $\Delta_M H_B^\infty$ = 14.2 J/(g polymer)

Polymer (B):	**poly(dimethylsiloxane)**	**1969BIA**
Characterization:	M_η/g.mol^{-1} = 170000	
Solvent (A):	**bromocyclohexane** **C$_6$H$_{11}$Br**	**108-85-0**

T/K = 303.15 $\Delta_M H_B^\infty$ = 753 J/(base mol polymer)

Polymer (B):	**poly(dimethylsiloxane)**	**1966BIA**
Characterization:	M_n/g.mol^{-1} = 30900, fractionated in the laboratory	
Solvent (A):	**2-butanone** **C$_4$H$_8$O**	**78-93-3**

T/K = 303.15 $\Delta_M H_B^\infty$ = 1054 J/(base mol polymer)

Comments: The final concentration is about 0.3 g polymer/100 ml solvent.

Polymer (B):	**poly(dimethylsiloxane)**	**1980SH1**
Characterization:	M_η/g.mol^{-1} = 80000, fractionated in the laboratory from a commercial PDMS, Shinetsu Chemicals Co., Japan	
Solvent (A):	**2-butanone** **C$_4$H$_8$O**	**78-93-3**

T/K	293.15	308.15	323.15
$\Delta_M H_B^\infty$/J(/g polymer)	14.4	14.3	14.3

Polymer (B): **poly(dimethylsiloxane)** **1969BIA**
Characterization: M_η/g.mol^{-1} = 170000
Solvent (A): **2-butanone** **C$_4$H$_8$O** **78-93-3**

T/K = 303.15 $\Delta_M H_B^\infty$ = 1092 J/(base mol polymer)

Polymer (B): **poly(dimethylsiloxane)** **1964PAT**
Characterization: M_η/g.mol^{-1} = 80000, commercial Dow Corning DC 200 silicone
Solvent (A): **butyl acetate** **C$_6$H$_{12}$O$_2$** **123-86-4**

T/K = 298.15 $\Delta_M H_B^\infty$ = 456 J/(base mol polymer)

Polymer (B): **poly(dimethylsiloxane)** **1964PAT**
Characterization: M_η/g.mol^{-1} = 80000, commercial Dow Corning DC 200 silicone
Solvent (A): **butyl propionate** **C$_7$H$_{14}$O$_2$** **590-01-2**

T/K = 298.15 $\Delta_M H_B^\infty$ = 360 J/(base mol polymer)

Polymer (B): **poly(dimethylsiloxane)** **1964DEL**
Characterization: M_η/g.mol^{-1} = 100000, commercial Dow Corning DC 200 silicone
Solvent (A): **chlorobenzene** **C$_6$H$_5$Cl** **108-90-7**

T/K = 298.15 $\Delta_M H_B^\infty$ = 7.53 J/(g polymer)

Polymer (B): **poly(dimethylsiloxane)** **1968MOR**
Characterization: M_n/g.mol^{-1} = 13000, η = 1.0 cSt, KF 96, Shinetsu Chemicals Co.
Solvent (A): **cyclohexane** **C$_6$H$_{12}$** **110-82-7**

T/K = 298.15 $\Delta_M H_B^\infty$ = 2.97 J/(g polymer)

Polymer (B): **poly(dimethylsiloxane)** **1979PH2**
Characterization: M_η/g.mol^{-1} = 20000, commercial Dow Corning DC 200 silicone
Solvent (A): **cyclohexane** **C$_6$H$_{12}$** **110-82-7**

T/K = 298.15 $\Delta_M H_B^\infty$ = 5.2 J/(g polymer)

Polymer (B): **poly(dimethylsiloxane)** **1964DEL**
Characterization: M_η/g.mol^{-1} = 100000, commercial Dow Corning DC 200 silicone
Solvent (A): **cyclohexane** **C$_6$H$_{12}$** **110-82-7**

T/K = 298.15 $\Delta_M H_B^\infty$ = 5.15 J/(g polymer)

Polymer (B):   **poly(dimethylsiloxane)**   **1979PH2**
Characterization: $M_\eta/\text{g.mol}^{-1} = 20000$, commercial Dow Corning DC 200 silicone
Solvent (A):   **cyclooctane**   **C$_8$H$_{16}$**   **292-64-8**

$T/\text{K} = 298.15$ $\Delta_M H_B{}^\infty = 6.8$ J/(g polymer)

Polymer (B):   **poly(dimethylsiloxane)**   **1979PH2**
Characterization: $M_\eta/\text{g.mol}^{-1} = 20000$, commercial Dow Corning DC 200 silicone
Solvent (A):   **cyclopentane**   **C$_5$H$_{10}$**   **287-92-3**

$T/\text{K} = 298.15$ $\Delta_M H_B{}^\infty = 1.0$ J/(g polymer)

Polymer (B):   **poly(dimethylsiloxane)**   **1979PH2**
Characterization: $M_\eta/\text{g.mol}^{-1} = 20000$, commercial Dow Corning DC 200 silicone
Solvent (A):   ***cis*-decahydronaphthalene**   **C$_{10}$H$_{18}$**   **493-01-6**

$T/\text{K} = 298.15$ $\Delta_M H_B{}^\infty = 7.1$ J/(g polymer)

Polymer (B):   **poly(dimethylsiloxane)**   **1979PH2**
Characterization: $M_\eta/\text{g.mol}^{-1} = 20000$, commercial Dow Corning DC 200 silicone
Solvent (A):   ***trans*-decahydronaphthalene**   **C$_{10}$H$_{18}$**   **493-02-7**

$T/\text{K} = 298.15$ $\Delta_M H_B{}^\infty = 4.3$ J/(g polymer)

Polymer (B):   **poly(dimethylsiloxane)**   **1962DE1**
Characterization: $M_\eta/\text{g.mol}^{-1} = 20000$, commercial Dow Corning DC 200 silicone
Solvent (A):   **decamethyltetrasiloxane**   **C$_{10}$H$_{30}$O$_3$Si$_4$**   **141-62-8**

$T/\text{K} = 297.65$ $\Delta_M H_B{}^\infty = -33$ J/(base mol polymer)

Polymer (B):   **poly(dimethylsiloxane)**   **1962DE1**
Characterization: $M_\eta/\text{g.mol}^{-1} = 20000$, commercial Dow Corning DC 200 silicone
Solvent (A):   **n-decane**   **C$_{10}$H$_{22}$**   **124-18-5**

$T/\text{K} = 297.65$ $\Delta_M H_B{}^\infty = 285$ J/(base mol polymer)

Polymer (B):   **poly(dimethylsiloxane)**   **1973CHA**
Characterization: $M_\eta/\text{g.mol}^{-1} = 20000$, SE-76, General Electric Co.
Solvent (A):   **n-decane**   **C$_{10}$H$_{22}$**   **124-18-5**

$T/\text{K} = 298.15$ $\Delta_M H_B{}^\infty = 3.85$ J/(g polymer)

Polymer (B):   **poly(dimethylsiloxane)**   **1964PAT**
Characterization: $M_\eta/\text{g.mol}^{-1} = 80000$, commercial Dow Corning DC 200 silicone
Solvent (A):   **n-decane**   **C$_{10}$H$_{22}$**   **124-18-5**

$T/\text{K} = 298.15$ $\Delta_M H_B{}^\infty = 284.5$ J/(base mol polymer)

Polymer (B):   **poly(dimethylsiloxane)**   **1964PAT**
Characterization: $M_\eta/\text{g.mol}^{-1} = 80000$, commercial Dow Corning DC 200 silicone
Solvent (A):   **decyl acetate** $C_{12}H_{24}O_2$ **112-17-4**

$T/\text{K} = 298.15$ $\Delta_M H_B^\infty = 335$ J/(base mol polymer)

Polymer (B):   **poly(dimethylsiloxane)**   **1964PAT**
Characterization: $M_\eta/\text{g.mol}^{-1} = 80000$, commercial Dow Corning DC 200 silicone
Solvent (A):   **dibutyl ether** $C_8H_{18}O$ **142-96-1**

$T/\text{K} = 298.15$ $\Delta_M H_B^\infty = 46$ J/(base mol polymer)

Polymer (B):   **poly(dimethylsiloxane)**   **1964PAT**
Characterization: $M_\eta/\text{g.mol}^{-1} = 80000$, commercial Dow Corning DC 200 silicone
Solvent (A):   **diethoxymethane** $C_5H_{12}O_2$ **462-95-3**

$T/\text{K} = 298.15$ $\Delta_M H_B^\infty = 134$ J/(base mol polymer)

Polymer (B):   **poly(dimethylsiloxane)**   **1964PAT**
Characterization: $M_\eta/\text{g.mol}^{-1} = 80000$, commercial Dow Corning DC 200 silicone
Solvent (A):   **diethyl ether** $C_4H_{10}O$ **60-29-7**

$T/\text{K} = 298.15$ $\Delta_M H_B^\infty = -96$ J/(base mol polymer)

Polymer (B):   **poly(dimethylsiloxane)**   **1979PH2**
Characterization: $M_\eta/\text{g.mol}^{-1} = 20000$, commercial Dow Corning DC 200 silicone
Solvent (A):   **3,3-diethylpentane** C_9H_{20} **4032-86-4**

$T/\text{K} = 298.15$ $\Delta_M H_B^\infty = 1.9$ J/(g polymer)

Polymer (B):   **poly(dimethylsiloxane)**   **1964PAT**
Characterization: $M_\eta/\text{g.mol}^{-1} = 80000$, commercial Dow Corning DC 200 silicone
Solvent (A):   **dihexyl ether** $C_{12}H_{26}O$ **112-58-3**

$T/\text{K} = 298.15$ $\Delta_M H_B^\infty = 226$ J/(base mol polymer)

Polymer (B):   **poly(dimethylsiloxane)**   **1964PAT**
Characterization: $M_\eta/\text{g.mol}^{-1} = 80000$, commercial Dow Corning DC 200 silicone
Solvent (A):   **1,2-dimethoxyethane** $C_4H_{10}O_2$ **110-71-4**

$T/\text{K} = 298.15$ $\Delta_M H_B^\infty = 904$ J/(base mol polymer)

Polymer (B):   **poly(dimethylsiloxane)**   **1964PAT**
Characterization: $M_\eta/\text{g.mol}^{-1} = 80000$, commercial Dow Corning DC 200 silicone
Solvent (A):   **dimethoxymethane** $C_3H_3O_2$ **109-87-5**

$T/\text{K} = 298.15$ $\Delta_M H_B^\infty = 552$ J/(base mol polymer)

Polymer (B):   **poly(dimethylsiloxane)**   **1968MOR**
Characterization: $M_n/\text{g.mol}^{-1} = 13000$, $\eta = 1.0$ cSt, KF 96, Shinetsu Chemicals Co.
Solvent (A):   **1,2-dimethylbenzene** $\mathbf{C_8H_{10}}$ **95-47-6**

$T/\text{K} = 298.15$ $\Delta_M H_B^\infty = 4.31$ J/(g polymer)

Polymer (B):   **poly(dimethylsiloxane)**   **1968MOR**
Characterization: $M_n/\text{g.mol}^{-1} = 13000$, $\eta = 1.0$ cSt, KF 96, Shinetsu Chemicals Co.
Solvent (A):   **1,3-dimethylbenzene** $\mathbf{C_8H_{10}}$ **108-38-3**

$T/\text{K} = 298.15$ $\Delta_M H_B^\infty = 3.0$ J/(g polymer)

Polymer (B):   **poly(dimethylsiloxane)**   **1968MOR**
Characterization: $M_n/\text{g.mol}^{-1} = 13000$, $\eta = 1.0$ cSt, KF 96, Shinetsu Chemicals Co.
Solvent (A):   **1,4-dimethylbenzene** $\mathbf{C_8H_{10}}$ **106-42-3**

$T/\text{K} = 298.15$ $\Delta_M H_B^\infty = 3.2$ J/(g polymer)

Polymer (B):   **poly(dimethylsiloxane)**   **1973CHA**
Characterization: $M_\eta/\text{g.mol}^{-1} = 20000$, SE-76, General Electric Co.
Solvent (A):   **1,4-dimethylbenzene** $\mathbf{C_8H_{10}}$ **106-42-3**

$T/\text{K} = 298.15$ $\Delta_M H_B^\infty = 4.23$ J/(g polymer)

Polymer (B):   **poly(dimethylsiloxane)**   **1975TAN**
Characterization: $M_n/\text{g.mol}^{-1} = 19000$, Dow Corning Silicones
Solvent (A):   **2,2-dimethylbutane** $\mathbf{C_6H_{14}}$ **75-83-2**

$T/\text{K} = 298.15$ $\Delta_M H_B^\infty = 0.0$ J/cm^3

Polymer (B):   **poly(dimethylsiloxane)**   **1980SH1**
Characterization: $M_\eta/\text{g.mol}^{-1} = 80000$, fractionated in the laboratory from a
                           commercial PDMS, Shinetsu Chemicals Co., Japan
Solvent (A):   **2,6-dimethyl-4-heptanone** $\mathbf{C_9H_{18}O}$ **108-83-8**

$T/\text{K} = 308.15$ $\Delta_M H_B^\infty = 6.1$ J/(g polymer)

Polymer (B):   **poly(dimethylsiloxane)**   **1979PH2**
Characterization: $M_\eta/\text{g.mol}^{-1} = 20000$, commercial Dow Corning DC 200 silicone
Solvent (A):   **2,2-dimethylpentane** $\mathbf{C_7H_{16}}$ **590-35-2**

$T/\text{K} = 298.15$ $\Delta_M H_B^\infty = 0.8$ J/(g polymer)

Polymer (B):   **poly(dimethylsiloxane)**   **1979PH2**
Characterization: M_η/g.mol^{-1} = 20000, commercial Dow Corning DC 200 silicone
Solvent (A):   **2,3-dimethylpentane**   **C$_7$H$_{16}$**   **565-59-3**

T/K = 298.15 $\Delta_M H_B^\infty$ = 1.4 J/(g polymer)

Polymer (B):   **poly(dimethylsiloxane)**   **1979PH2**
Characterization: M_η/g.mol^{-1} = 20000, commercial Dow Corning DC 200 silicone
Solvent (A):   **2,4-dimethylpentane**   **C$_7$H$_{16}$**   **108-08-7**

T/K = 298.15 $\Delta_M H_B^\infty$ = 1.6 J/(g polymer)

Polymer (B):   **poly(dimethylsiloxane)**   **1979PH2**
Characterization: M_η/g.mol^{-1} = 20000, commercial Dow Corning DC 200 silicone
Solvent (A):   **3,3-dimethylpentane**   **C$_7$H$_{16}$**   **562-49-2**

T/K = 298.15 $\Delta_M H_B^\infty$ = 0.5 J/(g polymer)

Polymer (B):   **poly(dimethylsiloxane)**   **1964PAT**
Characterization: M_η/g.mol^{-1} = 80000, commercial Dow Corning DC 200 silicone
Solvent (A):   **dipentyl ether**   **C$_{10}$H$_{22}$O**   **693-65-2**

T/K = 298.15 $\Delta_M H_B^\infty$ = 155 J/(base mol polymer)

Polymer (B):   **poly(dimethylsiloxane)**   **1964PAT**
Characterization: M_η/g.mol^{-1} = 80000, commercial Dow Corning DC 200 silicone
Solvent (A):   **dipropyl ether**   **C$_6$H$_{14}$O**   **111-43-3**

T/K = 298.15 $\Delta_M H_B^\infty$ = −87.9 J/(base mol polymer)

Polymer (B):   **poly(dimethylsiloxane)**   **1962DE1**
Characterization: M_η/g.mol^{-1} = 20000, commercial Dow Corning DC 200 silicone
Solvent (A):   **dodecamethylpentasiloxane**   **C$_{12}$H$_{36}$O$_4$Si$_5$**   **141-63-9**

T/K = 297.65 $\Delta_M H_B^\infty$ = −21 J/(base mol polymer)

Polymer (B):   **poly(dimethylsiloxane)**   **1975TAN**
Characterization: M_n/g.mol^{-1} = 19000, Dow Corning Silicones
Solvent (A):   **n-dodecane**   **C$_{12}$H$_{26}$**   **112-40-3**

T/K = 298.15 $\Delta_M H_B^\infty$ = 10.06 J/cm^3

Polymer (B):	**poly(dimethylsiloxane)**	**1962DE1**
Characterization:	$M_\eta/\text{g.mol}^{-1} = 20000$, commercial Dow Corning DC 200 silicone	
Solvent (A):	**n-dodecane** $\quad$ **C$_{12}$H$_{26}$**	**112-40-3**

$T/\text{K} = 297.65$ $\qquad \Delta_M H_B^\infty = 330$ J/(base mol polymer)

Polymer (B):	**poly(dimethylsiloxane)**	**1979PH1**
Characterization:	$M_\eta/\text{g.mol}^{-1} = 20000$, commercial Dow Corning DC 200 silicone	
Solvent (A):	**n-dodecane** $\quad$ **C$_{12}$H$_{26}$**	**112-40-3**

$T/\text{K} = 298.15$ $\qquad \Delta_M H_B^\infty = 4.4$ J/(g polymer)

Polymer (B):	**poly(dimethylsiloxane)**	**1964PAT**
Characterization:	$M_\eta/\text{g.mol}^{-1} = 80000$, commercial Dow Corning DC 200 silicone	
Solvent (A):	**n-dodecane** $\quad$ **C$_{12}$H$_{26}$**	**112-40-3**

$T/\text{K} = 298.15$ $\qquad \Delta_M H_B^\infty = 331$ J/(base mol polymer)

Polymer (B):	**poly(dimethylsiloxane)**	**1964PAT**
Characterization:	$M_\eta/\text{g.mol}^{-1} = 80000$, commercial Dow Corning DC 200 silicone	
Solvent (A):	**ethyl acetate** $\quad$ **C$_4$H$_8$O$_2$**	**141-78-6**

$T/\text{K} = 298.15$ $\qquad \Delta_M H_B^\infty = 941$ J/(base mol polymer)

Polymer (B):	**poly(dimethylsiloxane)**	**1969BIA**
Characterization:	$M_\eta/\text{g.mol}^{-1} = 170000$	
Solvent (A):	**ethyl acetate** $\quad$ **C$_4$H$_8$O$_2$**	**141-78-6**

$T/\text{K} = 303.15$ $\qquad \Delta_M H_B^\infty = 1017$ J/(base mol polymer)

Polymer (B):	**poly(dimethylsiloxane)**	**1968MOR**
Characterization:	$M_n/\text{g.mol}^{-1} = 13000$, $\eta = 1.0$ cSt, KF 96, Shinetsu Chemicals Co.	
Solvent (A):	**ethylbenzene** $\quad$ **C$_8$H$_{10}$**	**100-41-4**

$T/\text{K} = 298.15$ $\qquad \Delta_M H_B^\infty = 6.36$ J/(g polymer)

Polymer (B):	**poly(dimethylsiloxane)**	**1973CHA**
Characterization:	$M_\eta/\text{g.mol}^{-1} = 20000$, SE-76, General Electric Co.	
Solvent (A):	**ethylbenzene** $\quad$ **C$_8$H$_{10}$**	**100-41-4**

$T/\text{K} = 298.15$ $\qquad \Delta_M H_B^\infty = 6.24$ J/(g polymer)

Polymer (B):	**poly(dimethylsiloxane)**	**1964PAT**
Characterization:	$M_\eta/\text{g.mol}^{-1} = 80000$, commercial Dow Corning DC 200 silicone	
Solvent (A):	**ethyl butanoate** $\quad$ **C$_6$H$_{12}$O$_2$**	**105-54-4**

$T/\text{K} = 298.15$ $\qquad \Delta_M H_B^\infty = 443.5$ J/(base mol polymer)

Polymer (B): poly(dimethylsiloxane) **1964PAT**
Characterization: M_η/g.mol^{-1} = 80000, commercial Dow Corning DC 200 silicone
Solvent (A): **ethyl decanoate** $C_{12}H_{24}O_2$ **112-17-4**

T/K = 298.15 $\Delta_M H_B^\infty$ = 280 J/(base mol polymer)

Polymer (B): poly(dimethylsiloxane) **1964PAT**
Characterization: M_η/g.mol^{-1} = 80000, commercial Dow Corning DC 200 silicone
Solvent (A): **ethyl dodecanoate** $C_{14}H_{28}O_2$ **106-33-2**

T/K = 298.15 $\Delta_M H_B^\infty$ = 285 J/(base mol polymer)

Polymer (B): poly(dimethylsiloxane) **1964PAT**
Characterization: M_η/g.mol^{-1} = 80000, commercial Dow Corning DC 200 silicone
Solvent (A): **ethyl heptanoate** $C_9H_{18}O_2$ **106-30-9**

T/K = 298.15 $\Delta_M H_B^\infty$ = 301 J/(base mol polymer)

Polymer (B): poly(dimethylsiloxane) **1964PAT**
Characterization: M_η/g.mol^{-1} = 80000, commercial Dow Corning DC 200 silicone
Solvent (A): **ethyl hexanoate** $C_8H_{16}O_2$ **123-66-0**

T/K = 298.15 $\Delta_M H_B^\infty$ = 318 J/(base mol polymer)

Polymer (B): poly(dimethylsiloxane) **1964PAT**
Characterization: M_η/g.mol^{-1} = 80000, commercial Dow Corning DC 200 silicone
Solvent (A): **ethyl nonanoate** $C_{11}H_{22}O_2$ **123-29-5**

T/K = 298.15 $\Delta_M H_B^\infty$ = 276 J/(base mol polymer)

Polymer (B): poly(dimethylsiloxane) **1964PAT**
Characterization: M_η/g.mol^{-1} = 80000, commercial Dow Corning DC 200 silicone
Solvent (A): **ethyl octanoate** $C_{10}H_{20}O_2$ **106-32-1**

T/K = 298.15 $\Delta_M H_B^\infty$ = 285 J/(base mol polymer)

Polymer (B): poly(dimethylsiloxane) **1979PH2**
Characterization: M_η/g.mol^{-1} = 20000, commercial Dow Corning DC 200 silicone
Solvent (A): **3-ethylpentane** C_7H_{16} **617-78-7**

T/K = 298.15 $\Delta_M H_B^\infty$ = 0.6 J/(g polymer)

Polymer (B): poly(dimethylsiloxane) **1964PAT**
Characterization: M_η/g.mol^{-1} = 80000, commercial Dow Corning DC 200 silicone
Solvent (A): **ethyl propionate** $C_5H_{10}O_2$ **105-37-3**

T/K = 298.15 $\Delta_M H_B^\infty$ = 594 J/(base mol polymer)

Polymer (B):   **poly(dimethylsiloxane)**   **1975TAN**
Characterization: $M_n/\text{g.mol}^{-1} = 19000$, Dow Corning Silicones
Solvent (A):   **2,2,4,4,6,8,8-heptamethylnonane**   **$C_{16}H_{34}$**   **4390-04-9**

$T/\text{K} = 298.15$ $\Delta_M H_B^\infty = 4.81 \text{ J/cm}^3$

Polymer (B):   **poly(dimethylsiloxane)**   **1979PH2**
Characterization: $M_\eta/\text{g.mol}^{-1} = 20000$, commercial Dow Corning DC 200 silicone
Solvent (A):   **2,2,4,4,6,8,8-heptamethylnonane**   **$C_{16}H_{34}$**   **4390-04-9**

$T/\text{K} = 298.15$ $\Delta_M H_B^\infty = 3.5 \text{ J/(g polymer)}$

Polymer (B):   **poly(dimethylsiloxane)**   **1968MOR**
Characterization: $M_n/\text{g.mol}^{-1} = 13000$, $\eta = 1.0$ cSt, KF 96, Shinetsu Chemicals Co.
Solvent (A):   **n-heptane**   **C_7H_{16}**   **142-82-5**

$T/\text{K} = 298.15$ $\Delta_M H_B^\infty = 1.77 \text{ J/(g polymer)}$

Polymer (B):   **poly(dimethylsiloxane)**   **1962DE1**
Characterization: $M_\eta/\text{g.mol}^{-1} = 20000$, commercial Dow Corning DC 200 silicone
Solvent (A):   **n-heptane**   **C_7H_{16}**   **142-82-5**

$T/\text{K} = 297.65$ $\Delta_M H_B^\infty = 145 \text{ J/(base mol polymer)}$

Polymer (B):   **poly(dimethylsiloxane)**   **1973CHA**
Characterization: $M_\eta/\text{g.mol}^{-1} = 20000$, SE-76, General Electric Co.
Solvent (A):   **n-heptane**   **C_7H_{16}**   **142-82-5**

$T/\text{K} = 298.15$ $\Delta_M H_B^\infty = 2.01 \text{ J/(g polymer)}$

Polymer (B):   **poly(dimethylsiloxane)**   **1979PH1**
Characterization: $M_\eta/\text{g.mol}^{-1} = 20000$, commercial Dow Corning DC 200 silicone
Solvent (A):   **n-heptane**   **C_7H_{16}**   **142-82-5**

$T/\text{K} = 298.15$ $\Delta_M H_B^\infty = 1.9 \text{ J/(g polymer)}$

Polymer (B):   **poly(dimethylsiloxane)**   **1964PAT**
Characterization: $M_\eta/\text{g.mol}^{-1} = 80000$, commercial Dow Corning DC 200 silicone
Solvent (A):   **n-heptane**   **C_7H_{16}**   **142-82-5**

$T/\text{K} = 298.15$ $\Delta_M H_B^\infty = 151 \text{ J/(base mol polymer)}$

Polymer (B):   **poly(dimethylsiloxane)**   **1964DEL**
Characterization: $M_\eta/\text{g.mol}^{-1} = 100000$, commercial Dow Corning DC 200 silicone
Solvent (A):   **n-heptane**   **C_7H_{16}**   **142-82-5**

$T/\text{K} = 298.15$ $\Delta_M H_B^\infty = 2.10 \text{ J/(g polymer)}$

Polymer (B):	**poly(dimethylsiloxane)**		**1969BIA**
Characterization:	$M_\eta/\text{g.mol}^{-1} = 170000$		
Solvent (A):	**n-heptane**	**C$_7$H$_{16}$**	**142-82-5**

$T/\text{K} = 303.15$ $\Delta_M H_B^\infty = 138$ J/(base mol polymer)

Polymer (B):	**poly(dimethylsiloxane)**		**1980SH1**
Characterization:	$M_\eta/\text{g.mol}^{-1} = 80000$, fractionated in the laboratory from a commercial PDMS, Shinetsu Chemicals Co., Japan		
Solvent (A):	**3-heptanone**	**C$_7$H$_{14}$O**	**106-35-4**

T/K	308.15	323.15
$\Delta_M H_B^\infty/\text{J}/(\text{g polymer})$	8.80	8.75

Polymer (B):	**poly(dimethylsiloxane)**		**1975TAN**
Characterization:	$M_n/\text{g.mol}^{-1} = 19000$, Dow Corning Silicones		
Solvent (A):	**n-hexadecane**	**C$_{16}$H$_{34}$**	**544-76-3**

$T/\text{K} = 298.15$ $\Delta_M H_B^\infty = 12.64$ J/cm^3

Polymer (B):	**poly(dimethylsiloxane)**		**1962DE1**
Characterization:	$M_\eta/\text{g.mol}^{-1} = 20000$, commercial Dow Corning DC 200 silicone		
Solvent (A):	**n-hexadecane**	**C$_{16}$H$_{34}$**	**544-76-3**

$T/\text{K} = 297.65$ $\Delta_M H_B^\infty = 410$ J/(base mol polymer)

Polymer (B):	**poly(dimethylsiloxane)**		**1979PH1**
Characterization:	$M_\eta/\text{g.mol}^{-1} = 20000$, commercial Dow Corning DC 200 silicone		
Solvent (A):	**n-hexadecane**	**C$_{16}$H$_{34}$**	**544-76-3**

$T/\text{K} = 298.15$ $\Delta_M H_B^\infty = 5.5$ J/(g polymer)

Polymer (B):	**poly(dimethylsiloxane)**		**1962DE1**
Characterization:	$M_\eta/\text{g.mol}^{-1} = 20000$, commercial Dow Corning DC 200 silicone		
Solvent (A):	**hexamethyldisiloxane**	**C$_6$H$_{18}$OSi$_2$**	**107-46-0**

$T/\text{K} = 297.65$ $\Delta_M H_B^\infty = -90$ J/(base mol polymer)

Polymer (B):	**poly(dimethylsiloxane)**		**1973CHA**
Characterization:	$M_\eta/\text{g.mol}^{-1} = 20000$, SE-76, General Electric Co.		
Solvent (A):	**hexamethyldisiloxane**	**C$_6$H$_{18}$OSi$_2$**	**107-46-0**

$T/\text{K} = 298.15$ $\Delta_M H_B^\infty = -1.6$ J/(g polymer)

Polymer (B):   **poly(dimethylsiloxane)**   **1969BIA**
Characterization: $M_\eta/\text{g.mol}^{-1} = 170000$
Solvent (A):   **hexamethyldisiloxane** $C_6H_{18}OSi_2$ **107-46-0**

$T/\text{K} = 303.15$ $\Delta_M H_B^\infty = -109$ J/(base mol polymer)

Polymer (B):   **poly(dimethylsiloxane)**   **1975TAN**
Characterization: $M_n/\text{g.mol}^{-1} = 19000$, Dow Corning Silicones
Solvent (A):   **n-hexane** C_6H_{14} **110-54-3**

$T/\text{K} = 298.15$ $\Delta_M H_B^\infty = 2.8$ J/cm^3

Polymer (B):   **poly(dimethylsiloxane)**   **1973CHA**
Characterization: $M_\eta/\text{g.mol}^{-1} = 20000$, SE-76, General Electric Co.
Solvent (A):   **n-hexane** C_6H_{14} **110-54-3**

$T/\text{K} = 298.15$ $\Delta_M H_B^\infty = 0.67$ J/(g polymer)

Polymer (B):   **poly(dimethylsiloxane)**   **1962DE1**
Characterization: $M_\eta/\text{g.mol}^{-1} = 20000$, commercial Dow Corning DC 200 silicone
Solvent (A):   **n-hexane** C_6H_{14} **110-54-3**

$T/\text{K} = 297.65$ $\Delta_M H_B^\infty = 50$ J/(base mol polymer)

Polymer (B):   **poly(dimethylsiloxane)**   **1964PAT**
Characterization: $M_\eta/\text{g.mol}^{-1} = 80000$, commercial Dow Corning DC 200 silicone
Solvent (A):   **n-hexane** C_6H_{14} **110-54-3**

$T/\text{K} = 298.15$ $\Delta_M H_B^\infty = 50$ J/(base mol polymer)

Polymer (B):   **poly(dimethylsiloxane)**   **1969BIA**
Characterization: $M_\eta/\text{g.mol}^{-1} = 170000$
Solvent (A):   **n-hexane** C_6H_{14} **110-54-3**

$T/\text{K} = 303.15$ $\Delta_M H_B^\infty = 20.9$ J/(base mol polymer)

Polymer (B):   **poly(dimethylsiloxane)**   **1964PAT**
Characterization: $M_\eta/\text{g.mol}^{-1} = 80000$, commercial Dow Corning DC 200 silicone
Solvent (A):   **hexyl acetate** $C_8H_{16}O_2$ **142-92-7**

$T/\text{K} = 298.15$ $\Delta_M H_B^\infty = 368$ J/(base mol polymer)

Polymer (B):   **poly(dimethylsiloxane)**   **1968MOR**
Characterization: $M_n/\text{g.mol}^{-1} = 13000$, $\eta = 1.0$ cSt, KF 96, Shinetsu Chemicals Co.
Solvent (A):   **isopropylbenzene** C_9H_{12} **98-82-8**

$T/\text{K} = 298.15$ $\Delta_M H_B^\infty = 4.14$ J/(g polymer)

Polymer (B):   **poly(dimethylsiloxane)**   **1964PAT**
Characterization: M_η/g.mol^{-1} = 80000, commercial Dow Corning DC 200 silicone
Solvent (A):   **methyl butanoate**   **C$_5$H$_{10}$O$_2$**   **623-42-7**

T/K = 298.15 $\Delta_M H_B^\infty$ = 640 J/(base mol polymer)

Polymer (B):   **poly(dimethylsiloxane)**   **1968MOR**
Characterization: M_η/g.mol^{-1} = 13000, η = 1.0 cSt, KF 96, Shinetsu Chemicals Co.
Solvent (A):   **methylcyclohexane**   **C$_7$H$_{14}$**   **108-87-2**

T/K = 298.15 $\Delta_M H_B^\infty$ = 2.92 J/(g polymer)

Polymer (B):   **poly(dimethylsiloxane)**   **1964DEL**
Characterization: M_η/g.mol^{-1} = 100000, commercial Dow Corning DC 200 silicone
Solvent (A):   **methylcyclohexane**   **C$_7$H$_{14}$**   **108-87-2**

T/K = 298.15 $\Delta_M H_B^\infty$ = 1.90 J/(g polymer)

Polymer (B):   **poly(dimethylsiloxane)**   **1964PAT**
Characterization: M_η/g.mol^{-1} = 80000, commercial Dow Corning DC 200 silicone
Solvent (A):   **methyl decanoate**   **C$_{11}$H$_{22}$O$_2$**   **110-42-9**

T/K = 298.15 $\Delta_M H_B^\infty$ = 356 J/(base mol polymer)

Polymer (B):   **poly(dimethylsiloxane)**   **1979PH2**
Characterization: M_η/g.mol^{-1} = 20000, commercial Dow Corning DC 200 silicone
Solvent (A):   **3-methylhexane**   **C$_7$H$_{16}$**   **589-34-4**

T/K = 298.15 $\Delta_M H_B^\infty$ = 1.3 J/(g polymer)

Polymer (B):   **poly(dimethylsiloxane)**   **1964PAT**
Characterization: M_η/g.mol^{-1} = 80000, commercial Dow Corning DC 200 silicone
Solvent (A):   **methyl hexanoate**   **C$_7$H$_{14}$O$_2$**   **106-70-7**

T/K = 298.15 $\Delta_M H_B^\infty$ = 393 J/(base mol polymer)

Polymer (B):   **poly(dimethylsiloxane)**   **1964PAT**
Characterization: M_η/g.mol^{-1} = 80000, commercial Dow Corning DC 200 silicone
Solvent (A):   **methyl octanoate**   **C$_9$H$_{18}$O$_2$**   **111-11-5**

T/K = 298.15 $\Delta_M H_B^\infty$ = 368 J/(base mol polymer)

Polymer (B):   **poly(dimethylsiloxane)**   **1980SH1**
Characterization: M_η/g.mol^{-1} = 80000, fractionated in the laboratory from a
                           commercial PDMS, Shinetsu Chemicals Co., Japan
Solvent (A):   **4-methyl-2-pentanone   C$_6$H$_{12}$O**   **108-10-1**

T/K	293.15	308.15
$\Delta_M H_B^\infty$/J(/g polymer)	9.9	9.0

Polymer (B):   **poly(dimethylsiloxane)**   **1964PAT**
Characterization: M_η/g.mol^{-1} = 80000, commercial Dow Corning DC 200 silicone
Solvent (A):   **methyl propionate   C$_4$H$_8$O$_2$**   **554-12-1**

T/K = 298.15 $\Delta_M H_B^\infty$ = 895 J/(base mol polymer)

Polymer (B):   **poly(dimethylsiloxane)**   **1962DE1**
Characterization: M_η/g.mol^{-1} = 20000, commercial Dow Corning DC 200 silicone
Solvent (A):   **n-nonane   C$_9$H$_{20}$**   **111-84-2**

T/K = 297.65 $\Delta_M H_B^\infty$ = 250 J/(base mol polymer)

Polymer (B):   **poly(dimethylsiloxane)**   **1979PH1**
Characterization: M_η/g.mol^{-1} = 20000, commercial Dow Corning DC 200 silicone
Solvent (A):   **n-nonane   C$_9$H$_{20}$**   **111-84-2**

T/K = 298.15 $\Delta_M H_B^\infty$ = 3.3 J/(g polymer)

Polymer (B):   **poly(dimethylsiloxane)**   **1964PAT**
Characterization: M_η/g.mol^{-1} = 80000, commercial Dow Corning DC 200 silicone
Solvent (A):   **n-nonane   C$_9$H$_{20}$**   **111-84-2**

T/K = 298.15 $\Delta_M H_B^\infty$ = 251 J/(base mol polymer)

Polymer (B):   **poly(dimethylsiloxane)**   **1980SH2**
Characterization: M_η/g.mol^{-1} = 80000, fractionated in the laboratory from a
                           commercial PDMS, Shinetsu Chemicals Co., Japan
Solvent (A):   **octamethylcyclotetrasiloxane   C$_8$H$_{24}$O$_4$Si$_4$**   **556-67-2**

T/K = 298.15 $\Delta_M H_B^\infty$ = −0.4 J/(g polymer)

Polymer (B):   **poly(dimethylsiloxane)**   **1962DE1**
Characterization: M_η/g.mol^{-1} = 20000, commercial Dow Corning DC 200 silicone
Solvent (A):   **octamethyltrisiloxane   C$_8$H$_{24}$O$_2$Si$_3$**   **107-51-7**

T/K = 297.65 $\Delta_M H_B^\infty$ = −45 J/(base mol polymer)

Polymer (B):	poly(dimethylsiloxane)	**1973CHA**
Characterization:	M_η/g.mol^{-1} = 20000, SE-76, General Electric Co.	
Solvent (A):	**octamethyltrisiloxane** $C_8H_{24}O_2Si_3$	**107-51-7**

T/K = 298.15 $\Delta_M H_B^\infty$ = −0.8 J/(g polymer)

Polymer (B):	poly(dimethylsiloxane)	**1969BIA**
Characterization:	M_η/g.mol^{-1} = 170000	
Solvent (A):	**octamethyltrisiloxane** $C_8H_{24}O_2Si_3$	**107-51-7**

T/K = 303.15 $\Delta_M H_B^\infty$ = −71 J/(base mol polymer)

Polymer (B):	poly(dimethylsiloxane)	**1975TAN**
Characterization:	M_n/g.mol^{-1} = 19000, Dow Corning Silicones	
Solvent (A):	**n-octane** C_8H_{18}	**111-65-9**

T/K = 298.15 $\Delta_M H_B^\infty$ = 5.92 J/cm^3

Polymer (B):	poly(dimethylsiloxane)	**1962DE1**
Characterization:	M_η/g.mol^{-1} = 20000, commercial Dow Corning DC 200 silicone	
Solvent (A):	**n-octane** C_8H_{18}	**111-65-9**

T/K = 297.65 $\Delta_M H_B^\infty$ = 190 J/(base mol polymer)

Polymer (B):	poly(dimethylsiloxane)	**1973CHA**
Characterization:	M_η/g.mol^{-1} = 20000, SE-76, General Electric Co.	
Solvent (A):	**n-octane** C_8H_{18}	**111-65-9**

T/K = 298.15 $\Delta_M H_B^\infty$ = 2.6 J/(g polymer)

Polymer (B):	poly(dimethylsiloxane)	**1979PH1**
Characterization:	M_η/g.mol^{-1} = 20000, commercial Dow Corning DC 200 silicone	
Solvent (A):	**n-octane** C_8H_{18}	**111-65-9**

T/K = 298.15 $\Delta_M H_B^\infty$ = 2.4 J/(g polymer)

Polymer (B):	poly(dimethylsiloxane)	**1964PAT**
Characterization:	M_η/g.mol^{-1} = 80000, commercial Dow Corning DC 200 silicone	
Solvent (A):	**n-octane** C_8H_{18}	**111-65-9**

T/K = 298.15 $\Delta_M H_B^\infty$ = 192 J/(base mol polymer)

Polymer (B):	poly(dimethylsiloxane)	**1975TAN**
Characterization:	M_n/g.mol^{-1} = 19000, Dow Corning Silicones	
Solvent (A):	**2,2,4,6,6-pentamethylheptane** $C_{12}H_{26}$	**13475-82-6**

T/K = 298.15 $\Delta_M H_B^\infty$ = 4.47 J/cm^3

Polymer (B):   **poly(dimethylsiloxane)**   **1979PH1**
Characterization: M_η/g.mol^{-1} = 20000, commercial Dow Corning DC 200 silicone
Solvent (A):   **2,2,4,6,6-pentamethylheptane**   **$C_{12}H_{26}$**   **13475-82-6**

T/K = 298.15 $\Delta_M H_B^\infty$ = 2.7 J/(g polymer)

Polymer (B):   **poly(dimethylsiloxane)**   **1962DE1**
Characterization: M_η/g.mol^{-1} = 20000, commercial Dow Corning DC 200 silicone
Solvent (A):   **n-pentane**   **C_5H_{12}**   **109-66-0**

T/K = 297.65 $\Delta_M H_B^\infty$ = −70 J/(base mol polymer)

Polymer (B):   **poly(dimethylsiloxane)**   **1973CHA**
Characterization: M_η/g.mol^{-1} = 20000, SE-76, General Electric Co.
Solvent (A):   **n-pentane**   **C_5H_{12}**   **109-66-0**

T/K = 298.15 $\Delta_M H_B^\infty$ = −0.92 J/(g polymer)

Polymer (B):   **poly(dimethylsiloxane)**   **1964PAT**
Characterization: M_η/g.mol^{-1} = 80000, commercial Dow Corning DC 200 silicone
Solvent (A):   **n-pentane**   **C_5H_{12}**   **109-66-0**

T/K = 298.15 $\Delta_M H_B^\infty$ = −71 J/(base mol polymer)

Polymer (B):   **poly(dimethylsiloxane)**   **1964PAT**
Characterization: M_η/g.mol^{-1} = 80000, commercial Dow Corning DC 200 silicone
Solvent (A):   **pentyl acetate**   **$C_7H_{14}O_2$**   **628-63-7**

T/K = 298.15 $\Delta_M H_B^\infty$ = 431 J/(base mol polymer)

Polymer (B):   **poly(dimethylsiloxane)**   **1964PAT**
Characterization: M_η/g.mol^{-1} = 80000, commercial Dow Corning DC 200 silicone
Solvent (A):   **pentyl propionate**   **$C_8H_{16}O_2$**   **624-54-4**

T/K = 298.15 $\Delta_M H_B^\infty$ = 285 J/(base mol polymer)

Polymer (B):   **poly(dimethylsiloxane)**   **1964PAT**
Characterization: M_η/g.mol^{-1} = 80000, commercial Dow Corning DC 200 silicone
Solvent (A):   **propyl acetate**   **$C_5H_{10}O_2$**   **109-60-4**

T/K = 298.15 $\Delta_M H_B^\infty$ = 636 J/(base mol polymer)

Polymer (B):   **poly(dimethylsiloxane)**   **1969BIA**
Characterization: M_η/g.mol^{-1} = 170000
Solvent (A):   **propyl acetate**   **$C_5H_{10}O_2$**   **109-60-4**

T/K = 303.15 $\Delta_M H_B^\infty$ = 732 J/(base mol polymer)

Polymer (B):   **poly(dimethylsiloxane)**   **1964PAT**
Characterization: M_η/g.mol^{-1} = 80000, commercial Dow Corning DC 200 silicone
Solvent (A):   **propyl propionate** $C_6H_{12}O_2$ **106-36-5**

T/K = 298.15 $\Delta_M H_B^\infty$ = 414 J/(base mol polymer)

Polymer (B):   **poly(dimethylsiloxane)**   **1964DEL**
Characterization: M_η/g.mol^{-1} = 100000, commercial Dow Corning DC 200 silicone
Solvent (A):   **tetrachloromethane** CCl_4 **56-23-5**

T/K = 298.15 $\Delta_M H_B^\infty$ = 2.41 J/(g polymer)

Polymer (B):   **poly(dimethylsiloxane)**   **1962DE1**
Characterization: M_η/g.mol^{-1} = 20000, commercial Dow Corning DC 200 silicone
Solvent (A):   **n-tetradecane** $C_{14}H_{30}$ **629-59-4**

T/K = 297.65 $\Delta_M H_B^\infty$ = 375 J/(base mol polymer)

Polymer (B):   **poly(dimethylsiloxane)**   **1973CHA**
Characterization: M_η/g.mol^{-1} = 20000, SE-76, General Electric Co.
Solvent (A):   **n-tetradecane** $C_{14}H_{30}$ **629-59-4**

T/K = 298.15 $\Delta_M H_B^\infty$ = 5.11 J/(g polymer)

Polymer (B):   **poly(dimethylsiloxane)**   **1964PAT**
Characterization: M_η/g.mol^{-1} = 80000, commercial Dow Corning DC 200 silicone
Solvent (A):   **n-tetradecane** $C_{14}H_{30}$ **629-59-4**

T/K = 298.15 $\Delta_M H_B^\infty$ = 381 J/(base mol polymer)

Polymer (B):   **poly(dimethylsiloxane)**   **1979PH1**
Characterization: M_η/g.mol^{-1} = 20000, commercial Dow Corning DC 200 silicone
Solvent (A):   **2,2,4,4-tetramethylpentane** C_9H_{20} **1070-87-7**

T/K = 298.15 $\Delta_M H_B^\infty$ = 2.1 J/(g polymer)

Polymer (B):   **poly(dimethylsiloxane)**   **1979PH2**
Characterization: M_η/g.mol^{-1} = 20000, commercial Dow Corning DC 200 silicone
Solvent (A):   **2,2,4,4-tetramethylpentane** C_9H_{20} **1070-87-7**

T/K = 298.15 $\Delta_M H_B^\infty$ = 2.3 J/(g polymer)

Polymer (B):   **poly(dimethylsiloxane)**   **1979PH2**
Characterization: M_η/g.mol^{-1} = 20000, commercial Dow Corning DC 200 silicone
Solvent (A):   **2,3,3,4-tetramethylpentane** C_9H_{20} **16747-38-9**

T/K = 298.15 $\Delta_M H_B^\infty$ = 1.9 J/(g polymer)

Polymer (B):   **poly(dimethylsiloxane)**   **1968MOR**
Characterization: M_n/g.mol^{-1} = 13000, η = 1.0 cSt, KF 96, Shinetsu Chemicals Co.
Solvent (A):   **toluene**   **C$_7$H$_8$**   **108-88-3**

T/K = 298.15 $\Delta_M H_B^\infty$ = 5.48 J/(g polymer)

Polymer (B):   **poly(dimethylsiloxane)**   **1973CHA**
Characterization: M_η/g.mol^{-1} = 20000, SE-76, General Electric Co.
Solvent (A):   **toluene**   **C$_7$H$_8$**   **108-88-3**

T/K = 298.15 $\Delta_M H_B^\infty$ = 6.74 J/(g polymer)

Polymer (B):   **poly(dimethylsiloxane)**   **1962DE1**
Characterization: M_η/g.mol^{-1} = 20000, commercial Dow Corning DC 200 silicone
Solvent (A):   **n-tridecane**   **C$_{13}$H$_{28}$**   **629-50-5**

T/K = 297.65 $\Delta_M H_B^\infty$ = 355 J/(base mol polymer)

Polymer (B):   **poly(dimethylsiloxane)**   **1964PAT**
Characterization: M_η/g.mol^{-1} = 80000, commercial Dow Corning DC 200 silicone
Solvent (A):   **n-tridecane**   **C$_{13}$H$_{28}$**   **629-50-5**

T/K = 298.15 $\Delta_M H_B^\infty$ = 356 J/(base mol polymer)

Polymer (B):   **poly(dimethylsiloxane)**   **1968MOR**
Characterization: M_n/g.mol^{-1} = 13000, η = 1.0 cSt, KF 96, Shinetsu Chemicals Co.
Solvent (A):   **1,3,5-trimethylbenzene**   **C$_9$H$_{12}$**   **108-67-8**

T/K = 298.15 $\Delta_M H_B^\infty$ = 3.71 J/(g polymer)

Polymer (B):   **poly(dimethylsiloxane)**   **1975TAN**
Characterization: M_n/g.mol^{-1} = 19000, Dow Corning Silicones
Solvent (A):   **2,2,4-trimethylpentane**   **C$_8$H$_{18}$**   **540-84-1**

T/K = 298.15 $\Delta_M H_B^\infty$ = 3.54 J/cm^3

Polymer (B):   **poly(dimethylsiloxane)**   **1979PH1**
Characterization: M_η/g.mol^{-1} = 20000, commercial Dow Corning DC 200 silicone
Solvent (A):   **2,2,4-trimethylpentane**   **C$_8$H$_{18}$**   **540-84-1**

T/K = 298.15 $\Delta_M H_B^\infty$ = 1.4 J/(g polymer)

Polymer (B):   **poly(dimethylsiloxane)**   **1962DE1**
Characterization: M_η/g.mol^{-1} = 20000, commercial Dow Corning DC 200 silicone
Solvent (A):   **n-undecane**   **C$_{11}$H$_{24}$**   **1120-21-4**

T/K = 297.65 $\Delta_M H_B^\infty$ = 315 J/(base mol polymer)

Polymer (B):   **poly(dimethylsiloxane)**   **1964PAT**
Characterization: M_η/g.mol^{-1} = 80000, commercial Dow Corning DC 200 silicone
Solvent (A):   **n-undecane**   **C$_{11}$H$_{24}$**   **1120-21-4**

T/K = 298.15 $\Delta_M H_B^\infty$ = 318 J/(base mol polymer)

Polymer (B):   **polyethylene**   **1967SCH**
Characterization: M_w/g.mol^{-1} = 65000, polyethylene of intermediate density, MFI = 8.8
Solvent (A):   **1-chloronaphthalene**   **C$_{10}$H$_7$Cl**   **90-13-1**

T/K	353.15	363.15	373.15	383.25	393.15	403.05
$\Delta_{sol} H_B$/(J/base mol polymer)	1340	3430	21900	27400	22400	1380

Comments: The final concentration is about 0.3 g polymer/100 ml solvent; semicrystalline material below 400 K.

Polymer (B):   **polyethylene**   **1967SCH**
Characterization: M_w/g.mol^{-1} = 144000, low-pressure polyethylene, MFI = 2.0
Solvent (A):   **1-chloronaphthalene**   **C$_{10}$H$_7$Cl**   **90-13-1**

T/K	363.05	373.15	383.15	393.15	403.15	413.15	423.15
$\Delta_{sol} H_B$/(J/base mol polymer)	960	3680	25700	27800	19250	1880	2385

Comments: The final concentration is about 0.3 g polymer/100 ml solvent; semicrystalline material below 413 K.

Polymer (B):   **polyethylene**   **1967SCH**
Characterization: M_w/g.mol^{-1} = 670000, high-pressure polyethylene, MFI = 2.1
Solvent (A):   **1-chloronaphthalene**   **C$_{10}$H$_7$Cl**   **90-13-1**

T/K	353.05	363.15	373.15	383.25	393.15	403.15
$\Delta_{sol} H_B$/(J/base mol polymer)	8870	10540	12130	4560	1090	1000

Comments: The final concentration is about 0.3 g polymer/100 ml solvent; semicrystalline material below 393 K.

Polymer (B):   **polyethylene**   **1994PHU**
Characterization: M_w/g.mol^{-1} = 900000, ρ = 0.94 g/cm^3, crystallinity = 0.63, linear, Hostalen Gur 413, Hoechst AG, Germany
Solvent (A):   **1-chloronaphthalene**   **C$_{10}$H$_7$Cl**   **90-13-1**

T/K = 391.8 $\Delta_{sol} H_B^\infty$ = 245 J/(g polymer)

Comments: Semicrystalline material; the final concentration is 0.2%.

Polymer (B):   **polyethylene**   **1994PHU**
Characterization: M_w/g.mol^{-1} = 900000, ρ = 0.94 g/cm^3, crystallinity = 0.63,
                      linear, Hostalen Gur 413, Hoechst AG, Germany
Solvent (A):   **cyclohexane**   **C$_6$H$_{12}$**   **110-82-7**

T/K = 379.5 $\Delta_{sol} H_B^\infty$ = 205 J/(g polymer)

Comments: Semicrystalline material; the final concentration is 0.2%.

Polymer (B):   **polyethylene**   **1994PHU**
Characterization: M_w/g.mol^{-1} = 900000, ρ = 0.94 g/cm^3, crystallinity = 0.63,
                      linear, Hostalen Gur 413, Hoechst AG, Germany
Solvent (A):   **cyclopentane**   **C$_5$H$_{10}$**   **287-92-3**

T/K = 380.0 $\Delta_{sol} H_B^\infty$ = 190 J/(g polymer)

Comments: Semicrystalline material; the final concentration is 0.2%.

Polymer (B):   **polyethylene**   **1972BLA**
Characterization: LDPE, Alkathene
Solvent (A):   **decahydronaphthalene C$_{10}$H$_{18}$**   **91-17-8**

T/K = 349.85 $\Delta_{sol} H_B^\infty$ = 141.8 J/(g polymer)

Polymer (B):   **polyethylene**   **1972BLA**
Characterization: HDPE, Rigidex type 3
Solvent (A):   **decahydronaphthalene C$_{10}$H$_{18}$**   **91-17-8**

T/K = 366.65 $\Delta_{sol} H_B^\infty$ = 179.9 J/(g polymer)

Polymer (B):   **polyethylene**   **1972BLA**
Characterization: HDPE, Rigidex type 50
Solvent (A):   **decahydronaphthalene C$_{10}$H$_{18}$**   **91-17-8**

T/K = 366.65 $\Delta_{sol} H_B^\infty$ = 233.0 J/(g polymer)

Polymer (B):   **polyethylene**   **1994PHU**
Characterization: M_w/g.mol^{-1} = 900000, ρ = 0.94 g/cm^3, crystallinity = 0.63,
                      linear, Hostalen Gur 413, Hoechst AG, Germany
Solvent (A):   **decahydronaphthalene C$_{10}$H$_{18}$**   **91-17-8**

T/K = 384.0 $\Delta_{sol} H_B^\infty$ = 260 J/(g polymer)

Comments: Semicrystalline material; the final concentration is 0.2%.

Polymer (B):   **polyethylene**   **1958AKH**

Characterization: –

Solvent (A):   **1,2-dichloroethane**   **$C_2H_4Cl_2$**   **107-06-2**

$T/K =$	333.15	338.15	343.15	348.15
$\Delta_{sol}H_B$/(J/g polymer)	30.1	38.1	53.6	64.8

Comments: Semicrystalline material.

Polymer (B):   **polyethylene**   **1945RAI**

Characterization: M_η/g.mol^{-1} = 10000

Solvent (A):   **1,4-dimethylbenzene**   **C_8H_{10}**   **106-42-3**

T/K = 354.15 $\Delta_{sol}H_B$ = 138.9 J/(g polymer)

Comments: Semicrystalline material; the final concentration is 5%.

Polymer (B):   **polyethylene**   **1945RAI**

Characterization: M_η/g.mol^{-1} = 11800

Solvent (A):   **1,4-dimethylbenzene**   **C_8H_{10}**   **106-42-3**

T/K = 354.15 $\Delta_{sol}H_B$ = 139.3 J/(g polymer)

Comments: Semicrystalline material; the final concentration is 5%.

Polymer (B):   **polyethylene**   **1945RAI**

Characterization: M_η/g.mol^{-1} = 15600

Solvent (A):   **1,4-dimethylbenzene**   **C_8H_{10}**   **106-42-3**

T/K	353.65	363.65	368.15
$\Delta_{sol}H_B$/(J/g polymer)	154	113	104

Comments: Semicrystalline material; the final concentration is 5%.

Polymer (B):   **polyethylene**   **1994PHU**

Characterization: M_w/g.mol^{-1} = 900000, ρ = 0.94 g/cm^3, crystallinity = 0.63, linear, Hostalen Gur 413, Hoechst AG, Germany

Solvent (A):   **2,4-dimethylpentane**   **C_7H_{16}**   **108-08-7**

T/K = 393.0 $\Delta_{sol}H_B^\infty$ = 230 J/(g polymer)

Comments: Semicrystalline material; the final concentration is 0.2%.

Polymer (B):   **polyethylene**   **1994PHU**

Characterization: M_w/g.mol^{-1} = 900000, ρ = 0.94 g/cm^3, crystallinity = 0.63, linear, Hostalen Gur 413, Hoechst AG, Germany

Solvent (A):   **2,2,4,4,6,8,8-heptamethylnonane**   **$C_{16}H_{34}$**   **4390-04-9**

T/K = 399.5 $\Delta_{sol}H_B^\infty$ = 170 J/(g polymer)

Comments: Semicrystalline material; the final concentration is 0.2%.

Polymer (B):   **polyethylene**   **1994PHU**
Characterization: M_w/g.mol^{-1} = 900000, ρ = 0.94 g/cm^3, crystallinity = 0.63, linear, Hostalen Gur 413, Hoechst AG, Germany
Solvent (A):   **n-hexadecane**   **C$_{16}$H$_{34}$**   **544-76-3**

T/K = 399.5 $\Delta_{sol} H_B^\infty$ = 262 J/(g polymer)

Comments: Semicrystalline material; the final concentration is 0.2%.

Polymer (B):   **polyethylene**   **1994PHU**
Characterization: M_w/g.mol^{-1} = 900000, ρ = 0.94 g/cm^3, crystallinity = 0.63, linear, Hostalen Gur 413, Hoechst AG, Germany
Solvent (A):   **2-methylbutane**   **C$_5$H$_{12}$**   **78-78-4**

T/K = 394.2 $\Delta_{sol} H_B^\infty$ = 165 J/(g polymer)

Comments: Semicrystalline material; the final concentration is 0.2%.

Polymer (B):   **polyethylene**   **1967SCH**
Characterization: M_w/g.mol^{-1} = 65000, polyethylene of intermediate density, MFI = 8.8
Solvent (A):   **1,2,3,4-tetrahydronaphthalene**   **C$_{10}$H$_{12}$**   **119-64-2**

T/K	353.15	363.15	373.15	383.25	393.15	403.05
$\Delta_{sol} H_B$/(J/base mol polymer)	1841	12550	26230	27780	22200	1632

Comments: The final concentration is about 0.3 g polymer/100 ml solvent; semicrystalline material below 400 K.

Polymer (B):   **polyethylene**   **1967SCH**
Characterization: M_w/g.mol^{-1} = 84000, low-pressure polyethylene, MFI = 3.5
Solvent (A):   **1,2,3,4-tetrahydronaphthalene**   **C$_{10}$H$_{12}$**   **119-64-2**

T/K = 373.15 $\Delta_{sol} H_B$ = 23430 J/(base mol polymer)

Comments: The final concentration is about 0.3 g polymer/100 ml solvent; semicrystalline material.

Polymer (B):   **polyethylene**   **1967SCH**
Characterization: M_w/g.mol^{-1} = 130000, high-pressure polyethylene, MFI = 2.0
Solvent (A):   **1,2,3,4-tetrahydronaphthalene**   **C$_{10}$H$_{12}$**   **119-64-2**

T/K	353.15	373.15	393.15
$\Delta_{sol} H_B$/(J/base mol polymer)	17590	14520	1925

Comments: The final concentration is about 0.3 g polymer/100 ml solvent; semicrystalline material below 393 K.

Polymer (B):   **polyethylene**   **1967SCH**
Characterization: M_w/g.mol^{-1} = 144000, low-pressure polyethylene, MFI = 2.0
Solvent (A):   **1,2,3,4-tetrahydronaphthalene**   **C$_{10}$H$_{12}$**   **119-64-2**

T/K	363.05	373.15	383.15	393.15	403.15	413.15	423.15
$\Delta_{sol}H_B$/(J/base mol polymer)	880	17155	33680	31550	22470	3810	2470

Comments: The final concentration is about 0.3 g polymer/100 ml solvent; semicrystalline material below 413 K.

Polymer (B):   **polyethylene**   **1967SCH**
Characterization: M_w/g.mol^{-1} = 310000, high-pressure polyethylene, MFI = 1.5
Solvent (A):   **1,2,3,4-tetrahydronaphthalene**   **C$_{10}$H$_{12}$**   **119-64-2**

T/K = 343.15 $\Delta_{sol}H_B$ = 13600 J/(base mol polymer)

Comments: The final concentration is about 0.3 g polymer/100 ml solvent.

Polymer (B):   **polyethylene**   **1967SCH**
Characterization: M_w/g.mol^{-1} = 670000, high-pressure polyethylene, MFI = 2.1
Solvent (A):   **1,2,3,4-tetrahydronaphthalene**   **C$_{10}$H$_{12}$**   **119-64-2**

T/K	353.05	363.15	373.15	383.25	393.15	403.15
$\Delta_{sol}H_B$/(J/base mol polymer)	15650	15600	12970	4350	1883	1088

Comments: The final concentration is about 0.3 g polymer/100 ml solvent; semicrystalline material below 393 K.

Polymer (B):   **polyethylene**   **1956LIP**
Characterization: M_η/g.mol^{-1} = 16000
Solvent (A):   **toluene**   **C$_7$H$_8$**   **108-88-3**

T/K = 353.15 $\Delta_{sol}H_B^\infty$ = 106.3 J/(g polymer)   (non-oriented sample)

T/K = 353.15 $\Delta_{sol}H_B^\infty$ = 113.0 J/(g polymer)   (oriented sample, drawn 400%)

T/K = 353.15 $\Delta_{sol}H_B^\infty$ = 120.5 J/(g polymer)   (reoriented sample, drawn 800%)

Polymer (B):   **polyethylene**   **1956LIP**
Characterization: M_η/g.mol^{-1} = 22000
Solvent (A):   **toluene**   **C$_7$H$_8$**   **108-88-3**

T/K = 358.35 $\Delta_{sol}H_B^\infty$ = 118.0 J/(g polymer)   (non-oriented sample)

T/K = 358.35 $\Delta_{sol}H_B^\infty$ = 136.0 J/(g polymer)   (oriented sample, drawn 400%)

T/K = 367.85 $\Delta_{sol}H_B^\infty$ = 105.9 J/(g polymer)   (non-oriented sample)

T/K = 367.85 $\Delta_{sol}H_B^\infty$ = 126.8 J/(g polymer)   (oriented sample, drawn 400%)

Polymer (B):   **polyethylene**   **1994PHU**

Characterization: $M_w/\text{g.mol}^{-1} = 900000$, $\rho = 0.94$ g/cm^3, crystallinity = 0.63, linear, Hostalen Gur 413, Hoechst AG, Germany

Solvent (A):   **1,2,4-trichlorobenzene**   **C$_6$H$_3$Cl$_3$**   **120-82-1**

$T/\text{K} = 386.5$ $\Delta_{sol} H_B^\infty = 255$ J/(g polymer)

Comments: Semicrystalline material; the final concentration is 0.2%.

Polymer (B):   **poly(ethylene-*co*-propylene)**   **1979PH1, 1979PH2**

Characterization: $M_\eta/\text{g.mol}^{-1} = 145000$, 33 mol% ethylene

Solvent (A):   **cyclohexane**   **C$_6$H$_{12}$**   **110-82-7**

$T/\text{K} = 298.15$ $\Delta_M H_B^\infty = 1.4$ J/(g polymer)

Polymer (B):   **poly(ethylene-*co*-propylene)**   **1979PH1, 1979PH2**

Characterization: $M_\eta/\text{g.mol}^{-1} = 236000$, 63 mol% ethylene

Solvent (A):   **cyclohexane**   **C$_6$H$_{12}$**   **110-82-7**

$T/\text{K} = 298.15$ $\Delta_M H_B^\infty = 8.1$ J/(g polymer)

Polymer (B):   **poly(ethylene-*co*-propylene)**   **1979PH1, 1979PH2**

Characterization: $M_\eta/\text{g.mol}^{-1} = 109000$, 75 mol% ethylene

Solvent (A):   **cyclohexane**   **C$_6$H$_{12}$**   **110-82-7**

$T/\text{K} = 298.15$ $\Delta_M H_B^\infty = 11.8$ J/(g polymer)

Polymer (B):   **poly(ethylene-*co*-propylene)**   **1979PH1, 1979PH2**

Characterization: $M_\eta/\text{g.mol}^{-1} = 145000$, 33 mol% ethylene

Solvent (A):   **cyclooctane**   **C$_8$H$_{16}$**   **292-64-8**

$T/\text{K} = 298.15$ $\Delta_M H_B^\infty = 1.2$ J/(g polymer)

Polymer (B):   **poly(ethylene-*co*-propylene)**   **1979PH1, 1979PH2**

Characterization: $M_\eta/\text{g.mol}^{-1} = 236000$, 63 mol% ethylene

Solvent (A):   **cyclooctane**   **C$_8$H$_{16}$**   **292-64-8**

$T/\text{K} = 298.15$ $\Delta_M H_B^\infty = 6.9$ J/(g polymer)

Polymer (B):   **poly(ethylene-*co*-propylene)**   **1979PH1, 1979PH2**

Characterization: $M_\eta/\text{g.mol}^{-1} = 109000$, 75 mol% ethylene

Solvent (A):   **cyclooctane**   **C$_8$H$_{16}$**   **292-64-8**

$T/\text{K} = 298.15$ $\Delta_M H_B^\infty = 8.6$ J/(g polymer)

Polymer (B):   **poly(ethylene-*co*-propylene)**   **1979PH1, 1979PH2**
Characterization: M_η/g.mol^{-1} = 145000, 33 mol% ethylene
Solvent (A):   **cyclopentane**   **C$_5$H$_{10}$**   **287-92-3**

T/K = 298.15 $\Delta_M H_B^\infty$ = −3.5 J/(g polymer)

Polymer (B):   **poly(ethylene-*co*-propylene)**   **1979PH1, 1979PH2**
Characterization: M_η/g.mol^{-1} = 236000, 63 mol% ethylene
Solvent (A):   **cyclopentane**   **C$_5$H$_{10}$**   **287-92-3**

T/K = 298.15 $\Delta_M H_B^\infty$ = 1.1 J/(g polymer)

Polymer (B):   **poly(ethylene-*co*-propylene)**   **1979PH1, 1979PH2**
Characterization: M_η/g.mol^{-1} = 145000, 33 mol% ethylene
Solvent (A):   ***cis*-decahydronaphthalene**   **C$_{10}$H$_{18}$**   **493-01-6**

T/K = 298.15 $\Delta_M H_B^\infty$ = −2.4 J/(g polymer)

Polymer (B):   **poly(ethylene-*co*-propylene)**   **1979PH1, 1979PH2**
Characterization: M_η/g.mol^{-1} = 236000, 63 mol% ethylene
Solvent (A):   ***cis*-decahydronaphthalene**   **C$_{10}$H$_{18}$**   **493-01-6**

T/K = 298.15 $\Delta_M H_B^\infty$ = 2.4 J/(g polymer)

Polymer (B):   **poly(ethylene-*co*-propylene)**   **1979PH1, 1979PH2**
Characterization: M_η/g.mol^{-1} = 109000, 75 mol% ethylene
Solvent (A):   ***cis*-decahydronaphthalene**   **C$_{10}$H$_{18}$**   **493-01-6**

T/K = 298.15 $\Delta_M H_B^\infty$ = 3.9 J/(g polymer)

Polymer (B):   **poly(ethylene-*co*-propylene)**   **1979PH1, 1979PH2**
Characterization: M_η/g.mol^{-1} = 145000, 33 mol% ethylene
Solvent (A):   ***trans*-decahydronaphthalene**   **C$_{10}$H$_{18}$**   **493-02-7**

T/K = 298.15 $\Delta_M H_B^\infty$ = −4.8 J/(g polymer)

Polymer (B):   **poly(ethylene-*co*-propylene)**   **1979PH1, 1979PH2**
Characterization: M_η/g.mol^{-1} = 236000, 63 mol% ethylene
Solvent (A):   ***trans*-decahydronaphthalene**   **C$_{10}$H$_{18}$**   **493-02-7**

T/K = 298.15 $\Delta_M H_B^\infty$ = −1.3 J/(g polymer)

Polymer (B):   **poly(ethylene-*co*-propylene)**   **1979PH1, 1979PH2**
Characterization: M_η/g.mol^{-1} = 109000, 75 mol% ethylene
Solvent (A):   ***trans*-decahydronaphthalene**   **C$_{10}$H$_{18}$**   **493-02-7**

T/K = 298.15 $\Delta_M H_B^\infty$ = −0.3 J/(g polymer)

Polymer (B):   **poly(ethylene-*co*-propylene)**   **1979PH1, 1979PH2**
Characterization: M_η/g.mol^{-1} = 236000, 63 mol% ethylene
Solvent (A):   **3,3-diethylpentane**   **C$_9$H$_{20}$**   **4032-86-4**

T/K = 298.15 $\Delta_M H_B^\infty$ = −1.4 J/(g polymer)

Polymer (B):   **poly(ethylene-*co*-propylene)**   **1979PH1, 1979PH2**
Characterization: M_η/g.mol^{-1} = 109000, 75 mol% ethylene
Solvent (A):   **3,3-diethylpentane**   **C$_9$H$_{20}$**   **4032-86-4**

T/K = 298.15 $\Delta_M H_B^\infty$ = < 0.1 J/(g polymer)

Polymer (B):   **poly(ethylene-*co*-propylene)**   **1979PH1, 1979PH2**
Characterization: M_η/g.mol^{-1} = 236000, 63 mol% ethylene
Solvent (A):   **2,2-dimethylpentane**   **C$_7$H$_{16}$**   **590-35-2**

T/K = 298.15 $\Delta_M H_B^\infty$ = 5.3 J/(g polymer)

Polymer (B):   **poly(ethylene-*co*-propylene)**   **1979PH1, 1979PH2**
Characterization: M_η/g.mol^{-1} = 109000, 75 mol% ethylene
Solvent (A):   **2,2-dimethylpentane**   **C$_7$H$_{16}$**   **590-35-2**

T/K = 298.15 $\Delta_M H_B^\infty$ = 2.3 J/(g polymer)

Polymer (B):   **poly(ethylene-*co*-propylene)**   **1979PH1, 1979PH2**
Characterization: M_η/g.mol^{-1} = 236000, 63 mol% ethylene
Solvent (A):   **2,3-dimethylpentane**   **C$_7$H$_{16}$**   **565-59-3**

T/K = 298.15 $\Delta_M H_B^\infty$ = 0.7 J/(g polymer)

Polymer (B):   **poly(ethylene-*co*-propylene)**   **1979PH1, 1979PH2**
Characterization: M_η/g.mol^{-1} = 109000, 75 mol% ethylene
Solvent (A):   **2,3-dimethylpentane**   **C$_7$H$_{16}$**   **565-59-3**

T/K = 298.15 $\Delta_M H_B^\infty$ = 0.4 J/(g polymer)

Polymer (B):   **poly(ethylene-*co*-propylene)**   **1979PH1, 1979PH2**
Characterization: M_η/g.mol^{-1} = 145000, 33 mol% ethylene
Solvent (A):   **2,4-dimethylpentane**   **C$_7$H$_{16}$**   **108-08-7**

T/K = 298.15 $\Delta_M H_B^\infty$ = −1.2 J/(g polymer)

Polymer (B):   **poly(ethylene-*co*-propylene)**   **1979PH1, 1979PH2**
Characterization: M_η/g.mol^{-1} = 236000, 63 mol% ethylene
Solvent (A):   **2,4-dimethylpentane**   **C$_7$H$_{16}$**   **108-08-7**

T/K = 298.15 $\Delta_M H_B^\infty$ = 3.0 J/(g polymer)

Polymer (B):	**poly(ethylene-*co*-propylene)**	**1979PH1, 1979PH2**
Characterization:	M_η/g.mol^{-1} = 109000, 75 mol% ethylene	
Solvent (A):	**2,4-dimethylpentane C$_7$H$_{16}$**	**108-08-7**

T/K = 298.15 $\Delta_M H_B^\infty$ = 0.2 J/(g polymer)

Polymer (B):	**poly(ethylene-*co*-propylene)**	**1979PH1, 1979PH2**
Characterization:	M_η/g.mol^{-1} = 145000, 33 mol% ethylene	
Solvent (A):	**3,3-dimethylpentane C$_7$H$_{16}$**	**562-49-2**

T/K = 298.15 $\Delta_M H_B^\infty$ = −2.7 J/(g polymer)

Polymer (B):	**poly(ethylene-*co*-propylene)**	**1979PH1, 1979PH2**
Characterization:	M_η/g.mol^{-1} = 236000, 63 mol% ethylene	
Solvent (A):	**3,3-dimethylpentane C$_7$H$_{16}$**	**562-49-2**

T/K = 298.15 $\Delta_M H_B^\infty$ = 0.3 J/(g polymer)

Polymer (B):	**poly(ethylene-*co*-propylene)**	**1979PH1, 1979PH2**
Characterization:	M_η/g.mol^{-1} = 145000, 33 mol% ethylene	
Solvent (A):	**n-dodecane C$_{12}$H$_{26}$**	**112-40-3**

T/K = 298.15 $\Delta_M H_B^\infty$ = −0.1 J/(g polymer)

Polymer (B):	**poly(ethylene-*co*-propylene)**	**1979PH1, 1979PH2**
Characterization:	M_η/g.mol^{-1} = 236000, 63 mol% ethylene	
Solvent (A):	**n-dodecane C$_{12}$H$_{26}$**	**112-40-3**

T/K = 298.15 $\Delta_M H_B^\infty$ = 0.8 J/(g polymer)

Polymer (B):	**poly(ethylene-*co*-propylene)**	**1979PH1, 1979PH2**
Characterization:	M_η/g.mol^{-1} = 109000, 75 mol% ethylene	
Solvent (A):	**n-dodecane C$_{12}$H$_{26}$**	**112-40-3**

T/K = 298.15 $\Delta_M H_B^\infty$ = −4.0 J/(g polymer)

Polymer (B):	**poly(ethylene-*co*-propylene)**	**1979PH1, 1979PH2**
Characterization:	M_η/g.mol^{-1} = 236000, 63 mol% ethylene	
Solvent (A):	**3-ethylpentane C$_7$H$_{16}$**	**617-78-7**

T/K = 298.15 $\Delta_M H_B^\infty$ = 2.6 J/(g polymer)

Polymer (B):	**poly(ethylene-*co*-propylene)**	**1979PH1, 1979PH2**
Characterization:	M_η/g.mol^{-1} = 109000, 75 mol% ethylene	
Solvent (A):	**3-ethylpentane C$_7$H$_{16}$**	**617-78-7**

T/K = 298.15 $\Delta_M H_B^\infty$ = −0.6 J/(g polymer)

Polymer (B):   **poly(ethylene-*co*-propylene)**   **1979PH1, 1979PH2**
Characterization: M_η/g.mol^{-1} = 145000, 33 mol% ethylene
Solvent (A):   **2,2,4,4,6,8,8-heptamethylnonane**   **C$_{16}$H$_{34}$**   **4390-04-9**

T/K = 298.15 $\Delta_M H_B^\infty$ = −0.5 J/(g polymer)

Polymer (B):   **poly(ethylene-*co*-propylene)**   **1979PH1, 1979PH2**
Characterization: M_η/g.mol^{-1} = 236000, 63 mol% ethylene
Solvent (A):   **2,2,4,4,6,8,8-heptamethylnonane**   **C$_{16}$H$_{34}$**   **4390-04-9**

T/K = 298.15 $\Delta_M H_B^\infty$ = 2.2 J/(g polymer)

Polymer (B):   **poly(ethylene-*co*-propylene)**   **1979PH1, 1979PH2**
Characterization: M_η/g.mol^{-1} = 109000, 75 mol% ethylene
Solvent (A):   **2,2,4,4,6,8,8-heptamethylnonane**   **C$_{16}$H$_{34}$**   **4390-04-9**

T/K = 298.15 $\Delta_M H_B^\infty$ = −0.9 J/(g polymer)

Polymer (B):   **poly(ethylene-*co*-propylene)**   **1979PH1, 1979PH2**
Characterization: M_η/g.mol^{-1} = 145000, 33 mol% ethylene
Solvent (A):   **n-hexadecane**   **C$_{16}$H$_{34}$**   **544-76-3**

T/K = 298.15 $\Delta_M H_B^\infty$ = 0.7 J/(g polymer)

Polymer (B):   **poly(ethylene-*co*-propylene)**   **1979PH1, 1979PH2**
Characterization: M_η/g.mol^{-1} = 236000, 63 mol% ethylene
Solvent (A):   **n-hexadecane**   **C$_{16}$H$_{34}$**   **544-76-3**

T/K = 298.15 $\Delta_M H_B^\infty$ = −1.1 J/(g polymer)

Polymer (B):   **poly(ethylene-*co*-propylene)**   **1979PH1, 1979PH2**
Characterization: M_η/g.mol^{-1} = 109000, 75 mol% ethylene
Solvent (A):   **n-hexadecane**   **C$_{16}$H$_{34}$**   **544-76-3**

T/K = 298.15 $\Delta_M H_B^\infty$ = −4.6 J/(g polymer)

Comments: After subtracting the small heat of fusion of this polymer.

Polymer (B):   **poly(ethylene-*co*-propylene)**   **1979PH1, 1979PH2**
Characterization: M_η/g.mol^{-1} = 236000, 63 mol% ethylene
Solvent (A):   **3-methylhexane**   **C$_7$H$_{16}$**   **589-34-4**

T/K = 298.15 $\Delta_M H_B^\infty$ = 0.7 J/(g polymer)

Polymer (B):   poly(ethylene-*co*-propylene)   **1979PH1, 1979PH2**
Characterization: M_η/g.mol^{-1} = 109000, 75 mol% ethylene
Solvent (A):   **3-methylhexane**   **C$_7$H$_{16}$**   **589-34-4**

T/K = 298.15 $\Delta_M H_B^\infty$ = 1.7 J/(g polymer)

Polymer (B):   poly(ethylene-*co*-propylene)   **1979PH1, 1979PH2**
Characterization: M_η/g.mol^{-1} = 145000, 33 mol% ethylene
Solvent (A):   **n-octane**   **C$_8$H$_{18}$**   **111-65-9**

T/K = 298.15 $\Delta_M H_B^\infty$ = −1.6 J/(g polymer)

Polymer (B):   poly(ethylene-*co*-propylene)   **1979PH1, 1979PH2**
Characterization: M_η/g.mol^{-1} = 236000, 63 mol% ethylene
Solvent (A):   **n-octane**   **C$_8$H$_{18}$**   **111-65-9**

T/K = 298.15 $\Delta_M H_B^\infty$ = 3.6 J/(g polymer)

Polymer (B):   poly(ethylene-*co*-propylene)   **1979PH1, 1979PH2**
Characterization: M_η/g.mol^{-1} = 109000, 75 mol% ethylene
Solvent (A):   **n-octane**   **C$_8$H$_{18}$**   **111-65-9**

T/K = 298.15 $\Delta_M H_B^\infty$ = 0.3 J/(g polymer)

Comments: After subtracting the small heat of fusion of this polymer.

Polymer (B):   poly(ethylene-*co*-propylene)   **1979PH1, 1979PH2**
Characterization: M_η/g.mol^{-1} = 145000, 33 mol% ethylene
Solvent (A):   **2,2,4,6,6-pentamethylheptane**   **C$_{12}$H$_{26}$**   **13475-82-6**

T/K = 298.15 $\Delta_M H_B^\infty$ = −0.3 J/(g polymer)

Polymer (B):   poly(ethylene-*co*-propylene)   **1979PH1, 1979PH2**
Characterization: M_η/g.mol^{-1} = 236000, 63 mol% ethylene
Solvent (A):   **2,2,4,6,6-pentamethylheptane**   **C$_{12}$H$_{26}$**   **13475-82-6**

T/K = 298.15 $\Delta_M H_B^\infty$ = 3.6 J/(g polymer)

Polymer (B):   poly(ethylene-*co*-propylene)   **1979PH1, 1979PH2**
Characterization: M_η/g.mol^{-1} = 109000, 75 mol% ethylene
Solvent (A):   **2,2,4,6,6-pentamethylheptane**   **C$_{12}$H$_{26}$**   **13475-82-6**

T/K = 298.15 $\Delta_M H_B^\infty$ = 0.0 J/(g polymer)

Comments: After subtracting the small heat of fusion of this polymer.

Polymer (B):	**poly(ethylene-*co*-propylene)**	**1979PH1, 1979PH2**
Characterization:	M_η/g.mol^{-1} = 145000, 33 mol% ethylene	
Solvent (A):	**2,2,4,4-tetramethylpentane** $\quad$ **C$_9$H$_{20}$**	**1070-87-7**

T/K = 298.15 $\qquad$ $\Delta_M H_B^\infty$ = 3.2 J/(g polymer)

Polymer (B):	**poly(ethylene-*co*-propylene)**	**1979PH1, 1979PH2**
Characterization:	M_η/g.mol^{-1} = 236000, 63 mol% ethylene	
Solvent (A):	**2,2,4,4-tetramethylpentane** $\quad$ **C$_9$H$_{20}$**	**1070-87-7**

T/K = 298.15 $\qquad$ $\Delta_M H_B^\infty$ = 2.7 J/(g polymer)

Polymer (B):	**poly(ethylene-*co*-propylene)**	**1979PH1, 1979PH2**
Characterization:	M_η/g.mol^{-1} = 109000, 75 mol% ethylene	
Solvent (A):	**2,2,4,4-tetramethylpentane** $\quad$ **C$_9$H$_{20}$**	**1070-87-7**

T/K = 298.15 $\qquad$ $\Delta_M H_B^\infty$ = 3.1 J/(g polymer)

Polymer (B):	**poly(ethylene-*co*-propylene)**	**1979PH1, 1979PH2**
Characterization:	M_η/g.mol^{-1} = 145000, 33 mol% ethylene	
Solvent (A):	**2,2,4-trimethylpentane** $\quad$ **C$_8$H$_{18}$**	**540-84-1**

T/K = 298.15 $\qquad$ $\Delta_M H_B^\infty$ = −0.2 J/(g polymer)

Polymer (B):	**poly(ethylene-*co*-propylene)**	**1979PH1, 1979PH2**
Characterization:	M_η/g.mol^{-1} = 236000, 63 mol% ethylene	
Solvent (A):	**2,2,4-trimethylpentane** $\quad$ **C$_8$H$_{18}$**	**540-84-1**

T/K = 298.15 $\qquad$ $\Delta_M H_B^\infty$ = 1.9 J/(g polymer)

Polymer (B):	**poly(ethylene-*co*-propylene)**	**1979PH1, 1979PH2**
Characterization:	M_η/g.mol^{-1} = 109000, 75 mol% ethylene	
Solvent (A):	**2,2,4-trimethylpentane** $\quad$ **C$_8$H$_{18}$**	**540-84-1**

T/K = 298.15 $\qquad$ $\Delta_M H_B^\infty$ = 3.5 J/(g polymer)

Polymer (B):	**poly(ethylene-*co*-vinyl acetate)**	**2002RIG**
Characterization:	85.0 wt% vinyl acetate, Bayer AG	
Solvent (A):	**cyclopentanone** $\quad$ **C$_5$H$_8$O**	**120-92-3**

T/K = 298.15 $\qquad$ $\Delta_{sol} H_B^\infty$ = −0.50 J/(g polymer)

Polymer (B): **poly(ethylene-*co*-vinyl acetate)** **2002RIG**
Characterization: 85.0 wt% vinyl acetate, Bayer AG
Solvent (A): **cyclopentanone** **C_5H_8O** **120-92-3**
Polymer (C): **poly(vinyl chloride)**
Characterization: $M_n/\text{g.mol}^{-1} = 48000$, Fluka AG, Buchs, Switzerland

$T/\text{K} = 298.15$

w_B/w_C	0/100	5/95	10/90	20/80	50/50	80/20	90/20
$\Delta_{sol}H_{B+C}{}^\infty/(\text{J/g blend})$	−28.0	−19.8	−15.9	−12.2	−4.8	−2.2	−0.7

w_B/w_C	95/05	100/0
$\Delta_{sol}H_{B+C}{}^\infty/(\text{J/g blend})$	−0.2	−0.5

Polymer (B): **poly(ethylene-*co*-vinyl acetate)** **1990SHI**
Characterization: $M_w/\text{g.mol}^{-1} = 220000$, 70.0 wt% vinyl acetate, Bayer Japan
Solvent (A): **tetrahydrofuran** **C_4H_8O** **109-99-9**

$T/\text{K} = 304.65$ $\Delta_{sol}H_B{}^\infty = -1.33$ J/(g polymer)

Polymer (B): **poly(ethylene glycol)** **1979KOL, 1981KOL**
Characterization: $M_n/\text{g.mol}^{-1} = 180$, Hoechst AG, Germany
Solvent (A): **benzene** **C_6H_6** **71-43-2**

$T/\text{K} = 303.15$ $\Delta_{sol}H_B{}^\infty = 109$ J/(g polymer)

Comments: The final concentration is 0.13 polymer/100 cm^3 solvent.

Polymer (B): **poly(ethylene glycol)** **1979KOL, 1981KOL**
Characterization: $M_n/\text{g.mol}^{-1} = 385$, Hoechst AG, Germany
Solvent (A): **benzene** **C_6H_6** **71-43-2**

$T/\text{K} = 303.15$ $\Delta_{sol}H_B{}^\infty = 62$ J/(g polymer)

Comments: The final concentration is 0.13 polymer/100 cm^3 solvent.

Polymer (B): **poly(ethylene glycol)** **1979KOL, 1981KOL**
Characterization: $M_n/\text{g.mol}^{-1} = 560$, Hoechst AG, Germany
Solvent (A): **benzene** **C_6H_6** **71-43-2**

$T/\text{K} = 303.15$ $\Delta_{sol}H_B{}^\infty = 41$ J/(g polymer)

Comments: The final concentration is 0.13 polymer/100 cm^3 solvent.

Polymer (B): **poly(ethylene glycol)** **1979KOL, 1981KOL**
Characterization: $M_n/\text{g.mol}^{-1} = 1050$, Hoechst AG, Germany
Solvent (A): **benzene** **C_6H_6** **71-43-2**

$T/\text{K} = 303.15$ $\Delta_{sol}H_B{}^\infty = 89$ J/(g polymer)

Comments: The final concentration is 0.13 polymer/100 cm^3 solvent.

| **Polymer (B):** | **poly(ethylene glycol)** | | **1979KOL, 1981KOL** |

Polymer (B):   **poly(ethylene glycol)**   **1979KOL, 1981KOL**
Characterization: M_n/g.mol^{-1} = 1610, Hoechst AG, Germany
Solvent (A):   **benzene**   **C$_6$H$_6$**   **71-43-2**

T/K = 303.15 $\Delta_{sol} H_B^\infty$ = 142 J/(g polymer)

Comments: The final concentration is 0.13 polymer/100 cm^3 solvent.

Polymer (B):   **poly(ethylene glycol)**   **1979KOL, 1981KOL**
Characterization: M_n/g.mol^{-1} = 1940, Hoechst AG, Germany
Solvent (A):   **benzene**   **C$_6$H$_6$**   **71-43-2**

T/K = 303.15 $\Delta_{sol} H_B^\infty$ = 214 J/(g polymer)

Comments: The final concentration is 0.13 polymer/100 cm^3 solvent.

Polymer (B):   **poly(ethylene glycol)**   **1979KOL, 1981KOL**
Characterization: M_n/g.mol^{-1} = 3200, Hoechst AG, Germany
Solvent (A):   **benzene**   **C$_6$H$_6$**   **71-43-2**

T/K = 303.15 $\Delta_{sol} H_B^\infty$ = 197 J/(g polymer)

Comments: The final concentration is 0.13 polymer/100 cm^3 solvent.

Polymer (B):   **poly(ethylene glycol)**   **1979KOL, 1981KOL**
Characterization: M_n/g.mol^{-1} = 4330, Hoechst AG, Germany
Solvent (A):   **benzene**   **C$_6$H$_6$**   **71-43-2**

T/K = 303.15 $\Delta_{sol} H_B^\infty$ = 197 J/(g polymer)

Comments: The final concentration is 0.13 polymer/100 cm^3 solvent.

Polymer (B):   **poly(ethylene glycol)**   **1979KOL, 1981KOL**
Characterization: M_n/g.mol^{-1} = 5850, Hoechst AG, Germany
Solvent (A):   **benzene**   **C$_6$H$_6$**   **71-43-2**

T/K = 303.15 $\Delta_{sol} H_B^\infty$ = 192 J/(g polymer)

Comments: The final concentration is 0.13 polymer/100 cm^3 solvent.

Polymer (B):   **poly(ethylene glycol)**   **1979KOL, 1981KOL**
Characterization: M_n/g.mol^{-1} = 9950, Hoechst AG, Germany
Solvent (A):   **benzene**   **C$_6$H$_6$**   **71-43-2**

T/K = 303.15 $\Delta_{sol} H_B^\infty$ = 195 J/(g polymer)

Comments: The final concentration is 0.13 polymer/100 cm^3 solvent.

Polymer (B):   **poly(ethylene glycol)**   **1979KOL, 1981KOL**
Characterization: M_η/g.mol^{-1} = 43400, Hoechst AG, Germany
Solvent (A):   **benzene**   **C$_6$H$_6$**   **71-43-2**

T/K = 303.15 $\Delta_{sol}H_B^\infty$ = 192 J/(g polymer)

Comments: The final concentration is 0.13 polymer/100 cm^3 solvent.

Polymer (B):   **poly(ethylene glycol)**   **1993ZEL**
Characterization: M_n/g.mol^{-1} = 400, M_w/g.mol^{-1} = 420,
                       ρ = 1.1182 g/cm^3, Hoechst AG, Germany
Solvent (A):   **trichloromethane**   **CHCl$_3$**   **67-66-3**

T/K = 303.15 $\Delta_M H_B^\infty$ = −79.33 J/(g polymer)

Polymer (B):   **poly(ethylene glycol)**   **1993ZEL**
Characterization: M_n/g.mol^{-1} = 590, M_w/g.mol^{-1} = 615,
                       ρ = 1.1183 g/cm^3, Hoechst AG, Germany
Solvent (A):   **trichloromethane**   **CHCl$_3$**   **67-66-3**

T/K = 303.15 $\Delta_M H_B^\infty$ = −87.9 J/(g polymer)

Polymer (B):   **poly(ethylene glycol)**   **1979KOL, 1981KOL**
Characterization: M_n/g.mol^{-1} = 180, Hoechst AG, Germany
Solvent (A):   **water**   **H$_2$O**   **7732-18-5**

T/K = 303.15 $\Delta_M H_B^\infty$ = −136 J/(g polymer)

Comments: The final concentration is 0.13 polymer/100 cm^3 solvent.

Polymer (B):   **poly(ethylene glycol)**   **1983LAK**
Characterization: M_n/g.mol^{-1} = 200,
                       fractionated samples supplied by Union Carbide Corp.
Solvent (A):   **water**   **H$_2$O**   **7732-18-5**

T/K = 321.35 $\Delta_M H_B^\infty$ = −25 kJ/mol polymer

Polymer (B):   **poly(ethylene glycol)**   **1979KOL, 1981KOL**
Characterization: M_n/g.mol^{-1} = 385, Hoechst AG, Germany
Solvent (A):   **water**   **H$_2$O**   **7732-18-5**

T/K = 303.15 $\Delta_{sol}H_B^\infty$ = −159 J/(g polymer)

Comments: The final concentration is 0.13 polymer/100 cm^3 solvent.

Polymer (B):	**poly(ethylene glycol)**	**1983LAK**
Characterization:	M_n/g.mol^{-1} = 400,	
	fractionated samples supplied by Union Carbide Corp.	
Solvent (A):	**water** **H$_2$O**	**7732-18-5**

T/K = 321.35 $\Delta_M H_B^\infty$ = −60 kJ/mol polymer

Polymer (B):	**poly(ethylene glycol)**	**1979KOL, 1981KOL**
Characterization:	M_n/g.mol^{-1} = 560, Hoechst AG, Germany	
Solvent (A):	**water** **H$_2$O**	**7732-18-5**

T/K = 303.15 $\Delta_{sol} H_B^\infty$ = −150 J/(g polymer)

Comments: The final concentration is 0.13 polymer/100 cm^3 solvent.

Polymer (B):	**poly(ethylene glycol)**	**1983LAK**
Characterization:	M_n/g.mol^{-1} = 990,	
	fractionated samples supplied by Union Carbide Corp.	
Solvent (A):	**water** **H$_2$O**	**7732-18-5**

T/K = 321.35 $\Delta_{sol} H_B^\infty$ = −100 kJ/mol polymer

Polymer (B):	**poly(ethylene glycol)**	**1979KOL, 1981KOL**
Characterization:	M_n/g.mol^{-1} = 1050, Hoechst AG, Germany	
Solvent (A):	**water** **H$_2$O**	**7732-18-5**

T/K = 303.15 $\Delta_{sol} H_B^\infty$ = −106 J/(g polymer)

Comments: The final concentration is 0.13 polymer/100 cm^3 solvent.

Polymer (B):	**poly(ethylene glycol)**	**1983LAK**
Characterization:	M_n/g.mol^{-1} = 1460,	
	fractionated samples supplied by Union Carbide Corp.	
Solvent (A):	**water** **H$_2$O**	**7732-18-5**

T/K = 321.35 $\Delta_{sol} H_B^\infty$ = −2000 kJ/mol polymer

Polymer (B):	**poly(ethylene glycol)**	**1979KOL, 1981KOL**
Characterization:	M_n/g.mol^{-1} = 1610, Hoechst AG, Germany	
Solvent (A):	**water** **H$_2$O**	**7732-18-5**

T/K = 303.15 $\Delta_{sol} H_B^\infty$ = −6 J/(g polymer)

Polymer (B):	**poly(ethylene glycol)**	**1979KOL, 1981KOL**
Characterization:	M_n/g.mol^{-1} = 1940, Hoechst AG, Germany	
Solvent (A):	**water** **H$_2$O**	**7732-18-5**

T/K = 303.15 $\Delta_{sol} H_B^\infty$ = 57 J/(g polymer)

Comments: The final concentration is 0.13 polymer/100 cm^3 solvent.

Polymer (B):   **poly(ethylene glycol)**   **1979KOL, 1981KOL**
Characterization: M_n/g.mol^{-1} = 3200, Hoechst AG, Germany
Solvent (A):   **water**   **H$_2$O**   **7732-18-5**

T/K = 303.15 $\Delta_{sol}H_B^\infty$ = 58 J/(g polymer)

Comments: The final concentration is 0.13 polymer/100 cm^3 solvent.

Polymer (B):   **poly(ethylene glycol)**   **1979KOL, 1981KOL**
Characterization: M_n/g.mol^{-1} = 4330, Hoechst AG, Germany
Solvent (A):   **water**   **H$_2$O**   **7732-18-5**

T/K = 303.15 $\Delta_{sol}H_B^\infty$ = 28 J/(g polymer)

Comments: The final concentration is 0.13 polymer/100 cm^3 solvent.

Polymer (B):   **poly(ethylene glycol)**   **1979KOL, 1981KOL**
Characterization: M_n/g.mol^{-1} = 5850, Hoechst AG, Germany
Solvent (A):   **water**   **H$_2$O**   **7732-18-5**

T/K = 303.15 $\Delta_{sol}H_B^\infty$ = 39 J/(g polymer)

Comments: The final concentration is 0.13 polymer/100 cm^3 solvent.

Polymer (B):   **poly(ethylene glycol)**   **1979KOL, 1981KOL**
Characterization: M_n/g.mol^{-1} = 9950, Hoechst AG, Germany
Solvent (A):   **water**   **H$_2$O**   **7732-18-5**
T/K = 303.15 $\Delta_{sol}H_B^\infty$ = 30 J/(g polymer)

Comments: The final concentration is 0.13 polymer/100 cm^3 solvent.

Polymer (B):   **poly(ethylene glycol)**   **1970NAK**
Characterization: M_η/g.mol^{-1} = 14000, Wako Pure Chemicals Company, Japan
Solvent (A):   **water**   **H$_2$O**   **7732-18-5**

T/K = 303.15 $\Delta_{sol}H_B^\infty$ = 293 J/(base mol polymer)

T/K = 313.15 $\Delta_{sol}H_B^\infty$ = 1213 J/(base mol polymer)

Comments: Experimental data for mixtures of water and low-molecular glycols can be found in the original source.

Polymer (B):   **poly(ethylene glycol)**   **1979KOL, 1981KOL**
Characterization: M_η/g.mol^{-1} = 20300, Hoechst AG, Germany
Solvent (A):   **water**   **H$_2$O**   **7732-18-5**

T/K = 303.15 $\Delta_{sol}H_B^\infty$ = 45 J/(g polymer)

Comments: The final concentration is 0.13 polymer/100 cm^3 solvent.

Polymer (B):   poly(ethylene glycol)   **1979KOL, 1981KOL**
Characterization: M_η/g.mol^{-1} = 34500, Hoechst AG, Germany
Solvent (A):   **water**   **H$_2$O**   **7732-18-5**

T/K = 303.15 $\Delta_{sol} H_B^\infty$ = 34 J/(g polymer)

Comments: The final concentration is 0.13 polymer/100 cm^3 solvent.

Polymer (B):   poly(ethylene glycol)   **1979KOL, 1981KOL**
Characterization: M_η/g.mol^{-1} = 43400, Hoechst AG, Germany
Solvent (A):   **water**   **H$_2$O**   **7732-18-5**

T/K = 303.15 $\Delta_{sol} H_B^\infty$ = 40 J/(g polymer)

Polymer (B):   poly(ethylene glycol) dimethyl ether   **1993ZEL**
Characterization: M_n/g.mol^{-1} = 250
Solvent (A):   **tetrachloromethane**   **CCl$_4$**   **56-23-5**

T/K = 303.15 $\Delta_M H_B^\infty$ = −11.73 J/(g polymer)

T/K = 318.15 $\Delta_M H_B^\infty$ = −7.64 J/(g polymer)

Polymer (B):   poly(ethylene glycol) dimethyl ether   **1993ZEL**
Characterization: M_n/g.mol^{-1} = 398
Solvent (A):   **tetrachloromethane**   **CCl$_4$**   **56-23-5**

T/K = 303.15 $\Delta_M H_B^\infty$ = −12.31 J/(g polymer)

Polymer (B):   poly(ethylene glycol) dimethyl ether   **1993ZEL**
Characterization: M_n/g.mol^{-1} = 518, M_w/g.mol^{-1} = 550, ρ = 1.0712 g/cm^3
Solvent (A):   **tetrachloromethane**   **CCl$_4$**   **56-23-5**

T/K = 303.15 $\Delta_M H_B^\infty$ = −11.68 J/(g polymer)

T/K = 318.15 $\Delta_M H_B^\infty$ = −7.64 J/(g polymer)

Polymer (B):   poly(ethylene glycol) dimethyl ether   **1993ZEL**
Characterization: M_n/g.mol^{-1} = 250
Solvent (A):   **trichloromethane**   **CHCl$_3$**   **67-66-3**

T/K = 303.15 $\Delta_M H_B^\infty$ = −184 J/(g polymer)

Polymer (B):   poly(ethylene glycol) dimethyl ether   **1993ZEL**
Characterization: M_n/g.mol^{-1} = 518, M_w/g.mol^{-1} = 550, ρ = 1.0712 g/cm^3
Solvent (A):   **trichloromethane**   **CHCl$_3$**   **67-66-3**

T/K = 303.15 $\Delta_M H_B^\infty$ = −134.9 J/(g polymer)

Polymer (B):   **poly(ethylene glycol) monododecyl ether**   **1977MIU**
Characterization: M_n/g.mol^{-1} = 230, surfactant
Solvent (A):   **n-dodecane**   **C$_{12}$H$_{26}$**   **112-40-3**

T/K = 302.15 $\Delta_M H_B^\infty$ = 9620 J/mol

Polymer (B):   **poly(ethylene glycol) monododecyl ether**   **1977MIU**
Characterization: M_n/g.mol^{-1} = 274, surfactant
Solvent (A):   **n-dodecane**   **C$_{12}$H$_{26}$**   **112-40-3**

T/K = 302.15 $\Delta_M H_B^\infty$ = 6280 J/mol

Polymer (B):   **poly(ethylene glycol) monododecyl ether**   **1977MIU**
Characterization: M_n/g.mol^{-1} = 318, surfactant
Solvent (A):   **n-dodecane**   **C$_{12}$H$_{26}$**   **112-40-3**

T/K = 302.15 $\Delta_M H_B^\infty$ = 10900 J/mol

Polymer (B):   **poly(ethylene glycol) monododecyl ether**   **1977MIU**
Characterization: M_n/g.mol^{-1} = 362, surfactant
Solvent (A):   **n-dodecane**   **C$_{12}$H$_{26}$**   **112-40-3**

T/K = 302.15 $\Delta_M H_B^\infty$ = 13400 J/mol

Polymer (B):   **poly(ethylene glycol) monododecyl ether**   **1977MIU**
Characterization: M_n/g.mol^{-1} = 406, surfactant
Solvent (A):   **n-dodecane**   **C$_{12}$H$_{26}$**   **112-40-3**

T/K = 302.15 $\Delta_M H_B^\infty$ = 17200 J/mol

Polymer (B):   **poly(ethylene glycol) monomethyl ether**   **1993ZEL**
Characterization: M_n/g.mol^{-1} = 353, M_w/g.mol^{-1} = 377, ρ = 1.0845 g/cm^3
Solvent (A):   **trichloromethane**   **CHCl$_3$**   **67-66-3**

T/K = 303.15 $\Delta_M H_B^\infty$ = −125.1 J/(g polymer)

Polymer (B):   **poly(ethylene glycol) monomethyl ether**   **1993ZEL**
Characterization: M_n/g.mol^{-1} = 550, M_w/g.mol^{-1} = 580, ρ = 1.0991 g/cm^3
Solvent (A):   **trichloromethane**   **CHCl$_3$**   **67-66-3**

T/K = 303.15 $\Delta_M H_B^\infty$ = −117.2 J/(g polymer)

Polymer (B):	**poly(ethylene oxide)**	**1972MA2**
Characterization:	$M_n/\text{g.mol}^{-1} = 6000$, Carbowax 6000	
Solvent (A):	**dichloromethane** **CH_2Cl_2**	**75-09-2**

$T/K = 303.15$ $\Delta_{sol}H_B^\infty = 84$ J/(g polymer)

$T/K = 303.15$ $\Delta_M H_B^\infty = -160$ J/(g polymer)

Comments: $\Delta_M H_B^\infty$ is the value corrected for enthalpic contributions arising from destructing glassy and crystalline portions of the polymer.

Polymer (B):	**poly(ethylene oxide)**	**1972MA2**
Characterization:	$M_n/\text{g.mol}^{-1} = 6000$, Carbowax 6000	
Solvent (A):	**trichloromethane** **$CHCl_3$**	**67-66-3**

$T/K = 303.15$ $\Delta_{sol}H_B^\infty = 51$ J/(g polymer)

$T/K = 303.15$ $\Delta_M H_B^\infty = -185$ J/(g polymer)

Comments: $\Delta_M H_B^\infty$ is the value corrected for enthalpic contributions arising from destructing glassy and crystalline portions of the polymer.

Polymer (B):	**poly(ethylene oxide)**	**1975IKE**
Characterization:	$M_n/\text{g.mol}^{-1} = 1520$, $M_w/\text{g.mol}^{-1} = 1720$	
Solvent (A):	**water** **H_2O**	**7732-18-5**

$T/K = 293.15$ $\Delta_M H_B^\infty = 403.1$ J/(base mol polymer)   (quenched sample)

$T/K = 293.15$ $\Delta_M H_B^\infty = 392.2$ J/(base mol polymer)   (annealed sample)

$T/K = 298.15$ $\Delta_M H_B^\infty = 179.6$ J/(base mol polymer)   (quenched sample)

$T/K = 298.15$ $\Delta_M H_B^\infty = 150.1$ J/(base mol polymer)   (annealed sample)

$T/K = 303.15$ $\Delta_M H_B^\infty = 68.3$ J/(base mol polymer)   (quenched sample)

$T/K = 303.15$ $\Delta_M H_B^\infty = 109.0$ J/(base mol polymer)   (annealed sample)

Polymer (B):	**poly(ethylene oxide)**	**1972MA2**
Characterization:	$M_n/\text{g.mol}^{-1} = 6000$, Carbowax 6000	
Solvent (A):	**water** **H_2O**	**7732-18-5**

$T/K = 303.15$ $\Delta_{sol}H_B^\infty = 25$ J/(g polymer)

$T/K = 303.15$ $\Delta_M H_B^\infty = -50$ J/(g polymer)

Comments: $\Delta_M H_B^\infty$ is the value corrected for enthalpic contributions arising from destructing glassy and crystalline portions of the polymer.

Polymer (B):	**poly(ethylene oxide)**		**1975IKE**
Characterization:	M_n/g.mol^{-1} = 6840, M_w/g.mol^{-1} = 7590		
Solvent (A):	**water**	**H$_2$O**	**7732-18-5**

T/K = 293.15 $\Delta_M H_B^\infty$ = –28.0 J/(base mol polymer)   (quenched sample)

T/K = 293.15 $\Delta_M H_B^\infty$ = 208.9 J/(base mol polymer)   (annealed sample)

T/K = 298.15 $\Delta_M H_B^\infty$ = 241.0 J/(base mol polymer)   (quenched sample)

T/K = 298.15 $\Delta_M H_B^\infty$ = 539.8 J/(base mol polymer)   (annealed sample)

Polymer (B):	**poly(ethylene oxide)**		**1975IKE**
Characterization:	M_n/g.mol^{-1} = 19600, M_w/g.mol^{-1} = 23100		
Solvent (A):	**water**	**H$_2$O**	**7732-18-5**

T/K = 293.15 $\Delta_M H_B^\infty$ = –160.0 J/(base mol polymer)   (quenched sample)

T/K = 293.15 $\Delta_M H_B^\infty$ = –143.4 J/(base mol polymer)   (annealed sample)

T/K = 298.15 $\Delta_M H_B^\infty$ =   59.4 J/(base mol polymer)   (quenched sample)

T/K = 298.15 $\Delta_M H_B^\infty$ = 155.2 J/(base mol polymer)   (annealed sample)

T/K = 303.15 $\Delta_M H_B^\infty$ = 353.4 J/(base mol polymer)   (quenched sample)

T/K = 303.15 $\Delta_M H_B^\infty$ = 490.5 J/(base mol polymer)   (annealed sample)

Polymer (B):	**polyindene**		**1957VAN**
Characterization:	M_n/g.mol^{-1} = 756, M_w/g.mol^{-1} = 1023, ρ = 1.09 g/cm^3, commercial product, coumarone indene resin from Pennsylvania Ind. Chem. Corp.		
Solvent (A):	**anisole**	**C$_7$H$_8$O**	**100-66-3**

T/K = 299.15 $\Delta_{sol} H_B^\infty$ = 2.05 J/(g polymer)

Comments: The final concentration is about 1 g polymer/200 ml solvent.

Polymer (B):	**polyindene**		**1957VAN**
Characterization:	M_n/g.mol^{-1} = 756, M_w/g.mol^{-1} = 1023, ρ = 1.09 g/cm^3, commercial product, coumarone indene resin from Pennsylvania Ind. Chem. Corp.		
Solvent (A):	**benzene**	**C$_6$H$_6$**	**71-43-2**

T/K = 299.15 $\Delta_{sol} H_B^\infty$ = –0.04 J/(g polymer)

Comments: The final concentration is about 1 g polymer/200 ml solvent.

Polymer (B):	**polyindene**		**1957VAN**
Characterization:	M_n/g.mol^{-1} = 756, M_w/g.mol^{-1} = 1023, ρ = 1.09 g/cm^3, commercial product, coumarone indene resin from Pennsylvania Ind. Chem. Corp.		
Solvent (A):	**benzonitrile**	**C$_7$H$_5$N**	**100-47-0**

T/K = 299.15 $\Delta_{sol} H_B^\infty$ = –4.35 J/(g polymer)

Comments: The final concentration is about 1 g polymer/200 ml solvent.

Polymer (B):   **polyindene**   **1957VAN**
Characterization: M_n/g.mol^{-1} = 756, M_w/g.mol^{-1} = 1023, ρ = 1.09 g/cm^3, commercial
     product, coumarone indene resin from Pennsylvania Ind. Chem. Corp.
Solvent (A):   **bromobenzene**   **C$_6$H$_5$Br**   **108-86-1**

T/K = 299.15 $\Delta_{sol} H_B^\infty$ = −3.85 J/(g polymer)

Comments: The final concentration is about 1 g polymer/200 ml solvent.

Polymer (B):   **polyindene**   **1957VAN**
Characterization: M_n/g.mol^{-1} = 756, M_w/g.mol^{-1} = 1023, ρ = 1.09 g/cm^3, commercial
     product, coumarone indene resin from Pennsylvania Ind. Chem. Corp.
Solvent (A):   **2-butanone**   **C$_4$H$_8$O**   **78-93-3**

T/K = 299.15 $\Delta_{sol} H_B^\infty$ = 1.92 J/(g polymer)

Comments: The final concentration is about 1 g polymer/200 ml solvent.

Polymer (B):   **polyindene**   **1957VAN**
Characterization: M_n/g.mol^{-1} = 756, M_w/g.mol^{-1} = 1023, ρ = 1.09 g/cm^3, commercial
     product, coumarone indene resin from Pennsylvania Ind. Chem. Corp.
Solvent (A):   **chlorobenzene**   **C$_6$H$_5$Cl**   **108-90-7**

T/K = 299.15 $\Delta_{sol} H_B^\infty$ = −3.89 J/(g polymer)

Comments: The final concentration is about 1 g polymer/200 ml solvent.

Polymer (B):   **polyindene**   **1957VAN**
Characterization: M_n/g.mol^{-1} = 756, M_w/g.mol^{-1} = 1023, ρ = 1.09 g/cm^3, commercial
     product, coumarone indene resin from Pennsylvania Ind. Chem. Corp.
Solvent (A):   **1-chlorobutane**   **C$_4$H$_9$Cl**   **109-69-3**

T/K = 299.15 $\Delta_{sol} H_B^\infty$ = −4.02 J/(g polymer)

Comments: The final concentration is about 1 g polymer/200 ml solvent.

Polymer (B):   **polyindene**   **1957VAN**
Characterization: M_n/g.mol^{-1} = 756, M_w/g.mol^{-1} = 1023, ρ = 1.09 g/cm^3, commercial
     product, coumarone indene resin from Pennsylvania Ind. Chem. Corp.
Solvent (A):   **1-chloroheptane**   **C$_7$H$_{15}$Cl**   **629-06-1**

T/K = 299.15 $\Delta_{sol} H_B^\infty$ = 1.51 J/(g polymer)

Comments: The final concentration is about 1 g polymer/200 ml solvent.

Polymer (B):   **polyindene**   **1957VAN**
Characterization: $M_n/\text{g.mol}^{-1} = 756$, $M_w/\text{g.mol}^{-1} = 1023$, $\rho = 1.09$ g/cm³, commercial
                        product, coumarone indene resin from Pennsylvania Ind. Chem. Corp.
Solvent (A):   **cyclohexane**   **C₆H₁₂**   **110-82-7**

$T/\text{K} = 299.15$ $\Delta_{sol}H_B^\infty = 14.7$ J/(g polymer)

Comments: The final concentration is about 1 g polymer/200 ml solvent.

Polymer (B):   **polyindene**   **1957VAN**
Characterization: $M_n/\text{g.mol}^{-1} = 756$, $M_w/\text{g.mol}^{-1} = 1023$, $\rho = 1.09$ g/cm³, commercial
                        product, coumarone indene resin from Pennsylvania Ind. Chem. Corp.
Solvent (A):   **N,N-dimethylaniline**   **C₈H₁₁N**   **121-69-7**

$T/\text{K} = 299.15$ $\Delta_{sol}H_B^\infty = -8.16$ J/(g polymer)

Comments: The final concentration is about 1 g polymer/200 ml solvent.

Polymer (B):   **polyindene**   **1957VAN**
Characterization: $M_n/\text{g.mol}^{-1} = 756$, $M_w/\text{g.mol}^{-1} = 1023$, $\rho = 1.09$ g/cm³, commercial
                        product, coumarone indene resin from Pennsylvania Ind. Chem. Corp.
Solvent (A):   **ethyl acetate**   **C₄H₈O₂**   **141-78-6**

$T/\text{K} = 299.15$ $\Delta_{sol}H_B^\infty = 4.18$ J/(g polymer)

Comments: The final concentration is about 1 g polymer/200 ml solvent.

Polymer (B):   **polyindene**   **1957VAN**
Characterization: $M_n/\text{g.mol}^{-1} = 756$, $M_w/\text{g.mol}^{-1} = 1023$, $\rho = 1.09$ g/cm³, commercial
                        product, coumarone indene resin from Pennsylvania Ind. Chem. Corp.
Solvent (A):   **ethylbenzene**   **C₈H₁₀**   **100-41-4**

$T/\text{K} = 299.15$ $\Delta_{sol}H_B^\infty = -1.38$ J/(g polymer)

Comments: The final concentration is about 1 g polymer/200 ml solvent.

Polymer (B):   **polyindene**   **1957VAN**
Characterization: $M_n/\text{g.mol}^{-1} = 756$, $M_w/\text{g.mol}^{-1} = 1023$, $\rho = 1.09$ g/cm³, commercial
                        product, coumarone indene resin from Pennsylvania Ind. Chem. Corp.
Solvent (A):   **ethyl benzoate**   **C₉H₁₀O₂**   **93-89-0**

$T/\text{K} = 299.15$ $\Delta_{sol}H_B^\infty = -0.67$ J/(g polymer)

Comments: The final concentration is about 1 g polymer/200 ml solvent.

| **Polymer (B):** | **polyindene** | | **1957VAN** |

Characterization: M_n/g.mol^{-1} = 756, M_w/g.mol^{-1} = 1023, ρ = 1.09 g/cm^3, commercial product, coumarone indene resin from Pennsylvania Ind. Chem. Corp.

| **Solvent (A):** | **nitrobenzene** | **C$_6$H$_5$NO$_2$** | **98-95-3** |

T/K = 299.15 $\Delta_{sol} H_B^\infty$ = 4.56 J/(g polymer)

Comments: The final concentration is about 1 g polymer/200 ml solvent.

| **Polymer (B):** | **polyindene** | | **1957VAN** |

Characterization: M_n/g.mol^{-1} = 756, M_w/g.mol^{-1} = 1023, ρ = 1.09 g/cm^3, commercial product, coumarone indene resin from Pennsylvania Ind. Chem. Corp.

| **Solvent (A):** | **1-nitropropane** | **C$_3$H$_7$NO$_2$** | **108-03-2** |

T/K = 299.15 $\Delta_{sol} H_B^\infty$ = 8.41 J/(g polymer)

Comments: The final concentration is about 1 g polymer/200 ml solvent.

| **Polymer (B):** | **polyindene** | | **1957VAN** |

Characterization: M_n/g.mol^{-1} = 756, M_w/g.mol^{-1} = 1023, ρ = 1.09 g/cm^3, commercial product, coumarone indene resin from Pennsylvania Ind. Chem. Corp.

| **Solvent (A):** | **pyridine** | **C$_5$H$_5$N** | **110-86-1** |

T/K = 299.15 $\Delta_{sol} H_B^\infty$ = −6.74 J/(g polymer)

Comments: The final concentration is about 1 g polymer/200 ml solvent.

| **Polymer (B):** | **polyindene** | | **1957VAN** |

Characterization: M_n/g.mol^{-1} = 756, M_w/g.mol^{-1} = 1023, ρ = 1.09 g/cm^3, commercial product, coumarone indene resin from Pennsylvania Ind. Chem. Corp.

| **Solvent (A):** | **1,1,2,2-tetrachloroethane** | **C$_2$H$_2$Cl$_4$** | **79-34-5** |

T/K = 299.15 $\Delta_{sol} H_B^\infty$ = −18.7 J/(g polymer)

Comments: The final concentration is about 1 g polymer/200 ml solvent.

| **Polymer (B):** | **polyindene** | | **1957VAN** |

Characterization: M_n/g.mol^{-1} = 756, M_w/g.mol^{-1} = 1023, ρ = 1.09 g/cm^3, commercial product, coumarone indene resin from Pennsylvania Ind. Chem. Corp.

| **Solvent (A):** | **tetrachloromethane** | **CCl$_4$** | **56-23-5** |

T/K = 299.15 $\Delta_{sol} H_B^\infty$ = −2.47 J/(g polymer)

Comments: The final concentration is about 1 g polymer/200 ml solvent.

Polymer (B):   **polyindene**   **1957VAN**
Characterization: M_n/g.mol^{-1} = 756, M_w/g.mol^{-1} = 1023, ρ = 1.09 g/cm^3, commercial
                       product, coumarone indene resin from Pennsylvania Ind. Chem. Corp.
Solvent (A):   **1,1,1-trichloroethane**   **C$_2$H$_3$Cl$_3$**   **71-55-6**

T/K = 299.15 $\Delta_{sol}H_B^{\infty}$ = −1.76 J/(g polymer)

Comments: The final concentration is about 1 g polymer/200 ml solvent.

Polymer (B):   **polyindene**   **1957VAN**
Characterization: M_n/g.mol^{-1} = 756, M_w/g.mol^{-1} = 1023, ρ = 1.09 g/cm^3, commercial
                       product, coumarone indene resin from Pennsylvania Ind. Chem. Corp.
Solvent (A):   **trichloromethane**   **CHCl$_3$**   **67-66-3**

T/K = 299.15 $\Delta_{sol}H_B^{\infty}$ = −20.1 J/(g polymer)

Comments: The final concentration is about 1 g polymer/200 ml solvent.

Polymer (B):   **polyisobutylene**   **1977DES**
Characterization: M_n/g.mol^{-1} = 360, M_η/g.mol^{-1} = 700
Solvent (A):   **benzene**   **C$_6$H$_6$**   **71-43-2**

T/K = 298.15 $\Delta_M H_B^{\infty}$ = 30.2 J/(g polymer)

Polymer (B):   **polyisobutylene**   **1977DES**
Characterization: M_n/g.mol^{-1} = 1000, M_η/g.mol^{-1} = 2000
Solvent (A):   **benzene**   **C$_6$H$_6$**   **71-43-2**

T/K = 298.15 $\Delta_M H_B^{\infty}$ = 25.4 J/(g polymer)

Polymer (B):   **polyisobutylene**   **1977DES**
Characterization: M_n/g.mol^{-1} = 1300, M_η/g.mol^{-1} = 2500
Solvent (A):   **benzene**   **C$_6$H$_6$**   **71-43-2**

T/K = 298.15 $\Delta_M H_B^{\infty}$ = 23.0 J/(g polymer)

Polymer (B):   **polyisobutylene**   **1962DE2**
Characterization: M_η/g.mol^{-1} = 30000, fractionated in the laboratory from
                       Enjay Vistanex LM-MH-225
Solvent (A):   **benzene**   **C$_6$H$_6$**   **71-43-2**

T/K = 297.65 $\Delta_M H_B^{\infty}$ = 1088 J/(base mol polymer)

Polymer (B):	**polyisobutylene**		**1964DEL**
Characterization:	M_η/g.mol^{-1} = 30000, fractionated in the laboratory from Enjay Vistanex LM-MH-225		
Solvent (A):	**benzene**	**C$_6$H$_6$**	**71-43-2**

T/K = 298.15 $\Delta_M H_B^\infty$ = 19.37 J/(g polymer)

Polymer (B):	**polyisobutylene**		**1966BIA**
Characterization:	M_n/g.mol^{-1} = 44700, fractionated in the laboratory		
Solvent (A):	**benzene**	**C$_6$H$_6$**	**71-43-2**

T/K = 303.15 $\Delta_M H_B^\infty$ = 900 J/(base mol polymer)

Comments: The final concentration is about 0.3 g polymer/100 ml solvent.

Polymer (B):	**polyisobutylene**		**1979LEE**
Characterization:	M_η/g.mol^{-1} = 48000		
Solvent (A):	**benzene**	**C$_6$H$_6$**	**71-43-2**

T/K = 303.4 $\Delta_M H_B$ = 1082 J/(base mol polymer)

Comments: The final concentration is about 0.09 g polymer/7 g solvent.

Polymer (B):	**polyisobutylene**		**1966CUN**
Characterization:	M_η/g.mol^{-1} = 50000		
Solvent (A):	**benzene**	**C$_6$H$_6$**	**71-43-2**

T/K = 303.15 $\Delta_{sol} H_B^\infty$ = 920 J/(base mol polymer)

Polymer (B):	**polyisobutylene**		**1970LID**
Characterization:	M_η/g.mol^{-1} = 72000		
Solvent (A):	**benzene**	**C$_6$H$_6$**	**71-43-2**

T/K	300.15	323.15	343.15	375.15	394.15	423.15	437.15
$\Delta_M H_B^\infty$/(J/base mol polymer)	1076	1000	879	699	515	197	−26

T/K	453.15
$\Delta_M H_B^\infty$/(J/base mol polymer)	−264

Polymer (B):	**polyisobutylene**		**1958TA3**
Characterization:	M_η/g.mol^{-1} = 90000		
Solvent (A):	**benzene**	**C$_6$H$_6$**	**71-43-2**

T/K = 298.15 $\Delta_{sol} H_B^\infty$ = 6.7 J/(g polymer)

Polymer (B):   **polyisobutylene**   **1969BIA**
Characterization: $M_\eta/\text{g.mol}^{-1} = 160000$
Solvent (A):   **benzene**   **C₆H₆**   **71-43-2**

$T/\text{K} = 303.15$ $\Delta_{\text{sol}}H_B^\infty = 920$ J/(base mol polymer)

Polymer (B):   **polyisobutylene**   **1959HOR**
Characterization: $M_\eta/\text{g.mol}^{-1} = 560000$, fractionated in the laboratory
Solvent (A):   **benzene**   **C₆H₆**   **71-43-2**

$T/\text{K} = 298.15$ $\Delta_{\text{sol}}H_B^\infty = 18.4$ J/(g polymer)

Polymer (B):   **polyisobutylene**   **1950TAG, 1951TA1, 1951TA2**
Characterization: –
Solvent (A):   **benzene**   **C₆H₆**   **71-43-2**

$T/\text{K} = 298.15$ $\Delta_{\text{sol}}H_B^\infty = 6.78$ J/(g polymer)

Comments: The final concentration is about 1 g polymer/100 g solution.

Polymer (B):   **polyisobutylene**   **1962DE2**
Characterization: $M_\eta/\text{g.mol}^{-1} = 30000$, fractionated in the laboratory from
                          Enjay Vistanex LM-MH-225
Solvent (A):   **chlorobenzene**   **C₆H₅Cl**   **108-90-7**

$T/\text{K} = 297.65$ $\Delta_M H_B^\infty = 670$ J/(base mol polymer)

Polymer (B):   **polyisobutylene**   **1964DEL**
Characterization: $M_\eta/\text{g.mol}^{-1} = 30000$, fractionated in the laboratory from
                          Enjay Vistanex LM-MH-225
Solvent (A):   **chlorobenzene**   **C₆H₅Cl**   **108-90-7**

$T/\text{K} = 298.15$ $\Delta_M H_B^\infty = 12.6$ J/(g polymer)

Polymer (B):   **polyisobutylene**   **1966CUN**
Characterization: $M_\eta/\text{g.mol}^{-1} = 50000$
Solvent (A):   **chlorobenzene**   **C₆H₅Cl**   **108-90-7**

$T/\text{K} = 303.15$ $\Delta_{\text{sol}}H_B^\infty = 670$ J/(base mol polymer)

Polymer (B):   **polyisobutylene**   **1959HOR**
Characterization: $M_\eta/\text{g.mol}^{-1} = 560000$, fractionated in the laboratory
Solvent (A):   **chlorobenzene**   **C₆H₅Cl**   **108-90-7**

$T/\text{K} = 298.15$ $\Delta_{\text{sol}}H_B^\infty = 11.7$ J/(g polymer)

Polymer (B):   **polyisobutylene**   **1977DES**
Characterization: M_n/g.mol^{-1} = 360, M_η/g.mol^{-1} = 700
Solvent (A):   **cyclohexane** C_6H_{12} **110-82-7**

T/K = 298.15 $\Delta_M H_B^\infty$ = 3.78 J/(g polymer)

Polymer (B):   **polyisobutylene**   **1977DES**
Characterization: M_n/g.mol^{-1} = 1000, M_η/g.mol^{-1} = 2000
Solvent (A):   **cyclohexane** C_6H_{12} **110-82-7**

T/K = 298.15 $\Delta_M H_B^\infty$ = 1.21 J/(g polymer)

Polymer (B):   **polyisobutylene**   **1977DES**
Characterization: M_n/g.mol^{-1} = 1300, M_η/g.mol^{-1} = 2500
Solvent (A):   **cyclohexane** C_6H_{12} **110-82-7**

T/K = 298.15 $\Delta_M H_B^\infty$ = 1.13 J/(g polymer)

Polymer (B):   **polyisobutylene**   **1979PH2**
Characterization: M_η/g.mol^{-1} = 4500, sample Bh 505C, Polysar, Sarnia, Ontario
Solvent (A):   **cyclohexane** C_6H_{12} **110-82-7**

T/K = 298.15 $\Delta_M H_B^\infty$ = −0.6 J/(g polymer)

Polymer (B):   **polyisobutylene**   **1962DE2**
Characterization: M_η/g.mol^{-1} = 30000, fractionated in the laboratory from
                        Enjay Vistanex LM-MH-225
Solvent (A):   **cyclohexane** C_6H_{12} **110-82-7**

T/K = 297.65 $\Delta_M H_B^\infty$ = −40.2 J/(base mol polymer)

Polymer (B):   **polyisobutylene**   **1964DEL**
Characterization: M_η/g.mol^{-1} = 30000, fractionated in the laboratory from
                        Enjay Vistanex LM-MH-225
Solvent (A):   **cyclohexane** C_6H_{12} **110-82-7**

T/K = 298.15 $\Delta_M H_B^\infty$ = −0.63 J/(g polymer)

Polymer (B):   **polyisobutylene**   **1977DES**
Characterization: M_η/g.mol^{-1} = 40000
Solvent (A):   **cyclohexane** C_6H_{12} **110-82-7**

T/K = 298.15 $\Delta_M H_B^\infty$ = −0.72 J/(g polymer)

Polymer (B):   **polyisobutylene**   **1966CUN**
Characterization: M_η/g.mol^{-1} = 50000
Solvent (A):   **cyclohexane**   **C$_6$H$_{12}$**   **110-82-7**

T/K = 303.15 $\Delta_{sol}H_B^\infty$ = −40.2 J/(base mol polymer)

Polymer (B):   **polyisobutylene**   **1969BIA**
Characterization: M_η/g.mol^{-1} = 160000
Solvent (A):   **cyclohexane**   **C$_6$H$_{12}$**   **110-82-7**

T/K = 303.15 $\Delta_{sol}H_B^\infty$ = −33.5 J/(base mol polymer)

Polymer (B):   **polyisobutylene**   **1963TAG**
Characterization: M_η/g.mol^{-1} = 1990000
Solvent (A):   **cyclohexane**   **C$_6$H$_{12}$**   **110-82-7**

T/K = 298.15 $\Delta_{sol}H_B^\infty$ = −0.67 J/(g polymer)

Polymer (B):   **polyisobutylene**   **1979PH2**
Characterization: M_η/g.mol^{-1} = 4500, sample Bh 505C, Polysar, Sarnia, Ontario
Solvent (A):   **cyclooctane**   **C$_8$H$_{16}$**   **292-64-8**

T/K = 298.15 $\Delta_M H_B^\infty$ = 0.3 J/(g polymer)

Polymer (B):   **polyisobutylene**   **1979PH2**
Characterization: M_η/g.mol^{-1} = 4500, sample Bh 505C, Polysar, Sarnia, Ontario
Solvent (A):   **cyclopentane**   **C$_5$H$_{10}$**   **287-92-3**

T/K = 298.15 $\Delta_M H_B^\infty$ = −5.9 J/(g polymer)

Polymer (B):   **polyisobutylene**   **1979PH2**
Characterization: M_η/g.mol^{-1} = 4500, sample Bh 505C, Polysar, Sarnia, Ontario
Solvent (A):   ***cis*-decahydronaphthalene**   **C$_{10}$H$_{18}$**   **493-01-6**

T/K = 298.15 $\Delta_M H_B^\infty$ = 0.2 J/(g polymer)

Polymer (B):   **polyisobutylene**   **1979PH2**
Characterization: M_η/g.mol^{-1} = 4500, sample Bh 505C, Polysar, Sarnia, Ontario
Solvent (A):   ***trans*-decahydronaphthalene**   **C$_{10}$H$_{18}$**   **493-02-7**

T/K = 298.15 $\Delta_M H_B^\infty$ = −0.8 J/(g polymer)

Polymer (B):	**polyisobutylene**	**1962DE2**
Characterization:	M_η/g.mol^{-1} = 30000, fractionated in the laboratory from	
	Enjay Vistanex LM-MH-225	
Solvent (A):	**n-decane** $\quad$ $C_{10}H_{22}$	**124-18-5**

T/K = 297.65 $\qquad$ $\Delta_M H_B^\infty = -30.5$ J/(base mol polymer)

Polymer (B):	**polyisobutylene**	**1966CUN**
Characterization:	M_η/g.mol^{-1} = 50000	
Solvent (A):	**n-decane** $\quad$ $C_{10}H_{22}$	**124-18-5**

T/K = 303.15 $\qquad$ $\Delta_{sol} H_B^\infty = -30.5$ J/(base mol polymer)

Polymer (B):	**polyisobutylene**	**1962DE1**
Characterization:	M_η/g.mol^{-1} = 30000, fractionated in the laboratory from	
	Enjay Vistanex LM-MH-225	
Solvent (A):	**dibutyl ether** $\quad$ $C_8H_{18}O$	**142-96-1**

T/K = 297.65 $\qquad$ $\Delta_M H_B^\infty = 69$ J/(base mol polymer)

Polymer (B):	**polyisobutylene**	**1962DE1, 1962DE2**
Characterization:	M_η/g.mol^{-1} = 30000, fractionated in the laboratory from	
	Enjay Vistanex LM-MH-225	
Solvent (A):	**diethyl ether** $\quad$ $C_4H_{10}O$	**60-29-7**

T/K = 297.65 $\qquad$ $\Delta_M H_B^\infty = 155$ J/(base mol polymer)

Polymer (B):	**polyisobutylene**	**1979PH2**
Characterization:	M_η/g.mol^{-1} = 4500, sample Bh 505C, Polysar, Sarnia, Ontario	
Solvent (A):	**3,3-diethylpentane** $\quad$ C_9H_{20}	**4032-86-4**

T/K = 298.15 $\qquad$ $\Delta_M H_B^\infty = -1.4$ J/(g polymer)

Polymer (B):	**polyisobutylene**	**1962DE1**
Characterization:	M_η/g.mol^{-1} = 30000, fractionated in the laboratory from	
	Enjay Vistanex LM-MH-225	
Solvent (A):	**dihexyl ether** $\quad$ $C_{12}H_{26}O$	**112-58-3**

T/K = 297.65 $\qquad$ $\Delta_M H_B^\infty = 48$ J/(base mol polymer)

Polymer (B):	**polyisobutylene**	**1979PH2**
Characterization:	M_η/g.mol^{-1} = 4500, sample Bh 505C, Polysar, Sarnia, Ontario	
Solvent (A):	**2,2-dimethylpentane** $\quad$ C_7H_{16}	**590-35-2**

T/K = 298.15 $\qquad$ $\Delta_M H_B^\infty = -1.1$ J/(g polymer)

Polymer (B):   **polyisobutylene**   **1979PH2**
Characterization: M_η/g.mol^{-1} = 4500, sample Bh 505C, Polysar, Sarnia, Ontario
Solvent (A):   **2,3-dimethylpentane**   **C$_7$H$_{16}$**   **565-59-3**

T/K = 298.15 $\Delta_M H_B^\infty$ = –1.9 J/(g polymer)

Polymer (B):   **polyisobutylene**   **1979PH2**
Characterization: M_η/g.mol^{-1} = 4500, sample Bh 505C, Polysar, Sarnia, Ontario
Solvent (A):   **2,4-dimethylpentane**   **C$_7$H$_{16}$**   **108-08-7**

T/K = 298.15 $\Delta_M H_B^\infty$ = –1.1 J/(g polymer)

Polymer (B):   **polyisobutylene**   **1979PH2**
Characterization: M_η/g.mol^{-1} = 4500, sample Bh 505C, Polysar, Sarnia, Ontario
Solvent (A):   **3,3-dimethylpentane**   **C$_7$H$_{16}$**   **562-49-2**

T/K = 298.15 $\Delta_M H_B^\infty$ = –1.7 J/(g polymer)

Polymer (B):   **polyisobutylene**   **1962DE1**
Characterization: M_η/g.mol^{-1} = 30000, fractionated in the laboratory from
                       Enjay Vistanex LM-MH-225
Solvent (A):   **dipentyl ether**   **C$_{10}$H$_{22}$O**   **693-65-2**

T/K = 297.65 $\Delta_M H_B^\infty$ = 59 J/(base mol polymer)

Polymer (B):   **polyisobutylene**   **1962DE1**
Characterization: M_η/g.mol^{-1} = 30000, fractionated in the laboratory from
                       Enjay Vistanex LM-MH-225
Solvent (A):   **dipropyl ether**   **C$_6$H$_{14}$O**   **111-43-3**

T/K = 297.65 $\Delta_M H_B^\infty$ = 98 J/(base mol polymer)

Polymer (B):   **polyisobutylene**   **1977DES**
Characterization: M_n/g.mol^{-1} = 360, M_η/g.mol^{-1} = 700
Solvent (A):   **n-dodecane**   **C$_{12}$H$_{26}$**   **112-40-3**

T/K = 298.15 $\Delta_M H_B^\infty$ = 1.91 J/(g polymer)

Polymer (B):   **polyisobutylene**   **1977DES**
Characterization: M_n/g.mol^{-1} = 1000, M_η/g.mol^{-1} = 2000
Solvent (A):   **n-dodecane**   **C$_{12}$H$_{26}$**   **112-40-3**

T/K = 298.15 $\Delta_M H_B^\infty$ = 0.67 J/(g polymer)

Polymer (B):	**polyisobutylene**	**1977DES**
Characterization:	$M_n/\text{g.mol}^{-1} = 1300$, $M_\eta/\text{g.mol}^{-1} = 2500$	
Solvent (A):	**n-dodecane** $\quad$ **$C_{12}H_{26}$**	**112-40-3**

$T/\text{K} = 298.15$ $\qquad$ $\Delta_M H_B^\infty = 0.54$ J/(g polymer)

Polymer (B):	**polyisobutylene**	**1979PH1**
Characterization:	$M_\eta/\text{g.mol}^{-1} = 4500$, sample Bh 505C, Polysar, Sarnia, Ontario	
Solvent (A):	**n-dodecane** $\quad$ **$C_{12}H_{26}$**	**112-40-3**

$T/\text{K} = 298.15$ $\qquad$ $\Delta_M H_B^\infty = 0.2$ J/(g polymer)

Polymer (B):	**polyisobutylene**	**1962DE2**
Characterization:	$M_\eta/\text{g.mol}^{-1} = 30000$, fractionated in the laboratory from Enjay Vistanex LM-MH-225	
Solvent (A):	**n-dodecane** $\quad$ **$C_{12}H_{26}$**	**112-40-3**

$T/\text{K} = 297.65$ $\qquad$ $\Delta_M H_B^\infty = -6.3$ J/(base mol polymer)

Polymer (B):	**polyisobutylene**	**1964DEL**
Characterization:	$M_\eta/\text{g.mol}^{-1} = 30000$, fractionated in the laboratory from Enjay Vistanex LM-MH-225	
Solvent (A):	**n-dodecane** $\quad$ **$C_{12}H_{26}$**	**112-40-3**

$T/\text{K} = 298.15$ $\qquad$ $\Delta_M H_B^\infty = -0.08$ J/(g polymer)

Polymer (B):	**polyisobutylene**	**1979LEE**
Characterization:	$M_\eta/\text{g.mol}^{-1} = 48000$	
Solvent (A):	**ethylbenzene** $\quad$ **C_8H_{10}**	**100-41-4**

$T/\text{K} = 291.35$ $\qquad$ $\Delta_M H_B = 530$ J/(base mol polymer)

$T/\text{K} = 342.35$ $\qquad$ $\Delta_M H_B = 195$ J/(base mol polymer)

Comments: $\quad$ The final concentration is about 0.09 g polymer/7 g solvent. Additional data are given in a graph in the original source.

Polymer (B):	**polyisobutylene**	**1962DE1**
Characterization:	$M_\eta/\text{g.mol}^{-1} = 30000$, fractionated in the laboratory from Enjay Vistanex LM-MH-225	
Solvent (A):	**ethyl decanoate** $\quad$ **$C_{12}H_{24}O_2$**	**110-38-3**

$T/\text{K} = 297.65$ $\qquad$ $\Delta_M H_B^\infty = 170$ J/(base mol polymer)

Polymer (B):   **polyisobutylene**   **1962DE1**
Characterization: M_η/g.mol^{-1} = 30000, fractionated in the laboratory from
                          Enjay Vistanex LM-MH-225
Solvent (A):   **ethyl heptanoate**   **C$_9$H$_{18}$O$_2$**   **106-30-9**

T/K = 297.65 $\Delta_M H_B^\infty$ = 315 J/(base mol polymer)

Polymer (B):   **polyisobutylene**   **1962DE1**
Characterization: M_η/g.mol^{-1} = 30000, fractionated in the laboratory from
                          Enjay Vistanex LM-MH-225
Solvent (A):   **ethyl hexadecanoate**   **C$_{18}$H$_{36}$O$_2$**   **628-97-7**

T/K = 297.65 $\Delta_M H_B^\infty$ = 72 J/(base mol polymer)

Polymer (B):   **polyisobutylene**   **1962DE1**
Characterization: M_η/g.mol^{-1} = 30000, fractionated in the laboratory from
                          Enjay Vistanex LM-MH-225
Solvent (A):   **ethyl hexanoate**   **C$_8$H$_{16}$O$_2$**   **123-66-0**

T/K = 297.65 $\Delta_M H_B^\infty$ = 375 J/(base mol polymer)

Polymer (B):   **polyisobutylene**   **1962DE1**
Characterization: M_η/g.mol^{-1} = 30000, fractionated in the laboratory from
                          Enjay Vistanex LM-MH-225
Solvent (A):   **ethyl nonanoate**   **C$_{11}$H$_{22}$O$_2$**   **123-29-5**

T/K = 297.65 $\Delta_M H_B^\infty$ = 207 J/(base mol polymer)

Polymer (B):   **polyisobutylene**   **1962DE1**
Characterization: M_η/g.mol^{-1} = 30000, fractionated in the laboratory from
                          Enjay Vistanex LM-MH-225
Solvent (A):   **ethyl octanoate**   **C$_{10}$H$_{20}$O$_2$**   **106-32-1**

T/K = 297.65 $\Delta_M H_B^\infty$ = 255 J/(base mol polymer)

Polymer (B):   **polyisobutylene**   **1979PH2**
Characterization: M_η/g.mol^{-1} = 4500, sample Bh 505C, Polysar, Sarnia, Ontario
Solvent (A):   **3-ethylpentane**   **C$_7$H$_{16}$**   **617-78-7**

T/K = 298.15 $\Delta_M H_B^\infty$ = −2.0 J/(g polymer)

Polymer (B):   **polyisobutylene**   **1962DE1**
Characterization: M_η/g.mol^{-1} = 30000, fractionated in the laboratory from
                          Enjay Vistanex LM-MH-225
Solvent (A):   **ethyl tetradecanoate**   **C$_{16}$H$_{32}$O$_2$**   **124-06-1**

T/K = 297.65 $\Delta_M H_B^\infty$ = 99 J/(base mol polymer)

| Polymer (B): | polyisobutylene | | 1979PH2 |

Polymer (B): **polyisobutylene**   **1979PH2**
Characterization: M_η/g.mol^{-1} = 4500, sample Bh 505C, Polysar, Sarnia, Ontario
Solvent (A): **2,2,4,4,6,8,8-heptamethylnonane**   **C$_{16}$H$_{34}$**   **4390-04-9**

T/K = 298.15 $\Delta_M H_B^\infty$ = −0.5 J/(g polymer)

Polymer (B): **polyisobutylene**   **1950TAG, 1951TA1, 1951TA2**
Characterization: −
Solvent (A): **n-heptane**   **C$_7$H$_{16}$**   **142-82-5**

T/K = 298.15 $\Delta_{sol} H_B^\infty$ = −1.42 J/(g polymer)

Comments: The final concentration is about 1 g polymer/100 g solution.

Polymer (B): **polyisobutylene**   **1977DES**
Characterization: M_n/g.mol^{-1} = 360, M_η/g.mol^{-1} = 700
Solvent (A): **n-heptane**   **C$_7$H$_{16}$**   **142-82-5**

T/K = 298.15 $\Delta_M H_B^\infty$ = −0.48 J/(g polymer)

Polymer (B): **polyisobutylene**   **1977DES**
Characterization: M_n/g.mol^{-1} = 1000, M_η/g.mol^{-1} = 2000
Solvent (A): **n-heptane**   **C$_7$H$_{16}$**   **142-82-5**

T/K = 298.15 $\Delta_M H_B^\infty$ = −0.96 J/(g polymer)

Polymer (B): **polyisobutylene**   **1977DES**
Characterization: M_n/g.mol^{-1} = 1300, M_η/g.mol^{-1} = 2500
Solvent (A): **n-heptane**   **C$_7$H$_{16}$**   **142-82-5**

T/K = 298.15 $\Delta_M H_B^\infty$ = −1.38 J/(g polymer)

Polymer (B): **polyisobutylene**   **1979PH1**
Characterization: M_η/g.mol^{-1} = 4500, sample Bh 505C, Polysar, Sarnia, Ontario
Solvent (A): **n-heptane**   **C$_7$H$_{16}$**   **142-82-5**

T/K = 298.15 $\Delta_M H_B^\infty$ = −1.7 J/(g polymer)

Polymer (B): **polyisobutylene**   **1962DE2**
Characterization: M_η/g.mol^{-1} = 30000, fractionated in the laboratory from
                Enjay Vistanex LM-MH-225
Solvent (A): **n-heptane**   **C$_7$H$_{16}$**   **142-82-5**

T/K = 297.65 $\Delta_M H_B^\infty$ = −100 J/(base mol polymer)

Polymer (B):   **polyisobutylene**   **1964DEL**
Characterization: M_η/g.mol^{-1} = 30000, fractionated in the laboratory from
                          Enjay Vistanex LM-MH-225
Solvent (A):   **n-heptane**   **C$_7$H$_{16}$**   **142-82-5**

T/K = 298.15 $\Delta_M H_B^\infty$ = −2.01 J/(g polymer)

Polymer (B):   **polyisobutylene**   **1977DES**
Characterization: M_η/g.mol^{-1} = 40000
Solvent (A):   **n-heptane**   **C$_7$H$_{16}$**   **142-82-5**

T/K = 298.15 $\Delta_M H_B^\infty$ = −1.8 J/(g polymer)

Polymer (B):   **polyisobutylene**   **1966CUN**
Characterization: M_η/g.mol^{-1} = 50000
Solvent (A):   **n-heptane**   **C$_7$H$_{16}$**   **142-82-5**

T/K = 303.15 $\Delta_{sol} H_B^\infty$ = −100 J/(base mol polymer)

Polymer (B):   **polyisobutylene**   **1969BIA**
Characterization: M_η/g.mol^{-1} = 160000
Solvent (A):   **n-heptane**   **C$_7$H$_{16}$**   **142-82-5**

T/K = 303.15 $\Delta_{sol} H_B^\infty$ = −92.0 J/(base mol polymer)

Polymer (B):   **polyisobutylene**   **1977DES**
Characterization: M_n/g.mol^{-1} = 360, M_η/g.mol^{-1} = 700
Solvent (A):   **n-hexadecane**   **C$_{16}$H$_{34}$**   **544-76-3**

T/K = 298.15 $\Delta_M H_B^\infty$ = 4.47 J/(g polymer)

Polymer (B):   **polyisobutylene**   **1977DES**
Characterization: M_n/g.mol^{-1} = 1000, M_η/g.mol^{-1} = 2000
Solvent (A):   **n-hexadecane**   **C$_{16}$H$_{34}$**   **544-76-3**

T/K = 298.15 $\Delta_M H_B^\infty$ = 2.12 J/(g polymer)

Polymer (B):   **polyisobutylene**   **1977DES**
Characterization: M_n/g.mol^{-1} = 1300, M_η/g.mol^{-1} = 2500
Solvent (A):   **n-hexadecane**   **C$_{16}$H$_{34}$**   **544-76-3**

T/K = 298.15 $\Delta_M H_B^\infty$ = 0.98 J/(g polymer)

Polymer (B):	**polyisobutylene**	**1979PH1**
Characterization:	M_η/g.mol^{-1} = 4500, sample Bh 505C, Polysar, Sarnia, Ontario	
Solvent (A):	**n-hexadecane** $\quad$ $C_{16}H_{34}$	**544-76-3**

T/K = 298.15 $\qquad$ $\Delta_M H_B^\infty$ = 0.9 J/(g polymer)

Polymer (B):	**polyisobutylene**	**1962DE2**
Characterization:	M_η/g.mol^{-1} = 30000, fractionated in the laboratory from Enjay Vistanex LM-MH-225	
Solvent (A):	**n-hexadecane** $\quad$ $C_{16}H_{34}$	**544-76-3**

T/K = 297.65 $\qquad$ $\Delta_M H_B^\infty$ = 2.5 J/(base mol polymer)

Polymer (B):	**polyisobutylene**	**1977DES**
Characterization:	M_η/g.mol^{-1} = 40000	
Solvent (A):	**n-hexadecane** $\quad$ $C_{16}H_{34}$	**544-76-3**

T/K = 298.15 $\qquad$ $\Delta_M H_B^\infty$ = 0.45 J/(g polymer)

Polymer (B):	**polyisobutylene**	**1962DE2**
Characterization:	M_η/g.mol^{-1} = 30000, fractionated in the laboratory from Enjay Vistanex LM-MH-225	
Solvent (A):	**n-hexane** $\quad$ C_6H_{14}	**110-54-3**

T/K = 297.65 $\qquad$ $\Delta_M H_B^\infty$ = −142 J/(base mol polymer)

Polymer (B):	**polyisobutylene**	**1964DEL**
Characterization:	M_η/g.mol^{-1} = 30000, fractionated in the laboratory from Enjay Vistanex LM-MH-225	
Solvent (A):	**n-hexane** $\quad$ C_6H_{14}	**110-54-3**

T/K = 298.15 $\qquad$ $\Delta_M H_B^\infty$ = −2.59 J/(g polymer)

Polymer (B):	**polyisobutylene**	**1966CUN**
Characterization:	M_η/g.mol^{-1} = 50000	
Solvent (A):	**n-hexane** $\quad$ C_6H_{14}	**110-54-3**

T/K = 303.15 $\qquad$ $\Delta_{sol} H_B^\infty$ = −142 J/(base mol polymer)

Polymer (B):	**polyisobutylene**						**1970LID**
Characterization:	M_η/g.mol^{-1} = 72000						
Solvent (A):	**n-hexane**			C_6H_{14}			**110-54-3**

T/K	303.15	324.15	348.15	373.15	393.15	408.15	423.15
$\Delta_M H_B^\infty$/(J/base mol polymer)	−100	−130	−163	−205	−297	−373	−502

T/K	433.15
$\Delta_M H_B^\infty$/(J/base mol polymer)	−557

Polymer (B):   **polyisobutylene**   **1962DE2**
Characterization: M_η/g.mol^{-1} = 30000, fractionated in the laboratory from
     Enjay Vistanex LM-MH-225
Solvent (A):   **2-methylbutane**   **C$_5$H$_{12}$**   **78-78-4**

T/K = 297.65 $\Delta_M H_B^\infty$ = −175 J/(base mol polymer)

Polymer (B):   **polyisobutylene**   **1962DE2**
Characterization: M_η/g.mol^{-1} = 30000, fractionated in the laboratory from
     Enjay Vistanex LM-MH-225
Solvent (A):   **methylcyclohexane**   **C$_7$H$_{14}$**   **108-87-2**

T/K = 297.65 $\Delta_M H_B^\infty$ = −65.7 J/(base mol polymer)

Polymer (B):   **polyisobutylene**   **1966CUN**
Characterization: M_η/g.mol^{-1} = 50000
Solvent (A):   **methylcyclohexane**   **C$_7$H$_{14}$**   **108-87-2**

T/K = 303.15 $\Delta_{sol} H_B^\infty$ = −65.7 J/(base mol polymer)

Polymer (B):   **polyisobutylene**   **1979PH2**
Characterization: M_η/g.mol^{-1} = 4500, sample Bh 505C, Polysar, Sarnia, Ontario
Solvent (A):   **3-methylhexane**   **C$_7$H$_{16}$**   **589-34-4**

T/K = 298.15 $\Delta_M H_B^\infty$ = −1.0 J/(g polymer)

Polymer (B):   **polyisobutylene**   **1962DE2**
Characterization: M_η/g.mol^{-1} = 30000, fractionated in the laboratory from
     Enjay Vistanex LM-MH-225
Solvent (A):   **3-methylpentane**   **C$_6$H$_{14}$**   **96-14-0**

T/K = 297.65 $\Delta_M H_B^\infty$ = −159 J/(base mol polymer)

Polymer (B):   **polyisobutylene**   **1979PH1**
Characterization: M_η/g.mol^{-1} = 4500, sample Bh 505C, Polysar, Sarnia, Ontario
Solvent (A):   **n-nonane**   **C$_9$H$_{20}$**   **111-84-2**

T/K = 298.15 $\Delta_M H_B^\infty$ = −0.8 J/(g polymer)

Polymer (B):   **polyisobutylene**   **1962DE2**
Characterization: M_η/g.mol^{-1} = 30000, fractionated in the laboratory from
     Enjay Vistanex LM-MH-225
Solvent (A):   **n-nonane**   **C$_9$H$_{20}$**   **111-84-2**

T/K = 297.65 $\Delta_M H_B^\infty$ = −46 J/(base mol polymer)

Polymer (B):   **polyisobutylene**   **1979PH1**

Characterization: $M_\eta/\text{g.mol}^{-1} = 4500$, sample Bh 505C, Polysar, Sarnia, Ontario

Solvent (A):   **n-octane**   **C$_8$H$_{18}$**   **111-65-9**

$T/K = 298.15$ $\Delta_M H_B^\infty = -1.1$ J/(g polymer)

Polymer (B):   **polyisobutylene**   **1962DE2**

Characterization: $M_\eta/\text{g.mol}^{-1} = 30000$, fractionated in the laboratory from
Enjay Vistanex LM-MH-225

Solvent (A):   **n-octane**   **C$_8$H$_{18}$**   **111-65-9**

$T/K = 297.65$ $\Delta_M H_B^\infty = -67$ J/(base mol polymer)

Polymer (B):   **polyisobutylene**   **1970LID**

Characterization: $M_\eta/\text{g.mol}^{-1} = 72000$

Solvent (A):   **n-octane**   **C$_8$H$_{18}$**   **111-65-9**

T/K	303.15	324.15	348.15	373.15	393.15	423.15
$\Delta_M H_B^\infty$/(J/base mol polymer)	-17	-42	-50	-59	-75	-200

Polymer (B):   **polyisobutylene**   **1979PH1**

Characterization: $M_\eta/\text{g.mol}^{-1} = 4500$, sample Bh 505C, Polysar, Sarnia, Ontario

Solvent (A):   **2,2,4,6,6-pentamethylheptane**   **C$_{12}$H$_{26}$**   **13475-82-6**

$T/K = 298.15$ $\Delta_M H_B^\infty = -0.1$ J/(g polymer)

Polymer (B):   **polyisobutylene**   **1977DES**

Characterization: $M_n/\text{g.mol}^{-1} = 360$, $M_\eta/\text{g.mol}^{-1} = 700$

Solvent (A):   **n-pentane**   **C$_5$H$_{12}$**   **109-66-0**

$T/K = 298.15$ $\Delta_M H_B^\infty = -1.90$ J/(g polymer)

Polymer (B):   **polyisobutylene**   **1977DES**

Characterization: $M_n/\text{g.mol}^{-1} = 1000$, $M_\eta/\text{g.mol}^{-1} = 2000$

Solvent (A):   **n-pentane**   **C$_5$H$_{12}$**   **109-66-0**

$T/K = 298.15$ $\Delta_M H_B^\infty = -2.90$ J/(g polymer)

Polymer (B):   **polyisobutylene**   **1977DES**

Characterization: $M_n/\text{g.mol}^{-1} = 1300$, $M_\eta/\text{g.mol}^{-1} = 2500$

Solvent (A):   **n-pentane**   **C$_5$H$_{12}$**   **109-66-0**

$T/K = 298.15$ $\Delta_M H_B^\infty = -3.20$ J/(g polymer)

Polymer (B):   **polyisobutylene**   **1962DE2**
Characterization: $M_\eta/\text{g.mol}^{-1} = 30000$, fractionated in the laboratory from
                        Enjay Vistanex LM-MH-225
Solvent (A):   **n-pentane**   **C₅H₁₂**   **109-66-0**

$T/\text{K} = 297.65$ $\Delta_M H_B^\infty = -201$ J/(base mol polymer)

Polymer (B):   **polyisobutylene**   **1977DES**
Characterization: $M_\eta/\text{g.mol}^{-1} = 40000$
Solvent (A):   **n-pentane**   **C₅H₁₂**   **109-66-0**

$T/\text{K} = 298.15$ $\Delta_M H_B^\infty = -3.6$ J/(g polymer)

Polymer (B):   **polyisobutylene**   **1970LID**
Characterization: $M_\eta/\text{g.mol}^{-1} = 72000$
Solvent (A):   **n-pentane**   **C₅H₁₂**   **109-66-0**

T/K	303.15	333.15	352.15	365.15
$\Delta_M H_B^\infty/(\text{J/base mol polymer})$	−159	−193	−253	−306

Polymer (B):   **polyisobutylene**   **1977DES**
Characterization: $M_n/\text{g.mol}^{-1} = 360$, $M_\eta/\text{g.mol}^{-1} = 700$
Solvent (A):   **tetrachloromethane**   **CCl₄**   **56-23-5**

$T/\text{K} = 298.15$ $\Delta_M H_B^\infty = 5.90$ J/(g polymer)

Polymer (B):   **polyisobutylene**   **1977DES**
Characterization: $M_n/\text{g.mol}^{-1} = 1000$, $M_\eta/\text{g.mol}^{-1} = 2000$
Solvent (A):   **tetrachloromethane**   **CCl₄**   **56-23-5**

$T/\text{K} = 298.15$ $\Delta_M H_B^\infty = 5.75$ J/(g polymer)

Polymer (B):   **polyisobutylene**   **1977DES**
Characterization: $M_n/\text{g.mol}^{-1} = 1300$, $M_\eta/\text{g.mol}^{-1} = 2500$
Solvent (A):   **tetrachloromethane**   **CCl₄**   **56-23-5**

$T/\text{K} = 298.15$ $\Delta_M H_B^\infty = 5.00$ J/(g polymer)

Polymer (B):   **polyisobutylene**   **1963TAG**
Characterization: $M_\eta/\text{g.mol}^{-1} = 1990000$
Solvent (A):   **tetrachloromethane**   **CCl₄**   **56-23-5**

$T/\text{K} = 298.15$ $\Delta_{sol} H_B^\infty = 4.06$ J/(g polymer)

| Polymer (B): | polyisobutylene | | 1962DE2 |

Polymer (B): **polyisobutylene** 1962DE2
Characterization: $M_\eta/\text{g.mol}^{-1} = 30000$, fractionated in the laboratory from
Enjay Vistanex LM-MH-225
Solvent (A): **n-tetradecane** $C_{14}H_{30}$ **629-59-4**

$T/K = 297.65$ $\Delta_M H_B^\infty = 0.0$ J/(base mol polymer)

Polymer (B): **polyisobutylene** **1979PH1, 1979PH2**
Characterization: $M_\eta/\text{g.mol}^{-1} = 4500$, sample Bh 505C, Polysar, Sarnia, Ontario
Solvent (A): **2,2,4,4-tetramethylpentane** C_9H_{20} **1070-87-7**

$T/K = 298.15$ $\Delta_M H_B^\infty = -0.6$ J/(g polymer)

Polymer (B): **polyisobutylene** **1979PH2**
Characterization: $M_\eta/\text{g.mol}^{-1} = 4500$, sample Bh 505C, Polysar, Sarnia, Ontario
Solvent (A): **2,3,3,4-tetramethylpentane** C_9H_{20} **16747-38-9**

$T/K = 298.15$ $\Delta_M H_B^\infty = -2.3$ J/(g polymer)

Polymer (B): **polyisobutylene** **1950TAG, 1951TA1, 1951TA2**
Characterization: –
Solvent (A): **toluene** C_7H_8 **108-88-3**

$T/K = 298.15$ $\Delta_{sol} H_B^\infty = 1.84$ J/(g polymer)

Comments: The final concentration is about 1 g polymer/100 g solution.

Polymer (B): **polyisobutylene** **1966CUN**
Characterization: $M_\eta/\text{g.mol}^{-1} = 50000$
Solvent (A): **toluene** C_7H_8 **108-88-3**

$T/K = 303.15$ $\Delta_{sol} H_B^\infty = 415.5$ J/(base mol polymer)

Polymer (B): **polyisobutylene** **1969BIA**
Characterization: $M_\eta/\text{g.mol}^{-1} = 160000$
Solvent (A): **toluene** C_7H_8 **108-88-3**

$T/K = 303.15$ $\Delta_{sol} H_B^\infty = 415.5$ J/(base mol polymer)

Polymer (B): **polyisobutylene** **1963TAG**
Characterization: $M_\eta/\text{g.mol}^{-1} = 1990000$
Solvent (A): **toluene** C_7H_8 **108-88-3**

$T/K = 298.15$ $\Delta_{sol} H_B^\infty = 8.75$ J/(g polymer)

Polymer (B):	**polyisobutylene**	**1962DE2**
Characterization:	M_η/g.mol^{-1} = 30000, fractionated in the laboratory from	
	Enjay Vistanex LM-MH-225	
Solvent (A):	**n-tridecane** **C$_{13}$H$_{28}$**	**629-50-5**

T/K = 297.65 $\Delta_\mathrm{M}H_\mathrm{B}^\infty$ = –2.5 J/(base mol polymer)

Polymer (B):	**polyisobutylene**	**1979PH1**
Characterization:	M_η/g.mol^{-1} = 4500, sample Bh 505C, Polysar, Sarnia, Ontario	
Solvent (A):	**2,2,4-trimethylpentane C$_8$H$_{18}$**	**540-84-1**

T/K = 298.15 $\Delta_\mathrm{M}H_\mathrm{B}^\infty$ = –0.4 J/(g polymer)

Polymer (B):	**polyisobutylene**	**1962DE2**
Characterization:	M_η/g.mol^{-1} = 30000, fractionated in the laboratory from	
	Enjay Vistanex LM-MH-225	
Solvent (A):	**2,2,4-trimethylpentane C$_8$H$_{18}$**	**540-84-1**

T/K = 297.65 $\Delta_\mathrm{M}H_\mathrm{B}^\infty$ = –35.1 J/(base mol polymer)

Polymer (B):	**polyisobutylene**	**1950TAG, 1951TA1, 1951TA2**
Characterization:	–	
Solvent (A):	**2,2,4-trimethylpentane C$_8$H$_{18}$**	**540-84-1**

T/K = 298.15 $\Delta_\mathrm{sol}H_\mathrm{B}^\infty$ = 0.0 J/(g polymer)

Comments: The final concentration is about 1 g polymer/100 g solution.

Polymer (B):	**polyisobutylene**	**1963TAG**
Characterization:	M_η/g.mol^{-1} = 1990000	
Solvent (A):	**2,2,4-trimethylpentane C$_8$H$_{18}$**	**540-84-1**

T/K = 298.15 $\Delta_\mathrm{sol}H_\mathrm{B}^\infty$ = 0.0 J/(g polymer)

Polymer (B):	**polyisobutylene**	**1962DE2**
Characterization:	M_η/g.mol^{-1} = 30000, fractionated in the laboratory from	
	Enjay Vistanex LM-MH-225	
Solvent (A):	**n-undecane** **C$_{11}$H$_{24}$**	**1120-21-4**

T/K = 297.65 $\Delta_\mathrm{M}H_\mathrm{B}^\infty$ = –21.8 J/(base mol polymer)

Polymer (B): $\qquad$ **poly(isobutyl methacrylate)** $\qquad$ **1996SAT**
Characterization: $\qquad$ $M_w/\text{g.mol}^{-1} = 260000$, synthesized in the laboratory
Solvent (A): $\qquad$ **cyclohexanone** $\qquad$ **C₆H₁₀O** $\qquad$ **108-94-1**

$T/\text{K} = 303.15$ $\qquad$ $\Delta_{sol} H_B^\infty = -5.2$ J/(g polymer) $\qquad$ (glassy state polymer)

$T/\text{K} = 303.15$ $\qquad$ $\Delta_M H_B^\infty = +13.2$ J/(g polymer) $\qquad$ (liquid state polymer)

Comments: $\qquad$ $\Delta_M H_B^\infty$ is corrected for the enthalpy from the glass transition.

Polymer (B): $\qquad$ **poly(isobutyl methacrylate-*co*-methyl methacrylate)** $\qquad$ **1996SAT**
Characterization: $\qquad$ $M_w/\text{g.mol}^{-1} = 240000$, 49 wt% methyl methacrylate, synthesized in the laboratory
Solvent (A): $\qquad$ **cyclohexanone** $\qquad$ **C₆H₁₀O** $\qquad$ **108-94-1**

$T/\text{K} = 298.15$ $\qquad$ $\Delta_{sol} H_B^\infty = -10.8$ J/(g polymer) $\qquad$ (glassy state polymer)

$T/\text{K} = 298.15$ $\qquad$ $\Delta_M H_B^\infty = +14.5$ J/(g polymer) $\qquad$ (liquid state polymer)

Comments: $\qquad$ $\Delta_M H_B^\infty$ is corrected for the enthalpy from the glass transition.

Polymer (B): $\qquad$ **poly(*N*-isopropylacrylamide)** $\qquad$ **1968HES**
Characterization: $\qquad$ intrinsic viscosity = 2.6 dl/g in trichloromethane at 298 K
Solvent (A): $\qquad$ **water** $\qquad$ **H₂O** $\qquad$ **7732-18-5**

$T/\text{K} = 293.15$ $\qquad$ $\Delta_{sol} H_B^\infty = 0.213$ J/(g solution)

$T/\text{K} = 298.15$ $\qquad$ $\Delta_{sol} H_B^\infty = 0.201$ J/(g solution)

$T/\text{K} = 303.15$ $\qquad$ $\Delta_{sol} H_B^\infty = 0.163$ J/(g solution)

Comments: $\qquad$ The final concentration is 0.03 %.

Polymer (B): $\qquad$ **poly(methyl acrylate)** $\qquad$ **1956STR, 1958SLO**
Characterization: $\qquad$ intrinsic viscosity = 3.30
Solvent (A): $\qquad$ **2-propanone** $\qquad$ **C₃H₆O** $\qquad$ **67-64-1**

$T/\text{K} = 298.15$ $\qquad$ $\Delta_{sol} H_B^\infty = 0.0$ J/(g polymer)

Comments: $\qquad$ The final concentration is 1 g polymer/100 g solvent.

Polymer (B): $\qquad$ **poly(methyl methacrylate)** $\qquad$ **1996SHI**
Characterization: $\qquad$ $M_n/\text{g.mol}^{-1} = 73900$, $M_w/\text{g.mol}^{-1} = 170000$
Solvent (A): $\qquad$ **cyclohexanone** $\qquad$ **C₆H₁₀O** $\qquad$ **108-94-1**

$T/\text{K} = 304.15$ $\qquad$ $\Delta_{sol} H_B^\infty = -13.6$ J/(g polymer) $\qquad$ (glassy state polymer)

$T/\text{K} = 304.15$ $\qquad$ $\Delta_M H_B^\infty = +16.9$ J/(g polymer) $\qquad$ (liquid state polymer)

Comments: $\qquad$ $\Delta_M H_B^\infty$ is corrected for the enthalpy from the glass transition.

Polymer (B):   **poly(methyl methacrylate)**   **1958TA3**
Characterization: M_n/g.mol^{-1} = 1930
Solvent (A):   **1,2-dichloroethane**   **C$_2$H$_4$Cl$_2$**   **107-06-2**

T/K = 298.15 $\Delta_{sol}H_B^{\infty}$ = −19.7 J/(g polymer)

Polymer (B):   **poly(methyl methacrylate)**   **1958TA3**
Characterization: M_n/g.mol^{-1} = 240000
Solvent (A):   **1,2-dichloroethane**   **C$_2$H$_4$Cl$_2$**   **107-06-2**

T/K = 298.15 $\Delta_{sol}H_B^{\infty}$ = −27.2 J/(g polymer)

Polymer (B):   **poly(methyl methacrylate)**   **1958KAR**
Characterization: M_η/g.mol^{-1} = 53000, precipitated from a technical product
                         in the laboratory
Solvent (A):   **ethylbenzene**   **C$_8$H$_{10}$**   **100-41-4**

T/K = 298.15 $\Delta_{sol}H_B^{\infty}$ = −31.0 J/(g polymer)

Polymer (B):   **poly(methyl methacrylate)**   **1958KAR**
Characterization: M_η/g.mol^{-1} = 1800000, precipitated from a technical product
                         in the laboratory
Solvent (A):   **ethylbenzene**   **C$_8$H$_{10}$**   **100-41-4**

T/K = 298.15 $\Delta_{sol}H_B^{\infty}$ = −29.3 J/(g polymer)

Polymer (B):   **poly(methyl methacrylate)**   **1980GRA, 1984KIL**
Characterization: M_n/g.mol^{-1} = 28900, M_w/g.mol^{-1} = 35900, atactic,
                         Roehm GmbH, Darmstadt, Germany
Solvent (A):   **4-methyl-2-pentanone   C$_6$H$_{12}$O**   **108-10-1**

T/K = 303.15 $\Delta_{sol}H_B^{\infty}$ = −20.7 J/(g polymer)

Comments: The final concentration is 0.01 g/100 ml.

Polymer (B):   **poly(methyl methacrylate)**   **1980GRA, 1984KIL**
Characterization: M_n/g.mol^{-1} = 93940, M_w/g.mol^{-1} = 101000, atactic,
                         Roehm GmbH, Darmstadt, Germany
Solvent (A):   **4-methyl-2-pentanone   C$_6$H$_{12}$O**   **108-10-1**

T/K = 303.15 $\Delta_{sol}H_B^{\infty}$ = −23.5 J/(g polymer)

Comments: The final concentration is 0.04 g/100 ml.

| **Polymer (B):** | **poly(methyl methacrylate)** | | **1980GRA, 1984KIL** |

Polymer (B):   **poly(methyl methacrylate)**   **1980GRA, 1984KIL**
Characterization: M_n/g.mol^{-1} = 137000, M_w/g.mol^{-1} = 215000, atactic,
     Roehm GmbH, Darmstadt, Germany
Solvent (A):   **4-methyl-2-pentanone** **C$_6$H$_{12}$O**   **108-10-1**

T/K = 303.15 $\Delta_{sol} H_B^\infty$ = −28.0 J/(g polymer)

Comments: The final concentration is 0.005 g/100 ml.

Polymer (B):   **poly(methyl methacrylate)**   **1956STR, 1958SLO**
Characterization: intrinsic viscosity = 1.30
Solvent (A):   **2-propanone**   **C$_3$H$_6$O**   **67-64-1**

T/K = 298.15 $\Delta_{sol} H_B^\infty$ = −29.7 J/(g polymer)

Comments: The final concentration is 1 g polymer/100 g solvent.

Polymer (B):   **poly(methyl methacrylate)**   **1980GRA, 1984KIL**
Characterization: M_n/g.mol^{-1} = 93940, M_w/g.mol^{-1} = 101000, atactic,
     Roehm GmbH, Darmstadt, Germany
Solvent (A):   **toluene**   **C$_7$H$_8$**   **108-88-3**

T/K = 303.15 $\Delta_{sol} H_B^\infty$ = −22.0 J/(g polymer)

Comments: The final concentration is 0.02 g/100 ml.

Polymer (B):   **poly(methyl methacrylate)**   **1980GRA, 1984KIL**
Characterization: M_n/g.mol^{-1} = 689000, M_w/g.mol^{-1} = 782000, atactic,
     Roehm GmbH, Darmstadt, Germany
Solvent (A):   **toluene**   **C$_7$H$_8$**   **108-88-3**

T/K = 303.15 $\Delta_{sol} H_B^\infty$ = −24.0 J/(g polymer)

Comments: The final concentration is 0.002 g/100 ml.

Polymer (B):   **poly(methyl methacrylate)**   **1972MAS**
Characterization: M_η/g.mol^{-1} = 10000, synthesized in the laboratory
Solvent (A):   **trichloromethane**   **CHCl$_3$**   **67-66-3**

T/K = 298.15 $\Delta_M H_B^\infty$ = −46.95 J/cm^3

Polymer (B):   **poly(methyl methacrylate)**   **1970GER**
Characterization: M_η/g.mol^{-1} = 12000, atactic,
     Roehm & Haas GmbH, Darmstadt, Germany
Solvent (A):   **trichloromethane**   **CHCl$_3$**   **67-66-3**

T/K = 303.15 $\Delta_{sol} H_B^\infty$ = −64.52 J/(g polymer)

Comments: The final concentration is 0.003 base mol polymer/mol solvent.

Polymer (B):	**poly(methyl methacrylate)**	**1970GER**
Characterization:	$M_\eta/\text{g.mol}^{-1} = 54000$, atactic,	
	Roehm & Haas GmbH, Darmstadt, Germany	
Solvent (A):	**trichloromethane** **CHCl$_3$**	**67-66-3**

$T/\text{K} = 303.15$ $\Delta_{sol}H_B^\infty = -80.12$ J/(g polymer)

Comments: The final concentration is 0.003 base mol polymer/mol solvent.

Polymer (B):	**poly(methyl methacrylate)**	**1970GER**
Characterization:	$M_\eta/\text{g.mol}^{-1} = 80000$, atactic,	
	Roehm & Haas GmbH, Darmstadt, Germany	
Solvent (A):	**trichloromethane** **CHCl$_3$**	**67-66-3**

$T/\text{K} = 303.15$ $\Delta_{sol}H_B^\infty = -81.42$ J/(g polymer)

Comments: The final concentration is 0.003 base mol polymer/mol solvent.

Polymer (B):	**poly(methyl methacrylate)**	**1980GRA, 1984KIL**
Characterization:	$M_n/\text{g.mol}^{-1} = 93940$, $M_w/\text{g.mol}^{-1} = 101000$, atactic,	
	Roehm GmbH, Darmstadt, Germany	
Solvent (A):	**trichloromethane** **CHCl$_3$**	**67-66-3**

$T/\text{K} = 303.15$ $\Delta_{sol}H_B^\infty = -71$ J/(g polymer)

Comments: The final concentration is 0.04 g/100 ml.

Polymer (B):	**poly(methyl methacrylate)**	**1970GER**
Characterization:	$M_\eta/\text{g.mol}^{-1} = 100000$, atactic,	
	Roehm & Haas GmbH, Darmstadt, Germany	
Solvent (A):	**trichloromethane** **CHCl$_3$**	**67-66-3**

$T/\text{K} = 303.15$ $\Delta_{sol}H_B^\infty = -84.1$ J/(g polymer)

Comments: The final concentration is 0.003 base mol polymer/mol solvent.

Polymer (B):	**poly(methyl methacrylate)**	**2003CAR**
Characterization:	$M_w/\text{g.mol}^{-1} = 120000$, atactic, $T_g/\text{K} = 377$,	
	Aldrich Chem. Co., Inc., Milwaukee, WI	
Solvent (A):	**trichloromethane** **CHCl$_3$**	**67-66-3**

$T/\text{K} = 298.15$ $\Delta_{sol}H_B^\infty = -73.2$ J/(g polymer)   (glassy state polymer)

$T/\text{K} = 298.15$ $\Delta_{M}H_B^\infty = -46.3$ J/(g polymer)   (liquid state polymer)

Comments: The final concentration is between $0.005 < w_B < 0.020$.

Polymer (B):	**poly(methyl methacrylate)**	**1970GER**
Characterization:	$M_\eta/\text{g.mol}^{-1} = 320000$, atactic,	
	Roehm & Haas GmbH, Darmstadt, Germany	
Solvent (A):	**trichloromethane** **CHCl$_3$**	**67-66-3**

$T/\text{K} = 303.15$ $\Delta_{sol} H_B^\infty = -82.8$ J/(g polymer)

Comments: The final concentration is 0.003 base mol polymer/mol solvent.

Polymer (B):	**poly(methyl methacrylate)**	**1980GRA, 1984KIL**
Characterization:	$M_n/\text{g.mol}^{-1} = 689000$, $M_w/\text{g.mol}^{-1} = 782000$, atactic,	
	Roehm GmbH, Darmstadt, Germany	
Solvent (A):	**trichloromethane** **CHCl$_3$**	**67-66-3**

$T/\text{K} = 303.15$ $\Delta_{sol} H_B^\infty = -72$ J/(g polymer)

Comments: The final concentration is 0.04 g/100 ml.

Polymer (B):	**poly(methyl methacrylate)**	**1980GRA, 1984KIL**
Characterization:	$M_w/\text{g.mol}^{-1} = 2320000$, atactic,	
	Roehm GmbH, Darmstadt, Germany	
Solvent (A):	**trichloromethane** **CHCl$_3$**	**67-66-3**

$T/\text{K} = 303.15$ $\Delta_{sol} H_B^\infty = -73$ J/(g polymer)

Comments: The final concentration is 0.04 g/100 ml.

Polymer (B):	**poly(4-methyl-1-pentene)**	**1981AHA**
Characterization:	$M_\eta/\text{g.mol}^{-1} = 350000$	
Solvent (A):	**cyclohexane** **C$_6$H$_{12}$**	**110-82-7**

$T/\text{K} = 338.15$ $\Delta_{sol} H_B^\infty = 30$ J/(g polymer)

Comments: Semicrystalline material; the specific enthalpy of fusion is 40 J/g at 513 K.

Polymer (B):	**poly(α-methylstyrene)**	**1965COT**
Characterization:	$M_n/\text{g.mol}^{-1} = 1030$, $M_w/\text{g.mol}^{-1} = 1180$,	
	synthesized in the laboratory	
Solvent (A):	**toluene** **C$_7$H$_8$**	**108-88-3**

$T/\text{K} = 298.15$ $\Delta_{sol} H_B = -837$ J/(base mol polymer)

Comments: The final concentration is about 5%.

Polymer (B):	**poly(α-methylstyrene)**	**1965COT**
Characterization:	$M_n/\text{g.mol}^{-1} = 1430$, synthesized in the laboratory	
Solvent (A):	**toluene** **C$_7$H$_8$**	**108-88-3**

$T/\text{K} = 298.15$ $\Delta_{sol} H_B = -3556$ J/(base mol polymer)

Comments: The final concentration is about 5%.

Polymer (B):   **poly(α-methylstyrene)**   **1965COT**
Characterization: M_n/g.mol^{-1} = 1820, M_w/g.mol^{-1} = 2230,
                          synthesized in the laboratory
Solvent (A):   **toluene**   **C$_7$H$_8$**   **108-88-3**

T/K = 298.15 $\Delta_{sol}H_B$ = −3975 J/(base mol polymer)

Comments: The final concentration is about 5%.

Polymer (B):   **poly(α-methylstyrene)**   **1965COT**
Characterization: M_n/g.mol^{-1} = 1920, synthesized in the laboratory
Solvent (A):   **toluene**   **C$_7$H$_8$**   **108-88-3**

T/K = 298.15 $\Delta_{sol}H_B$ = −4350 J/(base mol polymer)

Comments: The final concentration is about 5%.

Polymer (B):   **poly(α-methylstyrene)**   **1965COT**
Characterization: M_n/g.mol^{-1} = 2700, M_w/g.mol^{-1} = 3300,
                          synthesized in the laboratory
Solvent (A):   **toluene**   **C$_7$H$_8$**   **108-88-3**

T/K = 298.15 $\Delta_{sol}H_B$ = −4560 J/(base mol polymer)

Comments: The final concentration is about 5%.

Polymer (B):   **poly(α-methylstyrene)**   **1965COT**
Characterization: M_n/g.mol^{-1} = 3280, synthesized in the laboratory
Solvent (A):   **toluene**   **C$_7$H$_8$**   **108-88-3**

T/K = 298.15 $\Delta_{sol}H_B$ = −5440 J/(base mol polymer)

Comments: The final concentration is about 5%.

Polymer (B):   **poly(α-methylstyrene)**   **1965COT**
Characterization: M_n/g.mol^{-1} = 5260, synthesized in the laboratory
Solvent (A):   **toluene**   **C$_7$H$_8$**   **108-88-3**

T/K = 298.15 $\Delta_{sol}H_B$ = −5440 J/(base mol polymer)

Comments: The final concentration is about 5%.

Polymer (B):   **poly(α-methylstyrene)**   **1965COT**
Characterization: M_n/g.mol^{-1} = 8600, synthesized in the laboratory
Solvent (A):   **toluene**   **C$_7$H$_8$**   **108-88-3**

T/K = 298.15 $\Delta_{sol}H_B$ = −5355 J/(base mol polymer)

Comments: The final concentration is about 5%.

Polymer (B):	**poly(α-methylstyrene)**	**1990PED**
Characterization:	M_n/g.mol^{-1} = 10500, M_w/M_n < 1.06,	
	Polymer Laboratories, Amherst, MA	
Solvent (A):	**toluene** $\quad\quad$ **C$_7$H$_8$**	**108-88-3**

T/K = 310.15 $\quad\quad$ $\Delta_{sol}H_B$ = −8.35 J/(g polymer)

Polymer (B):	**poly(α-methylstyrene)**	**1965COT**
Characterization:	M_n/g.mol^{-1} = 12170, synthesized in the laboratory	
Solvent (A):	**toluene** $\quad\quad$ **C$_7$H$_8$**	**108-88-3**

T/K = 298.15 $\quad\quad$ $\Delta_{sol}H_B$ = −5400 J/(base mol polymer)

Comments: $\quad$ The final concentration is about 5%.

Polymer (B):	**poly(α-methylstyrene)**	**1990PED**
Characterization:	M_n/g.mol^{-1} = 53000, M_w/M_n < 1.06,	
	Polymer Laboratories, Amherst, MA	
Solvent (A):	**toluene** $\quad\quad$ **C$_7$H$_8$**	**108-88-3**

T/K = 310.15 $\quad\quad$ $\Delta_{sol}H_B$ = −12.95 J/(g polymer)

Polymer (B):	**poly(α-methylstyrene)**	**1994BRU**
Characterization:	M_n/g.mol^{-1} = 55000, M_w/M_n < 1.06,	
	Polymer Laboratories, Amherst, MA	
Solvent (A):	**toluene** $\quad\quad$ **C$_7$H$_8$**	**108-88-3**

T/K = 333.15 $\quad\quad$ $\Delta_{sol}H_B$ = −15.5 J/(g polymer)

Comments: $\quad$ The final concentration is < 1.5%.

Polymer (B):	**poly(α-methylstyrene)**	**1988LAN**
Characterization:	M_n/g.mol^{-1} = 87000, M_w/M_n < 1.06,	
	Polymer Laboratories, Amherst, MA	
Solvent (A):	**toluene** $\quad\quad$ **C$_7$H$_8$**	**108-88-3**

T/K = 298.15 $\quad\quad$ $\Delta_{sol}H_B$ = −16.55 J/(g polymer)

T/K = 333.15 $\quad\quad$ $\Delta_{sol}H_B$ = −11.00 J/(g polymer)

Polymer (B):	**poly(α-methylstyrene)**	**1990PED**
Characterization:	M_n/g.mol^{-1} = 87000, M_w/M_n < 1.06,	
	Polymer Laboratories, Amherst, MA	
Solvent (A):	**toluene** $\quad\quad$ **C$_7$H$_8$**	**108-88-3**

T/K = 310.15 $\quad\quad$ $\Delta_{sol}H_B$ = −16.30 J/(g polymer)

Polymer (B):	poly(2-methyl-5-vinyltetrazole)		**1997KI2**
Characterization:	–		
Solvent (A):	**acetic acid**	$C_2H_4O_2$	**64-19-7**

$T/K = 298.15$ $\Delta_{sol}H_B^{\infty} = 47.2$ J/(g polymer)

Polymer (B):	poly(2-methyl-5-vinyltetrazole)		**1997KI2**
Characterization:	–		
Solvent (A):	**acetonitrile**	C_2H_3N	**75-05-8**

$T/K = 298.15$ $\Delta_{sol}H_B^{\infty} = 13.6$ J/(g polymer)

Polymer (B):	poly(2-methyl-5-vinyltetrazole)		**1997KI2**
Characterization:	–		
Solvent (A):	**1,2-dichloroethane**	$C_2H_4Cl_2$	**107-06-2**

$T/K = 298.15$ $\Delta_{sol}H_B^{\infty} = 17.0$ J/(g polymer)

Polymer (B):	poly(2-methyl-5-vinyltetrazole)		**1997KI2**
Characterization:	–		
Solvent (A):	**N,N-diethylacetamide**	$C_6H_{13}NO$	**685-91-6**

$T/K = 298.15$ $\Delta_{sol}H_B^{\infty} = 17.1$ J/(g polymer)

Polymer (B):	poly(2-methyl-5-vinyltetrazole)		**1997KI2**
Characterization:	–		
Solvent (A):	**N,N-dimethylformamide**	C_3H_7NO	**68-12-2**

$T/K = 298.15$ $\Delta_{sol}H_B^{\infty} = 33.0$ J/(g polymer)

Polymer (B):	poly(2-methyl-5-vinyltetrazole)		**1997KI2**
Characterization:	–		
Solvent (A):	**dimethylsulfoxide**	C_2H_6OS	**67-68-5**

$T/K = 298.15$ $\Delta_{sol}H_B^{\infty} = 10.2$ J/(g polymer)

Polymer (B):	poly(2-methyl-5-vinyltetrazole)		**1997KI2**
Characterization:	–		
Solvent (A):	**formamide**	CH_3NO	**75-12-7**

$T/K = 298.15$ $\Delta_{sol}H_B^{\infty} = 11.9$ J/(g polymer)

Polymer (B):	poly(2-methyl-5-vinyltetrazole)		**1997KI2**
Characterization:	–		
Solvent (A):	**formic acid**	CH_2O_2	**64-18-6**

$T/K = 298.15$ $\Delta_{sol}H_B^{\infty} = 110$ J/(g polymer)

| Polymer (B): | poly(2-methyl-5-vinyltetrazole) | | 1997KI2 |

Polymer (B): poly(2-methyl-5-vinyltetrazole)   **1997KI2**
Characterization: −
Solvent (A): nitromethane CH_3NO_2 **75-52-5**

$T/K = 298.15$ $\Delta_{sol}H_B^\infty = 9.6$ J/(g polymer)

Polymer (B): poly(2-methyl-5-vinyltetrazole)   **1997KI2**
Characterization: −
Solvent (A): pyridine C_5H_5N **110-86-1**

$T/K = 298.15$ $\Delta_{sol}H_B^\infty = 16.0$ J/(g polymer)

Polymer (B): poly(octamethylene oxide)   **1975IKE**
Characterization: M_n/g.mol^{-1} = 7000
Solvent (A): benzene C_6H_6 **71-43-2**

$T/K = 298.15$	$\Delta_M H_B^\infty = 19.74$	(sample crystallized from melt at constant $T/K = 303.15$)
$T/K = 303.15$	$\Delta_M H_B^\infty = 22.07$	(sample crystallized from melt at constant $T/K = 303.15$)
$T/K = 308.15$	$\Delta_M H_B^\infty = 24.66$	(sample crystallized from melt at constant $T/K = 303.15$)
$T/K = 308.15$	$\Delta_M H_B^\infty = 25.33$	(sample crystallized from melt at constant $T/K = 313.15$)
$T/K = 308.15$	$\Delta_M H_B^\infty = 26.78$	(sample crystallized from melt at constant $T/K = 323.15$)
$T/K = 308.15$	$\Delta_M H_B^\infty = 29.08$	(sample crystallized from melt at constant $T/K = 333.15$)

Polymer (B): polypentenamer   **1979PH2**
Characterization: M_η/g.mol^{-1} = 50000, Goodyear Tire & Rubber Company, Akron, OH
Solvent (A): cyclohexane C_6H_{12} **110-82-7**

$T/K = 298.15$ $\Delta_M H_B^\infty = 4.6$ J/(g polymer)

Polymer (B): polypentenamer   **1979PH2**
Characterization: M_η/g.mol^{-1} = 50000, Goodyear Tire & Rubber Company, Akron, OH
Solvent (A): cyclooctane C_8H_{16} **292-64-8**

$T/K = 298.15$ $\Delta_M H_B^\infty = 5.1$ J/(g polymer)

Polymer (B): polypentenamer   **1979PH2**
Characterization: M_η/g.mol^{-1} = 50000, Goodyear Tire & Rubber Company, Akron, OH
Solvent (A): cyclopentane C_5H_{10} **287-92-3**

$T/K = 298.15$ $\Delta_M H_B^\infty = -2.3$ J/(g polymer)

Polymer (B):	**polypentenamer**	**1979PH2**
Characterization:	M_η/g.mol^{-1} = 50000, Goodyear Tire & Rubber Company, Akron, OH	
Solvent (A):	*cis*-**decahydronaphthalene** **C$_{10}$H$_{18}$**	**493-01-6**

T/K = 298.15 $\Delta_M H_B^\infty$ = 2.6 J/(g polymer)

Polymer (B):	**polypentenamer**	**1979PH2**
Characterization:	M_η/g.mol^{-1} = 50000, Goodyear Tire & Rubber Company, Akron, OH	
Solvent (A):	*trans*-**decahydronaphthalene** **C$_{10}$H$_{18}$**	**493-02-7**

T/K = 298.15 $\Delta_M H_B^\infty$ = <0.1 J/(g polymer)

Polymer (B):	**polypentenamer**	**1979PH2**
Characterization:	M_η/g.mol^{-1} = 50000, Goodyear Tire & Rubber Company, Akron, OH	
Solvent (A):	**3,3-diethylpentane** **C$_9$H$_{20}$**	**4032-86-4**

T/K = 298.15 $\Delta_M H_B^\infty$ = 2.4 J/(g polymer)

Polymer (B):	**polypentenamer**	**1979PH2**
Characterization:	M_η/g.mol^{-1} = 50000, Goodyear Tire & Rubber Company, Akron, OH	
Solvent (A):	**2,2-dimethylpentane** **C$_7$H$_{16}$**	**590-35-2**

T/K = 298.15 $\Delta_M H_B^\infty$ = 3.3 J/(g polymer)

Polymer (B):	**polypentenamer**	**1979PH2**
Characterization:	M_η/g.mol^{-1} = 50000, Goodyear Tire & Rubber Company, Akron, OH	
Solvent (A):	**2,3-dimethylpentane** **C$_7$H$_{16}$**	**565-59-3**

T/K = 298.15 $\Delta_M H_B^\infty$ = 2.3 J/(g polymer)

Polymer (B):	**polypentenamer**	**1979PH2**
Characterization:	M_η/g.mol^{-1} = 50000, Goodyear Tire & Rubber Company, Akron, OH	
Solvent (A):	**2,4-dimethylpentane** **C$_7$H$_{16}$**	**108-08-7**

T/K = 298.15 $\Delta_M H_B^\infty$ = 3.3 J/(g polymer)

Polymer (B):	**polypentenamer**	**1979PH2**
Characterization:	M_η/g.mol^{-1} = 50000, Goodyear Tire & Rubber Company, Akron, OH	
Solvent (A):	**3,3-dimethylpentane** **C$_7$H$_{16}$**	**562-49-2**

T/K = 298.15 $\Delta_M H_B^\infty$ = 2.7 J/(g polymer)

Polymer (B):	**polypentenamer**	**1979PH2**
Characterization:	M_η/g.mol^{-1} = 50000, Goodyear Tire & Rubber Company, Akron, OH	
Solvent (A):	**n-dodecane** **C$_{12}$H$_{26}$**	**112-40-3**

T/K = 298.15 $\Delta_M H_B^\infty$ = 2.9 J/(g polymer)

Polymer (B):   **polypentenamer**   **1979PH2**
Characterization: M_η/g.mol^{-1} = 50000, Goodyear Tire & Rubber Company, Akron, OH
Solvent (A):   **3-ethylpentane**   **C$_7$H$_{16}$**   **617-78-7**

T/K = 298.15 $\Delta_M H_B^\infty$ = 2.1 J/(g polymer)

Polymer (B):   **polypentenamer**   **1979PH2**
Characterization: M_η/g.mol^{-1} = 50000, Goodyear Tire & Rubber Company, Akron, OH
Solvent (A):   **2,2,4,4,6,8,8-heptamethylnonane**   **C$_{16}$H$_{34}$**   **4390-04-9**

T/K = 298.15 $\Delta_M H_B^\infty$ = 3.2 J/(g polymer)

Polymer (B):   **polypentenamer**   **1979PH2**
Characterization: M_η/g.mol^{-1} = 50000, Goodyear Tire & Rubber Company, Akron, OH
Solvent (A):   **n-hexadecane**   **C$_{16}$H$_{34}$**   **544-76-3**

T/K = 298.15 $\Delta_M H_B^\infty$ = 2.6 J/(g polymer)

Polymer (B):   **polypentenamer**   **1979PH2**
Characterization: M_η/g.mol^{-1} = 50000, Goodyear Tire & Rubber Company, Akron, OH
Solvent (A):   **3-methylhexane**   **C$_7$H$_{16}$**   **589-34-4**

T/K = 298.15 $\Delta_M H_B^\infty$ = 2.4 J/(g polymer)

Polymer (B):   **polypentenamer**   **1979PH2**
Characterization: M_η/g.mol^{-1} = 50000, Goodyear Tire & Rubber Company, Akron, OH
Solvent (A):   **n-octane**   **C$_8$H$_{18}$**   **111-65-9**

T/K = 298.15 $\Delta_M H_B^\infty$ = 2.2 J/(g polymer)

Polymer (B):   **polypentenamer**   **1979PH2**
Characterization: M_η/g.mol^{-1} = 50000, Goodyear Tire & Rubber Company, Akron, OH
Solvent (A):   **2,2,4,6,6-pentamethylheptane**   **C$_{12}$H$_{26}$**   **13475-82-6**

T/K = 298.15 $\Delta_M H_B^\infty$ = 3.8 J/(g polymer)

Polymer (B):   **polypentenamer**   **1979PH1, 1979PH2**
Characterization: M_η/g.mol^{-1} = 50000, Goodyear Tire & Rubber Company, Akron, OH
Solvent (A):   **2,2,4,4-tetramethylpentane**   **C$_9$H$_{20}$**   **1070-87-7**

T/K = 298.15 $\Delta_M H_B^\infty$ = 4.5 J/(g polymer)

Polymer (B):   **polypentenamer**   **1979PH2**
Characterization: M_η/g.mol^{-1} = 50000, Goodyear Tire & Rubber Company, Akron, OH
Solvent (A):   **2,3,3,4-tetramethylpentane**   **C$_9$H$_{20}$**   **16747-38-9**

T/K = 298.15 $\Delta_M H_B^\infty$ = 2.4 J/(g polymer)

Polymer (B):   **polypentenamer**   **1979PH2**
Characterization: M_η/g.mol^{-1} = 50000, Goodyear Tire & Rubber Company, Akron, OH
Solvent (A):   **2,2,4-trimethylpentane C$_8$H$_{18}$**   **540-84-1**

T/K = 298.15 $\Delta_M H_B^\infty$ = 4.3 J/(g polymer)

Polymer (B):   **poly(m-phenylene)**   **1987KOJ**
Characterization: M_n/g.mol^{-1} = 382, m-quinquephenyl
Solvent (A):   **benzene**   **C$_6$H$_6$**   **71-43-2**

T/K = 293.15 $\Delta_{sol} H_B^\infty$ = (33.2 ± 0.2) kJ/mol

Polymer (B):   **poly(m-phenylene)**   **1987KOJ**
Characterization: M_n/g.mol^{-1} = 382, m-quinquephenyl
Solvent (A):   **N,N-dimethylacetamide   C$_4$H$_9$NO**   **127-19-5**

T/K = 293.15 $\Delta_{sol} H_B^\infty$ = (17.2 ± 0.3) kJ/mol

Polymer (B):   **poly(m-phenyleneisophthalamide)**   **1972SOK**
Characterization: synthesized in the laboratory
Solvent (A):   **N,N-dimethylacetamide   C$_4$H$_9$NO**   **127-19-5**

T/K = 298.15 $\Delta_{sol} H_B$ = −170.7 J/(g polymer)   (amorphous sample)

T/K = 298.15 $\Delta_{sol} H_B$ = −127.6 J/(g polymer)   (semicrystalline sample)

Comments: The final concentration is 0.1 g polymer/15 cm^3 solvent.

Polymer (B):   **poly(m-phenyleneisophthalamide)**   **1976KOC**
Characterization: –
Solvent (A):   **N,N-dimethylacetamide   C$_4$H$_9$NO**   **127-19-5**

T/K = 293.15 $\Delta_{sol} H_B^\infty$ = −23.0 kJ/(base mol polymer)

Polymer (B):   **poly(m-phenyleneisophthalamide)**   **1972SOK**
Characterization: synthesized in the laboratory
Solvent (A):   **N,N-dimethylformamide   C$_3$H$_7$NO**   **68-12-2**

T/K = 298.15 $\Delta_{sol} H_B$ = −148.5 J/(g polymer)   (amorphous sample)

T/K = 298.15 $\Delta_{sol} H_B$ = −124.7 J/(g polymer)   (semicrystalline sample)

Comments: The final concentration is 0.1 g polymer/15 cm^3 solvent.

Polymer (B):	**poly(m-phenyleneisophthalamide)**	**1972SOK**
Characterization:	synthesized in the laboratory	
Solvent (A):	**1-methyl-2-pyrrolidinone** **C$_5$H$_9$NO**	**872-50-4**

$T/K = 298.15$ $\Delta_{sol}H_B = -176.6$ J/(g polymer)   (amorphous sample)

$T/K = 298.15$ $\Delta_{sol}H_B = -117.6$ J/(g polymer)   (semicrystalline sample)

Comments: The final concentration is 0.1 g polymer/15 cm^3 solvent.

Polymer (B):	**poly(m-phenyleneisophthalamide-*co*-terephthalamide)**	**1982STA**
Characterization:	synthesized in the laboratory, iso/ter = 3/2	
Solvent (A):	**N,N-dimethylacetamide** **C$_4$H$_9$NO**	**127-19-5**

$T/K = 299.15$ $\Delta_{sol}H_B{}^\infty = -170$ J/(g polymer)

Polymer (B):	**poly(p-phenyleneterephthalamide)**	**1986BON**
Characterization:	–	
Solvent (A):	**sulfuric acid** **H$_2$SO$_4$**	**7664-93-9**

$T/K = 293.15$ $\Delta_{sol}H_B{}^\infty = -458$ J/(base mol polymer)

Polymer (B):	**polypropylene**	**1982OCH**
Characterization:	M_η/g.mol^{-1} = 18000, atactic, fractionated in the laboratory from a commercial sample, Sumitomo Chemical Industries Ltd., Japan	
Solvent (A):	**benzene** **C$_6$H$_6$**	**71-43-2**

$T/K - 298.15$ $\Delta_M H_B{}^\infty = 30.8$ J/(g polymer)

Polymer (B):	**polypropylene**	**1967SCH**
Characterization:	MFI = 0.4	
Solvent (A):	**1-chloronaphthalene** **C$_{10}$H$_7$Cl**	**90-13-1**

T/K	373.15	383.15	393.15	403.25	413.15	423.35
$\Delta_{sol}H_B$/(J/base mol polymer)	837	1088	7113	10380	11090	11590

Comments: The final concentration is about 0.3 g polymer/100 ml solvent; semicrystalline material.

Polymer (B):	**polypropylene**	**1979PH2**
Characterization:	M_η/g.mol^{-1} = 6000, atactic	
Solvent (A):	**cyclohexane** **C$_6$H$_{12}$**	**110-82-7**

$T/K = 298.15$ $\Delta_M H_B{}^\infty = 2.3$ J/(g polymer)

Polymer (B):	**polypropylene**	**1982OCH**
Characterization:	M_η/g.mol^{-1} = 18000, atactic, fractionated in the laboratory	
Solvent (A):	**cyclohexane** **C$_6$H$_{12}$**	**110-82-7**

$T/K = 298.15$ $\Delta_M H_B{}^\infty = 3.9$ J/(g polymer)

Polymer (B): **polypropylene** 1979PH2
Characterization: M_η/g.mol^{-1} = 6000, atactic
Solvent (A): **cyclooctane** **C$_8$H$_{16}$** 292-64-8

T/K = 298.15 $\Delta_M H_B^\infty$ = 3.0 J/(g polymer)

Polymer (B): **polypropylene** 1979PH2
Characterization: M_η/g.mol^{-1} = 6000, atactic
Solvent (A): **cyclopentane** **C$_5$H$_{10}$** 287-92-3

T/K = 298.15 $\Delta_M H_B^\infty$ = −2.3 J/(g polymer)

Polymer (B): **polypropylene** 1979PH2
Characterization: M_η/g.mol^{-1} = 6000, atactic
Solvent (A): ***cis*-decahydronaphthalene** **C$_{10}$H$_{18}$** 493-01-6

T/K = 298.15 $\Delta_M H_B^\infty$ = 0.5 J/(g polymer)

Polymer (B): **polypropylene** 1979PH2
Characterization: M_η/g.mol^{-1} = 6000, atactic
Solvent (A): ***trans*-decahydronaphthalene** **C$_{10}$H$_{18}$** 493-02-7

T/K = 298.15 $\Delta_M H_B^\infty$ = −2.4 J/(g polymer)

Polymer (B): **polypropylene** 1982OCH
Characterization: M_η/g.mol^{-1} = 18000, atactic, fractionated in the laboratory from
a commercial sample, Sumitomo Chemical Industries Ltd., Japan
Solvent (A): **n-decane** **C$_{10}$H$_{22}$** 124-18-5

T/K = 298.15 $\Delta_M H_B^\infty$ = 3.1 J/(g polymer)

Polymer (B): **polypropylene** 1979PH2
Characterization: M_η/g.mol^{-1} = 6000, atactic
Solvent (A): **3,3-diethylpentane** **C$_9$H$_{20}$** 4032-86-4

T/K = 298.15 $\Delta_M H_B^\infty$ = −3.9 J/(g polymer)

Polymer (B): **polypropylene** 1982OCH
Characterization: M_η/g.mol^{-1} = 18000, atactic, fractionated in the laboratory from
a commercial sample, Sumitomo Chemical Industries Ltd., Japan
Solvent (A): **1,2-dimethylbenzene** **C$_8$H$_{10}$** 95-47-6

T/K = 298.15 $\Delta_M H_B^\infty$ = 13.4 J/(g polymer)

Polymer (B):   **polypropylene**   **1982OCH**
Characterization: $M_\eta/\text{g.mol}^{-1} = 18000$, atactic, fractionated in the laboratory from a commercial sample, Sumitomo Chemical Industries Ltd., Japan
Solvent (A):   **1,3-dimethylbenzene**   **C$_8$H$_{10}$**   **108-38-3**

$T/\text{K} = 298.15$ $\Delta_M H_B^\infty = 12.3$ J/(g polymer)

Polymer (B):   **polypropylene**   **1982OCH**
Characterization: $M_\eta/\text{g.mol}^{-1} = 18000$, atactic, fractionated in the laboratory from a commercial sample, Sumitomo Chemical Industries Ltd., Japan
Solvent (A):   **1,4-dimethylbenzene**   **C$_8$H$_{10}$**   **106-42-3**

$T/\text{K} = 298.15$ $\Delta_M H_B^\infty = 10.3$ J/(g polymer)

Polymer (B):   **polypropylene**   **1979PH2**
Characterization: $M_\eta/\text{g.mol}^{-1} = 6000$, atactic
Solvent (A):   **2,2-dimethylpentane**   **C$_7$H$_{16}$**   **590-35-2**

$T/\text{K} = 298.15$ $\Delta_M H_B^\infty = -2.2$ J/(g polymer)

Polymer (B):   **polypropylene**   **1979PH2**
Characterization: $M_\eta/\text{g.mol}^{-1} = 6000$, atactic
Solvent (A):   **2,3-dimethylpentane**   **C$_7$H$_{16}$**   **565-59-3**

$T/\text{K} = 298.15$ $\Delta_M H_B^\infty = -2.5$ J/(g polymer)

Polymer (B):   **polypropylene**   **1979PH2**
Characterization: $M_\eta/\text{g.mol}^{-1} = 6000$, atactic
Solvent (A):   **2,4-dimethylpentane**   **C$_7$H$_{16}$**   **108-08-7**

$T/\text{K} = 298.15$ $\Delta_M H_B^\infty = -1.8$ J/(g polymer)

Polymer (B):   **polypropylene**   **1979PH2**
Characterization: $M_\eta/\text{g.mol}^{-1} = 6000$, atactic
Solvent (A):   **3,3-dimethylpentane**   **C$_7$H$_{16}$**   **562-49-2**

$T/\text{K} = 298.15$ $\Delta_M H_B^\infty = -3.0$ J/(g polymer)

Polymer (B):   **polypropylene**   **1979PH2**
Characterization: $M_\eta/\text{g.mol}^{-1} = 6000$, atactic
Solvent (A):   **n-dodecane**   **C$_{12}$H$_{26}$**   **112-40-3**

$T/\text{K} = 298.15$ $\Delta_M H_B^\infty = 1.7$ J/(g polymer)

Polymer (B):	**polypropylene**	**1982OCH**
Characterization:	$M_\eta/\text{g.mol}^{-1} = 18000$, atactic, fractionated in the laboratory from	
	a commercial sample, Sumitomo Chemical Industries Ltd., Japan	
Solvent (A):	**ethylbenzene** $\quad$ **C₈H₁₀**	**100-41-4**

$T/\text{K} = 298.15$ $\quad$ $\Delta_\text{M}H_\text{B}^\infty = 14.2$ J/(g polymer)

Polymer (B):	**polypropylene**	**1979PH2**
Characterization:	$M_\eta/\text{g.mol}^{-1} = 6000$, atactic	
Solvent (A):	**3-ethylpentane** $\quad$ **C₇H₁₆**	**617-78-7**

$T/\text{K} = 298.15$ $\quad$ $\Delta_\text{M}H_\text{B}^\infty = -2.5$ J/(g polymer)

Polymer (B):	**polypropylene**	**1979PH1**
Characterization:	$M_\eta/\text{g.mol}^{-1} = 6000$, atactic	
Solvent (A):	**2,2,4,4,6,8,8-heptamethylnonane** $\quad$ **C₁₆H₃₄**	**4390-04-9**

$T/\text{K} = 298.15$ $\quad$ $\Delta_\text{M}H_\text{B}^\infty = -0.7$ J/(g polymer)

Polymer (B):	**polypropylene**	**1979PH1**
Characterization:	$M_\eta/\text{g.mol}^{-1} = 6000$, atactic	
Solvent (A):	**n-heptane** $\quad$ **C₇H₁₆**	**142-82-5**

$T/\text{K} = 298.15$ $\quad$ $\Delta_\text{M}H_\text{B}^\infty = -1.6$ J/(g polymer)

Polymer (B):	**polypropylene**	**1982OCH**
Characterization:	$M_\eta/\text{g.mol}^{-1} = 18000$, atactic, fractionated in the laboratory from	
	a commercial sample, Sumitomo Chemical Industries Ltd., Japan	
Solvent (A):	**n-heptane** $\quad$ **C₇H₁₆**	**142-82-5**

$T/\text{K} = 298.15$ $\quad$ $\Delta_\text{M}H_\text{B}^\infty = 0.5$ J/(g polymer)

Polymer (B):	**polypropylene**	**1979PH1**
Characterization:	$M_\eta/\text{g.mol}^{-1} = 6000$, atactic	
Solvent (A):	**n-hexadecane** $\quad$ **C₁₆H₃₄**	**544-76-3**

$T/\text{K} = 298.15$ $\quad$ $\Delta_\text{M}H_\text{B}^\infty = 2.3$ J/(g polymer)

Polymer (B):	**polypropylene**	**1982OCH**
Characterization:	$M_\eta/\text{g.mol}^{-1} = 18000$, atactic, fractionated in the laboratory from	
	a commercial sample, Sumitomo Chemical Industries Ltd., Japan	
Solvent (A):	**n-hexane** $\quad$ **C₆H₁₄**	**110-54-3**

$T/\text{K} = 298.15$ $\quad$ $\Delta_\text{M}H_\text{B}^\infty = -1.4$ J/(g polymer)

Polymer (B):	**polypropylene**	**1979PH2**
Characterization:	$M_\eta/\text{g.mol}^{-1} = 6000$, atactic	
Solvent (A):	**3-methylhexane** C_7H_{16}	**589-34-4**

$T/\text{K} = 298.15$ $\Delta_M H_B^\infty = -1.8$ J/(g polymer)

Polymer (B):	**polypropylene**	**1979PH1**
Characterization:	$M_\eta/\text{g.mol}^{-1} = 6000$, atactic	
Solvent (A):	**n-nonane** C_9H_{20}	**111-84-2**

$T/\text{K} = 298.15$ $\Delta_M H_B^\infty = 0.8$ J/(g polymer)

Polymer (B):	**polypropylene**	**1982OCH**
Characterization:	$M_\eta/\text{g.mol}^{-1} = 18000$, atactic, fractionated in the laboratory from a commercial sample, Sumitomo Chemical Industries Ltd., Japan	
Solvent (A):	**n-nonane** C_9H_{20}	**111-84-2**

$T/\text{K} = 298.15$ $\Delta_M H_B^\infty = 2.4$ J/(g polymer)

Polymer (B):	**polypropylene**	**1979PH1**
Characterization:	$M_\eta/\text{g.mol}^{-1} = 6000$, atactic	
Solvent (A):	**n-octane** C_8H_{18}	**111-65-9**

$T/\text{K} = 298.15$ $\Delta_M H_B^\infty = -1.2$ J/(g polymer)

Polymer (B):	**polypropylene**	**1982OCH**
Characterization:	$M_\eta/\text{g.mol}^{-1} = 18000$, atactic, fractionated in the laboratory from a commercial sample, Sumitomo Chemical Industries Ltd., Japan	
Solvent (A):	**n-octane** C_8H_{18}	**111-65-9**

$T/\text{K} = 298.15$ $\Delta_M H_B^\infty = 1.0$ J/(g polymer)

Polymer (B):	**polypropylene**	**1979PH1**
Characterization:	$M_\eta/\text{g.mol}^{-1} = 6000$, atactic	
Solvent (A):	**2,2,4,6,6-pentamethylheptane** $C_{12}H_{26}$	**13475-82-6**

$T/\text{K} = 298.15$ $\Delta_M H_B^\infty = -0.2$ J/(g polymer)

Polymer (B):	**polypropylene**	**1982OCH**
Characterization:	$M_\eta/\text{g.mol}^{-1} = 18000$, atactic, fractionated in the laboratory from a commercial sample, Sumitomo Chemical Industries Ltd., Japan	
Solvent (A):	**n-pentane** C_5H_{12}	**109-66-0**

$T/\text{K} = 298.15$ $\Delta_M H_B^\infty = -4.7$ J/(g polymer)

Polymer (B):	**polypropylene**	**1977OCH**
Characterization:	M_η/g.mol^{-1} = 18000, atactic, fractionated in the laboratory from
	a commercial sample, Sumitomo Chemical Industries Ltd., Japan
Solvent (A):	**tetrachloromethane**	**CCl$_4$**	**56-23-5**

T/K = 298.15	$\Delta_M H_B^\infty$ = 6.44 J/(g polymer)

Polymer (B):	**polypropylene**	**1982OCH**
Characterization:	M_η/g.mol^{-1} = 18000, atactic, fractionated in the laboratory from
	a commercial sample, Sumitomo Chemical Industries Ltd., Japan
Solvent (A):	**tetrachloromethane**	**CCl$_4$**	**56-23-5**

T/K = 298.15	$\Delta_M H_B^\infty$ = 6.6 J/(g polymer)

Polymer (B):	**polypropylene**	**1967SCH**
Characterization:	MFI = 0.4
Solvent (A):	**1,2,3,4-tetrahydronaphthalene**	**C$_{10}$H$_{12}$**	**119-64-2**

T/K	373.15	383.15	393.15	403.25	413.15	423.35
$\Delta_{sol} H_B$/(J/base mol polymer)	5900	9120	13810	13765	14140	12220

Comments:	The final concentration is about 0.3 g polymer/100 ml solvent; semicrystalline material.

Polymer (B):	**polypropylene**	**1979PH2**
Characterization:	M_η/g.mol^{-1} = 6000, atactic
Solvent (A):	**2,2,4,4-tetramethylpentane**	**C$_9$H$_{20}$**	**1070-87-7**

T/K = 298.15	$\Delta_M H_B^\infty$ = –0.8 J/(g polymer)

Polymer (B):	**polypropylene**	**1979PH2**
Characterization:	M_η/g.mol^{-1} = 6000, atactic
Solvent (A):	**2,3,3,4-tetramethylpentane**	**C$_9$H$_{20}$**	**16747-38-9**

T/K = 298.15	$\Delta_M H_B^\infty$ = –3.1 J/(g polymer)

Polymer (B):	**polypropylene**	**1982OCH**
Characterization:	M_η/g.mol^{-1} = 18000, atactic, fractionated in the laboratory from
	a commercial sample, Sumitomo Chemical Industries Ltd., Japan
Solvent (A):	**toluene**	**C$_7$H$_8$**	**108-88-3**

T/K = 298.15	$\Delta_M H_B^\infty$ = 16.5 J/(g polymer)

Polymer (B):	**polypropylene**	**1977OCH**
Characterization:	M_η/g.mol^{-1} = 18000, atactic, fractionated in the laboratory from
	a commercial sample, Sumitomo Chemical Industries Ltd., Japan
Solvent (A):	**trichloromethane**	**CHCl$_3$**	**67-66-3**

T/K = 298.15	$\Delta_M H_B^\infty$ = 17.15 J/(g polymer)

Polymer (B):   **polypropylene**   **1982OCH**
Characterization: $M_\eta/\text{g.mol}^{-1} = 18000$, atactic, fractionated in the laboratory from
a commercial sample, Sumitomo Chemical Industries Ltd., Japan
Solvent (A):   **trichloromethane**   **CHCl$_3$**   **67-66-3**

$T/\text{K} = 298.15$ $\Delta_M H_B^\infty = 17.2$ J/(g polymer)

Polymer (B):   **polypropylene**   **1979PH2**
Characterization: $M_\eta/\text{g.mol}^{-1} = 6000$, atactic
Solvent (A):   **2,2,4-trimethylpentane**   **C$_8$H$_{18}$**   **540-84-1**

$T/\text{K} = 298.15$ $\Delta_M H_B^\infty = -1.0$ J/(g polymer)

Polymer (B):   **poly(propylene glycol)**   **1988PAR**
Characterization: $M_n/\text{g.mol}^{-1} = 150$,
fractionated samples supplied by Union Carbide Corp.
Solvent (A):   **benzene**   **C$_6$H$_6$**   **71-43-2**

$T/\text{K} = 321.35$ $\Delta_M H_B^\infty = 31$ kJ/(mol polymer)

Polymer (B):   **poly(propylene glycol)**   **1988PAR**
Characterization: $M_n/\text{g.mol}^{-1} = 425$,
fractionated samples supplied by Union Carbide Corp.
Solvent (A):   **benzene**   **C$_6$H$_6$**   **71-43-2**

$T/\text{K} = 321.35$ $\Delta_M H_B^\infty = 34$ kJ/(mol polymer)

Polymer (B):   **poly(propylene glycol)**   **1988PAR**
Characterization: $M_n/\text{g.mol}^{-1} = 2025$,
fractionated samples supplied by Union Carbide Corp.
Solvent (A):   **benzene**   **C$_6$H$_6$**   **71-43-2**

$T/\text{K} = 321.35$ $\Delta_M H_B^\infty = 88$ kJ/(mol polymer)

Polymer (B):   **poly(propylene glycol)**   **1988PAR**
Characterization: $M_n/\text{g.mol}^{-1} = 150$,
fractionated samples supplied by Union Carbide Corp.
Solvent (A):   **ethanol**   **C$_2$H$_6$O**   **64-17-5**

$T/\text{K} = 321.35$ $\Delta_M H_B^\infty = 6.5$ kJ/(mol polymer)

Polymer (B):   **poly(propylene glycol)**   **1988PAR**
Characterization: $M_n/\text{g.mol}^{-1} = 425$,
fractionated samples supplied by Union Carbide Corp.
Solvent (A):   **ethanol**   **C$_2$H$_6$O**   **64-17-5**

$T/\text{K} = 321.35$ $\Delta_M H_B^\infty = 27$ kJ/(mol polymer)

Polymer (B):   **poly(propylene glycol)**   **1988PAR**
Characterization: $M_n/\text{g.mol}^{-1} = 2025$,
                         fractionated samples supplied by Union Carbide Corp.
Solvent (A):   **ethanol**   **C_2H_6O**   **64-17-5**

$T/\text{K} = 321.35$ $\Delta_M H_B^\infty = 135$ kJ/(mol polymer)

Polymer (B):   **poly(propylene glycol)**   **1993ZEL**
Characterization: $M_n/\text{g.mol}^{-1} = 396$, $M_w/\text{g.mol}^{-1} = 412$, $\rho = 1.0042$ g/cm^3 (298 K)
Solvent (A):   **tetrachloromethane**   **CCl_4**   **56-23-5**

$T/\text{K} = 303.15$ $\Delta_M H_B^\infty = 4.68$ J/(g polymer)

$T/\text{K} = 318.15$ $\Delta_M H_B^\infty = 5.17$ J/(g polymer)

Polymer (B):   **poly(propylene glycol)**   **1993ZEL**
Characterization: $M_n/\text{g.mol}^{-1} = 1900$, $\rho = 1.0003$ g/cm^3 (298 K)
Solvent (A):   **tetrachloromethane**   **CCl_4**   **56-23-5**

$T/\text{K} = 303.15$ $\Delta_M H_B^\infty = -8.2$ J/(g polymer)

$T/\text{K} = 318.15$ $\Delta_M H_B^\infty = 11.22$ J/(g polymer)

Polymer (B):   **poly(propylene glycol)**   **1993ZEL**
Characterization: $M_n/\text{g.mol}^{-1} = 1900$, $\rho = 1.0003$ g/cm^3 (298 K)
Solvent (A):   **trichloromethane**   **$CHCl_3$**   **67-66-3**

$T/\text{K} = 303.15$ $\Delta_M H_B^\infty = -80.97$ J/(g polymer)

Polymer (B):   **poly(propylene glycol)**   **1988PAR**
Characterization: $M_n/\text{g.mol}^{-1} = 150$,
                         fractionated samples supplied by Union Carbide Corp.
Solvent (A):   **water**   **H_2O**   **7732-18-5**

$T/\text{K} = 321.35$ $\Delta_M H_B^\infty = -14$ kJ/(mol polymer)

Polymer (B):   **poly(propylene glycol)**   **1995CAR**
Characterization: $M_n/\text{g.mol}^{-1} = 400$, Fluka AG, Buchs, Switzerland
Solvent (A):   **water**   **H_2O**   **7732-18-5**

$T/\text{K} = 298.15$ $\Delta_M H_B^\infty = -165$ J/(g polymer)

Polymer (B):   **poly(propylene glycol)**   **1988PAR**
Characterization: $M_n/\text{g.mol}^{-1} = 425$,
                         fractionated samples supplied by Union Carbide Corp.
Solvent (A):   **water**   **H_2O**   **7732-18-5**

$T/\text{K} = 321.35$ $\Delta_M H_B^\infty = -40$ kJ/(mol polymer)

Polymer (B):	**polystyrene**		**1950TAG, 1951TA2**
Characterization:	–		
Solvent (A):	**benzene**	**C₆H₆**	**71-43-2**

$T/K = 298.15$ $\Delta_{sol}H_B^\infty = -10.3$ J/(g polymer)

Comments: The final concentration is about 1 g polymer/100 g solution.

Polymer (B):	**polystyrene**		**1970MO5**
Characterization:	M_w/g.mol⁻¹ = 600, $M_w/M_n \leq 1.10$,		
	Pressure Chemical Co., Pittsburgh, PA		
Solvent (A):	**benzene**	**C₆H₆**	**71-43-2**

$T/K = 298.15$ $\Delta_M H_B^\infty = -133$ J/(base mol polymer)

$T/K = 313.15$ $\Delta_M H_B^\infty = -265$ J/(base mol polymer)

Polymer (B):	**polystyrene**		**1955TA3**
Characterization:	M_η/g.mol⁻¹ = 784		
Solvent (A):	**benzene**	**C₆H₆**	**71-43-2**

$T/K = 298.15$ $\Delta_{sol}H_B^\infty = 0.0$ J/(g polymer)

Polymer (B):	**polystyrene**		**1970MO5**
Characterization:	M_w/g.mol⁻¹ = 900, $M_w/M_n \leq 1.10$,		
	Pressure Chemical Co., Pittsburgh, PA		
Solvent (A):	**benzene**	**C₆H₆**	**71-43-2**

$T/K = 291.15$ $\Delta_M H_B^\infty = -1038$ J/(base mol polymer)

$T/K = 318.15$ $\Delta_M H_B^\infty = -607$ J/(base mol polymer)

Polymer (B):	**polystyrene**		**1970MO5**
Characterization:	M_w/g.mol⁻¹ = 2000, $M_w/M_n \leq 1.10$,		
	Pressure Chemical Co., Pittsburgh, PA		
Solvent (A):	**benzene**	**C₆H₆**	**71-43-2**

$T/K = 291.15$ $\Delta_M H_B^\infty = -1720$ J/(base mol polymer)

$T/K = 318.15$ $\Delta_M H_B^\infty = -711$ J/(base mol polymer)

Polymer (B):	**polystyrene**		**1970MO5**
Characterization:	M_w/g.mol⁻¹ = 5000, $M_w/M_n \leq 1.10$,		
	Pressure Chemical Co., Pittsburgh, PA		
Solvent (A):	**benzene**	**C₆H₆**	**71-43-2**

$T/K = 291.15$ $\Delta_M H_B^\infty = -2372$ J/(base mol polymer)

$T/K = 318.15$ $\Delta_M H_B^\infty = -1226$ J/(base mol polymer)

Polymer (B):	**polystyrene**		**1970MO5**
Characterization:	M_w/g.mol^{-1} = 10300, $M_w/M_n \leq 1.10$,		
	Pressure Chemical Co., Pittsburgh, PA		
Solvent (A):	**benzene**	**C$_6$H$_6$**	**71-43-2**

T/K = 291.15 $\Delta_M H_B^\infty$ = −2674 J/(base mol polymer)

T/K = 318.15 $\Delta_M H_B^\infty$ = −1916 J/(base mol polymer)

Polymer (B):	**polystyrene**		**1955TA3**
Characterization:	M_η/g.mol^{-1} = 18000		
Solvent (A):	**benzene**	**C$_6$H$_6$**	**71-43-2**

T/K = 298.15 $\Delta_{sol}H_B^\infty$ = −4.06 J/(g polymer)

Polymer (B):	**polystyrene**		**1951HEL, 1952SC1**
Characterization:	M_n/g.mol^{-1} = 20000		
Solvent (A):	**benzene**	**C$_6$H$_6$**	**71-43-2**

T/K = 296.15 $\Delta_{sol}H_B^\infty$ = −15.44 J/(g polymer)

Polymer (B):	**polystyrene**		**1955TA3**
Characterization:	M_η/g.mol^{-1} = 29000		
Solvent (A):	**benzene**	**C$_6$H$_6$**	**71-43-2**

T/K = 298.15 $\Delta_{sol}H_B^\infty$ = −5.0 J/(g polymer)

Polymer (B):	**polystyrene**		**1955TA3**
Characterization:	M_η/g.mol^{-1} = 30000		
Solvent (A):	**benzene**	**C$_6$H$_6$**	**71-43-2**

T/K = 298.15 $\Delta_{sol}H_B^\infty$ = −7.45 J/(g polymer)

Polymer (B):	**polystyrene**		**1955TA3**
Characterization:	M_η/g.mol^{-1} = 59000		
Solvent (A):	**benzene**	**C$_6$H$_6$**	**71-43-2**

T/K = 298.15 $\Delta_{sol}H_B^\infty$ = −12.8 J/(g polymer)

Polymer (B):	**polystyrene**		**1955TA3**
Characterization:	M_η/g.mol^{-1} = 91000		
Solvent (A):	**benzene**	**C$_6$H$_6$**	**71-43-2**

T/K = 298.15 $\Delta_{sol}H_B^\infty$ = −14.6 J/(g polymer)

Polymer (B):	**polystyrene**	**1970MO5**
Characterization:	$M_w/\text{g.mol}^{-1} = 97200$, $M_w/M_n \leq 1.10$, Pressure Chemical Co., Pittsburgh, PA	
Solvent (A):	**benzene** $\quad$ **C$_6$H$_6$**	**71-43-2**

$T/\text{K} = 318.15$ $\quad$ $\Delta_M H_B^\infty = -2180$ J/(base mol polymer)

Polymer (B):	**polystyrene**	**1954OYA**
Characterization:	$M_n/\text{g.mol}^{-1} = 135000$	
Solvent (A):	**benzene** $\quad$ **C$_6$H$_6$**	**71-43-2**

$T/\text{K} = 296.15$ $\quad$ $\Delta_{sol} H_B^\infty = -23.4$ J/cm^3

Polymer (B):	**polystyrene**	**1955TA3**
Characterization:	$M_\eta/\text{g.mol}^{-1} = 142000$	
Solvent (A):	**benzene** $\quad$ **C$_6$H$_6$**	**71-43-2**

$T/\text{K} = 298.15$ $\quad$ $\Delta_{sol} H_B^\infty = -16.7$ J/(g polymer)

Polymer (B):	**polystyrene**	**1969BIA**
Characterization:	$M_\eta/\text{g.mol}^{-1} = 190000$	
Solvent (A):	**benzene** $\quad$ **C$_6$H$_6$**	**71-43-2**

$T/\text{K} = 303.15$ $\quad$ $\Delta_{sol} H_B^\infty = -1841$ J/(base mol polymer)

Polymer (B):	**polystyrene**	**1985AEL**
Characterization:	$M_n/\text{g.mol}^{-1} = 214000$	
Solvent (A):	**benzene** $\quad$ **C$_6$H$_6$**	**71-43-2**

$T/\text{K} = 300.15$ $\quad$ $\Delta_{sol} H_B^\infty = -16.21$ J/(g polymer)

Polymer (B):	**polystyrene**	**1955TA3**
Characterization:	$M_\eta/\text{g.mol}^{-1} = 216000$	
Solvent (A):	**benzene** $\quad$ **C$_6$H$_6$**	**71-43-2**

$T/\text{K} = 298.15$ $\quad$ $\Delta_{sol} H_B^\infty = -18.4$ J/(g polymer)

Polymer (B):	**polystyrene**	**1955TA3**
Characterization:	$M_\eta/\text{g.mol}^{-1} = 272000$	
Solvent (A):	**benzene** $\quad$ **C$_6$H$_6$**	**71-43-2**

$T/\text{K} = 298.15$ $\quad$ $\Delta_{sol} H_B^\infty = -21.3$ J/(g polymer)

Polymer (B):	**polystyrene**		**1953TAG**
Characterization:	M_η/g.mol^{-1} = 300000		
Solvent (A):	**benzene**	**C$_6$H$_6$**	**71-43-2**

T/K = 298.15 $\Delta_{sol}H_B^\infty$ = −21.3 J/(g polymer)

Polymer (B):	**polystyrene**		**1956STR, 1958SLO**
Characterization:	intrinsic viscosity = 2.90		
Solvent (A):	**benzene**	**C$_6$H$_6$**	**71-43-2**

T/K = 298.15 $\Delta_{sol}H_B^\infty$ = −27.4 J/(g polymer)

Comments: The final concentration is 1 g polymer/100 g solvent.

Polymer (B):	**polystyrene**		**1951HEL, 1952SC1**
Characterization:	M_n/g.mol^{-1} = 20000		
Solvent (A):	**2-butanone**	**C$_4$H$_8$O**	**78-93-3**

T/K = 296.15 $\Delta_{sol}H_B^\infty$ = −15.3 J/(g polymer)

Polymer (B):	**polystyrene**		**1958TA4**
Characterization:	M_η/g.mol^{-1} = 142000		
Solvent (A):	**2-butanone**	**C$_4$H$_8$O**	**78-93-3**

T/K = 298.15 $\Delta_{sol}H_B^\infty$ = −16.6 J/(g polymer)

Polymer (B):	**polystyrene**		**1951HEL, 1952SC1**
Characterization:	M_n/g.mol^{-1} = 20000		
Solvent (A):	**butyl acetate**	**C$_6$H$_{12}$O$_2$**	**123-86-4**

T/K = 296.15 $\Delta_{sol}H_B^\infty$ = −13.22 J/(g polymer)

Polymer (B):	**polystyrene**		**1969BIA**
Characterization:	M_η/g.mol^{-1} = 190000		
Solvent (A):	**butylbenzene**	**C$_{10}$H$_{14}$**	**104-51-8**

T/K = 303.15 $\Delta_{sol}H_B^\infty$ = −1423 J/(base mol polymer)

Polymer (B):	**polystyrene**		**1956JEN**
Characterization:	−		
Solvent (A):	**chlorobenzene**	**C$_6$H$_5$Cl**	**108-90-7**

T/K = 293.15 $\Delta_{sol}H_B^\infty$ = −4100 J/(base mol polymer)

Polymer (B):	**polystyrene**	**1968MA2**

Characterization: M_n/g.mol^{-1} = 150000

Solvent (A):   **chlorobenzene**   **C$_6$H$_5$Cl**   **108-90-7**

T/K = 293.15 $\Delta_{sol} H_B^\infty$ = -32.3 J/(g polymer)

Comments: $\Delta_{sol} H_B$ was found to be independent of concentration between φ_B = 0.0699 and 0.868.

Polymer (B):   **polystyrene**   **1959HOR**

Characterization: M_η/g.mol^{-1} = 266000, Monsanto Canada

Solvent (A):   **chlorobenzene**   **C$_6$H$_5$Cl**   **108-90-7**

T/K = 298.15 $\Delta_{sol} H_B^\infty$ = 5.4 J/(g polymer)

Comments: $\Delta_M H_B^\infty$ is corrected for the enthalpy from the glass transition.

Polymer (B):   **polystyrene**   **1956JEN**

Characterization: –

Solvent (A):   **cyclohexane**   **C$_6$H$_{12}$**   **110-82-7**

T/K = 293.15 $\Delta_{sol} H_B^\infty$ = -1464 J/(base mol polymer)

Polymer (B):   **polystyrene**   **1955SCH**

Characterization: M_n/g.mol^{-1} = 1260, T_g/K = 317.15

Solvent (A):   **cyclohexane**   **C$_6$H$_{12}$**   **110-82-7**

T/K = 298.15 $\Delta_{sol} H_B^\infty$ = 10.4 J/(g polymer)

Comments: The final concentration is 1 g polymer/100 cm^3 solvent.

Polymer (B):   **polystyrene**   **1955SCH**

Characterization: M_n/g.mol^{-1} = 1910, T_g/K = 329.15

Solvent (A):   **cyclohexane**   **C$_6$H$_{12}$**   **110-82-7**

T/K = 298.15 $\Delta_{sol} H_B^\infty$ = 5.4 J/(g polymer)

Comments: The final concentration is 1 g polymer/100 cm^3 solvent.

Polymer (B):   **polystyrene**   **1955SCH**

Characterization: M_n/g.mol^{-1} = 3160, T_g/K = 341.15

Solvent (A):   **cyclohexane**   **C$_6$H$_{12}$**   **110-82-7**

T/K = 298.15 $\Delta_{sol} H_B^\infty$ = -5.4 J/(g polymer)

Comments: The final concentration is 1 g polymer/100 cm^3 solvent.

Polymer (B):   **polystyrene**   **1955SCH**
Characterization: M_n/g.mol^{-1} = 3980, T_g/K = 353.15
Solvent (A):   **cyclohexane**   **C$_6$H$_{12}$**   **110-82-7**

T/K = 298.15 $\Delta_{sol}H_B^\infty$ = –6.86 J/(g polymer)

Comments: The final concentration is 1 g polymer/100 cm^3 solvent.

Polymer (B):   **polystyrene**   **1955SCH**
Characterization: M_n/g.mol^{-1} = 5630, T_g/K = 357.15
Solvent (A):   **cyclohexane**   **C$_6$H$_{12}$**   **110-82-7**

T/K = 298.15 $\Delta_{sol}H_B^\infty$ = –9.3 J/(g polymer)

Comments: The final concentration is 1 g polymer/100 cm^3 solvent.

Polymer (B):   **polystyrene**   **1955SCH**
Characterization: M_n/g.mol^{-1} = 9070, T_g/K = 361.15
Solvent (A):   **cyclohexane**   **C$_6$H$_{12}$**   **110-82-7**

T/K = 298.15 $\Delta_{sol}H_B^\infty$ = –10.8 J/(g polymer)

Comments: The final concentration is 1 g polymer/100 cm^3 solvent.

Polymer (B):   **polystyrene**   **1951HEL, 1952SC1**
Characterization: M_n/g.mol^{-1} = 20000
Solvent (A):   **cyclohexane**   **C$_6$H$_{12}$**   **110-82-7**

T/K = 296.15 $\Delta_{sol}H_B^\infty$ = 2.47 J/(g polymer)

Polymer (B):   **polystyrene**   **1969BIA**
Characterization: M_η/g.mol^{-1} = 190000
Solvent (A):   **cyclohexane**   **C$_6$H$_{12}$**   **110-82-7**

T/K = 303.15 $\Delta_{sol}H_B^\infty$ = –222 J/(base mol polymer)

Polymer (B):   **polystyrene**   **1951HEL, 1952SC1**
Characterization: M_n/g.mol^{-1} = 20000
Solvent (A):   **cyclohexene**   **C$_6$H$_{10}$**   **110-83-8**

T/K = 296.15 $\Delta_{sol}H_B^\infty$ = –9.37 J/(g polymer)

Polymer (B):   **polystyrene**   **1956STR, 1958SLO**
Characterization: M_n/g.mol^{-1} = 22400
Solvent (A):   **cyclohexanone**   **C$_6$H$_{10}$O**   **108-94-1**

T/K = 298.15 $\Delta_{sol}H_B^\infty$ = –28.8 J/(g polymer)

Comments: The final concentration is 1 g polymer/100 g solvent.

Polymer (B):   **polystyrene**   **1969BIA**
Characterization: $M_\eta/\text{g.mol}^{-1} = 190000$
Solvent (A):   **decahydronaphthalene** $C_{10}H_{18}$ **91-17-8**

$T/K = 303.15$ $\Delta_{sol} H_B^\infty = 397$ J/(base mol polymer)

Polymer (B):   **polystyrene**   **1988AUK**
Characterization: $M_w/\text{g.mol}^{-1} = 115000$, $M_w/M_n < 1.05$
Solvent (A):   **1,2-dichlorobenzene** $C_6H_4Cl_2$ **95-50-1**

$T/K = 303.05$ $\Delta_{sol} H_B = 25.9$ J/(g polymer)

Polymer (B):   **polystyrene**   **1951HEL, 1952SC1**
Characterization: $M_n/\text{g.mol}^{-1} = 20000$
Solvent (A):   **1,2-dimethylbenzene** C_8H_{10} **95-47-6**

$T/K = 296.15$ $\Delta_{sol} H_B^\infty = -13.4$ J/(g polymer)

Polymer (B):   **polystyrene**   **1951HEL, 1952SC1**
Characterization: $M_n/\text{g.mol}^{-1} = 20000$
Solvent (A):   **1,3-dimethylbenzene** C_8H_{10} **108-38-3**

$T/K = 296.15$ $\Delta_{sol} H_B^\infty = -12.05$ J/(g polymer)

Polymer (B):   **polystyrene**   **1969BIA**
Characterization: $M_\eta/\text{g.mol}^{-1} = 190000$
Solvent (A):   **1,3-dimethylbenzene** C_8H_{10} **108-38-3**

$T/K = 303.15$ $\Delta_{sol} H_B^\infty = -1570$ J/(base mol polymer)

Polymer (B):   **polystyrene**   **1954OYA**
Characterization: $M_n/\text{g.mol}^{-1} = 135000$
Solvent (A):   **1,4-dimethylbenzene** C_8H_{10} **106-42-3**

$T/K = 296.15$ $\Delta_{sol} H_B^\infty = -13.6$ J/cm^3

Polymer (B):   **polystyrene**   **1969BIA**
Characterization: $M_\eta/\text{g.mol}^{-1} = 190000$
Solvent (A):   **1,4-dioxane** $C_4H_8O_2$ **123-91-1**

$T/K = 303.15$ $\Delta_{sol} H_B^\infty = -1255$ J/(base mol polymer)

Polymer (B):   **polystyrene**   **1951HEL, 1952SC1**
Characterization: $M_n/\text{g.mol}^{-1} = 20000$
Solvent (A):   **ethyl acetate** $C_4H_8O_2$ **141-78-6**

$T/K = 296.15$ $\Delta_{sol} H_B^\infty = -10.97$ J/(g polymer)

Polymer (B):	**polystyrene**		**1958TA4**
Characterization:	$M_\eta/\text{g.mol}^{-1} = 142000$		
Solvent (A):	**ethyl acetate**	$C_4H_8O_2$	**141-78-6**

$T/\text{K} = 298.15$ $\Delta_{sol}H_B^\infty = -12.64$ J/(g polymer)

Polymer (B):	**polystyrene**		**1955TA3**
Characterization:	$M_\eta/\text{g.mol}^{-1} = 784$		
Solvent (A):	**ethylbenzene**	C_8H_{10}	**100-41-4**

$T/\text{K} = 298.15$ $\Delta_{sol}H_B^\infty = 0.0$ J/(g polymer)

Polymer (B):	**polystyrene**		**1955GAT**
Characterization:	$M_\eta/\text{g.mol}^{-1} = 784$		
Solvent (A):	**ethylbenzene**	C_8H_{10}	**100-41-4**

$T/\text{K} = 298.15$ $\Delta_{sol}H_B^\infty = 0.0$ J/(g polymer)

Polymer (B):	**polystyrene**		**1955GAT**
Characterization:	$M_\eta/\text{g.mol}^{-1} = 18000$		
Solvent (A):	**ethylbenzene**	C_8H_{10}	**100-41-4**

$T/\text{K} = 298.15$ $\Delta_{sol}H_B^\infty = -3.76$ J/(g polymer)

Polymer (B):	**polystyrene**		**1955TA3**
Characterization:	$M_\eta/\text{g.mol}^{-1} = 18000$		
Solvent (A):	**ethylbenzene**	C_8H_{10}	**100-41-4**

$T/\text{K} = 298.15$ $\Delta_{sol}H_B^\infty = -3.93$ J/(g polymer)

Polymer (B):	**polystyrene**		**1955TA3**
Characterization:	$M_\eta/\text{g.mol}^{-1} = 30000$		
Solvent (A):	**ethylbenzene**	C_8H_{10}	**100-41-4**

$T/\text{K} = 298.15$ $\Delta_{sol}H_B^\infty = -5.69$ J/(g polymer)

Polymer (B):	**polystyrene**		**1955TA3**
Characterization:	$M_\eta/\text{g.mol}^{-1} = 35000$		
Solvent (A):	**ethylbenzene**	C_8H_{10}	**100-41-4**

$T/\text{K} = 298.15$ $\Delta_{sol}H_B^\infty = -6.53$ J/(g polymer)

Polymer (B):	**polystyrene**		**1955TA3**
Characterization:	$M_\eta/\text{g.mol}^{-1} = 48000$		
Solvent (A):	**ethylbenzene**	C_8H_{10}	**100-41-4**

$T/\text{K} = 298.15$ $\Delta_{sol}H_B^\infty = -8.37$ J/(g polymer)

Polymer (B):	**polystyrene**		**1972MA1**
Characterization:	M_n/g.mol^{-1} = 60000		
Solvent (A):	**ethylbenzene**	**C$_8$H$_{10}$**	**100-41-4**

T/K = 303.15 $\qquad$ $\Delta_{sol} H_B^\infty$ = −21.75 J/(g polymer)

Polymer (B):	**polystyrene**		**1955TA3**
Characterization:	M_η/g.mol^{-1} = 91000		
Solvent (A):	**ethylbenzene**	**C$_8$H$_{10}$**	**100-41-4**

T/K = 298.15 $\qquad$ $\Delta_{sol} H_B^\infty$ = −11.17 J/(g polymer)

Polymer (B):	**polystyrene**		**1974FI2**
Characterization:	M_n/g.mol^{-1} = 113000, M_w/g.mol^{-1} = 122000		
Solvent (A):	**ethylbenzene**	**C$_8$H$_{10}$**	**100-41-4**

T/K	306.65	317.15	337.15	347.15	350.65	366.65	367.15
$\Delta_{sol} H_B^\infty$/(J/g polymer)	−24.48	−19.37	−10.54	−6.40	−4.94	−2.30	−2.09

T/K	368.65	372.15	378.15	385.15
$\Delta_{sol} H_B^\infty$/(J/g polymer)	−2.64	−4.18	−4.60	−5.65

Polymer (B):	**polystyrene**		**1955TA3**
Characterization:	M_η/g.mol^{-1} = 142000		
Solvent (A):	**ethylbenzene**	**C$_8$H$_{10}$**	**100-41-4**

T/K = 298.15 $\qquad$ $\Delta_{sol} H_B^\infty$ = −13.4 J/(g polymer)

Polymer (B):	**polystyrene**		**1968MA2**
Characterization:	M_n/g.mol^{-1} = 150000		
Solvent (A):	**ethylbenzene**	**C$_8$H$_{10}$**	**100-41-4**

T/K = 293.15 $\qquad$ $\Delta_{sol} H_B^\infty$ = −34.2 J/(g polymer)

Comments: $\qquad$ $\Delta_{sol} H_B$ was found to be independent of concentration between φ_B = 0.0699 and 0.868.

Polymer (B):	**polystyrene**		**1969BIA**
Characterization:	M_η/g.mol^{-1} = 190000		
Solvent (A):	**ethylbenzene**	**C$_8$H$_{10}$**	**100-41-4**

T/K = 303.15 $\qquad$ $\Delta_{sol} H_B^\infty$ = −1570 J/(base mol polymer)

Polymer (B):	**polystyrene**		**1955TA3**
Characterization:	M_η/g.mol^{-1} = 216000		
Solvent (A):	**ethylbenzene**	**C$_8$H$_{10}$**	**100-41-4**

T/K = 298.15 $\qquad$ $\Delta_{sol} H_B^\infty$ = −16.5 J/(g polymer)

Polymer (B):	**polystyrene**	**1952TAG**
Characterization:	–	
Solvent (A):	**ethylbenzene** C_8H_{10}	**100-41-4**

$T/K = 298.15$ $\Delta_{sol}H_B^\infty = -16.7$ J/(g polymer)

Comments: The final concentration is about 1%.

Polymer (B):	**polystyrene**	**1955TA3**
Characterization:	$M_\eta/\text{g.mol}^{-1} = 272000$	
Solvent (A):	**ethylbenzene** C_8H_{10}	**100-41-4**

$T/K = 298.15$ $\Delta_{sol}H_B^\infty = -18.2$ J/(g polymer)

Polymer (B):	**polystyrene**	**1955GAT**
Characterization:	$M_\eta/\text{g.mol}^{-1} = 272000$	
Solvent (A):	**ethylbenzene** C_8H_{10}	**100-41-4**

$T/K = 298.15$ $\Delta_{sol}H_B^\infty = -18.4$ J/(g polymer)

Polymer (B):	**polystyrene**	**1963TAG**
Characterization:	$M_\eta/\text{g.mol}^{-1} = 413000$	
Solvent (A):	**ethylbenzene** C_8H_{10}	**100-41-4**

$T/K = 298.15$ $\Delta_{sol}H_B^\infty = -24.1$ J/(g polymer)

Polymer (B):	**polystyrene**	**1956JEN**
Characterization:	–	
Solvent (A):	**ethylbenzene** C_8H_{10}	**100-41-4**

$T/K = 293.15$ $\Delta_{sol}H_B^\infty = -3556$ J/(base mol polymer)

Polymer (B):	**polystyrene**	**1958KAR**
Characterization:	–	
Solvent (A):	**ethylbenzene** C_8H_{10}	**100-41-4**

$T/K = 298.15$ $\Delta_{sol}H_B^\infty = -29.7$ J/(g polymer)

Polymer (B):	**polystyrene**	**1955SCH**
Characterization:	$M_n/\text{g.mol}^{-1} = 1260,\ T_g/K = 317.15$	
Solvent (A):	**2-propanone** C_3H_6O	**67-64-1**

$T/K = 298.15$ $\Delta_{sol}H_B^\infty = -0.59$ J/(g polymer)

Comments: The final concentration is 1 g polymer/100 cm^3 solvent.

Polymer (B):   **polystyrene**   **1955SCH**
Characterization: M_n/g.mol^{-1} = 1910, T_g/K = 329.15
Solvent (A):   **2-propanone**   **C$_3$H$_6$O**   **67-64-1**

T/K = 298.15 $\Delta_{sol} H_B^\infty$ = −7.74 J/(g polymer)

Comments: The final concentration is 1 g polymer/100 cm^3 solvent.

Polymer (B):   **polystyrene**   **1955SCH**
Characterization: M_n/g.mol^{-1} = 3160, T_g/K = 341.15
Solvent (A):   **2-propanone**   **C$_3$H$_6$O**   **67-64-1**

T/K = 298.15 $\Delta_{sol} H_B^\infty$ = −15.9 J/(g polymer)

Comments: The final concentration is 1 g polymer/100 cm^3 solvent.

Polymer (B):   **polystyrene**   **1955SCH**
Characterization: M_n/g.mol^{-1} = 3980, T_g/K = 353.15
Solvent (A):   **2-propanone**   **C$_3$H$_6$O**   **67-64-1**

T/K = 298.15 $\Delta_{sol} H_B^\infty$ = −17.4 J/(g polymer)

Comments: The final concentration is 1 g polymer/100 cm^3 solvent.

Polymer (B):   **polystyrene**   **1955SCH**
Characterization: M_n/g.mol^{-1} = 5630, T_g/K = 357.15
Solvent (A):   **2-propanone**   **C$_3$H$_6$O**   **67-64-1**

T/K = 298.15 $\Delta_{sol} H_B^\infty$ = −19.3 J/(g polymer)

Comments: The final concentration is 1 g polymer/100 cm^3 solvent.

Polymer (B):   **polystyrene**   **1955SCH**
Characterization: M_n/g.mol^{-1} = 9070, T_g/K = 361.15
Solvent (A):   **2-propanone**   **C$_3$H$_6$O**   **67-64-1**

T/K = 298.15 $\Delta_{sol} H_B^\infty$ = −21.4 J/(g polymer)

Comments: The final concentration is 1 g polymer/100 cm^3 solvent.

Polymer (B):   **polystyrene**   **1951HEL, 1952SC1**
Characterization: M_n/g.mol^{-1} = 20000
Solvent (A):   **2-propanone**   **C$_3$H$_6$O**   **67-64-1**

T/K = 296.15 $\Delta_{sol} H_B^\infty$ = −10.5 J/(g polymer)

Polymer (B):	polystyrene		**1969BIA**
Characterization:	M_η/g.mol^{-1} = 190000		
Solvent (A):	propylbenzene	C$_9$H$_{12}$	**103-65-1**

T/K = 303.15 $\Delta_{sol}H_B^\infty$ = −1420 J/(base mol polymer)

Polymer (B):	polystyrene		**1947ROB**
Characterization:	−		
Solvent (A):	styrene	C$_8$H$_8$	**100-42-5**

T/K = 298.15 $\Delta_{sol}H_B$ = −34.7 J/(g polymer)

Comments: The final concentration is 6.9 wt%.

Polymer (B):	polystyrene		**1951HEL, 1952SC1**
Characterization:	M_n/g.mol^{-1} = 20000		
Solvent (A):	styrene	C$_8$H$_8$	**100-42-5**

T/K = 296.15 $\Delta_{sol}H_B^\infty$ = −17.95 J/(g polymer)

Polymer (B):	polystyrene		**1963TAG**
Characterization:	M_η/g.mol^{-1} = 413000		
Solvent (A):	tetrachloromethane	CCl$_4$	**56-23-5**

T/K = 298.15 $\Delta_{sol}H_B^\infty$ = −22.13 J/(g polymer)

Polymer (B):	polystyrene		**1970MO5**
Characterization:	M_w/g.mol^{-1} = 600, $M_w/M_n \leq$ 1.10,		
	Pressure Chemical Co., Pittsburgh, PA		
Solvent (A):	toluene	C$_7$H$_8$	**108-88-3**

T/K = 298.15 $\Delta_M H_B^\infty$ = −150 J/(base mol polymer)

T/K = 313.15 $\Delta_M H_B^\infty$ = −333 J/(base mol polymer)

Polymer (B):	polystyrene		**1955SCH**
Characterization:	M_n/g.mol^{-1} = 800, T_g/K = 293.15		
Solvent (A):	toluene	C$_7$H$_8$	**108-88-3**

T/K	296.15	309.15	318.15
$\Delta_M H_B^\infty$/(J/g polymer)	−2.13	−1.84	−1.51

Polymer (B):	polystyrene			1970MO5
Characterization:	M_w/g.mol^{-1} = 900, $M_w/M_n \le 1.10$,			
	Pressure Chemical Co., Pittsburgh, PA			
Solvent (A):	**toluene**	**C_7H_8**		**108-88-3**

T/K = 291.15 $\Delta_M H_B^\infty$ = −761.5 J/(base mol polymer)

T/K = 318.15 $\Delta_M H_B^\infty$ = −686.2 J/(base mol polymer)

Polymer (B):	polystyrene				1955SCH
Characterization:	M_n/g.mol^{-1} = 1260, T_g/K = 317.15				
Solvent (A):	**toluene**	**C_7H_8**			**108-88-3**

T/K	296.15	303.15	309.15	
$\Delta_{sol} H_B^\infty$/(J/g polymer)	−10.75	−8.03	−5.86	

T/K	318.15	328.15	338.15	346.65
$\Delta_M H_B^\infty$/(J/g polymer)	−3.39	−2.34	−1.88	−1.25

Comments: The final concentration is 1 g polymer/100 cm^3 solvent.

Polymer (B):	polystyrene			1955SCH
Characterization:	M_n/g.mol^{-1} = 1910, T_g/K = 329.15			
Solvent (A):	**toluene**	**C_7H_8**		**108-88-3**

T/K	298.15	318.15
$\Delta_{sol} H_B^\infty$/(J/g polymer)	−15.82	−6.65

T/K	338.15	348.15
$\Delta_M H_B^\infty$/(J/g polymer)	−3.47	−2.51

Comments: The final concentration is 1 g polymer/100 cm^3 solvent.

Polymer (B):	polystyrene			1970MO5
Characterization:	M_w/g.mol^{-1} = 2000, $M_w/M_n \le 1.10$,			
	Pressure Chemical Co., Pittsburgh, PA			
Solvent (A):	**toluene**	**C_7H_8**		**108-88-3**

T/K = 291.15 $\Delta_M H_B^\infty$ = −1167 J/(base mol polymer)

T/K = 318.15 $\Delta_M H_B^\infty$ = −753 J/(base mol polymer)

Polymer (B):	polystyrene			1955SCH
Characterization:	M_n/g.mol^{-1} = 3160, T_g/K = 341.15			
Solvent (A):	**toluene**	**C_7H_8**		**108-88-3**

T/K = 298.15 $\Delta_{sol} H_B^\infty$ = −22.8 J/(g polymer)

Comments: The final concentration is 1 g polymer/100 cm^3 solvent.

Polymer (B):	**polystyrene**		**1955SCH**
Characterization:	M_n/g.mol^{-1} = 3980, T_g/K = 353.15		
Solvent (A):	**toluene**	**C$_7$H$_8$**	**108-88-3**

T/K	298.15	318.15	338.15
$\Delta_{sol}H_B^\infty$/(J/g polymer)	−24.4	−17.1	−9.83

Comments: The final concentration is 1 g polymer/100 cm^3 solvent.

Polymer (B):	**polystyrene**		**1970MO5**
Characterization:	M_w/g.mol^{-1} = 5000, M_w/M_n ≤ 1.10,		
	Pressure Chemical Co., Pittsburgh, PA		
Solvent (A):	**toluene**	**C$_7$H$_8$**	**108-88-3**

T/K = 291.15 $\Delta_M H_B^\infty$ = −2180 J/(base mol polymer)

T/K = 318.15 $\Delta_M H_B^\infty$ = −1138 J/(base mol polymer)

Polymer (B):	**polystyrene**		**1955SCH**
Characterization:	M_n/g.mol^{-1} = 5630, T_g/K = 357.15		
Solvent (A):	**toluene**	**C$_7$H$_8$**	**108-88-3**

T/K = 298.15 $\Delta_{sol}H_B^\infty$ = −25.9 J/(g polymer)

Comments: The final concentration is 1 g polymer/100 cm^3 solvent.

Polymer (B):	**polystyrene**		**1955SCH**
Characterization:	M_n/g.mol^{-1} = 9070, T_g/K = 361.15		
Solvent (A):	**toluene**	**C$_7$H$_8$**	**108-88-3**

T/K = 298.15 $\Delta_{sol}H_B^\infty$ = −28.45 J/(g polymer)

Comments: The final concentration is 1 g polymer/100 cm^3 solvent.

Polymer (B):	**polystyrene**		**1990PED**
Characterization:	M_n/g.mol^{-1} = 9000, M_w/M_n < 1.04,		
	Polymer Laboratories, Amherst, MA		
Solvent (A):	**toluene**	**C$_7$H$_8$**	**108-88-3**

T/K = 310.15 $\Delta_{sol}H_B$ = −9.20 J/(g polymer)

Polymer (B):	**polystyrene**		**1970MO5**
Characterization:	M_w/g.mol^{-1} = 10300, M_w/M_n ≤ 1.10,		
	Pressure Chemical Co., Pittsburgh, PA		
Solvent (A):	**toluene**	**C$_7$H$_8$**	**108-88-3**

T/K = 291.15 $\Delta_M H_B^\infty$ = −2515 J/(base mol polymer)

T/K = 318.15 $\Delta_M H_B^\infty$ = −1594 J/(base mol polymer)

Polymer (B):	**polystyrene**		**1951HEL, 1952SC1**
Characterization:	M_n/g.mol^{-1} = 20000		
Solvent (A):	**toluene**	**C$_7$H$_8$**	**108-88-3**

T/K = 296.15 $\Delta_{sol} H_B^\infty = -16.7$ J/(g polymer)

Polymer (B):	**polystyrene**		**1988LAN**
Characterization:	M_n/g.mol^{-1} = 20400, M_w/M_n < 1.07,		
	Polymer Laboratories, Amherst, MA		
Solvent (A):	**toluene**	**C$_7$H$_8$**	**108-88-3**

T/K = 298.15 $\Delta_{sol} H_B = -8.20$g polymer)

T/K = 333.15 $\Delta_{sol} H_B = -6.35$g polymer)

Polymer (B):	**polystyrene**		**1990PED**
Characterization:	M_n/g.mol^{-1} = 20400, M_w/M_n < 1.07,		
	Polymer Laboratories, Amherst, MA		
Solvent (A):	**toluene**	**C$_7$H$_8$**	**108-88-3**

T/K = 310.15 $\Delta_{sol} H_B = -8.35$ J/(g polymer)

Polymer (B):	**polystyrene**		**1990PED**
Characterization:	M_n/g.mol^{-1} = 47000, M_w/M_n < 1.06,		
	Polymer Laboratories, Amherst, MA		
Solvent (A):	**toluene**	**C$_7$H$_8$**	**108-88-3**

T/K = 310.15 $\Delta_{sol} H_B = -5.00$ J/(g polymer)

Polymer (B):	**polystyrene**		**1994BRU**
Characterization:	M_n/g.mol^{-1} = 50000, M_w/M_n < 1.06,		
	Polymer Laboratories, Amherst, MA		
Solvent (A):	**toluene**	**C$_7$H$_8$**	**108-88-3**

T/K = 333.15 $\Delta_{sol} H_B = -6.8$ J/(g polymer)

Comments: The final concentration is < 1.5%.

Polymer (B):	**polystyrene**		**1970MO5**
Characterization:	M_w/g.mol^{-1} = 97200, M_w/M_n ≤ 1.10,		
	Pressure Chemical Co., Pittsburgh, PA		
Solvent (A):	**toluene**	**C$_7$H$_8$**	**108-88-3**

T/K = 318.15 $\Delta_M H_B^\infty = -1770$ J/(base mol polymer)

Polymer (B):	**polystyrene**		**1972MA1**
Characterization:	$M_n/\text{g.mol}^{-1} = 60000$		
Solvent (A):	**toluene**	**C₇H₈**	**108-88-3**

$T/\text{K} = 303.15$ $\Delta_{sol}H_B^\infty = -21.4$ J/(g polymer)

Polymer (B):	**polystyrene**		**1974FI2**
Characterization:	$M_n/\text{g.mol}^{-1} = 113000$, $M_w/\text{g.mol}^{-1} = 122000$		
Solvent (A):	**toluene**	**C₇H₈**	**108-88-3**

T/K	304.15	306.15	306.65	316.15	333.15	337.15	346.15
$\Delta_{sol}H_B^\infty$/(J/g polymer)	−28.9	−26.8	−25.9	−22.5	−15.0	−12.4	−8.28

T/K	347.15	348.15	350.65	359.15	362.15	369.15	372.15
$\Delta_{sol}H_B^\infty$/(J/g polymer)	−7.82	−8.20	−6.53	−4.27	−2.72	−3.30	−2.76

Polymer (B):	**polystyrene**		**1990PED**
Characterization:	$M_n/\text{g.mol}^{-1} = 115000$, $M_w/M_n < 1.06$,		
	Polymer Laboratories, Amherst, MA		
Solvent (A):	**toluene**	**C₇H₈**	**108-88-3**

$T/\text{K} = 310.15$ $\Delta_{sol}H_B = -5.00$ J/(g polymer)

Polymer (B):	**polystyrene**		**1954OYA**
Characterization:	$M_n/\text{g.mol}^{-1} = 135000$		
Solvent (A):	**toluene**	**C₇H₈**	**108-88-3**

$T/\text{K} = 296.15$ $\Delta_{sol}H_B^\infty = -25.56$ J/cm³

Polymer (B):	**polystyrene**		**1968MA2**
Characterization:	$M_n/\text{g.mol}^{-1} = 150000$		
Solvent (A):	**toluene**	**C₇H₈**	**108-88-3**

$T/\text{K} = 293.15$ $\Delta_{sol}H_B^\infty = -34.05$ J/(g polymer)

Comments: $\Delta_{sol}H_B$ was found to be independent of concentration between $\varphi_B = 0.0699$ and 0.868.

Polymer (B):	**polystyrene**		**1969BIA**
Characterization:	$M_\eta/\text{g.mol}^{-1} = 190000$		
Solvent (A):	**toluene**	**C₇H₈**	**108-88-3**

$T/\text{K} = 303.15$ $\Delta_{sol}H_B^\infty = -1883$ J/(base mol polymer)

Polymer (B):	**polystyrene**		**1985AEL**
Characterization:	$M_n/\text{g.mol}^{-1} = 214000$		
Solvent (A):	**toluene**	**C₇H₈**	**108-88-3**

$T/\text{K} = 300.15$ $\Delta_{sol}H_B^\infty = -19.17$ J/(g polymer)

Polymer (B):   **polystyrene**   **1967BIA**
Characterization: M_w/g.mol^{-1} = 250000, fractionated in the laboratory
Solvent (A):   **toluene**   **C$_7$H$_8$**   **108-88-3**

T/K = 303.15 $\Delta_{sol} H_B^\infty$ = −1862 J/(base mol polymer)

Comments: Additional data for samples of different thermal histories are given in 1967BIA.

Polymer (B):   **polystyrene**   **1955SCH**
Characterization: M_n/g.mol^{-1} = 270000, T_g/K = 371.15
Solvent (A):   **toluene**   **C$_7$H$_8$**   **108-88-3**

T/K = 298.15 $\Delta_{sol} H_B^\infty$ = −32.64 J/(g polymer)

Comments: The final concentration is 1 g polymer/100 cm^3 solvent.

Polymer (B):   **polystyrene**   **1956JEN**
Characterization: −
Solvent (A):   **toluene**   **C$_7$H$_8$**   **108-88-3**

T/K = 293.15 $\Delta_{sol} H_B^\infty$ = −3580 J/(base mol polymer)

Polymer (B):   **polystyrene**   **1958AKH**
Characterization: −
Solvent (A):   **toluene**   **C$_7$H$_8$**   **108-88-3**

T/K =	298.65	308.15	318.15	333.15	343.15	353.15
$\Delta_{sol} H_B$/(J/g polymer)	−38.9	−33.9	−30.1	−23.0	−13.0	−12.6

Comments: Completely amorphous material.

Polymer (B):   **polystyrene**   **1968ALL**
Characterization: commercial product, ρ = 1.0496 g/cm^3
Solvent (A):   **toluene**   **C$_7$H$_8$**   **108-88-3**

T/K = 303.15 $\Delta_{sol} H_B$ = −26.8 J/(g polymer)

Comments: The final concentration is 15.6 g polymer/l.

Polymer (B):   **polystyrene**   **1970STO**
Characterization: −
Solvent (A):   **toluene**   **C$_7$H$_8$**   **108-88-3**

T/K = 298.15 $\Delta_{sol} H_B^\infty$ = −29.2 J/(g polymer)

T/K = 318.15 $\Delta_{sol} H_B^\infty$ = −21.4 J/(g polymer)

Comments: Completely amorphous material.

Polymer (B):   **polystyrene**   **1970MO5**
Characterization: M_w/g.mol^{-1} = 600, $M_w/M_n \leq 1.10$,
                             Pressure Chemical Co., Pittsburgh, PA
Solvent (A):   **trichloromethane   CHCl$_3$**   **67-66-3**

T/K = 298.15 $\Delta_M H_B^\infty$ = −1305 J/(base mol polymer)

T/K = 313.15 $\Delta_M H_B^\infty$ = −1025 J/(base mol polymer)

Polymer (B):   **polystyrene**   **1970MO5**
Characterization: M_w/g.mol^{-1} = 900, $M_w/M_n \leq 1.10$,
                             Pressure Chemical Co., Pittsburgh, PA
Solvent (A):   **trichloromethane   CHCl$_3$**   **67-66-3**

T/K = 291.15 $\Delta_M H_B^\infty$ = −2251 J/(base mol polymer)

T/K = 318.15 $\Delta_M H_B^\infty$ = −1523 J/(base mol polymer)

Polymer (B):   **polystyrene**   **1970MO5**
Characterization: M_w/g.mol^{-1} = 2000, $M_w/M_n \leq 1.10$,
                             Pressure Chemical Co., Pittsburgh, PA
Solvent (A):   **trichloromethane   CHCl$_3$**   **67-66-3**

T/K = 291.15 $\Delta_M H_B^\infty$ = −2883 J/(base mol polymer)

T/K = 318.15 $\Delta_M H_B^\infty$ = −1669 J/(base mol polymer)

Polymer (B):   **polystyrene**   **1970MO5**
Characterization: M_w/g.mol^{-1} = 5000, $M_w/M_n \leq 1.10$,
                             Pressure Chemical Co., Pittsburgh, PA
Solvent (A):   **trichloromethane   CHCl$_3$**   **67-66-3**

T/K = 291.15 $\Delta_M H_B^\infty$ = −3150 J/(base mol polymer)

T/K = 318.15 $\Delta_M H_B^\infty$ = −1920 J/(base mol polymer)

Polymer (B):   **polystyrene**   **1970MO5**
Characterization: M_w/g.mol^{-1} = 10300, $M_w/M_n \leq 1.10$,
                             Pressure Chemical Co., Pittsburgh, PA
Solvent (A):   **trichloromethane   CHCl$_3$**   **67-66-3**

T/K = 291.15 $\Delta_M H_B^\infty$ = −3410 J/(base mol polymer)

T/K = 318.15 $\Delta_M H_B^\infty$ = −2377 J/(base mol polymer)

Polymer (B):   **polystyrene**   **1952SC1**
Characterization: M_n/g.mol^{-1} = 20000
Solvent (A):   **trichloromethane   CHCl$_3$**   **67-66-3**

T/K = 296.15 $\Delta_{sol} H_B^\infty$ = −18.4 J/(g polymer)

Polymer (B):   **polystyrene**   **1956STR, 1958SLO**
Characterization: $M_n/\text{g.mol}^{-1} = 22400$
Solvent (A):   **trichloromethane**   **CHCl₃**   **67-66-3**

$T/\text{K} = 298.15$ $\Delta_{sol} H_B^\infty = -16.57$ J/(g polymer)

Comments: The final concentration is 1 g polymer/100 g solvent.

Polymer (B):   **polystyrene**   **1970MO5**
Characterization: $M_w/\text{g.mol}^{-1} = 97200$, $M_w/M_n \le 1.10$,
                     Pressure Chemical Co., Pittsburgh, PA
Solvent (A):   **trichloromethane**   **CHCl₃**   **67-66-3**

$T/\text{K} = 318.15$ $\Delta_M H_B^\infty = -2560$ J/(base mol polymer)

Polymer (B):   **polystyrene**   **1954OYA**
Characterization: $M_n/\text{g.mol}^{-1} = 135000$
Solvent (A):   **trichloromethane**   **CHCl₃**   **67-66-3**

$T/\text{K} = 296.15$ $\Delta_{sol} H_B^\infty = -28.28$ J/cm³

Polymer (B):   **polystyrene**   **2003CAR**
Characterization: $M_w/\text{g.mol}^{-1} = 250000$, $T_g/\text{K} = 375$,
                     synthesized in the laboratory by cationic polymerization
Solvent (A):   **trichloromethane**   **CHCl₃**   **67-66-3**

$T/\text{K} = 298.15$ $\Delta_{sol} H_B^\infty = -25.6$ J/(g polymer)   (glassy state polymer)

$T/\text{K} = 298.15$ $\Delta_M H_B^\infty = -2.5$ J/(g polymer)   (liquid state polymer)

Comments: The final concentration is between $0.005 < w_B < 0.020$.

Polymer (B):   **polystyrene**   **1951HEL, 1952SC1**
Characterization: $M_n/\text{g.mol}^{-1} = 20000$
Solvent (A):   **1,3,5-trimethylbenzene C₉H₁₂**   **108-67-8**

$T/\text{K} = 296.15$ $\Delta_{sol} H_B^\infty = -10.75$ J/(g polymer)

Polymer (B):   **polystyrene**   **1969BIA**
Characterization: $M_\eta/\text{g.mol}^{-1} = 190000$
Solvent (A):   **1,3,5-trimethylbenzene C₉H₁₂**   **108-67-8**

$T/\text{K} = 303.15$ $\Delta_{sol} H_B^\infty = -1400$ J/(base mol polymer)

Polymer (B):   **poly(styrene-*co*-acrylonitrile)**   **2003CAR**
Characterization: M_n/g.mol^{-1} = 95000, M_w/g.mol^{-1} = 153800, 4.5 wt% acrylonitrile,
                      T_g/K = 379.75, ENICHEM, Mantova, Italy
Solvent (A):   **trichloromethane**   **CHCl$_3$**   **67-66-3**

T/K = 298.15 $\Delta_{sol}H_B^\infty$ = -50.3 J/(g polymer)   (glassy state polymer)

T/K = 298.15 $\Delta_M H_B^\infty$ = -25.0 J/(g polymer)   (liquid state polymer)

Comments: The final concentration is between 0.005 < w_B < 0.020.

Polymer (B):   **poly(styrene-*co*-acrylonitrile)**   **2003CAR**
Characterization: M_n/g.mol^{-1} = 73500, M_w/g.mol^{-1} = 161600, 10.5 wt% acrylonitrile,
                      T_g/K = 382.25, ENICHEM, Mantova, Italy
Solvent (A):   **trichloromethane**   **CHCl$_3$**   **67-66-3**

T/K = 298.15 $\Delta_{sol}H_B^\infty$ = -58.7 J/(g polymer)   (glassy state polymer)

T/K = 298.15 $\Delta_M H_B^\infty$ = -30.1 J/(g polymer)   (liquid state polymer)

Comments: The final concentration is between 0.005 < w_B < 0.020.

Polymer (B):   **poly(styrene-*co*-acrylonitrile)**   **2003CAR**
Characterization: M_n/g.mol^{-1} = 86550, M_w/g.mol^{-1} = 142700, 15.6 wt% acrylonitrile,
                      T_g/K = 382.95, ENICHEM, Mantova, Italy
Solvent (A):   **trichloromethane**   **CHCl$_3$**   **67-66-3**

T/K = 298.15 $\Delta_{sol}H_B^\infty$ = -61.8 J/(g polymer)   (glassy state polymer)

T/K = 298.15 $\Delta_M H_B^\infty$ = -29.6 J/(g polymer)   (liquid state polymer)

Comments: The final concentration is between 0.005 < w_B < 0.020.

Polymer (B):   **poly(styrene-*co*-acrylonitrile)**   **2003CAR**
Characterization: M_n/g.mol^{-1} = 77950, M_w/g.mol^{-1} = 132400, 19.4 wt% acrylonitrile,
                      T_g/K = 384.85, ENICHEM, Mantova, Italy
Solvent (A):   **trichloromethane**   **CHCl$_3$**   **67-66-3**

T/K = 298.15 $\Delta_{sol}H_B^\infty$ = -67.4 J/(g polymer)   (glassy state polymer)

T/K = 298.15 $\Delta_M H_B^\infty$ = -32.7 J/(g polymer)   (liquid state polymer)

Comments: The final concentration is between 0.005 < w_B < 0.020.

Polymer (B):   **poly(styrene-*co*-acrylonitrile)**   **2003CAR**
Characterization: M_n/g.mol^{-1} = 48000, M_w/g.mol^{-1} = 78600, 25.0 wt% acrylonitrile,
                      T_g/K = 385.85, ENICHEM, Mantova, Italy
Solvent (A):   **trichloromethane**   **CHCl$_3$**   **67-66-3**

T/K = 298.15 $\Delta_{sol}H_B^\infty$ = -68.1 J/(g polymer)   (glassy state polymer)

T/K = 298.15 $\Delta_M H_B^\infty$ = -32.1 J/(g polymer)   (liquid state polymer)

Comments: The final concentration is between 0.005 < w_B < 0.020.

Polymer (B):   **poly(styrene-*co*-acrylonitrile)**   **2003CAR**
Characterization: M_n/g.mol^{-1} = 55500, M_w/g.mol^{-1} = 99900, 29.9 wt% acrylonitrile,
                               T_g/K = 386.15, ENICHEM, Mantova, Italy
Solvent (A):   **trichloromethane**   **CHCl$_3$**   **67-66-3**

T/K = 298.15 $\Delta_{sol} H_B^\infty$ = –66.8 J/(g polymer)   (glassy state polymer)

T/K = 298.15 $\Delta_M H_B^\infty$ = –30.7 J/(g polymer)   (liquid state polymer)

Comments: The final concentration is between 0.005 < w_B < 0.020.

Polymer (B):   **poly(styrene-*co*-acrylonitrile)**   **2003CAR**
Characterization: M_n/g.mol^{-1} = 46700, M_w/g.mol^{-1} = 78400, 33.8 wt% acrylonitrile,
                               T_g/K = 386.25, ENICHEM, Mantova, Italy
Solvent (A):   **trichloromethane**   **CHCl$_3$**   **67-66-3**

T/K = 298.15 $\Delta_{sol} H_B^\infty$ = –73.4 J/(g polymer)   (glassy state polymer)

T/K = 298.15 $\Delta_M H_B^\infty$ = –38.2 J/(g polymer)   (liquid state polymer)

Comments: The final concentration is between 0.005 < w_B < 0.020.

Polymer (B):   **poly(styrene-*co*-acrylonitrile)**   **2003CAR**
Characterization: M_n/g.mol^{-1} = 50700, M_w/g.mol^{-1} = 90300, 36.9 wt% acrylonitrile,
                               T_g/K = 386.35, ENICHEM, Mantova, Italy
Solvent (A):   **trichloromethane**   **CHCl$_3$**   **67-66-3**

T/K = 298.15 $\Delta_{sol} H_B^\infty$ = –71.8 J/(g polymer)   (glassy state polymer)

T/K = 298.15 $\Delta_M H_B^\infty$ = –36.5 J/(g polymer)   (liquid state polymer)

Comments: The final concentration is between 0.005 < w_B < 0.020.

Polymer (B):   **poly(tetramethylene oxide)**   **1985SHA**
Characterization: M_n/g.mol^{-1} = 650, Quaker Oats Corporation
Solvent (A):   **benzene**   **C$_6$H$_6$**   **71-43-2**

T/K = 313.15 $\Delta_M H_B^\infty$ = 2610 J/mol

Polymer (B):   **poly(tetramethylene oxide)**   **1985SHA**
Characterization: M_n/g.mol^{-1} = 1000, Quaker Oats Corporation
Solvent (A):   **benzene**   **C$_6$H$_6$**   **71-43-2**

T/K = 313.15 $\Delta_M H_B^\infty$ = 1975 J/mol

Polymer (B):   **poly(tetramethylene oxide)**   **1985SHA**
Characterization: $M_n/\text{g.mol}^{-1} = 2000$, Quaker Oats Corporation
Solvent (A):   **benzene**   **C$_6$H$_6$**   **71-43-2**

$T/\text{K} = 313.15$ $\Delta_M H_B^{\infty} = 2120$ J/mol

Polymer (B):   **poly(tetramethylene oxide)**   **1987SHA**
Characterization: $M_n/\text{g.mol}^{-1} = 650$, Quaker Oats Corporation
Solvent (A):   **1,2-dichloroethane**   **C$_2$H$_4$Cl$_2$**   **107-06-2**

$T/\text{K} = 313.15$ $\Delta_M H_B^{\infty} = 2000$ J/mol

Polymer (B):   **poly(tetramethylene oxide)**   **1987SHA**
Characterization: $M_n/\text{g.mol}^{-1} = 1000$, Quaker Oats Corporation
Solvent (A):   **1,2-dichloroethane**   **C$_2$H$_4$Cl$_2$**   **107-06-2**

$T/\text{K} = 313.15$ $\Delta_M H_B^{\infty} = 680$ J/mol

Polymer (B):   **poly(tetramethylene oxide)**   **1986SHA**
Characterization: $M_n/\text{g.mol}^{-1} = 650$, Quaker Oats Corporation
Solvent (A):   **1,2-dimethylbenzene**   **C$_8$H$_{10}$**   **95-47-6**

$T/\text{K} = 313.15$ $\Delta_M H_B^{\infty} = 3820$ J/mol

Polymer (B):   **poly(tetramethylene oxide)**   **1986SHA**
Characterization: $M_n/\text{g.mol}^{-1} = 1000$, Quaker Oats Corporation
Solvent (A):   **1,2-dimethylbenzene**   **C$_8$H$_{10}$**   **95-47-6**

$T/\text{K} = 313.15$ $\Delta_M H_B^{\infty} = 1820$ J/mol

Polymer (B):   **poly(tetramethylene oxide)**   **1986SHA**
Characterization: $M_n/\text{g.mol}^{-1} = 2000$, Quaker Oats Corporation
Solvent (A):   **1,2-dimethylbenzene**   **C$_8$H$_{10}$**   **95-47-6**

$T/\text{K} = 313.15$ $\Delta_M H_B^{\infty} = 1880$ J/mol

Polymer (B):   **poly(tetramethylene oxide)**   **1986SHA**
Characterization: $M_n/\text{g.mol}^{-1} = 650$, Quaker Oats Corporation
Solvent (A):   **1,3-dimethylbenzene**   **C$_8$H$_{10}$**   **108-38-3**

$T/\text{K} = 313.15$ $\Delta_M H_B^{\infty} = 4140$ J/mol

Polymer (B):   **poly(tetramethylene oxide)**   **1986SHA**
Characterization: $M_n/\text{g.mol}^{-1} = 1000$, Quaker Oats Corporation
Solvent (A):   **1,3-dimethylbenzene**   **C$_8$H$_{10}$**   **108-38-3**

$T/\text{K} = 313.15$ $\Delta_M H_B^{\infty} = 585$ J/mol

Polymer (B):   **poly(tetramethylene oxide)**   **1986SHA**
Characterization: M_n/g.mol^{-1} = 2000, Quaker Oats Corporation
Solvent (A):   **1,3-dimethylbenzene**   **C$_8$H$_{10}$**   **108-38-3**

T/K = 313.15 $\Delta_M H_B^\infty$ = 1615 J/mol

Polymer (B):   **poly(tetramethylene oxide)**   **1986SHA**
Characterization: M_n/g.mol^{-1} = 650, Quaker Oats Corporation
Solvent (A):   **1,4-dimethylbenzene**   **C$_8$H$_{10}$**   **106-42-3**

T/K = 313.15 $\Delta_M H_B^\infty$ = 2785 J/mol

Polymer (B):   **poly(tetramethylene oxide)**   **1986SHA**
Characterization: M_n/g.mol^{-1} = 1000, Quaker Oats Corporation
Solvent (A):   **1,4-dimethylbenzene**   **C$_8$H$_{10}$**   **106-42-3**

T/K = 313.15 $\Delta_M H_B^\infty$ = 1840 J/mol

Polymer (B):   **poly(tetramethylene oxide)**   **1986SHA**
Characterization: M_n/g.mol^{-1} = 2000, Quaker Oats Corporation
Solvent (A):   **1,4-dimethylbenzene**   **C$_8$H$_{10}$**   **106-42-3**

T/K = 313.15 $\Delta_M H_B^\infty$ = 1345 J/mol

Polymer (B):   **poly(tetramethylene oxide)**   **1982SH1**
Characterization: M_n/g.mol^{-1} = 650, Quaker Oats Corporation
Solvent (A):   **1,4-dioxane**   **C$_4$H$_8$O$_2$**   **123-91-1**

T/K = 321.35 $\Delta_M H_B^\infty$ = 2600 J/mol

Polymer (B):   **poly(tetramethylene oxide)**   **1982SH1**
Characterization: M_n/g.mol^{-1} = 1000, Quaker Oats Corporation
Solvent (A):   **1,4-dioxane**   **C$_4$H$_8$O$_2$**   **123-91-1**

T/K = 321.35 $\Delta_M H_B^\infty$ = 2300 J/mol

Polymer (B):   **poly(tetramethylene oxide)**   **1982SH1**
Characterization: M_n/g.mol^{-1} = 2000, Quaker Oats Corporation
Solvent (A):   **1,4-dioxane**   **C$_4$H$_8$O$_2$**   **123-91-1**

T/K = 321.35 $\Delta_M H_B^\infty$ = 2000 J/mol

Polymer (B):   **poly(tetramethylene oxide)**   **1985SHA**
Characterization: M_n/g.mol^{-1} = 650, Quaker Oats Corporation
Solvent (A):   **ethylbenzene**   **C$_8$H$_{10}$**   **100-41-4**

T/K = 313.15 $\Delta_M H_B^\infty$ = 4480 J/mol

Polymer (B):   **poly(tetramethylene oxide)**   **1985SHA**
Characterization: M_n/g.mol^{-1} = 1000, Quaker Oats Corporation
Solvent (A):   **ethylbenzene**   **C$_8$H$_{10}$**   **100-41-4**

T/K = 313.15 $\Delta_M H_B^\infty$ = 3440 J/mol

Polymer (B):   **poly(tetramethylene oxide)**   **1985SHA**
Characterization: M_n/g.mol^{-1} = 2000, Quaker Oats Corporation
Solvent (A):   **ethylbenzene**   **C$_8$H$_{10}$**   **100-41-4**

T/K = 313.15 $\Delta_M H_B^\infty$ = −95.6 J/mol

Polymer (B):   **poly(tetramethylene oxide)**   **1985SHA**
Characterization: M_n/g.mol^{-1} = 650, Quaker Oats Corporation
Solvent (A):   **propylbenzene**   **C$_9$H$_{12}$**   **103-65-1**

T/K = 313.15 $\Delta_M H_B^\infty$ = 3770 J/mol

Polymer (B):   **poly(tetramethylene oxide)**   **1985SHA**
Characterization: M_n/g.mol^{-1} = 1000, Quaker Oats Corporation
Solvent (A):   **propylbenzene**   **C$_9$H$_{12}$**   **103-65-1**

T/K = 313.15 $\Delta_M H_B^\infty$ = 1300 J/mol

Polymer (B):   **poly(tetramethylene oxide)**   **1985SHA**
Characterization: M_n/g.mol^{-1} = 2000, Quaker Oats Corporation
Solvent (A):   **propylbenzene**   **C$_9$H$_{12}$**   **103-65-1**

T/K = 313.15 $\Delta_M H_B^\infty$ = 1820 J/mol

Polymer (B):   **poly(tetramethylene oxide)**   **1987SHA**
Characterization: M_n/g.mol^{-1} = 650, Quaker Oats Corporation
Solvent (A):   **tetrachloromethane**   **CCl$_4$**   **56-23-5**

T/K = 313.15 $\Delta_M H_B^\infty$ = 2140 J/mol

Polymer (B):   **poly(tetramethylene oxide)**   **1982SH2**
Characterization: M_n/g.mol^{-1} = 1000, Quaker Oats Corporation
Solvent (A):   **tetrachloromethane**   **CCl$_4$**   **56-23-5**

T/K = 321.35 $\Delta_M H_B^\infty$ = 1400 J/mol

Polymer (B):   **poly(tetramethylene oxide)**   **1982SH2**
Characterization: M_n/g.mol^{-1} = 2000, Quaker Oats Corporation
Solvent (A):   **tetrachloromethane**   **CCl$_4$**   **56-23-5**

T/K = 321.35 $\Delta_M H_B^\infty$ = 1300 J/mol

Polymer (B):	**poly(tetramethylene oxide)**	**1985SHA**
Characterization:	$M_n/\text{g.mol}^{-1} = 650$, Quaker Oats Corporation	
Solvent (A):	**toluene** $\quad$ **C₇H₈**	**108-88-3**

$T/\text{K} = 313.15 \qquad \Delta_M H_B^\infty = 2765$ J/mol

Polymer (B):	**poly(tetramethylene oxide)**	**1985SHA**
Characterization:	$M_n/\text{g.mol}^{-1} = 1000$, Quaker Oats Corporation	
Solvent (A):	**toluene** $\quad$ **C₇H₈**	**108-88-3**

$T/\text{K} = 313.15 \qquad \Delta_M H_B^\infty = 2000$ J/mol

Polymer (B):	**poly(tetramethylene oxide)**	**1985SHA**
Characterization:	$M_n/\text{g.mol}^{-1} = 2000$, Quaker Oats Corporation	
Solvent (A):	**toluene** $\quad$ **C₇H₈**	**108-88-3**

$T/\text{K} = 313.15 \qquad \Delta_M H_B^\infty = 1820$ J/mol

Polymer (B):	**poly(tetramethylene oxide)**	**1986SHA**
Characterization:	$M_n/\text{g.mol}^{-1} = 650$, Quaker Oats Corporation	
Solvent (A):	**1,3,5-trimethylbenzene C₉H₁₂**	**108-67-8**

$T/\text{K} = 313.15 \qquad \Delta_M H_B^\infty = 4375$ J/mol

Polymer (B):	**poly(tetramethylene oxide)**	**1986SHA**
Characterization:	$M_n/\text{g.mol}^{-1} = 1000$, Quaker Oats Corporation	
Solvent (A):	**1,3,5-trimethylbenzene C₉H₁₂**	**108-67-8**

$T/\text{K} = 313.15 \qquad \Delta_M H_B^\infty = 2720$ J/mol

Polymer (B):	**poly(tetramethylene oxide)**	**1986SHA**
Characterization:	$M_n/\text{g.mol}^{-1} = 2000$, Quaker Oats Corporation	
Solvent (A):	**1,3,5-trimethylbenzene C₉H₁₂**	**108-67-8**

$T/\text{K} = 313.15 \qquad \Delta_M H_B^\infty = 1120$ J/mol

Polymer (B):	**poly(vinyl acetate)**	**1969BIA**
Characterization:	$M_\eta/\text{g.mol}^{-1} = 350000$	
Solvent (A):	**2-butanone** $\quad$ **C₄H₈O**	**78-93-3**

$T/\text{K} = 303.15 \qquad \Delta_{sol} H_B^\infty = -150.6$ J/(base mol polymer)

Polymer (B):	**poly(vinyl acetate)**	**1969BIA**
Characterization:	$M_\eta/\text{g.mol}^{-1} = 350000$	
Solvent (A):	**butyl acetate** $\quad$ **C₆H₁₂O₂**	**123-86-4**

$T/\text{K} = 303.15 \qquad \Delta_{sol} H_B^\infty = 83.7$ J/(base mol polymer)

Polymer (B):	**poly(vinyl acetate)**	**1959HOR**
Characterization:	$M_\eta/\text{g.mol}^{-1} = 135000$, fractionated in the laboratory	
Solvent (A):	**chlorobenzene** $\quad$ **C₆H₅Cl**	**108-90-7**

$T/\text{K} = 298.15 \qquad \Delta_{\text{sol}}H_B^\infty = 5.0 \text{ J/(g polymer)}$

Polymer (B):	**poly(vinyl acetate)**	**1952TAG**
Characterization:	–	
Solvent (A):	**ethyl acetate** $\quad$ **C₄H₈O₂**	**141-78-6**

$T/\text{K} = 298.15 \qquad \Delta_{\text{sol}}H_B^\infty = 0.0 \text{ J/(g polymer)}$

Comments: $\quad$ The final concentration is about 1.7%.

Polymer (B):	**poly(vinyl acetate)**	**1969BIA**
Characterization:	$M_\eta/\text{g.mol}^{-1} = 350000$	
Solvent (A):	**ethyl acetate** $\quad$ **C₄H₈O₂**	**141-78-6**

$T/\text{K} = 303.15 \qquad \Delta_{\text{sol}}H_B^\infty = -577.4 \text{ J/(base mol polymer)}$

Polymer (B):	**poly(vinyl acetate)**	**1966BIA**
Characterization:	$M_\eta/\text{g.mol}^{-1} = 26000$, fractionated in the laboratory	
Solvent (A):	**3-heptanone** $\quad$ **C₇H₁₄O**	**106-35-4**

$T/\text{K} = 303.15 \qquad \Delta_M H_B^\infty = 602.5 \text{ J/(base mol polymer)}$

Comments: $\quad$ The final concentration is about 0.3 g polymer/100 ml solvent.

Polymer (B):	**poly(vinyl acetate)**	**1969BIA**
Characterization:	$M_\eta/\text{g.mol}^{-1} = 350000$	
Solvent (A):	**3-heptanone** $\quad$ **C₇H₁₄O**	**106-35-4**

$T/\text{K} = 303.15 \qquad \Delta_{\text{sol}}H_B^\infty = 418 \text{ J/(base mol polymer)}$

Polymer (B):	**poly(vinyl acetate)**	**1969BIA**
Characterization:	$M_\eta/\text{g.mol}^{-1} = 350000$	
Solvent (A):	**methyl acetate** $\quad$ **C₃H₆O₂**	**79-20-9**

$T/\text{K} = 303.15 \qquad \Delta_{\text{sol}}H_B^\infty = -837 \text{ J/(base mol polymer)}$

Polymer (B):	**poly(vinyl acetate)**	**1969BIA**
Characterization:	$M_\eta/\text{g.mol}^{-1} = 350000$	
Solvent (A):	**2-pentanone** $\quad$ **C₅H₁₀O**	**107-87-9**

$T/\text{K} = 303.15 \qquad \Delta_{\text{sol}}H_B^\infty = 0.0 \text{ J/(base mol polymer)}$

Polymer (B):	**poly(vinyl acetate)**	**1956STR, 1958SLO**
Characterization:	intrinsic viscosity = 1.77	
Solvent (A):	**2-propanone** C_3H_6O	**67-64-1**

$T/K = 298.15$ $\Delta_{sol} H_B^\infty = -2.85$ J/(g polymer)

Comments: The final concentration is 1 g polymer/100 g solvent.

Polymer (B):	**poly(vinyl acetate)**	**1958TA3**
Characterization:	$M_\eta/\text{g.mol}^{-1} = 93000$	
Solvent (A):	**2-propanone** C_3H_6O	**67-64-1**

$T/K = 298.15$ $\Delta_{sol} H_B^\infty = -0.4$ J/(g polymer)

Polymer (B):	**poly(vinyl acetate)**	**1969BIA**
Characterization:	$M_\eta/\text{g.mol}^{-1} = 350000$	
Solvent (A):	**2-propanone** C_3H_6O	**67-64-1**

$T/K = 303.15$ $\Delta_{sol} H_B^\infty = -335$ J/(base mol polymer)

Polymer (B):	**poly(vinyl acetate)**	**1969BIA**
Characterization:	$M_\eta/\text{g.mol}^{-1} = 350000$	
Solvent (A):	**propyl acetate** $C_5H_{10}O_2$	**109-60-4**

$T/K = 303.15$ $\Delta_{sol} H_B^\infty = -230$ J/(base mol polymer)

Polymer (B):	**poly(vinyl acetate)**	**1990SHI**
Characterization:	$M_w/\text{g.mol}^{-1} = 150000$	
Solvent (A):	**tetrahydrofuran** C_4H_8O	**109-99-9**

$T/K = 304.65$ $\Delta_{sol} H_B^\infty = 4.52$ J/(g polymer)

Polymer (B):	**poly(vinyl acetate-*co*-vinyl alcohol)**	**1958TA2**
Characterization:	9.0 wt% vinyl acetate, synthesized in the laboratory	
Solvent (A):	**ethyl acetate** $C_4H_8O_2$	**141-78-6**
Solvent (C):	**ethanol** C_2H_6O	**64-17-5**

$T/K = 298.15$ $\Delta_{sol} H_B^\infty = 1.67$ J/(g polymer)

Comments: The solvent mixture concentration is equal to the comonomer contents, i.e., 9.0 wt% ethyl acetate/91.0 wt% ethanol.

Polymer (B):	poly(vinyl acetate-*co*-vinyl alcohol)		1958TA2
Characterization:	44.0 wt% vinyl acetate, synthesized in the laboratory		
Solvent (A):	ethyl acetate	$C_4H_8O_2$	141-78-6
Solvent (C):	ethanol	C_2H_6O	64-17-5

$T/K = 298.15$ $\Delta_{sol}H_B^\infty = 0.84$ J/(g polymer)

Comments: The solvent mixture concentration is equal to the comonomer contents, i.e., 44.0 wt% ethyl acetate/56.0 wt% ethanol.

Polymer (B):	poly(vinyl acetate-*co*-vinyl alcohol)		1958TA2
Characterization:	57.0 wt% vinyl acetate, synthesized in the laboratory		
Solvent (A):	ethyl acetate	$C_4H_8O_2$	141-78-6
Solvent (C):	ethanol	C_2H_6O	64-17-5

$T/K = 298.15$ $\Delta_{sol}H_B^\infty = 0.84$ J/(g polymer)

Comments: The solvent mixture concentration is equal to the comonomer contents, i.e., 57.0 wt% ethyl acetate/43.0 wt% ethanol.

Polymer (B):	poly(vinyl acetate-*co*-vinyl alcohol)		1958TA2
Characterization:	67.0 wt% vinyl acetate, synthesized in the laboratory		
Solvent (A):	ethyl acetate	$C_4H_8O_2$	141-78-6
Solvent (C):	ethanol	C_2H_6O	64-17-5

$T/K = 298.15$ $\Delta_{sol}H_B^\infty = 0.0$ J/(g polymer)

Comments: The solvent mixture concentration is equal to the comonomer contents, i.e., 67.0 wt% ethyl acetate/33.0 wt% ethanol.

Polymer (B):	poly(vinyl acetate-*co*-vinyl alcohol)		1958TA2
Characterization:	68.6 wt% vinyl acetate, synthesized in the laboratory		
Solvent (A):	ethyl acetate	$C_4H_8O_2$	141-78-6
Solvent (C):	ethanol	C_2H_6O	64-17-5

$T/K = 298.15$ $\Delta_{sol}H_B^\infty = 0.0$ J/(g polymer)

Comments: The solvent mixture concentration is equal to the comonomer contents, i.e., 68.6 wt% ethyl acetate/31.4 wt% ethanol.

Polymer (B):	poly(vinyl acetate-*co*-vinyl alcohol)		1958TA1
Characterization:	9.0 wt% vinyl acetate, synthesized in the laboratory		
Solvent (A):	2-propanone	C_3H_6O	67-64-1

$T/K = 298.15$ $\Delta_{sol}H_B^\infty = 6.3$ J/(g polymer)

Polymer (B):	poly(vinyl acetate-*co*-vinyl alcohol)	1958TA1
Characterization:	44.0 wt% vinyl acetate, synthesized in the laboratory	
Solvent (A):	2-propanone C_3H_6O	67-64-1

$T/K = 298.15$ $\Delta_{sol} H_B^\infty = 4.6$ J/(g polymer)

Polymer (B):	poly(vinyl acetate-*co*-vinyl alcohol)	1958TA1
Characterization:	57.0 wt% vinyl acetate, synthesized in the laboratory	
Solvent (A):	2-propanone C_3H_6O	67-64-1

$T/K = 298.15$ $\Delta_{sol} H_B^\infty = 0.0$ J/(g polymer)

Polymer (B):	poly(vinyl acetate-*co*-vinyl alcohol)	1958TA1
Characterization:	67.0 wt% vinyl acetate, synthesized in the laboratory	
Solvent (A):	2-propanone C_3H_6O	67-64-1

$T/K = 298.15$ $\Delta_{sol} H_B^\infty = -1.26$ J/(g polymer)

Polymer (B):	poly(vinyl acetate-*co*-vinyl alcohol)	1955OYA
Characterization:	M_n/g.mol^{-1} = 7560, 4.2 mol% vinyl acetate	
Solvent (A):	water H_2O	7732-18-5

$T/K = 303.15$ $\Delta_{sol} H_B^\infty = -41.3$ J/(g polymer)

Comments: The final concentration is < 2 g polymer/200 cm^3 solvent.

Polymer (B):	poly(vinyl acetate-*co*-vinyl alcohol)	1955OYA
Characterization:	M_n/g.mol^{-1} = 64300, 4.3 mol% vinyl acetate	
Solvent (A):	water H_2O	7732-18-5

$T/K = 303.15$ $\Delta_{sol} H_B^\infty = -48.6$ J/(g polymer)

Comments: The final concentration is < 2 g polymer/200 cm^3 solvent.

Polymer (B):	poly(vinyl acetate-*co*-vinyl alcohol)	1955OYA
Characterization:	M_n/g.mol^{-1} = 66900, 9.0 mol% vinyl acetate	
Solvent (A):	water H_2O	7732-18-5

$T/K = 303.15$ $\Delta_{sol} H_B^\infty = -54.6$ J/(g polymer)

Comments: The final concentration is < 2 g polymer/200 cm^3 solvent.

Polymer (B):	poly(vinyl acetate-*co*-vinyl alcohol)	1955OYA
Characterization:	M_n/g.mol^{-1} = 7970, 10.3 mol% vinyl acetate	
Solvent (A):	water H_2O	7732-18-5

$T/K = 303.15$ $\Delta_{sol} H_B^\infty = -46.8$ J/(g polymer)

Comments: The final concentration is < 2 g polymer/200 cm^3 solvent.

Polymer (B):   **poly(vinyl acetate-*co*-vinyl alcohol)**   **1955OYA**
Characterization: M_n/g.mol^{-1} = 8300, 15.3 mol% vinyl acetate
Solvent (A):   **water**   **H$_2$O**   **7732-18-5**

T/K = 303.15 $\Delta_{sol}H_B^{\infty}$ = −60.2 J/(g polymer)

Comments: The final concentration is < 2 g polymer/200 cm^3 solvent.

Polymer (B):   **poly(vinyl acetate-*co*-vinyl alcohol)**   **1955OYA**
Characterization: M_n/g.mol^{-1} = 70700, 15.4 mol% vinyl acetate
Solvent (A):   **water**   **H$_2$O**   **7732-18-5**

T/K = 303.15 $\Delta_{sol}H_B^{\infty}$ = −64.7 J/(g polymer)

Comments: The final concentration is < 2 g polymer/200 cm^3 solvent.

Polymer (B):   **poly(vinyl acetate-*co*-vinyl alcohol)**   **1955OYA**
Characterization: M_n/g.mol^{-1} = 73100, 19.5 mol% vinyl acetate
Solvent (A):   **water**   **H$_2$O**   **7732-18-5**

T/K = 303.15 $\Delta_{sol}H_B^{\infty}$ = −66.4 J/(g polymer)

Comments: The final concentration is < 2 g polymer/200 cm^3 solvent.

Polymer (B):   **poly(vinyl acetate-*co*-vinyl alcohol)**   **1955OYA**
Characterization: M_n/g.mol^{-1} = 8800, 22.1 mol% vinyl acetate
Solvent (A):   **water**   **H$_2$O**   **7732-18-5**

T/K = 303.15 $\Delta_{sol}H_B^{\infty}$ = −60.1 J/(g polymer)

Comments: The final concentration is < 2 g polymer/200 cm^3 solvent.

Polymer (B):   **poly(vinyl acetate-*co*-vinyl alcohol)**   **1955OYA**
Characterization: M_n/g.mol^{-1} = 77000, 26.2 mol% vinyl acetate
Solvent (A):   **water**   **H$_2$O**   **7732-18-5**

T/K = 303.15 $\Delta_{sol}H_B^{\infty}$ = −64.0 J/(g polymer)

Comments: The final concentration is < 2 g polymer/200 cm^3 solvent.

Polymer (B):   **poly(vinyl acetate-*co*-vinyl alcohol)**   **1955OYA**
Characterization: M_n/g.mol^{-1} = 9370, 30.6 mol% vinyl acetate
Solvent (A):   **water**   **H$_2$O**   **7732-18-5**

T/K = 303.15 $\Delta_{sol}H_B^{\infty}$ = −53.2 J/(g polymer)

Comments: The final concentration is < 2 g polymer/200 cm^3 solvent.

Polymer (B):	**poly(vinyl acetate-*co*-vinyl alcohol)**	**1955OYA**
Characterization:	M_n/g.mol^{-1} = 81600, 34.0 mol% vinyl acetate	
Solvent (A):	**water** $\quad$ **H$_2$O**	**7732-18-5**

T/K = 303.15 $\qquad$ $\Delta_{sol}H_B^\infty$ = –60.4 J/(g polymer)

Comments: $\quad$ The final concentration is < 2 g polymer/200 cm^3 solvent.

Polymer (B):	**poly(vinyl acetate-*co*-vinyl alcohol)**	**1955OYA**
Characterization:	M_n/g.mol^{-1} = 9670, 34.7 mol% vinyl acetate	
Solvent (A):	**water** $\quad$ **H$_2$O**	**7732-18-5**

T/K = 303.15 $\qquad$ $\Delta_{sol}H_B^\infty$ = –44.4 J/(g polymer)

Comments: $\quad$ The final concentration is < 2 g polymer/200 cm^3 solvent.

Polymer (B):	**poly(vinyl acetate-*co*-vinyl chloride)**	**1997SAT**
Characterization:	M_w/g.mol^{-1} = 12400, M_w/g.mol^{-1} = 26000, 10.0 wt% vinyl acetate,	
	Scientific Polymer Products, Inc., Ontario, NY	
Solvent (A):	**cyclohexanone** $\quad$ **C$_6$H$_{10}$O**	**108-94-1**

T/K = 304.15 $\qquad$ $\Delta_{sol}H_B^\infty$ = –37.2 J/(g polymer) $\qquad\qquad$ (glassy state polymer)

T/K = 304.15 $\qquad$ $\Delta_M H_B^\infty$ = –16.5 J/(g polymer) $\qquad\qquad$ (liquid state polymer)

Comments: $\quad$ $\Delta_M H_B^\infty$ is corrected for the enthalpy from the glass transition.

Polymer (B):	**poly(vinyl alcohol)**	**1952TAG**
Characterization:	–	
Solvent (A):	**ethanol** $\quad$ **C$_2$H$_6$O**	**64-17-5**

T/K = 298.15 $\qquad$ $\Delta_{sol}H_B^\infty$ = 3.8 J/(g polymer)

Comments: $\quad$ The final concentration is about 1.7%.

Polymer (B):	**poly(vinyl alcohol)**	**1958TA2**
Characterization:	–	
Solvent (A):	**ethanol** $\quad$ **C$_2$H$_6$O**	**64-17-5**

T/K = 298.15 $\qquad$ $\Delta_{sol}H_B^\infty$ = 9.6 J/(g polymer)

Polymer (B):	**poly(vinyl alcohol)**	**1955OYA**
Characterization:	M_n/g.mol^{-1} = 7260	
Solvent (A):	**water** $\quad$ **H$_2$O**	**7732-18-5**

T/K = 303.15 $\qquad$ $\Delta_{sol}H_B^\infty$ = –33.9 J/(g polymer)

Polymer (B):	poly(vinyl alcohol)		**1958TA3**
Characterization:	M_η/g.mol^{-1} = 17000		
Solvent (A):	**water**	**H$_2$O**	**7732-18-5**

T/K = 298.15 $\Delta_{sol}H_B^\infty$ = −17.6 J/(g polymer)

Polymer (B):	poly(vinyl alcohol)		**1955OYA**
Characterization:	M_n/g.mol^{-1} = 61600		
Solvent (A):	**water**	**H$_2$O**	**7732-18-5**

T/K = 303.15 $\Delta_{sol}H_B^\infty$ = −40.9 J/(g polymer)

Comments: The final concentration is < 2 g polymer/200 cm^3 solvent.

Polymer (B):	poly(vinyl alcohol)		**1952TAG**
Characterization:	–		
Solvent (A):	**water**	**H$_2$O**	**7732-18-5**

T/K = 298.15 $\Delta_{sol}H_B^\infty$ = −8.4 J/(g polymer)

Comments: The final concentration is about 1%.

Polymer (B):	poly(vinyl chloride)		**1960MUE**
Characterization:	–		
Solvent (A):	**chlorobenzene**	**C$_6$H$_5$Cl**	**108-90-7**

T/K = room temperature $\Delta_{sol}H_B$ = −16.5 J/(g polymer)

Comments: The final concentration is 0.1 g polymer in 1 g solvent.

Polymer (B):	poly(vinyl chloride)		**1972MA3**
Characterization:	M_n/g.mol^{-1} = 23200, B.F. Goodrich Chemical Company		
Solvent (A):	**cyclohexanone**	**C$_6$H$_{10}$O**	**108-94-1**

T/K = 303.15 $\Delta_{sol}H_B^\infty$ = −27.0 J/(g polymer)

T/K = 303.15 $\Delta_M H_B^\infty$ = −7.5 J/(g polymer)

Comments: $\Delta_M H_B^\infty$ is the value corrected for enthalpic contributions arising from destructing glassy and crystalline portions of the polymer.

Polymer (B):	poly(vinyl chloride)		**1972MA3**
Characterization:	M_n/g.mol^{-1} = 38700, B.F. Goodrich Chemical Company		
Solvent (A):	**cyclohexanone**	**C$_6$H$_{10}$O**	**108-94-1**

T/K = 303.15 $\Delta_{sol}H_B^\infty$ = −28.6 J/(g polymer)

T/K = 303.15 $\Delta_M H_B^\infty$ = −6.5 J/(g polymer)

Comments: $\Delta_M H_B^\infty$ is the value corrected for enthalpic contributions arising from destructing glassy and crystalline portions of the polymer.

Polymer (B):   **poly(vinyl chloride)**   1972MA3
Characterization: $M_n/\text{g.mol}^{-1} = 53500$, B.F. Goodrich Chemical Company
Solvent (A):   **cyclohexanone** $C_6H_{10}O$ 108-94-1

$T/\text{K} = 303.15$ $\Delta_{sol} H_B^{\infty} = -28.2$ J/(g polymer)

$T/\text{K} = 303.15$ $\Delta_M H_B^{\infty} = -6.3$ J/(g polymer)

Comments: $\Delta_M H_B^{\infty}$ is the value corrected for enthalpic contributions arising from destructing glassy and crystalline portions of the polymer.

Polymer (B):   **poly(vinyl chloride)**   1972MA3
Characterization: $M_n/\text{g.mol}^{-1} = 66700$, B.F. Goodrich Chemical Company
Solvent (A):   **cyclohexanone** $C_6H_{10}O$ 108-94-1

$T/\text{K} = 303.15$ $\Delta_{sol} H_B^{\infty} = -28.9$ J/(g polymer)

$T/\text{K} = 303.15$ $\Delta_M H_B^{\infty} = -6.1$ J/(g polymer)

Comments: $\Delta_M H_B^{\infty}$ is the value corrected for enthalpic contributions arising from destructing glassy and crystalline portions of the polymer.

Polymer (B):   **poly(vinyl chloride)**   1972MA3
Characterization: $M_n/\text{g.mol}^{-1} = 136000$, B.F. Goodrich Chemical Company
Solvent (A):   **cyclohexanone** $C_6H_{10}O$ 108-94-1

$T/\text{K} = 303.15$ $\Delta_{sol} H_B^{\infty} = -30.8$ J/(g polymer)

$T/\text{K} = 303.15$ $\Delta_M H_B^{\infty} = -5.8$ J/(g polymer)

Comments: $\Delta_M H_B^{\infty}$ is the value corrected for enthalpic contributions arising from destructing glassy and crystalline portions of the polymer.

Polymer (B):   **poly(vinyl chloride)**   1972MA3
Characterization: $M_n/\text{g.mol}^{-1} = 155400$, B.F. Goodrich Chemical Company
Solvent (A):   **cyclohexanone** $C_6H_{10}O$ 108-94-1

$T/\text{K} = 303.15$ $\Delta_{sol} H_B^{\infty} = -32.4$ J/(g polymer)

$T/\text{K} = 303.15$ $\Delta_M H_B^{\infty} = -5.8$ J/(g polymer)

Comments: $\Delta_M H_B^{\infty}$ is the value corrected for enthalpic contributions arising from destructing glassy and crystalline portions of the polymer.

Polymer (B):   **poly(vinyl chloride)**   2002RIG
Characterization: $M_n/\text{g.mol}^{-1} = 48000$, Fluka AG, Buchs, Switzerland
Solvent (A):   **cyclopentanone** C_5H_8O 120-92-3

$T/\text{K} = 298.15$ $\Delta_{sol} H_B^{\infty} = -28.0$ J/(g polymer)

Polymer (B):   **poly(vinyl chloride)**   **1958AKH**
Characterization: –
Solvent (A):   **1,2-dichloroethane**   **C$_2$H$_4$Cl$_2$**   **107-06-2**

$T/K =$	323.15	328.15	333.15	368.15	373.15	378.15
$\Delta_{sol}H_B$/(J/g polymer)	24.3	33.5	37.7	43.9	46.4	46.4

Comments: Completely amorphous material.

Polymer (B):   **poly(vinyl chloride)**   **1959ZEL**
Characterization: –
Solvent (A):   **N,N-dimethylformamide**   **C$_3$H$_7$NO**   **68-12-2**

T/K	295.15	295.15	308.15	308.15	323.15	323.15	338.15
$\Delta_{sol}H_B^{\infty}$/(J/g polymer)	−29.3	−27.1	−20.5	−17.6	−14.4	−13.4	−7.82

T/K	338.15	353.15	353.15
$\Delta_{sol}H_B^{\infty}$/(J/g polymer)	−7.11	+2.93	+1.80

Comments: The final concentration is about 1 g polymer/165 g solvent.

Polymer (B):   **poly(vinyl chloride)**   **1972MA3**
Characterization: M_n/g.mol^{-1} = 23200, B.F. Goodrich Chemical Company
Solvent (A):   **tetrahydrofuran**   **C$_4$H$_8$O**   **109-99-9**

$T/K = 303.15$ $\Delta_{sol}H_B^{\infty} = -33.7$ J/(g polymer)

$T/K = 303.15$ $\Delta_{M}H_B^{\infty} = -14.3$ J/(g polymer)

Comments: $\Delta_{M}H_B^{\infty}$ is the value corrected for enthalpic contributions arising from destructing glassy and crystalline portions of the polymer.

Polymer (B):   **poly(vinyl chloride)**   **1972MA3**
Characterization: M_n/g.mol^{-1} = 38700, B.F. Goodrich Chemical Company
Solvent (A):   **tetrahydrofuran**   **C$_4$H$_8$O**   **109-99-9**

$T/K = 303.15$ $\Delta_{sol}H_B^{\infty} = -35.4$ J/(g polymer)

$T/K = 303.15$ $\Delta_{M}H_B^{\infty} = -14.3$ J/(g polymer)

Comments: $\Delta_{M}H_B^{\infty}$ is the value corrected for enthalpic contributions arising from destructing glassy and crystalline portions of the polymer.

Polymer (B):   **poly(vinyl chloride)**   **1972MA3**
Characterization: M_n/g.mol^{-1} = 53500, B.F. Goodrich Chemical Company
Solvent (A):   **tetrahydrofuran**   **C$_4$H$_8$O**   **109-99-9**

$T/K = 303.15$ $\Delta_{sol}H_B^{\infty} = -38.5$ J/(g polymer)

$T/K = 303.15$ $\Delta_{M}H_B^{\infty} = -14.2$ J/(g polymer)

Comments: $\Delta_{M}H_B^{\infty}$ is the value corrected for enthalpic contributions arising from destructing glassy and crystalline portions of the polymer.

Polymer (B):   **poly(vinyl chloride)**   1972MA3
Characterization: M_n/g.mol^{-1} = 66700, B.F. Goodrich Chemical Company
Solvent (A):   **tetrahydrofuran**   **C$_4$H$_8$O**   109-99-9

T/K = 303.15 $\Delta_{sol} H_B^\infty$ = −36.0 J/(g polymer)

T/K = 303.15 $\Delta_M H_B^\infty$ = −14.2 J/(g polymer)

Comments: $\Delta_M H_B^\infty$ is the value corrected for enthalpic contributions arising from destructing glassy and crystalline portions of the polymer.

Polymer (B):   **poly(vinyl chloride)**   1972MA3
Characterization: M_n/g.mol^{-1} = 136000, B.F. Goodrich Chemical Company
Solvent (A):   **tetrahydrofuran**   **C$_4$H$_8$O**   109-99-9

T/K = 303.15 $\Delta_{sol} H_B^\infty$ = −38.5 J/(g polymer)

T/K = 303.15 $\Delta_M H_B^\infty$ = −14.2 J/(g polymer)

Comments: $\Delta_M H_B^\infty$ is the value corrected for enthalpic contributions arising from destructing glassy and crystalline portions of the polymer.

Polymer (B):   **poly(vinyl chloride)**   1972MA3
Characterization: M_n/g.mol^{-1} = 155400, B.F. Goodrich Chemical Company
Solvent (A):   **tetrahydrofuran**   **C$_4$H$_8$O**   109-99-9

T/K = 303.15 $\Delta_{sol} H_B^\infty$ = −38.9 J/(g polymer)

T/K = 303.15 $\Delta_M H_B^\infty$ = −14.2 J/(g polymer)

Comments: $\Delta_M H_B^\infty$ is the value corrected for enthalpic contributions arising from destructing glassy and crystalline portions of the polymer.

Polymer (B):   **poly(vinyl chloride-*co*-vinylidene chloride)**   1956LIP
Characterization: −
Solvent (A):   **trichloromethane**   **CHCl$_3$**   67-66-3

T/K = 297.15 $\Delta_{sol} H_B^\infty$ = −17.15 J/(g polymer)   (non-oriented sample)

T/K = 297.15 $\Delta_{sol} H_B^\infty$ = −15.9 J/(g polymer)   (oriented sample, drawn 100%)

Polymer (B):   **poly(1-vinyl-3,5-dimethyl-1,2,4-triazole)**   1991TA1
Characterization: −
Solvent (A):   **N,N-dimethylformamide**   **C$_3$H$_7$NO**   68-12-2

T/K = 298.15 $\Delta_M H_B^\infty$ = −28.0 J/(g polymer)

Polymer (B):   **poly(1-vinyl-3,5-dimethyl-1,2,4-triazole)**   1991TA1
Characterization: −
Solvent (A):   **water**   **H$_2$O**   7732-18-5

T/K = 298.15 $\Delta_M H_B^\infty$ = −139 J/(g polymer)

Polymer (B):   **poly(1-vinylimidazole)**   **1991TA1**
Characterization: M_n/g.mol^{-1} = 20700
Solvent (A):   **acetic acid** $C_2H_4O_2$ **64-19-7**

T/K = 298.15 $\Delta_M H_B^\infty$ = −37000 J/(base mol polymer)

Polymer (B):   **poly(1-vinylimidazole)**   **1991TA1**
Characterization: M_n/g.mol^{-1} = 20700
Solvent (A):   **butanoic acid** $C_4H_8O_2$ **107-92-6**

T/K = 298.15 $\Delta_M H_B^\infty$ = −30300 J/(base mol polymer)

Polymer (B):   **poly(1-vinylimidazole)**   **1988TAG, 1991TA1**
Characterization: M_n/g.mol^{-1} = 20700
Solvent (A):   **N,N-dimethylacetamide** C_4H_9NO **127-19-5**

T/K = 298.15 $\Delta_M H_B^\infty$ = −4510 J/(base mol polymer)

Polymer (B):   **poly(1-vinylimidazole)**   **1988TAG, 1991TA1**
Characterization: M_n/g.mol^{-1} = 20700
Solvent (A):   **N,N-dimethylformamide** C_3H_7NO **68-12-2**

T/K = 298.15 $\Delta_M H_B^\infty$ = −4470 J/(base mol polymer)

Polymer (B):   **poly(1-vinylimidazole)**   **1988TAG, 1991TA1**
Characterization: M_n/g.mol^{-1} = 20700
Solvent (A):   **1-methyl-2-pyrrolidinone** C_5H_9NO **872-50-4**

T/K = 298.15 $\Delta_M H_B^\infty$ = −5090 J/(base mol polymer)

Polymer (B):   **poly(1-vinylimidazole)**   **1991TA1**
Characterization: M_n/g.mol^{-1} = 20700
Solvent (A):   **pentanoic acid** $C_5H_{10}O_2$ **109-52-4**

T/K = 298.15 $\Delta_M H_B^\infty$ = −30600 J/(base mol polymer)

Polymer (B):   **poly(1-vinylimidazole)**   **1991TA1**
Characterization: M_n/g.mol^{-1} = 20700
Solvent (A):   **propanoic acid** $C_3H_6O_2$ **79-09-4**

T/K = 298.15 $\Delta_M H_B^\infty$ = −26200 J/(base mol polymer)

Polymer (B):   **poly(1-vinylimidazole)**   **1988TAG, 1989SAF, 1991TA1**
Characterization: M_n/g.mol^{-1} = 20700
Solvent (A):   **water** H_2O **7732-18-5**

T/K = 298.15 $\Delta_M H_B^\infty$ = −119 J/(g polymer) = −11200 J/(base mol polymer)

Polymer (B):	**poly(1-vinylpyrazole)**	**1991TA1**
Characterization:	$M_n/\text{g.mol}^{-1} = 18900$	
Solvent (A):	**acetic acid** $C_2H_4O_2$	**64-19-7**

$T/\text{K} = 298.15$ $\Delta_M H_B^\infty = -8400$ J/(base mol polymer)

Polymer (B):	**poly(1-vinylpyrazole)**	**1991TA1**
Characterization:	$M_n/\text{g.mol}^{-1} = 18900$	
Solvent (A):	**butanoic acid** $C_4H_8O_2$	**107-92-6**

$T/\text{K} = 298.15$ $\Delta_M H_B^\infty = -5700$ J/(base mol polymer)

Polymer (B):	**poly(1-vinylpyrazole)**	**1991TA1**
Characterization:	$M_n/\text{g.mol}^{-1} = 18900$	
Solvent (A):	**N,N-dimethylacetamide** C_4H_9NO	**127-19-5**

$T/\text{K} = 298.15$ $\Delta_M H_B^\infty = -2430$ J/(base mol polymer)

Polymer (B):	**poly(1-vinylpyrazole)**	**1991TA1**
Characterization:	$M_n/\text{g.mol}^{-1} = 18900$	
Solvent (A):	**N,N-dimethylformamide** C_3H_7NO	**68-12-2**

$T/\text{K} = 298.15$ $\Delta_M H_B^\infty = -2700$ J/(base mol polymer)

Polymer (B):	**poly(1-vinylpyrazole)**	**1991TA1**
Characterization:	$M_n/\text{g.mol}^{-1} = 18900$	
Solvent (A):	**pentanoic acid** $C_5H_{10}O_2$	**109-52-4**

$T/\text{K} = 298.15$ $\Delta_M H_B^\infty = -4900$ J/(base mol polymer)

Polymer (B):	**poly(1-vinylpyrazole)**	**1991TA1**
Characterization:	$M_n/\text{g.mol}^{-1} = 18900$	
Solvent (A):	**propanoic acid** $C_3H_6O_2$	**79-09-4**

$T/\text{K} = 298.15$ $\Delta_M H_B^\infty = -3400$ J/(base mol polymer)

Polymer (B):	**poly(1-vinyl-2-pyrrolidinone)**	**1968GOL**
Characterization:	$M_\eta/\text{g.mol}^{-1} = 32000$	
Solvent (A):	**water** H_2O	**7732-18-5**

$T/\text{K} = 298.15$ $\Delta_{sol} H_B^\infty = -149.6$ J/(g polymer) $= -16610$ J/(base mol polymer)

Polymer (B):	**poly(1-vinyl-1,2,4-triazole)**	**1991TA1**
Characterization:	$M_n/\text{g.mol}^{-1} = 69500$	
Solvent (A):	**acetic acid** $C_2H_4O_2$	**64-19-7**

$T/\text{K} = 298.15$ $\Delta_M H_B^\infty = -8200$ J/(base mol polymer)

Polymer (B):   **poly(1-vinyl-1,2,4-triazole)**   **1991TA1**
Characterization: M_n/g.mol^{-1} = 69500
Solvent (A):   **butanoic acid** $C_4H_8O_2$ **107-92-6**

T/K = 298.15 $\Delta_M H_B^\infty$ = −7100 J/(base mol polymer)

Polymer (B):   **poly(1-vinyl-1,2,4-triazole)**   **1988TAG, 1991TA1**
Characterization: M_n/g.mol^{-1} = 69500
Solvent (A):   **N,N-dimethylacetamide** C_4H_9NO **127-19-5**

T/K = 298.15 $\Delta_M H_B^\infty$ = −4700 J/(base mol polymer)

Polymer (B):   **poly(1-vinyl-1,2,4-triazole)**   **1988TAG, 1991TA1**
Characterization: M_n/g.mol^{-1} = 69500
Solvent (A):   **N,N-dimethylformamide** C_3H_7NO **68-12-2**

T/K = 298.15 $\Delta_{sol} H_B^\infty$ = −48.0 J/(g polymer) = −4540 J/(base mol polymer)

Polymer (B):   **poly(1-vinyl-1,2,4-triazole)**   **1988TAG, 1991TA1**
Characterization: M_n/g.mol^{-1} = 69500
Solvent (A):   **1-methyl-2-pyrrolidinone** C_5H_9NO **872-50-4**

T/K = 298.15 $\Delta_M H_B^\infty$ = −5270 J/(base mol polymer)

Polymer (B):   **poly(1-vinyl-1,2,4-triazole)**   **1991TA1**
Characterization: M_n/g.mol^{-1} = 69500
Solvent (A):   **pentanoic acid** $C_5H_{10}O_2$ **109-52-4**

T/K = 298.15 $\Delta_M H_B^\infty$ = −7200 J/(base mol polymer)

Polymer (B):   **poly(1-vinyl-1,2,4-triazole)**   **1991TA1**
Characterization: M_n/g.mol^{-1} = 69500
Solvent (A):   **propanoic acid** $C_3H_6O_2$ **79-09-4**

T/K = 298.15 $\Delta_M H_B^\infty$ = −6900 J/(base mol polymer)

Polymer (B):   **poly(1-vinyl-1,2,4-triazole)**   **1988TAG, 1989SAF, 1991TA1**
Characterization: M_n/g.mol^{-1} = 69500
Solvent (A):   **water** H_2O **7732-18-5**

T/K = 298.15 $\Delta_{sol} H_B^\infty$ = −68.9 J/(g polymer) = −6550 J/(base mol polymer)

Polymer (B):   **tetra(ethylene glycol)**   **1993ZEL**
Characterization: M/g.mol^{-1} = 194.23
Solvent (A):   **trichloromethane** $CHCl_3$ **67-66-3**

T/K = 303.15 $\Delta_M H_B^\infty$ = −76.83 J/(g polymer)

Polymer (B):   **tetra(ethylene glycol)**   **1998OHT**
Characterization: M/g.mol^{-1} = 194.23
Solvent (A):   **water**   **H$_2$O**   **7732-18-5**

T/K = 283.15 $\Delta_M H_B^\infty$ = −32.70 ± 0.033 kJ/mol

T/K = 288.15 $\Delta_M H_B^\infty$ = −32.10 ± 0.028 kJ/mol

T/K = 293.15 $\Delta_M H_B^\infty$ = −31.33 ± 0.014 kJ/mol

T/K = 298.15 $\Delta_M H_B^\infty$ = −30.47 ± 0.016 kJ/mol

T/K = 303.15 $\Delta_M H_B^\infty$ = −29.51 ± 0.040 kJ/mol

T/K = 308.15 $\Delta_M H_B^\infty$ = −28.80 ± 0.029 kJ/mol

T/K = 313.15 $\Delta_M H_B^\infty$ = −27.73 ± 0.034 kJ/mol

Polymer (B):   **tetra(ethylene glycol) dimethyl ether**   **1987CIF**
Characterization: M/g.mol^{-1} = 222.28
Solvent (A):   **N,N-dimethylformamide**   **C$_3$H$_7$NO**   **68-12-2**

T/K = 298.15 $\Delta_{sol} H_B^\infty$ = −1.20 kJ/mol polymer

Polymer (B):   **tetra(ethylene glycol) dimethyl ether**   **1993ZEL**
Characterization: M/g.mol^{-1} = 222.28
Solvent (A):   **tetrachloromethane**   **CCl$_4$**   **56-23-5**

T/K = 303.15 $\Delta_M H_B^\infty$ = −12.31 J/(g polymer)

T/K = 318.15 $\Delta_M H_B^\infty$ = −8.12 J/(g polymer)

T/K = 333.15 $\Delta_M H_B^\infty$ = −8.45 J/(g polymer)

Polymer (B):   **tetra(ethylene glycol) dimethyl ether**   **1993ZEL**
Characterization: M/g.mol^{-1} = 222.28
Solvent (A):   **trichloromethane**   **CHCl$_3$**   **67-66-3**

T/K = 303.15 $\Delta_M H_B^\infty$ = −104.16 J/(g polymer)

Polymer (B):   **tetra(ethylene glycol) dimethyl ether**   **1987CIF**
Characterization: M/g.mol^{-1} = 222.28
Solvent (A):   **water**   **H$_2$O**   **7732-18-5**

T/K = 298.15 $\Delta_M H_B^\infty$ = −42.60 kJ/mol polymer

3.2. References

1934LIE   Liepatoff, S. and Preobagenskaja, S., Zur Lehre von den lyophilen Kolloiden, *Kolloid Z. Z. Polym.*, 68, 324, 1934.

1935KAR   Kargin, V. and Papkov, S., Über die Wärmeentwicklung bei Einwirkung von Lösungs-mitteln auf Nitrozellulose, *Acta Physico-Chim. URSS*, 3, 839, 1935.

1937PAP   Papkov, S. and Kargin, V., Über die Wärmeentwicklung bei der Einwirkung von Lösungs-mitteln auf Nitrozellulose-II, *Acta Physico-Chim. URSS*, 7, 667, 1937.

1941TAG   Tager, A. and Kargin, V., Lösungs- und Quellungsvorgang von Zelluloseestern, *Acta Physico-Chim. URSS*, 14, 713, 1941.

1945RAI   Raine, H.C., Richards, R.B., and Ryder, H., The heat capacity, heat of solution, and crystallinity of polythene, *Trans. Faraday Soc.*, 41, 56, 1945.

1947ROB   Roberts, D.E., Walton, W.W., and Jessup, R.S., Heats of combustion and solution of liquid styrene and solid polystyrene and heat of polymerization of styrene, *J. Polym. Sci.*, 2, 420, 1947.

1950TAG   Tager, A. and Sanatina, V., Heats of solution and swelling of some synthetic high-molecular compounds (Russ.), *Kolloidn. Zh.*, 12, 474, 1950.

1951GLI   Glikman, S.A. and Root, L.A., Kharakteristika energeticheskikh effektov razbavleniya rastvorov vysokopolimerov dilatometricheskim metodom, *Zh. Obshch. Khim.*, 21, 58, 1951.

1951HEL   Hellfritz, H., Über die Lösungswärme von Polystyrol in verschiedenen Lösungsmitteln, *Makromol. Chem.*, 7, 191, 1951.

1951TA1   Tager, A. and Vershkain, R., Rastvoryayushchaya sposobnost' rastvoritelya i vsyazkost' rastvorov vysokomolularnykh soedinenii, *Kolloidn. Zh.*, 13, 123, 1951.

1951TA2   Tager, A. and Sanatina, V., Heats of solution and swelling of some synthetic high-molecular compounds, *Rubber Chem. Technol.*, 24, 773, 1951.

1952SC1   Schulz, G.V., Gute und schlechte Lösungsmittel für hochpolymere Stoffe, *Angew. Chem.*, 64, 553, 1952.

1952TAG   Tager, A.A. and Kargin, V.A., Thermodynamic study of systems of polymers and hydrogenated monomers (Russ.), *Kolloidn. Zh.*, 14, 367, 1952.

1953TAG   Tager, A.A. and Dombek, Zh.S., Thermodynamic study of polystyrene solutions (Russ.), *Kolloidn. Zh.*, 15, 69, 1953.

1954OYA   Oya, S., Heat of solution of synthetic high polymers. I. Heat of solution and cohesive energy densities of polystyrene, *Kobunshi Kagaku*, 11, 95, 1954.

1955GAT   Gatovskaya, T.V., Kargin, V.A., and Tager, A.A., The effect of the molecular weight of polystyrene on the packing density of the molecular chain (Russ.), *Zh. Fiz. Khim.*, 29, 883, 1955.

1955OYA   Oya, S., Heat of solution of synthetic high polymers. III. Heat of solution of partially acetylated poly(vinyl alcohol), *Chem. High Polym. Japan*, 12, 122, 1955.

1955SCH   Schulz, G.V., Günner, K. von, and Gerrens, H., Thermodynamische Untersuchung der Beziehungen zwischen Glas, Schmelze und Lösung bei Polystyrol, *Z. Phys. Chem., N.F.*, 4, 192, 1955.

1955TA1   Tager, A.A. and Kosova, L.K., Thermodynamic studies of copolymer solutions 2. Thermodynamic study of butadiene-acrylonitrile copolymer solutions (Russ.), *Kolloidn. Zh.*, 17, 391, 1955.

1955TA2   Tager, A.A., Kosova, L.K., Karlinskaya, D.Yu., and Yurina, I.A., Thermodynamic studies of copolymer solutions 1. Thermodynamic study of butadiene-styrene copolymer solutions (Russ.), *Kolloidn. Zh.*, 17, 315, 1955.

1955TA3   Tager, A.A., Krivokorytova, R.V., and Khodorov, P.M., The heats of solution of polystyrenes of various molecular weights and the packing density of rigid chains (Russ.), *Dokl. Akad. Nauk SSSR*, 100, 741, 1955.

1955ZAZ   Zazulina, Z.A. and Rogovin, Z.A., Studies on properties of synthetic carbochain fibres. 2. Mechanism of the thermorelaxation behavior of carbochain fibres (Russ.), *Kolloidn. Zh.*, 17, 343, 1955.

1956GLI   Glikman, S.A. and Root, L.A., Ob ob'emnykh effektakh razbavleniya rastvorov vysoko-polimerov, *Kolloidn. Zh.*, 18, 523, 1956.

1956JEN   Jenckel, E. and Gorke, K., Zur Kalorimetrie von Hochpolymeren. II. Die integralen Verdünnungswärmen der Lösungen des Polystyrols in Ethylbenzol, Toluol, Chlorbenzol und Cyclohexan, *Z. Elektrochem.*, 60, 579, 1956.

1956LIP   Lipatov, Yu.S., Kargin, V.A., and Slonimskii, G.L., Study on orientation in high polymers II. Crystalline polymers (Russ.), *Zh. Fiz. Khim.*, 30, 1202, 1956.

1956MEE   Meerson, S.I. and Lipatov, S.M., Dependence of the heat of solution of polymers on their physical state (Russ.), *Kolloidn. Zh.*, 18, 447, 1956.

1956MIK   Mikhailov, N.V. and Fainberg, E.Z., Issledovanie struktury sinteticheskikh poliamidnykh volokon 6. Integral'naya teplota rastvoreniya kapronovogo volokna v muravinoi kislote, *Kolloidn. Zh.*, 18, 44, 1956.

1956STR   Struminskii, G.V. and Slonimskii, G.L., Mutual solubility of polymers III. Heats of mixing (Russ.), *Zh. Fiz. Khim.*, 30, 1941, 1956.

1957GAL   Gal'perin, D.I. and Moseev, L.I., On the heats of swelling and of solution of nitrocellulose (Russ.), *Kolloidn. Zh.*, 19, 167, 1957.

1957VAN   Vanderryn, J. and Zettlemoyer, A.C., Heats of solution of polyindene, *Ind. Eng. Chem., Chem. Eng. Data Ser.*, 2, 56, 1958.

1958AKH   Akhmedov, K.S., Heats of swelling and solution of polyethylene, poly(vinyl chloride), perchlorovinyl resin, and polystyrene in 1,2-dichloroethane or toluene (Russ.), *Uzb. Khim. Zh.*, (1), 19, 1958.

1958KAR   Kargin, V.A. and Lipatov, Yu.S., Orientation studies on high polymers IV. (Russ.), *Zh. Fiz. Khim.*, 32, 326, 1958.

1958MEE   Meerson, S.I. and Lipatov, S.M., Melting of gelatin gels (Russ.), *Kolloidn. Zh.*, 20, 353, 1958.

1958SLO   Slonimskii, G.L., Mutual solubility of polymers and properties of their mixtures, *J. Polym. Sci.*, 30, 625, 1958.

1958TA1   Tager, A.A. and Iovleva, M., Thermodynamic studies of copolymer solutions III. Saponified poly(vinyl acetate)s (Russ.), *Zh. Fiz. Khim.*, 32, 1774, 1958.

1958TA2   Tager, A.A. and Kargin, V.A., Thermodynamic studies of polymer-hydrogenated monomer systems II. Enthalpies of solution of copolymers in mixtures of hydrogenated monomers (Russ.), *Zh. Fiz. Khim.*, 32, 1362, 1958.

1958TA3   Tager, A.A. and Kargin, V.A., The heat of solution of polymers and their hydrated monomers in the same liquid (Russ.), *Zh. Fiz. Khim.*, 32, 2694, 1958.

1958TA4   Tager, A.A. and Galkina, L.A., Thermodynamic study of the process of solution of polystyrene in methyl ethyl ketone and ethyl acetate (Russ.), *Nauchn. Dokl. Vyssh. Shkol., Khim. Khim. Tekhnol.*, (2), 357, 1958.

1959HOR   Horth, A., Patterson, D., and Rinfret, M., The apparent specific volume of polymers in dilute solutions, *J. Polym. Sci.*, 39, 189, 1959.

1959ZEL   Zelikman, S.G. and Mikhailov, N.V., Studies on the structure and properties of carbochain polymers in dilute solutions. IV. The integral and differential heats of solution and densities of the polymers (Russ.), *Vysokomol. Soedin.*, 1, 1077, 1959.

1960MUE   Müller, F.H. and Engelter, A., Über den Einfluss der Lagerung auf die innere Energie von kaltverstrecktem Polyvinylchlorid, *Kolloid Z. Z. Polym.*, 171, 152, 1960.

1962DE1   Delmas, G., Patterson, D., and Böhme, A., Application of the Prigogine solution theory to the heats of mixing of some polymer-solvent systems, *Trans. Faraday Soc.*, 58, 2116, 1962.

1962DE2   Delmas, G., Patterson, D., and Somcynsky, T., Thermodynamics of polyisobutylene-n-alkane systems, *J. Polym. Sci.*, 57, 79, 1962.

1963TAG   Tager, A.A. and Podlesnyak, A.I., Concentrated polymer solutions. I. Determination of the integral and differential heats of solution and dilution of polyisobutylene and polystyrene (Russ.), *Vysokomol. Soedin., Ser. A*, 5, 87, 1963.

1964DEL   Delmas, G., Patterson, D., and Bhattacharyya, S.N., Heats of mixing of polymers with mixed-solvent media, *J. Phys. Chem.*, 68, 1468, 1964.

1964DIE   Diemusch, J. and Banderet, A., Chaleur de dissolution de differentes fibres acryliques, *C. R. Acad. Sci. Paris*, 258, 4719, 1964.

1964PAT   Patterson, D., Heats of mixing of polymers with ester and ether solvents, *J. Polym. Sci.: Part A*, 2, 5177, 1964.

1964ZVE   Zverev, M.P., Barash, A.N., and Zubov, P.I., The heat of precipitation of polyacrylonitrile from solution (Russ.), *Vysokomol. Soedin., Ser. A*, 6, 1012, 1964.

1965COT   Cottam, B.J., Cowie, J.M.G., and Bywater, S., Some solution properties of poly(α-methyl-styrene)s of low molecular weight, *Makromol. Chem.*, 86, 116, 1965.

1965NAU   Naumova, S.F. and Vasil'eva, V.V., Thermal effects of polymerization of 1,3-cyclohexadiene under the action of $TiCl_4$ in various solvents (Beloruss.), *Vestsi Akad. Navuk BSSR, Ser. Khim. Navuk*, (4), 26, 1965.

1966BIA   Bianchi, U., Pedemonte, E., and Rossi, C., Heat of solution of amorphous polymers in θ-solvents, *Makromol. Chem.*, 92, 114, 1966.

1966CUN   Cuniberti, C. and Bianchi, U., Thermodynamic of mixing: excess volumes in polyiso-butylene-solvent systems, *Polymer*, 7, 151, 1966.

1966GIA   Giacommeti, G. and Turolla, A., Heat of solution of poly(γ-benzyl-L-glutamate) in dichloroethane-dichloroacetic acid mixtures, *Z. Phys. Chem. N. F.*, 51, 108, 1966.

1967BIA   Bianchi, U., Cuniberti, C., Pedemonte, E., and Rossi, C., Energy contents of amorphous polystyrene with different thermal histories, *J. Polym. Sci.: Part A-2*, 5, 743, 1967.

1967SCH   Schreiber, H.P. and Waldman, M.H., Isothermal calorimeter for studies of polymer solution processes, *J. Polym. Sci.: Part A-2*, 5, 555, 1967.

1968ALL   Allen, G., Ayerst, R.C., Cleveland, J.R., Gee, G., and Price, C., Effect of conditions of glass formation on the density and energy of polystyrene, *J. Polym. Sci.: Part C*, 23, 127, 1968.

1968GOL   Goldfarb, J. and Rodriguez, S., Aqueous solutions of polyvinylpyrrolidone, *Makromol. Chem.*, 116, 96, 1968.

1968HES   Heskins, M. and Guillet, J.E., Solution properties of poly(*N*-isopropylacrylamide), *J. Macromol. Sci.-Chem. A*, 2, 1441, 1968.

1968MA2   Maron, S.H. and Daniels, C.A., Thermodynamics of polystyrene solutions. II. Glass contributions and thermal behavior, *J. Macromol. Sci.-Phys. B*, 2, 769, 1968.

1968MOR   Morimoto, S., On the heat of solution of polymer with solvent: polydimethylsiloxane-solvent systems, *J. Polym. Sci.: Part A-1*, 6, 1547, 1968.

1969BIA   Bianchi, U., Cuniberti, C., Pedemonte, E., and Rossi, C., Conformational energy contribution to the heat of solution in polymer-solvent systems. Part II. Experimental results, *J. Polym. Sci.: Part A-2*, 7, 855, 1969.

1970GER   Gerth, Ch. and Müller, F.H., Folgerungen auf Nahordnungsstrukturen in isotaktischen Polymethylmethacrylaten aus Lösungswärmen, *Kolloid Z. Z. Polym.*, 241, 1071, 1970.

1970GIA   Giacommeti, G., Turolla, A., and Boni, R., Enthalpy of helix-coil transitions from heats of solution. II. Poly(ϵ-carbobenzoxy-L-lysine) and copolymers with L-phenylalanine, *Biopolymers*, 9, 979, 1970.

1970LID   Liddell, A.H. and Swinton, F.L., Thermodynamic properties of some polymer solutions at elevated temperatures, *Discuss. Faraday Soc.*, 49, 115, 1970.

1970MO5   Morimoto, S., Molecular weight dependence of the heat of solution of polystyrene (Jap.), *Nippon Kagaku Zasshi*, 91, 31, 1970.

1970NAK   Nakayama, H., Temperature dependence of the heat of solution of poly(ethylene glycol) and of related compounds, *Bull. Chem. Soc. Japan*, 43, 1683, 1970.

1970STO   Stölting, J. and Müller, F.H., Energie-Elastizität von Polystyrol bei der Warmverstreckung. Teil 1: Experimentelle Ergebnisse - Lösungswärmen in Abhängigkeit von der Verstreckung, *Kolloid Z. Z. Polym.*, 238, 459, 1970.

1972BLA   Blackadder, D.A. and Roberts, T.L., The application of a new DTA technique to the measurement of the heats of fusion of bulk crystallized polyethylenes and polyethylene single crystals, *Angew. Makromol. Chem.*, 27, 165, 1972.

1972MA1   Maron, S.H. and Filisko, F.E., A modified Tian-Calvet microcalorimeter for polymer solution measurements, *J. Macromol. Sci.-Phys. B*, 6, 57, 1972.

1972MA2   Maron, S.H. and Filisko, F.E., Heats of solution and dilution for polyethylene oxide in several solvents, *J. Macromol. Sci.-Phys. B*, 6, 79, 1972.

1972MA3   Maron, S.H. and Filisko, F.E., Heats of solution and dilution for polyvinyl chloride in cyclohexanone and tetrahydrofuran, *J. Macromol. Sci.-Phys. B*, 6, 413, 1972.

1972MAS   Masa, Z., Biros, J., Trekoval, J., and Pouchly, J., Specific interactions in solutions of polymers. IV. A calorimetric study of butyl methacrylate oligomers in chloroform, *J. Polym. Sci.: Part C*, 39, 219, 1972.

1972SOK   Sokolova, D.F., Sokolov, L.B., and Gerasimov, V.D., Teplota rastvoreniya poli-m-fenilen-izoftalamida razlichnoi struktury v amidno-solevykh rastvoritelyakh, *Vysokomol. Soedin., Ser. B*, 14, 580, 1972.

1973CHA   Chahal, R.S., Kao, W.-P., and Patterson, D., Thermodynamics of polydimethylsiloxane solutions, *J. Chem. Soc., Faraday Trans. I*, 69, 1834, 1973.

1973DEL   Delmas, G. and Tancrede, P., Heats of mixing of atactic polybutene-1 and of a model molecule with normal alkanes: order in long alkanes, *Eur. Polym. J.*, 9, 199, 1973.

1974FI1   Filisko, F.E. and Raghava, R.S., Amorphous structure heat: molecular mechanisms from solution heats of poly(2,6-dimethyl-p-phenylene oxide) in orthodichlorobenzene, *J. Appl. Phys.*, 45, 4151, 1974.

1974FI2   Filisko, F.E., Raghava, R.S., and Yeh, G.S.Y., Amorphous structure heat. Temperature dependence of heats of solution for polystyrene in toluene and ethylbenzene, *J. Macromol. Sci.-Phys. B*, 10, 371, 1974.

1975BAB   Baba, Y., Fujimoto, K., and Kagemoto, A., Heats of solution of mixture of poly-γ-benzyl-L- and - D-glutamate, *Mem. Osaka Inst. Technol., Ser. A Sci. Technol.*, 20, 99, 1975.

1975IKE   Ikeda, M., Suga, H., and Seki, S., Thermodynamic studies of solid polyethers: 5. Crystalline-amorphous interfacial thermal properties, *Polymer*, 16, 634, 1975.

1975TAN   Tancrede, P., Patterson, D., and Lam, V.-T., Thermodynamic effects of orientation order in chain-molecule mixtures, *J. Chem. Soc., Faraday Trans. II*, 71, 985, 1975.

1976KIS   Kiselev, V.P., Shakhova, E.M., Fainberg, E.Z., Virnik, A.D, and Rogovin, Z.A., Measure-ment of density and heat of solution of various dextran samples (Russ.), *Vysokomol. Soedin., Ser. B*, 18, 847, 1976.

1976KOC   Koch, W., Heats of solution and reactions of aromatic amides, nylons, and acid chlorides in the N,N-dimethylacetamide/lithium chloride system, *Lenzinger Berichte*, 40, 14, 1976.

1976PET   Petrosyan, V.A., Gabrielyan, G.A., and Rogovin, Z.A., Investigation of some properties of fiber-forming copolymers of acrylonitrile and isoprene, *Arm. Khim. Zh.*, 29, 516, 1976.

1977DES   Deshpande, D.D. and Prabhu, C.S., Contact interactions of polyisobutylene polymers, *Macromolecules*, 10, 433, 1977.

1977MIU   Miura, T. and Nakamura, M., The heat of solution of poly(oxyethylene) dodecyl ethers in dodecane, *Bull. Chem. Soc. Japan*, 50, 2528, 1977.

1977OCH   Ochiai, H., Nishihara, Y., Yamaguchi, S., and Murakami, I., Thermodynamic properties of polypropylene–chlorinated methane systems, *J. Sci. Hiroshima Univ., Ser. A*, 41, 157, 1977.

1978SOK   Sokolova, D.F., Kudim, T.V., Sokolov, L.B., Zhegalova, N.I., and Zhuravlev, N.D., Thermochemical evaluation of the orderliness of copolymer polyarylates (Russ.), *Vysokomol. Soedin., Ser. B*, 20, 596, 1978.

1979BAS   Basedow, A.M. and Ebert, K.H., Production, characterization, and solution properties of dextran fractions of narrow molecular weight distributions, *J. Polym. Sci. Polym. Symp.*, 66, 101, 1979.

1979LEE   Lee, J.-O., Ono, M., Hamada, F., and Nakajima, A., Thermodynamic study on heat of mixing of polyisobutylene with ethylbenzene, *Polym. Bull.*, 1, 763, 1979.

1979PH1   Phuong-Nguyen, H. and Delmas, G., Heats of mixing at infinite dilution of atactic polymers. 1. Effect of correlations of orientations in the solvents or in the polymers, *Macromolecules*, 12, 740, 1979.

1979PH2   Phuong-Nguyen, H. and Delmas, G., Heats of mixing at infinite dilution of atactic polymers. 2. Effect of steric hindrance and of the shape of the solvent, *Macromolecules*, 12, 746, 1979.

1979KOL   Koller, J., Kalorimetrische Untersuchungen der inter- und intramolekularen Wechselwirkungen von Polyethylenglycolen in wässriger und benzolischer Lösung, *Dissertation*, TU München, 1979.

1980BAS   Basedow, A.M., Ebert, K.H., and Feigenbutz, W., Polymer-solvent interactions. Dextrans in water and DMSO, *Makromol. Chem.*, 181, 1071, 1980.

1980GRA   Graun, K., Kalorimetrische Untersuchungen der Wechselwirkungen von taktischen und ataktischen Polymethylmethacrylaten in verdünnter Lösung, *Dissertation*, TU München, 1980.

1980SH1   Shiomi, T., Izumi, Z., Hamada, F., and Nakajima, A., Thermodynamics of solutions of poly(dimethylsiloxane). 1. Solutions of poly(dimethylsiloxane) in methyl ethyl ketone, methyl isobutyl ketone, and diisobutyl ketone, *Macromolecules*, 13, 1149, 1980.

1980SH2   Shiomi, T., Kohra, Y., Hamada, F., and Nakajima, A., Thermodynamics of solutions of poly(dimethylsiloxane). 2. Solution of poly(dimethylsiloxane) in octamethylcyclotetrasiloxane, *Macromolecules*, 13, 1154, 1980.

1981AHA   Aharoni, S.M., Charlet, G., and Delmas, G., Investigation of solutions and gels of poly(4-methyl-1-pentene) in cyclohexane and decalin by viscometry, calorimetry, and X-ray diffraction. A new crystalline form of poly(4-methyl-1-pentene) from gels, *Macromolecules*, 14, 1390, 1981.

1981KOL   Koller, J. and Killmann, E., Lösungs- und Verdünnungswärmen von Polyethylenglykolen in Wasser und Benzol, 1. Experimentelle Ergebnisse, *Makromol. Chem.*, 182, 3579, 1981.

1982MEE   Meerson, S.I., Shakhova, E.M., Prokof'eva, M.V., and Khin, N.N., Distribution of electrolytes and phase equilibria in cellulose ethers-water-electrolyte systems (Russ.), *Vysokomol. Soedin., Ser. A*, 24, 634, 1982.

1982OCH   Ochiai, H., Ohashi, T., Tadokoro, Y., and Murakami, I., Thermodynamic properties of atactic polypropylene solutions, *Polym. J.*, 14, 457, 1982.

1982SH1   Sharma, S.C., Mahajan, R., Sharma, V.K., and Lakhanpal, M.L., Enthalpies of mixing of poly(tetramethylene oxides) with dioxane and cyclohexane, *Indian J. Chem.*, 21A, 682, 1982.

1982SH2   Sharma, S.C., Mahajan, R., Sharma, V.K., and Lakhanpal, M.L., Enthalpies of mixing of poly(tetramethylene oxides) with benzene and carbon tetrachloride, *Indian J. Chem.*, 21A, 685, 1982.

1982STA   Starikova, N.A., Medved', Z.N., Mayatskii, V.A., and Terasimov, V.D., Kalorimetricheskoe izuchenie aromaticheskikh sopoliamidov, *Vysokomol. Soedin., Ser. B*, 24, 709, 1982.

1983LAK   Lakhanpal, M.L. and Parashar, R.N., Enthalpies of mixing for aqueous solutions of polyoxyethylene glycols, *Indian J. Chem.*, 22A, 48, 1983.

1984KIL   Killmann, E. and Graun, K., Microcalorimetric studies on enthalpic interactions of poly(methyl methacrylate) of varying tacticity in solution (Ger.), *Makromol. Chem.*, 185, 1199, 1984.

1985AEL   Aeleni, N., Studiul dizolvarii polistirenului in benzen si toluen, *Mater. Plast. (Bucharest)*, 22, 92, 1985.

1985SHA   Sharma, S.C. and Sharma, V.K., Enthalpies of mixing of poly(tetramethylene oxide) with benzene, methylbenzene, ethylbenzene and propylbenzene at 313.15 K, *Indian J. Chem.*, 24A, 292, 1985.

1986BON   Bondarenko, O.A., Polyak, B.A., Shablygin, M.V., Papkov, S.P., and Okromechedlidze, N.P., Thermochemistry of the dissolution of poly(N-phenyleneterephthalamide) in concentrated sulfuric acid (Russ.), *Khim. Volokna*, (5), 45, 1986.

1986SHA   Sharma, S.C., Bhalla, S., and Sharma, V.K., Enthalpies of mixing of poly(tetramethylene oxide) with o-, m-, and p-xylenes and mesitylene, *Indian J. Chem.*, 25A, 131, 1986.

1987CIF   Cifra, P. and Romanov, A., Enthalpies of solution of methoxy and ethoxy end group oligomers of poly(ethylene glycol) in binary solvent water-dimethylformamide. Study of hydrophobic hydration, *Chem. Pap.*, 41, 395, 1987.

1987KOJ   Kojima, T., Hamada, M., Miyahara, M., and Hosaka, Y., Solute-solvent interactions of oligo(m-benzamide)s in N,N-dimethylacetamide, *J. Polym. Sci.: Part B: Polym. Phys.*, 25, 1481, 1987.

1987SHA   Sharma, S.C., Syngal, M., and Sharma, V.K., Enthalpies and excess volumes of mixing of poly(tetramethylene oxide) fractions with tetrachloromethane, 1,2-dichloroethane and 1,1,1-trichloroethane at 313.15 K, *Indian J. Chem.*, 26A, 285, 1987.

1988AUK   Aukett, P.N. and Brown, C.S., The measurement of heats of mixing of polymer alloys by a heat of solution method, *J. Therm. Anal.*, 33, 1079, 1988.

1988LAN   Lanzavecchia, L. and Pedemonte, E., Evaluation of the interaction parameter for polystyrene-poly(α-methylstyrene) mixtures from heat of solution measurements, *Thermochim. Acta*, 137, 123, 1988.

1988PAR   Parashar, R. and Sharma, S.C., Enthalpies of mixing of polyoxypropyleneglycols with benzene, ethanol and water, *Indian J. Chem.*, 27A, 1092, 1988.

1988TAG   Tager, A.A., Safronov, A.P., Voit, V.V., Lopyrev, V.A., Ermakova, T.G., Tatarova, L.A., and Shagelaeva, N.S., Heats of solution of poly(1-vinylpyrazole), poly(1-vinylimidazole), and poly(1-vinyl-1,2,4-triazole) in organic donor solvents (Russ.), *Vysokomol. Soedin., Ser. A*, 30, 2360, 1988.

1989SAF   Safronov, A.P., Tager, A.A., Sharina, S.V., Lopyrev, V.A., Ermakova, T.G., Tatarova, L.A., and Kashik, T.N., Nature of hydration in aqueous solutions of poly(1-vinylazoles) (Russ.), *Vysokomol. Soedin., Ser. A*, 31, 2662, 1989.

1990PED   Pedemonte, E. and Lanzavecchia, L., Pair interaction parameter for compatible polystyrene-poly(α-methylstyrene) mixtures, *Thermochim. Acta*, 162, 223, 1990.

1990SHI   Shiomi, T., Ishimatsu, H., Eguchi, T., and Imai, K., Application of equation-of-state theory to random copolymer systems. 1. Copolymer solutions in solvent, *Macromolecules*, 23, 4970, 1990.

1991TA1   Tager, A.A. and Safronov, A.P., Complex formation in aqueous and nonaqueous solutions of poly(vinylazoles) (Russ.), *Vysokomol. Soedin., Ser. A*, 33, 67, 1991.

1993ZEL   Zellner, H., Mischungsthermodynamik und Energetik der Wechselwirkungen von Polyethern in Lösungsmitteln unterschiedlicher Polarität, *Dissertation*, TU München, 1993.

1994BRU   Brunacci, A., Pedemonte, E., Cowie, J.M.G., and McEwen, I.J., The thermodynamics of mixing of polystyrene and poly(α-methylstyrene) from a calorimetric viewpoint, *Polymer*, 35, 2893, 1994.

1994PHU   Phuong-Nguyen, H. and Delmas, G., Chain dynamics of high molecular weight polyethylene as observed from heats of dissolution in slow calorimetry, *J. Solution Chem.*, 23, 249, 1994.

1995CAR   Carlsson, M., Hallen, D., and Linse, P., Mixing enthalpy and phase separation in a poly(propylene oxide)-water system, *J. Chem. Soc., Faraday Trans.*, 91, 2081, 1995.

1996SAT   Sato, T., Tohyama, M., Suzuki, M., Shiomi, T., and Imai, K., Application of equation-of-state theory to random copolymer blends with upper critical solution temperature type miscibility, *Macromolecules*, 29, 8231, 1996.

1996SHI   Shiomi, T., Tohyama, M., Endo, M., Sato, T., and Imai, K., Dependence of Flory-Huggins χ-parameters on the copolymer composition for solutions of poly(methyl methacrylate-*ran*-n-butyl methacrylate) in cyclohexanone, *J. Polym. Sci., Part B: Polym. Phys.*, 34, 2599, 1996.

1997KI1   Kizhnyaev, V.N., Gorkovenko, O.P., Safronov, A.P., and Adamova, L.V., Thermodynamics of the interaction between tetrazole-containing polyelectrolytes and water (Russ.), *Vysokomol. Soedin., Ser. A*, 39, 527, 1997.

1997KI2   Kizhnyaev, V.N., Gorkovenko, O.P., Bazhenov, D.N., and Smirnov, A.I., The solubilities and enthalpies of a solution of polyvinyltetrazoles in organic solvents (Russ.), *Vysokomol. Soedin., Ser. A*, 39, 856, 1997.

1997SAT   Sato, T., Suzuki, M., Tohyama, M., Endo, M., Shiomi, T., and Imai, K., Behavior of the temperature dependence of χ-parameter in random copolymer blends showing an immiscibility window, *Polym. J.*, 29, 417, 1997.

1998OHT   Ohta, A., Takiue, T., Ikeda, N., and Aratono, M., Calorimetric study of dilute aqueous solutions of ethylene glycol oligomers, *J. Phys. Chem. B*, 102, 4809, 1998.

1999NOV   Novoselov, N.P., Sahsina, E.S., Tager, E., Kozlov, I.L., Kurlykin, M.P., Polythermal study of the heat of dissolution of cellulose in N-methylmorpholine-N-oxide (Russ.), *Zh. Prikl. Khim.*, 72, 1192, 1999.

2001NOV   Novoselov, N.P., Sahsina, E.S., Kozlov, I.L., Solubility of cellulose in mixtures of N-methylmorpholine-N-oxide monohydrate and aprotic solvents (Russ.), *Zh. Fiz. Khim.*, 75, 1254, 2001.

2002RIG   Righetti, M.C., Cardelli, C., Scalari, M., Tombari, E., and Conti, G., Thermodynamics of mixing poly(vinyl chloride) and poly(ethylene-*co*-vinyl acetate), *Polymer*, 43, 5035, 2002.

2003CAR   Cardelli, C., Conti, G., Gianni, P., and Porta, R., Blend formation between homo- and co-polymers at 298.15 K. PMMA-SAN blends, *J. Therm. Anal. Calorim.*, 71, 353, 2003.

4. SOLVENT PARTIAL ENTHALPIES OF MIXING MEASURED BY CALORIMETRY

4.1. Experimental data

Polymer (B):	hexamethyldisiloxane		1983NEF
Characterization:	$M/\text{g.mol}^{-1} = 162.4$		
Solvent (A):	benzene	C_6H_6	71-43-2

$T/\text{K} = 298.15$

φ_B	0.412	0.600	0.726	0.927
$\Delta_M H_A/(RT\varphi_B^2)$	0.873	0.912	0.923	0.974

Comments: The originally measured intermediary enthalpies of dilution were reduced by the authors to $\Delta_M H_A/(RT\varphi_B^2)$ to get data which depend less on concentration (see also the next data sets below).

Polymer (B):	hexamethyldisiloxane		1983NEF
Characterization:	$M/\text{g.mol}^{-1} = 162.4$		
Solvent (A):	n-heptane	C_7H_{16}	142-82-5

$T/\text{K} = 298.15$

φ_B	0.426	0.527	0.695	0.770	0.857	0.955
$\Delta_M H_A/(RT\varphi_B^2)$	0.223	0.227	0.234	0.237	0.244	0.245

$T/\text{K} = 313.15$

φ_B	0.349	0.390	0.573	0.625	0.858	0.955
$\Delta_M H_A/(RT\varphi_B^2)$	0.187	0.194	0.204	0.210	0.228	0.244

Polymer (B):	hexamethyldisiloxane		1983NEF
Characterization:	$M/\text{g.mol}^{-1} = 162.4$		
Solvent (A):	n-nonane	C_9H_{20}	111-84-2

$T/\text{K} = 298.15$

φ_B	0.327	0.361	0.445	0.623	0.681	0.749	0.837	0.946
$\Delta_M H_A/(RT\varphi_B^2)$	0.355	0.352	0.362	0.373	0.375	0.374	0.382	0.402

$T/\text{K} = 313.15$

φ_B	0.359	0.404	0.695	0.770	0.858	0.955
$\Delta_M H_A/(RT\varphi_B^2)$	0.308	0.294	0.322	0.328	0.330	0.332

Polymer (B):	**hexamethyldisiloxane**						**1983NEF**
Characterization:	$M/\text{g.mol}^{-1} = 162.4$						
Solvent (A):	**n-octane**		C_8H_{18}				**111-65-9**

$T/\text{K} = 298.15$

φ_B	0.375	0.525	0.610	0.659	0.715	0.772	0.860	0.953
$\Delta_M H_A/(RT\varphi_B{}^2)$	0.279	0.282	0.285	0.291	0.289	0.283	0.304	0.318

$T/\text{K} = 313.15$

φ_B	0.394	0.481	0.528	0.655	0.718	0.780	0.855	0.956
$\Delta_M H_A/(RT\varphi_B{}^2)$	0.245	0.243	0.245	0.242	0.242	0.247	0.257	0.280

Polymer (B):	**hexamethyldisiloxane**						**1983NEF**
Characterization:	$M/\text{g.mol}^{-1} = 162.4$						
Solvent (A):	**toluene**		C_7H_8				**108-88-3**

$T/\text{K} = 298.15$

φ_B	0.291	0.402	0.598	0.708	0.924
$\Delta_M H_A/(RT\varphi_B{}^2)$	0.535	0.528	0.599	0.629	0.730

$T/\text{K} = 313.15$

φ_B	0.268	0.287	0.319	0.360	0.400	0.610	0.655	0.725
$\Delta_M H_A/(RT\varphi_B{}^2)$	0.520	0.530	0.521	0.542	0.546	0.573	0.605	0.610

φ_B	0.813	0.919
$\Delta_M H_A/(RT\varphi_B{}^2)$	0.638	0.656

Polymer (B):	**hexamethyldisiloxane**						**1983NEF**
Characterization:	$M/\text{g.mol}^{-1} = 162.4$						
Solvent (A):	**2,2,4-trimethylpentane**		C_8H_{18}				**540-84-1**

$T/\text{K} = 298.15$

φ_B	0.362	0.406	0.676	0.746	0.830	0.943
$\Delta_M H_A/(RT\varphi_B{}^2)$	0.205	0.203	0.209	0.213	0.213	0.217

Polymer (B):	**maltopentaose**						**2002COO**
Characterization:	$M/\text{g.mol}^{-1} = 828.7$						
Solvent (A):	**water**		H_2O				**7732-18-5**

$T/\text{K} = 318.15$

x_B	0.00084	0.00113	0.00159	0.00229
$\Delta_M H_A/\text{J.mol}^{-1}$	0.05	−0.02	−0.10	−0.14

Polymer (B):	**maltotetraose**			**2002COO**
Characterization:	$M/\text{g.mol}^{-1} = 666.6$			
Solvent (A):	**water**	H_2O		**7732-18-5**

$T/K = 318.15$

x_B	0.00123	0.00171	0.00250
$\Delta_M H_A/\text{J.mol}^{-1}$	−0.01	−0.10	−0.22

Polymer (B):	**maltotriose**			**2002COO**
Characterization:	$M/\text{g.mol}^{-1} = 504.4$			
Solvent (A):	**water**	H_2O		**7732-18-5**

$T/K = 318.15$

x_B	0.00128	0.00195	0.00303	0.00345	0.00482	0.00541	0.01259
$\Delta_M H_A/\text{J.mol}^{-1}$	0.02	0.01	−0.18	−0.54	−0.64	−1.77	−5.07

Polymer (B):	**octamethylcyclotetrasiloxane**			**1983NEF**
Characterization:	$M/\text{g.mol}^{-1} = 296.6$			
Solvent (A):	**benzene**	C_6H_6		**71-43-2**

$T/K = 298.15$

φ_B	0.220	0.245	0.305	0.305	0.425	0.425
$\Delta_M H_A/(RT\varphi_B^2)$	0.633	0.643	0.627	0.633	0.634	0.640

Comments: The originally measured intermediary enthalpies of dilution were reduced by the authors to $\Delta_M H_A/(RT\varphi_B^2)$ to get data which depend less on concentration (see also the next data sets below).

Polymer (B):	**octamethylcyclotetrasiloxane**			**1983NEF**
Characterization:	$M/\text{g.mol}^{-1} = 296.6$			
Solvent (A):	**n-nonane**	C_9H_{20}		**111-84-2**

$T/K = 298.15$

φ_B	0.413	0.636	0.700	0.772	0.861	0.958
$\Delta_M H_A/(RT\varphi_B^2)$	0.615	0.664	0.683	0.690	0.682	0.713

Polymer (B):	**octamethylcyclotetrasiloxane**			**1983NEF**
Characterization:	$M/\text{g.mol}^{-1} = 296.6$			
Solvent (A):	**toluene**	C_7H_8		**108-88-3**

$T/K = 298.15$

φ_B	0.415	0.635	0.858	0.957
$\Delta_M H_A/(RT\varphi_B^2)$	0.381	0.483	0.584	0.657

Polymer (B):   **octamethylcyclotetrasiloxane**   **1983NEF**
Characterization: $M/g.mol^{-1} = 296.6$
Solvent (A):   **2,2,4-trimethylpentane**   **C_8H_{18}**   **540-84-1**

$T/K = 298.15$

φ_B	0.309	0.420	0.599	0.768	0.851
$\Delta_M H_A/(RT\varphi_B^2)$	0.460	0.468	0.477	0.461	0.469

Polymer (B):   **poly(dimethylsiloxane)**   **1983NEF**
Characterization: $M_n/g.mol^{-1} = 594$, $M_w/g.mol^{-1} = 665$, PDMS2, Goldschmidt AG
Solvent (A):   **benzene**   **C_6H_6**   **71-43-2**

$T/K = 298.15$

φ_B	0.190	0.256	0.343	0.350	0.420	0.440	0.610	0.637
$\Delta_M H_A/(RT\varphi_B^2)$	0.575	0.606	0.644	0.655	0.662	0.665	0.732	0.740

φ_B	0.662	0.698	0.730	0.765	0.812	0.860	0.911	0.965
$\Delta_M H_A/(RT\varphi_B^2)$	0.752	0.760	0.787	0.794	0.801	0.823	0.844	0.869

Comments: The originally measured intermediary enthalpies of dilution were reduced by the authors to $\Delta_M H_A/(RT\varphi_B^2)$ to get data which depend less on concentration (see also the next data sets below).

Polymer (B):   **poly(dimethylsiloxane)**   **1983NEF**
Characterization: $M_n/g.mol^{-1} = 958$, $M_w/g.mol^{-1} = 1418$, PDMS10, Goldschmidt AG
Solvent (A):   **benzene**   **C_6H_6**   **71-43-2**

$T/K = 298.15$

φ_B	0.182	0.237	0.275	0.308	0.339	0.376	0.428	0.462
$\Delta_M H_A/(RT\varphi_B^2)$	0.436	0.459	0.477	0.491	0.503	0.528	0.542	0.573

φ_B	0.465	0.535	0.655	0.671	0.692
$\Delta_M H_A/(RT\varphi_B^2)$	0.551	0.582	0.650	0.649	0.672

Polymer (B):   **poly(dimethylsiloxane)**   **1950NEW**
Characterization: $M_n/g.mol^{-1} = 1140$, $\eta = 10$ cSt, Dow Corning Corporation, MI
Solvent (A):   **benzene**   **C_6H_6**   **71-43-2**

$T/K = 298.15$

x_A	0.1	0.2	0.3	0.4	0.5	0.6	0.7	0.8
φ_A	0.0085	0.0220	0.0318	0.0487	0.0714	0.103	0.151	0.234
$\Delta_M H_A/(J/mol\ solvent)$	2109	2029	1975	1841	1770	1578	1356	1059

x_A	0.85	0.9	0.925	0.95	0.975	0.985	0.990
φ_A	0.302	0.408	0.485	0.592	0.748	0.835	0.885
$\Delta_M H_A/(J/mol\ solvent)$	816	531	393	232	87	36	17

Polymer (B):	poly(dimethylsiloxane)						**1983NEF**
Characterization:	M_n/g.mol^{-1} = 1540, M_w/g.mol^{-1} = 1880, PDMS20, Goldschmidt AG						
Solvent (A):	**benzene**		**C$_6$H$_6$**				**71-43-2**

T/K = 298.15

φ_B	0.252	0.301	0.335	0.378	0.432	0.587	0.612	0.648
$\Delta_M H_A/(RT\varphi_B{}^2)$	0.405	0.419	0.433	0.449	0.467	0.567	0.580	0.598

φ_B	0.668	0.706	0.746	0.798	0.846	0.904	0.968	
$\Delta_M H_A/(RT\varphi_B{}^2)$	0.616	0.635	0.646	0.680	0.707	0.766	0.792	

Polymer (B):	poly(dimethylsiloxane)						**1950NEW**
Characterization:	M_η/g.mol^{-1} = 3850, η = 50 cSt, Dow Corning Corporation, MI						
Solvent (A):	**benzene**		**C$_6$H$_6$**				**71-43-2**

T/K = 298.15

x_A	0.1	0.2	0.3	0.4	0.5	0.6	0.7	0.8
φ_A	0.00255	0.00574	0.00981	0.0151	0.0225	0.0335	0.0511	0.0843
$\Delta_M H_A/$(J/mol solvent)	1929	1921	1900	1870	1833	1778	1686	1557

x_A	0.9	0.925	0.94	0.95	0.96	0.97	0.98	0.99
φ_A	0.171	0.221	0.266	0.305	0.356	0.427	0.530	0.695
$\Delta_M H_A/$(J/mol solvent)	1226	1050	912	791	653	494	301	115

x_A	0.995
φ_A	0.820
$\Delta_M H_A/$(J/mol solvent)	33.5

Polymer (B):	poly(dimethylsiloxane)						**1983NEF**
Characterization:	M_n/g.mol^{-1} = 4170, M_w/g.mol^{-1} = 5920, PDMS100, Goldschmidt AG						
Solvent (A):	**benzene**		**C$_6$H$_6$**				**71-43-2**

T/K = 298.15

φ_B	0.196	0.200	0.248	0.266	0.328	0.369	0.418	0.538
$\Delta_M H_A/(RT\varphi_B{}^2)$	0.333	0.373	0.357	0.375	0.393	0.406	0.453	0.529

φ_B	0.567	0.621	0.676	0.706	0.737	0.761	0.841	
$\Delta_M H_A/(RT\varphi_B{}^2)$	0.544	0.587	0.613	0.630	0.641	0.671	0.710	

Polymer (B):	poly(dimethylsiloxane)						**1983NEF**
Characterization:	M_n/g.mol^{-1} = 6560, M_w/g.mol^{-1} = 11220, PDMS350, Goldschmidt AG						
Solvent (A):	**benzene**		**C$_6$H$_6$**				**71-43-2**

T/K = 298.15

φ_B	0.246	0.284	0.308	0.320	0.338	0.369	0.404	0.420
$\Delta_M H_A/(RT\varphi_B{}^2)$	0.324	0.367	0.384	0.365	0.409	0.427	0.433	0.469

φ_B	0.440	0.462	0.491	0.506	0.544	0.555	0.603	0.608
$\Delta_M H_A/(RT\varphi_B{}^2)$	0.458	0.467	0.452	0.504	0.498	0.548	0.545	0.551

φ_B	0.647	0.677
$\Delta_M H_A/(RT\varphi_B{}^2)$	0.584	0.608

Polymer (B):	**poly(dimethylsiloxane)**					**1983NEF**

Characterization: M_n/g.mol^{-1} = 7860, M_w/g.mol^{-1} = 17060, PDMS1000, Goldschmidt AG

Solvent (A):   **benzene**   **C$_6$H$_6$**   **71-43-2**

T/K = 298.15

φ_B	0.348	0.384	0.425	0.470	0.530
$\Delta_M H_A/(RT\varphi_B^2)$	0.406	0.415	0.424	0.458	0.513

Polymer (B):	**poly(dimethylsiloxane)**							**1950NEW**

Characterization: M_η/g.mol^{-1} = 15700, η = 350 cSt, Dow Corning Corporation, MI

Solvent (A):   **benzene**   **C$_6$H$_6$**   **71-43-2**

T/K = 298.15

x_A	0.1	0.2	0.3	0.4	0.5	0.6	0.7	0.8
φ_A	0.00062	0.00139	0.00256	0.00368	0.0055	0.00825	0.0127	0.0217
$\Delta_M H_A$/(J/mol solvent)	1841	1837	1833	1828	1820	1808	1770	1753

x_A	0.9	0.95	0.975	0.985	0.990	0.995	0.997	0.999
φ_A	0.0478	0.096	0.178	0.267	0.356	0.525	0.650	0.848
$\Delta_M H_A$/(J/mol solvent)	1632	1431	1138	874	598	280	138	18

Polymer (B):	**poly(dimethylsiloxane)**							**1983NEF**

Characterization: M_n/g.mol^{-1} = 594, M_w/g.mol^{-1} = 665, PDMS2, Goldschmidt AG

Solvent (A):   **n-heptane**   **C$_7$H$_{16}$**   **142-82-5**

T/K = 298.15

φ_B	0.390	0.429	0.471	0.520	0.585	0.615	0.671	0.744
$\Delta_M H_A/(RT\varphi_B^2)$	0.162	0.168	0.178	0.189	0.198	0.202	0.212	0.226

φ_B	0.827	0.936
$\Delta_M H_A/(RT\varphi_B^2)$	0.244	0.273

T/K = 313.15

φ_B	0.459	0.507	0.560	0.617	0.685	0.689	0.830	0.945
$\Delta_M H_A/(RT\varphi_B^2)$	0.168	0.178	0.196	0.204	0.214	0.199	0.248	0.275

Comments: The originally measured intermediary enthalpies of dilution were reduced by the authors to $\Delta_M H_A/(RT\varphi_B^2)$ to get data which depend less on concentration (see also the next data sets below).

Polymer (B):	**poly(dimethylsiloxane)**							**1983NEF**

Characterization: M_n/g.mol^{-1} = 958, M_w/g.mol^{-1} = 1418, PDMS10, Goldschmidt AG

Solvent (A):   **n-heptane**   **C$_7$H$_{16}$**   **142-82-5**

T/K = 298.15

φ_B	0.427	0.465	0.475	0.517	0.528	0.595	0.595	0.667
$\Delta_M H_A/(RT\varphi_B^2)$	0.139	0.146	0.148	0.158	0.160	0.172	0.175	0.184

φ_B	0.748	0.862	0.942
$\Delta_M H_A/(RT\varphi_B^2)$	0.201	0.226	0.268

Polymer (B):   **poly(dimethylsiloxane)**   **1983NEF**
Characterization: M_n/g.mol^{-1} = 1540, M_w/g.mol^{-1} = 1880, PDMS20, Goldschmidt AG
Solvent (A):   **n-heptane**   **C$_7$H$_{16}$**   **142-82-5**

T/K = 298.15

φ_B	0.410	0.460	0.512	0.516	0.559	0.573	0.573	0.623
$\Delta_M H_A/(RT\varphi_B^2)$	0.107	0.118	0.128	0.133	0.145	0.144	0.148	0.148

φ_B	0.626	0.677	0.708	0.749	0.830	0.942
$\Delta_M H_A/(RT\varphi_B^2)$	0.159	0.162	0.184	0.179	0.212	0.254

T/K = 313.15

φ_B	0.410	0.494	0.557	0.656	0.785	0.875
$\Delta_M H_A/(RT\varphi_B^2)$	0.104	0.125	0.150	0.149	0.188	0.186

Polymer (B):   **poly(dimethylsiloxane)**   **1983NEF**
Characterization: M_n/g.mol^{-1} = 4170, M_w/g.mol^{-1} = 5920, PDMS100, Goldschmidt AG
Solvent (A):   **n-heptane**   **C$_7$H$_{16}$**   **142-82-5**

T/K = 298.15

φ_B	0.408	0.450	0.499	0.560	0.560	0.610	0.672
$\Delta_M H_A/(RT\varphi_B^2)$	0.100	0.110	0.124	0.132	0.141	0.142	0.157

T/K = 313.15

φ_B	0.550	0.608	0.667	0.717
$\Delta_M H_A/(RT\varphi_B^2)$	0.112	0.129	0.146	0.165

Polymer (B):   **poly(dimethylsiloxane)**   **1983NEF**
Characterization: M_n/g.mol^{-1} = 6560, M_w/g.mol^{-1} = 11220, PDMS350, Goldschmidt AG
Solvent (A):   **n-heptane**   **C$_7$H$_{16}$**   **142-82-5**

T/K = 298.15

φ_B	0.406	0.444	0.486	0.543	0.606
$\Delta_M H_A/(RT\varphi_B^2)$	0.101	0.108	0.121	0.124	0.130

T/K = 313.15

φ_B	0.450	0.508	0.556
$\Delta_M H_A/(RT\varphi_B^2)$	0.093	0.098	0.107

Polymer (B):   **poly(dimethylsiloxane)**   **1983NEF**
Characterization: M_n/g.mol^{-1} = 594, M_w/g.mol^{-1} = 665, PDMS2, Goldschmidt AG
Solvent (A):   **n-nonane**   **C$_9$H$_{20}$**   **111-84-2**

T/K = 298.15

φ_B	0.297	0.355	0.438	0.649	0.716	0.788	0.872	0.958
$\Delta_M H_A/(RT\varphi_B^2)$	0.298	0.304	0.321	0.373	0.392	0.415	0.445	0.475

Polymer (B):	**poly(dimethylsiloxane)**						**1983NEF**
Characterization:	M_n/g.mol^{-1} = 1540, M_w/g.mol^{-1} = 1880, PDMS20, Goldschmidt AG						
Solvent (A):	**n-nonane**		**C$_9$H$_{20}$**				**111-84-2**

T/K = 298.15

φ_B	0.364	0.397	0.442	0.477	0.535	0.603	0.642	0.706
$\Delta_M H_A/(RT\varphi_B^2)$	0.207	0.226	0.233	0.267	0.270	0.304	0.314	0.345

φ_B	0.774	0.861	0.957
$\Delta_M H_A/(RT\varphi_B^2)$	0.382	0.437	0.506

T/K = 313.15

φ_B	0.355	0.400	0.725	0.800	0.874	0.955
$\Delta_M H_A/(RT\varphi_B^2)$	0.208	0.223	0.321	0.343	0.379	0.433

Polymer (B):	**poly(dimethylsiloxane)**			**1983NEF**
Characterization:	M_n/g.mol^{-1} = 6560, M_w/g.mol^{-1} = 11220, PDMS350, Goldschmidt AG			
Solvent (A):	**n-nonane**	**C$_9$H$_{20}$**		**111-84-2**

T/K = 298.15

φ_B	0.365	0.401	0.442	0.481
$\Delta_M H_A/(RT\varphi_B^2)$	0.192	0.210	0.219	0.236

T/K = 313.15

φ_B	0.408	0.502	0.561
$\Delta_M H_A/(RT\varphi_B^2)$	0.200	0.220	0.245

Polymer (B):	**poly(dimethylsiloxane)**			**1983NEF**
Characterization:	M_n/g.mol^{-1} = 594, M_w/g.mol^{-1} = 665, PDMS2, Goldschmidt AG			
Solvent (A):	**octamethylcyclotetrasiloxane**	**C$_8$H$_{24}$O$_4$Si$_4$**		**556-67-2**

T/K = 298.15

φ_B	0.465	0.497	0.590	0.616	0.676	0.745	0.831	0.939
$\Delta_M H_A/(RT\varphi_B^2)$	0.114	0.110	0.116	0.115	0.118	0.128	0.129	0.138

Comments: The originally measured intermediary enthalpies of dilution were reduced by the authors to $\Delta_M H_A/(RT\varphi_B^2)$ to get data which depend less on concentration (see also the next data sets below).

Polymer (B):	**poly(dimethylsiloxane)**			**1983NEF**
Characterization:	M_n/g.mol^{-1} = 958, M_w/g.mol^{-1} = 1418, PDMS10, Goldschmidt AG			
Solvent (A):	**octamethylcyclotetrasiloxane**	**C$_8$H$_{24}$O$_4$Si$_4$**		**556-67-2**

T/K = 298.15

φ_B	0.625	0.681	0.752	0.840	0.942
$\Delta_M H_A/(RT\varphi_B^2)$	0.078	0.083	0.084	0.093	0.096

Polymer (B):   **poly(dimethylsiloxane)**   **1983NEF**
Characterization: $M_n/\text{g.mol}^{-1} = 1540$, $M_w/\text{g.mol}^{-1} = 1880$, PDMS20, Goldschmidt AG
Solvent (A):   **octamethylcyclotetrasiloxane**   **C$_8$H$_{24}$O$_4$Si$_4$**   **556-67-2**

$T/\text{K} = 298.15$

φ_B	0.625	0.682	0.751	0.839	0.944
$\Delta_M H_A/(RT\varphi_B^2)$	0.055	0.055	0.057	0.064	0.066

Polymer (B):   **poly(dimethylsiloxane)**   **1983NEF**
Characterization: $M_n/\text{g.mol}^{-1} = 4170$, $M_w/\text{g.mol}^{-1} = 5920$, PDMS100, Goldschmidt AG
Solvent (A):   **octamethylcyclotetrasiloxane**   **C$_8$H$_{24}$O$_4$Si$_4$**   **556-67-2**

$T/\text{K} = 298.15$

φ_B	0.613	0.685
$\Delta_M H_A/(RT\varphi_B^2)$	0.041	0.045

Polymer (B):   **poly(dimethylsiloxane)**   **1983NEF**
Characterization: $M_n/\text{g.mol}^{-1} = 594$, $M_w/\text{g.mol}^{-1} = 665$, PDMS2, Goldschmidt AG
Solvent (A):   **n-octane**   **C$_8$H$_{18}$**   **111-65-9**

$T/\text{K} = 298.15$

φ_B	0.332	0.369	0.408	0.460	0.635	0.750	0.840	0.942
$\Delta_M H_A/(RT\varphi_B^2)$	0.229	0.234	0.248	0.252	0.298	0.325	0.343	0.389

Comments: The originally measured intermediary enthalpies of dilution were reduced by the authors to $\Delta_M H_A/(RT\varphi_B^2)$ to get data which depend less on concentration (see also the next data sets below).

Polymer (B):   **poly(dimethylsiloxane)**   **1983NEF**
Characterization: $M_n/\text{g.mol}^{-1} = 958$, $M_w/\text{g.mol}^{-1} = 1418$, PDMS10, Goldschmidt AG
Solvent (A):   **n-octane**   **C$_8$H$_{18}$**   **111-65-9**

$T/\text{K} = 303.15$

φ_B	0.349	0.380	0.424	0.465	0.643	0.707	0.879	0.958
$\Delta_M H_A/(RT\varphi_B^2)$	0.205	0.206	0.220	0.236	0.257	0.270	0.306	0.347

Polymer (B):   **poly(dimethylsiloxane)**   **1983NEF**
Characterization: $M_n/\text{g.mol}^{-1} = 1540$, $M_w/\text{g.mol}^{-1} = 1880$, PDMS20, Goldschmidt AG
Solvent (A):   **n-octane**   **C$_8$H$_{18}$**   **111-65-9**

$T/\text{K} = 298.15$

φ_B	0.371	0.410	0.450	0.495	0.557	0.621	0.678	0.750
$\Delta_M H_A/(RT\varphi_B^2)$	0.155	0.161	0.170	0.189	0.206	0.219	0.236	0.267

φ_B	0.832
$\Delta_M H_A/(RT\varphi_B^2)$	0.306

continued

continued

$T/\text{K} = 313.15$

φ_B	0.445	0.490	0.550	0.564	0.650	0.689	0.744	0.780
$\Delta_M H_A/(RT\varphi_B^2)$	0.172	0.180	0.192	0.190	0.204	0.225	0.244	0.255

φ_B	0.865	0.960
$\Delta_M H_A/(RT\varphi_B^2)$	0.284	0.332

Polymer (B):   **poly(dimethylsiloxane)**   **1983NEF**
Characterization: $M_n/\text{g.mol}^{-1} = 6560$, $M_w/\text{g.mol}^{-1} = 11220$, PDMS350, Goldschmidt AG
Solvent (A):   **n-octane**   **C$_8$H$_{18}$**   **111-65-9**

$T/\text{K} = 298.15$

φ_B	0.380	0.416	0.455	0.505	0.564
$\Delta_M H_A/(RT\varphi_B^2)$	0.146	0.157	0.173	0.195	0.192

$T/\text{K} = 313.15$

φ_B	0.411	0.450	0.500	0.550
$\Delta_M H_A/(RT\varphi_B^2)$	0.137	0.148	0.154	0.182

Polymer (B):   **poly(dimethylsiloxane)**   **1983NEF**
Characterization: $M_n/\text{g.mol}^{-1} = 594$, $M_w/\text{g.mol}^{-1} = 665$, PDMS2, Goldschmidt AG
Solvent (A):   **toluene**   **C$_7$H$_8$**   **108-88-3**

$T/\text{K} = 298.15$

φ_B	0.242	0.285	0.398	0.437	0.650	0.714	0.780	0.856
$\Delta_M H_A/(RT\varphi_B^2)$	0.370	0.374	0.390	0.403	0.483	0.509	0.538	0.591

φ_B	0.955
$\Delta_M H_A/(RT\varphi_B^2)$	0.645

$T/\text{K} = 313.15$

φ_B	0.339	0.365	0.411	0.462	0.610	0.650	0.715	0.780
$\Delta_M H_A/(RT\varphi_B^2)$	0.370	0.381	0.389	0.418	0.427	0.464	0.485	0.511

φ_B	0.861	0.955
$\Delta_M H_A/(RT\varphi_B^2)$	0.541	0.618

Polymer (B):   **poly(dimethylsiloxane)**   **1983NEF**
Characterization: $M_n/\text{g.mol}^{-1} = 958$, $M_w/\text{g.mol}^{-1} = 1418$, PDMS10, Goldschmidt AG
Solvent (A):   **toluene**   **C$_7$H$_8$**   **108-88-3**

$T/\text{K} = 298.15$

φ_B	0.275	0.300	0.329	0.369	0.415	0.620	0.676	0.721
$\Delta_M H_A/(RT\varphi_B^2)$	0.305	0.308	0.319	0.328	0.345	0.413	0.440	0.453

continued

continued

φ_B	0.775	0.858	0.950
$\Delta_M H_A/(RT\varphi_B^2)$	0.489	0.524	0.595

Polymer (B):	**poly(dimethylsiloxane)**							**1983NEF**
Characterization:	M_n/g.mol^{-1} = 1540, M_w/g.mol^{-1} = 1880, PDMS20, Goldschmidt AG							
Solvent (A):	**toluene**		**C$_7$H$_8$**					**108-88-3**

T/K = 298.15

φ_B	0.211	0.290	0.421	0.465	0.518	0.573	0.638	0.698
$\Delta_M H_A/(RT\varphi_B^2)$	0.243	0.260	0.299	0.325	0.350	0.348	0.394	0.427

φ_B	0.767	0.769	0.770
$\Delta_M H_A/(RT\varphi_B^2)$	0.444	0.463	0.479

T/K = 313.15

φ_B	0.350	0.378	0.415	0.461	0.505	0.554	0.661
$\Delta_M H_A/(RT\varphi_B^2)$	0.277	0.291	0.300	0.312	0.333	0.358	0.375

Polymer (B):	**poly(dimethylsiloxane)**							**1983NEF**
Characterization:	M_n/g.mol^{-1} = 4170, M_w/g.mol^{-1} = 5920, PDMS100, Goldschmidt AG							
Solvent (A):	**toluene**		**C$_7$H$_8$**					**108-88-3**

T/K = 298.15

φ_B	0.222	0.306	0.382	0.417	0.465	0.513	0.570	0.627
$\Delta_M H_A/(RT\varphi_B^2)$	0.210	0.237	0.270	0.282	0.300	0.317	0.339	0.389

φ_B	0.686	0.758	0.837
$\Delta_M H_A/(RT\varphi_B^2)$	0.410	0.449	0.482

Polymer (B):	**poly(dimethylsiloxane)**							**1983NEF**
Characterization:	M_n/g.mol^{-1} = 6560, M_w/g.mol^{-1} = 11220, PDMS350, Goldschmidt AG							
Solvent (A):	**toluene**		**C$_7$H$_8$**					**108-88-3**

T/K = 298.15

φ_B	0.322	0.349	0.390	0.472	0.480
$\Delta_M H_A/(RT\varphi_B^2)$	0.233	0.243	0.255	0.267	0.283

T/K = 313.15

φ_B	0.235	0.262	0.475	0.514	0.559	0.610	0.670	0.725
$\Delta_M H_A/(RT\varphi_B^2)$	0.200	0.206	0.277	0.300	0.315	0.344	0.368	0.416

Polymer (B):	**poly(dimethylsiloxane)**							**1983NEF**
Characterization:	M_n/g.mol^{-1} = 7860, M_w/g.mol^{-1} = 17060, PDMS1000, Goldschmidt AG							
Solvent (A):	**toluene**		**C$_7$H$_8$**					**108-88-3**

continued

continued

$T/K = 298.15$

φ_B	0.312	0.336	0.367	0.408	0.449
$\Delta_M H_A/(RT\varphi_B^2)$	0.220	0.234	0.245	0.253	0.281

Polymer (B):	**poly(dimethylsiloxane)**		**1983NEF**
Characterization:	$M_n/\text{g.mol}^{-1} = 1540$, $M_w/\text{g.mol}^{-1} = 1880$, PDMS20, Goldschmidt AG		
Solvent (A):	**2,2,4-trimethylpentane**	**C$_8$H$_{18}$**	**540-84-1**

$T/K = 298.15$

φ_B	0.396	0.443	0.748	0.818	0.875
$\Delta_M H_A/(RT\varphi_B^2)$	0.122	0.128	0.180	0.187	0.195

Polymer (B):	**poly(dimethylsiloxane)**		**1983NEF**
Characterization:	$M_n/\text{g.mol}^{-1} = 6560$, $M_w/\text{g.mol}^{-1} = 11220$, PDMS350, Goldschmidt AG		
Solvent (A):	**2,2,4-trimethylpentane**	**C$_8$H$_{18}$**	**540-84-1**

$T/K = 298.15$

φ_B	0.548
$\Delta_M H_A/(RT\varphi_B^2)$	0.145

Polymer (B):	**poly(ethylene glycol)**		**1989AND**
Characterization:	$M/\text{g.mol}^{-1} = 400$		
Solvent (A):	**water**	**H$_2$O**	**7732-18-5**

$T/K = 278.15$

w_B	0.0000	0.0241	0.0589	0.1750	0.3805	0.6188	0.7850	0.9060
$\Delta_M H_A/(\text{J/g solvent})$	+3	−9	−6	−53	−412	−1529	−2638	−3726

w_B	0.9540	0.9763
$\Delta_M H_A/(\text{J/g solvent})$	−4383	−4963

$T/K = 298.15$

w_B	0.0000	0.0241	0.0589	0.1750	0.3805	0.6188	0.7850	0.9060
$\Delta_M H_A/(\text{J/g solvent})$	+0.3	−0.3	−12	−52	−373	−1462	−2507	−3452

w_B	0.9540	0.9763	0.9917	1.0000
$\Delta_M H_A/(\text{J/g solvent})$	−3724	−4215	−4561	−4753

$T/K = 317.15$

w_B	0.0000	0.0241	0.0589	0.1750	0.3805	0.6188	0.7850	0.9060
$\Delta_M H_A/(\text{J/g solvent})$	+4	−0.6	−5	−48	−379	−1353	−2288	−2963

w_B	0.9540	0.9763	0.9917	1.0000
$\Delta_M H_A/(\text{J/g solvent})$	−3286	−3826	−4299	−4583

Polymer (B):   **poly(propylene glycol)**   **1995CAR**
Characterization: M_n/g.mol^{-1} = 400, PPO400, Fluka AG, Buchs, Switzerland
Solvent (A):   **water**   **H$_2$O**   **7732-18-5**

T/K = 298.15

w_B	0.02496	0.05003	0.1701	0.2494	0.3801	0.5899	0.7486	0.8949
$\Delta_M H_A$/J.g^{-1}	−0.094	−0.365	−5.19	−12.30	−32.08	−60.6	−83.4	−103.6

w_B	0.9997
$\Delta_M H_A$/J.g^{-1}	−136.4

Polymer (B):   **poly(vinyl acetate)**   **1989BEN**
Characterization: M_w/g.mol^{-1} = 194800, Aldrich Chem. Co., Inc., Milwaukee, WI
Solvent (A):   **methanol**   **CH$_4$O**   **67-56-1**

T/K = 298.15

φ_B	0.009	0.025	0.041	0.054	0.060	0.066	0.086	0.094
$\Delta_M H_A$/(RTφ_B^2)	0.17	0.17	0.14	0.15	0.15	0.20	0.18	0.15

φ_B	0.112	0.123	0.135	0.161	0.177
$\Delta_M H_A$/(RTφ_B^2)	0.14	0.18	0.17	0.18	0.10

Comments: The average value of $\Delta_M H_A$/(RTφ_B^2) is 0.16. Its standard deviation is 0.03.
The originally measured intermediary enthalpies of dilution were reduced by the
authors to $\Delta_M H_A$/(RTφ_B^2) to get data which depend much less on concentration.

Polymer (B):   **poly(vinyl acetate)**   **1989BEN**
Characterization: M_w/g.mol^{-1} = 194800, Aldrich Chem. Co., Inc., Milwaukee, WI
Solvent (A):   **2-propanone**   **C$_3$H$_6$O**   **67-64-1**

T/K = 298.15

φ_B	0.004	0.012	0.037	0.044	0.051	0.070	0.091	0.101
$\Delta_M H_A$/(RTφ_B^2)	−0.26	−0.13	−0.15	−0.17	−0.19	−0.21	−0.08	−0.05

φ_B	0.132	0.146	0.158	0.176
$\Delta_M H_A$/(RTφ_B^2)	−0.24	−0.15	−0.14	−0.19

Polymer (B):   **poly(vinyl acetate)**   **1989BEN**
Characterization: M_w/g.mol^{-1} = 194800, Aldrich Chem. Co., Inc., Milwaukee, WI
Solvent (A):   **trichloromethane**   **CHCl$_3$**   **67-66-3**

T/K = 298.15

φ_B	0.008	0.014	0.021	0.036	0.061	0.071	0.078	0.086
$\Delta_M H_A$/(RTφ_B^2)	−0.62	−1.17	−0.95	−1.32	−1.35	−1.40	−1.21	−1.45

φ_B	0.093	0.065	0.071	0.078	0.098	0.107	0.118	0.130
$\Delta_M H_A$/(RTφ_B^2)	−1.39	−0.47	−0.17	−0.87	−0.66	−0.88	−0.83	−0.78

continued

continued

φ_B	0.143	0.157
$\Delta_M H_A/(RT\varphi_B{}^2)$	-0.92	-0.82

Comments: The average value of $\Delta_M H_A/(RT\varphi_B{}^2)$ is -0.96. Its standard deviation is 0.36. The originally measured intermediary enthalpies of dilution were reduced by the authors to $\Delta_M H_A/(RT\varphi_B{}^2)$ to get data which depend much less on concentration.

Polymer (B):   **poly(1-vinylimidazole)**   **1988TAG, 1991TA1**
Characterization: $M_n/\text{g.mol}^{-1} = 20700$
Solvent (A):   **N,N-dimethylacetamide**   **C$_4$H$_9$NO**   **127-19-5**

$T/\text{K} = 298.15$ $\Delta_M H_A{}^\infty = -23900$ J/mol solvent

Polymer (B):   **poly(1-vinylimidazole)**   **1988TAG, 1991TA1**
Characterization: $M_n/\text{g.mol}^{-1} = 20700$
Solvent (A):   **N,N-dimethylformamide**   **C$_3$H$_7$NO**   **68-12-2**

$T/\text{K} = 298.15$ $\Delta_M H_A{}^\infty = -12800$ J/mol solvent

Polymer (B):   **poly(1-vinylimidazole)**   **1988TAG, 1991TA1**
Characterization: $M_n/\text{g.mol}^{-1} = 20700$
Solvent (A):   **1-methyl-2-pyrrolidinone**   **C$_5$H$_9$NO**   **872-50-4**

$T/\text{K} = 298.15$ $\Delta_M H_A{}^\infty = -26300$ J/mol solvent

Polymer (B):   **poly(1-vinylimidazole)**   **1988TAG, 1989SAF, 1991TA1**
Characterization: $M_n/\text{g.mol}^{-1} = 20700$
Solvent (A):   **water**   **H$_2$O**   **7732-18-5**

$T/\text{K} = 298.15$ $\Delta_M H_A{}^\infty = -13500$ J/mol solvent

Polymer (B):   **poly(1-vinylpyrazole)**   **1991TA1**
Characterization: $M_n/\text{g.mol}^{-1} = 18900$
Solvent (A):   **N,N-dimethylacetamide**   **C$_4$H$_9$NO**   **127-19-5**

$T/\text{K} = 298.15$ $\Delta_M H_A{}^\infty = -19600$ J/mol solvent

Polymer (B):   **poly(1-vinylpyrazole)**   **1991TA1**
Characterization: $M_n/\text{g.mol}^{-1} = 18900$
Solvent (A):   **N,N-dimethylformamide**   **C$_3$H$_7$NO**   **68-12-2**

$T/\text{K} = 298.15$ $\Delta_M H_A{}^\infty = -12000$ J/mol solvent

Polymer (B):   **poly(1-vinyl-1,2,4-triazole)**   **1988TAG, 1991TA1**
Characterization: $M_n/\text{g.mol}^{-1} = 69500$
Solvent (A):   **N,N-dimethylacetamide**   **C$_4$H$_9$NO**   **127-19-5**

$T/\text{K} = 298.15$ $\Delta_M H_A{}^\infty = -14400$ J/mol solvent

Polymer (B):	poly(1-vinyl-1,2,4-triazole)		1988TAG, 1991TA1
Characterization:	M_n/g.mol^{-1} = 69500		
Solvent (A):	N,N-dimethylformamide	C$_3$H$_7$NO	68-12-2

T/K = 298.15 $\Delta_M H_A^{\infty}$ = −14200 J/mol solvent

Polymer (B):	poly(1-vinyl-1,2,4-triazole)		1988TAG, 1991TA1
Characterization:	M_n/g.mol^{-1} = 69500		
Solvent (A):	1-methyl-2-pyrrolidinone	C$_5$H$_9$NO	872-50-4

T/K = 298.15 $\Delta_M H_A^{\infty}$ = −15800 J/mol solvent

Polymer (B):	poly(1-vinyl-1,2,4-triazole)		1988TAG, 1991TA1
Characterization:	M_n/g.mol^{-1} = 69500		
Solvent (A):	water	H$_2$O	7732-18-5

T/K = 298.15 $\Delta_M H_A^{\infty}$ = −14300 J/mol solvent

Polymer (B):	starch		1981BER
Characterization:	native potato starch		
Solvent (A):	water	H$_2$O	7732-18-5

T/K = 298.15 $\Delta_M H_A^{\infty}$ = −12.54 kJ/mol solvent

Polymer (B):	tetra(ethylene glycol) monooctyl ether		1989AND
Characterization:	M/g.mol^{-1} = 306.44		
Solvent (A):	water	H$_2$O	7732-18-5

T/K = 283.15

w_B	0.0000	0.0164	0.0943	0.2600	0.5020	0.7820	0.9000	0.9800
$\Delta_M H_A$/(J/g solvent)	−5.0	−5.0	−29	−76	−357	−1457	−2340	−3509

w_B	0.9900	1.0000
$\Delta_M H_A$/(J/g solvent)	−3958	−4407

T/K = 298.15

w_B	0.0000	0.0216	0.0520	0.1053	0.2600	0.5020	0.7820	0.9000
$\Delta_M H_A$/(J/g solvent)	−3.6	−3.6	−3.9	−10	−30	−282	−1362	−2225

w_B	0.9700	0.9800	0.9900	1.0000
$\Delta_M H_A$/(J/g solvent)	−2950	−3240	−3704	−4179

T/K = 313.15

w_B	0.0000	0.0323	0.0992	0.2600	0.5020	0.7820	0.9000	0.9700
$\Delta_M H_A$/(J/g solvent)	−2.3	−2.3	−4	−23	−129	−1106	−1929	−2368

w_B	1.0000
$\Delta_M H_A$/(J/g solvent)	−2885

4.2. References

1950NEW   Newing, M.J., Thermodynamic studies of silicones in benzene solution, *Trans. Faraday Soc.*, 46, 613, 1950.

1981BER   Berg, C. van den, *Dissertation*, University Wageningen, 1981.

1983NEF   Neff, B., Heintz, A., and Lichtenthaler, R.N., Thermodynamics of polydimethylsiloxane solutions. I. Calorimetric results of the enthalpy of dilution with seven organic solvents, *Ber. Bunsenges. Phys. Chem.*, 87, 1164, 1983.

1988TAG   Tager, A.A., Safronov, A.P., Voit, V.V., Lopyrev, V.A., Ermakova, T.G., Tatarova, L.A., and Shagelaeva, N.S., Heats of solution of poly(1-vinylpyrazole), poly(1-vinylimidazole), and poly(1-vinyl-1,2,4-triazole) in organic donor solvents (Russ.), *Vysokomol. Soedin., Ser. A*, 30, 2360, 1988.

1989AND   Andersson, B. and Olofsson, G., Calorimetric study of binary systems of tetraethyleneglycol octylether and polyethyleneglycol with water, *J. Solution Chem.*, 18, 1019, 1989.

1989BEN   Bender, M. and Heintz, A., Enthalpies of solution of polyvinylacetate in chloroform, acetone, and methanol using titration calorimetry, *J. Solution Chem.*, 18, 1, 1989.

1989SAF   Safronov, A.P., Tager, A.A., Sharina, S.V., Lopyrev, V.A., Ermakova, T.G., Tatarova, L.A., and Kashik, T.N., Nature of hydration in aqueous solutions of poly(1-vinylazoles) (Russ.), *Vysokomol. Soedin., Ser. A*, 31, 2662, 1989.

1991TA1   Tager, A.A. and Safronov, A.P., Complex formation in aqueous and nonaqueous solutions of poly(vinylazoles) (Russ.), *Vysokomol. Soedin., Ser. A*, 33, 67, 1991.

1995CAR   Carlsson, M., Hallen, D., and Linse, P., Mixing enthalpy and phase separation in a poly(propylene oxide)-water system, *J. Chem. Soc., Faraday Trans.*, 91, 2081, 1995.

2002COO   Cooke, S.A., Jonsdottir, S.O., and Westh, P., A thermodynamic study of glucose and related oligomers in aqueous solution: vapor pressures and enthalpies of mixing, *J. Chem. Eng. Data*, 47, 1185, 2002.

5. PARTIAL MOLAR ENTHALPIES OF MIXING AT INFINITE DILUTION OF SOLVENTS AND ENTHALPIES OF SOLUTION OF GASES/VAPORS OF SOLVENTS IN MOLTEN POLYMERS FROM INVERSE GAS-LIQUID CHROMATOGRAPHY (IGC)

5.1. Experimental data

Polymer (B):	cellulose acetate		1983ASP
Characterization:	D.S. = 2.45, Eastman Kodak Company		

Solvent (A)	T-range/ K	$\Delta_M H_A^\infty/$ kJ/mol	
water	333.15-358.15	-9.5 ± 1	

Polymer (B):	cellulose monoacetate		1983ASP
Characterization:	D.S. = 0.89, prepared in the laboratory by partial saponification of cellulose triacetate		

Solvent (A)	T-range/ K	$\Delta_M H_A^\infty/$ kJ/mol	
water	333.15-358.15	-16.0 ± 1	

Polymer (B):	cellulose triacetate		1983ASP
Characterization:	D.S. = 3.0 , Eastman Kodak Company		

Solvent (A)	T-range/ K	$\Delta_M H_A^\infty/$ kJ/mol	
water	333.15-358.15	-5.0 ± 1	

Polymer (B): **α,ω-dihydroxy poly[di(oxyethylene)oxyadipoyl]** **1989ED1**
Characterization: $M_n/\text{g.mol}^{-1} = 1910$

Solvent (A)	T-range/ K	$\Delta_M H_A^{\infty}/$ kJ/mol	$\Delta_{sol} H_{A(vap)}^{\infty}/$ kJ/mol
triacetin	402.1-431.1	−0.5	−69.2

Polymer (B): **α,ω-dihydroxy poly[di(oxyethylene)oxysuccinyl]** **1989ED1**
Characterization: $M_n/\text{g.mol}^{-1} = 890$

Solvent (A)	T-range/ K	$\Delta_M H_A^{\infty}/$ kJ/mol	$\Delta_{sol} H_{A(vap)}^{\infty}/$ kJ/mol
triacetin	402.1-431.1	+0.5	−68.2

Polymer (B): **α,ω-dihydroxy poly(hexamethylene carbonate)** **1989ED1**
Characterization: $M_n/\text{g.mol}^{-1} = 2410$

Solvent (A)	T-range/ K	$\Delta_M H_A^{\infty}/$ kJ/mol	$\Delta_{sol} H_{A(vap)}^{\infty}/$ kJ/mol
triacetin	402.1-431.1	+2.0	−66.8

Polymer (B): **α,ω-dihydroxy poly(oxyethyleneoxysuccinyl)- b-poly(oxyethylene) diblock copolymer** **1989ED1**
Characterization: $M_n/\text{g.mol}^{-1} = 1810$

Solvent (A)	T-range/ K	$\Delta_M H_A^{\infty}/$ kJ/mol	$\Delta_{sol} H_{A(vap)}^{\infty}/$ kJ/mol
triacetin	402.1-431.1	−0.8	−69.5

Polymer (B): α,ω-**dihydroxy poly[oxy-3-(2-methoxyethoxy)**   **1989ED1**
 propyleneoxysuccinyl]
Characterization: $M_n/\text{g.mol}^{-1} = 1820$

Solvent (A)	T-range/ K	$\Delta_M H_A^\infty/$ kJ/mol	$\Delta_{sol} H_{A(vap)}^\infty/$ kJ/mol
triacetin	402.1-431.1	+2.0	−66.3

Polymer (B): α,ω-**dihydroxy poly(tetramethylene carbonate)**   **1989ED1**
Characterization: $M_n/\text{g.mol}^{-1} = 2210$

Solvent (A)	T-range/ K	$\Delta_M H_A^\infty/$ kJ/mol	$\Delta_{sol} H_{A(vap)}^\infty/$ kJ/mol
triacetin	402.1-431.1	+3.5	−65.4

Polymer (B): α,ω-**dihydroxy poly(tetramethylene carbonate)**   **1989ED2**
Characterization: $M_n/\text{g.mol}^{-1} = 2210$

Solvent (A)	T-range/ K	$\Delta_M H_A^\infty/$ kJ/mol	$\Delta_{sol} H_{A(vap)}^\infty/$ kJ/mol
1,2,4-butanetriyl trinitrate	402.1-431.1		−78.0
triacetin	402.1-431.1	3.442	−65.4
1,2,3-propanetriyl trinitrate	402.1-431.1	10.68	−73.1

Polymer (B): α,ω-**dihydroxy poly[tri(oxyethylene)oxysuccinyl]**   **1989ED1**
Characterization: $M_n/\text{g.mol}^{-1} = 1250$

Solvent (A)	T-range/ K	$\Delta_M H_A^\infty/$ kJ/mol	$\Delta_{sol} H_{A(vap)}^\infty/$ kJ/mol
triacetin	402.1-431.1	−0.2	−69.1

Polymer (B):	**ethylcellulose**		**1983ASP**
Characterization:	D.S. = 1.4, Dow Ethocel		

Solvent (A)	T-range/ K	$\Delta_M H_A^\infty$/ kJ/mol	
water	333.15-358.15	−12.0	

Polymer (B):	**ethylcellulose**		**1998BLO**
Characterization:	M_n/g.mol^{-1} = 107000, 46% C_2H_5O groups		

Solvent (A)	T-range/ K	$\Delta_M H_A^\infty$/ kJ/mol	$\Delta_{sol} H_{A(vap)}^\infty$/ kJ/mol
dichloromethane	333.15-368.15	−19.1	−46.8
1,4-dioxane	333.15-368.15	−26.6	−59.9
ethanol	333.15-368.15	−10.5	−51.2
ethyl acetate	333.15-368.15	−19.1	−52.2
2-propanone	333.15-368.15	−22.8	−53.2
tetrachloromethane	333.15-368.15	−23.0	−55.2

Polymer (B):	**hydroxypropylcellulose**	**1979ASP**
Characterization:	M_w/g.mol^{-1} = 100000, MS = 4	
	Klucel L, Hercules Inc., Wilmington, DE	

Solvent (A)	T/ K	$\Delta_M H_A^\infty$/ kJ/mol
water	288.15	−2.6
water	292.85	−3.2
water	298.15	−3.8
water	302.45	−4.3
water	307.35	−4.9
water	313.45	−5.6
water	318.25	−6.1
water	323.05	−6.6

continued

continued

Solvent (A)	$T/$ K	$\Delta_M H_A{}^\infty/$ kJ/mol
water	327.95	−7.2
water	332.75	−7.7
water	338.05	−8.2
water	342.15	−8.6
water	347.85	−9.2
water	352.65	−9.7
water	357.65	−10.2

Polymer (B):	**hydroxypropylcellulose**		**1982ASP**
Characterization:	$M_w/\text{g.mol}^{-1} = 100000$, MS = 4		
	Klucel L, Hercules Inc., Wilmington, DE		

Solvent (A)	T-range/ K	$\Delta_M H_A{}^\infty/$ kJ/mol
1-butanol	313.15-348.15	−1.9
n-decane	313.15-343.15	5.0
1,4-dioxane	313.15-358.15	−0.7
ethanol	313.15-348.15	−1.4
n-heptane	318.15-358.15	2.6
methanol	313.15-348.15	−4.4
1-propanol	313.15-348.15	−1.8
2-propanone	313.15-333.15	−3.1
tetrahydrofuran	313.15-373.15	0.2
toluene	313.15-358.15	0.2
trichloromethane	313.15-343.15	−4.8
water	288.15-358.15	−5.7

Polymer (B):	**hydroxypropylcellulose**		**1983ASP**
Characterization:	$M_w/\text{g.mol}^{-1} = 100000$, MS = 4		
	Klucel L, Hercules Inc., Wilmington, DE		

Solvent (A)	T-range/ K	$\Delta_M H_A{}^\infty/$ kJ/mol
water	288.15-358.15	−5.7

| **Polymer (B):** | **methylcellulose** | | **1983ASP** |
| *Characterization:* | D.S. = 1.9, Dow Methocel | | |

Solvent (A)	T-range/ K	$\Delta_M H_A^\infty$/ kJ/mol
water	333.15-358.15	−18.5

| **Polymer (B):** | **natural rubber** | | **2003VOZ** |
| *Characterization:* | NR-RSS-1, Malaysia | | |

Solvent (A)	T-range/ K	$\Delta_M H_A^\infty$/ kJ/mol
benzene	303.15-413.15	1.94
cyclohexane	303.15-413.15	−1.04
n-decane	303.15-413.15	2.31
n-heptane	303.15-413.15	1.83
n-hexane	303.15-413.15	1.83
n-nonane	303.15-413.15	1.36
n-octane	303.15-413.15	0.89
toluene	303.15-413.15	1.31

| **Polymer (B):** | **nylon 6** | | **1992BON** |
| *Characterization:* | M_n/g.mol^{-1} = 18000 | | |

Solvent (A)	T-range/ K	$\Delta_M H_A^\infty$/ kJ/mol	$\Delta_{sol} H_{A(vap)}^\infty$/ kJ/mol	
ε-caprolactam	523.15-553.15	−8.37	−55.9	(packed column)
ε-caprolactam	523.15-553.15	−7.36	−58.4	(capillary column)

| **Polymer (B):** | **oxypropylcellulose** | | **1996BLO** |
| *Characterization:* | M_n/g.mol^{-1} = 130000, Klucel J, Hercules Inc., Wilmington, DE | | |

Solvent (A)	T-range/ K	$\Delta_M H_A^\infty$/ kJ/mol	$\Delta_{sol} H_{A(vap)}^\infty$/ kJ/mol
dichloromethane	333.15-373.15	11.9	−13.2
1,4-dioxane	333.15-373.15	11.7	−20.9
ethanol	333.15-373.15	11.7	−26.1
2-propanone	333.15-373.15	17.6	−9.4

Polymer (B):   **phenoxy (bisphenol-A polyhydroxyether)**   **1989IRI, 1992ETX**

Characterization: $M_n/\text{g.mol}^{-1} = 18000$, $M_w/\text{g.mol}^{-1} = 50700$, $M_\eta/\text{g.mol}^{-1} = 43000$,

                 $T_g/\text{K} = 369$, PKHH resin, Quimidroga, Barcelona, Spain

Solvent (A)	T-range/ K	$\Delta_M H_A^\infty/$ kJ/mol
acetonitrile	363.15-423.15	0.79
benzene	363.15-423.15	5.48
2-butanone	363.15-423.15	2.26
chlorobenzene	363.15-423.15	3.14
1,2-dichloroethane	363.15-423.15	2.22
N,N-dimethylformamide	363.15-423.15	−9.05
1,4-dioxane	363.15-423.15	−0.54
ethyl acetate	363.15-423.15	−1.30
bis(2-methoxyethyl) ether	363.15-423.15	−6.11
1-propanol	363.15-423.15	3.52
toluene	363.15-423.15	4.90

Polymer (B):   **poly(acrylonitrile-*co*-butadiene)**   **1993ISS**

Characterization: $M_n/\text{g.mol}^{-1} = 60000$, 34.0 wt% acrylonitrile,

                 Perbunan 3307, Bayer AG, Germany

Solvent (A)	T-range/ K	$\Delta_{sol} H_{A(vap)}^\infty/$ kJ/mol
benzene	323.15-393.15	−31.80
n-heptane	323.15-393.15	−25.94
n-nonane	323.15-393.15	−35.62
n-octane	323.15-393.15	−30.23
toluene	323.15-393.15	−35.30

Polymer (B):   **poly(acrylonitrile-*co*-butadiene)**   **1997SCH**

Characterization: $M_n/\text{g.mol}^{-1} = 60000$, 34.0 wt% acrylonitrile,

                 Perbunan 3307, Bayer AG, Germany

Solvent (A)	T-range/ K	$\Delta_{sol} H_{A(vap)}^\infty/$ kJ/mol
benzene	323.15-393.15	−31.82
cyclohexane	323.15-393.15	−31.65
n-heptane	323.15-393.15	−35.41
n-hexane	323.15-393.15	−30.50
toluene	323.15-393.15	−35.60

Polymer (B):   **poly(acrylonitrile-*co*-methyl acrylate)**   **1994COS**
Characterization: 91.9 wt% acrylonitrile, Yalova Fibre Co., Turkey

Solvent (A)	T-range/ K	$\Delta_M H_A^\infty$/ kJ/mol
1-butanol	383.15-413.15	31.74
ethanol	383.15-413.15	33.24
1-hexanol	383.15-413.15	45.05
1-pentanol	383.15-413.15	31.11
1-propanol	383.15-413.15	32.41

Polymer (B):   **poly(acrylonitrile-*co*-[2-(3-methyl-**   **2000KAY**
                     3-phenylcyclobutyl)-2-hydroxyethyl methacrylate]
Characterization: 45.0 mol% acrylonitrile, ρ_B = 1.075 g/cm^3, T_g/K = 377,
                     synthesized in the laboratory

Solvent (A)	T-range/ K	$\Delta_M H_A^\infty$/ kJ/mol
benzene	443.15-453.15	24.9
2-butanone	443.15-453.15	8.3
n-decane	443.15-453.15	23.3
1,2-dimethylbenzene	443.15-453.15	23.3
n-dodecane	443.15-453.15	26.6
ethanol	443.15-453.15	38.3
ethyl acetate	443.15-453.15	15.0
methanol	443.15-453.15	24.9
methyl acetate	443.15-453.15	16.6
n-nonane	443.15-453.15	23.3
n-octane	443.15-453.15	16.6
2-propanone	443.15-453.15	18.3
toluene	443.15-453.15	24.9
n-undecane	443.15-453.15	20.0

Polymer (B):   **poly(acrylonitrile-*co*-α-methylstyrene)**   **1984SIO, 1986SIO**
Characterization: M_w/g.mol^{-1} = 160000, 30 wt% acrylonitrile,
            Luran KR 2556U, BASF

Solvent (A)	T-range/ K	$\Delta_M H_A^\infty$/ kJ/mol	$\Delta_{sol} H_{A(vap)}^\infty$/ kJ/mol
benzene	438.15-468.15	8.46	−18.46
1-butanol	438.15-468.15	14.57	−24.12
2-butanone	438.15-468.15	6.95	−20.10
butyl acetate	438.15-468.15	11.35	−21.94
butylbenzene	438.15-468.15	10.59	−30.60
chlorobenzene	438.15-468.15	4.31	−29.14
cyclohexanol	438.15-468.15	15.78	−25.79
dichloromethane	438.15-468.15	−2.51	−24.91
1,4-dioxane	438.15-468.15	4.48	−25.87
n-dodecane	438.15-468.15	31.36	−16.66
n-tetradecane	438.15-468.15	26.08	−29.98
tetrahydrofuran	438.15-468.15	11.68	−13.19
toluene	438.15-468.15	6.82	−23.57
trichloromethane	438.15-468.15	7.41	−17.08

Polymer (B):   **poly(5,5'-bibenzimidazol-2,2'-diyl-1,3-phenylene)**   **1981DAN**
Characterization: –

Solvent (A)	T/ K	$\Delta_{sol} H_{A(vap)}^\infty$/ kJ/mol
acetaldehyde	462.9	−64.42
acetic acid	462.9	−67.93
acetonitrile	462.9	−71.89
acrylonitrile	462.9	−80.57
allyl alcohol	462.9	−96.63
1-bromobutane	462.9	−60.11
1-butanol	462.9	−79.70
2-butanone	462.9	−63.95
butyraldehyde	462.9	−58.70
n-decane	462.9	−70.01
diethylamine	462.9	−44.06
n-dodecane	462.9	−82.21

continued

continued

Solvent (A)	$T/$ K	$\Delta_{sol}H_{A(vap)}^{\infty}/$ kJ/mol
ethanol	462.9	−94.05
ethyl acetate	462.9	−67.07
formamide	462.9	−69.63
formic acid	462.9	−62.07
n-heptane	462.9	−47.51
nitroethane	462.9	−74.70
nitromethane	462.9	−80.20
1-nitropropane	462.9	−78.25
n-nonane	462.9	−64.17
n-octane	462.9	−55.64
1-pentanol	462.9	−74.33
1,2-propanediamine	462.9	−60.64
1-propanol	462.9	−93.33
2-propanone	462.9	−73.63
propionitrile	462.9	−77.65
1,1,3,3-tetramethylurea	462.9	−93.66

Polymer (B):	**poly[4,4'-bis(hexamethyleneoxy) biphenyl allylmalonate]**	**1992RO1, 1994ROM**
Characterization:	M_n/g.mol^{-1} = 21100, synthesized in the laboratory	

Solvent (A)	T-range/ K	$\Delta_M H_A^{\infty}/$ kJ/mol	$\Delta_{sol}H_{A(vap)}^{\infty}/$ kJ/mol
isotropic melt phase:			
n-decane	423.15-433.15	11.0	−31.7
1,2-dimethylbenzene	423.15-433.15	4.3	−33.1
1,3-dimethylbenzene	423.15-433.15	4.8	−31.6
1,4-dimethylbenzene	423.15-433.15	5.3	−31.4
toluene	423.15-433.15	5.1	−27.8
smectic A phase:			
n-decane	375.15-393.15	23.4	−21.7
1,2-dimethylbenzene	375.15-393.15	19.7	−20.9
1,3-dimethylbenzene	375.15-393.15	17.9	−21.5
1,4-dimethylbenzene	375.15-393.15	16.3	−21.3
toluene	375.15-393.15	18.5	−15.3

continued

continued

Solvent (A)	T-range/ K	$\Delta_M H_A{}^\infty$/ kJ/mol	$\Delta_{sol}H_{A(vap)}{}^\infty$/ kJ/mol
smectic B phase:			
n-decane	352.15-373.15	8.3	−38.9
1,2-dimethylbenzene	352.15-373.15	3.9	−33.5
1,3-dimethylbenzene	352.15-373.15	3.1	−35.7
1,4-dimethylbenzene	352.15-373.15	3.6	−32.5
toluene	352.15-373.15	3.8	−31.1

Polymer (B):   **1,2-polybutadiene**   **1997SCH**
Characterization: M_w/g.mol^{-1} = 215000, 69% 1,2-*vinyl*, 18% 1,4-*cis*, 13% 1,4-*trans*, Buna VII969, Buna-Werke Huels/Marl, Germany

Solvent (A)	T-range/ K	$\Delta_{sol}H_{A(vap)}{}^\infty$/ kJ/mol
benzene	323.15-393.15	−31.59
cyclohexane	323.15-393.15	−30.49
n-heptane	323.15-393.15	−34.18
n-hexane	323.15-393.15	−29.24
toluene	323.15-393.15	−36.66

Polymer (B):   **1,4-*cis*-polybutadiene**   **1997SCH**
Characterization: −

Solvent (A)	T-range/ K	$\Delta_{sol}H_{A(vap)}{}^\infty$/ kJ/mol
benzene	323.15-393.15	−31.88
cyclohexane	323.15-393.15	−31.76
n-heptane	323.15-393.15	−34.65
n-hexane	323.15-393.15	−29.09
toluene	323.15-393.15	−36.11

| Polymer (B): | **1,4-*cis*-polybutadiene** | **1985ITU** |

Characterization: M_n/g.mol^{-1} = 200000-300000, 98% 1,4-*cis*
Scientific Polymer Products, Inc., Ontario, NY

Solvent (A)	T-range/ K	$\Delta_{sol}H_{A(vap)}{}^{\infty}$/ kJ/mol
benzene	423.15-498.15	−28
1,3-butadiene	423.15-498.15	−22
n-butane	423.15-498.15	−24
cyclohexane	423.15-498.15	−28
cyclopentane	423.15-498.15	−29
n-decane	423.15-498.15	−43
1,3-dimethylbenzene	423.15-498.15	−37
2,2-dimethylbutane	423.15-498.15	−27
n-dodecane	423.15-498.15	−51
ethylbenzene	423.15-498.15	−36
n-heptane	423.15-498.15	−31
n-hexane	423.15-498.15	−28
2-methylbutane	423.15-498.15	−26
2-methylpropane	423.15-498.15	−20
n-nonane	423.15-498.15	−38
n-octane	423.15-498.15	−34
n-pentane	423.15-498.15	−27
propane	423.15-498.15	−17
propene	423.15-498.15	−17
toluene	423.15-498.15	−31
1,3,5-trimethylbenzene	423.15-498.15	−41

| Polymer (B): | **polybutadiene** | **1991ROM** |

Characterization: M_n/g.mol^{-1} = 22600, M_w/g.mol^{-1} = 24000,
8% 1,2-*vinyl*, 40% 1,4-*cis*, 52% 1,4-*trans*,
Pressure Chemical Co., Pittsburgh, PA

Solvent (A)	T-range/ K	$\Delta_{sol}H_{A(vap)}{}^{\infty}$/ kJ/mol
acetonitrile	353.15-373.15	−31.7
aniline	353.15-373.15	−43.5

continued

continued

Solvent (A)	T-range/ K	$\Delta_{sol}H_{A(vap)}^\infty$/ kJ/mol
benzaldehyde	353.15-373.15	−42.7
benzene	353.15-373.15	−31.2
benzyl alcohol	353.15-373.15	−46.9
1-butanol	353.15-373.15	−34.1
2-butanone	353.15-373.15	−29.5
butylbenzene	353.15-373.15	−48.4
cyclohexane	353.15-373.15	−30.3
cyclohexanone	353.15-373.15	−38.8
1,2-dichloroethane	353.15-373.15	−30.8
dichloromethane	353.15-373.15	−26.1
1,4-dimethylbenzene	353.15-373.15	−39.9
N,N-dimethylformamide	353.15-373.15	−37.2
ethyl acetate	353.15-373.15	−31.7
ethylbenzene	353.15-373.15	−38.9
n-hexane	353.15-373.15	−28.8
methanol	353.15-373.15	−43.5
4-methyl-2-pentanone	353.15-373.15	−35.5
2-propanol	353.15-373.15	−26.7
2-propanone	353.15-373.15	−27.7
tetrachloromethane	353.15-373.15	−31.0
tetrahydrofuran	353.15-373.15	−30.8
toluene	353.15-373.15	−35.6
trichloromethane	353.15-373.15	−30.0

Polymer (B):   **polybutadiene**   **1993ALE**

Characterization: M_n/g.mol^{-1} = 11900, M_w/g.mol^{-1} = 12100,
8% 1,2-*vinyl*, 46.5% 1,4-*cis*, 45.5% 1,4-*trans*

Solvent (A)	T-range/ K	$\Delta_{sol}H_{A(vap)}^\infty$/ kJ/mol
acetonitrile	333.15-373.15	−24
benzene	333.15-373.15	−32
1-butanol	333.15-373.15	−34
2-butanone	333.15-373.15	−29

continued

continued

Solvent (A)	T-range/ K	$\Delta_{sol}H_{A(vap)}{}^\infty$/ kJ/mol
butyl acetate	333.15-373.15	−39
butyronitrile	333.15-373.15	−32
1-chlorobutane	333.15-373.15	−31
1-chloropentane	333.15-373.15	−35
1-chloropropane	333.15-373.15	−27
cyclohexane	333.15-373.15	−31
dibutyl ether	333.15-373.15	−41
diethyl ether	333.15-373.15	−26
dipropyl ether	333.15-373.15	−33
ethanol	333.15-373.15	−23
ethyl acetate	333.15-373.15	−33
ethylcyclohexane	333.15-373.15	−38
n-heptane	333.15-373.15	−33
1-heptene	333.15-373.15	−34
n-hexane	333.15-373.15	−28
1-hexene	333.15-373.15	−30
methanol	333.15-373.15	−18
methyl acetate	333.15-373.15	−42
methylcyclohexane	333.15-373.15	−33
4-methyl-2-pentanone	333.15-373.15	−36
n-octane	333.15-373.15	−38
1-octene	333.15-373.15	−37
n-pentane	333.15-373.15	−24
3-pentanone	333.15-373.15	−34
1-propanol	333.15-373.15	−29
2-propanol	333.15-373.15	−23
2-propanone	333.15-373.15	−25
propionitrile	333.15-373.15	−28
propyl acetate	333.15-373.15	−35
tetrachloromethane	333.15-373.15	−31
tetrahydrofuran	333.15-373.15	−31
toluene	333.15-373.15	−36
trichloromethane	333.15-373.15	−31

Polymer (B):	**polybutadiene**	1993ALE
Characterization:	M_n/g.mol^{-1} = 8800, M_w/g.mol^{-1} = 10800, 12.9% 1,2-*vinyl*, 40.6% 1,4-*cis*, 46.5% 1,4-*trans*	

Solvent (A)	T-range/ K	$\Delta_{sol}H_{A(vap)}^\infty$/ kJ/mol
benzene	333.15-373.15	−31
1-butanol	333.15-373.15	−41
2-butanone	333.15-373.15	−27
1-chlorobutane	333.15-373.15	−31
1-chloropropane	333.15-373.15	−27
cyclohexane	333.15-373.15	−30
ethyl acetate	333.15-373.15	−32
n-heptane	333.15-373.15	−29
1-heptene	333.15-373.15	−32
4-methyl-2-pentanone	333.15-373.15	−31
1-octene	333.15-373.15	−36
n-octane	333.15-373.15	−37
3-pentanone	333.15-373.15	−33
1-propanol	333.15-373.15	−36
propionitrile	333.15-373.15	−29
propyl acetate	333.15-373.15	−35
tetrachloromethane	333.15-373.15	−30
toluene	333.15-373.15	−35
trichloromethane	333.15-373.15	−33

Polymer (B):	**polybutadiene**	1993ALE
Characterization:	M_n/g.mol^{-1} = 11800, M_w/g.mol^{-1} = 13100, 69% 1,2-*vinyl*, 20.8% 1,4-*cis*, 10.2% 1,4-*trans*	

Solvent (A)	T-range/ K	$\Delta_{sol}H_{A(vap)}^\infty$/ kJ/mol
acetonitrile	333.15-373.15	−24
benzene	333.15-373.15	−31
1-butanol	333.15-373.15	−32
2-butanone	333.15-373.15	−30
butyl acetate	333.15-373.15	−40
butyronitrile	333.15-373.15	−33

continued

continued

Solvent (A)	T-range/ K	$\Delta_{sol}H_{A(vap)}^{\infty}$/ kJ/mol
1-chlorobutane	333.15-373.15	−30
1-chloropentane	333.15-373.15	−35
1-chloropropane	333.15-373.15	−27
cyclohexane	333.15-373.15	−30
diethyl ether	333.15-373.15	−26
ethanol	333.15-373.15	−23
ethyl acetate	333.15-373.15	−31
ethylcyclohexane	333.15-373.15	−38
n-heptane	333.15-373.15	−34
1-heptene	333.15-373.15	−33
n-hexane	333.15-373.15	−30
1-hexene	333.15-373.15	−29
methanol	333.15-373.15	−18
methyl acetate	333.15-373.15	−27
methylcyclohexane	333.15-373.15	−33
4-methyl-2-pentanone	333.15-373.15	−35
n-octane	333.15-373.15	−38
1-octene	333.15-373.15	−38
n-pentane	333.15-373.15	−25
3-pentanone	333.15-373.15	−34
1-propanol	333.15-373.15	−29
2-propanol	333.15-373.15	−25
2-propanone	333.15-373.15	−25
propionitrile	333.15-373.15	−28
propyl acetate	333.15-373.15	−35
tetrachloromethane	333.15-373.15	−32
tetrahydrofuran	333.15-373.15	−30
toluene	333.15-373.15	−36
trichloromethane	333.15-373.15	−30

Polymer (B):	**polybutadiene**		**1998YAM**
Characterization:	SKD type		

Solvent (A)	T-range/ K	$\Delta_{M}H_{A}^{\infty}$/ kJ/mol	$\Delta_{sol}H_{A(vap)}^{\infty}$/ kJ/mol
dichloromethane	298.15-383.15	−0.10	−29.7
tetrachloromethane	298.15-383.15	−0.08	−32.2
trichloromethane	298.15-383.15	−0.08	−31.0

Polymer (B):   **poly(butadiene-co-styrene)**   **1997SCH**
Characterization: $M_n/\text{g.mol}^{-1} = 240000$, 15.0 wt% styrene,
                         Buna EM BT98, Buna Werke Huels/Marl, Germany

Solvent (A)	T-range/ K	$\Delta_{sol}H_{A(vap)}^\infty/$ kJ/mol
benzene	323.15-393.15	−32.25
cyclohexane	323.15-393.15	−30.75
n-heptane	323.15-393.15	−33.67
n-hexane	323.15-393.15	−29.44
toluene	323.15-393.15	−35.96

Polymer (B):   **poly(butadiene-co-styrene)**   **1997SCH**
Characterization: $M_n/\text{g.mol}^{-1} = 210000$, $M_w/\text{g.mol}^{-1} = 400000$, 23.0 wt% styrene,
                         Buna EM 1500, Buna Werke Huels/Marl, Germany

Solvent (A)	T-range/ K	$\Delta_{sol}H_{A(vap)}^\infty/$ kJ/mol
benzene	323.15-393.15	−31.84
cyclohexane	323.15-393.15	−30.58
n-heptane	323.15-393.15	−34.93
n-hexane	323.15-393.15	−29.00
toluene	323.15-393.15	−36.41

Polymer (B):   **poly(butadiene-co-styrene)**   **1997SCH**
Characterization: $M_n/\text{g.mol}^{-1} = 190000$, 40.0 wt% styrene,
                         Buna EM 1516, Buna Werke Huels/Marl, Germany

Solvent (A)	T-range/ K	$\Delta_{sol}H_{A(vap)}^\infty/$ kJ/mol
benzene	323.15-393.15	−32.47
cyclohexane	323.15-393.15	−30.87
n-heptane	323.15-393.15	−33.48
n-hexane	323.15-393.15	−29.22
toluene	323.15-393.15	−35.55

Polymer (B):	poly(1-butene)		1978DIP1
Characterization:	–		

Solvent (A)	*T*-range/ K	$\Delta_M H_A^\infty$/ kJ/mol
n-tetradecane	453.15-483.15	0.61

Polymer (B):	poly(butyl methacrylate)		1991WOH
Characterization:	M_n/g.mol^{-1} = 36800, M_w/g.mol^{-1} = 73600, synthesized in the laboratory		

Solvent (A)	*T*-range/ K	$\Delta_M H_A^\infty$/ kJ/mol	$\Delta_{sol} H_{A(vap)}^\infty$/ kJ/mol
benzene	343.15-423.15	0.42	−30.8
2-butanone	343.15-423.15	2.39	−29.1
1,2-dichloroethane	343.15-423.15	0.93	−32.4
dichloromethane	343.15-423.15	−2.54	−27.4
1,2-dimethylbenzene	343.15-423.15	1.56	−38.0
1,4-dimethylbenzene	343.15-423.15	0.83	−38.1
ethylbenzene	343.15-423.15	1.58	−36.7
2-pentanone	343.15-423.15	2.08	−33.3
2-propanone	343.15-423.15	0.89	−23.7
propylbenzene	343.15-423.15	0.94	−40.7
tetrachloromethane	343.15-423.15	0.79	−28.3
toluene	343.15-423.15	0.98	−33.5
trichloromethane	343.15-423.15	−6.03	−32.4

Polymer (B):	poly(butyl methacrylate)		1987TYA
Characterization:	M_n/g.mol^{-1} = 91000, M_η/g.mol^{-1} = 372000, T_g/K = 291, synthesized in the laboratory		

Solvent (A)	*T*-range/ K	$\Delta_M H_A^\infty$/ kJ/mol	$\Delta_{sol} H_{A(vap)}^\infty$/ kJ/mol
benzene	343.15-423.15	−0.51	−30.36
cyclohexane	343.15-423.15		−27.04

continued

continued

Solvent (A)	T-range/ K	$\Delta_M H_A^\infty/$ kJ/mol	$\Delta_{sol} H_{A(vap)}^\infty/$ kJ/mol
n-decane	343.15-423.15	3.86	
dichloromethane	343.15-383.15	−1.13	−29.30
1,4-dimethylbenzene	343.15-423.15	−0.34	−38.33
n-heptane	343.15-423.15	2.34	−29.93
n-hexane	343.15-413.15	2.89	−25.72
n-nonane	343.15-423.15	3.49	−37.39
n-octane	343.15-423.15	2.78	−33.65
n-pentane	343.15-373.15	4.73	−22.17
tetrachloromethane	343.15-423.15	−0.05	−28.86
toluene	343.15-423.15	−0.66	−34.61
trichloroethene	343.15-423.15	−1.87	−32.92
trichloromethane	343.15-423.15	−4.95	−32.29
2,2,4-trimethylpentane	343.15-423.15		−26.63

Polymer (B):	**poly(butyl methacrylate)**	**1994TIA**
Characterization:	T_g/K = 293, ρ = 1.055 g/cm^3,	
	Scientific Polymer Products, Inc., Ontario, NY	

Solvent (A)	T-range/ K	$\Delta_{sol} H_{A(vap)}^\infty/$ kJ/mol
benzene	343.15-383.15	−31.59
n-butane	343.15-383.15	−18.35
1-butanol	343.15-383.15	−39.04
2-butanone	343.15-383.15	−30.38
butyl acetate	343.15-383.15	−38.41
chlorobenzene	343.15-383.15	−38.16
1-chlorohexane	343.15-383.15	−37.87
1-chlorooctane	343.15-383.15	−46.02
cycloheptane	343.15-383.15	−32.13
cyclohexadiene	343.15-383.15	−29.87
cyclohexane	343.15-383.15	−26.69
cyclohexene	343.15-383.15	−28.70
cyclooctane	343.15-383.15	−36.44
cyclopentane	343.15-383.15	−25.36
n-decane	343.15-383.15	−42.63

continued

continued

Solvent (A)	T-range/ K	$\Delta_{sol}H_{A(vap)}{}^{\infty}/$ kJ/mol
1,1-dichloroethane	343.15-383.15	−29.58
1,2-dichloroethane	343.15-383.15	−32.97
dichloromethane	343.15-383.15	−29.33
1,4-dioxane	343.15-383.15	−33.60
ethanol	343.15-383.15	−31.97
ethyl acetate	343.15-383.15	−31.34
ethylbenzene	343.15-383.15	−39.20
n-heptane	343.15-383.15	−30.88
n-hexane	343.15-383.15	−26.94
methyl acetate	343.15-383.15	−28.74
n-nonane	343.15-383.15	−38.37
n-octane	343.15-383.15	−34.35
1-pentanol	343.15-383.15	−43.10
1-propanol	343.15-383.15	−35.10
2-propanone	343.15-383.15	−27.03
propyl acetate	343.15-383.15	−34.69
toluene	343.15-383.15	−35.56
1,1,1-trichloroethane	343.15-383.15	−29.41
trichloroethene	343.15-383.15	−33.43
trichloromethane	343.15-383.15	−33.01
n-undecane	343.15-383.15	−46.82

Polymer (B):   **poly(4-*tert*-butylstyrene)**   1999KA5
Characterization: M_w/g.mol^{-1} = 50000-100000
    T_g/K = 399-404, Aldrich Chem. Co., Inc., Milwaukee, WI

Solvent (A)	T-range/ K	$\Delta_M H_A{}^{\infty}/$ kJ/mol
benzene	453.15-463.15	19.34
2-butanone	453.15-463.15	18.17
n-decane	453.15-463.15	2.05
1,2-dimethylbenzene	453.15-463.15	4.19
n-dodecane	453.15-463.15	3.56
ethanol	453.15-463.15	3.56
ethyl acetate	453.15-463.15	7.58
methanol	453.15-463.15	21.98

continued

continued

Solvent (A)	T-range/ K	$\Delta_M H_A^\infty$/ kJ/mol
methyl acetate	453.15-463.15	10.47
n-nonane	453.15-463.15	13.98
n-octane	453.15-463.15	11.26
2-propanone	453.15-463.15	15.41
toluene	453.15-463.15	25.16
n-undecane	453.15-463.15	3.27

Polymer (B):	**poly(ε-caprolactone)**	**1994TIA**
Characterization:	T_g/K = 213, ρ = 1.095 g/cm³,	
	Scientific Polymer Products, Inc., Ontario, NY	

Solvent (A)	T-range/ K	$\Delta_{sol} H_{A(vap)}^\infty$/ kJ/mol
benzene	343.15-383.15	−32.55
1-butanol	343.15-383.15	−41.25
2-butanone	343.15-383.15	−32.26
butyl acetate	343.15-383.15	−40.29
chlorobenzene	343.15-383.15	−40.33
1-chlorobutane	343.15-383.15	−31.34
1-chlorohexane	343.15-383.15	−38.83
1-chlorooctane	343.15-383.15	−46.78
1-chloropentane	343.15-383.15	−35.19
cycloheptane	343.15-383.15	−31.92
cyclohexadiene	343.15-383.15	−30.75
cyclohexane	343.15-383.15	−27.32
cyclohexene	343.15-383.15	−29.58
cyclooctane	343.15-383.15	−36.48
cyclopentane	343.15-383.15	−24.85
n-decane	343.15-383.15	−41.30
1,1-dichloroethane	343.15-383.15	−31.25
1,2-dichloroethane	343.15-383.15	−35.10
dichloromethane	343.15-383.15	−31.21
1,4-dioxane	343.15-383.15	−36.74
ethanol	343.15-383.15	−34.22
ethyl acetate	343.15-383.15	−33.43
ethylbenzene	343.15-383.15	−39.83

continued

continued

Solvent (A)	T-range/ K	$\Delta_{sol}H_{A(vap)}^{\infty}/$ kJ/mol
n-heptane	343.15-383.15	−29.54
n-hexane	343.15-383.15	−25.98
methyl acetate	343.15-383.15	−30.92
n-nonane	343.15-383.15	−37.28
n-octane	343.15-383.15	−33.35
1-pentanol	343.15-383.15	−45.23
1-propanol	343.15-383.15	−37.53
2-propanone	343.15-383.15	−28.45
propyl acetate	343.15-383.15	−36.69
tetrachloromethane	343.15-383.15	−31.76
tetrahydrofuran	343.15-383.15	−30.79
toluene	343.15-383.15	−36.53
1,1,1-trichloroethane	343.15-383.15	−31.71
trichloroethene	343.15-383.15	−34.98
trichloromethane	343.15-383.15	−35.02
n-undecane	343.15-383.15	−45.15

Polymer (B):   **poly(ε-caprolactone)**   **2001ALG**
Characterization: $M_n/\text{g.mol}^{-1}$ = high, Aldrich Chem. Co., Inc., Milwaukee, WI

Solvent (A)	T K	$\Delta_{M}H_{A}^{\infty}/$ kJ/mol	$\Delta_{sol}H_{A(vap)}^{\infty}/$ kJ/mol
1-butanol	383.15		−15.5
butyl acetate	373.15		−32.0
n-decane	373.15	19.2	−24.7
n-dodecane	383.15	17.4	−28.9
ethanol	373.15		−15.4
ethyl acetate	373.15		−28.1
n-heptane	373.15	17.0	−29.7
n-hexane	373.15	16.7	−32.6
methanol	373.15		−12.9
methyl acetate	373.15		−42.4
n-nonane	373.15	19.0	−24.3
n-octane	373.15	18.2	−21.3
n-pentane	373.15	15.8	−45.7
1-propanol	373.15		−15.3
propyl acetate	373.15		−28.1
n-undecane	373.15	19.5	−28.0

Polymer (B):   **poly(ε-caprolactone)**   **2002SAR**

Characterization: M_n/g.mol^{-1} = 33000, Polysciences, Inc., Warrington, PA

Solvent (A)	T-range/ K	$\Delta_M H_A^\infty$/ kJ/mol	$\Delta_{sol} H_{A(vap)}^\infty$/ kJ/mol
tert-butyl acetate	343.15-413.15	2.51	−33.1
n-decane	343.15-413.15	5.02	−40.6
n-heptane	343.15-413.15	2.93	−28.9
n-hexane	343.15-413.15	1.67	−25.9
isobutyl acetate	343.15-413.15	0.0	−36.8
isopropyl acetate	343.15-413.15	−0.0	−32.6
n-nonane	343.15-413.15	4.18	−36.8
n-octane	343.15-413.15	4.18	−32.6
n-pentane	343.15-413.15	2.93	−20.1

Polymer (B):   **poly(ε-caprolactone)**   **2005SAR**

Characterization: M_n/g.mol^{-1} = 33000, Polysciences, Inc., Warrington, PA

Solvent (A)	T-range/ K	$\Delta_M H_A^\infty$/ kJ/mol	$\Delta_{sol} H_{A(vap)}^\infty$/ kJ/mol
benzene	343.15-413.15	−2.05	−32.2
butyl acetate	343.15-413.15	−0.80	−39.3
chlorobenzene	343.15-413.15	−2.67	−39.3
ethyl acetate	343.15-413.15	−1.17	−31.8
ethylbenzene	343.15-413.15	−1.46	−39.8
isopentyl acetate	343.15-413.15	+1.55	−42.4
isopropylbenzene	343.15-413.15	−1.34	−40.6
methyl acetate	343.15-413.15	−1.76	−29.7
propyl acetate	343.15-413.15	−0.17	−35.2
propylbenzene	343.15-413.15	−0.26	−41.0
toluene	343.15-413.15	−2.47	−36.4

Polymer (B):	**polycarbonate bisphenol-A**	**1992DIP**

Characterization: M_n/g.mol^{-1} = 83000, M_w/g.mol^{-1} = 163000, ρ = 1.202 g/cm^3,
T_g/K = 429, Makrolon 5705, Bayer AG, Ludwigshafen, Germany

Solvent (A)	*T*-range/ K	$\Delta_M H_A^\infty$/ kJ/mol
benzene	473.15-523.15	-0.88
butyl acetate	473.15-523.15	-2.26
butylbenzene	473.15-523.15	0.63
tert-butylbenzene	473.15-523.15	0.13
butylcyclohexane	473.15-523.15	4.69
chlorobenzene	473.15-523.15	-1.67
cyclohexanol	473.15-523.15	14.24
cyclohexanone	473.15-523.15	-0.08
n-decane	473.15-523.15	4.65
dichloromethane	473.15-503.15	-4.98
n-dodecane	473.15-523.15	7.95
ethylbenzene	473.15-523.15	-0.63
methylcyclohexane	473.15-523.15	4.27
2-pentanone	473.15-523.15	-3.06
1-propanol	473.15-523.15	7.41
n-tetradecane	473.15-523.15	7.41
toluene	473.15-523.15	-0.80
trichloromethane	473.15-523.15	-3.22

Polymer (B):	**polycarbonate 4,4'-(1-phenylethylidene)bisphenol**	**1992DIP**

Characterization: M_n/g.mol^{-1} = 23000, M_w/g.mol^{-1} = 57000, ρ = 1.084 g/cm^3,
T_g/K = 456, Fuji Xerox

Solvent (A)	*T*-range/ K	$\Delta_M H_A^\infty$/ kJ/mol
benzene	503.15-533.15	-2.05
butyl acetate	503.15-533.15	-3.43
butylbenzene	503.15-533.15	-0.25
tert-butylbenzene	503.15-533.15	1.51
butylcyclohexane	503.15-533.15	5.90
chlorobenzene	503.15-533.15	2.43
cyclohexanol	503.15-533.15	15.74
cyclohexanone	503.15-533.15	0.84
n-decane	503.15-533.15	4.48

continued

continued

Solvent (A)	T-range/ K	$\Delta_M H_A^\infty/$ kJ/mol
dichloromethane	503.15-533.15	−8.33
n-dodecane	503.15-533.15	7.91
ethylbenzene	503.15-533.15	−0.80
methylcyclohexane	503.15-533.15	2.43
2-pentanone	503.15-533.15	−3.81
1-propanol	503.15-533.15	13.52
n-tetradecane	503.15-533.15	9.25
toluene	503.15-533.15	−6.28
trichloromethane	503.15-533.15	4.06

Polymer (B):   **polycarbonate 4,4'-diphenylmethylidenebisphenol**   **1992DIP**

Characterization: M_n/g.mol^{-1} = 44000, M_w/g.mol^{-1} = 110000, ρ = 1.219 g/cm^3, T_g/K = 477, Fuji Xerox

Solvent (A)	T-range/ K	$\Delta_M H_A^\infty/$ kJ/mol
benzene	513.15-533.15	−1.55
butyl acetate	513.15-533.15	−2.01
butylbenzene	513.15-533.15	−1.09
tert-butylbenzene	513.15-533.15	3.77
butylcyclohexane	513.15-533.15	9.38
chlorobenzene	513.15-533.15	−1.13
cyclohexanol	513.15-533.15	15.16
cyclohexanone	513.15-533.15	1.51
n-decane	513.15-533.15	6.91
dichloromethane	513.15-533.15	−13.36
n-dodecane	513.15-533.15	6.24
ethylbenzene	513.15-533.15	−0.80
methylcyclohexane	513.15-533.15	9.29
2-pentanone	513.15-533.15	−0.04
1-propanol	513.15-533.15	6.74
n-tetradecane	513.15-533.15	11.05
toluene	513.15-533.15	−2.43
trichloromethane	513.15-533.15	−5.32

Polymer (B):	**polycarbonate 4,4'-cyclohexylidenebisphenol**	**1992DIP**

Characterization: M_n/g.mol^{-1} = 67000, M_w/g.mol^{-1} = 170000, ρ = 1.206 g/cm^3, T_g/K = 454, Fuji Xerox

Solvent (A)	T-range/ K	$\Delta_M H_A^\infty$/ kJ/mol
benzene	483.15-523.15	−0.59
butyl acetate	483.15-523.15	−3.64
butylbenzene	483.15-523.15	−0.04
tert-butylbenzene	483.15-523.15	1.88
butylcyclohexane	483.15-523.15	4.98
chlorobenzene	483.15-523.15	−2.51
cyclohexanol	483.15-523.15	14.82
cyclohexanone	483.15-523.15	1.47
n-decane	483.15-523.15	2.93
dichloromethane	483.15-503.15	−6.15
n-dodecane	483.15-523.15	6.36
ethylbenzene	483.15-523.15	−1.80
2-pentanone	483.15-523.15	−2.72
1-propanol	483.15-523.15	1.88
n-tetradecane	483.15-523.15	7.66
toluene	483.15-523.15	−2.13
trichloromethane	483.15-523.15	−5.36

Polymer (B):	**polycarbonate 4,4'-(4-*tert*-butylcyclohexylidene)- bisphenol**	**1992DIP**

Characterization: M_n/g.mol^{-1} = 20000, M_w/g.mol^{-1} = 57000, ρ = 1.132 g/cm^3, T_g/K = 476, synthesized in the laboratory

Solvent (A)	T-range/ K	$\Delta_M H_A^\infty$/ kJ/mol
butyl acetate	513.15-533.15	−4.01
butylbenzene	513.15-533.15	−0.42
tert-butylbenzene	513.15-533.15	4.40
butylcyclohexane	513.15-533.15	4.98
chlorobenzene	513.15-533.15	−1.46
cyclohexanol	513.15-533.15	15.90
cyclohexanone	513.15-533.15	0.21
n-decane	513.15-533.15	3.48
n-dodecane	513.15-533.15	0.63
methylcyclohexane	513.15-533.15	10.17
n-tetradecane	513.15-533.15	4.10
toluene	513.15-533.15	−1.93

Polymer (B):   **polycarbonate 4,4'-cyclohexylidene-**   **1992DIP**
                          2,2'-dimethylbisphenol

Characterization: $M_n/\text{g.mol}^{-1} = 22000$, $M_w/\text{g.mol}^{-1} = 62000$, $\rho = 1.182$ g/cm^3,
                         $T_g/\text{K} = 411$, synthesized in the laboratory

Solvent (A)	T-range/ K	$\Delta_M H_A^\infty/$ kJ/mol
benzene	473.15-523.15	−1.30
butyl acetate	473.15-523.15	−1.17
butylbenzene	473.15-523.15	0.33
tert-butylbenzene	473.15-523.15	2.55
butylcyclohexane	473.15-523.15	5.99
chlorobenzene	473.15-523.15	−1.80
cyclohexanol	473.15-523.15	15.49
cyclohexanone	473.15-523.15	0.50
n-decane	473.15-523.15	5.44
dichloromethane	473.15-503.15	−5.61
n-dodecane	473.15-523.15	5.73
ethylbenzene	473.15-523.15	0.50
methylcyclohexane	473.15-523.15	7.03
2-pentanone	473.15-523.15	7.91
1-propanol	473.15-523.15	−2.26
n-tetradecane	473.15-523.15	6.28
toluene	473.15-523.15	−1.21
trichloromethane	473.15-523.15	−4.14

Polymer (B):   **polychloroprene**   **1998YAM**

Characterization: −

Solvent (A)	T-range/ K	$\Delta_M H_A^\infty/$ kJ/mol	$\Delta_{sol} H_{A(vap)}^\infty/$ kJ/mol
dichloromethane	298.15-383.15	−0.22	−26.7
tetrachloromethane	298.15-383.15	−0.98	−30.1
trichloromethane	298.15-383.15	−0.30	−30.6

| Polymer (B): | poly(chloroprene-*co*-methyl methacrylate) | | | 1998YAM |
| Characterization: | 15.0 mol% methyl methacrylate, synthesized in the laboratory | | | |

Solvent (A)	T-range/ K	$\Delta_M H_A^\infty$/ kJ/mol	$\Delta_{sol} H_{A(vap)}^\infty$/ kJ/mol
dichloromethane	298.15-383.15	−1.68	−22.2
tetrachloromethane	298.15-383.15	−4.64	−20.1
trichloromethane	298.15-383.15	−1.47	−22.2

| Polymer (B): | poly(chloroprene-*co*-methyl methacrylate-*co*-methacrylic acid) | | | 1998YAM |
| Characterization: | 13.8 mol% methyl methacrylate and 78.2 mol% chloroprene, synthesized in the laboratory | | | |

Solvent (A)	T-range/ K	$\Delta_M H_A^\infty$/ kJ/mol	$\Delta_{sol} H_{A(vap)}^\infty$/ kJ/mol
dichloromethane	298.15-383.15	−1.30	−18.8
tetrachloromethane	298.15-383.15	−3.75	−29.7
trichloromethane	298.15-383.15	−0.52	−30.6

| Polymer (B): | poly(chloroprene-*co*-methyl methacrylate-*co*-methacrylic acid) | | | 1998YAM |
| Characterization: | 14.8 mol% methyl methacrylate and 83.7 mol% chloroprene, synthesized in the laboratory | | | |

Solvent (A)	T-range/ K	$\Delta_M H_A^\infty$/ kJ/mol	$\Delta_{sol} H_{A(vap)}^\infty$/ kJ/mol
dichloromethane	298.15-383.15	−0.77	−28.9
tetrachloromethane	298.15-383.15	−3.36	−27.6
trichloromethane	298.15-383.15	−1.67	−29.7

| Polymer (B): | poly(p-chlorostyrene) | | 1992YIL |

Characterization: M_η/g.mol^{-1} = 250000, Polysciences, Inc., Warrington, PA

Solvent (A)	T-range/ K	$\Delta_M H_A^\infty$/ kJ/mol	$\Delta_{sol} H_{A(vap)}^\infty$/ kJ/mol
benzene	423.15-443.15	3.60	−23.4
n-heptane	423.15-443.15	1.88	−27.7
n-hexane	423.15-443.15	3.98	−20.6
isopropylbenzene	423.15-443.15	0.71	−36.2
n-pentane	423.15-443.15	1.38	−17.9
propylbenzene	423.15-443.15	3.10	−35.0
toluene	423.15-443.15	2.93	−29.5

| Polymer (B): | poly(3,4-dichlorobenzyl methacrylate-*co*-ethyl methacrylate) | 2001DEM |

Characterization: 13 mol% ethyl methacrylate, synthesized in the laboratory

Solvent (A)	T-range/ K	$\Delta_M H_A^\infty$/ kJ/mol
benzene	403.15-423.15	19.40
2-butanone	403.15-423.15	18.01
n-decane	403.15-423.15	16.63
1,2-dimethylbenzene	403.15-423.15	13.10
n-dodecane	403.15-423.15	21.48
ethanol	403.15-423.15	54.04
ethyl acetate	403.15-423.15	20.79
methanol	403.15-423.15	22.86
methyl acetate	403.15-423.15	18.71
n-nonane	403.15-423.15	29.10
n-octane	403.15-423.15	27.71
2-propanone	403.15-423.15	20.09
toluene	403.15-423.15	14.90
n-undecane	403.15-423.15	27.02

Polymer (B):	**poly(3,4-dichlorobenzyl methacrylate-*co*-ethyl methacrylate)**	2001KA1
Characterization:	27 mol% ethyl methacrylate, $T_g/K = 336.2$, $\rho = 1.09$ g/cm^3, synthesized in the laboratory	

Solvent (A)	T-range/ K	$\Delta_M H_A^\infty/$ kJ/mol
benzene	403.15-423.15	19.40
2-butanone	403.15-423.15	19.40
n-decane	403.15-423.15	11.09
1,2-dimethylbenzene	403.15-423.15	21.48
n-dodecane	403.15-423.15	21.48
ethanol	403.15-423.15	61.66
ethyl acetate	403.15-423.15	22.86
methanol	403.15-423.15	24.94
methyl acetate	403.15-423.15	20.09
n-nonane	403.15-423.15	28.41
n-octane	403.15-423.15	31.87
2-propanone	403.15-423.15	18.71
toluene	403.15-423.15	22.17
n-undecane	403.15-423.15	28.41

Polymer (B):	**poly(3,4-dichlorobenzyl methacrylate-*co*-ethyl methacrylate)**	2001KA1
Characterization:	60 mol% ethyl methacrylate, $T_g/K = 334.2$, $\rho = 1.14$ g/cm^3, synthesized in the laboratory	

Solvent (A)	T-range/ K	$\Delta_M H_A^\infty/$ kJ/mol
benzene	403.15-423.15	22.17
2-butanone	403.15-423.15	21.48
n-decane	403.15-423.15	31.87
1,2-dimethylbenzene	403.15-423.15	20.09
n-dodecane	403.15-423.15	26.33
ethanol	403.15-423.15	65.82
ethyl acetate	403.15-423.15	22.86
methanol	403.15-423.15	22.17
methyl acetate	403.15-423.15	19.40
n-nonane	403.15-423.15	29.79
n-octane	403.15-423.15	25.64
2-propanone	403.15-423.15	19.70
toluene	403.15-423.15	23.56
n-undecane	403.15-423.15	29.79

Polymer (B):	**poly(3,4-dichlorobenzyl methacrylate-*co*-ethyl methacrylate)**	**2001KA1**

Characterization: 70 mol% ethyl methacrylate, $T_g/K = 332.2$, $\rho = 1.14$ g/cm^3, synthesized in the laboratory

Solvent (A)	T-range/ K	$\Delta_M H_A^\infty$/ kJ/mol
benzene	403.15-423.15	21.48
2-butanone	403.15-423.15	18.71
n-decane	403.15-423.15	29.79
1,2-dimethylbenzene	403.15-423.15	18.71
n-dodecane	403.15-423.15	24.25
ethanol	403.15-423.15	67.21
ethyl acetate	403.15-423.15	20.09
methanol	403.15-423.15	27.71
methyl acetate	403.15-423.15	20.09
n-nonane	403.15-423.15	28.41
n-octane	403.15-423.15	27.02
2-propanone	403.15-423.15	16.63
toluene	403.15-423.15	23.56
n-undecane	403.15-423.15	28.41

Polymer (B):	**poly(3,4-dichlorobenzyl methacrylate-*co*-ethyl methacrylate)**	**2001KA1**

Characterization: 82 mol% ethyl methacrylate, $T_g/K = 331.2$, $\rho = 1.17$ g/cm^3, synthesized in the laboratory

Solvent (A)	T-range/ K	$\Delta_M H_A^\infty$/ kJ/mol
benzene	403.15-423.15	17.32
2-butanone	403.15-423.15	22.17
n-decane	403.15-423.15	27.02
1,2-dimethylbenzene	403.15-423.15	13.86
n-dodecane	403.15-423.15	9.70
ethanol	403.15-423.15	65.82
ethyl acetate	403.15-423.15	18.01
methanol	403.15-423.15	29.79
methyl acetate	403.15-423.15	22.86
n-nonane	403.15-423.15	31.18
n-octane	403.15-423.15	27.71
2-propanone	403.15-423.15	35.34
toluene	403.15-423.15	14.55
n-undecane	403.15-423.15	13.86

Polymer (B):	poly(3,4-dichlorobenzyl methacrylate-*co*-ethyl methacrylate)		2001DEM
Characterization:	93 mol% ethyl methacrylate, synthesized in the laboratory		

Solvent (A)	*T*-range/ K	$\Delta_M H_A^\infty$/ kJ/mol
benzene	403.15-423.15	24.94
2-butanone	403.15-423.15	23.56
n-decane	403.15-423.15	33.26
1,2-dimethylbenzene	403.15-423.15	18.71
n-dodecane	403.15-423.15	7.62
ethanol	403.15-423.15	73.44
ethyl acetate	403.15-423.15	21.48
methanol	403.15-423.15	28.41
methyl acetate	403.15-423.15	29.79
n-nonane	403.15-423.15	32.56
n-octane	403.15-423.15	29.79
2-propanone	403.15-423.15	27.71
toluene	403.15-423.15	14.55
n-undecane	403.15-423.15	13.86

Polymer (B):	poly(diethyl maleate-*co*-vinyl acetate)		1995NEM
Characterization:	M_n/g.mol^{-1} = 51200, 50 mol% vinyl acetate synthesized in the laboratory		

Solvent (A)	*T*-range/ K	$\Delta_M H_A^\infty$/ kJ/mol	$\Delta_{sol} H_{A(vap)}^\infty$/ kJ/mol
n-heptane	359.96-378.36	12.75	−19.15
n-hexane	359.96-378.36	3.95	−24.07
n-nonane	359.96-378.36	29.69	−11.01
n-octane	359.96-378.36	21.21	−14.89

Polymer (B):   **poly(dimethylsilmethylene)**   **2002SOL**

Characterization: $M_n/\text{g.mol}^{-1} = 135000$, $M_w/\text{g.mol}^{-1} = 357000$,

                              $T_g/\text{K} = 181$, $\rho = 0.906$ g/cm^3

Solvent (A)	T-range/ K	$\Delta_{sol}H_{A(vap)}{}^\infty$/ kJ/mol
n-butane	313.15-393.15	−19.89
n-heptane	313.15-393.15	−31.83
n-hexane	313.15-393.15	−28.45
n-pentane	313.15-393.15	−24.38
propane	313.15-393.15	−13.38

Polymer (B):   **poly(dimethylsiltrimethylene)**   **2002SOL**

Characterization: $M_n/\text{g.mol}^{-1} = 786000$, $M_w/\text{g.mol}^{-1} = 1100000$,

                              $T_g/\text{K} = 197$, $\rho = 0.908$ g/cm^3

Solvent (A)	T-range/ K	$\Delta_{sol}H_{A(vap)}{}^\infty$/ kJ/mol
n-butane	313.15-393.15	−22.48
n-heptane	313.15-393.15	−34.55
n-hexane	313.15-393.15	−30.29
n-pentane	313.15-393.15	−26.08
propane	313.15-393.15	−20.51

Polymer (B):   **poly(dimethylsiloxane)**   **1994TIA**

Characterization: $T_g/\text{K} = 146$, $\rho = 0.970$ g/cm^3,

                              Scientific Polymer Products, Inc., Ontario, NY

Solvent (A)	T-range/ K	$\Delta_{sol}H_{A(vap)}{}^\infty$/ kJ/mol
benzene	333.15-383.15	−29.16
n-butane	333.15-383.15	−20.38
1-butanol	333.15-383.15	−31.84
2-butanone	333.15-383.15	−27.45
butyl acetate	333.15-383.15	−37.78

continued

continued

Solvent (A)	*T*-range/ K	$\Delta_{sol}H_{A(vap)}{}^{\infty}$/ kJ/mol
chlorobenzene	333.15-383.15	−36.11
1-chlorobutane	333.15-383.15	−28.95
1-chlorohexane	333.15-383.15	−37.53
1-chlorooctane	333.15-383.15	−46.15
1-chloropentane	333.15-383.15	−33.30
cycloheptane	333.15-383.15	−34.10
cyclohexadiene	333.15-383.15	−29.04
cyclohexane	333.15-383.15	−29.29
cyclohexene	333.15-383.15	−29.58
cyclooctane	333.15-383.15	−38.79
cyclopentane	333.15-383.15	−26.11
n-decane	333.15-383.15	−45.90
1,1-dichloroethane	333.15-383.15	−25.94
1,2-dichloroethane	333.15-383.15	−28.45
dichloromethane	333.15-383.15	−24.60
1,4-dioxane	333.15-383.15	−31.71
ethyl acetate	333.15-383.15	−29.54
ethylbenzene	333.15-383.15	−37.23
n-heptane	333.15-383.15	−32.34
n-hexane	333.15-383.15	−28.20
methyl acetate	333.15-383.15	−25.69
n-nonane	333.15-383.15	−40.79
n-octane	333.15-383.15	−36.61
n-pentane	333.15-383.15	−24.06
1-pentanol	333.15-383.15	−36.28
1-propanol	333.15-383.15	−27.20
2-propanone	333.15-383.15	−21.80
propyl acetate	333.15-383.15	−33.64
tetrachloromethane	333.15-383.15	−29.33
tetrahydrofuran	333.15-383.15	−28.70
toluene	333.15-383.15	−33.43
1,1,1-trichloroethane	333.15-383.15	−28.35
trichloroethene	333.15-383.15	−30.92
trichloromethane	333.15-383.15	−27.74
n-undecane	333.15-383.15	−48.12

Polymer (B):	**poly(dimethylsiloxane)**	**1986ROT**
Characterization:	$M_n/\text{g.mol}^{-1} = 2410$, $M_w/\text{g.mol}^{-1} = 18000$	
	MS 200 Midland Silicones, Barry, Great Britain	

Solvent (A)	$T/$ K	$\Delta_M H_A^\infty/$ J/mol
n-heptane	303.15	563
n-heptane	333.15	335
n-heptane	363.15	107
n-hexane	303.15	+362
n-hexane	333.15	−40
n-hexane	363.15	−442
n-octane	303.15	1057
n-octane	333.15	797
n-octane	363.15	536
n-pentane	303.15	−127
n-pentane	333.15	−478
n-pentane	363.15	−829

Polymer (B):	**poly(dimethylsiloxane)**	**1986ROT**
Characterization:	$M_n/\text{g.mol}^{-1} = 3480$, $M_w/\text{g.mol}^{-1} = 8810$	
	SF 96, Hewlett-Packard Co., Avondale, PA	

Solvent (A)	$T/$ K	$\Delta_M H_A^\infty/$ J/mol
n-heptane	303.15	674
n-heptane	333.15	242
n-heptane	363.15	−190
n-hexane	303.15	328
n-hexane	333.15	−107
n-hexane	363.15	−542
n-octane	303.15	1122
n-octane	333.15	708
n-octane	363.15	284
n-pentane	303.15	−112
n-pentane	333.15	−607
n-pentane	363.15	−1102

Polymer (B): *Characterization:*	**poly(dimethylsiloxane)** $M_n/\text{g.mol}^{-1} = 15100$, $M_w/\text{g.mol}^{-1} = 28700$ OV-101, Supelco Inc., Bellefonte, PA	**1986ROT**

Solvent (A)	$T/$ K	$\Delta_M H_A^\infty/$ J/mol
n-heptane	303.15	424
n-heptane	333.15	199
n-heptane	363.15	−26
n-hexane	303.15	189
n-hexane	333.15	−162
n-hexane	363.15	−513
n-octane	303.15	941
n-octane	333.15	668
n-octane	363.15	395
n-pentane	303.15	−278
n-pentane	333.15	−557
n-pentane	363.15	−836

Polymer (B): *Characterization:*	**poly(dimethylsiloxane)** $M_n/\text{g.mol}^{-1} = 20700$, $M_w/\text{g.mol}^{-1} = 95300$ DC 200, Hewlett-Packard Co., Avondale, PA	**1986ROT**

Solvent (A)	$T/$ K	$\Delta_M H_A^\infty/$ J/mol
n-heptane	303.15	509
n-heptane	333.15	200
n-heptane	363.15	−109
n-hexane	303.15	152
n-hexane	333.15	−172
n-hexane	363.15	−496
n-octane	303.15	929
n-octane	333.15	683
n-octane	363.15	437
n-pentane	303.15	−303
n-pentane	333.15	−591
n-pentane	363.15	−879

Polymer (B):	poly(dimethylsiloxane)		2002PRI
Characterization:	M_n/g.mol^{-1} = 24100, M_w/M_n = 3.8, fractionated from DC12500 fluid, Dow Corning		

Solvent (A)	$T/$ K	$\Delta_{sol}H_{A(vap)}{}^\infty/$ kJ/mol
benzene	363.15	−29.4
cyclohexane	363.15	−29.1
1,2-dimethylbenzene	363.15	−35.8
1,3-dimethylbenzene	363.15	−35.3
1,4-dimethylbenzene	363.15	−35.5
2,3-dimethylpentane	363.15	−31.4
2,4-dimethylpentane	363.15	−30.2
ethylbenzene	363.15	−37.9
n-heptane	363.15	−32.8
n-hexane	363.15	−28.8
2-methylhexane	363.15	−31.5
3-methylhexane	363.15	−31.8
n-nonane	363.15	−40.3
n-octane	363.15	−36.4
n-pentane	363.15	−24.5
toluene	363.15	−33.5
2,2,3-trimethylbutane	363.15	−29.8

Polymer (B):	poly(dimethylsiloxane)		1986ROT
Characterization:	M_n/g.mol^{-1} = 218000, M_w/g.mol^{-1} = 480000 SE 30, Carlo Erba Strumentazione, SpA., Milano, Italy		

Solvent (A)	$T/$ K	$\Delta_M H_A{}^\infty/$ J/mol
n-heptane	303.15	499
n-heptane	333.15	277
n-heptane	363.15	55
n-hexane	303.15	179
n-hexane	333.15	−112
n-hexane	363.15	−403
n-octane	303.15	907
n-octane	333.15	685
n-octane	363.15	463
n-pentane	303.15	−152
n-pentane	333.15	−515
n-pentane	363.15	−878

Polymer (B):	**poly(dimethylsiloxane)**		**1991BEC**
Characterization:	type OV-101		

Solvent (A)	T-range/ K	$\Delta_M H_A^\infty$/ kJ/mol	$\Delta_{sol} H_{A(vap)}^\infty$/ kJ/mol
benzene	363.15-473.15	4.27	−27.42
1-butanol	363.15-473.15	15.45	−29.02
n-decane	363.15-473.15	4.31	−40.57
N,N-dimethylaniline	363.15-473.15	−1.88	−41.62
n-dodecane	363.15-473.15	3.98	−48.52
ethyl acetate	363.15-473.15	6.49	−26.71
n-heptane	363.15-473.15	4.15	−29.98
n-hexane	363.15-473.15	3.68	−26.80
n-nonane	363.15-473.15	1.72	−39.48
n-octane	363.15-473.15	2.22	−35.25
1-octanol	363.15-473.15	14.70	−42.12
2-pentanone	363.15-473.15	5.40	−29.35
pyridine	363.15-473.15	6.16	−30.35
toluene	363.15-473.15	3.43	−31.95
n-undecane	363.15-473.15	4.27	−44.38

Polymer (B):	**poly(dimethylsiloxane)**		**1992BEC**
Characterization:	type PS-255		

Solvent (A)	T/ K	$\Delta_M H_A^\infty$/ kJ/mol	$\Delta_{sol} H_{A(vap)}^\infty$/ kJ/mol
benzene	423.15	4.94	−26.9
1-butanol	423.15	17.7	−27.4
2-butanone	423.15	4.39	−26.9
n-decane	423.15	3.01	−42.1
n-dodecane	423.15	2.93	−50.0
n-heptane	423.15	4.52	−29.7
n-hexane	423.15	4.80	−25.8
n-nonane	423.15	3.22	−38.2
n-octane	423.15	3.47	−34.2
1-octanol	423.15	14.9	−43.3
2-octanone	423.15	7.03	−10.28
n-pentane	423.15	4.48	−43.0
1-pentanol	423.15	13.4	−32.6
2-pentanone	423.15	5.69	−29.4
pentylbenzene	423.15	1.05	−47.2
2-propanone	423.15	6.50	−24.5
propylbenzene	423.15	5.45	−39.5
toluene	423.15	4.85	−30.6

Polymer (B):   **poly(dimethylsiloxane)**   **1993BAL**

Characterization: M_n/g.mol^{-1} = 464000, General Electric Co.

Solvent (A)	T-range/ K	$\Delta_M H_A^\infty$/ kJ/mol
decamethylcyclopentasiloxane	423.15-473.15	0.92
dodecamethylcyclohexasiloxane	423.15-473.15	0.71
octamethylcyclotetrasiloxane	423.15-473.15	1.34

Polymer (B):   **poly[dimethylsiloxane-*co*-methyl(4-cyanobiphenoxy)**   **2002PRI**
butylsiloxane]

Characterization: 50 mol% dimethylsiloxane, 40 repeat units, Merck Ltd., UK

Solvent (A)	T/ K	$\Delta_{sol} H_{A(vap)}^\infty$/ kJ/mol
benzene	358.15	−25.3
cyclohexane	358.15	−29.0
1,2-dimethylbenzene	358.15	−32.2
1,3-dimethylbenzene	358.15	−31.8
1,4-dimethylbenzene	358.15	−31.5
2,3-dimethylpentane	358.15	−20.2
2,4-dimethylpentane	358.15	−18.3
ethylbenzene	358.15	−30.2
n-heptane	358.15	−22.5
n-hexane	358.15	−19.9
2-methylhexane	358.15	−17.8
3-methylhexane	358.15	−19.4
n-nonane	358.15	−29.6
n-octane	358.15	−25.3
n-pentane	358.15	−14.9
toluene	358.15	−28.5
2,2,3-trimethylbutane	358.15	−17.6

Polymer (B):   **poly(epichlorohydrin)**   **1994TIA**

Characterization: $T_g/K = 251$, $\rho = 1.860$ g/cm³,

                          Scientific Polymer Products, Inc., Ontario, NY

Solvent (A)	T-range/ K	$\Delta_{sol}H_{A(vap)}^{\infty}/$ kJ/mol
benzene	333.15-383.15	−31.13
1-butanol	333.15-383.15	−37.57
2-butanone	333.15-383.15	−32.89
butyl acetate	333.15-383.15	−40.17
chlorobenzene	333.15-383.15	−38.16
1-chlorobutane	333.15-383.15	−29.16
1-chlorohexane	333.15-383.15	−36.48
1-chlorooctane	333.15-383.15	−43.76
1-chloropentane	333.15-383.15	−32.76
cycloheptane	333.15-383.15	−29.12
cyclohexadiene	333.15-383.15	−28.87
cyclohexane	333.15-383.15	−24.27
cyclohexene	333.15-383.15	−27.24
cyclooctane	333.15-383.15	−33.47
cyclopentane	333.15-383.15	−22.01
n-decane	333.15-383.15	−36.40
1,1-dichloroethane	333.15-383.15	−28.41
1,2-dichloroethane	333.15-383.15	−33.26
dichloromethane	333.15-383.15	−27.91
1,4-dioxane	333.15-383.15	−37.74
ethanol	333.15-383.15	−31.34
ethyl acetate	333.15-383.15	−33.35
ethylbenzene	333.15-383.15	−38.07
n-heptane	333.15-383.15	−25.23
n-hexane	333.15-383.15	−21.71
methyl acetate	333.15-383.15	−30.75
n-nonane	333.15-383.15	−32.72
n-octane	333.15-383.15	−28.95
1-pentanol	333.15-383.15	−40.92
1-propanol	333.15-383.15	−33.68
2-propanone	333.15-383.15	−29.37
propyl acetate	333.15-383.15	−36.74
tetrachloromethane	333.15-383.15	−27.70
tetrahydrofuran	333.15-383.15	−31.34
toluene	333.15-383.15	−35.02
1,1,1-trichloroethane	333.15-383.15	−28.45
trichloroethene	333.15-383.15	−31.34
trichloromethane	333.15-383.15	−30.38
n-undecane	333.15-383.15	−40.17

Polymer (B):	poly(ethyl acrylate)		1992COS
Characterization:	synthesized in the laboratory		

Solvent (A)	T-range/ K	$\Delta_{sol} H_{A(vap)}^\infty/$ kJ/mol	
n-decane	373.15-423.15	−42.53	
n-hexane	373.15-423.15	−31.71	
n-nonane	373.15-423.15	−37.53	

Polymer (B):	poly(ethyl acrylate)		1994TIA
Characterization:	T_g/K = 249, ρ = 1.12 g/cm^3, Scientific Polymer Products, Inc., Ontario, NY		

Solvent (A)	T-range/ K	$\Delta_{sol} H_{A(vap)}^\infty/$ kJ/mol
benzene	343.15-383.15	−31.71
chlorobenzene	343.15-383.15	−39.33
1-chlorobutane	343.15-383.15	−29.46
1-chloropentane	343.15-383.15	−32.89
cycloheptane	343.15-383.15	−30.25
cyclohexane	343.15-383.15	−25.98
cyclooctane	343.15-383.15	−35.98
cyclopentane	343.15-383.15	−23.89
n-decane	343.15-383.15	−39.08
1,1-dichloroethane	343.15-383.15	−30.12
1,2-dichloroethane	343.15-383.15	−34.48
dichloromethane	343.15-383.15	−29.91
1,4-dioxane	343.15-383.15	−36.36
ethyl acetate	343.15-383.15	−32.26
ethylbenzene	343.15-383.15	−37.32
n-heptane	343.15-383.15	−27.24
n-hexane	343.15-383.15	−24.35
n-nonane	343.15-383.15	−35.40
n-octane	343.15-383.15	−31.30
tetrachloromethane	343.15-383.15	−29.83
tetrahydrofuran	343.15-383.15	−30.21
toluene	343.15-383.15	−36.19
1,1,1-trichloroethane	343.15-383.15	−30.29
trichloroethene	343.15-383.15	−33.68
trichloromethane	343.15-383.15	−33.51
n-undecane	343.15-383.15	−43.10

Polymer (B):   **polyethylene**   **1989GAL**
Characterization: M_n/g.mol^{-1} = 11200, M_w/g.mol^{-1} = 55000, T_m/K = 400,
                       linear PE, DuPont Corp.

Solvent (A)	T-range/ K	$\Delta_M H_A^\infty$/ kJ/mol
benzene	408.15-448.15	1.46
2-butanone	408.15-448.15	4.95
butyl acetate	408.15-448.15	4.44
chlorobenzene	408.15-448.15	2.73
1-chlorobutane	408.15-448.15	2.80
1-chlorohexane	408.15-448.15	2.47
cyclohexane	408.15-448.15	0.90
cyclohexanone	408.15-448.15	7.41
n-decane	408.15-448.15	0.80
1,5-dichloropentane	408.15-448.15	3.91
N,N-dimethylacetamide	408.15-448.15	8.90
N,N-dimethylformamide	408.15-448.15	9.80
dimethylsulfoxide	408.15-448.15	9.55
dipropyl ether	408.15-448.15	3.41
dipropyl thioether	408.15-448.15	1.60
n-dodecane	408.15-448.15	0.60
ethyl propionate	408.15-448.15	5.11
n-heptane	408.15-448.15	1.05
n-octane	408.15-448.15	1.02
2-pentanone	408.15-448.15	4.10
1,1,2,2-tetrachloroethane	408.15-448.15	2.94
tetrachloromethane	408.15-448.15	1.59
toluene	408.15-448.15	1.84
1,1,2-trichloroethane	408.15-448.15	4.14
trichloromethane	408.15-448.15	2.70

Polymer (B):   **polyethylene**   **1980DIP**
Characterization: ρ = 0.917 g/cm^3, LDPE, Tenite 800 E, Eastman Kodak

Solvent (A)	T-range/ K	$\Delta_M H_A^\infty$/ kJ/mol	$\Delta_{sol} H_{A(vap)}^\infty$/ kJ/mol
benzene	393.15-423.15	1.46	−27.05
1-butanol	393.15-423.15	11.43	−31.15
chlorobenzene	398.15-418.15	1.38	−34.16
1-chlorobutane	398.15-418.15	0.84	−26.46

continued

continued

Solvent (A)	T-range/ K	$\Delta_M H_A^\infty$/ kJ/mol	$\Delta_{sol} H_{A(vap)}^\infty$/ kJ/mol
chlorocyclohexane	393.15-418.15		−35.59
cyclohexane	398.15-418.15	1.46	−26.38
cyclohexanol	393.15-418.15	13.27	−35.34
cyclohexanone	393.15-418.15		−35.17
n-decane	398.15-418.15	0.34	−42.58
n-nonane	398.15-418.15	1.00	−37.97
1-octene	393.15-423.15	0.08	−33.87
2-pentanone	398.15-418.15	4.10	−28.64
phenol	398.15-418.15	6.57	−43.21
tetrachloromethane	398.15-418.15	1.59	−26.04
trichloromethane	398.15-418.15	0.71	−25.33

Polymer (B): **polyethylene** 1978DIP1
Characterization: LDPE

Solvent (A)	T-range/ K	$\Delta_M H_A^\infty$/ kJ/mol
n-octane	383.15-416.15	1.36

Polymer (B): **polyethylene** 1978DIP1
Characterization: HDPE, linear Marlex 6050

Solvent (A)	T-range/ K	$\Delta_M H_A^\infty$/ kJ/mol
n-tetradecane	453.15-483.15	1.36

Polymer (B): **poly(ethylene-*co*-carbon monoxide)** 1978DIP1
Characterization: 10.5 wt% carbon monoxide

Solvent (A)	T-range/ K	$\Delta_M H_A^\infty$/ kJ/mol
n-octane	381.23-411.23	1.08

Polymer (B):	poly(ethylene-*co*-propylene)	1979ITO
Characterization:	40 wt% propylene, Exxon Vistalon 404	

Solvent (A)	T-range/ K	$\Delta_M H_A^\infty$/ kJ/mol
benzene	303.15-346.15	2.39
tert-butylbenzene	303.15-346.15	1.93
cyclohexane	303.15-346.15	0.92
n-decane	303.15-346.15	1.00
ethylbenzene	303.15-346.15	2.13
n-hexane	303.15-346.15	1.05
n-octane	303.15-346.15	1.37
2,2,4-trimethylpentane	303.15-346.15	1.30

Polymer (B):	poly(ethylene-*co*-propylene-*co*-diene)	1993ISS
Characterization:	M_n/g.mol^{-1} = 90000, M_w/g.mol^{-1} = 200000, 50 mol% propylene, Buna AP341, Buna-Werke Huels/Marl, Germany	

Solvent (A)	T-range/ K	$\Delta_{sol} H_{A(vap)}^\infty$/ kJ/mol
benzene	333.15-393.15	−29.38
n-heptane	333.15-393.15	−33.91
n-hexane	333.15-393.15	−29.05
n-nonane	333.15-393.15	−43.25
n-octane	333.15-393.15	−38.71
toluene	333.15-393.15	−34.51

Polymer (B):	poly(ethylene-*co*-propylene-*co*-diene)	1997SCH
Characterization:	M_n/g.mol^{-1} = 90000, M_w/g.mol^{-1} = 200000, 50 mol% propylene, Buna AP341, Buna-Werke Huels/Marl, Germany	

Solvent (A)	T-range/ K	$\Delta_{sol} H_{A(vap)}^\infty$/ kJ/mol
benzene	323.15-393.15	−29.35
n-heptane	323.15-393.15	−33.89
n-hexane	323.15-393.15	−29.41
toluene	323.15-393.15	−34.50

Polymer (B):	poly(ethylene-*co*-vinyl acetate)		1978DIP1
Characterization:	18.0 wt% vinyl acetate		

Solvent (A)	T-range/ K	$\Delta_M H_A^\infty$/ kJ/mol
n-octane	365.15-393.15	1.63

Polymer (B):	poly(ethylene-*co*-vinyl acetate)		1980DIP
Characterization:	18.0 wt% vinyl acetate, $MI = 2.5$, DuPont Elvax 460		

Solvent (A)	T-range/ K	$\Delta_M H_A^\infty$/ kJ/mol	$\Delta_{sol} H_{A(vap)}^\infty$/ kJ/mol
benzene	388.15-408.15	0.08	−28.93
1-butanol	398.15-418.15	10.47	−32.07
chlorobenzene	388.15-408.15	0.88	−35.13
1-chlorobutane	388.15-408.15	2.01	−26.08
chlorocyclohexane	388.15-408.15		−36.13
cyclohexane	388.15-408.15	1.72	−26.59
cyclohexanol	388.15-408.15	9.59	−39.94
cyclohexanone	388.15-408.15		−37.30
n-hexane	388.15-408.15	1.09	−25.67
2-pentanone	388.15-408.15	3.01	−30.02
phenol	398.15-418.15	−1.42	−51.16
tetrachloromethane	388.15-408.15	2.60	−25.50
trichloromethane	388.15-408.15	−2.47	−28.81

Polymer (B):	poly(ethylene-*co*-vinyl acetate)		1998KIM
Characterization:	M_n/g.mol^{-1} = 37000, M_w/g.mol^{-1} = 49000, T_g/K = 312.3, 20.0 wt% vinyl acetate, Polymer Laboratories		

Solvent (A)	T-range/ K	$\Delta_M H_A^\infty$/ kJ/mol	$\Delta_{sol} H_{A(vap)}^\infty$/ kJ/mol
benzene	423.15-463.15	−14.07	−41.88
1-butanol	423.15-463.15	−11.56	−48.85
2-butanone	423.15-463.15	−11.79	−38.79

continued

continued

Solvent (A)	T-range/ K	$\Delta_M H_A^\infty$/ kJ/mol	$\Delta_{sol} H_{A(vap)}^\infty$/ kJ/mol
butyl acetate	423.15-463.15	−11.55	−45.30
cyclohexane	423.15-463.15	−12.97	−38.91
n-decane	423.15-463.15	−12.61	−51.39
1,4-dimethylbenzene	423.15-463.15	−13.14	−47.60
ethanol	423.15-463.15	− 9.72	−41.50
ethyl acetate	423.15-463.15	−10.42	−37.10
methanol	423.15-463.15	−12.01	−41.78
n-octane	423.15-463.15	−12.29	−44.55
2-propanone	423.15-463.15	−12.19	−36.52
tetrachloromethane	423.15-463.15	−15.11	−46.88
toluene	423.15-463.15	−11.02	−38.79
trichloromethane	423.15-463.15	−14.62	−39.19

Polymer (B):   **poly(ethylene-*co*-vinyl acetate)**   **1978DIN**
Characterization: M_n/g.mol^{-1} = 43200, 29.0 wt% vinyl acetate, ρ = 0.9513 g/cm^3, T_m/K < 373, MI = 15.0, DX-31034, Union Carbide

Solvent (A)	T-range/ K	$\Delta_{sol} H_{A(vap)}^\infty$/ kJ/mol
acetaldehyde	423.61-433.68	−23.08
acetic acid	423.61-433.68	−24.91
acetonitrile	423.61-433.68	−28.21
acrylonitrile	423.61-433.68	−30.46
benzene	423.61-433.68	−31.12
1-bromobutane	423.61-433.68	−37.10
2-bromobutane	423.61-433.68	−33.76
1-butanol	423.61-433.68	−36.23
2-butanol	423.61-433.68	−26.99
2-butanone	423.61-433.68	−27.27
chlorobenzene	423.61-433.68	−41.02
cyclohexane	423.61-433.68	−26.03
1,2-dichloroethane	423.61-433.68	−41.30
dichloromethane	423.61-433.68	−22.64
diethyl ether	423.61-433.68	−34.21
1,3-dimethylbenzene	423.61-433.68	−44.36
1,4-dioxane	423.61-433.68	−33.08

continued

continued

Solvent (A)	T-range/ K	$\Delta_{sol}H_{A(vap)}{}^\infty$/ kJ/mol
dipropyl ether	423.61-433.68	−34.44
ethanol	423.61-433.68	−21.32
ethyl acetate	423.61-433.68	−35.34
formic acid	423.61-433.68	−30.09
furan	423.61-433.68	−21.97
n-heptane	423.61-433.68	−31.79
n-hexane	423.61-433.68	−24.63
methanol	423.61-433.68	−14.48
methylcyclohexane	423.61-433.68	−30.94
nitroethane	423.61-433.68	−32.98
nitromethane	423.61-433.68	−31.78
1-nitropropane	423.61-433.68	−33.52
2-nitropropane	423.61-433.68	−35.36
n-octane	423.61-433.68	−38.49
1-octene	423.61-433.68	−37.04
n-pentane	423.61-433.68	−16.92
3-pentanone	423.61-433.68	−36.17
2-propanol	423.61-433.68	−17.65
propionitrile	423.61-433.68	−30.06
propyl acetate	423.61-433.68	−35.46
tetrachloromethane	423.61-433.68	−29.65
tetrahydrofuran	423.61-433.68	−29.94
2,2,2-trifluoroethanol	423.61-433.68	−35.11
water	423.61-433.68	−28.00

Polymer (B):   **poly(ethylene-*co*-vinyl alcohol)**   **1996GAV**
Characterization: 71.0 wt% vinyl alcohol, CERDATO, Elf Atochem, France

Solvent (A)	T-range/ K	$\Delta_{sol}H_{A(vap)}{}^\infty$/ kJ/mol
1-butanol	333.15-363.15	−71.18
2-butanol	333.15-363.15	−19.26
ethanol	333.15-363.15	−74.52
methanol	333.15-363.15	−15.49
2-methyl-1-propanol	333.15-363.15	−17.58

continued

continued

Solvent (A)	T-range/ K	$\Delta_{sol}H_{A(vap)}{}^{\infty}$/ kJ/mol	
2-methyl-2-propanol	333.15-363.15	−25.12	
1-propanol	333.15-363.15	−81.22	
2-propanol	333.15-363.15	−67.41	

Polymer (B):   **poly(ethylene glycol)**   **1969AND**
Characterization: M_n/g.mol^{-1} = 600

Solvent (A)	T/ K	$\Delta_M H_A{}^{\infty}$/ kJ/mol	$\Delta_{sol}H_{A(vap)}{}^{\infty}$/ kJ/mol
2-aminoethanol	423.15	−12.1	−57.7
aziridine	423.15	−7.1	−31.8
diethylamine	423.15	−7.95	−29.3
2,5-dimethylpyrazine	423.15	5.0	−38.9
1,2-ethanediamine	423.15	−10.0	−52.3
N-ethylpiperazine	423.15	0.0	−43.5
N-methylpiperazine	423.15	−2.5	−42.7
2-methylpyrazine	423.15	3.8	−36.0
N-methylpyrrole	423.15	2.1	−33.1
piperazine	423.15	−4.6	−48.5
pyrazine	423.15	2.5	−35.1
pyrrole	423.15	−7.5	−50.2
pyrrolidine	423.15	−11.3	−41.0
triethylene diamine	423.15	18.0	−41.8

Polymer (B):   **poly(ethylene glycol)**   **1989ED1**
Characterization: M_n/g.mol^{-1} = 1570

Solvent (A)	T-range/ K	$\Delta_M H_A{}^{\infty}$/ kJ/mol	$\Delta_{sol}H_{A(vap)}{}^{\infty}$/ kJ/mol
triacetin	402.1-431.1	−0.60	−69.3

Polymer (B):	poly(ethylene glycol)		1989ED2
Characterization:	M_n/g.mol^{-1} = 1570		

Solvent (A)	T-range/ K	$\Delta_M H_A^\infty$/ kJ/mol	$\Delta_{sol} H_{A(vap)}^\infty$/ kJ/mol
triacetin	402.1-431.1	-0.62	-69.3
1,2,3-propanetriyl trinitrate	402.1-431.1	4.27	-79.7
1,2,4-butanetriyl trinitrate	402.1-431.1		-84.4

Polymer (B):	poly(ethylene glycol)		1969AND
Characterization:	M_n/g.mol^{-1} = 2000		

Solvent (A)	T/ K	$\Delta_M H_A^\infty$/ kJ/mol	$\Delta_{sol} H_{A(vap)}^\infty$/ kJ/mol
2-aminoethanol	423.15	-11.7	-55.7
aziridine	423.15	-18.4	-43.1
diethylamine	423.15	-11.3	-32.7
1,2-ethanediamine	423.15	-10.9	-53.2
N-ethylpiperazine	423.15	-4.2	-47.7
N-methylpiperazine	423.15	5.4	-34.7
2-methylpyrazine	423.15	-3.4	-43.1
N-methylpyrrole	423.15	8.4	-26.8
piperazine	423.15	-9.2	-53.2
pyrazine	423.15	-1.7	-39.4
pyrrole	423.15	-7.5	-50.2
pyrrolidine	423.15	-12.6	-42.3
triethylene diamine	423.15	12.1	-47.7

Polymer (B):	poly(ethylene glycol)		1998KIM
Characterization:	M_n/g.mol^{-1} = 31000, M_w/g.mol^{-1} = 82000, T_g/K = 311.6, Aldrich Chem. Co., Inc., Milwaukee, WI		

Solvent (A)	T-range/ K	$\Delta_M H_A^\infty$/ kJ/mol	$\Delta_{sol} H_{A(vap)}^\infty$/ kJ/mol
benzene	353.15-393.15	-3.27	-33.50
1-butanol	353.15-393.15	2.99	-48.89

continued

continued

Solvent (A)	T-range/ K	$\Delta_M H_A^\infty$/ kJ/mol	$\Delta_{sol} H_{A(vap)}^\infty$/ kJ/mol
2-butanone	353.15-393.15	−2.22	−33.14
butyl acetate	353.15-393.15	−0.70	−39.47
cyclohexane	353.15-393.15	3.37	−26.16
n-decane	353.15-393.15	8.83	−36.91
1,4-dimethylbenzene	353.15-393.15	−0.64	−40.83
ethanol	353.15-393.15	0.34	−38.76
ethyl acetate	353.15-393.15	−1.95	−33.11
methanol	353.15-393.15	−0.86	−34.04
n-octane	353.15-393.15	6.04	−30.84
2-propanone	353.15-393.15	−2.45	−30.35
tetrachloromethane	353.15-393.15	−3.09	−32.11
toluene	353.15-393.15	−3.27	−36.55
trichloromethane	353.15-393.15	−10.45	−38.07

Polymer (B):   **poly(ethylene glycol) dibutyl ether**   **1998BRU**
Characterization: M_n/g.mol^{-1} = 270, Genosorb 1843, ρ = 0.927 g/cm^3 (293 K)

Solvent (A)	T-range/ K	$\Delta_M H_A^\infty$/ kJ/mol
1,2-dichloroethane	315.15-363.15	−2.4
dichloromethane	315.15-363.15	−4.4
ethyl acetate	315.15-363.15	2.1
tetrachloroethene	315.15-363.15	−2.8
tetrachloromethane	315.15-363.15	0.2
1,1,1-trichloroethane	315.15-363.15	0.0
trichloroethene	315.15-363.15	−5.8
trichloromethane	315.15-363.15	−8.9

Polymer (B): **poly(ethylene glycol) dimethyl ether** **1998BRU**

Characterization: M_n/g.mol^{-1} = 280, Genosorb 300, ρ = 1.02-1.04 g/cm^3 (293 K)

Solvent (A)	T-range/ K	$\Delta_M H_A^\infty$/ kJ/mol
1,2-dichloroethane	315.15-363.15	−6.8
dichloromethane	315.15-363.15	−8.0
ethyl acetate	315.15-363.15	−1.2
tetrachloroethene	315.15-363.15	−0.5
tetrachloromethane	315.15-363.15	−2.3
1,1,1-trichloroethane	315.15-363.15	−1.6
trichloroethene	315.15-363.15	−5.3
trichloromethane	315.15-363.15	−10.5

Polymer (B): **poly(ethylene oxide)** **1984GAL**

Characterization: M_n/g.mol^{-1} = 10000

Solvent (A)	T-range/ K	$\Delta_{sol} H_{A(vap)}^\infty$/ kJ/mol
acetonitrile	343.15-393.15	−34.63
benzene	343.15-393.15	−32.36
2-butanone	343.15-393.15	−34.54
cyclohexane	343.15-393.15	−22.86
n-decane	343.15-393.15	−36.01
1,2-dimethoxyethane	343.15-393.15	−33.37
1,4-dioxane	343.15-393.15	−35.25
n-dodecane	343.15-393.15	−45.64
ethyl acetate	343.15-393.15	−31.82
ethylbenzene	343.15-393.15	−37.17
n-heptane	343.15-393.15	−26.63
bis(2-methoxyethyl) ether	343.15-393.15	−43.96
1-propanol	343.15-393.15	−38.94
2-propanol	343.15-393.15	−35.67
pyridine	343.15-393.15	−38.85
toluene	343.15-393.15	−35.09
trichloromethane	343.15-393.15	−36.84
2,2,2-trifluoroethanol	343.15-393.15	−48.06

Polymer (B):   **poly(ethylene oxide)**   **1982FER**
Characterization: $M_w/\text{g.mol}^{-1} = 300000$, Polysciences, Inc., Warrington, PA

Solvent (A)	T-range/ K	$\Delta_{sol}H_{A(vap)}{}^{\infty}/$ kJ/mol
benzene	353.15-393.15	−30.97
chlorobenzene	353.15-393.15	−39.19
cyclohexane	353.15-393.15	−21.84
ethanol	353.15-393.15	−33.35
bis(2-ethoxyethyl) ether	353.15-393.15	−50.01
ethyl acetate	353.15-393.15	−32.31
n-hexane	353.15-393.15	−23.87
methanol	353.15-393.15	−31.28
n-octane	353.15-393.15	−40.53
1-propanol	353.15-393.15	−36.67
2-propanone	353.15-393.15	−28.07
toluene	353.15-393.15	−34.77

Comments: Partial molar enthalpies of mixing given in the original source for three temperature intervals (343-363 K, 363-383 K, 383-403 K) differ too much and are not included here.

Polymer (B):   **poly(ethylene oxide)**   **1999ALS**
Characterization: $M_w/\text{g.mol}^{-1} = 1100000$, $T_m/\text{K} = 346.2$,
                          Union Carbide, South Charleston, WV

Solvent (A)	T-range/ K	$\Delta_{M}H_{A}{}^{\infty}/$ kJ/mol	$\Delta_{sol}H_{A(vap)}{}^{\infty}/$ kJ/mol
1-butanol	380.15-400.15		−40.8
butyl acetate	360.15-400.15		−34.1
n-decane	360.15-400.15	20.8	−25.0
n-dodecane	380.15-400.15	13.0	−40.0
ethyl acetate	360.15-400.15		−28.1
n-heptane	340.15-400.15	19.7	−16.0
n-hexane	340.15-400.15	12.4	−18.3
methanol	360.15-400.15		−33.8
methyl acetate	360.15-400.15		−29.2
n-nonane	360.15-400.15	18.0	−21.0
n-octane	340.15-400.15	17.5	−21.4
n-pentane	340.15-400.15	12.2	−18.2
1-propanol	360.15-400.15		−34.5
propyl acetate	360.15-400.15		−25.0
n-undecane	360.15-400.15	14.1	−34.9

Polymer (B):   **poly(ethylethylene)**   **1994TIA**

Characterization: $T_g/K = 269$, $\rho = 0.872$ g/cm^3,

                         Scientific Polymer Products, Inc., Ontario, NY

Solvent (A)	T-range/ K	$\Delta_{sol}H_{A(vap)}^\infty$/ kJ/mol
benzene	333.15-383.15	−28.95
n-butane	333.15-383.15	−19.66
1-butanol	333.15-383.15	−28.28
2-butanone	333.15-383.15	−25.48
butyl acetate	333.15-383.15	−36.11
chlorobenzene	333.15-383.15	−35.94
1-chlorobutane	333.15-383.15	−29.12
1-chlorohexane	333.15-383.15	−38.12
1-chlorooctane	333.15-383.15	−46.86
1-chloropentane	333.15-383.15	−33.64
cycloheptane	333.15-383.15	−35.52
cyclohexadiene	333.15-383.15	−29.75
cyclohexane	333.15-383.15	−30.38
cyclohexene	333.15-383.15	−30.63
cyclooctane	333.15-383.15	−40.29
cyclopentane	333.15-383.15	−26.74
n-decane	333.15-383.15	−46.69
1,1-dichloroethane	333.15-383.15	−25.31
1,2-dichloroethane	333.15-383.15	−27.78
dichloromethane	333.15-383.15	−22.76
1,4-dioxane	333.15-383.15	−30.04
ethyl acetate	333.15-383.15	−27.36
ethylbenzene	333.15-383.15	−37.45
n-heptane	333.15-383.15	−33.60
n-hexane	333.15-383.15	−29.12
methyl acetate	333.15-383.15	−23.47
n-nonane	333.15-383.15	−42.47
n-octane	333.15-383.15	−37.99
n-pentane	333.15-383.15	−24.64
1-pentanol	333.15-383.15	−32.13
1-propanol	333.15-383.15	−23.60
2-propanone	333.15-383.15	−18.87
propyl acetate	333.15-383.15	−31.71
tetrachloromethane	333.15-383.15	−29.25
tetrahydrofuran	333.15-383.15	−27.66
toluene	333.15-383.15	−33.56
1,1,1-trichloroethane	333.15-383.15	−28.16
trichloroethene	333.15-383.15	−31.09
trichloromethane	333.15-383.15	−26.78

Polymer (B):   **poly(ethyl methacrylate)**   **1978KAR**
Characterization: M_n/g.mol^{-1} = 144000

Solvent (A)	T-range/ K	$\Delta_{sol}H_{A(vap)}^{\infty}$/ kJ/mol
acetonitrile	417.74-427.55	−38.24
acrylonitrile	417.74-427.55	−36.99
benzene	417.74-427.55	−32.71
bromobenzene	417.74-427.55	−45.75
1-bromobutane	417.74-427.55	−35.76
1-butanol	417.74-427.55	−38.03
2-butanone	417.74-427.55	−33.68
butyl acetate	417.74-427.55	−36.76
butylamine	417.74-427.55	−30.37
butyraldehyde	417.74-427.55	−33.93
chlorobenzene	417.74-427.55	−45.45
cyclohexane	417.74-427.55	−12.23
n-decane	417.74-427.55	−35.66
dipropyl ether	417.74-427.55	−25.50
ethylbenzene	417.74-427.55	−40.73
ethylcyclohexane	417.74-427.55	−18.03
n-heptane	417.74-427.55	−20.22
methylcyclohexane	417.74-427.55	−15.60
nitroethane	417.74-427.55	−40.13
nitromethane	417.74-427.55	−36.71
1-nitropropane	417.74-427.55	−43.69
n-nonane	417.74-427.55	−29.73
n-octane	417.74-427.55	−24.82
1-octene	417.74-427.55	−29.09
1-pentanol	417.74-427.55	−38.70
propanoic acid	417.74-427.55	−32.91
1-propanol	417.74-427.55	−37.27
2-propanone	417.74-427.55	−34.35
propionitrile	417.74-427.55	−39.12
propyl acetate	417.74-427.55	−35.04
tetrachloromethane	417.74-427.55	−5.94
toluene	417.74-427.55	−37.38
trichloromethane	417.74-427.55	−34.29

Polymer (B):	poly(ethyl methacrylate)	1990CHE
Characterization:	$M_\eta/\text{g.mol}^{-1} = 215000$, $T_g/\text{K} = 338$,	
	Aldrich Chem. Co., Inc., Milwaukee, WI	

Solvent (A)	T-range/ K	$\Delta_{sol} H_{A(vap)}^\infty/$ kJ/mol
1-butanol	448.15-468.15	−42.3
butyl acetate	448.15-468.15	−41.5
n-decane	448.15-468.15	−40.5
n-dodecane	448.15-468.15	−46.7
ethanol	448.15-468.15	−32.8
ethyl acetate	448.15-468.15	−33.8
n-heptane	448.15-468.15	−29.6
methanol	448.15-468.15	−28.6
methyl acetate	448.15-468.15	−30.2
n-nonane	448.15-468.15	−37.1
n-octane	448.15-468.15	−33.8
2-pentanol	448.15-468.15	−45.1
2-propanol	448.15-468.15	−37.2
propyl acetate	448.15-468.15	−37.8
n-undecane	448.15-468.15	−44.0

Polymer (B):	poly(ethyl methacrylate)	1991ALS
Characterization:	$M_\eta/\text{g.mol}^{-1} = 215000$, $T_g/\text{K} = 338$,	
	Aldrich Chem. Co., Inc., Milwaukee, WI	

Solvent (A)	T-range/ K	$\Delta_{sol} H_{A(vap)}^\infty/$ kJ/mol
1-butanol	380.15-413.15	−28.11
butyl acetate	380.15-413.15	−26.90
n-decane	380.15-413.15	−26.23
ethanol	380.15-413.15	−26.48
ethyl acetate	380.15-413.15	−23.51
n-heptane	380.15-413.15	−21.24
methanol	380.15-413.15	−26.98
methyl acetate	380.15-413.15	−23.13
n-nonane	380.15-413.15	−23.30
n-octane	380.15-413.15	−22.54
2-pentanol	380.15-413.15	−26.19
2-propanol	380.15-413.15	−27.03
propyl acetate	380.15-413.15	−23.92
n-undecane	380.15-413.15	−28.37

Polymer (B):   **poly(glycidyl methacrylate-*co*-butyl methacrylate)**   **2002KAY**
Characterization: M_n/g.mol^{-1} = 530000, M_w/g.mol^{-1} = 738000, T_g/K = 334,
                          ρ (298 K) = 1.178 g/cm^3, 41 mol% butyl methacrylate

Solvent (A)	T-range/ K	$\Delta_M H_A^\infty$/ kJ/mol
benzene	403.15-423.15	5.85
1-chlorobutane	403.15-423.15	3.68
1-chloropropane	403.15-423.15	3.05
n-decane	403.15-423.15	10.7
1,4-dimethylbenzene	403.15-423.15	7.73
n-heptane	403.15-423.15	6.90
n-hexane	403.15-423.15	5.56
n-nonane	403.15-423.15	9.61
n-octane	403.15-423.15	8.07
n-pentane	403.15-423.15	3.93
tetrachloromethane	403.15-423.15	0.63
toluene	403.15-423.15	6.19

Polymer (B):   **poly(glycidyl methacrylate-*co*-ethyl methacrylate)**   **2002KAY**
Characterization: M_n/g.mol^{-1} = 441000, M_w/g.mol^{-1} = 700000, T_g/K = 355,
                          ρ (298 K) = 1.186 g/cm^3, 44 mol% ethyl methacrylate

Solvent (A)	T-range/ K	$\Delta_M H_A^\infty$/ kJ/mol
benzene	403.15-423.15	5.52
1-chlorobutane	403.15-423.15	4.39
1-chloropropane	403.15-423.15	2.84
n-decane	403.15-423.15	9.99
1,4-dimethylbenzene	403.15-423.15	6.98
n-heptane	403.15-423.15	4.89
n-hexane	403.15-423.15	4.89
n-nonane	403.15-423.15	9.20
n-octane	403.15-423.15	7.73
n-pentane	403.15-423.15	3.47
tetrachloromethane	403.15-423.15	1.05
toluene	403.15-423.15	5.48

Polymer (B):	poly(glycidyl methacrylate-*co*-methyl methacrylate)	2002KAY
Characterization:	M_n/g.mol^{-1} = 555000, M_w/g.mol^{-1} = 710000, T_g/K = 373, ρ (298 K) = 1.204 g/cm^3, 38 mol% methyl methacrylate	

Solvent (A)	*T*-range/ K	$\Delta_M H_A^\infty$/ kJ/mol
benzene	403.15-423.15	5.23
1-chlorobutane	403.15-423.15	2.63
1-chloropropane	403.15-423.15	2.63
n-decane	403.15-423.15	10.4
1,4-dimethylbenzene	403.15-423.15	7.23
1,4-dioxane	403.15-423.15	4.10
n-heptane	403.15-423.15	6.40
n-hexane	403.15-423.15	5.10
n-nonane	403.15-423.15	9.28
n-octane	403.15-423.15	7.90
n-pentane	403.15-423.15	3.76
tetrachloromethane	403.15-423.15	0.59
toluene	403.15-423.15	5.60

Polymer (B):	poly(2-hydroxyethyl acrylate)	1994TIA
Characterization:	T_g/K = 258, ρ = 1.310 g/cm^3, Scientific Polymer Products, Inc., Ontario, NY	

Solvent (A)	*T*-range/ K	$\Delta_{sol} H_{A(vap)}^\infty$/ kJ/mol
1-butanol	333.15-383.15	−43.81
2-butanone	333.15-383.15	−29.78
butyl acetate	333.15-383.15	−34.85
chlorobenzene	333.15-383.15	−35.15
1-chlorohexane	333.15-383.15	−29.04
1-chlorooctane	333.15-383.15	−39.54
1,2-dichloroethane	333.15-383.15	−31.00
dichloromethane	333.15-383.15	−27.07
1,4-dioxane	333.15-383.15	−35.48
ethanol	333.15-383.15	−38.45
ethyl acetate	333.15-383.15	−29.33
ethylbenzene	333.15-383.15	−32.26
methyl acetate	333.15-383.15	−28.24
1-pentanol	333.15-383.15	−46.82
1-propanol	333.15-383.15	−40.50

continued

continued

Solvent (A)	T-range/ K	$\Delta_{sol}H_{A(vap)}^{\infty}/$ kJ/mol
2-propanone	333.15-383.15	-27.49
propyl acetate	333.15-383.15	-31.59
tetrahydrofuran	333.15-383.15	-28.24
toluene	333.15-383.15	-28.99
trichloroethene	333.15-383.15	-28.37
trichloromethane	333.15-383.15	-28.45

Polymer (B):	**poly(2-hydroxyethyl methacrylate)**	**2000DEM**
Characterization:	$T_g/K = 339$, $\rho = 1.10$ g/cm^3,	

Solvent (A)	T-range/ K	$\Delta_M H_A^{\infty}/$ kJ/mol
benzene	403.15-423.15	27.71
2-butanone	403.15-423.15	23.56
n-decane	403.15-423.15	33.85
1,2-dimethylbenzene	403.15-423.15	35.34
n-dodecane	403.15-423.15	25.64
ethanol	403.15-423.15	65.13
ethyl acetate	403.15-423.15	26.33
methanol	403.15-423.15	19.40
methyl acetate	403.15-423.15	19.40
n-nonane	403.15-423.15	29.10
n-octane	403.15-423.15	28.41
2-propanone	403.15-423.15	22.17
toluene	403.15-423.15	33.26
n-undecane	403.15-423.15	38.11

Polymer (B):	**poly(4-hydroxystyrene)**	**1995LEZ**
Characterization:	$M_n/\text{g.mol}^{-1} = 1500$, $M_w/\text{g.mol}^{-1} = 3000$	

Solvent (A)	T/ K	$\Delta_M H_A^{\infty}/$ kJ/mol	$\Delta_{sol}H_{A(vap)}^{\infty}/$ kJ/mol
2-butanone	443.15-463.15	-6.16	-37.9
butyl acetate	443.15-463.15	22.1	-11.9
ethyl acetate	443.15-463.15	-4.44	-30.4

continued

continued

Solvent (A)	T-range/ K	$\Delta_M H_A^\infty/$ kJ/mol	$\Delta_{sol} H_{A(vap)}^\infty/$ kJ/mol
3-pentanone	443.15-463.15	24.3	−5.61
2-propanol	443.15-463.15	2.47	−25.6
2-propanone	443.15-463.15	−31.0	−54.6
propyl acetate	443.15-463.15	8.37	−20.8
tetrahydrofuran	443.15-463.15	−7.16	−32.2

Polymer (B):	**poly(isobornyl methacrylate)**	**1999KA2**
Characterization:	Aldrich Chem. Co., Inc., Milwaukee, WI	

Solvent (A)	T-range/ K	$\Delta_M H_A^\infty/$ kJ/mol
benzene	423.15-473.15	18.2
2-butanone	423.15-473.15	19.3
n-decane	423.15-473.15	37.2
1,2-dimethylbenzene	423.15-473.15	27.0
n-dodecane	423.15-473.15	42.1
ethyl acetate	423.15-473.15	19.0
methanol	423.15-473.15	27.0
methyl acetate	423.15-473.15	17.3
n-nonane	423.15-473.15	33.8
n-octane	423.15-473.15	28.6
2-propanone	423.15-473.15	19.1
tetrahydrofuran	423.15-473.15	19.7
n-undecane	423.15-473.15	38.5

Polymer (B):	**polyisobutylene**	**1978TAG**
Characterization:	$M_n/\text{g.mol}^{-1} = 20000$	

Solvent (A)	T-range/ K	$\Delta_M H_A^\infty/$ kJ/mol
benzene	298.15-321.95	3.56
2,2,4-trimethylpentane	298.15-321.95	4.61

Polymer (B):   **polyisobutylene**   **1978TAG**
Characterization: M_n/g.mol^{-1} = 85000

Solvent (A)	T-range/ K	$\Delta_M H_A^\infty$/ kJ/mol
benzene	298.15-321.95	3.56
2,2,4-trimethylpentane	298.15-321.95	4.61

Polymer (B):   **polyisobutylene**   **1978TAG**
Characterization: M_n/g.mol^{-1} = 280000

Solvent (A)	T-range/ K	$\Delta_M H_A^\infty$/ kJ/mol
benzene	298.15-321.95	3.56
2,2,4-trimethylpentane	298.15-321.95	4.61

Polymer (B):   **1,4-*cis*-polyisoprene**   **2003VOZ**
Characterization: 98.5 % 1,4-*cis*, SKI-5, industrial sample

Solvent (A)	T-range/ K	$\Delta_M H_A^\infty$/ kJ/mol
benzene	303.15-413.15	1.40
cyclohexane	303.15-413.15	0.47
n-decane	303.15-413.15	1.45
n-heptane	303.15-413.15	1.40
n-hexane	303.15-413.15	1.34
hexafluorobenzene	303.15-413.15	3.92
n-nonane	303.15-413.15	1.49
n-octane	303.15-413.15	1.43
toluene	303.15-413.15	0.94

Polymer (B):	poly(N-isopropylacrylamide)	1978DIP1
Characterization:	–	

Solvent (A)	T-range/ K	$\Delta_M H_A^\infty/$ kJ/mol
1-butanol	412.15-449.15	7.53

Polymer (B):	poly(isopropyl methacrylate)	2000FEN
Characterization:	$\rho = 1.033$ g/cm^3	

Solvent (A)	T-range/ K	$\Delta_{sol} H_{A(vap)}^\infty/$ kJ/mol
benzene	393.15-433.15	−23.98
1-butanol	393.15-433.15	−33.80
butyl acetate	393.15-433.15	−28.68
1,2-dichloroethane	393.15-433.15	−31.90
ethanol	393.15-433.15	−32.34
ethyl acetate	393.15-433.15	−29.89
ethylbenzene	393.15-433.15	−26.03
methyl acetate	393.15-433.15	−34.16
2-methyl-1-propanol	393.15-433.15	−30.38
2-propanol	393.15-433.15	−31.51
propyl acetate	393.15-433.15	−28.78
toluene	393.15-433.15	−25.27
trichloromethane	393.15-433.15	−24.70

Polymer (B):	poly[2-(3-mesityl-3-methylcyclobutyl)-2-hydroxyethyl methacrylate]	1999KA4
Characterization:	M_n/g.mol^{-1} = 84000, M_w/g.mol^{-1} = 200000, T_g/K = 410, ρ (298 K) = 1.042 g/cm^3, synthesized in the laboratory	

Solvent (A)	T-range/ K	$\Delta_M H_A^\infty/$ kJ/mol
benzene	453.15-473.15	19.97
2-butanone	453.15-473.15	11.63
n-decane	453.15-473.15	11.64

continued

continued

Solvent (A)	T-range/ K	$\Delta_M H_A^\infty$/ kJ/mol
1,2-dimethylbenzene	453.15-473.15	4.19
n-dodecane	453.15-473.15	14.15
ethanol	453.15-473.15	24.12
ethyl acetate	453.15-473.15	16.62
methanol	453.15-473.15	24.12
methyl acetate	453.15-473.15	20.81
n-nonane	453.15-473.15	25.79
n-octane	453.15-473.15	19.13
2-propanone	453.15-473.15	14.99
toluene	453.15-473.15	24.95
n-undecane	453.15-473.15	15.83

Polymer (B):   **poly(methyl acrylate)**   **1978DIP1, 1978DIP2**
Characterization: M_n/g.mol^{-1} = 63200, M_w/g.mol^{-1} = 200000,
                     Aldrich Chem. Co., Inc., Milwaukee, WI

Solvent (A)	T/ K	$\Delta_M H_A^\infty$/ kJ/mol	$\Delta_{sol} H_{A(vap)}^\infty$/ kJ/mol
benzene	361.15-385.15	1.42	−28.78
butylbenzene	361.15-385.15	4.95	−41.10
tert-butylbenzene	361.15-385.15	4.19	−39.05
butylcyclohexane	361.15-385.15	11.47	−33.30
cyclohexane	361.15-385.15	10.43	−19.08
n-decane	351.15-381.15	13.45	−32.83
cis-decahydronaphthalene	361.15-385.15	12.11	−34.71
trans-decahydronaphthalene	361.15-385.15	12.93	−32.41
n-dodecane	361.15-385.15	14.15	−40.52
ethylbenzene	361.15-385.15	3.89	−34.25
naphthalene	372.15-395.15	1.44	−49.39
n-octane	361.15-385.15	11.20	−25.61
n-tetradecane	361.15-385.15	17.25	−46.61
1,2,3,4-tetrahydronaphthalene	361.15-385.15	4.88	−44.64
3,3,4,4-tetramethylhexane	361.15-385.15	12.91	−29.47
toluene	361.15-385.15	1.74	−32.58
2,2,5-trimethylhexane	361.15-385.15	14.97	−20.67
3,4,5-trimethylheptane	361.15-385.15	12.84	−28.70
2,2,4-trimethylpentane	361.15-385.15	16.09	−15.19

Polymer (B):	**poly(methyl acrylate)**	**1994TIA**
Characterization:	$T_g/K = 283$, $\rho = 1.220$ g/cm^3,	
	Scientific Polymer Products, Inc., Ontario, NY	

Solvent (A)	T-range/ K	$\Delta_{sol}H_{A(vap)}^\infty$/ kJ/mol
benzene	333.15-383.15	−29.62
1-butanol	333.15-383.15	−38.70
2-butanone	333.15-383.15	−30.92
butyl acetate	333.15-383.15	−37.99
chlorobenzene	333.15-383.15	−36.94
1-chlorobutane	333.15-383.15	−27.07
1-chlorohexane	333.15-383.15	−34.43
1-chlorooctane	333.15-383.15	−41.76
1-chloropentane	333.15-383.15	−30.67
cyclohexadiene	333.15-383.15	−26.40
cyclooctane	333.15-383.15	−29.08
n-decane	333.15-383.15	−33.18
1,1-dichloroethane	333.15-383.15	−28.11
1,2-dichloroethane	333.15-383.15	−33.68
dichloromethane	333.15-383.15	−28.95
1,4-dioxane	333.15-383.15	−35.56
ethanol	333.15-383.15	−32.68
ethyl acetate	333.15-383.15	−31.88
ethylbenzene	333.15-383.15	−36.28
methyl acetate	333.15-383.15	−29.29
n-nonane	333.15-383.15	−29.08
1-pentanol	333.15-383.15	−41.76
1-propanol	333.15-383.15	−35.48
2-propanone	333.15-383.15	−27.87
propyl acetate	333.15-383.15	−34.77
tetrachloromethane	333.15-383.15	−25.65
tetrahydrofuran	333.15-383.15	−28.49
toluene	333.15-383.15	−33.22
1,1,1-trichloroethane	333.15-383.15	−26.19
trichloroethene	333.15-383.15	−31.63
trichloromethane	333.15-383.15	−31.59
n-undecane	333.15-383.15	−36.94

| Polymer (B): | **poly(methylcyanopropylsiloxane)** | | **1991BEC** |
| *Characterization:* | 5 % cyanopropyl groups, type OV-105 | | |

Solvent (A)	T-range/ K	$\Delta_M H_A^\infty$/ kJ/mol	$\Delta_{sol} H_{A(vap)}^\infty$/ kJ/mol
benzene	372.90-473.15	4.90	−26.75
1-butanol	372.90-413.50	14.44	−29.60
n-decane	372.90-473.15	3.73	−40.95
n-dodecane	372.90-473.15	3.10	−48.86
n-heptane	372.90-473.15	4.61	−29.39
n-hexane	372.90-473.15	5.44	−25.16
n-nonane	372.90-473.15	3.94	−37.05
n-octane	372.90-473.15	4.23	−33.12
2-pentanone	372.90-473.15	5.02	−29.56
pyridine	372.90-473.15	6.11	−30.27
n-undecane	372.90-473.15	3.39	−44.92

| Polymer (B): | **poly(methyl methacrylate)** | | **1998KIM** |
| *Characterization:* | M_n/g.mol^{-1} = 945, M_w/g.mol^{-1} = 1000, T_g/K = 313.6, Aldrich Chem. Co., Inc., Milwaukee, WI | | |

Solvent (A)	T-range/ K	$\Delta_M H_A^\infty$/ kJ/mol	$\Delta_{sol} H_{A(vap)}^\infty$/ kJ/mol
benzene	403.15-423.15	3.11	−28.64
1-butanol	403.15-423.15	12.55	−29.06
2-butanone	403.15-423.15	−8.94	−19.93
butyl acetate	403.15-423.15	−1.77	−37.90
n-decane	403.15-423.15	25.85	−19.27
1,2-dimethylbenzene	403.15-423.15	−1.89	−38.01
ethanol	403.15-423.15	−12.66	−47.62
ethyl acetate	403.15-423.15	−6.14	−36.84
ethylbenzene	403.15-423.15	3.02	−32.51
methanol	403.15-423.15	−8.94	−41.48
4-methyl-2-pentanone	403.15-423.15	11.23	−25.01
toluene	403.15-423.15	−7.12	−39.53
trichloromethane	403.15-423.15	−6.10	−34.48

Polymer (B):   **poly(methyl methacrylate)**   **1996KAY**

Characterization: $M_w/\text{g.mol}^{-1} = 120000$,
                       Aldrich Chem. Co., Inc., Milwaukee, WI

Solvent (A)	T-range/ K	$\Delta_{sol} H_{A(vap)}^\infty/$ kJ/mol
2-butanone	433.15-473.15	−8.75
n-decane	433.15-473.15	−26.4
1,2-dimethylbenzene	433.15-473.15	−13.1
n-dodecane	433.15-473.15	−40.9
ethyl acetate	433.15-473.15	−7.49
methanol	433.15-473.15	−8.58
methyl acetate	433.15-473.15	−7.41
n-nonane	433.15-473.15	−16.1
n-octane	433.15-473.15	−9.80
2-propanone	433.15-473.15	−8.50
n-undecane	433.15-473.15	−34.8

Polymer (B):   **poly(methyl methacrylate)**   **1999KA1**

Characterization: $M_w/\text{g.mol}^{-1} = 120000$,
                       Aldrich Chem. Co., Inc., Milwaukee, WI

Solvent (A)	T-range/ K	$\Delta_M H_A^\infty/$ kJ/mol	$\Delta_{sol} H_{A(vap)}^\infty/$ kJ/mol
benzene	403.15-433.15	2.35	−8.29
2-butanone	403.15-433.15	11.01	−8.75
n-decane	403.15-433.15	7.54	−26.4
1,2-dimethylbenzene	403.15-433.15	2.51	−13.1
1,4-dioxane	403.15-433.15	18.63	−11.1
n-dodecane	403.15-433.15	5.86	−36.6
ethanol	403.15-433.15	21.02	−9.10
ethyl acetate	403.15-433.15	12.23	−7.49
methanol	403.15-433.15	21.65	−8.58
methyl acetate	403.15-433.15	13.44	−7.41
n-nonane	403.15-433.15	9.80	−16.1
n-octane	403.15-433.15	11.68	−9.80
2-propanone	403.15-433.15	13.98	−8.50
tetrachloromethane	403.15-433.15	9.13	−9.80
tetrahydrofuran	403.15-433.15	20.06	−10.6
toluene	403.15-433.15	2.51	−9.46
trichloromethane	403.15-433.15	17.79	−8.29
n-undecane	403.15-433.15	5.53	−36.5

Polymer (B):　　　**poly[2-(3-methyl-3-phenylcyclobutyl)-**　　　　　　　　**2000KAY**
　　　　　　　　　　2-hydroxyethyl methacrylate]

Characterization:　$M_n/\text{g.mol}^{-1} = 47700$, $M_w/\text{g.mol}^{-1} = 125000$, $\rho_B = 1.053 \text{ g/cm}^3$,
　　　　　　　　　synthesized in the laboratory

Solvent (A)	T-range/ K	$\Delta_M H_A^\infty/$ kJ/mol
benzene	443.15-453.15	29.9
2-butanone	443.15-453.15	16.6
n-decane	443.15-453.15	23.3
1,2-dimethylbenzene	443.15-453.15	21.6
n-dodecane	443.15-453.15	23.3
ethanol	443.15-453.15	39.9
ethyl acetate	443.15-453.15	23.3
methanol	443.15-453.15	34.9
methyl acetate	443.15-453.15	24.9
n-nonane	443.15-453.15	28.3
n-octane	443.15-453.15	31.6
2-propanone	443.15-453.15	21.6
toluene	443.15-453.15	26.6
n-undecane	443.15-453.15	13.3

Polymer (B):　　　**poly(methyl trifluoropropyl siloxane)**　　　　　　**1992BEC**
Characterization:　　50/50 methyl/3,3,3-trifluorpropyl, type OV-215

Solvent (A)	T-range/ K	$\Delta_M H_A^\infty/$ kJ/mol	$\Delta_{sol} H_{A(vap)}^\infty/$ kJ/mol
benzene	363.15-453.15	3.64	−28.2
1-butanol	363.15-453.15	16.16	−28.9
2-butanone	363.15-453.15	−2.39	−33.7
butyronitrile	363.15-453.15	2.01	−34.5
N,N-dimethylaniline	363.15-453.15	−5.65	−45.7
ethyl acetate	363.15-453.15	3.27	−30.1
n-decane	363.15-453.15	8.46	−36.7
n-dodecane	363.15-453.15	8.96	−44.1
n-heptane	363.15-453.15	7.12	−27.1
n-hexane	363.15-453.15	8.25	−22.4
n-nonane	363.15-453.15	7.87	−33.6
n-octane	363.15-453.15	7.66	−30.1
1-octanol	363.15-453.15	14.03	−44.1
2-octanone	363.15-453.15	3.31	−46.8

continued

continued

Solvent (A)	T-range/ K	$\Delta_M H_A^\infty$/ kJ/mol	$\Delta_{sol} H_{A(vap)}^\infty$/ kJ/mol
1-pentanol	363.15-453.15	13.94	−32.1
2-pentanone	363.15-453.15	0.96	−34.1
pentylbenzene	363.15-453.15	3.31	−44.8
2-propanone	363.15-453.15	0.25	−30.8
propylbenzene	363.15-453.15	2.97	−37.8
pyridine	363.15-453.15	4.10	−32.6
toluene	363.15-453.15	3.06	−32.4
n-undecane	363.15-453.15	8.54	−40.5
valeronitrile	363.15-453.15	0.84	

Polymer (B): **poly[1-methyl-1-(trimethylsilyl-methyl)-1-silacyclobutane]**   1994YAM

Characterization: M_n/g.mol^{-1} = 1556000, M_w/g.mol^{-1} = 1650000

Solvent (A)	T-range/ K	$\Delta_M H_A^\infty$/ kJ/mol	$\Delta_{sol} H_{A(vap)}^\infty$/ kJ/mol
benzene	333.15-393.15	−11.7	−34.3
ethylbenzene	353.15-433.15	−3.8	−40.2
n-hexane	343.15-393.15	−1.3	−31.0
n-heptane	323.15-393.15	−4.2	−31.8
toluene	343.15-403.15	−9.2	−41.4

Polymer (B): **poly[4-(2-naphthalenesulfonyl)-styrene]**   1999KA3
Characterization: sulfonation of polystyrene (M_η/g.mol^{-1} = 200000) with 2-naphthalenesulfonyl chloride in the laboratory

Solvent (A)	T-range/ K	$\Delta_{sol} H_{A(vap)}^\infty$/ kJ/mol
benzene	453.15-473.15	−14.15
2-butanone	453.15-473.15	−11.64
n-decane	453.15-473.15	−22.48
1,2-dimethylbenzene	453.15-473.15	−17.46
n-dodecane	453.15-473.15	−28.30

continued

continued

Solvent (A)	T-range/ K	$\Delta_{sol}H_{A(vap)}{}^{\infty}/$ kJ/mol
ethanol	453.15-473.15	−22.48
ethyl acetate	453.15-473.15	−11.22
methanol	453.15-473.15	−14.15
methyl acetate	453.15-473.15	−9.17
n-nonane	453.15-473.15	−15.83
n-octane	453.15-473.15	−10.80
2-propanone	453.15-473.15	−10.43
toluene	453.15-473.15	−19.13
n-undecane	453.15-473.15	−24.95

Polymer (B):   **poly(2-phenyl-1,3-dioxolane-4-yl-methyl methacrylate)**   **2002ILT**
Characterization: $M_n/\text{g.mol}^{-1} = 60000$, $M_w/\text{g.mol}^{-1} = 259000$,
                         ρ (298 K) = 1.252 g/cm^3, T_g/K = 403

Solvent (A)	T-range/ K	$\Delta_{M}H_{A}{}^{\infty}/$ kJ/mol
benzene	433.15-453.15	12.8
1-butanol	433.15-453.15	31.1
butylamine	433.15-453.15	14.9
n-decane	433.15-453.15	14.9
ethanol	433.15-453.15	29.7
n-heptane	433.15-453.15	11.8
n-hexane	433.15-453.15	13.1
n-octane	433.15-453.15	17.7
1-pentanol	433.15-453.15	28.8
1-propanol	433.15-453.15	32.0
propylamine	433.15-453.15	14.4

Polymer (B): **poly(2-phenyl-1,3-dioxolane-4-yl-methyl** **2003ACI**
methacrylate-*co*-butyl methacrylate)

Characterization: M_n/g.mol^{-1} = 301100, M_w/g.mol^{-1} = 880200,
ρ (298 K) = 1.217 g/cm^3, 55.0 mol% butyl methacrylate

Solvent (A)	T-range/ K	$\Delta_M H_A^\infty$/ kJ/mol
1-butanol	413.15-453.15	31.06
ethanol	413.15-453.15	29.68
methanol	413.15-453.15	26.79
1-pentanol	413.15-453.15	28.80
1-propanol	413.15-453.15	31.98

Polymer (B): **poly(2-phenyl-1,3-dioxolane-4-yl-methyl** **2004ILT**
methacrylate-*co*-glycidyl methacrylate)

Characterization: M_n/g.mol^{-1} = 244050, M_w/g.mol^{-1} = 623430, T_g/K = 367,
ρ (298 K) = 1.229 g/cm^3, 60.0 mol% glycidyl methacrylate

Solvent (A)	T-range/ K	$\Delta_M H_A^\infty$/ kJ/mol
1-butanol	413.15-463.15	28.44
n-decane	413.15-463.15	28.10
ethanol	413.15-463.15	24.27
n-heptane	413.15-463.15	20.37
n-hexane	413.15-463.15	15.79
n-octane	413.15-463.15	22.40
1-pentanol	413.15-463.15	25.16
1-propanol	413.15-463.15	26.64

Polymer (B): **poly(phenyl ether)** **1982LAN**
Characterization: 5-ring ether from Anspec

Solvent (A)	T-range/ K	$\Delta_M H_A^\infty$/ kJ/mol
benzene	353.15-373.15	−0.41
butyl *tert*-butyl ether	353.15-373.15	1.26
butyl acetate	353.15-373.15	−2.09

continued

continued

Solvent (A)	T-range/ K	$\Delta_M H_A^\infty$/ kJ/mol
butyl pentafluoropropionate	353.15-373.15	3.27
butyl trifluoroacetate	353.15-373.15	2.81
butyl trimethylsilyl ether	353.15-373.15	0.92
cyclohexyl acetate	353.15-373.15	−1.84
cyclohexyl pentafluoropropionate	353.15-373.15	4.40
cyclohexyl trifluoroacetate	353.15-373.15	1.88
cyclohexyl trimethylsilyl ether	353.15-373.15	1.38
1,2-dimethylbenzene	353.15-373.15	0.38
n-heptane	353.15-373.15	3.98
1-methyl-3-ethylbenzene	353.15-373.15	0.46
n-nonane	353.15-373.15	5.11
n-octane	353.15-373.15	4.44
toluene	353.15-373.15	−0.07
m-tolyl pentafluoropropionate	353.15-373.15	3.52
p-tolyl pentafluoropropionate	353.15-373.15	3.48
m-tolyl trifluoroacetate	353.15-373.15	2.01
p-tolyl trifluoroacetate	353.15-373.15	3.06
m-tolyl trimethylsilyl ether	353.15-373.15	−1.26
p-tolyl trimethylsilyl ether	353.15-373.15	−1.55

Polymer (B):	**poly(2-(3-phenyl-3-methylcyclobutyl)-2-hydroxyethyl methacrylate-*co*-methacrylic acid)**	**2001KA2**
Characterization:	M_n/g.mol^{-1} = 17200, M_w/g.mol^{-1} = 45000, 45 mol% methacrylic acid, synthesized in the laboratory	

Solvent (A)	T-range/ K	$\Delta_M H_A^\infty$/ kJ/mol
benzene	423.15-453.15	12.64
2-butanone	423.15-453.15	27.42
1,2-dimethylbenzene	423.15-453.15	14.82
n-dodecane	423.15-453.15	18.30
ethanol	423.15-453.15	29.73
ethyl acetate	423.15-453.15	15.16
methanol	423.15-453.15	26.63
methyl acetate	423.15-453.15	10.47
2-propanone	423.15-453.15	24.95
toluene	423.15-453.15	15.83

Polymer (B):	**polypropylene**	**1978DIP1**
Characterization:	–	

Solvent (A)	T-range/ K	$\Delta_M H_A^\infty$/ kJ/mol
n-decane	448.15-483.15	−0.61
n-dodecane	448.15-483.15	−0.61
n-tetradecane	453.15-483.15	0.82

Polymer (B):	**polypropylene**	**1994TIA**
Characterization:	T_g/K = 257, ρ = 0.865 g/cm^3, sample PP1, Scientific Polymer Products, Inc., Ontario, NY	

Solvent (A)	T-range/ K	$\Delta_{sol} H_{A(vap)}^\infty$/ kJ/mol
benzene	343.15-383.15	−28.37
1-butanol	343.15-383.15	−26.94
butyl acetate	343.15-383.15	−34.89
chlorobenzene	343.15-383.15	−33.89
1-chlorohexane	343.15-383.15	−36.40
1-chlorooctane	343.15-383.15	−45.48
1-chloropentane	343.15-383.15	−32.93
cycloheptane	343.15-383.15	−34.56
cyclohexadiene	343.15-383.15	−29.12
cyclohexane	343.15-383.15	−29.92
cyclohexene	343.15-383.15	−30.00
cyclooctane	343.15-383.15	−38.16
cyclopentane	343.15-383.15	−27.07
1,1-dichloroethane	343.15-383.15	−24.81
1,2-dichloroethane	343.15-383.15	−27.49
1,4-dioxane	343.15-383.15	−28.74
ethylbenzene	343.15-383.15	−35.98
n-hexane	343.15-383.15	−28.83
n-nonane	343.15-383.15	−38.99
n-octane	343.15-383.15	−36.86
n-pentane	343.15-383.15	−25.27
1-pentanol	343.15-383.15	−31.25
propyl acetate	343.15-383.15	−29.62
tetrachloromethane	343.15-383.15	−27.53

continued

continued

Solvent (A)	T-range/ K	$\Delta_{sol}H_{A(vap)}^{\infty}/$ kJ/mol
tetrahydrofuran	343.15-383.15	−26.78
toluene	343.15-383.15	−32.13
1,1,1-trichloroethane	343.15-383.15	−26.99
trichloroethene	343.15-383.15	−29.79
trichloromethane	343.15-383.15	−24.98
n-undecane	343.15-383.15	−50.29

Polymer (B): **poly(propylene glycol)**   **1991WOH**
Characterization: $M_n/\text{g.mol}^{-1} = 1700$

Solvent (A)	T-range/ K	$\Delta_{M}H_{A}^{\infty}/$ kJ/mol	$\Delta_{sol}H_{A(vap)}^{\infty}/$ kJ/mol
benzene	343.15-423.15	1.45	−30.1
2-butanone	343.15-423.15	1.25	−31.0
1,2-dichloroethane	343.15-423.15	−0.43	−33.5
dichloromethane	343.15-423.15	2.00	−30.4
1,2-dimethylbenzene	343.15-423.15	0.68	−37.4
1,4-dimethylbenzene	343.15-423.15	0.10	−37.4
ethylbenzene	343.15-423.15	−0.33	−37.6
2-pentanone	343.15-423.15	1.35	−33.4
2-propanone	343.15-423.15	2.16	−22.9
propylbenzene	343.15-423.15	−0.36	−40.7
tetrachloromethane	343.15-423.15	2.64	−30.8
toluene	343.15-423.15	−1.60	−34.3
trichloromethane	343.15-423.15	−6.77	−33.3

Polymer (B): **poly(propylene glycol)**   **1989ED1**
Characterization: $M_n/\text{g.mol}^{-1} = 2000$

Solvent (A)	T-range/ K	$\Delta_{M}H_{A}^{\infty}/$ kJ/mol	$\Delta_{sol}H_{A(vap)}^{\infty}/$ kJ/mol
triacetin	402.1-431.1	+2.5	−65.8

| Polymer (B): | **poly(propylene oxide)** | | **1996MOR** |
| *Characterization:* | Parel 58, B.F. Goodrich Company | | |

Solvent (A)	T-range/ K	$\Delta_M H_A^\infty/$ kJ/mol	$\Delta_{sol} H_{A(vap)}^\infty/$ kJ/mol
benzene	354.1-363.9	29.68	−7.20
N,N-dimethylformamide	354.1-363.9	−4.35	−51.92
ethanol	354.1-363.9	8.00	−25.58
ethyl acetate	354.1-363.9	−40.95	−62.22
n-heptane	354.1-363.9	20.10	−9.88
n-nonane	354.1-363.9	5.07	−31.15
n-octane	354.1-363.9	16.33	−8.88
1-propanol	354.1-363.9	15.16	−14.11
toluene	354.1-363.9	11.47	−24.70

| Polymer (B): | **polystyrene** | **1986INO** |
| *Characterization:* | $M_w/\text{g.mol}^{-1} = 2200$, Toyo Soda Manufacturing Co., Japan | |

Solvent (A)	T-range/ K	$\Delta_M H_A^\infty/$ kJ/mol
anisole	443.15-463.15	2.16
benzene	443.15-463.15	0.09
ethylbenzene	443.15-463.15	0.98
pyridine	443.15-463.15	−1.32
toluene	443.15-463.15	−0.71

| Polymer (B): | **polystyrene** | **1986INO** |
| *Characterization:* | $M_w/\text{g.mol}^{-1} = 4000$, Toyo Soda Manufacturing Co., Japan | |

Solvent (A)	T-range/ K	$\Delta_M H_A^\infty/$ kJ/mol
anisole	443.15-463.15	2.12
benzene	443.15-463.15	−0.36
ethylbenzene	443.15-463.15	2.23
pyridine	443.15-463.15	−0.76
toluene	443.15-463.15	−0.13

Polymer (B):	polystyrene			1995BOG
Characterization:	$M_n/\text{g.mol}^{-1} = 14500$, $T_g/\text{K} = 392$, synthesized in the laboratory			

Solvent (A)	T-range/ K	$\Delta_M H_A^{\infty}/$ kJ/mol	$\Delta_{sol} H_{A(vap)}^{\infty}/$ kJ/mol
1-butanol	408.15-433.15	−6.0	−29.0
n-decane	408.15-433.15	−23.0	−16.0
dibutyl ether	408.15-433.15	−15.0	−19.0
1,4-dioxane	408.15-433.15	−8.0	−23.0
n-heptane	408.15-433.15	−22.0	−5.0
n-nonane	408.15-433.15	−22.0	−13.0
n-octane	408.15-433.15	−19.0	−11.0

Polymer (B):	polystyrene		1986INO
Characterization:	$M_w/\text{g.mol}^{-1} = 17500$, Toyo Soda Manufacturing Co., Japan		

Solvent (A)	T-range/ K	$\Delta_M H_A^{\infty}/$ kJ/mol
anisole	443.15-463.15	1.25
benzene	443.15-463.15	−2.06
ethylbenzene	443.15-463.15	0.63
pyridine	443.15-463.15	−0.50
toluene	443.15-463.15	−0.85

Polymer (B):	polystyrene		1980GUN
Characterization:	$M_n/\text{g.mol}^{-1} = 20000 \pm 5000$		
	PS-K500, Petrokimya A.S., Yarimca, Turkey		

Solvent (A)	T-range/ K	$\Delta_{sol} H_{A(vap)}^{\infty}/$ kJ/mol
acetic acid	435.46-502.58	−13.52
acetonitrile	435.46-502.58	3.73
aniline	435.46-502.58	−42.20
benzaldehyde	435.46-502.58	−35.21
benzene	435.46-502.58	−23.45

continued

continued

Solvent (A)	T-range/ K	$\Delta_{sol} H_{A(vap)}^\infty/$ kJ/mol
benzyl alcohol	435.46-502.58	−39.36
1-butanol	435.46-502.58	−41.32
2-butanone	435.46-502.58	−7.95
butyl acetate	435.46-502.58	−17.84
chlorobenzene	435.46-502.58	−23.86
cyclohexane	435.46-502.58	−17.84
cyclohexanone	435.46-502.58	−24.07
1,2-dichloroethane	435.46-502.58	−14.28
dichloromethane	435.46-502.58	−8.29
diethyl ether	435.46-502.58	−18.88
diisopropyl ether	435.46-502.58	35.21
1,2-dimethylbenzene	435.46-502.58	−30.35
1,4-dioxane	435.46-502.58	−17.96
1,2-ethanediol	435.46-502.58	−62.93
ethanol	435.46-502.58	6.45
ethyl acetate	435.46-502.58	−3.89
formamide	435.46-502.58	−33.12
n-heptane	435.46-502.58	−1.13
n-hexane	435.46-502.58	14.19
methanol	435.46-502.58	−0.54
2-methyl-1-propanol	435.46-502.58	−14.32
nitrobenzene	435.46-502.58	−33.70
n-octane	435.46-502.58	8.04
1-octanol	435.46-502.58	−26.88
n-pentane	435.46-502.58	19.05
1-pentanol	435.46-502.58	−25.46
1-propanol	435.46-502.58	5.74
2-propanol	435.46-502.58	16.29
2-propanone	435.46-502.58	−9.38
pyridine	435.46-502.58	26.54
tetrachloromethane	435.46-502.58	−14.53
tetrahydrofuran	435.46-502.58	−14.15
toluene	435.46-502.58	−17.38
trichloroethene	435.46-502.58	−17.25
trichloromethane	435.46-502.58	−30.52
2,2,4-trimethylpentane	435.46-502.58 -	50.95
water	435.46-502.58	18.90

Polymer (B): **polystyrene** **1984SC2**
Characterization: $M_n/\text{g.mol}^{-1} = 50700$, $M_w/\text{g.mol}^{-1} = 53700$

Solvent (A)	T-range/ K	$\Delta_M H_A^\infty/$ kJ/mol	$\Delta_{sol} H_{A(vap)}^\infty/$ kJ/mol
cyclohexane	433.15-463.15	1.13	−25.6
n-heptane	433.15-463.15	5.42	
n-hexane	433.15-463.15	6.01	−20.2
2-methylheptane	433.15-463.15	4.66	−27.1
2-methylpentane	433.15-463.15	4.37	−18.5
3-methylpentane	433.15-463.15	4.84	−19.2
n-nonane	433.15-463.15	3.80	
n-octane	433.15-463.15	4.49	−28.1
1-octene	433.15-463.15	3.34	−29.6
cis-2-octene	433.15-463.15	1.86	−30.4
2,2,4-trimethylpentane	433.15-463.15	4.00	−24.3

Polymer (B): **polystyrene** **1990GAL**
Characterization: $M_n/\text{g.mol}^{-1} = 62700$, $M_w/\text{g.mol}^{-1} = 69000$, $T_g/\text{K} = 368.5$

Solvent (A)	T-range/ K	$\Delta_M H_A^\infty/$ kJ/mol
acetonitrile	423.15-463.15	−0.48
benzene	423.15-463.15	−0.56
2-butanone	423.15-463.15	0.12
chlorobenzene	423.15-463.15	−0.65
1-chlorohexane	423.15-463.15	0.60
cyclohexane	423.15-463.15	2.60
cyclohexanone	423.15-463.15	−0.60
1,2-dichloroethane	423.15-463.15	−0.87
1,5-dichloropentane	423.15-463.15	0.20
N,N-dimethylformamide	423.15-463.15	−0.29
1,4-dioxane	423.15-463.15	1.15
dipropyl ether	423.15-463.15	3.20
dipropyl thioether	423.15-463.15	0.60
ethylbenzene	423.15-463.15	0.30
n-heptane	423.15-463.15	4.04
n-hexane	423.15-463.15	3.40
pyridine	423.15-463.15	0.10
toluene	423.15-463.15	−0.20
trichloromethane	423.15-463.15	−1.60

Polymer (B):   **polystyrene**   **1986INO**

Characterization: M_w/g.mol^{-1} = 107000, Toyo Soda Manufacturing Co., Japan

Solvent (A)	T-range/ K	$\Delta_M H_A^{\infty}$/ kJ/mol
anisole	443.15-463.15	1.33
benzene	443.15-463.15	−1.42
ethylbenzene	443.15-463.15	0.47
toluene	443.15-463.15	−0.10

Polymer (B):   **polystyrene**   **1978DIP1**

Characterization: M_w/g.mol^{-1} = 120000, Polymer Corporation

Solvent (A)	T-range/ K	$\Delta_M H_A^{\infty}$/ kJ/mol
n-decane	441.15-491.15	4.04
n-dodecane	441.15-491.15	4.18
n-hexadecane	441.15-491.15	5.73
n-tetradecane	441.15-491.15	4.78

Polymer (B):   **polystyrene**   **1978DIP2**

Characterization: M_w/g.mol^{-1} = 120000, Polymer Corporation

Solvent (A)	T/ K	$\Delta_M H_A^{\infty}$/ kJ/mol	$\Delta_{sol} H_{A(vap)}^{\infty}$/ kJ/mol
benzene	433.15-453.15	−0.04	−27.17
butylbenzene	456.15-476.15	0.70	−39.70
butylcyclohexane	433.15-453.15	4.14	−36.56
cyclohexane	433.15-453.15	2.60	−23.87
n-decane	456.15-476.15	4.04	−35.09
n-dodecane	456.15-476.15	4.19	−42.79
cis-decahydronaphthalene	456.15-476.15	3.84	−37.47
trans-decahydronaphthalene	456.15-476.15	4.33	−35.61
n-hexadecane	456.15-476.15	5.74	−56.84
naphthalene	456.15-476.15	0.51	−44.81
n-tetradecane	456.15-476.15	4.79	−50.07
1,2,3,4-tetrahydronaphthalene	456.15-476.15	1.12	−42.76
3,3,4,4-tetramethylhexane	433.15-453.15	10.78	−27.60

| Polymer (B): | **polystyrene** | **1984SC1** |
| Characterization: | $M_n/\text{g.mol}^{-1} = 158000$, $M_w/\text{g.mol}^{-1} = 160000$, Pressure Chemical Co., Pittsburgh, PA | |

Solvent (A)	T-range/ K	$\Delta_{sol}H_{A(vap)}{}^{\infty}/$ kJ/mol
tert-butyl acetate	416.25-456.15	−25.57

| Polymer (B): | **polystyrene** | **1991OZD** |
| Characterization: | $M_{\eta}/\text{g.mol}^{-1} = 145000$ | |

Solvent (A)	$T/$ K	$\Delta_{M}H_{A}{}^{\infty}/$ kJ/mol	$\Delta_{sol}H_{A(vap)}{}^{\infty}/$ kJ/mol
n-decane	413.15-453.15	15.2	−52.85
n-heptane	413.15-453.15	4.67	−50.52
n-hexane	413.15-453.15	0.42	−31.49
n-nonane	413.15-453.15	14.1	−53.23

| Polymer (B): | **polystyrene** | **1980IWA** |
| Characterization: | $M_{\eta}/\text{g.mol}^{-1} = 309000$ | |

Solvent (A)	T-range/ K	$\Delta_{sol}H_{A(vap)}{}^{\infty}/$ kJ/mol
benzene	423.15-498.15	−26.4
n-decane	423.15-498.15	−32.7
1,3-dimethylbenzene	423.15-498.15	−31.4
n-nonane	423.15-498.15	−28.1
n-octane	423.15-498.15	−23.4
toluene	423.15-498.15	−30.1
1,3,5-trimethylbenzene	423.15-498.15	−33.1

| Polymer (B): | **polystyrene** | | **1986INO** |

Characterization: $M_w/\text{g.mol}^{-1} = 1800000$, Toyo Soda Manufacturing Co., Japan

Solvent (A)	T-range/ K	$\Delta_M H_A^\infty$/ kJ/mol
anisole	443.15-463.15	0.87
benzene	443.15-463.15	−1.70
ethylbenzene	443.15-463.15	0.55
toluene	443.15-463.15	−1.45

| Polymer (B): | **polystyrene** | | **1992COS** |

Characterization: synthesized in the laboratory

Solvent (A)	T-range/ K	$\Delta_{sol} H_{A(vap)}^\infty$/ kJ/mol
n-decane	373.15-423.15	−21.24
n-hexane	373.15-423.15	−11.43
n-nonane	373.15-423.15	−17.51

| Polymer (B): | **polystyrene-b-polybutadiene-b-polystyrene triblock copolymer** | | **1992RO2** |

Characterization: $M_n/\text{g.mol}^{-1} = 114800$, $M_w/\text{g.mol}^{-1} = 188300$,
17.0 wt% styrene, Kraton D-1301X

Solvent (A)	T-range/ K	$\Delta_M H_A^\infty$/ kJ/mol	$\Delta_{sol} H_{A(vap)}^\infty$/ kJ/mol
benzene	308.15-348.15	−0.252	−32.7
2-butanone	308.15-348.15	0.302	−32.8
cyclohexane	308.15-348.15	0.727	−30.8
1,4-dimethylbenzene	308.15-348.15	0.270	−40.4
ethylbenzene	308.15-348.15	0.397	−40.2
n-heptane	308.15-348.15	0.464	−34.4
n-hexane	308.15-348.15	−0.228	−30.0
toluene	308.15-348.15	−0.106	−36.7
trichloromethane	308.15-348.15	−2.467	−32.4

| Polymer (B): | **polystyrene-b-polybutadiene-b-polystyrene triblock copolymer** | | | **1992RO2** |

Characterization: M_n/g.mol^{-1} = 110800, M_w/g.mol^{-1} = 139600,
31.0 wt% styrene, Kraton D-1101

Solvent (A)	T-range/ K	$\Delta_M H_A^{\infty}$/ kJ/mol	$\Delta_{sol} H_{A(vap)}^{\infty}$/ kJ/mol
benzene	308.15-348.15	−0.745	−33.2
2-butanone	308.15-348.15	−1.11	−34.2
cyclohexane	308.15-348.15	1.09	−30.4
1,4-dimethylbenzene	308.15-348.15	0.158	−40.5
ethylbenzene	308.15-348.15	−0.130	−40.7
n-heptane	308.15-348.15	0.459	−34.4
n-hexane	308.15-348.15	−0.765	−30.6
toluene	308.15-348.15	−0.134	−36.7
trichloromethane	308.15-348.15	−4.49	−34.4

| Polymer (B): | **poly(styrene-*co*-butyl methacrylate)** | | **1981DIP** |

Characterization: M_w/g.mol^{-1} = 70000-75000, M_w/M_n = 2.0-2.3, MI = 15.5
58.0 wt% styrene, Xerox Corporation

Solvent (A)	T-range/ K	$\Delta_M H_A^{\infty}$/ kJ/mol
benzene	393.15-423.15	−0.25
1-butanol	393.15-423.15	5.11
butyl acetate	393.15-423.15	−0.46
butylbenzene	393.15-423.15	0.25
tert-butylbenzene	393.15-423.15	0.84
butylcyclohexane	393.15-423.15	3.85
chlorobenzene	393.15-423.15	−0.75
1-chlorobutane	393.15-423.15	−1.38
cyclohexane	393.15-423.15	2.39
cyclohexanol	393.15-423.15	5.74
n-decane	393.15-423.15	4.73
dichloromethane	393.15-423.15	−2.22
n-dodecane	393.15-423.15	5.15
ethylbenzene	393.15-423.15	0.13
methylcyclohexane	393.15-423.15	3.22
n-octane	393.15-423.15	3.81
2-pentanone	393.15-423.15	−1.17
tetrachloromethane	393.15-423.15	0.08
trichloromethane	393.15-423.15	−3.94
2,2,4-trimethylpentane	393.15-423.15	5.73
3,4,5-trimethylheptane	393.15-423.15	5.31

Polymer (B):	**poly(styrene-*co*-divinylbenzene)**		**1987SAN, 1989SAN**
Characterization:	5 wt% divinylbenzene, network, synthesized in the laboratory		

Solvent (A)	T-range/ K	$\Delta_M H_A^\infty$/ kJ/mol	$\Delta_{sol} H_{A(vap)}^\infty$/ kJ/mol
1-butanol	445-465	22.0	−5.7
ethanol	445-465	17.0	−16.0
methanol	445-465	11.9	−12.7
1-propanol	445-465	22.6	− 8.8

Polymer (B):	**poly(styrene-*co*-divinylbenzene)**		**1987SAN, 1989SAN**
Characterization:	10 wt% divinylbenzene, network, synthesized in the laboratory		

Solvent (A)	T-range/ K	$\Delta_M H_A^\infty$/ kJ/mol	$\Delta_{sol} H_{A(vap)}^\infty$/ kJ/mol
1-butanol	455-480	17.0	−16.0
methanol	455-480	10.3	−17.9
1-pentanol	455-480	16.0	−15.1
1-propanol	455-480	10.2	−16.9

Polymer (B):	**poly(styrene-*co*-divinylbenzene)**		**1987SAN, 1989SAN**
Characterization:	20 wt% divinylbenzene, network, synthesized in the laboratory		

Solvent (A)	T-range/ K	$\Delta_M H_A^\infty$/ kJ/mol	$\Delta_{sol} H_{A(vap)}^\infty$/ kJ/mol
1-butanol	460-480	25.9	−11.7
ethanol	460-480	13.8	−16.8
methanol	460-480	14.6	−13.4
1-pentanol	460-480	22.2	−8.5
1-propanol	460-480	17.2	−11.9

Polymer (B):	**poly(styrene-*co*-ethyl acrylate)**		**1992COS**
Characterization:	18.5 mol% ethyl acrylate, synthesized in the laboratory		

Solvent (A)	T-range/ K	$\Delta_{sol}H_{A(vap)}{}^{\infty}$/ kJ/mol
n-decane	373.15-423.15	−21.20
n-hexane	373.15-423.15	−2.66
n-nonane	373.15-423.15	−21.28

Polymer (B):	**poly(styrene-*co*-ethyl acrylate)**		**1992COS**
Characterization:	37.8 mol% ethyl acrylate, synthesized in the laboratory		

Solvent (A)	T-range/ K	$\Delta_{sol}H_{A(vap)}{}^{\infty}$/ kJ/mol
n-decane	373.15-423.15	−23.82
n-hexane	373.15-423.15	−6.47
n-nonane	373.15-423.15	−15.49

Polymer (B):	**poly(styrene-*co*-ethyl acrylate)**		**1992COS**
Characterization:	63.4 mol% ethyl acrylate, synthesized in the laboratory		

Solvent (A)	T-range/ K	$\Delta_{sol}H_{A(vap)}{}^{\infty}$/ kJ/mol
n-decane	373.15-423.15	−13.63
n-hexane	373.15-423.15	−14.21
n-nonane	373.15-423.15	−28.50

Polymer (B):	**poly(styrene-*co*-ethyl acrylate)**		**1992COS**
Characterization:	81.0 mol% ethyl acrylate, synthesized in the laboratory		

Solvent (A)	T-range/ K	$\Delta_{sol}H_{A(vap)}{}^{\infty}$/ kJ/mol
n-decane	373.15-423.15	−26.15
n-hexane	373.15-423.15	−19.05
n-nonane	373.15-423.15	−10.91

Polymer (B):	poly(styrene-*co*-maleic anhydride-*co*-methacrylic acid)		1995BOG
Characterization:	M_n/g.mol^{-1} = 28000, 33 mol% styrene, (1:1:1), T_g/K = 510, synthesized in the laboratory		

Solvent (A)	T-range/ K	$\Delta_M H_A^\infty$/ kJ/mol	$\Delta_{sol} H_{A(vap)}^\infty$/ kJ/mol
1-butanol	526.6-556.6	−23.0	−11.0
n-decane	526.6-556.6	−13.0	−18.0
dibutyl ether	526.6-556.6	−16.0	−15.0
1,4-dioxane	526.6-556.6	−13.0	−15.0
n-heptane	526.6-556.6	−16.0	−15.0
n-nonane	526.6-556.6	−15.0	−15.0
n-octane	526.6-556.6	−13.0	−17.0

Polymer (B):	poly(styrene-*co*-methyl methacrylate)		1995NEM
Characterization:	M_n/g.mol^{-1} = 57800, 42 mol% styrene, synthesized in the laboratory		

Solvent (A)	T-range/ K	$\Delta_M H_A^\infty$/ kJ/mol	$\Delta_{sol} H_{A(vap)}^\infty$/ kJ/mol
n-heptane	443.2-471.7	8.15	−19.15
n-hexane	443.2-471.7	10.10	−12.67
n-nonane	443.2-471.7	6.65	−25.53
n-octane	443.2-471.7	8.31	−28.72

Polymer (B):	poly[styrene-*co*-2-(3-methyl-3-phenylcyclobutyl)-2-hydroxyethyl methacrylate]	2000KAY
Characterization:	40.0 mol% styrene, ρ_B = 1.045 g/cm^3, T_g/K = 375, synthesized in the laboratory	

Solvent (A)	T-range/ K	$\Delta_M H_A^\infty$/ kJ/mol
benzene	443.15-453.15	11.6
2-butanone	443.15-453.15	23.3
n-decane	443.15-453.15	21.6

continued

continued

Solvent (A)	T-range/ K	$\Delta_M H_A^\infty$/ kJ/mol
1,2-dimethylbenzene	443.15-453.15	10.0
n-dodecane	443.15-453.15	16.6
ethanol	443.15-453.15	41.6
ethyl acetate	443.15-453.15	29.9
methanol	443.15-453.15	21.6
methyl acetate	443.15-453.15	21.6
n-nonane	443.15-453.15	20.0
n-octane	443.15-453.15	21.6
2-propanone	443.15-453.15	16.6
toluene	443.15-453.15	43.2
n-undecane	443.15-453.15	5.0

Polymer (B): **poly(styrene-*co*-nonyl methacrylate)** **1995BOG**
Characterization: M_n/g.mol^{-1} = 29000, 80 mol% styrene, T_g/K = 340, synthesized in the laboratory

Solvent (A)	T-range/ K	$\Delta_M H_A^\infty$/ kJ/mol	$\Delta_{sol} H_{A(vap)}^\infty$/ kJ/mol
1-butanol	358.15-383.15	− 5.0	−38.0
n-decane	358.15-383.15	− 4.0	−41.0
dibutyl ether	358.15-383.15	8.0	−46.0
1,4-dioxane	358.15-383.15	0.0	−34.0
n-heptane	358.15-383.15	−14.0	−16.0
n-nonane	358.15-383.15	0.0	−40.0
n-octane	358.15-383.15	− 9.0	−25.0

Polymer (B): **poly(tetrafluoroethylene-*co*-2,2-bis(trifluoromethyl)-4,5-difluoro-1,3-dioxole)** **1999BON**
Characterization: 13 mol% tetrafluoroethylene, AF2400, DuPont, Wilmington, DE

Solvent (A)	T-range/ K	$\Delta_{sol} H_{A(vap)}^\infty$/ kJ/mol
benzene	325.15-505.15	−30.0
n-butane	325.15-505.15	−21.1
n-decane	325.15-505.15	−44.4

continued

continued

Solvent (A)	T-range/ K	$\Delta_{sol}H_{A(vap)}^\infty/$ kJ/mol
n-dodecane	325.15-505.15	−44.4
n-heptane	325.15-505.15	−36.3
hexafluorobenzene	325.15-505.15	−37.7
n-hexane	325.15-505.15	−31.7
2-methylpropane	325.15-505.15	−25.0
n-nonane	325.15-505.15	−38.5
octafluorotoluene	325.15-505.15	−44.4
n-pentane	325.15-505.15	−27.1
n-tridecane	325.15-505.15	−41.0
n-undecane	325.15-505.15	−47.3

Polymer (B): **poly(tetramethyl-p-silphenylene siloxane)** **1986WUY**
Characterization: $M_n/\text{g.mol}^{-1} = 105000$

Solvent (A)	T-range/ K	$\Delta_M H_A^\infty/$ kJ/mol	$\Delta_{sol}H_{A(vap)}^\infty/$ kJ/mol
benzene	335.15-406.15	−0.95	−31.21
1-butanol	335.15-406.15	11.04	−31.91
1,4-dioxane	335.15-406.15	−2.15	−35.85
n-dodecane	335.15-406.15	4.94	−51.31
n-octane	335.15-406.15	−1.90	−35.71
1-octanol	335.15-406.15	14.5	−50.25
toluene	335.15-406.15	−3.46	−35.15
trichloromethane	335.15-406.15	0.11	−28.63

Polymer (B): **poly(vinyl acetate)** **1978DIP3**
Characterization: $M_n/\text{g.mol}^{-1} = 86000$, B1000, Wacker Chemicals, Munich, Germany

Solvent (A)	T-range/ K	$\Delta_M H_A^\infty/$ kJ/mol	$\Delta_{sol}H_{A(vap)}^\infty/$ kJ/mol
benzene	398.15-418.15	−0.17	−28.72
butylbenzene	398.15-418.15	4.27	−39.61
butylcyclohexane	398.15-418.15	10.38	−32.32
cyclohexane	398.15-418.15	9.17	−19.13

continued

continued

Solvent (A)	T-range/ K	$\Delta_M H_A^\infty$/ kJ/mol	$\Delta_{sol} H_{A(vap)}^\infty$/ kJ/mol
cis-decahydronaphthalene	398.15-418.15	10.17	−34.50
n-decane	398.15-418.15	11.39	−31.86
n-dodecane	398.15-418.15	13.52	−38.56
n-nonane	398.15-418.15	10.76	−28.26
1,2,3,4-tetrahydronaphthalene	398.15-418.15	3.98	−43.17
3,3,4,4-tetramethylhexane	398.15-418.15	9.97	−30.40
n-undecane	398.15-418.15	12.35	−35.09

Polymer (B):   **poly(vinyl acetate)**   **1980DIP**
Characterization: M_n/g.mol^{-1} = 86000, B1000, Wacker Chemicals, Munich, Germany

Solvent (A)	T-range/ K	$\Delta_M H_A^\infty$/ kJ/mol	$\Delta_{sol} H_{A(vap)}^\infty$/ kJ/mol
benzene	398.15-418.15	−0.167	−28.72
1-butanol	393.15-423.15	7.12	−35.42
chlorobenzene	398.15-418.15	0.042	−35.50
1-chlorobutane	398.15-418.15	0.96	−26.34
chlorocyclohexane	398.15-418.15		−33.03
cyclohexane	388.15-408.15	9.17	−19.13
cyclohexanol	398.15-418.15	5.74	−42.25
cyclohexanone	398.15-418.15		−38.90
n-decane	393.15-423.15	11.39	−31.86
n-nonane	393.15-423.15	10.76	−28.26
1-octene	393.15-423.15	6.99	−26.92
2-pentanone	393.15-423.15	−0.084	−32.87
tetrachloromethane	398.15-418.15	2.05	−25.58
trichloromethane	398.15-418.15	−4.15	−30.19

Polymer (B):	poly(vinyl acetate)	1990GAL

Characterization: $M_n/\text{g.mol}^{-1} = 113000$, $M_w/\text{g.mol}^{-1} = 192000$, $T_g/\text{K} = 305.15$

Solvent (A)	T-range/ K	$\Delta_M H_A^\infty/$ kJ/mol
acetonitrile	383.15-423.15	−2.17
benzene	383.15-423.15	0.40
2-butanone	383.15-423.15	−0.73
chlorobenzene	383.15-423.15	0.10
1-chlorohexane	383.15-423.15	2.40
cyclohexane	383.15-423.15	6.15
1,2-dichloroethane	383.15-423.15	−2.07
1,5-dichloropentane	383.15-423.15	−1.00
N,N-dimethylacetamide	383.15-423.15	0.50
N,N-dimethylformamide	383.15-423.15	−0.51
dimethylsulfoxide	383.15-423.15	−0.70
1,4-dioxane	383.15-423.15	2.50
dipropyl ether	383.15-423.15	4.37
dipropyl thioether	383.15-423.15	3.23
ethyl acetate	383.15-423.15	−0.90
n-heptane	383.15-423.15	8.10
n-hexane	383.15-423.15	6.70
2-propanone	383.15-423.15	−1.10
tetrachloromethane	383.15-423.15	1.00
tetrahydrofuran	383.15-423.15	3.00
toluene	383.15-423.15	1.50
trichloromethane	383.15-423.15	−3.40

Polymer (B):	poly(vinyl acetate)	1985CAS

Characterization: $M_w/\text{g.mol}^{-1} = 500000$, Polysciences, Inc., Warrington, PA

Solvent (A)	T-range/ K	$\Delta_{sol}H_{A(vap)}^\infty/$ kJ/mol
benzene	393.15-423.15	−27.72
1-butanol	393.15-423.15	−36.09
2-butanol	393.15-423.15	−33.70
cyclohexanol	393.15-423.15	−44.38
n-decane	393.15-423.15	−32.41
1-decanol	393.15-423.15	−55.18
n-dodecane	393.15-423.15	−40.11
ethylbenzene	393.15-423.15	−35.71

continued

continued

Solvent (A)	T-range/ K	$\Delta_{sol}H_{A(vap)}{}^{\infty}/$ kJ/mol
1-heptanol	393.15-423.15	−46.85
n-hexadecane	393.15-423.15	−51.29
1-hexanol	393.15-423.15	−44.09
2-methyl-1-propanol	393.15-423.15	−36.51
n-nonane	393.15-423.15	−29.10
n-octane	393.15-423.15	−24.99
1-octanol	393.15-423.15	−49.66
1-pentanol	393.15-423.15	−41.95
2-pentanol	393.15-423.15	−47.27
1-propanol	393.15-423.15	−33.66
2-propanol	393.15-423.15	−31.40
n-tetradecane	393.15-423.15	−45.43
toluene	393.15-423.15	−32.57
n-undecane	393.15-423.15	−36.84

Polymer (B):   **poly(vinyl chloride)**   1978DIP1
Characterization: −

Solvent (A)	T-range/ K	$\Delta_{M}H_{A}{}^{\infty}/$ kJ/mol
n-dodecane	403.15-433.15	14.2

Polymer (B):   **poly(vinyl chloride)**   1995KAY
Characterization: M_{η}/g.mol^{-1} = 86000, Aliaga Petkim Co., Izmir, Turkey

Solvent (A)	T-range/ K	$\Delta_{M}H_{A}{}^{\infty}/$ kJ/mol
benzene	393.15-433.15	5.16
n-hexane	393.15-433.15	7.75
methyl acetate	393.15-433.15	16.05
toluene	393.15-433.15	13.65

| Polymer (B): | **poly(vinyl chloride)** | | | **1999KA1** |
| Characterization: | M_η/g.mol^{-1} = 86000, Aliaga Petkim Co., Izmir, Turkey | | | |

Solvent (A)	T-range/ K	$\Delta_M H_A^\infty$/ kJ/mol	$\Delta_{sol} H_{A(vap)}^\infty$/ kJ/mol
benzene	403.15-433.15	5.15	−19.30
2-butanone	403.15-433.15	14.36	−10.05
n-decane	403.15-433.15	8.79	−32.32
1,2-dimethylbenzene	403.15-433.15	13.90	−28.89
1,4-dioxane	403.15-433.15	11.43	−16.66
n-dodecane	403.15-433.15	6.57	−45.30
ethanol	403.15-433.15	22.48	−10.38
ethyl acetate	403.15-433.15	16.41	−11.93
methanol	403.15-433.15	23.11	−9.50
methyl acetate	403.15-433.15	16.04	−11.56
n-nonane	403.15-433.15	10.72	−19.76
n-octane	403.15-433.15	12.48	−14.44
2-propanone	403.15-433.15	15.99	−9.42
tetrachloromethane	403.15-433.15	13.61	−9.63
tetrahydrofuran	403.15-433.15	10.80	−16.08
toluene	403.15-433.15	13.65	−24.37
trichloromethane	403.15-433.15	18.92	−8.88
n-undecane	403.15-433.15	6.91	−40.03

| Polymer (B): | **poly(vinyl chloride)** | | **1992KAL** |
| Characterization: | M_w/g.mol^{-1} = 100000, Polysciences, Inc., Warrington, PA | | |

Solvent (A)	T-range/ K	$\Delta_{sol} H_{A(vap)}^\infty$/ kJ/mol
water	298.15-323.15	−43.66

| Polymer (B): | **poly(vinyl chloride-*co*-vinylidene chloride)** | | **1992KAL** |
| Characterization: | M_w/g.mol^{-1} = 90000, 20 wt% vinyl chloride, Polysciences, Inc., Warrington, PA | | |

Solvent (A)	T-range/ K	$\Delta_{sol} H_{A(vap)}^\infty$/ kJ/mol
water	298.15-323.15	−42.70

Polymer (B):	poly(vinylidene fluoride)	1985GAL

Characterization: M_n/g.mol^{-1} = 131000, M_w/g.mol^{-1} = 275000, T_m/K = 433.7, Kynar 461, Pennwalt Corp., Philadelphia, PA

Solvent (A)	T-range/ K	$\Delta_M H_A^\infty$/ kJ/mol
2-butanone	433.15-483.15	-1.68
γ-butyrolactone	433.15-483.15	1.68
1-chlorodecane	433.15-483.15	10.89
1-chlorooctane	433.15-483.15	7.96
cyclohexanone	433.15-483.15	-0.84
1,2-dichloroethane	433.15-483.15	5.02
1,2-dimethoxyethane	433.15-483.15	0.42
N,N-dimethylacetamide	433.15-483.15	-5.44
N,N-dimethylformamide	433.15-483.15	-7.12
dimethylsulfoxide	433.15-483.15	-6.70
n-dodecane	433.15-483.15	16.33
ethyl acetate	433.15-483.15	-2.51
n-hexadecane	433.15-483.15	22.61
bis(2-methoxyethyl) ether	433.15-483.15	2.51
1-methyl-2-pyrrolidinone	433.15-483.15	-7.54
pentachloroethane	433.15-483.15	9.63
propylene carbonate	433.15-483.15	0.84
sulfolane	433.15-483.15	-1.26
1,1,2,2-tetrachloroethane	433.15-483.15	7.12
n-tetradecane	433.15-483.15	18.72

Polymer (B):	poly(vinylidene fluoride)	1990GAL

Characterization: M_n/g.mol^{-1} = 131000, M_w/g.mol^{-1} = 275000, T_m/K = 433.7, Kynar 461, Pennwalt Corp., Philadelphia, PA

Solvent (A)	T-range/ K	$\Delta_M H_A^\infty$/ kJ/mol
acetophenone	433.15-483.15	1.13
2-butanone	433.15-483.15	-2.85
butyl acetate	433.15-483.15	-0.83
butylbenzene	433.15-483.15	5.69
γ-butyrolactone	433.15-483.15	0.80
chlorobenzene	433.15-483.15	3.22
1-chlorodecane	433.15-483.15	10.76
1-chlorohexane	433.15-483.15	6.13

continued

continued

Solvent (A)	T-range/ K	$\Delta_M H_A^\infty$/ kJ/mol
1-chlorooctane	433.15-483.15	8.04
cyclohexanone	433.15-483.15	−1.25
cis-decahydronaphthalene	433.15-483.15	13.0
1,2-dichlorobenzene	433.15-483.15	6.07
2,2'-dichlorodiethyl ether	433.15-483.15	4.04
1,2-dichloroethane	433.15-483.15	2.55
1,5-dichloropentane	433.15-483.15	3.25
N,N-dimethylacetamide	433.15-483.15	−5.40
1,3-dimethylbenzene	433.15-483.15	3.77
dimethylethyleneurea	433.15-483.15	−4.40
N,N-dimethylformamide	433.15-483.15	−6.90
dimethylpropyleneurea	433.15-483.15	−3.60
dimethylsulfoxide	433.15-483.15	−6.36
dipropyl thioether	433.15-483.15	5.49
n-dodecane	433.15-483.15	15.75
ethyl acetate	433.15-483.15	−2.08
ethylbenzene	433.15-483.15	3.06
ethyl propionate	433.15-483.15	−0.90
n-hexadecane	433.15-483.15	22.40
hexamethylphosphoramide	433.15-483.15	−1.30
1-methyl-2-pyrrolidinone	433.15-483.15	−5.05
pentachloroethane	433.15-483.15	8.60
propylene carbonate	433.15-483.15	−1.97
pyridine	433.15-483.15	0.10
sulfolane	433.15-483.15	−1.10
1,1,2,2-tetrachloroethane	433.15-483.15	4.60
n-tetradecane	433.15-483.15	18.70
1,2,3,4-tetrahydronaphthalene	433.15-483.15	6.57
1,1,3,3-tetramethylurea	433.15-483.15	−3.27
toluene	433.15-483.15	2.34
1,1,2-trichloroethane	433.15-483.15	4.52

Polymer (B):   **poly(vinylidene fluoride)**   **1989CHE, 2000ALS**
Characterization: M_w/g.mol^{-1} = 250000, Aldrich Chem. Co., Inc., Milwaukee, WI

Solvent (A)	T-range/ K	$\Delta_M H_A^\infty$/ kJ/mol	$\Delta_{sol} H_{A(vap)}^\infty$/ kJ/mol
butyl acetate	443.15-493.15		−35.92
n-decane	443.15-493.15	20.98	−13.61

continued

continued

Solvent (A)	T-range/ K	$\Delta_M H_A^\infty$/ kJ/mol	$\Delta_{sol} H_{A(vap)}^\infty$/ kJ/mol
n-dodecane	443.15-493.15	24.74	-17.21
ethyl acetate	443.15-493.15		-34.79
methyl acetate	443.15-493.15		-28.51
n-nonane	443.15-493.15	15.95	-15.45
n-octane	443.15-493.15	15.74	-11.01
propyl acetate	443.15-493.15		-35.88
n-undecane	443.15-493.15	20.14	-18.13

Polymer (B):   **poly(vinylidene fluoride)**   **2000TAZ**
Characterization: $M_w/\text{g.mol}^{-1} = 470000$, $T_m/\text{K} = 450$, crystallinity = 59.5%, Kureha Chemical Co., Japan

Solvent (A)	T-range/ K	$\Delta_{sol} H_{A(vap)}^\infty$/ kJ/mol
γ-butyrolactone	463.15-493.15	-29.1
cycloheptanone	463.15-493.15	-29.5
cyclohexanone	463.15-493.15	-29.7
N,N-dimethylformamide	463.15-493.15	-29.1
ethyl acetate	463.15-493.15	-34.5
3-heptanone	463.15-493.15	-25.5
n-hexane	463.15-493.15	-14.5
3-hexanone	463.15-493.15	-25.5
n-octane	463.15-493.15	-15.3
3-octanone	463.15-493.15	-33.8
3-pentanone	463.15-493.15	-29.6
2-propanone	463.15-493.15	-26.1

Polymer (B):   **poly(vinylidene fluoride)**   **1982DIP**
Characterization: –

Solvent (A)	T-range/ K	$\Delta_M H_A^\infty$/ kJ/mol
acetophenone	463.15-503.15	-0.59
butylbenzene	463.15-503.15	6.62

continued

continued

Solvent (A)	T-range/ K	$\Delta_M H_A^\infty$/ kJ/mol
1-chlorodecane*	463.15-503.15	10.8
1-chlorooctane*	463.15-493.15	8.04
cyclohexanol	463.15-503.15	18.2
cyclohexanone	463.15-503.15	0.96
1,2-dichlorobenzene	463.15-503.15	6.07
N,N-dimethylformamide*	463.15-503.15	−8.96
n-tetradecane	463.15-488.15	19.8
1,2,3,4-tetrahydronaphthalene	463.15-503.15	6.57

* These values were not corrected for vapor phase nonideality of the solvent probe.

Polymer (B):	**poly(vinyl isobutyl ether)**	**1987DES**
Characterization:	atactic, M_n/g.mol^{-1} = 2340, M_η/g.mol^{-1} = 8600	

Solvent (A)	T-range/ K	$\Delta_M H_A^\infty$/ kJ/mol	$\Delta_{sol} H_{A(vap)}^\infty$/ kJ/mol
benzene	323.15-363.15	−1.358	−32.64
cyclohexane	323.15-363.15	1.341	−30.73
dichloromethane	323.15-363.15	−2.208	−28.63
n-heptane	323.15-363.15	1.364	−32.79
n-hexane	323.15-363.15	0.911	−28.57
n-octane	323.15-363.15	1.522	−38.00
n-pentane	323.15-363.15	0.840	−23.31
tetrachloromethane	323.15-363.15	−1.047	−31.53
toluene	323.15-363.15	−2.107	−37.94
trichloroethene	323.15-363.15	−3.241	−35.70
trichloromethane	323.15-363.15	−6.024	−34.97
2,2,4-trimethylpentane	323.15-363.15	1.351	−31.38

Polymer (B):	**poly(vinyl isobutyl ether)**	**1987DES**
Characterization:	isotactic, M_n/g.mol^{-1} = 2500, M_η/g.mol^{-1} = 12400	

Solvent (A)	T-range/ K	$\Delta_{sol} H_{A(vap)}^\infty$/ kJ/mol
benzene	323.15-363.15	−32.64
cyclohexane	323.15-363.15	−30.73

continued

continued

Solvent (A)	T-range/ K	$\Delta_{sol}H_{A(vap)}^{\infty}$/ kJ/mol
dichloromethane	323.15-363.15	−28.63
n-heptane	323.15-363.15	−32.79
n-hexane	323.15-363.15	−28.57
n-octane	323.15-363.15	−38.00
n-pentane	323.15-363.15	−23.31
tetrachloromethane	323.15-363.15	−31.53
toluene	323.15-363.15	−37.94
trichloroethene	323.15-363.15	−35.70
trichloromethane	323.15-363.15	−34.97
2,2,4-trimethylpentane	323.15-363.15	−31.38

Polymer (B):	**poly(vinyl methyl ketone)**	**2000ALS**
Characterization:	M_w/g.mol^{-1} = 500000, T_m/K = 433, Aldrich Chem. Co., Inc., Milwaukee, WI	

Solvent (A)	T-range/ K	$\Delta_{sol}H_{A(vap)}^{\infty}$/ kJ/mol
n-decane	353.15-443.15	−43.46
n-dodecane	353.15-443.15	−52.21
ethanol	353.15-443.15	−39.61
ethyl acetate	353.15-443.15	−18.97
n-heptane	353.15-443.15	−34.12
methanol	353.15-443.15	−38.64
methyl acetate	353.15-443.15	−25.08
n-nonane	353.15-443.15	−41.99
n-octane	353.15-443.15	−35.88
1-propanol	353.15-443.15	−16.75
propyl acetate	353.15-443.15	−18.67
n-undecane	353.15-443.15	−46.77

Polymer (B):	**starch**	**1998BEN**
Characterization:	amorphous sample	

Solvent (A)	T-range/ K	$\Delta_M H_A^{\infty}$/ kJ/mol
water	378.15-438.15	−12.7

5.2. References

1969AND   Anderson, A.A. and Shimanskaya, M.B., Gas-liquid chromatography of some aliphatic and heterocyclic polyfunctional amines II. Solution thermodynamics of amines in the stationary phases (Russ.), *Latv. PSR Zinat. Akad. Vest. Kim. Ser.*, 5, 537, 1969.

1978DIN   Dincer, S. and Bonner, D.C., Thermodynamic analysis of ethylene and vinyl acetate copolymer with various solvents by gas chromatography, *Macromolecules*, 11, 107, 1978.

1978DIP1   DiPaola-Baranyi, G., Braun, J.-M., and Guillet, J.E., Partial molar heats of mixing of small molecules with polymers by gas chromatography, *Macromolecules*, 11, 224, 1978.

1978DIP2   DiPaola-Baranyi, G. and Guillet, J.E., Estimation of polymer solubility parameters by gas chromatography, *Macromolecules*, 11, 228, 1978.

1978DIP3   DiPaola-Baranyi, G., Guillet, J.E., Klein, J., and Jeberien, H.-E., Estimation of solubility parameters for poly(vinyl acetate) by inverse gas chromatography, *J. Chromatogr.*, 166, 349, 1978.

1978KAR   Karim, K.A. and Bonner, D.C., Thermodynamic interpretation of solute-polymer interactions at infinite dilution, *J. Appl. Polym. Sci.*, 22, 1277, 1978.

1978TAG   Tager, A.A., Kirillova, T.I., and Ikanina, T.V., The possibility to use the method of inverse gas chromatography for calculating thermodynamic parameters of the affinity between a polymer and solvent (Russ.), *Vysokomol. Soedin., Ser. A*, 20, 2543, 1978.

1979ASP   Aspler, J.S. and Gray, D.G., Gas chromatographic and static measurements of solute activity for a polymeric liquid-crystalline phase, *Macromolecules*, 12, 562, 1979.

1979ITO   Ito, K. and Guillet, J.E., Estimation of solubility parameters for some olefin polymers and copolymers by inverse gas chromatography, *Macromolecules*, 12, 1163, 1979.

1980DIP   DiPaola-Baranyi, G., Guillet, J.E., Jeberien, H.-E., and Klein, J., Thermodynamics of hydrogen bonding polymer-solute interactions by inverse gas chromatography, *Makromol. Chem.*, 181, 215, 1980.

1980GUN   Gunduz, S. and Dincer, S., Solubility behaviour of polystyrene. Thermodynamic studies using gas chromatography, *Polymer*, 21, 1041, 1980.

1980IWA   Iwai, Y., Nagafuji, M., and Arai, Y., Solubilities of volatile hydrocarbons in molten polystyrene, *J. Jap. Petrol. Inst.*, 23, 215, 1980.

1981DAN   Dangayach, K.C.B., Karim, K.A., and Bonner, D.C., Interaction of organic solvents with aromatic heterocyclic polymers. I. m-Phenylene polybenzimidazole, *J. Appl. Polym. Sci.*, 26, 559, 1981.

1981DIP   DiPaola-Baranyi, G., Thermodynamic miscibility of various solutes with styrene-butyl methacrylate polymers and copolymers, *Macromolecules*, 14, 683, 1981.

1982ASP   Aspler, J.S. and Gray, D.G., Interaction of organic vapours with hydroxypropyl cellulose, *Polymer*, 23, 43, 1982.

1982DIP   DiPaola-Baranyi, G., Fletcher, S.J., and Degre, P., Gas chromatographic investigation of poly(vinylidene fluoride)-poly(methyl methacrylate) blends, *Macromolecules*, 15, 885, 1982.

1982FER   Fernandez-Berridi, M.J., Otero, T.F., Guzman, G.M., and Elorza, J.M., Determination of the solubility parameter of poly(ethylene oxide) at 25°C by gas-liquid chromatography, *Polymer*, 23, 1361, 1982.

1982LAN   Langer, S.H., Sheehan, R.J., and Huang, J.-C., Gas-chromatographic study of the solution thermodynamics of hydroxylic derivatives and related compounds, *J. Phys. Chem.*, 86, 4605, 1982.

1983ASP   Aspler, J.S. and Gray, D.G., An inverse gas-chromatography study of the interactions of water with some cellulose derivatives, *J. Polym. Sci. Polym. Phys. Ed.* 21, 1675, 1983.

1984GAL   Galin, M., Gas-liquid chromatography study of poly(ethylene oxide)-solvent interactions. A molecular approach to solvation mechanisms, *Polymer*, 25, 1784, 1984.

1984SC1   Schotsch, K. and Wolf, B.A., Concentration dependence of the Flory-Huggins parameter at different thermodynamic conditions, *Makromol. Chem.*, 185, 2169, 1984.

1984SC2   Schuster, R.H., Gräter, H., and Cantow, H.J., Thermodynamic studies on polystyrene-solvent systems by gas chromatography, *Macromolecules*, 17, 619, 1984.

1984SIO   Siow, K.S., Goh, S.H., and Yap, K.S., Solubility parameters of poly(α-methylstyrene-*co*-acrylnitrile) from gas-liquid chromatography, *Polym. Mater. Sci. Eng.*, 51, 532, 1984.

1985CAS   Castells, R.C., Mazza, G.D., and Arancibia, E.L., Estudio de interacciones polimero-solvente por chromatografia gaseosa. Sistemas constituidos por hidrocarburos y alcoholes con poli(acetato de vinylo), *An. Asoc. Quim. Argent.*, 73, 519, 1985.

1985GAL   Galin, M. and Maslinko, L., Gas-liquid chromatography study of poly(vinylidene fluoride)-solvent ineractions. Correlation analysis of the partial molar enthalpy of mixing with probe polarity, *Macromolecules*, 18, 2192, 1985.

1985ITU   Ituno, S., Ohzono, M., Iwai, Y., and Arai, Y., Solubilities of hydrocarbon gases in polybutadiene, *Kobunshi Ronbunshu*, 42, 73, 1985.

1986INO   Inoue, K., Fujii, R., Baba, Y., and Kagemoto, A., Heat interaction parameter of polystrene solutions determined by inverse gas-liquid chromatography, *Makromol. Chem.*, 187, 923, 1986.

1986ROT   Roth, H. and Novak, J., Thermodynamics of poly(dimethylsiloxane)-alkane systems by gas-liquid chromatography, *Macromolecules*, 19, 364, 1986.

1986SIO   Siow, K.S., Goh, S.H., and Yap, K.S., Solubility parameters of poly(α-methylstyrene-*co*-acrylonitrile) from gas-liquid chromatography, *J. Chromatogr.*, 354, 75, 1986.

1986WUY   Wu, Y.-M., Direct determinationof thermodynamic quantities and solubility parameter for poly(tetramethylene p-silphenylene siloxane) by gas liquid chromatography (Chin.), *Sepu*, 4, 232, 1986.

1987DES   Deshpande, D.D. and Tyagi, O.S., Inverse gas chromatography of poly(vinyl isobutyl ether)s and polymer-solute thermodynamic interactions, *J. Appl. Polym. Sci.*, 33, 715, 1987.

1987SAN   Sanetra, R., Kolarz, B., and Wlochowicz, A., Determination of thermodynamic data for the interaction of aliphatic alcohols with poly(styrene-*co*-divinylbenzene) using inverse gas chromatography, *Polymer*, 28, 1753, 1987.

1987TYA   Tyagi, O.S. and Deshpande, D.D., Thermodynamic studies on poly(n-butyl methacrylate)-solute systems using gas chromatography, *Polym. J.*, 11, 1231, 1987.

1989CHE   Chen, C.-T. and Al-Saigh, Z.Y., Characterization of semicrystalline polymers by inverse gas chromatography. 1. Poly(vinylidene fluoride), *Macromolecules*, 22, 2974, 1989.

1989ED1   Edelmann, A. and Fradet, A., Inverse gas chromatography study of some triacetin-polymer systems, *Polymer*, 30, 317, 1989.

1989ED2   Edelmann, A. and Fradet, A., Inverse gas chromatography study of some alkyl trinitrate-polymer systems, *Polymer*, 30, 324, 1989.

1989GAL   Galin, M., Gas-liquid chromatography study of the thermodynamics of interactions between linear polyethylene and non-polar and polar solutes, *Polymer*, 30, 2074, 1989.

1989IRI   Iribarren, J.I., Iriarte, M., Uriarte, C., and Iruin, J.J., Phenoxy resin: characterization, solution properties, and inverse gas chromatography investigation of its potential miscibility with other polymers, *J. Appl. Polym. Sci.*, 37, 3459, 1989.

1989SAN   Sanetra, R., Kolarz, B.N., and Wlochowicz, A., Badanie metoda inwesyjnej chromatografii gazowej struktury usieciowanych kopolimerow na przykladzie kopolimerow styren/diwinylobenzen, *Polimery*, 34, 490, 1989.

1990CHE   Chen, C.-T. and Al-Saigh, Z.Y., Characterization of poly(ethyl methacrylate) by inverse gas chromatography, *Polymer*, 31, 1170, 1990.

1990GAL   Galin, M., Gas-liquid chromatography study of polar polymer-solvent systems. Correlation analysis of the partial molar enthalpy of mixing with solvent polarity, *Macromolecules*, 23, 3006, 1990.

1991ALS   Al-Saigh, Z.Y., Thermodynamics of poly(ethyl methacrylate)-solute systems using inverse gas chromatography, *Polym. Commun.*, 32, 459, 1991.

1991BEC   Becerra, M.R., Fernandez-Sanchez, E., Fernandez-Torres, A., Garcia-Dominguze, J.A., and Santiuste, J.M., Evaluation of the effect of the cyanopropyl radical on the interaction of the methylene group with silicone stationary phases, *J. Chromatogr.*, 547, 269, 1991.

1991OZD   Ozdemir, E., Coskun, M., Acikses, A., Estimation of thermodynamic parameters of polystyrene + n-hydrocarbons systems using inverse gas chromatography, *Macromol. Rep., Suppl. 2*, A28, 129, 1991.

1991ROM   Romdhane, I.H. and Danner, R.P., Solvent volatilities from polymer solutions by gas-liquid chromatography, *J. Chem. Eng. Data*, 36, 15, 1991.

1991WOH   Wohlfarth, C., Personalcomputergestütztes dezentrales Datenbankprojekt Thermodynamische Eigenschaften von Polymersystemen 3. Aktivitätskoeffizienten, χ-Funktion, Lösungs- und Exzessenthalpien, Henry-Konstanten (Daten aus der IGC), *Plaste Kautschuk*, 38, 228, 1991.

1992BEC   Becerra, M.R., Fernandez-Sanchez, E., Fernandez-Torres, A., Garcia-Dominguez, J.A., and Santiuste, J.M., Thermodynamic characterization by inverse gas chromatography of a 50% methyl, 50% trifluoropropyl polysiloxane, *Macromolecules*, 25, 4665, 1992.

1992BON   Bonifaci, L. and Ravanetti, G.P., Measurement of infinite dilution diffusion coefficients of ε-caprolactam in nylon-6/solvent at elevated temperatures by inverse gas chromatography, *J. Chromatogr.*, 607, 145, 1992.

1992COS   Coskun, M., Özdemir, E., Benzen, R., and Pulat, E., Stiren-etilakrilat kopolimerlerinin termodinamik oezellikerinin invers gaz kromatografisi ile incelenmesi, *Doga - Turk Kimya Dergisi*, 16, 76, 1992.

1992DIP   DiPaola-Baranyi, G., Hsiao, C.K., Spes, M., Odell, P.G., and Burt, R.A., Inverse gas chromatography of polycarbonates: I. Determination of polymer-solvent interaction parameters, *J. Appl. Polym. Sci., Appl. Polym. Symp.*, 51 (1992) 195-208

1992ETX   Etxeberria, A., Alfageme, J., Uriarte, C., and Iruin, J.J., Inverse gas chromatography in the characterization of polymeric materials, *J. Chromatogr.*, 607, 227, 1992.

1992KAL   Kalaouzis, P.J. and Demertzis, P.G., Water sorption and water vapour diffusion in food-grade plastics packaging materials: effect of a polymeric plasticizer, *Packag. Technol. Sci.*, 5, 133, 1992.

1992RO1   Romansky, M. and Guillet, J.E., Characterization of a thermotropic liquid crystal polymer by inverse gas chromatography, *Polym. Mater. Sci. Eng.*, 67, 441, 1992.

1992RO2   Romdhane, I.H., Plana, A., Hwang, S., and Danner, R.P., Thermodynamic interactions of solvents with styrene-butadiene-styrene triblock copolymers, *J. Appl. Polym. Sci.*, 45, 2049, 1992.

1992YIL   Yilmaz, F., Cankurtaran, O.G., and Baysal, B.M., Thermodynamic interactions and characterization of poly(p-chlororstyrene) with some aliphatic and aromatic probes by inverse gas chromatography, *Polymer*, 33, 4563, 1992.

1993ALE   Alessi, P., Cortesi, A., Sacomani, P., and Valles, E., Solvent-polymer interactions in polybutadiene, *Macromolecules*, 26, 6175, 1993.

1993BAL   Baltus, R.E., Alger, M.M., and Stanley, T.J., Solubility and diffusivity of cyclic oligomers in poly(dimethylsiloxane) using capillary column inverse gas chromatography, *Macromolecules*, 26, 5651, 1993.

1993ISS   Issel, H.-M., Thermodynamische und rheologische Steuerung der Materialeigenschaften von Elastomeren durch *trans*-Poly(octenylen), *Dissertation*, Universität Hannover, 1993.

1994COS   Coskun, M., Özdemir, E., Celik, S., and Cansiz, A., Inverse gas chromatography of n-alcohols–poly(acrylonitrile-*co*-methyl acrylate) systems, *J. Macromol. Sci., Macromol. Rep. A, (Suppl. 1 & 2)*, 31, 63, 1994.

1994ROM   Romansky, M. and Guillet, J.E., The use of inverse gas chromatography to study liquid crystalline polymers, *Polymer*, 35, 584, 1994.

1994TIA   Tian, M. and Munk, P., Characterization of polymer-solvent interactions and their temperature dependence using inverse gas chromatography, *J. Chem. Eng. Data*, 39, 742, 1994.

1994YAM   Yampol'skii, Y., Pavlova, A., Ushakov, N., and Finkelshtein, E., On the unusually high solubility of a trimethylsilyl derivate of poly(dimethylsilacyclobutane), *Macromol. Rapid Commun.*, 15, 917, 1994.

1995BOG   Bogillo, V.I. and Voelkel, A., Solution properties of amorphous co- and terpolymers of styrene as examined by inverse gas chromatography, *J. Chromatogr. A*, 715, 127, 1995.

1995KAY   Kaya, I., Ceylan, K., and Özdemir, E., Determination of thermodynamic parameters of poly(vinyl chloride) using inverse gas chromatography, *Turk. J. Chem.*, 19, 94, 1995.

1995LEZ   Lezcano, E.G., Salom, C., Masegosa, R.M., and Prolongo, M.G., Polymer-solvent interaction parameter for poly(4-hydroxystyrene) by IGC, *Polym. Bull.*, 34, 677, 1995.

1995NEM   Nemtoi, G. and Beldie, C., Thermodynamic characterization of some copolymers by gas-chromatographic measurements, *Rev. Roum. Chim.*, 40, 335, 1995.

1996BLO   Blokihna, S.V., Alekseeva, O.V., Ol'khovich, M.V., and Lokhanova, A.V., Issledovanie termodinamiki rastvoreniya nizkomolekulyarnykh soedinenii v oksipropiltsellyuloze metodom orashchennoi gazovoi khromatografii, *Zh. Fiz. Khim.*, 70, 2073, 1996.

1996KAY   Kaya, I., Özdemir, E., and Coskun, M., Thermodynamic parameters and characterization of poly(methyl methacrylate) with some probes using inverse gas chromatography, *Macromol. Rep., Suppl. 1*, A33, 37, 1996.

1996MOR   Morales, E. and Acosta, J.L., Polymer solubility parameters of poly(propylene oxide) rubber from inverse gas chromatography measurements, *Polym. J.*, 28, 127, 1996.

1997SCH   Schuster, R.H., Issel, H.M., and Peterseim, V., Charakterisierung der Kautschuk-Lösungsmittel Wechselwirkung mittels inverser Gaschromatographie, *Kautschuk Gummi Kunststoffe*, 50, 890, 1997.

1998BEN   Benczedi, D., Tomka, I., and Escher, F., Thermodynamics of amorphous starch-water systems. 2. Concentration fluctuations, *Macromolecules*, 31, 3062, 1998.

1998BLO   Blokhina, S.V., Alekseeva, O.V., Prusov, A.N., Ol'khovich, M.V., and Lokhanova, A.V., Gas-chromatographic study of the thermodynamic parameters of interaction between ethyl cellulose and low-molecular-mass compounds (Russ.), *Vysokomol. Soedin., Ser. B*, 40, 2099, 1998.

1998BRU   Brunazzi, E., Paglianti, A., and Petraca, L., Measurement of activity coefficients at infinite dilution of chlorinated solvents in commercial polyethylene glycol ethers, *J. Chem. Eng. Data*, 43, 443, 1998.

1998KIM   Kim, N.H., Won, Y.S., and Choi, J.S., Partial molar heat of mixing at infinite dilution in solvent/polymer (PEG, PMMA, EVA) solutions, *Fluid Phase Equil.*, 146, 223, 1998.

1998YAM   Yampolskii, Yu.P. and Bondarenko, G.N., Evidence of hydrogen bonding during sorption of chloromethanes in copolymers of chloroprene with methyl methacrylate and methacrylic acid, *Polymer*, 39, 2241, 1998.

1999ALS   Al-Saigh, Z.Y., Inverse gas chromatographic characterization of poly(ethylene oxide), *Polymer*, 40, 3479, 1999.

1999BON   Bondar, V.I., Freeman, B.D., and Yampolskii, Yu.P., Sorption of gases and vapors in an amorphous glassy perfluorodioxol copolymer, *Macromolecules*, 32, 6163, 1999.

1999KA1   Kaya, I., Determination of thermodynamic interactions of the PVC-PMMA blend system by inverse gas chromatography, *Polym-Plast. Technol. Eng.*, 38, 385, 1999.

1999KA2   Kaya, I. and Özdemir, E., Thermodynamic interactions and characterisation of poly(iso-bornyl methacrylate) by inverse gas chromatography at various temperatures, *Polymer*, 40, 2405, 1999.

1999KA3   Kaya, I. and Demirelli, K., Study of thermodynamic interaction parameters of poly(p-naphthalene-2(β)-sulfonylstyrene) using inverse gas chromatography, *J. Polym. Eng.*, 19, 61, 1999.

1999KA4   Kaya, I. and Demirelli, K., A study of the thermodynamic properties of poly[2-(3-mesityl-3-methylcyclobutyl)-2-hydroxyethyl methacrylate] at infinite dilution using inverse gas chromatography, *Turk. J. Chem.*, 23, 171, 1999.

1999KA5   Kaya, I. and Mart, H., Determination of thermodynamic properties of poly(4-*tert*-butyl-styrene) by inverse gas chromatography, *J. Polym. Eng.*, 19, 197, 1999.

2000ALS   Al-Saigh, Z.Y. and Al-Ghamdi, A., Miscibility and surface characterization of a poly(vinylidene fluoride)-poly(vinyl methyl ketone) blend by inverse gas chromatography, *J. Polym. Sci.: Part B: Polym. Phys.*, 38, 1155, 2000.

2000DEM   Demirelli, K., Kaya, I., Oezdemir, E., Determination of thermodynamic porperties of poly(2-hydroxyethyl methacrylate) at infinite dilution by using inverse gas chromatography, *J. Polym. Eng.*, 20, 351, 2000.

2000FEN   Feng, Y., Ye, R., and Liu, H., Measurement of infinite diluted activity coefficient of solvents in polymer by inverse gas chromatography method (Chin.), *Chin. J. Chem. Eng.*, 8, 167, 2000.

2000KAY   Kaya, I. and Demirelli, K., Determination of thermodynamic properties of poly[2-(3-methyl-3-phenyl-cyclobutyl)-2-hydroxyethylmethacrylate] and its copolymers at infinite dilution using inverse gas chromatography, *Polymer*, 41, 2855, 2000.

2000TAZ   Tazaki, M., Wada, R., Okabe, M., and Omma, T., Inverse gas chromatographic observation of thermodynamic interaction between poly(vinylidene fluoride) and organic solvents, *Polym. Bull.*, 44, 93, 2000.

2001ALG   Al-Ghamdi, A.M.S., Inverse gas chromatographic characterization of polycaprolactone, *Arab. J. Sci. Eng.*, 26, 11, 2001.

2001DEM   Demirelli, K., Kaya, I., and Coskun, M., 3,4-Dichlorobenzyl methacrylate and ethyl methacrylate system. Monomer reactivity ratios and determination of thermodynamic properties at infinite dilution using inverse gas chromatography, *Polymer*, 42, 5181, 2001.

2001KA1   Kaya, I. and Demirelli, K., Study of some thermodynamic properties of poly(3,4-di-chlorobenzyl methacrylate-*co*-ethyl methacrylate) using inverse gas chromatography, *J. Polym. Eng.*, 21, 1, 2001.

2001KA2   Kaya, I. and Demirelli, K., Determination of the thermodynamic properties of poly[2-(3-phenyl-3-methylcyclobutyl)-2-hydroxyethyl methacrylate-*co*-methacrylic acid] at infinite dilution by inverse gas chromatography, *Turk. J. Chem.*, 25, 11, 2001.

2002ILT   Ilter, Z., Kaya, I., and Acikses, A., Determination of thermodynamic properties of poly[(2-phenyl-1,3-dioxolane-4-yl)methyl methacrylate] by inverse gas chromatography, *J. Polym. Eng.*, 22, 45, 2002.

2002KAY   Kaya, I., Ilter, Z., and Senol, D., Thermodynamic interactions and characterisation of poly[(glycidyl methacrylate-co-methyl, ethyl, butyl) methacrylate] by inverse gas chromatography, *Polymer*, 43, 6455, 2002.

2002PRI   Price, G.J. and Shillcock, I.M., Inverse gas chromatography study of the thermodynamic behaviour of thermotropic low molar mass and polymeric liquid crystals (supplementary material), *Phys. Chem. Chem. Phys.*, 4, 5307, 2002.

2002SAR   Sarac, A., Tuere, A., Cankurtaran, O., and Yilmaz, F., Determination of some thermo-dynamical interaction parameters of polycaprolactone with some solvents by inverse gas chromatography, *Polym. Int.*, 51, 1285, 2002.

2002SOL   Solovev, S.A., Yampolskii, Yu.P., Economou, I.G., Ushakov, N.V., and Finkelshtein, E.Sh., Thermodynamic parameters of hydrocarbon sorption by poly(silmethylenes) (Russ.), *Vysokomol. Soedin., Ser. A*, 44, 465, 2002.

2003ACI   Acikses, A.. Kaya, I., and Ilter, Z., Study of some thermodynamic properties of poly[(2-phenyl-1,3-dioxolane-4-yl)methyl methacrylate-*co*-butyl methacrylate] by inverse gas chromatography, *Polym.-Plast. Technol. Eng.*, 42, 431, 2003.

2003VOZ   Voznyakovskii, A.P., Comparative study of sorption behavior of natural and synthetic polyisoprenes (Russ.), *Vysokomol. Soedin., Ser. A*, 45, 262, 2003.

2004ILT   Ilter, Z., Kaya, I., and Acikses, A., Determination of poly[(2-phenyl-1,3-dioxolane-4-yl)-methyl methacrylate-*co*-glycidyl methacrylate]-probe interactions by inverse gas chromatography, *Polym.-Plast. Technol. Eng.*, 43, 229, 2004.

2005SAR   Sarac, A., Sakar, D., Cankurtaran, O., and Karaman, F.Y., The ratio of crystallinity and thermodynamical interactions of polycaprolactone with some aliphatic esters and aromatic solvents by inverse gas chromatography, *Polym. Bull.*, 53, 349, 2005.

6. TABLE OF SYSTEMS FOR ADDITIONAL INFORMATION ON ENTHALPY EFFECTS IN POLYMER SOLUTIONS

6.1. List of systems

Polymer (B)	Solvent (A)	Enthalpy	T-range	Ref.
Butyl rubber				
	carbon dioxide	$\Delta_{sol}H_{A(vap)}^{\infty}$		1950AME
	ethene	$\Delta_{sol}H_{A(vap)}^{\infty}$	333-353 K	1948BAR
	ethyne	$\Delta_{sol}H_{A(vap)}^{\infty}$		1950AME
	helium	$\Delta_{sol}H_{A(vap)}^{\infty}$		1950AME
	hydrogen	$\Delta_{sol}H_{A(vap)}^{\infty}$		1950AME
	nitrogen	$\Delta_{sol}H_{A(vap)}^{\infty}$		1950AME
	nitrogen	$\Delta_{sol}H_{A(vap)}^{\infty}$	313-353 K	1948BAR
	oxygen	$\Delta_{sol}H_{A(vap)}^{\infty}$		1950AME
	popane	$\Delta_{sol}H_{A(vap)}^{\infty}$	333-353 K	1948BAR
Carboxymethylcellulose sodium salt				
	water	$\Delta_{dil}H^{12}$	298-313 K	1974MIT
	water	$\Delta_{sol}H_B^{\infty}$	298 K	2000SUV
	water	$\Delta_{sol}H_B^{\infty}$	298, 353 K	2002SUV
Cellulose				
	acetic acid	$\Delta_M H$	298 K	1992IV1
	1,2-dichloroethane	$\Delta_{sol}H_B$	330 K	1990ZAV
	dichloromethane	$\Delta_{sol}H_B$	330 K	1990ZAV
	N,N-dimethylacetamide	$\Delta_{sol}H_B$	330 K	1990ZAV
	N,N-dimethylformamide	$\Delta_{sol}H_B$	330 K	1990ZAV
	dimethylsulfoxide	$\Delta_{sol}H_B$	330 K	1990ZAV
	N-methyl-morpholine-N-oxide	$\Delta_{sol}H_B$	355 K	1992IV2
	N-methyl-morpholine-N-oxide	$\Delta_{sol}H_B$	358 K	1986GOL
	nitric acid	$\Delta_M H$	298 K	1992IV1
	nitric acid	$\Delta_M H$	298 K	1993IVA
	phosphoric acid	$\Delta_M H$	298 K	1992IV1
	sulfuric acid	$\Delta_M H$	298 K	1992IV1
	sulfuric acid	$\Delta_M H$	298 K	1993IVA
	trichloromethane	$\Delta_{sol}H_B$	330 K	1990ZAV
	trifluoroacetic acid	$\Delta_M H$	298 K	1992IV1
	trifluoroacetic acid	$\Delta_M H$	298 K	1993IVA
	trifluoroacetic acid	$\Delta_{sol}H_B$	330 K	1990ZAV
	water	$\Delta_M H_A$	323 K	1940ME2

Polymer (B)	Solvent (A)	Enthalpy	T-range	Ref.
Cellulose (*continued*)				
	water	$\Delta_M H$	298 K	1985IOE
	water	$\Delta_{sol} H_A$	298 K	1994KHA
	water/NaOH	$\Delta_{sol} H_B$	298 K	1966KAR
Cellulose acetate				
	acetic acid	$\Delta_M H_A^\infty$	303-313 K	1958MOO
	aniline	$\Delta_M H_A^\infty$	298 K	1957MOO
	aniline	$\Delta_M H_A^\infty$	303-313 K	1958MOO
	carbon dioxide	$\Delta_{sol} H_{A(vap)}^\infty$	283-343 K	1978STE
	1,2-dichloroethane	$\Delta_{sol} H_B$	330 K	1990ZAV
	dichloromethane	$\Delta_{sol} H_B$	330 K	1990ZAV
	N,N-dimethylacetamide	$\Delta_{sol} H_B$		1973KHA
	N,N-dimethylacetamide	$\Delta_{sol} H_B$	330 K	1990ZAV
	N,N-dimethylformamide	$\Delta_{sol} H_B$		1973KHA
	N,N-dimethylformamide	$\Delta_{sol} H_B$	330 K	1990ZAV
	dimethylsulfoxide	$\Delta_{sol} H_B$		1973KHA
	dimethylsulfoxide	$\Delta_{sol} H_B$	330 K	1990ZAV
	1,4-dioxane	$\Delta_M H_A^\infty$	298 K	1957MOO
	1,4-dioxane	$\Delta_M H_A$	303-308 K	1963MO2
	1,4-dioxane	$\Delta_M H_A^\infty$	303-313 K	1958MOO
	formic acid	$\Delta_{sol} H$		1958KHA
	methyl acetate	$\Delta_M H_A^\infty$	298 K	1957MOO
	methyl acetate	$\Delta_M H_A$	303-308 K	1963MO2
	methyl acetate	$\Delta_M H_A^\infty$	303-313 K	1958MOO
	2-methylpyridine	$\Delta_M H_A^\infty$	298 K	1957MOO
	3-methylpyridine	$\Delta_M H_A^\infty$	298 K	1957MOO
	4-methylpyridine	$\Delta_M H_A^\infty$	298 K	1957MOO
	nitromethane	$\Delta_M H_A^\infty$	298 K	1957MOO
	nitromethane	$\Delta_M H_A^\infty$	303-313 K	1958MOO
	phenol	$\Delta_M H_A^\infty$	298-368 K	1957JEF
	2-propanone	$\Delta_{sol} H$		1958KHA
	2-propanone	$\Delta_M H_A^\infty$	298 K	1957MOO
	2-propanone	$\Delta_M H$	298 K	1975TA3
	2-propanone	$\Delta_M H_A^\infty$	298-368 K	1957JEF
	2-propanone	$\Delta_M H_A$	303-308 K	1963MO2
	2-propanone	$\Delta_M H_A^\infty$	303-313 K	1958MOO
	pyridine	$\Delta_{sol} H$		1958KHA
	pyridine	$\Delta_M H_A^\infty$	298 K	1957MOO
	pyridine	$\Delta_M H_A$	303-308 K	1963MO2
	pyridine	$\Delta_M H_A^\infty$	303-313 K	1958MOO
	1,1,2,2-tetrachloroethane	$\Delta_M H_A$	298 K	1941TAG
	trichloromethane	$\Delta_{sol} H_B$	330 K	1990ZAV
	trifluoroacetic acid	$\Delta_{sol} H_B$	330 K	1990ZAV
	water	$\Delta_{sol} H_A$	298 K	1994KHA

Polymer (B)	Solvent (A)	Enthalpy	T-range	Ref.
Cellulose diacetate				
	1,2-dichloroethane	$\Delta_{sol}H_B$	330 K	1990ZAV
	dichloromethane	$\Delta_{sol}H_B$	330 K	1990ZAV
	N,N-dimethylacetamide	$\Delta_{sol}H_B$	330 K	1990ZAV
	N,N-dimethylformamide	$\Delta_{sol}H_B$	330 K	1990ZAV
	N,N-dimethylformamide	$\Delta_M H$	293-333 K	1991AZI
	dimethyl phthalate	$\Delta_M H$	240-300 K	1985ZAR
	dimethylsulfoxide	$\Delta_{sol}H_B$	330 K	1990ZAV
	ethanol	$\Delta_{sol}H_B$	298 K	1993BRU
	2-propanone	$\Delta_{sol}H_B$	298 K	1993BRU
	2-propanone	$\Delta_M H_A$	285-323 K	1980SUZ
	triacetin	$\Delta_M H$	240-300 K	1985RA3
	trichloromethane	$\Delta_{sol}H_B$	330 K	1990ZAV
	trifluoroacetic acid	$\Delta_{sol}H_B$	330 K	1990ZAV
Cellulose triacetate				
	1,2-dichloroethane	$\Delta_{sol}H_B$	330 K	1990ZAV
	dichloromethane	$\Delta_{sol}H_B$	330 K	1990ZAV
	dichloromethane	$\Delta_M H_A$	293-298 K	1963MO3
	N,N-dimethylacetamide	$\Delta_{sol}H_B$	330 K	1990ZAV
	N,N-dimethylformamide	$\Delta_{sol}H_B$	330 K	1990ZAV
	dimethylsulfoxide	$\Delta_{sol}H_B$	330 K	1990ZAV
	1,4-dioxane	$\Delta_M H_A^{\infty}$		1941KUN
	2-propanone	$\Delta_M H$		1959TA1
	1,1,2,2-tetrachloroethane	$\Delta_M H_A^{\infty}$	303 K	1940HAG
	trichloromethane	$\Delta_M H$		1959TA1
	trichloromethane	$\Delta_M H_A$	303-308 K	1963MO3
	trichloromethane	$\Delta_{sol}H_B$	330 K	1990ZAV
	trifluoroacetic acid	$\Delta_{sol}H_B$	330 K	1990ZAV
Cellulose tribenzoate				
	dichloromethane	$\Delta_{sol}H_B$	373-493 K	1963KOZ
Cellulose tributyrate				
	benzophenone	$\Delta_M H_A^{\infty}$	438 K	1951MAN
	dimethyl phthalate	$\Delta_M H_A^{\infty}$	438 K	1951MAN
	ethyl benzoate	$\Delta_M H_A^{\infty}$	438 K	1951MAN
	ethyl laurate	$\Delta_M H_A^{\infty}$	438 K	1951MAN
	hydrochinone			
	monomethyl ether	$\Delta_M H_A^{\infty}$	438 K	1951MAN
	tributyrin	$\Delta_M H_A^{\infty}$	438 K	1951MAN
Cellulose xanthogenate				
	water/NaOH	$\Delta_{dil}H^{12}$	298 K	1949LAU

Polymer (B)	Solvent (A)	Enthalpy	T-range	Ref.
Chitosan				
	water/acetic acid	$\Delta_{dil}H^{12}$	298 K	2002SA1
Chloropolyethylene				
	2-octyl acetate	$\Delta_M H$	338-358 K	1983WAL
Dextran				
	dimethylsulfoxide	$\Delta_M H$	298 K	1980BAS
	water	$\Delta_{dil}H^{12}$	279-306 K	1978COM
	water	$\Delta_M H$	298 K	1980BAS
Epoxy resin				
	water	$\Delta_{sol}H_A$	310-333K	1982GAR
Ethylcellulose				
	benzene	$\Delta_M H_A^{\infty}$	298 K	1957MOO
	benzene	$\Delta_M H_A$	313-353 K	1957BAR
	benzene	$\Delta_{sol}H_A$	313-353 K	1982SAX
	n-butane	$\Delta_M H_A$	303-348 K	1958BAR
	2-butanone	$\Delta_M H_A^{\infty}$	298 K	1957MOO
	butyl acetate	$\Delta_M H_A^{\infty}$	298 K	1957MOO
	2,2-dimethylpropane	$\Delta_M H_A$	303-348 K	1958BAR
	ethyl acetate	$\Delta_M H_A^{\infty}$	298 K	1957MOO
	2-hexanone	$\Delta_M H_A^{\infty}$	298 K	1957MOO
	methanol	$\Delta_M H_A$	313-353 K	1957BAR
	methanol	$\Delta_{sol}H_A$	313-353 K	1982SAX
	methyl acetate	$\Delta_M H_A^{\infty}$	298 K	1957MOO
	3-methylbutane	$\Delta_M H_A$	303-348 K	1958BAR
	2-methylpropane	$\Delta_M H_A$	303-348 K	1958BAR
	n-pentane	$\Delta_M H_A$	303-348 K	1958BAR
	2-pentanone	$\Delta_M H_A^{\infty}$	298 K	1957MOO
	pentyl acetate	$\Delta_M H_A^{\infty}$	298 K	1957MOO
	2-propanone	$\Delta_M H_A^{\infty}$	298 K	1957MOO
	2-propanone	$\Delta_M H_A$	313-353 K	1957BAR
	2-propanone	$\Delta_{sol}H_A$	313-353 K	1982SAX
	propyl acetate	$\Delta_M H_A^{\infty}$	298 K	1957MOO
	tetrachloromethane	$\Delta_M H_A^{\infty}$	298 K	1957MOO
	toluene	$\Delta_M H_A^{\infty}$	298 K	1957MOO
	trichloromethane	$\Delta_M H_A^{\infty}$	298 K	1957MOO
Hemoglobin				
	water	$\Delta_{dil}H^{12}$	298 K	1981FUJ

Polymer (B)	Solvent (A)	Enthalpy	T-range	Ref.
Gelatine				
	water	$\Delta_{dil}H^{12}$	318 K	1999CES
Guttapercha				
	toluene	$\Delta_{M}H_{A}{}^{\infty}$	297-309 K	1940WOL
Hexa(ethylene glycol) monododecyl ether				
	water	$\Delta_{M}H$	298 K	1967CLU
	water	$\Delta_{M}H$	298 K	1969CLU
Hydroxypropylcellulose				
	water	$\Delta_{M}H_{A}$	298 K	1981ASP
Maltodextrin				
	water	$\Delta_{dil}H^{12}$	318 K	1999CES
Methylcellulose				
	water	$\Delta_{sol}H_{B}$	298 K	1997RAS
	water	$\Delta_{sol}H_{B}$	298 K	2000SUV
	water/NaOH	$\Delta_{sol}H_{B}$	298 K	1959TSU
Natural rubber				
	benzene	$\Delta_{M}H_{A}$		1945FLO
	benzene	$\Delta_{M}H_{A}$		1947GEE
	benzene	$\Delta_{M}H_{A}$		1947ORR
	benzene	$\Delta_{M}H$	295 K	1945FER
	benzene	$\Delta_{M}H$	298 K	1959MAR
	benzene	$\Delta_{M}H_{A}$	298 K	1942GEE
	benzene	$\Delta_{M}H_{A}$	298 K	1951MIN
	benzene	$\Delta_{M}H_{A}$	298-323 K	1946GEE
	n-butane	$\Delta_{sol}H_{A(vap)}{}^{\infty}$	313-353 K	1948BAR
	2-butanone	$\Delta_{M}H_{A}$	298 K	1964BO1
	carbon dioxide	$\Delta_{sol}H_{A(vap)}{}^{\infty}$		1950AME
	carbon disulfide	$\Delta_{M}H$	295 K	1945FER
	ethane	$\Delta_{sol}H_{A(vap)}{}^{\infty}$	313-353 K	1948BAR
	ethene	$\Delta_{sol}H_{A(vap)}{}^{\infty}$	313-353 K	1948BAR
	ethyl acetate	$\Delta_{M}H_{A}$	298 K	1964BO1
	ethyl acetate	$\Delta_{M}H_{A}$	298-323 K	1957BOO
	ethyne	$\Delta_{sol}H_{A(vap)}{}^{\infty}$		1950AME
	helium	$\Delta_{sol}H_{A(vap)}{}^{\infty}$		1950AME
	n-heptane	$\Delta_{M}H$	295 K	1945FER
	hydrogen	$\Delta_{sol}H_{A(vap)}{}^{\infty}$		1950AME
	hydrogen	$\Delta_{sol}H_{A(vap)}{}^{\infty}$	303-313 K	1947CAR

Polymer (B)	Solvent (A)	Enthalpy	T-range	Ref.
Natural rubber (*continued*)				
	methane	$\Delta_{sol}H_{A(vap)}{}^{\infty}$	313-353 K	1948BAR
	methanol	$\Delta_M H_A$	303 K	1945FER
	methyl acetate	$\Delta_M H_A$	298 K	1964BO1
	nitrogen	$\Delta_{sol}H_{A(vap)}{}^{\infty}$		1950AME
	nitrogen	$\Delta_{sol}H_{A(vap)}{}^{\infty}$	298-323 K	1947CAR
	nitrogen	$\Delta_{sol}H_{A(vap)}{}^{\infty}$	313-353 K	1948BAR
	oxygen	$\Delta_{sol}H_{A(vap)}{}^{\infty}$	303-318 K	1947CAR
	2-pentanone	$\Delta_M H_A$	298 K	1964BO1
	propane	$\Delta_{sol}H_{A(vap)}{}^{\infty}$	313-353 K	1948BAR
	2-propanone	$\Delta_M H$	295 K	1945FER
	2-propanone	$\Delta_M H_A$	298 K	1964BO1
	2-propanone	$\Delta_{sol}H_B$	303 K	1969ICH
	oxygen	$\Delta_{sol}H_{A(vap)}{}^{\infty}$		1950AME
	tetrachloromethane	$\Delta_M H$	295 K	1945FER
	toluene	$\Delta_M H_A$	298-308 K	1939MEY
	toluene	$\Delta_M H$	295 K	1945FER
	toluene	$\Delta_M H_A{}^{\infty}$	297-309 K	1940ME1
	trichloromethane	$\Delta_M H$	295 K	1945FER
Nitrocellulose				
	2-butanone	$\Delta_{sol}H_B$	298 K	1941TAG
	2-butanone	$\Delta_M H_A{}^{\infty}$	298 K	1957MOO
	butyl acetate	$\Delta_{sol}H_A, \Delta_{sol}H_B$		1945CA2
	butyl acetate	$\Delta_M H_A{}^{\infty}$	298 K	1957MOO
	butyl acetate	$\Delta_{sol}H_B$	291-323 K	1950LI1
	cyclohexanone	$\Delta_M H_A$	298-309 K	1937BOI
	cyclohexanone	$\Delta_M H_A$	298-309 K	1938BOI
	dibutyl phthalate	$\Delta_M H$		1985RA1
	dibutyl phthalate	$\Delta_M H$	298 K	1984KRY
	diethyl ether	$\Delta_{sol}H_A, \Delta_{sol}H_B$		1937PAP
	diethyl ether	$\Delta_M H_A, \Delta_M H_B$		1937PAP
	diethyl ether	$\Delta_{sol}H_B$		1945CA4
	1,3-diethyl-1,3-diphenyl-urea	$\Delta_M H$	348 K	2003KSI
	diphenyl 4-*tert*-butyl-phenyl phosphate	$\Delta_M H$	298 K	1984KRY
	diphenyl 2-ethylhexyl phosphate	$\Delta_M H$	298 K	1984KRY
	ethanol	$\Delta_{sol}H_B$		1945CA4
	ethyl acetate	$\Delta_{sol}H_A, \Delta_{sol}H_B$		1945CA2
	ethyl acetate	$\Delta_{dil}H^{12}$		1949GLI
	ethyl acetate	$\Delta_M H_A{}^{\infty}$	298 K	1957MOO
	ethyl acetate	$\Delta_{sol}H_B$	288-323 K	1950LI2
	2-heptanone	$\Delta_M H_A{}^{\infty}$	298 K	1957MOO

Polymer (B)	Solvent (A)	Enthalpy	T-range	Ref.
Nitrocellulose (*continued*)				
	2-hexanone	$\Delta_M H_A^\infty$	298 K	1957MOO
	methanol	$\Delta_{sol} H_B$	288-323 K	1950LI2
	methanol	$\Delta_{sol} H_B$	291-323 K	1950LI1
	methyl acetate	$\Delta_{sol} H_A$, $\Delta_{sol} H_B$		1945CA2
	methyl acetate	$\Delta_M H_A^\infty$	298 K	1957MOO
	methyl acetate	$\Delta_M H_A$	303-308 K	1963MO2
	methyl nitrate	$\Delta_{sol} H_A$, $\Delta_{sol} H_B$		1945CA3
	nitric acid esters	$\Delta_M H_A$	293-333 K	1965BRO
	2,4-pentanedione	$\Delta_{sol} H_B$	298 K	1941TAG
	2-pentanone	$\Delta_{sol} H_B$	298 K	1941TAG
	2-pentanone	$\Delta_M H_A^\infty$	298 K	1957MOO
	pentyl acetate	$\Delta_M H_A^\infty$	298 K	1957MOO
	2-propanone	$\Delta_{sol} H_A$, $\Delta_{sol} H_B$		1945CA1
	2-propanone	$\Delta_{sol} H_A$, $\Delta_{sol} H_B$		1945CA2
	2-propanone	$\Delta_M H_A$		1945CA5
	2-propanone	$\Delta_M H_A$		1947CHE
	2-propanone	$\Delta_{dil} H^{12}$		1949GLI
	2-propanone	$\Delta_{dil} H^{12}$		1951GLI
	2-propanone	$\Delta_M H_A$	283-303 K	1937SCH
	2-propanone	$\Delta_M H_A$	283-303 K	1938SCH
	2-propanone	$\Delta_M H_A$	288-313 K	1970OIK
	2-propanone	$\Delta_{sol} H_B$	288-323 K	1950LI2
	2-propanone	$\Delta_M H_A$	290 K	1941CA1
	2-propanone	$\Delta_{sol} H_B$	291-323 K	1950LI1
	2-propanone	$\Delta_{sol} H_B$	298 K	1941TAG
	2-propanone	$\Delta_M H_A^\infty$	298 K	1957MOO
	2-propanone	$\Delta_M H$	298 K	1975TA3
	2-propanone	$\Delta_M H_A$	303-308 K	1963MO2
	2-propanone	$\Delta_M H_A$	313 K	1942CAL
	propyl acetate	$\Delta_{sol} H_A$, $\Delta_{sol} H_B$		1945CA2
	propyl acetate	$\Delta_M H_A^\infty$	298 K	1957MOO
	triacetin	$\Delta_M H$	220-300 K	1985RA2
	triacetin	$\Delta_M H$	298-304 K	1983KIR
	trichloropropyl phosphate	$\Delta_M H$	298 K	1984KRY
	tricresyl phosphate	$\Delta_{sol} H_B$	291-323 K	1950LI1
Nylon 6				
	2-chlorophenol	$\Delta_M H_A^\infty$	500 K	1963GEC
	hydrochloric acid	$\Delta_{sol} H_B$		1957POK
	formic acid	$\Delta_{sol} H_B$		1957POK
	4-methoxyphenol	$\Delta_M H_A^\infty$	500 K	1963GEC
	3-methylphenol	$\Delta_M H_A^\infty$	500 K	1963GEC
	3-nitrotoluene	$\Delta_M H_A^\infty$	500 K	1963GEC

Polymer (B)	Solvent (A)	Enthalpy	T-range	Ref.
Nylon 6 (*continued*)				
	phenol	$\Delta_{sol}H_B$		1957BYK
	phenol	$\Delta_{sol}H_B$		1957POK
	phenol	$\Delta_M H_A^{\infty}$	500 K	1963GEC
	sulfuric acid	$\Delta_{sol}H_B$		1957POK
	thymol	$\Delta_M H_A^{\infty}$	500 K	1963GEC
Nylon 6,6				
	formic acid	$\Delta_M H_A^{\infty}$	298 K	1953LIQ
	sulfuric acid	$\Delta_M H_A^{\infty}$	298 K	1953LIQ
	3-methylphenol	$\Delta_M H_A^{\infty}$	293-368 K	1953LIQ
Nylon 11				
	argon	$\Delta_{sol}H_{A(vap)}^{\infty}$		1970ASH
	carbon dioxide	$\Delta_{sol}H_{A(vap)}^{\infty}$		1970ASH
	helium	$\Delta_{sol}H_{A(vap)}^{\infty}$		1970ASH
	hydrogen	$\Delta_{sol}H_{A(vap)}^{\infty}$		1970ASH
	neon	$\Delta_{sol}H_{A(vap)}^{\infty}$		1970ASH
Octa(ethylene glycol) monododecyl ether				
	water	$\Delta_M H$	298 K	1967CLU
Octamethylcyclotetrasiloxane				
	benzene	$\Delta_M H$, $\Delta_M H_A$	315 K	1968MAR
Penta(ethylene glycol)				
	water	$\Delta_M H$	283-313K	1998OHT
Penta(ethylene glycol) monooctyl ether				
	water	$\Delta_M H$, $\Delta_M H_B$	283-313K	2001OHT
	water/surfactant	$\Delta_{sol}H$	298 K	2004DER
Perchlorovinylic resin				
	cyclohexanone	$\Delta_M H_A^{\infty}$	283-303 K	1953BYK
	cyclohexanone	$\Delta_{sol}H_B^{\infty}$	293 K	1953BYK
	1,2-dichloroethane	$\Delta_M H_A^{\infty}$	283-303 K	1953BYK
	1,2-dichloroethane	$\Delta_{sol}H_B^{\infty}$	293 K	1953BYK
	2-propanone	$\Delta_M H_A^{\infty}$	283-303 K	1953BYK
	2-propanone	$\Delta_{sol}H_B^{\infty}$	293 K	1953BYK

Polymer (B)	Solvent (A)	Enthalpy	T-range	Ref.
Poly(acrylamide)				
	dimethylsulfoxide	$\Delta_{sol}H_B$	303 K	1970GRO
	formamide	$\Delta_{sol}H_B$	303 K	1970GRO
	water	$\Delta_M H_A^\infty$	305-315 K	1957SIL
	water	$\Delta_{sol}H_B$	303 K	1970GRO
	water	$\Delta_{sol}H_B$, $\Delta_{dil}H^{12}$	298 K	1978SCH
	water	$\Delta_{dil}H^{12}$	298 K	1981DAY
Poly(2-acrylamido-2-methyl-propanesulfonic acid)				
	water	$\Delta_{dil}H^{12}$	298 K	1993KEL
	water	$\Delta_{dil}H^{12}$	298 K	1996KEL
Poly(acrylic acid)				
	1-butanol	$\Delta_M H$	298-305 K	1996SAF
	N,N-dimethylformamide	$\Delta_M H$	298-313 K	1996SAF
	1,4-dioxane	$\Delta_{dil}H^{12}$	298-316 K	1990COW
	1,2-ethanediol	$\Delta_{sol}H_B$	298 K	1979KLE
	1,2-ethanediol	$\Delta_{sol}H_B$	298 K	1978SCH
	1,2-ethanediol	$\Delta_{dil}H^{12}$	298 K	1978SCH
	formamide	$\Delta_{sol}H_B$	298 K	1978SCH
	formamide	$\Delta_{dil}H^{12}$	298 K	1978SCH
	formamide	$\Delta_{sol}H_B$	298 K	1979KLE
	hydrochloric acid	$\Delta_M H_A^\infty$	287-307 K	1957SIL
	water	$\Delta_{dil}H^{12}$	298 K	1972CRE
	water	$\Delta_{dil}H^{12}$	298 K	1974SKE
	water	$\Delta_{sol}H_B$, $\Delta_{dil}H^{12}$	298 K	1978SCH
	water	$\Delta_{sol}H_B$	298 K	1979KLE
	water	$\Delta_M H$	298 K	1983TA2
	water	$\Delta_{dil}H^{12}$	273-313 K	1986SK1
	water	$\Delta_{dil}H^{12}$	298 K	1990DAO
	water	$\Delta_M H$	298 K	1992TAG
	water	$\Delta_{dil}H^{12}$	298 K	1993SAF
	water	$\Delta_M H$	290-313 K	1996SAF
	water/penta(ethylene glycol) monooctyl ether	$\Delta_{sol}H$	298 K	2004DER
	water/sodium dodecyl sulfate	ΔH	298 K	2005WAN

Polymer (B)	Solvent (A)	Enthalpy	T-range	Ref.
Poly(acrylic acid) salts				
	water	$\Delta_{dil}H^{12}$	298 K	1973ISE
	water	$\Delta_{dil}H^{12}$	298-313 K	1974MIT
	water	$\Delta_{dil}H^{12}$	298 K	1974SKE
	water	$\Delta_{dil}H^{12}$	298 K	1980DAO
	water	$\Delta_{dil}H^{12}$	273-313 K	1986SK1
	water	$\Delta_{sol}H$	298 K	2004DER
Poly(acrylic acid-*co*-methyl acrylate)				
	water	$\Delta_{M}H$	298 K	1993ADA
	water	$\Delta_{M}H$	298 K	1993TA1
Poly(acrylic acid-*co*-methyl methacrylate)				
	water	$\Delta_{M}H$	298 K	1993ADA
Poly(acrylonitrile)				
	N,N-dimethylformamide	$\Delta_{sol}H_{B}^{\infty}$	298 K	1991ZVE
	N,N-dimethylformamide	$\Delta_{M}H_{A}^{\infty}$	303 K	1964KAW
	N,N-dimethylformamide	$\Delta_{sol}H_{B}$	303-353 K	1963ME1
	N,N-dimethylformamide	$\Delta_{sol}H_{B}$		1967ZVE
	dimethylsulfoxide	$\Delta_{sol}H_{B}$		1967ZVE
	dimethylsulfoxide	$\Delta_{M}H_{A}^{\infty}$	303 K	1964KAW
	water	$\Delta_{sol}H_{B}^{\infty}$	298 K	1991ZVE
Poly(acrylonitrile-*co*-butadiene)				
	2-butanone	$\Delta_{M}H_{A}$	311 K	1969POD
	benzene	$\Delta_{M}H_{A}^{\infty}$	298 K	1955TA1
	benzene	$\Delta_{M}H$	298 K	1955TA1
	cyclohexane	$\Delta_{M}H_{A}$	311 K	1969POD
	hydrogen	$\Delta_{sol}H_{A(vap)}^{\infty}$	293-343 K	1939BAR
	nitrogen	$\Delta_{sol}H_{A(vap)}^{\infty}$	293-343 K	1939BAR
Poly(acrylonitrile-*co*-isoprene)				
	carbon dioxide	$\Delta_{sol}H_{A(vap)}^{\infty}$		1950AME
	helium	$\Delta_{sol}H_{A(vap)}^{\infty}$		1950AME
	hydrogen	$\Delta_{sol}H_{A(vap)}^{\infty}$		1950AME
	nitrogen	$\Delta_{sol}H_{A(vap)}^{\infty}$		1950AME
	oxygen	$\Delta_{sol}H_{A(vap)}^{\infty}$		1950AME

Polymer (B)	Solvent (A)	Enthalpy	T-range	Ref.
Poly(acrylonitrile-*co*-2-methyl-furan)				
	N,N-dimethylformamide	$\Delta_{sol}H_B^\infty$		1964USM
Poly(acrylonitrile-*co*-styrene)				
	acrylonitrile	$\Delta_{sol}H_B^\infty$	298 K	1989EGO
	N,N-dimethylformamide	$\Delta_{sol}H_B^\infty$	298 K	1991ZVE
	styrene	$\Delta_{sol}H_B^\infty$	298 K	1989EGO
	trichloromethane	$\Delta_{sol}H_B^\infty$	303-323 K	1993FRE
	trichloromethane	$\Delta_{sol}H_B^\infty$	303-323 K	1994FRE
	water	$\Delta_{sol}H_B^\infty$	298 K	1991ZVE
Poly(acrylonitrile-*co*-styrene-sulfonate)				
	N,N-dimethylformamide	$\Delta_{sol}H_B^\infty$	298 K	1991ZVE
	water	$\Delta_{sol}H_B^\infty$	298 K	1991ZVE
Poly(acrylonitrile-*co*-vinyl chloride)				
	N,N-dimethylformamide	$\Delta_{sol}H_B^\infty$	298 K	1991ZVE
	water	$\Delta_{sol}H_B^\infty$	298 K	1991ZVE
Polyamide				
	argon	$\Delta_{sol}H_{A(vap)}^\infty$	298-368 K	1960TIK
	helium	$\Delta_{sol}H_{A(vap)}^\infty$	298-368 K	1960TIK
	phenol	$\Delta_{sol}H_B$		1957BYK
Poly(γ-benzyl-L-glutamate)				
	1,2-dichloroethane	$\Delta_{sol}H_B$	303 K	1968GIA
	dichloroacetic acid	$\Delta_{sol}H_B$	303 K	1968GIA
Poly(benzyl methacrylate)				
	carbon dioxide	$\Delta_{sol}H_{A(vap)}^\infty$	298-348 K	1996WAN
Polybutadiene				
	benzene	$\Delta_M H$	298 K	1976POU
	benzene	$\Delta_M H_A$	300 K	1958JES
	cyclohexane	$\Delta_M H$	298 K	1976POU
	trichloromethane	$\Delta_M H_A$	273-298 K	1964BO2

Polymer (B)	Solvent (A)	Enthalpy	T-range	Ref.
Poly(butadiene-*co*-methyl methacrylate)				
	argon	$\Delta_{sol}H_{A(vap)}{}^{\infty}$	293-343 K	1939BAR
	hydrogen	$\Delta_{sol}H_{A(vap)}{}^{\infty}$	293-343 K	1939BAR
	nitrogen	$\Delta_{sol}H_{A(vap)}{}^{\infty}$	293-343 K	1939BAR
	oxygen	$\Delta_{sol}H_{A(vap)}{}^{\infty}$	293-343 K	1939BAR
Poly(butadiene-*co*-styrene)				
	argon	$\Delta_{sol}H_{A(vap)}{}^{\infty}$	293-343 K	1939BAR
	argon	$\Delta_{sol}H_{A(vap)}{}^{\infty}$	298 K	1960TIK
	benzene	$\Delta_M H_A$	293 K	1955TA2
	hydrogen	$\Delta_{sol}H_{A(vap)}{}^{\infty}$	293-343 K	1939BAR
	nitrogen	$\Delta_{sol}H_{A(vap)}{}^{\infty}$	293-343 K	1939BAR
Poly(butyl methacrylate)				
	butyl acetate	$\Delta_{sol}H_B{}^{\infty}$	273-343 K	1956MEE
	ethanol	$\Delta_M H$	313-354 K	1987NUN
	2-propanol	ΔH	274-332 K	1963BLA
	2-propanol	$\Delta_M H$	300-359 K	1987NUN
	trichloromethane	$\Delta_M H$	298 K	1972MAS
Poly(ε-caprolactone)				
	1,4-dioxane	$\Delta_{dil}H^{12}$	298 K	1973MAN
Polycarbonate bisphenol-A				
	carbon dioxide	$\Delta_{sol}H_{A(vap)}{}^{\infty}$	293-313 K	1987SAD
	carbon dioxide	$\Delta_{sol}H_{A(vap)}{}^{\infty}$	308 K	1995BAN
	N,N-dimethylformamide	$\Delta_{dil}H^{12}, \Delta_M H$	298 K	2002SA3
	methane	$\Delta_{sol}H_{A(vap)}{}^{\infty}$	293-313 K	1987SAD
	1-propanol	$\Delta_M H$	298 K	1987JAN
	1,1,2,2-tetrachloroethane	$\Delta_M H$	298 K	1987JAN
	trichloromethane	$\Delta_{dil}H^{12}, \Delta_M H$	298 K	2002SA3
Poly(3,3-bis-chloromethyl-oxycyclobutane)				
	bromobenzene	$\Delta_{sol}H$	423 K	1966HEL
	chlorobenzene	$\Delta_{sol}H$	423 K	1966HEL
	cyclohexanone	$\Delta_{sol}H$	423 K	1966HEL
	1,4-dimethylbenzene	$\Delta_{sol}H$	423 K	1966HEL
	1,3,5-trimethylbenzene	$\Delta_{sol}H$	423 K	1966HEL
	pentyl acetate	$\Delta_{sol}H$	423 K	1966HEL
	tetrachloromethane	$\Delta_{sol}H$	423 K	1966HEL

Polymer (B)	Solvent (A)	Enthalpy	T-range	Ref.
Polychloroprene				
	argon	$\Delta_{sol}H_{A(vap)}^{\infty}$	273-373 K	1939BAR
	helium	$\Delta_{sol}H_{A(vap)}^{\infty}$	273-373 K	1939BAR
	hydrogen	$\Delta_{sol}H_{A(vap)}^{\infty}$	293-343 K	1939BAR
	nitrogen	$\Delta_{sol}H_{A(vap)}^{\infty}$	293-343 K	1939BAR
	oxygen	$\Delta_{sol}H_{A(vap)}^{\infty}$	293-343 K	1939BAR
Poly(4-chlorostyrene)				
	tert-butyl acetate	$\Delta_{M}H_{A}$	303 K	1972IZU
	2-(2-butoxyethoxy)-ethanol	$\Delta_{M}H_{A}$	303 K	1972IZU
	chlorobenzene	$\Delta_{M}H_{A}$	303-333 K	1966TAK
	ethylbenzene	$\Delta_{M}H_{A}$	303 K	1972IZU
	2-(2-ethoxyethoxy)-ethanol	$\Delta_{M}H_{A}$	303 K	1972IZU
	ethyl chloroacetate	$\Delta_{M}H_{A}$	303 K	1972IZU
	isopropyl acetate	$\Delta_{M}H_{A}$	303 K	1972IZU
	isopropylbenzene	$\Delta_{M}H_{A}$	303 K	1972IZU
	isopropyl chloroacetate	$\Delta_{M}H_{A}$	303 K	1972IZU
	methyl chloroacetate	$\Delta_{M}H_{A}$	303 K	1972IZU
	tetrachloroethene	$\Delta_{M}H_{A}$	303 K	1972IZU
	tetrachloromethane	$\Delta_{M}H_{A}$	303 K	1972IZU
	toluene	$\Delta_{M}H$	298-353 K	1977TA3
	toluene	$\Delta_{M}H_{A}$	303-333 K	1966TAK
Poly(cyclopentene-*co*-maleic acid)				
	water	$\Delta_{dil}H^{12}$	298 K	1981PAO
Poly(*N,N*-dimethylacrylamide)				
	water	$\Delta_{dil}H^{12}$	298 K	1981DAY
Poly(2,6-dimethyl-1,4-phenylene oxide)				
	1,2-dichlorobenzene	$\Delta_{sol}H_{B}$	298-423 K	1974FI1
Poly(dimethylsilmethylene)				
	carbon dioxide	$\Delta_{sol}H_{A(vap)}^{\infty}$	283-328 K	1993SHA
	methane	$\Delta_{sol}H_{A(vap)}^{\infty}$	283-328 K	1993SHA
	propane	$\Delta_{sol}H_{A(vap)}^{\infty}$	283-328 K	1993SHA

Polymer (B)	Solvent (A)	Enthalpy	T-range	Ref.
Poly(dimethylsiloxane)				
	benzene	$\Delta_M H$		1950NEW
	benzene	$\Delta_M H_A$	270 K	1961KUW
	benzene/chlorobenzene	$\Delta_M H_B^\infty$	298 K	1964DEL
	benzene/cyclohexane	$\Delta_M H_B^\infty$	298 K	1964DEL
	benzene/n-heptane	$\Delta_M H_B^\infty$	298 K	1964DEL
	benzene/tetrachloro-methane	$\Delta_M H_B^\infty$	298 K	1964DEL
	benzene	$\Delta_M H$		1970MO2
	benzene	$\Delta_{sol} H_B^\infty$	298 K	1970MO3
	benzene	$\Delta_M H_A$	306 K	1970MUR
	benzene	$\Delta_M H_A$	281-333 K	1968SEE
	benzene	$\Delta_M H_A$	298 K	1972FLO
	benzene	$\Delta_M H$	298 K	1976POU
	benzene	$\Delta_M H_B^\infty$		1984PAN
	benzene	$\Delta_{sol} H_{A(vap)}^\infty$	323-423 K	2005KLO
	methane	$\Delta_M H_B^\infty$	298 K	1964DEL
	n-butane	$\Delta_M H_A$	303-343K	1962BAR
	n-butane	$\Delta_{sol} H_A$	303-343K	1962BAR
	2-butanone	$\Delta_M H_B^\infty$		1984PAN
	2-butanone	$\Delta_M H_A$	281-333 K	1968SEE
	2-butanone	$\Delta_M H, \Delta_M H_A$	293 K	1996VSH
	2-butanone	$\Delta_M H_B^\infty$	298 K	1946SCO
	2-butanone	$\Delta_{sol} H_B^\infty$	298 K	1970MO3
	2-butanone	$\Delta_M H_A$	303 K	1962MUR
	butyl acetate	$\Delta_M H$		1970MO2
	carbon dioxide	$\Delta_{sol} H_{A(vap)}^\infty$	283-328 K	1986SH2
	carbon dioxide	$\Delta_{sol} H_{A(vap)}^\infty$	308 K	1995BAN
	chlorobenzene	$\Delta_M H_A$	298 K	1972FLO
	chlorobenzene	$\Delta_M H$	298 K	1976POU
	chlorobenzene	$\Delta_M H_B^\infty$		1984PAN
	chlorobenzene	$\Delta_{sol} H_{A(vap)}^\infty$	323-423 K	2005KLO
	cumene	$\Delta_{sol} H_B^\infty$	298 K	1970MO3
	cyclohexane	$\Delta_M H_B^\infty$		1984PAN
	cyclohexane/tetra-chloromethane	$\Delta_M H_B^\infty$	298 K	1964DEL
	cyclohexane	$\Delta_{sol} H_B^\infty$	298 K	1970MO3
	cyclohexane	$\Delta_M H_A$	298 K	1972FLO
	cyclohexane	$\Delta_M H$	298 K	1976POU
	n-decane	$\Delta_{sol} H_B^\infty$	298 K	1970MO3
	n-decane	$\Delta_M H_B^\infty$	303 K-T_b	1971DRE
	n-decane	$\Delta_{sol} H_{A(vap)}^\infty$	323-423 K	2005KLO
	1,2-dichlorobenzene	$\Delta_{sol} H_{A(vap)}^\infty$	323-423 K	2005KLO
	1,3-dichlorobenzene	$\Delta_{sol} H_{A(vap)}^\infty$	323-423 K	2005KLO
	diethyl ether	$\Delta_M H$		1970MO2

Polymer (B)	Solvent (A)	Enthalpy	T-range	Ref.
Poly(dimethylsiloxane) (*continued*)				
	diethyl ether	$\Delta_{sol}H_B^{\infty}$	298 K	1970MO3
	1,2-dimethylbenzene	$\Delta_{sol}H_B^{\infty}$	298 K	1970MO3
	1,2-dimethylbenzene	$\Delta_{sol}H_{A(vap)}^{\infty}$	323-423 K	2005KLO
	1,3-dimethylbenzene	$\Delta_M H$		1970MO2
	1,3-dimethylbenzene	$\Delta_{sol}H_B^{\infty}$	298 K	1970MO3
	1,4-dimethylbenzene	$\Delta_M H_A$	281-333 K	1968SEE
	1,4-dimethylbenzene	$\Delta_{sol}H_B^{\infty}$	298 K	1970MO3
	1,4-dimethylbenzene	$\Delta_{sol}H_{A(vap)}^{\infty}$	323-423 K	2005KLO
	2,2-dimethylpropane	$\Delta_M H_A$	303-343K	1962BAR
	2,2-dimethylpropane	$\Delta_{sol}H_A$	303-343K	1962BAR
	dipropyl ether	$\Delta_{sol}H_B^{\infty}$	298 K	1970MO3
	n-dodecane	$\Delta_{sol}H_B^{\infty}$	298 K	1970MO3
	ethylbenzene	$\Delta_M H$		1970MO2
	ethylbenzene	$\Delta_{sol}H_B^{\infty}$	298 K	1970MO3
	ethylbenzene	$\Delta_M H_A$	306 K	1970MUR
	ethylbenzene	$\Delta_{sol}H_{A(vap)}^{\infty}$	323-423 K	2005KLO
	n-heptane	$\Delta_M H$		1970MO2
	n-heptane	$\Delta_{sol}H_B^{\infty}$	298 K	1970MO3
	n-heptane	$\Delta_M H$	298 K	1976POU
	n-heptane	$\Delta_M H_B^{\infty}$	303 K-T_b	1971DRE
	n-heptane	$\Delta_M H_A$	306 K	1970MUR
	n-heptane	$\Delta_{sol}H_{A(vap)}^{\infty}$	323-423 K	2005KLO
	n-hexadecane	$\Delta_M H_B^{\infty}$	303 K-T_b	1971DRE
	hexamethyldisiloxane	$\Delta_M H$	298 K	1968PAT
	n-hexane	$\Delta_{sol}H_B^{\infty}$	298 K	1970MO3
	n-hexane	$\Delta_M H_B^{\infty}$	303 K-T_b	1971DRE
	n-hexane	$\Delta_M H_A$	306 K	1970MUR
	n-hexane	$\Delta_{sol}H_{A(vap)}^{\infty}$	323-423 K	2005KLO
	1-hexanol	$\Delta_M H_A$	281-333 K	1968SEE
	methane	$\Delta_{sol}H_{A(vap)}^{\infty}$	283-328 K	1986SH2
	methanol	$\Delta_M H_A$	281-333 K	1968SEE
	methylcyclohexane/			
	tetrachloromethane	$\Delta_M H_B^{\infty}$	298 K	1964DEL
	methylcyclohexane	$\Delta_{sol}H_B^{\infty}$	298 K	1970MO3
	methylcyclohexane	$\Delta_M H$	298 K	1976POU
	2-methylpropane	$\Delta_M H_A$	303-343K	1962BAR
	2-methylpropane	$\Delta_{sol}H_A$	303-343K	1962BAR
	n-nonane	$\Delta_{sol}H_B^{\infty}$	298 K	1970MO3
	octamethylcyclotetra-			
	siloxane	$\Delta_M H_A^{\infty}$	298 K	1946SCO
	octamethyltrisiloxane	$\Delta_M H_A$	298 K	1968PAT
	n-octane	$\Delta_{sol}H_B^{\infty}$	298 K	1970MO3
	n-octane	$\Delta_M H_B^{\infty}$	303 K-T_b	1971DRE
	n-octane	$\Delta_M H_A$	306 K	1970MUR

Polymer (B)	Solvent (A)	Enthalpy	T-range	Ref.
Poly(dimethylsiloxane) (*continued*)				
	n-octane	$\Delta_{sol}H_{A(vap)}^{\infty}$	323-423 K	2005KLO
	octafluoropropane	$\Delta_{sol}H_{A(vap)}^{\infty}$	298-328 K	2005PRA
	oligodimethylsiloxane	$\Delta_{M}H_{B}^{\infty}$	303-363 K	1971DRE
	n-pentane	$\Delta_{sol}H_{B}^{\infty}$	298 K	1970MO3
	n-pentane	$\Delta_{M}H_{A}$	303-343K	1962BAR
	n-pentane	$\Delta_{sol}H_{A}$	303-343K	1962BAR
	n-pentane	$\Delta_{M}H_{B}^{\infty}$	303 K-T_{b}	1971DRE
	n-pentane	$\Delta_{sol}H_{A(vap)}^{\infty}$	323-423 K	2005KLO
	2-pentanone	$\Delta_{M}H_{A}$	281-333 K	1968SEE
	propane	$\Delta_{sol}H_{A(vap)}^{\infty}$	283-328 K	1986SH2
	propane	$\Delta_{sol}H_{A(vap)}^{\infty}$	298-328 K	2005PRA
	1-propanol	$\Delta_{M}H_{A}$	281-333 K	1968SEE
	2-propanone	$\Delta_{M}H_{A}$	281-333 K	1968SEE
	tetrachloromethane	$\Delta_{M}H$	298 K	1976POU
	n-tetradecane	$\Delta_{sol}H_{B}^{\infty}$	298 K	1970MO3
	toluene	$\Delta_{M}H_{A}$	281-333 K	1968SEE
	toluene	$\Delta_{sol}H_{B}^{\infty}$	298 K	1970MO3
	toluene	$\Delta_{M}H_{A}$	306 K	1970MUR
	toluene	$\Delta_{sol}H_{A(vap)}^{\infty}$	323-423 K	2005KLO
	n-tridecane	$\Delta_{M}H$		1970MO2
	n-tridecane	$\Delta_{sol}H_{B}^{\infty}$	298 K	1970MO3
	1,3,5-trimethylbenzene	$\Delta_{sol}H_{B}^{\infty}$	298 K	1970MO3
	n-undecane	$\Delta_{sol}H_{B}^{\infty}$	298 K	1970MO3
	n-undecane	$\Delta_{sol}H_{A(vap)}^{\infty}$	323-423 K	2005KLO
Poly(3-dodecylthiophene)				
	1,2-dimethylbenzene	$\Delta_{M}H$		2004MAL
	1,3-dimethylbenzene	$\Delta_{M}H$		2004MAL
	1,4-dimethylbenzene	$\Delta_{M}H$		2004MAL
Polyester-siloxane block copolymer				
	carbon dioxide	$\Delta_{sol}H_{A(vap)}$		1988KAR
Poly(ethacrylic acid) salts				
	water	$\Delta_{dil}H^{12}$	298 K	1980DAO
Poly(ether urethane)				
	ethanol	$\Delta_{sol}H_{A(vap)}$	298 K	2004SMI
	water	$\Delta_{sol}H_{A(vap)}$	298 K	2004SMI

Polymer (B)	Solvent (A)	Enthalpy	T-range	Ref.
Poly(ethyl acrylate)				
	1-butanol	$\Delta_M H_A$	298 K	1967LLO
	ethanol	$\Delta_M H_A$	298 K	1967LLO
	methanol	$\Delta_M H_A$	298 K	1967LLO
	1-propanol	$\Delta_M H_A$	298 K	1967LLO
Poly(ethyl acrylate-*co*-tetraethylene glycol dimethacrylate)				
	argon	$\Delta_{sol} H_{A(vap)}{}^{\infty}$	333 K	1968BAR
	carbon dioxide	$\Delta_{sol} H_{A(vap)}{}^{\infty}$	333 K	1968BAR
	hydrogen	$\Delta_{sol} H_{A(vap)}{}^{\infty}$	333 K	1968BAR
	krypton	$\Delta_{sol} H_{A(vap)}{}^{\infty}$	333 K	1968BAR
	methane	$\Delta_{sol} H_{A(vap)}{}^{\infty}$	333 K	1968BAR
	neon	$\Delta_{sol} H_{A(vap)}{}^{\infty}$	333 K	1968BAR
	nitrogen	$\Delta_{sol} H_{A(vap)}{}^{\infty}$	333 K	1968BAR
	oxygen	$\Delta_{sol} H_{A(vap)}{}^{\infty}$	333 K	1968BAR
Polyethylene				
	argon	$\Delta_{sol} H_{A(vap)}{}^{\infty}$	298-368 K	1960TIK
	argon	$\Delta_{sol} H_{A(vap)}{}^{\infty}$	278-328 K	1961MIC
	argon	$\Delta_{sol} H_{A(vap)}{}^{\infty}$		1970ASH
	benzyl phenyl ether	$\Delta_M H_A$	393-473 K	1966NAK
	carbon dioxide	$\Delta_{sol} H_{A(vap)}{}^{\infty}$	278-328 K	1961MIC
	carbon dioxide	$\Delta_{sol} H_{A(vap)}{}^{\infty}$		1970ASH
	carbon dioxide	$\Delta_{sol} H_{A(vap)}{}^{\infty}$		2002ARE
	carbon monoxide	$\Delta_{sol} H_{A(vap)}{}^{\infty}$	278-328 K	1961MIC
	1-decanol	$\Delta_M H_A$	393-473 K	1966NAK
	dichlorodifluoromethane	$\Delta_{sol} H_{A(vap)}{}^{\infty}$	293-453 K	1975HOR
	dichlorodifluoromethane	$\Delta_M H_A{}^{\infty}$	293-453 K	1975HOR
	dichlorotetrafluoroethane	$\Delta_{sol} H_{A(vap)}{}^{\infty}$	293-453 K	1975HOR
	dichlorotetrafluoroethane	$\Delta_M H_A{}^{\infty}$	293-453 K	1975HOR
	diphenyl	$\Delta_M H_A$	393-473 K	1966NAK
	diphenyl ether	$\Delta_M H_A$	393-473 K	1966NAK
	diphenylmethane	$\Delta_M H_A$	393-473 K	1966NAK
	ethane	$\Delta_{sol} H_{A(vap)}{}^{\infty}$	278-328 K	1961MIC
	fluorotrichloromethane	$\Delta_{sol} H_{A(vap)}{}^{\infty}$	293-453 K	1975HOR
	fluorotrichloromethane	$\Delta_M H_A{}^{\infty}$	293-453 K	1975HOR
	helium	$\Delta_{sol} H_{A(vap)}{}^{\infty}$	298-368 K	1960TIK
	helium	$\Delta_{sol} H_{A(vap)}{}^{\infty}$	278-328 K	1961MIC
	helium	$\Delta_{sol} H_{A(vap)}{}^{\infty}$		1970ASH
	n-heptane	$\Delta_{sol} H_{A(vap)}{}^{\infty}$	293-453 K	1975HOR
	n-heptane	$\Delta_M H_A{}^{\infty}$	293-453 K	1975HOR
	n-hexane	$\Delta_{sol} H_{A(vap)}{}^{\infty}$	293-453 K	1975HOR
	n-hexane	$\Delta_M H_A{}^{\infty}$	293-453 K	1975HOR

Polymer (B)	Solvent (A)	Enthalpy	T-range	Ref.
Polyethylene (*continued*)				
	n-hexane	$\Delta_{sol}H_{A(vap)}^{\infty}$	273-313 K	1987CAS
	hydrogen	$\Delta_{sol}H_{A(vap)}^{\infty}$		1970ASH
	methane	$\Delta_{sol}H_{A(vap)}^{\infty}$	278-328 K	1961MIC
	methoxybenzene	$\Delta_{M}H_{A}$	393-473 K	1966NAK
	2-methylpropane	$\Delta_{sol}H_{A(vap)}^{\infty}$	293-453 K	1975HOR
	2-methylpropane	$\Delta_{M}H_{A}^{\infty}$	293-453 K	1975HOR
	neon	$\Delta_{sol}H_{A(vap)}^{\infty}$		1970ASH
	nitrogen	$\Delta_{sol}H_{A(vap)}^{\infty}$	278-328 K	1961MIC
	nitrogen	$\Delta_{sol}H_{A(vap)}^{\infty}$		1970ASH
	4-nonylphenol	$\Delta_{M}H_{A}$	393-473 K	1966NAK
	n-octane	$\Delta_{sol}H_{A(vap)}^{\infty}$	293-453 K	1975HOR
	n-octane	$\Delta_{M}H_{A}^{\infty}$	293-453 K	1975HOR
	1-octanol	$\Delta_{M}H_{A}$	393-473 K	1966NAK
	4-octylphenol	$\Delta_{M}H_{A}$	393-473 K	1966NAK
	oxygen	$\Delta_{sol}H_{A(vap)}^{\infty}$	278-328 K	1961MIC
	n-pentane	$\Delta_{sol}H_{A(vap)}^{\infty}$	293-453 K	1975HOR
	n-pentane	$\Delta_{M}H_{A}^{\infty}$	293-453 K	1975HOR
	n-pentane	$\Delta_{sol}H_{A(vap)}^{\infty}$	263-308 K	1987CAS
	propane	$\Delta_{sol}H_{A(vap)}^{\infty}$	278-328 K	1961MIC
	propene	$\Delta_{sol}H_{A(vap)}^{\infty}$	278-328 K	1961MIC
	propyne	$\Delta_{sol}H_{A(vap)}^{\infty}$	278-328 K	1961MIC
	sulfur hexafluoride	$\Delta_{sol}H_{A(vap)}^{\infty}$	278-328 K	1961MIC
	trichlorotrifluoroethane	$\Delta_{sol}H_{A(vap)}^{\infty}$	293-453 K	1975HOR
	trichlorotrifluoroethane	$\Delta_{M}H_{A}^{\infty}$	293-453 K	1975HOR
Poly(ethylene-*co*-ethyl acrylate)				
	carbon dioxide	$\Delta_{sol}H_{A(vap)}^{\infty}$	423-473 K	2004ARE
Poly(ethylene-*co*-maleic acid)				
	water	$\Delta_{dil}H^{12}$	298 K	1981PAO
Poly(ethylene-*co*-propylene)				
	benzene	$\Delta_{sol}H_{A(vap)}^{\infty}$	283-313 K	1964FRE
	benzene	$\Delta_{M}H_{A}$	298-373 K	1964HOL
	chlorobenzene	$\Delta_{M}H_{A}$	298-373 K	1964HOL
	cyclohexane	$\Delta_{sol}H_{A(vap)}^{\infty}$	283-313 K	1964FRE
	cyclohexane	$\Delta_{M}H_{A}$	298-373 K	1964HOL
	dichloromethane	$\Delta_{sol}H_{A(vap)}^{\infty}$	283-313 K	1964FRE
	ethane	$\Delta_{sol}H_{A(vap)}^{\infty}$	287-363 K	2001TSU
	ethene	$\Delta_{sol}H_{A(vap)}^{\infty}$	287-363 K	2001TSU
	n-heptane	$\Delta_{M}H_{A}$	298-373 K	1964HOL
	n-hexane	$\Delta_{sol}H_{A(vap)}^{\infty}$	283-313 K	1964FRE
	n-pentane	$\Delta_{sol}H_{A(vap)}^{\infty}$	283-313 K	1964FRE

Polymer (B)	Solvent (A)	Enthalpy	T-range	Ref.
Poly(ethylene-*co*-propylene) (*continued*)	propane	$\Delta_{sol}H_{A(vap)}{}^\infty$	287-363 K	2001TSU
	propene	$\Delta_{sol}H_{A(vap)}{}^\infty$	287-363 K	2001TSU
	tetrachloromethane	$\Delta_{sol}H_{A(vap)}{}^\infty$	283-313 K	1964FRE
	tetrachloromethane	$\Delta_M H_A$	298-373 K	1964HOL
	1,1,2-trichloro-trifluoroethane	$\Delta_{sol}H_{A(vap)}{}^\infty$	283-313 K	1964FRE
Poly(ethylene-*co*-vinyl acetate)				
	n-dodecane	$\Delta_M H$	383-433 K	1981DRU
	n-octadecane	$\Delta_M H$	383-433 K	1981DRU
Poly(ethylene adipate)				
	1,4-dioxane	$\Delta_M H$		1965TAG
	1,4-dioxane	$\Delta_{dil}H^{12}$	298 K	1975MAN
Poly(ethylene glycol)				
	benzene	$\Delta_M H$	328 K	1976LA2
	butanol	$\Delta_M H$	298-313 K	1982RAE
	chlorobenzene	$\Delta_{sol}H_B$	298 K	2002SP1
	dichloromethane	$\Delta_{sol}H_B$	298 K	2002SP1
	1,4-dioxane/water	$\Delta_M H$	328 K	1976LA2
	ethanol	$\Delta_M H$	298-313 K	1982RAE
	ethanol	$\Delta_M H$	298-313 K	2005BIG
	ethanol/water	$\Delta_M H$	328 K	1976LA2
	1-propanol	$\Delta_M H$	298-313 K	1982RAE
	tetrachloromethane	$\Delta_M H_A$	303-333 K	1993ZEL
	trichloromethane	$\Delta_M H_A$	303 K	1993ZEL
	trichloromethane	$\Delta_{sol}H_B$	298 K	2002SP1
	water	$\Delta_M H_A{}^\infty$	324 K	1965SAT
	water	$\Delta_M H_A$	298-308 K	1968LA1
	water	$\Delta_M H$	328 K	1976LA2
	water	$\Delta_M H$	298 K	1981MED
	water	$\Delta_M H$	298 K	1982MED
	water	$\Delta_M H, \Delta_{sol}H_B$	278-317 K	1989AND
	water	$\Delta_{dil}H^{12}$	298 K	1990DAO
	water	$\Delta_{dil}H^{12}$	298 K	1993AUC
	water	$\Delta_{dil}H^{12}$	298 K	1994AUC
	water	$\Delta_M H_A$	278-308 K	1994NAG
	water	$\Delta_{sol}H_B$	298 K	1995GRO
	water	$\Delta_{sol}H_B$	298 K	2002SP1
	water	$\Delta_M H$	298 K	2003JAB

Polymer (B)	Solvent (A)	Enthalpy	T-range	Ref.
Poly(ethylene glycol) dibutyl ether				
	n-dodecane	$\Delta_M H$	304 K	1985ALK
	n-hexadecane	$\Delta_M H$	304 K	1985ALK
	n-tetradecane	$\Delta_M H$	304 K	1985ALK
Poly(ethylene glycol) didodecyl ether				
	water	$\Delta_{dil} H^{12}$	293 K	1994PER
Poly(ethylene glycol) dimethyl ether				
	n-decane	$\Delta_M H$	304 K	1983ALK
	n-dodecane	$\Delta_M H$	304 K	1983ALK
	n-hexadecane	$\Delta_M H$	304 K	1983ALK
	tetrachloromethane	$\Delta_M H_A$	303-333 K	1993ZEL
	trichloromethane	$\Delta_M H_A$	303 K	1993ZEL
	n-tetradecane	$\Delta_M H$	304 K	1983ALK
	n-undecane	$\Delta_M H$	304 K	1983ALK
Poly(ethylene glycol) mono-methyl ether				
	tetrachloromethane	$\Delta_M H_A$	303-318 K	1993ZEL
	trichloromethane	$\Delta_M H_A$	303 K	1993ZEL
Poly(ethylene imine) hydrochloride				
	water	$\Delta_{dil} H^{12}$	298 K	1975MIT
Poly(ethylene oxide)				
	benzene	$\Delta_{sol} H_B^{\infty}$	298 K	1975IKE
	benzene	$\Delta_{dil} H^{12}$	303 K	1979MOR
	benzene	$\Delta_M H_A$	333 K	1971BO1
	cyclohexanone	$\Delta_{sol} H_B$	298 K	1993BRU
	dideuterium oxide	$\Delta_M H$	264 K	1991HEY
	formamide	$\Delta_{sol} H_B$	298-313 K	1991OLO
	ethanol	$\Delta_M H, \Delta_{dil} H^{12}$	298 K	1984DAO
	toluene	$\Delta_{dil} H^{12}$	303 K	1979MOR
	toluene	$\Delta_M H_A$	333 K	1981ANS
	toluene	$\Delta_{sol} H_B$	298 K	1993BRU
	toluene	$\Delta_M H, \Delta_{dil} H^{12}$	298 K	1999SAF
	trichloromethane	$\Delta_M H, \Delta_{dil} H^{12}$	298 K	1999SAF
	trichloromethane	$\Delta_{sol} H_B^{\infty}$	298 K	1975IKE
	water	$\Delta_M H$	298 K	1983TA2

Polymer (B)	Solvent (A)	Enthalpy	T-range	Ref.
Poly(ethylene oxide) (*continued*)				
	water	$\Delta_M H$, $\Delta_{dil} H^{12}$	298 K	1984DAO
	water	$\Delta_M H$	263 K	1987BOG
	water	$\Delta_M H$	264 K	1991HEY
	water	$\Delta_{dil} H^{12}$	288-293 K	1993EAG
	water	$\Delta_M H$, $\Delta_{dil} H^{12}$	298 K	1999SAF
	water/sodium dodecylsulfate	ΔH		2004SIL
Poly(ethylene oxide)-b-poly (propylene oxide)-b-poly (ethylene oxide) triblock copolymer				
	water	$\Delta_M H$	298 K	2002THU
	water/surfactant	ΔH		2004DEL
	water/surfactant	ΔH		2004JAN
Poly(ethylene oxide-*co*-tetra-hydrofuran)				
	2-propanol	$\Delta_M H$	303 K	1998STU
	tetrachloromethane	$\Delta_M H$	303 K	1998STU
	trichloromethane	$\Delta_M H$	303 K	1998STU
Poly(ethylene oxide)-b-poly(tetra-hydrofuran) diblock copolymer				
	2-propanol	$\Delta_M H$	303 K	1999STU
	tetrachloromethane	$\Delta_M H$	303 K	1999STU
	trichloromethane	$\Delta_M H$	303 K	1999STU
Poly(ethylene sebacate)				
	1,4-dioxane	$\Delta_M H$		1965TAG
	furfurol	$\Delta_{sol} H_B$	318 K	1959KOZ
	3-methylphenol	$\Delta_{sol} H_B$	318 K	1959KOZ
Poly(ethylene succinate)				
	1,4-dioxane	$\Delta_M H$		1965TAG
Poly(ethylenesulfonic acid) sodium salt				
	water	$\Delta_{dil} H^{12}$	298 K	1976BO1

Polymer (B)	Solvent (A)	Enthalpy	T-range	Ref.
Poly(ethylene terephthalate)				
	furfurol	$\Delta_{sol}H_B$	318 K	1959KOZ
	hydrogen	$\Delta_{sol}H_{A(vap)}^{\infty}$	293-373 K	1962DRA
	3-methylphenol	$\Delta_{sol}H_B$	318 K	1959KOZ
Poly(γ-ethyl-L-glutamate)				
	1,2-dichloroethane	$\Delta_{sol}H_B$	303 K	1968GIA
	dichloroacetic acid	$\Delta_{sol}H_B$	303 K	1968GIA
Poly(ethyl methacrylate)				
	carbon dioxide	$\Delta_{sol}H_{A(vap)}^{\infty}$	303-328 K	1981KOR
	carbon dioxide	$\Delta_{sol}H_{A(vap)}^{\infty}$	288-318 K	1989CHI
	methane	$\Delta_{sol}H_{A(vap)}^{\infty}$	298-318 K	1989CHI
	propane	$\Delta_{sol}H_{A(vap)}^{\infty}$		1987GOR
Poly(L-glutamic acid)				
	water	$\Delta_{dil}H^{12}$	298 K	2005GOD
Poly(hexamethylene adipate)				
	1,4-dioxane	$\Delta_{dil}H^{12}$	298 K	1975MAN
Poly(2-(2-hydroxyethoxy)ethyl methacrylate)				
	water	$\Delta_M H$	298 K	1982POU
Poly(2-hydroxyethyl methacrylate)				
	water	$\Delta_M H$	298 K	1982POU
Poly(9,9-bis(4-hydroxyphenyl)-fluorene-*co*-terephthaloyl chloride)				
	1,1,2,2-tetrachloroethane	$\Delta_M H$	298K	1977TA2
	trichloromethane	$\Delta_M H$	298K	1977TA2
Poly(4-hydroxystyrene)				
	2-propanone	$\Delta_M H$	298-323 K	1994COM
Polyimide				
	carbon dioxide	$\Delta_{sol}H_{A(vap)}^{\infty}$	293-313 K	1988SAD
Polyionene				
	water	$\Delta_{sol}H_B$, $\Delta_{dil}H^{12}$	286-343 K	1991BUR

Polymer (B)	Solvent (A)	Enthalpy	T-range	Ref.
Polyisobutylene				
	n-alkanes	$\Delta_M H$	298-423 K	1968FL1
	n-alkanes	$\Delta_{sol} H_B, \Delta_{dil} H^{12}$		1968MA3
	benzene	$\Delta_M H_A$	298 K	1946GEE
	benzene	$\Delta_M H_A$	298-338 K	1956BA1
	benzene	$\Delta_M H_A{}^\infty$	298-308 K	1957SIL
	benzene	$\Delta_M H_A$	298 K	1961DAO
	benzene	$\Delta_M H_A$	298 K	1961KUW
	benzene	$\Delta_{dil} H^{12}$	298 K	1962SEN
	benzene	$\Delta_M H_A$	298-313 K	1964BO2
	benzene	$\Delta_{sol} H_B, \Delta_M H$	298 K	1968MA1
	benzene	$\Delta_M H_A$	298 K	1968TAG
	benzene	$\Delta_M H_B{}^\infty$	298 K	1973NEK
	benzene	$\Delta_M H$	298 K	1976POU
	benzene	$\Delta_M H$	298 K	1977HUA
	benzene/cyclohexane	$\Delta_M H_B{}^\infty$	298 K	1964DEL
	benzene/n-heptane	$\Delta_M H_B{}^\infty$	298 K	1964DEL
	n-butane	$\Delta_M H_A$	428-463 K	1983ALI
	chlorobenzene	$\Delta_M H_A$	298 K	1961DAO
	chlorobenzene	$\Delta_{dil} H^{12}$	298 K	1962SEN
	chlorobenzene	$\Delta_{sol} H_B$	298 K	1968MA1
	chlorobenzene	$\Delta_M H_B{}^\infty$	298 K	1973NEK
	chlorobenzene	$\Delta_M H$	298 K	1976POU
	chlorobenzene/n-heptane	$\Delta_M H_B{}^\infty$	298 K	1964DEL
	cyclohexane	$\Delta_{dil} H^{12}$	298 K	1973GAE
	cyclohexane	$\Delta_M H_B{}^\infty$	298 K	1973NEK
	cyclohexane	$\Delta_M H$	298 K	1976POU
	cyclohexane	$\Delta_M H_B{}^\infty$		1984PAN
	n-decane	$\Delta_M H_B{}^\infty$	298 K	1968FL2
	n-decane	$\Delta_M H_B{}^\infty$	303 K-T_b	1971DRE
	n-decane	$\Delta_M H_B{}^\infty$	298 K	1973NEK
	n-dodecane/n-hexane	$\Delta_M H_B{}^\infty$	298 K	1964DEL
	ethylbenzene	$\Delta_{sol} H_B$	298 K	1968MA1
	n-heptane	$\Delta_M H_B{}^\infty$	298 K	1968FL2
	n-heptane	$\Delta_M H_A$	298 K	1968TAG
	n-heptane	$\Delta_{sol} H_B$	298 K	1968MA1
	n-heptane	$\Delta_M H_B{}^\infty$	303 K-T_b	1971DRE
	n-heptane	$\Delta_M H$	298 K	1976POU
	n-hexadecane	$\Delta_M H_B{}^\infty$	298 K	1968FL2
	n-hexadecane	$\Delta_M H_B{}^\infty$	303 K-T_b	1971DRE
	n-hexane	$\Delta_M H_B{}^\infty$	298 K	1968FL2
	n-hexane	$\Delta_M H_B{}^\infty$	303 K-T_b	1971DRE
	n-hexane	$\Delta_M H_B{}^\infty$	298 K	1973NEK
	n-octane	$\Delta_M H_B{}^\infty$	298 K	1968FL2
	n-octane	$\Delta_M H_B{}^\infty$	303 K-T_b	1971DRE

Polymer (B)	Solvent (A)	Enthalpy	T-range	Ref.
Polyisobutylene (*continued*)				
	n-octane	$\Delta_M H_B^\infty$	298 K	1973NEK
	oligodimethylsiloxane	$\Delta_M H_B^\infty$	303-367 K	1971DRE
	n-pentane	$\Delta_M H_A$	298-328 K	1962BAK
	n-pentane	$\Delta_M H_B^\infty$	298 K	1968FL2
	n-pentane	$\Delta_M H_B^\infty$	303 K-T_b	1971DRE
	n-pentane	$\Delta_M H_A$	343 K	1972LEC
	n-pentane	$\Delta_{dil} H^{12}$	298 K	1973GAE
	n-pentane	$\Delta_M H_B^\infty$	298 K	1973NEK
	tetrachloromethane	$\Delta_{dil} H^{12}, \Delta_M H$	298 K	1963TAG
	toluene	$\Delta_M H_A$	298 K	1961DAO
	toluene	$\Delta_{dil} H^{12}$	298 K	1962SEN
	toluene	$\Delta_{dil} H^{12}, \Delta_M H$	298 K	1963TAG
	toluene	$\Delta_{sol} H_B$	298 K	1968MA1
	2,2,4-trimethylpentane	$\Delta_M H_A$	297 K	1948TAG
	2,2,4-trimethylpentane	$\Delta_M H$	298 K	1968TAG
	2,2,4-trimethylpentane	$\Delta_M H_A$	298 K	1972LEC
Poly(isobutylene-*co*-maleic acid)				
	water	$\Delta_{dil} H^{12}$	298 K	1981PAO
Polyisoprene				
	helium	$\Delta_{sol} H_{A(vap)}^\infty$	273-373 K	1939BAR
Polyisoprene-b-polystyrene diblock copolymer				
	cyclohexane	$\Delta_M H_A$	293-333 K	1972GIR
	4-methyl-2-pentanone	$\Delta_M H_A$	293-333 K	1972GIR
	toluene	$\Delta_M H_A$	293-333 K	1972GIR
Poly(5-isopropenyltetrazole)				
	N,N-dimethylformamide	$\Delta_M H$	298 K	1997KI2
	water	$\Delta_{sol} H_B, \Delta_{dil} H^{12}$	298 K	1997KI1
Poly(*N*-isopropylacrylamide)				
	water	$\Delta_M H_A$	278-308 K	1994NAG
	water	$\Delta_M H_A$	283-323 K	1996NAG
	water	ΔH		2003CHO
	water/surfactant	ΔH		2004LOH

Polymer (B)	Solvent (A)	Enthalpy	T-range	Ref.
Poly(*N*-isopropylacrylamide-*co*-*N*-octadecylacrylamide)				
	water	$\Delta_{dil}H^{12}$	303 K	1996FAE
Poly(L-lysine)				
	N,N-dimethylformamide	$\Delta_{dil}H^{12}$	298 K	1982TAN
	ethanol	$\Delta_{dil}H^{12}$	298 K	1982TAN
	2-propanol	$\Delta_{dil}H^{12}$	298 K	1982TAN
	water	$\Delta_{dil}H^{12}$	298 K	1982TAN
Poly(methacrylic acid)				
	water	$\Delta_{M}H_{A}^{\infty}$	309-329 K	1957SIL
	water	$\Delta_{dil}H^{12}$	298 K	1972CRE
	water	$\Delta_{M}H$	294 K	1981HOR
	water	$\Delta_{M}H$	298-313 K	1996SAF
Poly(methacrylic acid) salts				
	water	$\Delta_{dil}H^{12}$	298 K	1980DAO
Poly(*N*-methacryloyl-L-alanine)				
	water	$\Delta_{dil}H^{12}$	298 K	1982MOR
Poly(*N*-methacryloyl-L-alanine-*co*-*N*-phenylmethacrylamide)				
	water	$\Delta_{dil}H^{12}$	298 K	1982MOR
Poly(3-methacryloyloxypropane-1-sulfonic acid) sodium salt				
	water	$\Delta_{dil}H^{12}$	298 K	1976BO1
Poly(*N*-methylacrylamide)				
	water	$\Delta_{M}H_{A}^{\infty}$	279-299 K	1957SIL
	water	$\Delta_{dil}H^{12}$	298 K	1981DAY
Poly(γ-methyl-L-glutamate)				
	1,2-dichloroethane	$\Delta_{sol}H_{B}$	303 K	1968GIA
	dichloroacetic acid	$\Delta_{sol}H_{B}$	303 K	1968GIA

Polymer (B)	Solvent (A)	Enthalpy	T-range	Ref.
Poly(methyl methacrylate)				
	acetonitrile	$\Delta_M H_A$		1972DAN
	acetonitrile	$\Delta_{dil} H^{12}$, $\Delta_M H$	298 K	1997SAF
	benzene	$\Delta_M H_A$	293-313 K	1948BRE
	benzene	$\Delta_M H_A$	273-303 K	1952SC1
	benzene	$\Delta_M H_A$		1952SC2
	benzene	$\Delta_M H_A$		1953SC1
	benzene	$\Delta_M H$, $\Delta_M H_A$	298 K	1953TAG
	benzene	$\Delta_M H_A^\infty$	300 K	1962FOX
	benzene	$\Delta_M H_A^\infty$	298 K	1963MO1
	benzene	$\Delta_M H_A$	303 K	1968KUW
	benzene	$\Delta_M H_A$		1972DAN
	1-butanol	$\Delta_M H_A$	298 K	1967LLO
	2-butanone	$\Delta_M H_A$	293-313 K	1948BRE
	2-butanone	$\Delta_M H_A$	273-303 K	1952SC1
	2-butanone	$\Delta_M H_A^\infty$	303-338 K	1961VAR
	2-butanone	$\Delta_M H_A^\infty$	300 K	1962FOX
	2-butanone	$\Delta_M H_A$		1972DAN
	butyl acetate	$\Delta_M H_A$		1953SC1
	butyl acetate	$\Delta_M H_A$		1972DAN
	tert-butyl acetate	$\Delta_M H$	283-323 K	1987NUN
	carbon dioxide	$\Delta_{sol} H_{A(vap)}^\infty$	308-353 K	1981KOR
	chlorobenzene	$\Delta_M H_A$		1972DAN
	1-chlorobutane	$\Delta_M H_A$		1972DAN
	cyclohexanone	$\Delta_{sol} H_B$	298 K	1993BRU
	1,2-dichloroethane	$\Delta_M H$		1959TA2
	1,2-dichloroethane	$\Delta_M H_A^\infty$	300 K	1962FOX
	1,2-dichlorobenzene	$\Delta_M H_A$		1972DAN
	1,2-dichloroethane	$\Delta_M H_A$		1972DAN
	1,2-dichlorobenzene	$\Delta_{sol} H_B$	313-433 K	1974RAG
	1,3-dimethylbenzene	$\Delta_M H_A$	273-303 K	1952SC1
	1,3-dimethylbenzene	$\Delta_M H_A$		1952SC2
	1,3-dimethylbenzene	$\Delta_M H_A$		1953SC1
	N,N-dimethylformamide	$\Delta_{dil} H^{12}$, $\Delta_M H$	298 K	1997SAF
	1,4-dioxane	$\Delta_M H_A$	273-303 K	1952SC1
	1,4-dioxane	$\Delta_M H_A$		1952SC2
	1,4-dioxane	$\Delta_M H_A$		1953SC1
	1,4-dioxane	$\Delta_M H_A$		1972DAN
	ethyl acetate	$\Delta_M H_A$		1953SC1
	ethyl acetate	$\Delta_M H_A^\infty$	303-338 K	1961VAR
	ethyl acetate	$\Delta_M H_A^\infty$	300 K	1962FOX
	ethyl acetate	$\Delta_{dil} H^{12}$, $\Delta_M H$	298 K	1997SAF
	2-heptanone	$\Delta_M H_A^\infty$	300 K	1962FOX
	3-heptanone	$\Delta_M H_A^\infty$	300 K	1962FOX
	4-heptanone	$\Delta_M H_A^\infty$	300 K	1962FOX

Polymer (B)	Solvent (A)	Enthalpy	T-range	Ref.
Poly(methyl methacrylate) *(continued)*	4-heptanone	$\Delta_M H_A$		1972DAN
	methyl isobutyrate	$\Delta_M H_A^\infty$	300 K	1962FOX
	methyl methacrylate	$\Delta_M H_A^\infty$	300 K	1962FOX
	nitroethane	$\Delta_M H_A$		1972DAN
	2-octanone	$\Delta_M H_A^\infty$	300 K	1962FOX
	3-pentanone	$\Delta_M H_A$	273-303 K	1952SC1
	3-pentanone	$\Delta_M H_A$		1952SC2
	3-pentanone	$\Delta_M H_A$		1953SC1
	pentyl acetate	$\Delta_M H_A^\infty$	303-338 K	1961VAR
	pentyl acetate	$\Delta_M H_A^\infty$	300 K	1962FOX
	1-propanol	$\Delta_M H_A$	298 K	1967LLO
	2-propanone	$\Delta_M H_A$		1952SC2
	2-propanone	$\Delta_M H_A$		1953SC1
	2-propanone	$\Delta_M H_A^\infty$	300 K	1962FOX
	2-propanone	$\Delta_M H_A$	303 K	1968KUW
	2-propanone	$\Delta_{sol} H_B$	303 K	1969ICH
	2-propanone	$\Delta_M H_A$		1972DAN
	2-propanone	$\Delta_{dil} H^{12}, \Delta_M H$	298 K	1997SAF
	tetrachloromethane	$\Delta_M H_A^\infty$	300 K	1962FOX
	tetrahydrofuran	$\Delta_M H_A$	273-303 K	1952SC1
	tetrahydrofuran	$\Delta_M H_A$		1952SC2
	tetrahydrofuran	$\Delta_M H_A$		1953SC1
	toluene	$\Delta_M H_A$	293-333 K	1948BRE
	toluene	$\Delta_M H_A$	273-303 K	1952SC1
	toluene	$\Delta_M H_A$		1952SC2
	toluene	$\Delta_M H_A$		1953SC1
	toluene	$\Delta_M H_A^\infty$	303-338 K	1961VAR
	toluene	$\Delta_M H_A^\infty$	300 K	1962FOX
	toluene	$\Delta_M H_A$		1972DAN
	toluene	$\Delta_{dil} H^{12}, \Delta_M H$	298 K	1997SAF
	trichloromethane	$\Delta_M H_A$		1952SC2
	trichloromethane	$\Delta_M H_A$	273-303 K	1952SC1
	trichloromethane	$\Delta_M H_A$		1953SC1
	trichloromethane	$\Delta_M H_A^\infty$	300 K	1962FOX
	trichloromethane	$\Delta_M H$	298 K	1972MAS
	trichloromethane	$\Delta_{sol} H_B$	303-323 K	1993FRE
	trichloromethane	$\Delta_{sol} H_B$	303-323 K	1994FRE
	trichloromethane	$\Delta_{dil} H^{12}, \Delta_M H$	298 K	1997SAF

Polymer (B)	Solvent (A)	Enthalpy	T-range	Ref.
Poly(methyl methacrylate-*co*-methacrylic acid)				
	dibutyl phthalate	$\Delta_M H$	298 K	1991TA3
	didodecyl phthalate	$\Delta_M H$	298 K	1991TA3
	dioctyl phthalate	$\Delta_M H$	298 K	1991TA3
	dioctyl sebacate	$\Delta_M H$	298 K	1991TA3
	tetramethyl pyrromellitate	$\Delta_M H$	298 K	1991TA3
	tricresyl phosphate	$\Delta_M H$	298 K	1991TA3
Poly(methyloctylsiloxane)				
	carbon dioxide	$\Delta_{sol} H_{A(vap)}{}^{\infty}$	283-328 K	1986SH2
	methane	$\Delta_{sol} H_{A(vap)}{}^{\infty}$	283-328 K	1986SH2
	propane	$\Delta_{sol} H_{A(vap)}{}^{\infty}$	283-328 K	1986SH2
Poly(methylphenylsiloxane)				
	carbon dioxide	$\Delta_{sol} H_{A(vap)}{}^{\infty}$	283-328 K	1986SH2
	methane	$\Delta_{sol} H_{A(vap)}{}^{\infty}$	283-328 K	1986SH2
	propane	$\Delta_{sol} H_{A(vap)}{}^{\infty}$	283-328 K	1986SH2
Poly(4-methyl-1-pentene)				
	cyclohexane	$\Delta_M H$	338 K	1981AHA
Poly(4-methyl-1-pentene-*co*-maleic acid)				
	water	$\Delta_{dil} H^{12}$	298 K	1981PAO
Poly(α-methylstyrene)				
	benzene	$\Delta_{sol} H_B$	303 K	1971IC2
	cumene	$\Delta_M H_A$	338 K	1966CAN
	cyclohexane	$\Delta_M H$	298 K	1965COT
	α-methylstyrene	$\Delta_M H_A$	338 K	1966CAN
	tetrahydrofuran	$\Delta_M H$	293 K	1969LEO
	tetrahydrofuran	$\Delta_M H$	298 K	1965COT
	toluene	$\Delta_M H$	298 K	1965COT
	toluene	$\Delta_{sol} H_B$	310 K	1987PED
	toluene	$\Delta_{sol} H_B$	333 K	1993BRU
Poly(methyltrifluoropropyl siloxane)				
	carbon dioxide	$\Delta_{sol} H_{A(vap)}{}^{\infty}$	283-328 K	1986SH2
	methane	$\Delta_{sol} H_{A(vap)}{}^{\infty}$	283-328 K	1986SH2
	propane	$\Delta_{sol} H_{A(vap)}{}^{\infty}$	283-328 K	1986SH2

Polymer (B)	Solvent (A)	Enthalpy	T-range	Ref.
Poly(2-methyl-1-vinylimidazole chloride)				
	water	$\Delta_{dil}H^{12}$	298 K	1993KEL
	water	$\Delta_{dil}H^{12}$	298 K	1996KEL
Poly(2-methyl-5-vinyl-pyridine)				
	N,N-dimethylformamide	$\Delta_M H$	293-333 K	1991AZI
Poly(2-methyl-5-vinyltetrazole)				
	acetic acid	$\Delta_{sol}H_B^{\infty}$	298 K	1994KIZ
	N,N-dimethylformamide	$\Delta_{dil}H^{12}$	298 K	1994KIZ
	N,N-dimethylformamide	$\Delta_{sol}H_B^{\infty}$	298 K	1994KIZ
	N,N-dimethylformamide	$\Delta_{sol}H_B$	298 K	1995KIZ
	N,N-dimethylformamide	$\Delta_M H$	298 K	1997KI2
	2-methyl-5-ethyltetrazole	$\Delta_M H$	298 K	1997KI2
	water	$\Delta_{sol}H_B$	298 K	1995KIZ
Poly(phenolphthalein-*co*-terephthaloyl chloride)				
	1,1,2,2-tetrachloroethane	$\Delta_M H$	298 K	1977TA2
	trichloromethane	$\Delta_M H$	298 K	1977TA1
	trichloromethane	$\Delta_M H$	298 K	1977TA2
Poly(m-phenyleneisophthalamide)				
	N,N-dimethylacetamide	$\Delta_M H$	298 K	1974TS1
	N,N-dimethylformamide	$\Delta_M H$	298 K	1974TS2
Poly(phenylmethylsiloxane)				
	carbon dioxide	$\Delta_{sol}H_{A(vap)}^{\infty}$	283-328 K	1986SH2
	methane	$\Delta_{sol}H_{A(vap)}^{\infty}$	283-328 K	1986SH2
	propane	$\Delta_{sol}H_{A(vap)}^{\infty}$	283-328 K	1986SH2
Poly(β-propiolactone)				
	1,4-dioxane	$\Delta_{dil}H^{12}$	298 K	1973MAN
Polypropylene				
	carbon dioxide	$\Delta_{sol}H_{A(vap)}^{\infty}$	423-473 K	2004ARE
	2,4-dimethyl-3-pentanone	$\Delta_M H$	298-318 K	1970DAN
	3-pentanone	$\Delta_M H$	298-318 K	1970DAN
	propane	$\Delta_{sol}H_{A(vap)}^{\infty}$	433-473 K	1994HOR
	propene	$\Delta_{sol}H_{A(vap)}^{\infty}$	287-363 K	2001TSU

Polymer (B)	Solvent (A)	Enthalpy	T-range	Ref.
Poly(propylene-*co*-maleic acid)				
	water	$\Delta_{dil}H^{12}$	298 K	1981PAO
Poly(propylene glycol)				
	benzene	$\Delta_M H$	300 K	1989PAR
	ethanol	$\Delta_M H$	300 K	1989PAR
	n-hexane	$\Delta_M H$	316-338	1972TAG
	pyridine	$\Delta_M H$	300 K	1989PAR
	tetrachloromethane	$\Delta_M H_A$	303-318 K	1993ZEL
	trichloromethane	$\Delta_M H_A$	303 K	1993ZEL
	water	$\Delta_M H$	298-313 K	1972TAG
	water	$\Delta_M H$	298-329 K	1974BES
	water	$\Delta_M H$	298-329 K	1975TA2
	water	$\Delta_M H$	298 K	1979SIM
	water	$\Delta_{sol}H_A^{\infty}, \Delta_{sol}H_B^{\infty}$	298 K	1980BIL
	water	$\Delta_M H$	298 K	1981MED
	water	$\Delta_M H$	298 K	1982MED
	water	$\Delta_M H$	298 K	1983ME1
	water	$\Delta_M H$	298 K	1983ME2
Poly(propylene glycol adipate)				
	dioctyl sebacate	$\Delta_M H$	340 K	1986OVC
	triacetin	$\Delta_M H$	298 K	1992SUV
Poly(propylene oxide)				
	benzene	$\Delta_M H_A$	333 K	1971BO2
	1-butanol	$\Delta_M H$	298 K	1964CON
	2-butanol	$\Delta_M H$	298 K	1964CON
	tert-butanol	$\Delta_M H$	298 K	1964CON
	cyclohexane	$\Delta_M H$	298 K	1964CON
	ethanol	$\Delta_M H$	298 K	1964CON
	n-hexane	$\Delta_M H$	298 K	1964CON
	methanol	$\Delta_M H$	298 K	1964CON
	3-methyl-1-butanol	$\Delta_M H$	298 K	1964CON
	2-methyl-1-propanol	$\Delta_M H$	298 K	1964CON
	1-pentanol	$\Delta_M H$	298 K	1964CON
	1-propanol	$\Delta_M H$	298 K	1964CON
	2-propanol	$\Delta_M H$	298 K	1964CON
	2-propanol	ΔH	274-332 K	1963BLA
	tetrachloromethane	$\Delta_M H$	298 K	1964CON

Polymer (B)	Solvent (A)	Enthalpy	T-range	Ref.
Polystyrene				
	acetophenone	$\Delta_{dil}H^{12}$	303 K	1977BA1
	argon	$\Delta_{sol}H_{A(vap)}{}^{\infty}$	293-363 K	1992REI
	benzene	$\Delta_{M}H_{A}$		1953SC2
	benzene	$\Delta_{sol}H_{B}{}^{\infty}$	283-353 K	1956MEE
	benzene	$\Delta_{M}H$	293 K	1961MIZ
	benzene	$\Delta_{dil}H^{12}$	298 K	1964BIR
	benzene	$\Delta_{dil}H^{12}$	298 K	1967POU
	benzene	$\Delta_{sol}H_{B}, \Delta_{M}H$	293 K	1968MA2
	benzene	$\Delta_{M}H_{B}{}^{\infty}$	298 K	1970MO1
	benzene	$\Delta_{M}H_{A}$	298 K	1970MO4
	benzene	$\Delta_{M}H$	300 K	1970TAG
	benzene	$\Delta_{sol}H_{B}$	303 K	1971IC1
	benzene	$\Delta_{M}H_{B}{}^{\infty}$	288-318 K	1971MOR
	n-butane	$\Delta_{M}H_{A}$	428-463 K	1983ALI
	2-butanone	$\Delta_{M}H_{A}$		1950BAW
	2-butanone	$\Delta_{M}H_{A}{}^{\infty}$	300-323 K	1950SCH
	2-butanone	$\Delta_{M}H_{A}$		1953SC2
	2-butanone	$\Delta_{M}H_{A}$	298-328 K	1954BAW
	2-butanone	$\Delta_{M}H_{A}$	298 K	1955ROS
	2-butanone	$\Delta_{M}H_{A}$	280-313 K	1956CAN
	2-butanone	$\Delta_{M}H$	298 K	1958TA2
	2-butanone	$\Delta_{M}H_{A}$	298-323 K	1964BO2
	2-butanone	$\Delta_{sol}H_{B}, \Delta_{M}H$	293 K	1968MA2
	2-butanone	$\Delta_{M}H_{A}$	298 K	1971HO1
	2-butanone	$\Delta_{dil}H^{12}$	298 K	1973GAE
	2-butanone	$\Delta_{M}H_{A}$	298-313 K	1974NAK
	2-butanone	$\Delta_{dil}H^{12}$	298 K	1979FUJ
	butyl acetate	$\Delta_{M}H_{A}$		1953SC2
	butyl acetate	$\Delta_{M}H$	298 K	1977HUA
	carbon dioxide	$\Delta_{sol}H_{A(vap)}{}^{\infty}$	293-313 K	1987SAD
	carbon dioxide	$\Delta_{sol}H_{A(vap)}{}^{\infty}$	293-363 K	1992REI
	carbon dioxide	$\Delta_{sol}H_{A(vap)}{}^{\infty}$	423-473 K	2004ARE
	cyclohexane	$\Delta_{M}H_{A}{}^{\infty}$	300-343 K	1950SCH
	cyclohexane	$\Delta_{sol}H_{B}$	298 K	1953GUE
	cyclohexane	$\Delta_{M}H_{A}$		1953SC2
	cyclohexane	$\Delta_{M}H_{A}$	298 K	1955ROS
	cyclohexane	$\Delta_{M}H_{A}$	296-318 K	1956CAN
	cyclohexane	$\Delta_{M}H_{A}$	300 K	1956SCH
	cyclohexane	$\Delta_{M}H_{A}$	300 K	1959KRI
	cyclohexane	$\Delta_{M}H_{A}$	298 K	1961KUW
	cyclohexane	$\Delta_{M}H$	293 K	1961MIZ
	cyclohexane	$\Delta_{M}H_{A}$	307-317 K	1964BO2
	cyclohexane	$\Delta_{sol}H_{B}, \Delta_{M}H$	293 K	1968MA2
	cyclohexane	$\Delta_{M}H$	300 K	1970TAG

Polymer (B)	Solvent (A)	Enthalpy	T-range	Ref.
Polystyrene (*continued*)				
	cyclohexane	$\Delta_M H_A$	298 K	1971HO3
	cyclohexane	$\Delta_M H$	298 K	1977HUA
	cyclohexane	$\Delta_{dil} H^{12}$	298-318 K	1978OKU
	cyclohexane	$\Delta_M H$	303-343 K	1987NUN
	cyclohexane	$\Delta_M H_A$	308 K	1988KAM
	cyclohexane	$\Delta_M H, \Delta_{dil} H^{12}$	306 K	1991SAF
	cyclohexane	$\Delta_M H, \Delta_{dil} H^{12}$	306 K	2002SA2
	cyclohexanone	$\Delta_{dil} H^{12}$	298 K	1979FUJ
	cyclohexene	$\Delta_M H$	300 K	1970TAG
	decahydronaphthalene	$\Delta_M H, \Delta_{dil} H^{12}$	308 K	1991SAF
	decahydronaphthalene	$\Delta_M H, \Delta_{dil} H^{12}$	308 K	2002SA2
	trans-decahydro-naphthalene	$\Delta_M H_A$	273-293 K	1977WOL
	trans-decahydro-naphthalene	$\Delta_M H_A$	297 K	1988KAM
	dibenzyl ether	$\Delta_{dil} H^{12}$	303 K	1977BA1
	diethyl ether	$\Delta_M H_A^{\infty}$	243-298 K	1976WOL
	dimethoxymethane	$\Delta_M H_A^{\infty}$	243-298 K	1976WOL
	1,2-dimethylbenzene	$\Delta_M H_A$		1953SC2
	1,3-dimethylbenzene	$\Delta_M H_A$		1953SC2
	1,4-dimethylbenzene	$\Delta_M H$	293 K	1961MIZ
	1,4-dioxane	$\Delta_{sol} H_B^{\infty}$	283-353 K	1956MEE
	1,4-dioxane	$\Delta_{sol} H_B, \Delta_M H$	293 K	1968MA2
	1,3-diphenylbutane	$\Delta_{dil} H^{12}$	298 K	1964BIR
	1,3-diphenylbutane	$\Delta_{dil} H^{12}$	298 K	1967POU
	ethyl acetate	$\Delta_M H_A^{\infty}$	310-323 K	1950SCH
	ethyl acetate	$\Delta_M H_A$		1953SC2
	ethyl acetate	$\Delta_M H$	298 K	1958TA2
	ethyl acetate	$\Delta_{sol} H_B, \Delta_M H$	293 K	1968MA2
	ethylbenzene	$\Delta_M H_A$	298 K	1955GAT
	ethylbenzene	$\Delta_{sol} H_B$	273-373 K	1958TA1
	ethylbenzene	$\Delta_M H$	293 K	1961MIZ
	ethylbenzene	$\Delta_{dil} H^{12}$	298 K	1964BIR
	ethylbenzene	$\Delta_{dil} H^{12}$	298 K	1967POU
	ethylbenzene	$\Delta_M H$	300 K	1970TAG
	ethylbenzene	$\Delta_M H_A$	298 K	1971HO2
	ethylbenzene	$\Delta_{dil} H^{12}$	298 K	1973GAE
	ethylbenzene	$\Delta_M H, \Delta_{dil} H^{12}$	298 K	2002SA2
	methane	$\Delta_{sol} H_{A(vap)}^{\infty}$	293-313 K	1987SAD
	methane	$\Delta_{sol} H_{A(vap)}^{\infty}$	293-363 K	1992REI
	methoxybenzene	$\Delta_{sol} H_B, \Delta_M H$	293 K	1968MA2
	methoxybenzene	$\Delta_{dil} H^{12}$	298 K	1977BA2
	methoxybenzene	$\Delta_{dil} H^{12}$	303 K	1977BA1
	3-pentanone	$\Delta_{dil} H^{12}$	298 K	1983FUJ

Polymer (B)	Solvent (A)	Enthalpy	T-range	Ref.
Polystyrene (*continued*)				
	2-propanone	$\Delta_{sol}H_B$	298 K	1953GUE
	2-propanone	$\Delta_M H_A$	298-323 K	1964BO2
	2-propanone	$\Delta_M H_A$	298-323 K	1956BA2
	propyl acetate	$\Delta_M H_A$	298-323 K	1956BA2
	styrene	$\Delta_M H$	293 K	1961MIZ
	tetrachloromethane	$\Delta_M H_A$		1953SC2
	tetrachloromethane	$\Delta_{dil}H^{12}, \Delta_M H$	298 K	1963TAG
	tetrachloromethane	$\Delta_{sol}H_B$	298 K	1972DAV
	1,2,3,4-tetrahydro-naphthalene	$\Delta_M H$	300 K	1970TAG
	1,2,3,4-tetrahydro-naphthalene	$\Delta_M H_A$	298 K	1972LEC
	toluene	$\Delta_M H_A$	283-303 K	1937SCH
	toluene	$\Delta_M H_A$	283-303 K	1938SCH
	toluene	$\Delta_M H_A$		1950BAW
	toluene	$\Delta_M H_A^{\infty}$	310-333 K	1950SCH
	toluene	$\Delta_{sol}H_B$	298-353 K	1953GUE
	toluene	$\Delta_M H_A$		1953SC2
	toluene	$\Delta_M H_A$	298-323 K	1954BAW
	toluene	$\Delta_M H_A$	298 K	1955ROS
	toluene	$\Delta_{sol}H_B^{\infty}$	283-353 K	1956MEE
	toluene	$\Delta_M H_A^{\infty}$	308 K	1957SIL
	toluene	$\Delta_{sol}H_B$	273-373 K	1958TA1
	toluene	$\Delta_{sol}H_B^{\infty}$	283-353 K	1959MEE
	toluene	$\Delta_M H_A$	298 K	1959SCH
	toluene	$\Delta_M H$	293 K	1961MIZ
	toluene	$\Delta_{sol}H_B^{\infty}$		1965BIA
	toluene	$\Delta_{dil}H^{12}$	303 K	1969LEW
	toluene	$\Delta_M H_B^{\infty}$	298 K	1970MO1
	toluene	$\Delta_M H_A$	298 K	1970MO4
	toluene	$\Delta_M H$	300 K	1970TAG
	toluene	$\Delta_{sol}H_B$	298-373 K	1971BRU
	toluene	$\Delta_M H_B^{\infty}$	288-318 K	1971MOR
	toluene	$\Delta_{sol}H_B$	298 K	1972DAV
	toluene	$\Delta_{dil}H^{12}$	298 K	1973GAE
	toluene	$\Delta_{sol}H_B$	303 K	1974TAK
	toluene	$\Delta_{dil}H^{12}$	298 K	1977BA2
	toluene	$\Delta_M H$	298 K	1977HUA
	toluene	$\Delta_{sol}H_B$	310 K	1987PED
	toluene	$\Delta_{sol}H_B$	298 K	1993BRU
	toluene/poly(vinyl methyl ether)	$\Delta_M H$	323 K	2001CAS
	trichloromethane	$\Delta_M H_A$		1953SC2
	trichloromethane	$\Delta_M H_A$	298 K	1955ROS
	trichloromethane	$\Delta_M H_A$	298-323 K	1956BA2

Polymer (B)	Solvent (A)	Enthalpy	T-range	Ref.
Polystyrene (*continued*)				
	trichloromethane	$\Delta_M H_B^{\infty}$	298 K	1970MO1
	trichloromethane	$\Delta_M H_A$	298 K	1970MO4
	trichloromethane	$\Delta_M H_B^{\infty}$	288-318 K	1971MOR
	trichloromethane	$\Delta_M H, \Delta_{dil} H^{12}$	298 K	2002SA2
	trichloromethane	$\Delta_{sol} H_{A(vap)}^{\infty}$	298-328 K	2005SAN
Poly(styrene oxide)-b-poly(ethylene oxide) diblock copolymer				
	water/sodium dodecyl sulfate	ΔH	298 K	2005CAS
Poly(styrenesulfonic acid)				
	water	$\Delta_{dil} H^{12}$	298 K	1974DOL
	water	$\Delta_{dil} H^{12}$	298 K	1977SKE
	water	$\Delta_{dil} H^{12}$	298 K	1984VE2
Poly(styrenesulfonic acid) salts				
	1,4-dioxane	$\Delta_{dil} H^{12}$	298 K	1980SKE
	N-methylformamide	$\Delta_{dil} H^{12}$	298 K	1986SK2
	water	$\Delta_{dil} H^{12}$	298 K	1967SK2
	water	$\Delta_{dil} H^{12}$	298 K	1974DOL
	water	$\Delta_{dil} H^{12}$	298 K	1976BO1
	water	$\Delta_M H$	298 K	1976BO2
	water	$\Delta_{dil} H^{12}$	298 K	1976MIT
	water	$\Delta_{dil} H^{12}$	298 K	1977SKE
	water	$\Delta_{dil} H^{12}$	298 K	1980SKE
	water	$\Delta_{dil} H^{12}$	273-313 K	1984VE1
	water	$\Delta_{dil} H^{12}$	298 K	1984VE2
	water	$\Delta_{dil} H^{12}$	273-313 K	1987VES
Polysulfone				
	1,2-dichlorobenzene	$\Delta_{sol} H_B$	298-423 K	1976FIL
Poly(tetrafluoroethylene)				
	argon	$\Delta_{sol} H_{A(vap)}^{\infty}$	298-368 K	1960TIK
	helium	$\Delta_{sol} H_{A(vap)}^{\infty}$	298-368 K	1960TIK
Poly(tetrafluoroethylene-*co*-2,2-bistrifluoromethyl-4,5-difluoro-1,3-dioxole)				
	n-butane	$\Delta_{sol} H_{A(vap)}^{\infty}$	298-318 K	1999BON
	n-butane	$\Delta_{sol} H_{A(vap)}^{\infty}$	298-318 K	2002DEA

Polymer (B)	Solvent (A)	Enthalpy	T-range	Ref.
Poly(tetrafluoroethylene-*co*-2,2-bistrifluoromethyl-4,5-difluoro-1,3-dioxole) (*continued*)				
	carbon dioxide	$\Delta_{sol}H_{A(vap)}^{\infty}$	298-318 K	1999BON
	carbon dioxide	$\Delta_{sol}H_{A(vap)}^{\infty}$	298-318 K	2002DEA
	ethane	$\Delta_{sol}H_{A(vap)}^{\infty}$	298-318 K	1999BON
	ethane	$\Delta_{sol}H_{A(vap)}^{\infty}$	298-318 K	2002DEA
	helium	$\Delta_{sol}H_{A(vap)}^{\infty}$	298-318 K	1999BON
	methane	$\Delta_{sol}H_{A(vap)}^{\infty}$	298-318 K	1999BON
	methane	$\Delta_{sol}H_{A(vap)}^{\infty}$	298-318 K	2002DEA
	nitrogen	$\Delta_{sol}H_{A(vap)}^{\infty}$	298-318 K	1999BON
	nitrogen	$\Delta_{sol}H_{A(vap)}^{\infty}$	298-318 K	2002DEA
	oxygen	$\Delta_{sol}H_{A(vap)}^{\infty}$	298-318 K	1999BON
	oxygen	$\Delta_{sol}H_{A(vap)}^{\infty}$	298-318 K	2002DEA
	propane	$\Delta_{sol}H_{A(vap)}^{\infty}$	298-318 K	1999BON
	propane	$\Delta_{sol}H_{A(vap)}^{\infty}$	298-318 K	2002DEA
	tetrafluoroethane	$\Delta_{sol}H_{A(vap)}^{\infty}$	298-318 K	1999BON
	tetrafluoroethane	$\Delta_{sol}H_{A(vap)}^{\infty}$	298-318 K	2002DEA
	tetrafluoromethane	$\Delta_{sol}H_{A(vap)}^{\infty}$	298-318 K	1999BON
	tetrafluoromethane	$\Delta_{sol}H_{A(vap)}^{\infty}$	298-318 K	2002DEA
Poly(tetramethylene oxide)				
	benzene	$\Delta_M H$	321 K	1982SH2
	benzene	$\Delta_M H$	313 K	1985SHA
	cyclohexane	$\Delta_M H$	321 K	1982SH1
	1,2-dichloroethane	$\Delta_M H$	313 K	1987SHA
	1,2-dimethylbenzene	$\Delta_M H$	313 K	1986SH1
	1,3-dimethylbenzene	$\Delta_M H$	313 K	1986SH1
	1,4-dimethylbenzene	$\Delta_M H$	313 K	1986SH1
	1,4-dioxane	$\Delta_M H$	321 K	1982SH1
	ethylbenzene	$\Delta_M H$	313 K	1985SHA
	2-propanol	$\Delta_M H$	303 K	1998STU
	2-propanol	$\Delta_M H$	303 K	1999STU
	propylbenzene	$\Delta_M H$	313 K	1985SHA
	tetrachloromethane	$\Delta_M H$	321 K	1982SH2
	tetrachloromethane	$\Delta_M H$	313 K	1987SHA
	tetrachloromethane	$\Delta_M H$	303 K	1998STU
	tetrachloromethane	$\Delta_M H$	303 K	1999STU
	toluene	$\Delta_M H$	313 K	1985SHA
	1,1,1-trichloroethane	$\Delta_M H$	313 K	1987SHA
	trichloromethane	$\Delta_M H$	303 K	1998STU
	trichloromethane	$\Delta_M H$	303 K	1999STU
	1,3,5-trimethylbenzene	$\Delta_M H$	313 K	1986SH1

Polymer (B)	Solvent (A)	Enthalpy	T-range	Ref.
Poly(tetramethylsilhexylene siloxane)				
	carbon dioxide	$\Delta_{sol}H_{A(vap)}{}^{\infty}$	283-328 K	1993SHA
	methane	$\Delta_{sol}H_{A(vap)}{}^{\infty}$	283-328 K	1993SHA
	propane	$\Delta_{sol}H_{A(vap)}{}^{\infty}$	283-328 K	1993SHA
Poly(1-trimethylsilyl-1-propyne)				
	octafluoropropane	$\Delta_{sol}H_{A(vap)}{}^{\infty}$	298-328 K	2005PRA
	propane	$\Delta_{sol}H_{A(vap)}{}^{\infty}$	298-328 K	2005PRA
Poly(trimethyl-1-pentene-*co*-maleic acid)				
	water	$\Delta_{dil}H^{12}$	298 K	1981PAO
Polyurethane				
	benzene	$\Delta_{sol}H_{A(vap)}{}^{\infty}$	303-323 K	1974HUN
	2-butanone	$\Delta_{sol}H_{A(vap)}{}^{\infty}$	303-323 K	1974HUN
	chlorobenzene	$\Delta_{sol}H_{A(vap)}{}^{\infty}$	303-323 K	1974HUN
	1,4-dioxane	$\Delta_{M}H$		1965TAG
	1,4-dioxane	$\Delta_{sol}H_{A(vap)}{}^{\infty}$	298 K	1968LIP
	ethyl acetate	$\Delta_{sol}H_{A(vap)}{}^{\infty}$	303-323 K	1974HUN
	n-hexane	$\Delta_{sol}H_{A(vap)}{}^{\infty}$	303-323 K	1974HUN
	1-heptanol	$\Delta_{sol}H_{A(vap)}{}^{\infty}$	303-323 K	1974HUN
	1-hexanol	$\Delta_{sol}H_{A(vap)}{}^{\infty}$	303-323 K	1974HUN
	methyl acetate	$\Delta_{sol}H_{A(vap)}{}^{\infty}$	303-323 K	1974HUN
	3-methyl-1-butanol	$\Delta_{sol}H_{A(vap)}{}^{\infty}$	303-323 K	1974HUN
	1-octanol	$\Delta_{sol}H_{A(vap)}{}^{\infty}$	303-323 K	1974HUN
	toluene	$\Delta_{sol}H_{A(vap)}{}^{\infty}$	303-323 K	1974HUN
Poly(δ-valerolactone)				
	1,4-dioxane	$\Delta_{dil}H^{12}$	298 K	1975MAN
Poly(vinyl acetate)				
	argon	$\Delta_{sol}H_{A(vap)}{}^{\infty}$	277-317 K	1954MEA
	2-butanone	$\Delta_{M}H_{A}{}^{\infty}$	283-318 K	1949BRO
	cyclohexanone	$\Delta_{sol}H_{B}$	398 K	1993BRU
	1,2-dichloroethane	$\Delta_{M}H$	300 K	1971TAG
	1,4-dioxane	$\Delta_{dil}H^{12}$	298 K	1987SCH
	helium	$\Delta_{sol}H_{A(vap)}{}^{\infty}$	277-317 K	1954MEA
	hydrogen	$\Delta_{sol}H_{A(vap)}{}^{\infty}$	277-317 K	1954MEA
	neon	$\Delta_{sol}H_{A(vap)}{}^{\infty}$	277-317 K	1954MEA
	oxygen	$\Delta_{sol}H_{A(vap)}{}^{\infty}$	277-317 K	1954MEA
	1-propanol	$\Delta_{M}H_{A}$	303-323 K	1953KOK
	2-propanone	$\Delta_{M}H$	298 K	1975TA3

Polymer (B)	Solvent (A)	Enthalpy	T-range	Ref.
Poly(vinyl acetate) (*continued*)				
	2-propanone	$\Delta_{sol}H_B$	303 K	1969ICH
	2-propanone	$\Delta_M H_A$	303-323 K	1953KOK
	tetrachloromethane	$\Delta_M H, \Delta_{dil}H^{12}$	300 K	1991SAF
	tetrachloromethane	$\Delta_M H, \Delta_{dil}H^{12}$	300 K	2002SA2
	toluene	$\Delta_M H$	300 K	1971TAG
	toluene	$\Delta_{sol}H_B$	333 K	1993BRU
	1,2,3-trichloropropane	$\Delta_M H_A^{\infty}$	288-323 K	1949BRO
Poly(vinyl acetate-*co*-1-vinyl-2-pyrrolidinone)				
	vinyl acetate	$\Delta_{sol}H_B^{\infty}$	298 K	1989EGO
	1-vinyl-2-pyrrolidinone	$\Delta_{sol}H_B^{\infty}$	298 K	1989EGO
Poly(vinyl alcohol)				
	1,4-dioxane	$\Delta_M H_A$	298 K	1969NAP
	water	$\Delta_{sol}H_B$	313-353 K	1963ME1
	water	$\Delta_M H_A$	298 K	1969NAP
	water	$\Delta_M H$	294 K	1981HOR
	water	$\Delta_M H$	294 K	1983TA2
	water	$\Delta_{dil}H^{12}$	298 K	1987SCH
Poly(vinyl amine-*co*-vinyl caprolactam)				
	water	$\Delta_M H$	298-308 K	1994TAG
Poly(1-vinyl-5-aminotetrazole)				
	dimethylsulfoxide	$\Delta_M H$	298 K	1997KI2
Poly(vinylbenzyltrimethyl-ammonium) halides				
	water	$\Delta_{dil}H^{12}$	298-318 K	1993PER
Poly(vinylbenzyltrimethyl-ammonium) hydroxide				
	water	$\Delta_{dil}H^{12}$	298-318 K	1993PER
Poly(*N*-vinylcaprolactam)				
	water	$\Delta_M H, \Delta_{dil}H^{12}$	298 K	1990TAG
	water	$\Delta_M H_A, \Delta_{sol}H_B$	298 K	1990TAG
	water	$\Delta_{dil}H^{12}$	298 K	1993TA2
	water	$\Delta_M H$	298-308 K	1994TAG

Polymer (B)	Solvent (A)	Enthalpy	T-range	Ref.
Poly(vinyl chloride)				
	argon	$\Delta_{sol}H_{A(vap)}{}^{\infty}$	298-368 K	1960TIK
	bromobenzene	$\Delta_M H_A$	293-353 K	1969BOH
	2-butanone	$\Delta_M H_A$	293-353 K	1969BOH
	butyl acetate	$\Delta_M H_A$	293-353 K	1969BOH
	chlorobenzene	$\Delta_M H_A$	293-353 K	1969BOH
	cyclohexanone	$\Delta_M H_A{}^{\infty}$	313 K	1957SIL
	cyclohexanone	$\Delta_M H_A$	298 K	1959MEN
	cyclohexanone	$\Delta_{sol}H_B$	283-363 K	1963ME2
	cyclohexanone	$\Delta_M H$	358 K	1968ZUB
	cyclohexanone	$\Delta_M H_A$	293-353 K	1969BOH
	cyclohexanone	$\Delta_M H, \Delta_{dil}H^{12}$	303-373 K	1973MA1
	cyclohexanone	$\Delta_M H$	298-353 K	1996ABD
	cyclopentanone	$\Delta_M H_A$	298-313 K	1959MEN
	cyclopentanone	$\Delta_M H_A$	293-353 K	1969BOH
	cyclopentanone	$\Delta_M H, \Delta_{dil}H^{12}$	303-363 K	1973MA2
	dibutyl phthalate	$\Delta_M H$	298 K	1982YUS
	dibutyl phthalate	$\Delta_M H$	293-373 K	1983TA1
	1,2-dichloroethane	$\Delta_{sol}H_B{}^{\infty}$	298-333 K	1957AKH
	1,2-dichloroethane	$\Delta_{sol}H_B$	283-363 K	1963ME2
	1,2-dichloroethane	$\Delta_M H_A$	293-353 K	1969BOH
	didodecyl phthalate	$\Delta_M H$	298 K	1982YUS
	diisododecyl phthalate	$\Delta_M H$	304 K	1985KIR
	N,N-dimethylformamide	$\Delta_{sol}H_B$	283-363 K	1963ME2
	N,N-dimethylformamide	$\Delta_M H$	358 K	1968ZUB
	N,N-dimethylformamide	$\Delta_M H_A$	293-353 K	1969BOH
	dimethyl phthalate	$\Delta_M H$	293-373 K	1983TA1
	dioctyl phthalate	$\Delta_M H$	298 K	1982YUS
	dioctyl phthalate	$\Delta_M H$	293-373 K	1983TA1
	dioctyl phthalate	$\Delta_M H$	413-443 K	1985EZH
	dioctyl sebacate	$\Delta_M H$	298 K	1982YUS
	dioctyl sebacate	$\Delta_M H$	413-443 K	1985EZH
	1,4-dioxane	$\Delta_M H_A$	293-353 K	1969BOH
	diphenyl 4-*tert*-butyl-phenyl phosphate	$\Delta_M H$	293-373 K	1983TA1
	ethyl acetate	$\Delta_M H_A$	293-353 K	1969BOH
	ethyl acetoacetate	$\Delta_M H_A$	293-353 K	1969BOH
	bis(2-ethylhexyl) adipate	$\Delta_M H$	304 K	1985KIR
	bis(2-ethylhexyl) sebacate	$\Delta_M H$	304 K	1985KIR
	fumarates	$\Delta_M H$	>373 K	1962ANA
	furfurol	$\Delta_M H$	298-353 K	1996ABD
	helium	$\Delta_{sol}H_{A(vap)}{}^{\infty}$	298-368 K	1960TIK
	maleates	$\Delta_M H$	>373 K	1962ANA
	methoxybenzene	$\Delta_M H_A$	293-353 K	1969BOH
	nitrobenzene	$\Delta_{sol}H_B{}^{\infty}$	293-363 K	1957AKH
	nitrobenzene	$\Delta_M H_A$	293-353 K	1969BOH

Polymer (B)	Solvent (A)	Enthalpy	T-range	Ref.
Poly(vinyl chloride) (*continued*)				
	phthalates	$\Delta_M H$	>373 K	1962ANA
	2-propanone	$\Delta_M H_A$	293-353 K	1969BOH
	tetrachloroethane	$\Delta_M H_A$	293-353 K	1969BOH
	tetrahydrofuran	$\Delta_M H_A$	298 K	1959MEN
	tetrahydrofuran	$\Delta_M H_A$	293-353 K	1969BOH
	tetrahydrofuran	$\Delta_M H$	298 K	1969POU
	tetrahydrofuran	$\Delta_M H, \Delta_{dil} H^{12}$	293-323 K	1973MA3
	tetrahydrofuran	$\Delta_M H$	413-443 K	1985EZH
	tetrahydrofuran	$\Delta_M H$	298 K	1994YUS
	tributoxyethyl phosphate	$\Delta_M H$	293-373 K	1983TA1
	trichloromethane	$\Delta_M H_A$	293-353 K	1969BOH
	tricresyl phosphate	$\Delta_M H$	293-373 K	1983TA1
Poly(vinyl ethyl ether)				
	water	$\Delta_{sol} H_B$	288-313 K	1966NAG
Poly(vinylidene fluoride)				
	ε-caprolactam	$\Delta_M H_A^\infty$	450 K	1996LIU
Poly(1-vinylimidazole)				
	acetic acid	$\Delta_{sol} H_B^\infty$	298 K	1994KIZ
	N,N-dimethylformamide	$\Delta_{sol} H_B^\infty$	298 K	1994KIZ
	water	$\Delta_M H, \Delta_{dil} H^{12}$	298 K	1987TAG
	water	$\Delta_M H_A, \Delta_{sol} H_B$	298 K	1987TAG
	water	$\Delta_M H_B^\infty$	298 K	1989SAF
Poly(vinyl methyl ether)				
	toluene/polystyrene	$\Delta_M H$	323 K	2001CAS
Poly(*N*-vinyl-*N*-propylacetamide)				
	water	$\Delta_M H$	298-308 K	1991TA2
Poly(1-vinylpyrazole)				
	acetic acid	$\Delta_{sol} H_B^\infty$	298 K	1994KIZ
	N,N-dimethylformamide	$\Delta_{sol} H_B^\infty$	298 K	1994KIZ
	toluene	$\Delta_M H, \Delta_{dil} H^{12}$	307 K	1991SAF
	toluene	$\Delta_M H, \Delta_{dil} H^{12}$	307 K	2002SA2

Polymer (B)	Solvent (A)	Enthalpy	T-range	Ref.
Poly(1-vinyl-2-pyrrolidinone)				
	ethanol	$\Delta_{sol}H_B$	298 K	1993BRU
	methanol	$\Delta_{dil}H^{12}$	298 K	1975MUR
	2-propanone	$\Delta_{sol}H_B$	298 K	1993BRU
	trichloromethane	$\Delta_{dil}H^{12}$	298 K	1985KIL
	water	$\Delta_{dil}H^{12}$	298 K	1975MUR
	water/methanol	$\Delta_{dil}H^{12}$	298 K	1975MUR
	water	$\Delta_{dil}H^{12}$	298 K	1979GAR
	water	$\Delta_{dil}H^{12}$	298 K	1985SAR
	water	$\Delta_{sol}H_B$	298 K	1997RAS
Poly(1-vinyltetrazole)				
	acetic acid	$\Delta_{sol}H_B^{\infty}$	298 K	1994KIZ
	N,N-dimethylformamide	$\Delta_M H, \Delta_{dil}H^{12}$	298 K	1994KIZ
	N,N-dimethylformamide	$\Delta_M H_A, \Delta_M H_B$	298 K	1994KIZ
	N,N-dimethylformamide	$\Delta_{sol}H_B^{\infty}$	298 K	1994KIZ
	N,N-dimethylformamide	$\Delta_{sol}H_B^{\infty}$	298 K	1995KIZ
	N,N-dimethylformamide	$\Delta_M H$	298 K	1997KI2
	dimethylsulfoxide	$\Delta_M H$	298 K	1997KI2
	water	$\Delta_{sol}H_B$	298 K	1995KIZ
Poly(5-vinyltetrazole)				
	N,N-dimethylformamide	$\Delta_M H$	298 K	1997KI2
	dimethylsulfoxide	$\Delta_M H$	298 K	1997KI2
	water	$\Delta_{sol}H_B, \Delta_{dil}H^{12}$	298 K	1997KI1
Poly(5-vinyltetrazole-*co*-2-methyl-5-vinyltetrazole)				
	N,N-dimethylformamide	$\Delta_{sol}H_B^{\infty}$	298 K	1995KIZ
	water	$\Delta_{sol}H_B^{\infty}, \Delta_{dil}H^{12}$	298 K	1997KI1
	water	$\Delta_M H$	298 K	1997KI1
Poly(1-vinyl-1,2,4-triazole)				
	acetic acid	$\Delta_{sol}H_B^{\infty}$	298 K	1994KIZ
	N,N-dimethylformamide	$\Delta_{sol}H_B^{\infty}$	298 K	1994KIZ
	water	$\Delta_M H, \Delta_{dil}H^{12}$	298 K	1987TAG
	water	$\Delta_M H_A, \Delta_{sol}H_B$	298 K	1987TAG
	water	$\Delta_M H_B^{\infty}$	298 K	1989SAF
Starch				
	water	$\Delta_{sol}H_B$	298 K	2000SUV
	water	$\Delta_{sol}H_B$	298, 353 K	2002SUV

Polymer (B)	Solvent (A)	Enthalpy	T-range	Ref.
Tetra(ethylene glycol)				
	water	$\Delta_M H$	283-313K	1998OHT
Tetra(ethylene glycol) monooctyl ether				
	water	$\Delta_M H$	283-308 K	1997ARA
	water	$\Delta_M H, \Delta_M H_B$	283-313 K	2001OHT
Tri(ethylene glycol) monooctyl ether				
	water	$\Delta_M H, \Delta_M H_B$	283-313 K	2001OHT

6.2. References

1937BOI   Boissonnas, Ch.G. and Meyer, K.H., Proprietes des polymeres en solution VI. Energie libre et chaleur de solution. Systeme nitrocellulose-cyclohexanone, *Helv. Chim. Acta*, 20, 783, 1937.

1937PAP   Papkov, S. and Kargin, V., Über die Wärmeentwicklung bei der Einwirkung von Lösungsmitteln auf Nitrozellulose II., *Acta Physicochim. URSS*, 7, 667, 1937.

1937SCH   Schulz, G.V., Über die Temperaturabhängigkeit des osmotischen Druckes und den Molekularzustand in hochmolekularen Lösungen, *Z. Phys. Chem., Abt. A*, 180, 1, 1937.

1938BOI   Boissonnas, Ch.G. and Meyer, K.H., Osmotischer Druck, Verdünnungswärme und Verdünnungsentropie in den Systemen Nitrocellulose-Cyclohexanon und Nitrocellulose-Aceton, *Z. Phys. Chem., Abt. B*, 40, 108, 1938.

1938SCH   Schulz, G.V., Temperaturabhängigkeit des osmotischen Druckes, Verdünnungswärme und Verdünnungsentropie hochmolekularer Lösungen, *Z. Phys. Chem., Abt. B*, 40, 319, 1938.

1939BAR   Barrer, R.M., Permeation, diffusion and solution of gases in organic polymers, *Trans. Faraday Soc.*, 35, 628, 1939.

1939MEY   Meyer, K.H., Wolff, E., and Boissonnas, Ch.G., Heat of dilution in the system: rubber-toluene, *Rubber Chem. Technol.*, 12, 504, 1939.

1940HAG   Hagger, O. and Van der Wyk, A.J.A., Proprietes des polymeres en solution XIV. Systeme triacetylcellulose-tetrachloroethane, *Helv. Chim. Acta*, 23, 484, 1940.

1940ME1   Meyer, K.H., Wolff, E., and Boissonnas, Ch.G., Proprietes des polymeres en solution. XII. Energie libre et chaleur de dilution. Systeme caoutchouc-toluene, *Helv. Chim. Acta*, 23, 430, 1940.

1940ME2   Meyer, K.H. and Van der Wyk, A.J.A., Proprietes de polymeres en solution. Resume de propriete thermodynamique de systeme binaires liquides, *Helv. Chim. Acta*, 23, 488, 1940.

1940WOL   Wolff, E., Proprietes des polymeres en solution. XIII. Energie libre et chaleur de dilution. Systeme guttapercha-toluene, *Helv. Chim. Acta*, 23, 439, 1940.

1941CA1   Calvet, E., Effects thermiques produits au cours de l'adsorption d'acetone par les nitrocelluloses, *C. R. Hebd. Seanc. Acad. Sci.*, 213, 126, 1941.

1941CA2   Calvet, E. and Izac, H., Calorimetric study of the absorption of acetone by nitrocellulose, *Compt. Rend. Trav. Fac. Sci. Marseille*, 1, 4, 1941.

1941KUN   Kunze, F., Mischungsentropie und Verdünnungswärme verdünnter Lösungen von Triacetylcellulose in Dioxan, *Z. Phys. Chem., Abt. A*, 188, 90, 1941.

1941TAG   Tager, A. and Kargin, V., Lösungs- und Quellungsvorgang von Zelluloseestern, *Acta Physicochim. URSS*, 14, 713, 1941.

1942CAL   Calvet, E., Tensions de vapeur des gels de nitrocellulose a l'acetone, *C. R. Hebd. Seanc. Acad. Sci.*, 214, 767, 1942.

1942GEE   Gee, G. and Treloar, L.R.G., The interaction between rubber and liquids. 1. A thermodynamical study of the system rubber-benzene, *Trans. Faraday Soc.*, 38, 147, 1942.

1945CA1   Calvet, E. and Maurizot, A., The mechanism of solution studied by the microcalorimeter. I. Application to the study of the gelatinization of nitrocellulose, *Mem. Services Chim. Etat (Paris)*, 32, 168, 1945.

1945CA2   Calvet, E. and Coutelle, J., The mechanism of solution studied by the microcalorimeter. III. Calorimetric study of the gelatinization of nitrocellulose by methyl nitrate, *Mem. Services Chim. Etat (Paris)*, 32, 200, 1945.

1945CA3   Calvet, E. and Coutelle, J., The mechanism of solution studied by the microcalorimeter. IV. Calorimetric study of the gelatinization of nitrocellulose by acetic esters, *Mem. Services Chim. Etat (Paris)*, 32, 204, 1945.

1945CA4   Calvet, E, The mechanism of solution studied by the microcalorimeter. V. Calorimetric study of the gelatinization of nitrocellulose by alcohol-ether mixture, *Mem. Services Chim. Etat (Paris)*, 32, 211, 1945.

1945CA5   Calvet, E., Sur les volumes specifiques des systemes nitrocellulose-acetone, les chaleurs de dilution et l'etat' semi-ideal des solutions acetoniques de nitrocellulose, *Bull. Soc. Chim. Belg.*, 12, 553, 1945.

1945FER   Ferry, J., Gee, G., and Treloar, L.R.G., The interaction between rubber and liquids. VII. The heats and entropies of dilution of natural rubber by various liquids, *Trans. Faraday Soc.*, 41, 340, 1945.

1945FLO   Flory, P.J., Thermodynamics of dilute solutions of high polymers, *J. Chem. Phys.*, 11, 453, 1945.

1946SCO   Scott, D.W., Osmotic pressure measurements with polydimethylsilicone fractions, *J. Amer. Chem. Soc.*, 68, 1877, 1946.

1946GEE   Gee, G. and Orr, W.J.C., The interaction between rubber and liquids. VIII. A new examination of the thermodynamic properties of the system rubber + benzene, *Trans. Faraday Soc.*, 42, 507, 1946.

1947CHE   Chedin, J. and Vandoni, R., Contribution a l'etude de la gelatinisation des hauts polymeres. Cas des nitrocelluloses, *Mem. Services Chim. Etat (Paris)*, 33, 205, 1947.

1947CAR   Carpenter, A.S., Crystallinity in solid colloids. Solution and diffusion in high polymers, *Trans. Faraday Soc.*, 43, 529, 1947.

1947GEE   Gee, G., Equilibrium properties of high polymer solutions and gels, *J. Chem. Soc.*, 280, 1947.

1947ORR   Orr, W.J.C., Statistical treatment of polymer solutions at infinite dilution, *Trans. Faraday Soc.*, 43, 12, 1947.

1947ROB   Roberts, D.E., Walton, W.W., and Jessup, R.S., Heats of combustion and solution of liquid styrene and solid polystyrene and heat of polymerization of styrene, *J. Polym. Sci.*, 2, 420, 1947.

1948BAR   Barrer, R.M. and Skirrow, G., Transport and equilibrium phenomena in gas-elastomer systems. II. Equilibrium phenomena, *J. Polym. Sci.*, 3, 564, 1948.

1948BRE   Breitenbach, J.W. and Frank, H.P., Über osmotische und Quellungsgleichgewichte von Polystyrol, *Monatsh. Chem.*, 79, 531, 1948.

1948TAG   Tager, A. and Kargin, V.A., The mechanism of swelling of rubber-like high-molecular substances. High-molecular isoparaffins and isooctane (Russ.), *Kolloidn. Zh.*, 10, 455, 1948.

1949BRO   Browning, G.V. and Ferry, J.D., Thermodynamic studies of polyvinyl acetate solutions in the dilute and moderately concentrated range, *J. Chem. Phys.*, 17, 1107, 1949.

1949GLI   Glikman, S.A. and Root, L.A., Kharakteristika energeticheskikh effektov razbavleniya rastvorov vysokopolimerov dilatometricheskim metodom, *Dokl. Akad. Nauk SSR*, 65, 701, 1949.

1949LAU   Lauer, K., Wilde, O., and Dobberstein, O., Zur Kenntnis des Lösungszustandes des Zellulosexanthogenats in Viskosen, *Kolloid Z. Z. Polymere*, 112, 16, 1949.

1950AME   Amerongen, G.J. van, Influence of structure of elastomers on their permeability to gases, *J. Polym. Sci.*, 5, 307, 1950.

1950BAW   Bawn, C.E.H., Freeman, R.F.J., and Kamaliddin, A.R., Vapor pressure of polystyrene solutions, *Trans. Faraday Soc.*, 46, 677, 1950.

1950LI1   Lipatov, S.M. and Meerson, S.I., Thermodynamic properties of polymer solutions. Effect of temperature on the heat of solution of polymers in various liquids (Russ), *Kolloidn. Zh.*, 12, 122, 1950.

1950LI2   Lipatov, S.M. and Meerson, S.I., Heat of solution of nitrocellulose films obtained from solutions of various concentrations (Russ.), *Kolloidn. Zh.*, 12, 427, 1950.

1950NEW   Newing, M.J., Thermodynamic studies of silicones in benzene solution, *Trans. Faraday Soc.*, 46, 613, 1950.

1950SCH   Schick, M.J., Doty, P., and Zimm, B.H., Thermodynamic properties of concentrated polystyrene solutions, *J. Amer. Chem. Soc.*, 72, 530, 1950.

1951GLI   Glikman, S.A. and Root, L.A., Kharakteristika energeticheskikh effektov razbavleniya rastvorov vysokopolimerov dilatometricheskim metodom, *Zh. Obshch. Khim.*, 21, 58, 1951.

1951MAN   Mandelkern, L. and Flory, P.J., Melting and glassy state transitions in cellulose esters and their mixtures with diluents, *J. Amer. Chem. Soc.*, 73, 3206, 1951.

1951MIN   Minassian, L. der and Magat, M., Contribution a la thermodynamique des solutions de hauts polymeres. VI. Thermodynamique du systeme polyisobutylene-cyclohexane, *J. Chim. Phys.*, 48, 574, 1951.

1952SC1   Schulz, G.V., Gute und schlechte Lösungsmittel für hochpolymere Stoffe, *Angew. Chem.*, 64, 553, 1952.

1952SC2   Schulz, G.V. and Doll, H., Thermodynamische Analyse der Lösungen von Polymethacryl-säureestern in verschiedenen Lösungsmitteln, *Z. Elektrochem.*, 56, 248, 1952.

1953BYK   Bykov, A.N. and Pakshver, A.B., Investigation of solutions of perchlorovinylic resins (Russ.), *Kolloind. Zh.*, 15, 321, 1953.

1953GUE   Guenner, K. [von] and Schulz, G.V., Lösungswärme und Phasenumwandlung von Polystyrolen, *Naturwiss.*, 40, 164, 1953.

1953KOK   Kokes, R.J., DiPietro, A.R., and Long, F.A., Equilibrium sorption of several organic diluents in polyvinyl acetate, *J. Amer. Chem. Soc.*, 75, 6319, 1953.

1953LIQ   Liquori, M. and Mele, A., Influenza del solvente e della temperatura sulla forma e le dimensioni di molecole poliammidiche in soluzione, *Gazz. Chim. Ital.*, 83, 941, 1953.

1953SC1   Schulz, G.V. and Doll, H., Thermodynamische Analyse der Lösungen von Polymethacrylsäuremethylester in verschiedenen Lösungsmitteln II., *Z. Elektrochem.*, 57, 841, 1953.

1953SC2   Schulz, G.V. and Hellfritz, H., Thermodynamische Analyse der Lösungen von Polystyrol in verschiedenen Lösungsmitteln, *Z. Elektrochem.*, 57, 835, 1953.

1953TAG   Tager, A.A. and Dombek, Zh.S., Thermodynamic study of polystyrene solutions (Russ.), *Kolloidn. Zh.*, 15, 69, 1953.

1954BAW   Bawn, C.E.H. and Wajid, M.A., Thermodynamic properties of polystyrene solutions, *J. Polym. Sci.*, 12, 109, 1954.

1954MEA   Meares, P., The diffusion of gases through polyvinyl acetate, *J. Amer. Chem. Soc.*, 76, 3415, 1954.

1955GAT   Gatovskaya, T.V., Kargin, V.A., and Tager, A.A., The effect of the molecular weight of polystyrene on the packing density of the molecular chain (Russ.), *Zh. Fiz. Khim.*, 29, 883, 1955.

1955ROS   Rosen, B., Polymer/solvent interactions in dilute solution, *J. Polym. Sci.*, 17, 559, 1955.

1955TA1   Tager, A.A. and Kosova, L.K., Thermodynamic studies of copolymer solutions 2. Thermodynamic study of butadiene-acrylonitrile copolymer solutions (Russ.), *Kolloidn. Zh.*, 17, 391, 1955.

1955TA2   Tager, A.A., Kosova, L.K., Karlinskaya, D.Yu., and Yurina, I.A., Thermodynamic studies of copolymer solutions 1. Thermodynamic study of butadiene-styrene copolymer solutions (Russ.), *Kolloidn. Zh.*, 17, 315, 1955.

1955TA3   Tager, A.A., Krivokorytova, R.V., and Khodorov, P.M., The heats of solution of polystyrenes of various molecular weights and the packing density of rigid chains (Russ.), *Dokl. Akad. Nauk SSSR*, 100, 741, 1955.

1956BA1   Bawn, C.E.H. and Patel, R.D., High polymer solutions. Part 8. The vapour pressure of solutions of polyisobutylene in toluene and cyclohexane, *Trans.Faraday Soc.*, 52, 1664, 1956.

1956BA2   Bawn, C.E.H. and Wajid, M.A., High polymer solutions. Part 7. Vapour pressure of polystyrene solutions in acetone, chloroform and propyl acetate, *Trans.Faraday Soc.*, 52, 1658, 1956.

1956CAN   Cantow, H.-J., Cantow, I., and Lindner, K., Determination of thermodynamic data from the temperature dependence of the light dispersion of high-molecular substances in solution (Ger.), *Z. Phys. Chem., N.F.*, 7, 58, 1956.

1956MEE   Meerson, S.I. and Lipatov, S.M., Dependence of the heat of solution of polymers on their physical state (Russ.), *Kolloidn. Zh.*, 18, 447, 1956.

1956SCH   Schmoll, K. and Jenckel, E., Über den Dampfdruck von Polystyrol-Lösungen in Toluol und Cyclohexan im ganzen Konzentrationsbereich, *Z. Elektrochemie*, 60, 756, 1965.

1957AKH   Akhmedov, K.S. and Lipatov, S.M., The heats of swelling and of solution of polymers in relation to molecular weights and temperatures (Russ.), *Kolloidn. Zh.*, 19, 257, 1957.

1957BAR   Barrer, R.M. and Barrie, J.A., Sorption and diffusion in ethyl cellulose. Part II. Quantitative examination of settled isotherms and permeation rates, *J. Polym. Sci.*, 23, 331, 1957.

1957BOO   Booth, C., Gee, G., and Williamson, G.R., Departures from random mixing in polyethylene-liquid systems, *J. Polym. Sci.*, 23, 3, 1957.

1957BYK   Bykov, A.N., Ivanova, M.I., and Pakshver, A.B., Altering the properties of polyamide by the inclusion method (Russ.), *Kolloidn. Zh.*, 19, 542, 1957.

1957JEF   Jeffries, R., The thermodynamic properties of mixtures of secondary cellulose acetate and its solvents, *Trans. Faraday Soc.*, 53, 1592, 1957.

1957MOO   Moore, W.R., Epstein, J.A., Brown, A.M., and Tidswell, B.M., Cellulose derivative-solvent interaction, *J. Polym. Sci.*, 23, 23, 1957.

1957SIL   Silberberg, A., Eliassaf, J., and Katchalsky, A., Temperature-dependence of light scattering and intrinsic viscosity of hydrogen bonding polymers, *J. Polym. Sci.*, 23, 259, 1957.

1957POK   Pokrovskii, L.I. and Pakshver, A.B., Izmenenie mezhmolekulyarnoi struktury kapronovogo volokna v rezul'tate teplovoi obrabotki, *Kolloidn. Zh.*, 19, 478, 1957.

1958BAR   Barrer, R.M., Barrie, J.A., and Slater, J., Sorption and diffusion in ethyl cellulose. Part III. Comparison between ethyl cellulose and rubber, *J. Polym. Sci.*, 27, 177, 1958.

1958JES   Jessup, R.S., Some thermodynamic properties of the systems polybutadiene-benzene and polyisobutylene-benzene, *J. Res. Natnl. Bur. Stand.*, 60, 47, 1958.

1958KHA   Kharitonova, V.P. and Pakshver, A.B., Influence of acetyl groups in acetylcellulose on the properties of its solutions (Russ.), *Kolloidn. Zh.*, 20, 110, 1958.

1958MOO   Moore, W.R. and Tidswell, B.M., Thermodynamic properties of solutions of cellulose derivatives. I. Dilute solutions of secondary cellulose acetate, *J. Polym. Sci.*, 27, 459, 1958.

1958TA1   Tager, A.A. and Gur'yanova, N.M., Heat of solution and packing of polymer molecules in various physical states (Russ.), *Zh. Fiz. Khim.*, 32, 1958, 1958.

1958TA2   Tager, A.A. and Galkina, L.A., Thermodynamic study of the process of solution of polystyrene in methyl ethyl ketone and ethyl acetate (Russ.), *Nauchn. Dokl. Vyssh. Shkol., Khim. Khim. Tekhnol.*, (2), 357, 1958.

1959KOZ   Kozlov, P.V., Iovleva, M.M., and Shiryaeva, L.L., Thermodynamic study of copolymer solutions of ethylene glycol and terephthalic and sebacic acids (Russ.), *Vysokomol. Soedin.*, 1, 1106, 1959.

1959KRI   Krigbaum, W.R. and Geymer, D.O., Thermodynamics of polymer solutions. The polystyrene-cyclohexane system near the Flory theta temperature, *J. Amer. Chem. Soc.*, 81, 1859, 1959.

1959MAR   Maron, S.H. and Nakajima, N., A theory of the thermodynamic behavior of nonelectrolyte solutions. II. Application to the system rubber-benzene, *J. Polym. Sci.*, 40, 59, 1959.

1959MEE   Meerson, S.I. and Lipatov, S.M., Temperature dependence of the heat of solution of polystyrene and its packing density (Russ.), *Kolloidn. Zh.*, 21, 531, 1959.

1959MEN   Mencik, Z., Interaction between polyvinylchloride and solvent, *Coll. Czech. Chem. Commun.*, 24, 3291, 1959.

1959SCH   Schulz, G.V. and Horbach, A., Kalorimetrische Messungen zur Thermodynamik von Polystyrollösungen, *Z. Phys. Chem., N.F.*, 22, 377, 1959.

1959TA1   Tager, A. A. and Popova, O., The effect of the molecular weight of vitreous polymers on the packing density of their chains. III. Triacetylcellulose (Russ.), *Zh. Fiz. Khim.*, 33, 593, 1959.

1959TA2   Tager, A.A., Tsilipotkina, M.V., and Doronina, B.K., The effect of the molecular weight of vitreous polymers on the packing density of their chains. II. Poly(methyl methacrylate) (Russ.), *Zh. Fiz. Khim.*, 33, 335, 1959.

1959TSU   Tsuda, Y., Heat of solution of methanolized cellulose, *Bull. Chem. Soc. Japan*, 32, 372, 1959.

1960TIK   Tikhomirova, N.S., Malinksii, Yu.M., and Karpov, V.L., Studies on diffusion processes in polymers. I. Diffusion of monoatomic gases through polymer films of different structures (Russ.), *Vysokomol. Soedin.*, 2, 221, 1960.

1961DAO   Daoust, H. and Senez, M., Viscometric study of the polyisobutylene-chlorobenzene system, *Polymer*, 2, 393, 1961.

1961KUW   Kuwahara, N. and Miyake, Y., Second virial coefficient of polymer solutions, *Kobunshi Kagaku*, 18, 153, 1961.

1961MIC   Michaels, A.S. and Bixler, H.J., Solubility of gases in polyethylene, *J. Polym. Sci.*, 50, 393, 1961.

1961MIZ   Mizutani, H., Extension of high polymer molecule and energy change by mixing, *J. Phys. Soc. Japan*, 16, 282, 1961.

1961VAR   Varadaiah, V.V. and Rao, V.S.R., Intrinsic viscosity-temperature studies of polymethyl methacrylate, *J. Sci. Ind. Res. India*, 20B, 280, 1961.

1962ANA   Anagnostopoulos, C.E. and Coran, A.Y., Polymer-diluent interactions. II. Poly(vinyl chloride)-diluent interactions, *J. Polym. Sci.*, 57, 1, 1962.

1962BAK   Baker, C.H., Brown, W.B., Gee, G., Rowlinson, J.S., Stubley, D., and Yeadon, R.E., A study of the thermodynamic properties and phase equilibria of solutions of polyisobutylene in n-pentane, *Polymer*, 3, 215, 1962.

1962BAR   Barrer, R.M., Barrie, J.A., and Raman, N.K., Solution and diffusion in silicone rubber. I. A comparison with natural rubber, *Polymer*, 3, 595, 1962.

1962DRA   Draisbach, H.-Ch., Jeschke, D., and Stuart, H.A., Zur Sorption und Diffusion von Gasen in Poly-Äthylenglykol-Terephthalsäure-Ester, *Z. Naturfosch.*, 17A, 447, 1962.

1962FOX   Fox, T.G., Properties of dilute polymer solutions III. Intrinsic viscosity-temperature relationships for conventional polymethyl methacrylate, *Polymer*, 3, 111, 1962.

1962MUR   Muramoto, A., Dependence of interaction parameter on molecular weight and concentration for solutions of poly(dimethylsiloxane) in methylethylketone, *Polymer*, 23, 1311, 1962.

1962SEN   Senez, M. and Daoust, H., Heats of dilution of polyisobutylene solutions, *Can. J. Chem.*, 40, 734, 1962.

1963BLA   Blanks, R.F. and Prausnitz, J.M., Infrared studies of hydrogen bonding in isopropanol and in mixtures of isopropanol with poly(n-butyl methacrylate) and with poly(propylene oxide), *J. Chem. Phys.*, 38, 1500, 1963.

1963GEC   Gechele, G.B. and Crescentini, L., Melting temperatures and polymer-solvent interaction for polycaprolactam, *J. Appl. Polym. Sci.*, 7, 1349, 1963.

1963KOZ   Kozlov, P.V., Fainberg, E.Z., and Andreeva, I.N., Effect of thermal treatment on the structure of cellulose tribenzoate (Russ.), *Vysokomol. Soedin., Tsellyuloza Proisvodn., Sb. Stat.*, 192, 1962.

1963ME1   Meerson, S.I. and Zagraevskaya, I.M., Thermochemical and dilatometric investigations of crystallizable polymer gels (Russ.), *Kolloidn. Zh.*, 25, 197, 1963.

1963ME2   Meerson, S.I. and Zagraevskaya, I.M., Thermochemical and dilatometric investigations of amorphous polymer gels (Russ.), *Kolloidn. Zh.*, 25, 202, 1963.

1963MO1   Moore, W.R. and Fort, R.J., Viscosities of dilute solutions of polymethyl methacrylate, *J. Polym. Sci.: Part A*, 1, 929, 1963.

1963MO2   Moore, W.R. and Shuttleworth, R., Thermodynamic properties of solutions of cellulose acetate and cellulose nitrate, *J. Polym. Sci.: Part A*, 1, 733, 1963.

1963MO3   Moore, W.R. and Shuttleworth, R., Thermodynamic properties of solutions of cellulose triacetates, *J. Polym. Sci.: Part A*, 1, 1985, 1963.

1963TAG   Tager, A.A. and Podlesnyak, A.I., Concentrated polymer solutions. I. Determination of the integral and differential heats of solution and dilution of polyisobutylene and polystyrene (Russ.), *Vysokomol. Soedin., Ser. A*, 5, 87, 1963.

1964BIR   Biros, J., Solc, K., and Pouchly, J., Ein Beitrag zur thermodynamischen Wechselwirkung Polymer-Lösungsmittel, *Faserforsch. Textiltechn.*, 15, 608, 1964.

1964BO1   Booth, C., Gee, G., Holden, G., and Williamson, G.R., Studies in the thermodynamics of polymer-liquid systems. Part I - Natural rubber and polar liquids, *Polymer*, 5, 343, 1964.

1964BO2   Booth, C., Gee, G., Jones, M.N., Taylor, W.D., Studies in the thermodynamics of polymer-liquid systems. Part II - A reassessment of published data, *Polymer*, 5, 353, 1964.

1964CON   Conway, B.E. and Nicholson, J.P., Some experimental studies on enthalpy and entropy effects in equilibrium swelling of polyoxypropylene elastomers, *Polymer*, 5, 387, 1964.

1964DEL   Delmas, G., Patterson, D., and Bhattacharyya, S.N., Heats of mixing of polymers with mixed-solvent media, *J. Phys. Chem.*, 68, 1468, 1964.

1964FRE   Frensdorff, H.K., Diffusion and sorption of vapors in ethylene-propylene copolymers. I. Equilibrium sorption, *J. Polym. Sci., Pt. A*, 2, 333, 1964.

1964HOL   Holly, E.D., Interaction parameters and heats of dilution for ethylene-propylene rubber in various solvents, *J. Polym. Sci., Pt. A, 2*, 5267, 1964.

1964KAW   Kawai, T. and Ida, E., Analysis of intrinsic viscosity data on polyacrylonitrile in various solvents, *Kolloid Z. Z. Polym.*, 194, 40, 1964.

1964USM   Usmanov, Kh.U., Tillaev, R.S., and Musaev, Y.N., Radiation copolymerization of acrylonitrile with sylvane (Russ.), *Nauchn. Tr. Tashkent. Gos. Univ.*, 257, 3, 1964.

1965BIA   Bianchi, U., Pedemonte, E., and LoGiudice, M., Heat of solution of polymeric glasses. Atactic polystyrene, *Ric. Sci. Rend. Chim.*, 8, 1083, 1965.

1965BRO   Brooks, R.L. and Lawrence, A.R., Polymer-diluent interactions of some nitrocellulose-nitrate ester systems, *J. Appl. Polym. Sci.*, 9, 707, 1965.

1965COT   Cottam, B.J., Cowie, J.M.G., and Bywater, S., Some solution properties of poly(α-methylstyrene)s of low molecular weight, *Makromol. Chem.*, 86, 116, 1965.

1965SAT   Sato, H. and Nakamura, R., Thermodynamic properties of solutions of poly(ethylene glycol), *Nippon Kagaku Zasshi*, 86, 168, 1965.

1965TAG   Tager, A.A. and Karas, L.Ya., Thermodynamics of the swelling of cross-linked polyurethanes (Russ.), *Dokl. Akad. Nauk SSSR*, 165, 1122, 1965.

1966CAN   Canagaratna, S.G., Margerison, D., and Newport, J.P., Thermodynamics of solutions of polymers of α-methyl styrene in α-methyl styrene and cumene, *Trans. Faraday Soc.*, 62, 3058, 1966.

1966HEL   Hellwege, K.-H., Johnsen, U., and Lehmann, J., Der Einfluss des Lösungsmittels auf Eigenschaften von Einkristallen des Poly-3,3-(bischloromethyl)oxacyclobutans (Penton), *Kolloid Z. Z. Polymere*, 209, 1, 1966.

1966KAR   Karasev, N.E., Dymarchuk, N.P., and Mishchenko, K.P., Determination of integral heat of interaction between cellulose and sodium hydroxide from heat of dilution (Russ.), *Zh. Prikl. Khim.*, 39, 2301, 1966.

1966NAG   Nagasawa, M., Asai, Y., and Sugiura, I., Determination of molecular weight distribution of linear polymers by measuring specific heat of solution (Jap.), *Kogyo Kagaku Zasshi*, 69, 1759, 1966.

1966NAK   Nakajima, A., Fujiwara, H., and Hamada, F., Phase relationships and thermodynamic interactions in linear polyethylene-diluent systems, *J. Polym. Sci.: Part A-2*, 4, 507, 1966.

1966TAK   Takamizawa, K., Osmotic pressure measurements for poly-p-chlorostyrene solutions, *Bull. Chem. Soc. Japan*, 39, 1186, 1966.

1967CLU   Clunie, J.S., Corkill, J.M., Goodman, J.F., and Symons, P.C., Thermodynamics of nonionic surface-active agent + water systems, *Trans. Faraday Soc.*, 63, 2839, 1967.

1967LLO   Llopis, J., Albert, A., and Usobiaga, P., Studies of poly(ethyl acrylate) in theta solvents, *Eur. Polym. J.*, 3, 259, 1967.

1967POU   Pouchly, J., Biros, J., Solc, K., and Vondrejsova, J., The thermodynamic properties of solutions of polystyrene and of low molecular weight model compounds, *J. Polym. Sci.: Part C*, 16, 679, 1967.

1967SK1   Skerjanc, J., Dolar, D., and Leskovsek, D., Heats of dilution of polyelectrolyte solutions. I. Polystyrenesulphonic acid and its sodium salt, *Z. Phys. Chem., N. F.*, 56, 207, 1967.

1967SK2   Skerjanc, J., Dolar, D., and Leskovsek, D., Heats of dilution of polyelectrolyte solutions. II. Zinc polystyrenesulphonate, *Z. Phys. Chem., N. F.*, 56, 218, 1967.

1967ZVE   Zverev, M.P., Barash, A.N., Nikonorova, L.I., Ivanova, L.V., and Zubov, P.I., Effect of solvent nature and precipitant composition on polyacrylonitrile behavior (Russ.), *Vysokomol. Soedin., Ser. A*, 9, 927, 1967.

1968BAR   Barrer, R.M., Barrie, J.A., and Wong, P.S.-L., The diffusion and solution of gases in highly crosslinked copolymers, *Polymer*, 9, 609, 1968.

1968FL1   Flory, P.J., Eichinger, B.E., and Orwoll, R.A., Thermodynamics of mixing polymethylene and polyisobutylene, *Macromolecules*, 1, 287, 1968.

1968FL2   Flory, P.J., Ellenson, J.L., and Eichinger, B.E., Thermodynamics of mixing of n-alkanes with polyisobutylene, *Macromolecules*, 1, 279, 1968.

1968GIA   Giacommeti, G., Turolla, A., and Boni, R., Enthalpy of helix-coil transitions from heats of solution. I. Polyglutamates, *Biopolymers*, 6, 441, 1968.

1968KUW   Kuwahara, N., Oikawa, T., and Kaneko, M., Excluded-volume effect of poly(methyl methacrylate) solutions, *J. Chem. Phys.*, 49, 4972, 1968.

1968LA1   Lakhanpal, M.L., Chhina, K.S., and Sharma, S.C., Thermodynamic properties of aqueous solutions of polyoxethyleneglycols, *Indian J. Chem.*, 6, 505, 1968.

1968LA2   Lakhanpal, M.L., Singh, H.G., Singh, H., and Sharma, S.C., A comparative study of heats of mixing of polyoxypropylene glycol-water and polyoxypropylene glycol-water systems, *Indian J. Chem.*, 6, 95, 1968.

1968LIP   Lipatov, Yu.S., Sergeeva, L.M., and Kovalenko, G.F., Investigation of sorption of vapors into polyurethane (Russ.), *Vysokomol. Soedin., Ser. B*, 10, 205, 1968.

1968MA1   Maron, S.H. and Daniels, C.A., Thermodynamics of polyisobutylene solutions. III. Thermal behavior and polymer order, *J. Macromol. Sci.-Phys. B*, 2, 591, 1968.

1968MA2   Maron, S.H. and Daniels, C.A., Thermodynamics of polystyrene solutions. II. Glass contributions and thermal behavior, *J. Macromol. Sci.-Phys. B*, 2, 769, 1968.

1968MA3   Maron, S.H. and Filisko, F., Use of microcalorimetry for determination of order in polymers, *Anal. Calorimetry, Proc. Amer. Chem. Soc. Symp.*, 155, 153, 1968.

1968MAR   Marsh, K.N., Thermodynamics of octamethylcyclotetrasiloxane mixtures, *Trans. Faraday Soc.*, 64, 883, 1986.

1968PAT   Patterson, D., Bhattacharyya, S.N., and Picker, P., Thermodynamics of chain-molecule mixtures: heats of mixing linear methylsiloxanes, *Trans. Faraday Soc.*, 64, 648, 1968.

1968SEE   Seeley, R.D., Thermodynamic properties and polymer solvent interaction parameters for silicone rubber networks, *Rubber Chem. Technol.*, 41, 608, 1968.

1968TAG   Tager, A.A., Podlesnyak, A.I., and Demidova, L.V., Temperature dependence of the thermodynamic parameters of polyisobutylene dissolution (Russ.), *Vysokomol. Soedin., Ser. B*, 10, 601, 1968.

1968ZUB   Zubov, L.N., Pakshver, A.B., Fikhman, V.D., and Meerson, S.I., Effect of spinning solution characteristics of the properties of poly(vinyl chloride) fibers (Russ.), *Khim. Volokna*, (1), 26 1968.

1969BOH   Bohdanecky, M., On the structure and properties of vinyl polymers and their models. XIII. Viscometrically determined polyvinyl chloride-solvent interaction parameters, *Coll. Czech. Chem. Commun.*, 34, 2065, 1969.

1969CLU   Clunie, J.S., Goodman, J.F., and Symons, P.C., Phase equilibria of dodecylhexaoxyethylene glycol monoether in water, *Trans. Faraday Soc.*, 65, 287, 1969.

1969ICH   Ichihara, S. and Hata, T., Studies on miscibility of polymer II. Heat of mixing of polymers, *Kobunshi Kagaku*, 26, 249, 1969.

1969LEO   Leonard, J., Thermodynamics of equilibrium polymerization in solution. Effect of polymer concentration on the equilibrium monomer concentration, *Macromolecules*, 2, 661, 1969.

1969LEW   Lewis, G. and Johnson, A.F., Heat of dilution of polymer solutions. Part I. Polystyrene-toluene system, *J. Chem. Soc. A*, 1816, 1969.

1969NAP   Napper, D.H., A reversible transformation from enthalpic to entropic stabilization, *Kolloid Z. Z. Polym.*, 234, 1149, 1969.

1969POD   Poddubnyi, I.Ya. and Podalinskii, A.V., Possible production of thermodynamically ideal polymer-solvent systems (Russ.), *Dokl. Akad. Nauk SSSR*, 185, 401, 1969.

1969POU   Pouchly, J. and Biros, J., On the site of specific solvation in the poly(vinyl chloride) molecule, *J. Polym. Sci.: Polym. Lett.*, 7, 463, 1969.

1970ASH   Ash, R., Barrer, R.M., and Palmer, D.G., Solubility and transport of gases in nylon and polyethylene, *Polymer*, 11, 421, 1970.

1970DAN   Daniels, C.A. and Maron, S.H., Effect of polymer order on the vapor pressures of polypropylene solutions, *J. Macromol. Sci.-Phys. B*, 4, 227, 1970.

1970GRO   Gromov, V.F., Khomikovskii, P.M., and Abkin, A.D., Medium influence on radical polymerization of acrylamide (Russ.), *Vysokomol. Soedin., Ser. B*, 12, 767, 1970.

1970MO1   Morimoto, S., Calorimetric investigations on polymer solutions. IV. Heats of dilution of polystyrene solutions and their molecular weight dependence, *Bull. Res. Inst. Polym. Textil.*, 90(3), 38, 1970.

1970MO2   Morimoto, S., Contact interactions of polydimethylsiloxane with solvents, *Rep. Progr. Polym. Phys. Japan*, 13, 29, 1970.

1970MO3   Morimoto, S., Contact interactions of polydimethylsiloxane and its low oligomers with solvents, *Makromol. Chem.*, 133, 197, 1970.

1970MO4   Morimoto, S., Heats of dilution of polystyrene solutions and their molecular weight dependence (Jap.), *Nippon Kagaku Zasshi*, 91, 117, 1970.

1970MUR   Muramoto, A., Studies on the interaction parameter in polysiloxane solutions, *Polym.J.*, 1, 450, 1970.

1970OIK   Oikawa, T., Kuwahara, N., and Kaneko, M., Thermodynamic properties of cellulose-trinitrate-acetone solution, *Rep. Progr. Polym. Phys. Japan*, 13, 9, 1970.

1970TAG   Tager, A.A., Podlesnyak, A.I., Tsilipotkina, M.V., Adamova, L.V., Bakhareva, A.A., and Demidova, L.V., Temperature and concentration functions of enthalpy, free energy, and entropy of mixing of polystyrene solutions with upper and lower critical solution temperatures (Russ.), *Vysokomol. Soedin., Ser. A*, 12, 1320, 1970.

1971BO1   Booth, C. and Devoy, C.J., Thermodynamics of mixtures of poly(ethylene oxide) and benzene, *Polymer*, 12, 309, 1971.

1971BO2   Booth, C. and Devoy, C.J., Thermodynamics of mixtures of poly(propylene oxide) and benzene, *Polymer*, 12, 320, 1971.

1971BRU   Bruns, W., Mehdorn, F., and Ueberreiter, K., Ein Beitrag zur Thermodynamik der Glasumwandlung. Lösungs- und Verdünnungswärme des Systems Polystyrol-Toluen oberhalb und unterhalb der Glastemperatur, *Kolloid-Z. Z. Polym.*, 244, 202, 1971.

1971HO1   Höcker, H. and Flory, P.J., Thermodynamics of polystyrene solutions Part 1. Polystyrene and methyl ethyl ketone, *Trans. Faraday Soc.*, 67, 2258, 1971.

1971HO2   Höcker, H. and Flory, P.J., Thermodynamics of polystyrene solutions Part 2. Polystyrene and ethylbenzene, *Trans. Faraday Soc.*, 67, 2270, 1971.

1971HO3   Höcker, H., Shih, H., and Flory, P.J., Thermodynamics of polystyrene solutions Part 3. Polystyrene and cyclohexane, *Trans. Faraday Soc.*, 67, 2275, 1971.

1971DRE   Dreifus, D.W. and Patterson, D., Heats of mixing of polymer + solvent systems and corresponding states theories, *Trans. Faraday Soc.*, 67, 631, 1971.

1971IC1   Ichihara, S., Komatsu, A., and Hata, T., Thermodynamic studies on the glass transition and the glassy state of polymers. II. Enthalpies and specific heats of polystyrene glasses of different thermal histories, *Polym. J.*, 2, 644, 1971.

1971IC2   Ichihara, S., Komatsu, A., and Hata, T., Thermodynamic studies on the glass transition and the glassy state of polymers. III. Poly(α-methylstyrene), *Polym. J.*, 2, 650, 1971.

1971MOR   Morimoto, S., Calorimetric study of polystyrenes of low molecular weight, *Bull. Chem. Soc. Japan*, 44, 879, 1971.

1971TAG   Tager, A.A., Suvorova, A.I., Bessonov, Yu.S., Podlesnyak, A.I., Koroleva, I.A., Adamova, L.V., and Tsilipotkina, M.V., Orientation order in poly(vinyl acetate) solutions (Russ.), *Vysokomol. Soedin., Ser. A*, 13, 2454, 1971.

1972CRE   Crescenzi, V., Quadrifolio, F., and Delben, F., Calorimetric investigation of poly(methacrylic acid) and poly(acrylic acid) in aqueous solution, *J. Polym. Sci.: Part A-2*, 10, 357, 1972.

1972DAN   Daniels, C.A. and Maron, S.H., Thermodynamic behavior of polymethyl methacrylate solutions, *J. Macromol. Sci.-Phys. B*, 6, 1, 1972.

1972DAV   Davalloo, P., Gainer, J.L., and Hall, K.R., Enthalpies of solution in complex systems. Albumin + KCl(aq), polystyrene + toluene, and polystyrene + carbon tetrachloride, *J. Chem. Thermodyn.*, 4, 691, 1972.

1972FLO   Flory, P.J. and Shih, H., Thermodynamics of solutions of poly(dimethylsiloxane) in benzene, cyclohexane, and chlorobenzene, *Macromolecules*, 5, 761, 1972.

1972GIR   Girolamo, M. and Urwin, J.R., Thermodynamic parameters from osmotic studies on solutions of block copolymers of polyisoprene and polystyrene, *Eur. Polym. J.*, 8, 299, 1972.

1972IZU   Izumi, Y. and Miyake, Y., Study of linear poly(p-chlorostyrene)-diluent systems. I. Solubilities, phase relationships, and thermodynamic interactions, *Polym. J.*, 3, 647, 1972.

1972LEC   Lechner, M.D., Schulz, G.V., and Wolf, B.A., Thermodynamics of polymer solutions as functions of pressure and temperature, *J. Coll. Interface Sci.*, 39, 462, 1972.

1972MAS   Masa, Z., Biros, J., Trekoval, J., and Pouchly, J., Specific interactions in solutions of polymers. IV. A calorimetric study of butyl methacrylate oligomers in chloroform, *J. Polym. Sci.: Part C*, 39, 219, 2972.

1972TAG   Tager, A.A., Adamova, L.V., Bessonov, Yu.S., Kuznetsov, V.N., Plyusnina, T.A., Soldatov, V.V., and Tsilipotkina, M.V., Thermodynamic study of oligomeric poly(oxypropylene)diol solutions in water and n-hexane in the precritical region (Russ.), *Vysokomol. Soedin., Ser. A*, 14, 1991, 1972.

1973GAE   Gaeckle, D., Kao, W.-P., Patterson, D., and Rinfret, M., Exothermic heats of dilution in non-polar polymer solutions, *J. Chem. Soc., Faraday Trans. I*, 69, 1849, 1973.

1973ISE   Ise, N., Mita, K., and Okubo, T., Heat of dilution of aqueous solutions of polyacrylates, *J. Chem. Soc., Faraday Trans. I*, 69, 106, 1973.

1973KHA   Khalik, E.M., Gal'braikh, L.S., Ilieva, N.I., Meerson, S.I., and Rogovin, Z.A., Synthesis and study of some properties of partially substituted cellulose acetates (Russ.), *Vysokomol. Soedin., Ser. B*, 15, 452, 1973.

1973MAN   Manzini, G. and Crescenzi, V., Heat of dilution and density data for poly(β-propiolactone) and poly(ε-caprolactone), *Polymer*, 14, 343, 1973.

1973MA1   Maron, S.H. and Lee, M.-S., Thermodynamics of polyvinyl chloride solutions. I. Solutions in cyclohexanone, *J. Macromol. Sci.-Phys. B*, 7, 29, 1973.

1973MA2   Maron, S.H. and Lee, M.-S., Thermodynamics of polyvinyl chloride solutions. II. Solutions in cyclopentanone, *J. Macromol. Sci.-Phys. B*, 7, 47, 1973.

1973MA3   Maron, S.H. and Lee, M.-S., Thermodynamics of polyvinyl chloride solutions. III. Solutions in tetrahydrofuran, *J. Macromol. Sci.-Phys. B*, 7, 61, 1973.

1973NEK   Nekrasova, T.N. and Eskin, V.E., Partial specific volume of polyisobutylene in various solvents (Russ.), *Vysokomol. Soedin., Ser. A*, 15, 2429, 1973.

1974BES   Bessonov, Yu.S. and Tager, A.A., Thermodynamic investigation of oligomeric polyoxypropylenediol-water system in the demixing region (Russ.), *Tr. Khim. Khim. Tekhnol.*, (1), 150, 1974.

1974DOL   Dolar, D. and Skerjanc, J., Studies of polyelectrolyte solutions containing mixtures of mono- and divalent counterions. II. Heat of dilution, *J. Phys. Chem.*, 61, 4106, 1974.

1974FI1   Filisko, F.E. and Raghava, R.S., Amorphous structure heat: molecular mechanisms from solution heats of poly(2,6-dimethyl-p-phenylene oxide) in orthodichlorobenzene, *J. Appl. Phys.*, 45, 4151, 1974.

1974FI2   Filisko, F.E., Raghava, R.S., and Yeh, G.S.Y., Amorphous structure heat. Temperature dependence of heats of solution for polystyrene in toluene and ethylbenzene, *J. Macromol. Sci.-Phys. B*, 10, 371, 1974.

1974HUN   Hung, G.W.C., Solvent-polymer interaction. I. Molecular transport of some selected organic liquids in polymer membrane, *Microchem. J.*, 19, 130, 1974.

1974MIT   Mita, K. and Okubo, T., Heat of dilution of aqueous solutions of sodium carboxymethyl-cellulose and sodium polyacrylate, *J. Chem. Soc., Faraday Trans. I*, 70, 1546, 1974.

1974NAK   Nakajima, A., Hamada, F., Yasue, K., Fujisawa, K., and Shiomi, T., Thermodynamic studies based on corresponding states theory for solutions of polystyrene in methyl ethyl ketone, *Makromol. Chem.*, 175, 197, 1974.

1974RAG   Raghava, R.S. and Filisko, F.E., Amorphous structure heat: Molecular structure from solution heats of polymethylmethacrylate in o-dichlorobenzene, *J. Appl. Phys.*, 45, 4155, 1974.

1974TAK   Takashima, Y., Misa, K., Miyata, S., and Sakaoku, K., Construction of a microcalorimeter and measurement of heats of solution of stretched glassy polystyrene, *Anal. Calorim.*, (3), 1, 1974.

1974TS1   Tsiperman, R.F., Pavlinov, L.I., and Krasnov, E.P., Thermodynamic properties of the system poly(m-phenyleneisophthalamide)-dimethylacetamide (Russ.), *Khim. Volokna*, (4), 16, 1974.

1974TS2   Tsiperman, R.F., Pavlinov, L.I., and Krasnov, E.P., Thermodynamic properties of the poly(m-phenyleneisophthalamide)-dimethylformamide system (Russ.), *Vysokomol. Soedin., Ser. B*, 16, 460, 1974.

1975HOR   Horacek, H., Gleichgewichtsdrücke, Löslichkeit und Mischbarkeit des Systems Polyethylen niedriger Dichte und Kohlenwasserstoffen bzw. halogenierten Kohlenwasserstoffen, *Makromol. Chem., Suppl.*, 1, 415, 1975.

1975IKE   Ikeda, M., Suga, H., and Seki, S., Thermodynamic studies of solid polyethers. 5. Crystalline-amorphous interfacial thermal properties, *Polymer*, 16, 634, 1975.

1975MAN   Manzini, G., Crescenzi, V., and Furlanetto, R., Thermodynamics of aliphatic diester and polyester solutions, *Macromolecules*, 8, 198, 1975.

1975MIT   Mita, K., Okubo, T., and Ise, N., Heats of dilution of aqueous solutions of polyethyleneimine hydrochloride and sodium polyphosphate and their low molecular weight analogues, *J. Chem. Soc., Faraday Trans. I*, 71, 1932, 1975.

1975MUR   Murakami, S., Kimura, F., and Fujishiro, R., Enthalpies of dilution of poly(1-vinyl-2-pyrrolidone) + aqueous methanol solution, *Makromol. Chem.*, 176, 3425, 1975.

1975TA1   Tager, A.A. and Bessonov, Yu.S., Thermodynamic study of solutions of poly(vinyl acetate) and cellulose tricarbanilate in the precritical region (Russ.), *Vysokomol. Soedin., Ser. A*, 17, 2377, 1975.

1975TA2   Tager, A.A. and Bessonov, Yu.S., Thermochemistry of solutions of polymers and polymer compositions in the phase separation region (Russ.), *Vysokomol. Soedin., Ser. A*, 17, 2383, 1975.

1975TA3   Tager, A.A., Sholokhovich, T.I., and Bessonov, Yu.S., Thermodynamics of mixing of polymers, *Eur. Polym. J.*, 11, 321, 1975.

1976BO1   Boyd, G.E. and Wilson, D.P., Enthalpies of dilution of strong polyelectrolyte solutions. Comparisons with the cell and line charge theories, *J. Phys. Chem.*, 80, 805, 1976.

1976BO2   Boyd, G.E., Wilson, D.P., and Mannig, G.S., Enthalpies of mixing of polyelectrolytes with simple aqueous electrolyte solutions, *J. Phys. Chem.*, 80, 808, 1976.

1976FIL   Filisko, F.E. and Raghava, R.S., Amorphous structure heat: heats of solutions versus temperature for polysulfone in o-dichlorobenzene, *J. Macromol. Sci.-Phys. B*, 12, 317, 1976.

1976LA1   Lakhanpal, M.L., Sharma, S.C., Krishan, B., and Parashar, R.N., Enthalpies of mixing of polyoxyethylene glycols in benzene, carbon tetrachloride, methyl alcohol and ethyl alcohol, *Indian J. Chem.*, 14A, 642, 1976.

1976LA2   Lakhanpal, M.L., Chaturvedi, L.K., Puri, T., and Sharma, S.C., Thermochemical studies on ternary systems. Part I - Enthalpies of mixing of polyoxyethylene glycols in aqueous dioxane and ethyl alcohol, *Indian J. Chem.*, 14A, 645, 1976.

1976MIT   Mita, K., Okubo, T., and Ise, N., Heat of dilution of aqueous solutions of sodium salts of partially sulphonated polystyrenes, *J. Chem. Soc., Faraday Trans. I*, 72, 504, 1976.

1976POU   Pouchly, J. and Patterson, D., Polymers in mixed solvents, *Macromolecules*, 9, 574, 1976.

1976WOL   Wolf, B,A., Bieringer, H.F., and Breitenbach, J.W., Comparison of the unperturbed dimension of polystyrene for endothermal and exothermal heats of dilution within the same system, *Polymer*, 17, 605, 1976.

1977BA1   Baba, Y., Fujimoto, K., Kagemoto, A., and Fujishiro, R., The heats of dilution of polystyrene solutions, *Makromol. Chem.*, 178, 1439, 1977.

1977BA2   Baba, Y. and Kagemoto, A., The heats of dilution of isotactic polystyrene solution in toluene and in anisole, *Netsusokutei*, 4(2), 57, 1977.

1977HUA   Huang, Y.-C. and Eichinger, B.E., Flow microcalorimetry on dilute polymer solutions, *Polymer*, 18, 55, 1977.

1977TA1   Tager, A.A., Kolmakova, L.K., Bessonov, Yu.S., Salazkin, S.N., and Trofimova, N.M., Effect of the molecular mass and porous structure of cardo polyarylate on the thermodynamic parameters of dissolution (Russ.), *Vysokomol. Soedin., Ser. A*, 19, 1475, 1977.

1977TA2   Tager, A.A., Kolmakova, L.K., Anufriev, V.A., Bessonov, Yu.S., Zhigunova, O.A., Vinogradova, S.V., Salazkin, S.N., and Tsilipotkina, M.V., Thermodynamics of the dissolution of cardo polyarylates in chloroform and tetrachloroethane (Russ.), *Vysokomol. Soedin., Ser. A*, 19, 2367, 1977.

1977TA3   Tager, A.A. and Ikanina, T.V., Thermodynamic study of dilute and concentrated solutions of poly(p-chlorostyrene) (Russ.), *Vysokomol. Soedin., Ser. B*, 19, 192, 1977.

1977WOL   Wolf, B.A. and Jend, R., Über die Möglichkeiten zur Bestimmung von Mischungs-enthalpien und -volumina aus der Molekulargewichtsabhängigkeit der kritischen Entmischungstemperaturen und -drücke am Beispiel des Systems *trans*-Decahydronaphthalin/Polystyrol, *Makromol. Chem.*, 178, 1811, 1977.

1978COM   Comper, W.D. and Laurent, T.C., An estimate of the enthalpic contribution to the interaction between dextran and albumin, *Biochem. J.*, 175, 703, 1978.

1978OKU   Okubo, M. and Ueberreiter, K., Surface tension of polymer solutions 7. *Colloid Polym. Sci.*, 256, 941, 1978.

1978SCH   Scholz, W., Kalorimetrische Untersuchungen zum Lösungsverhalten von Polyacrylsäure und Polyacrylamid in Wasser, Formamid und Ethylenglykol, *Dissertation*, TU Braunschweig, 1978.

1978STE   Stern, S.A. and DeMeringo, A.H., Solubility of carbon dioxide in cellulose acetate at elevated pressure, *J. Polym. Sci.: Polym. Phys. Ed.*, 16, 735, 1978.

1979FUJ   Fujihara, I., Tamura, K., Murakami, S., and Fujishiro, R., Enthalpies of dilution of polymer solutions, *Polym. J.*, 11, 153, 1979.

1979GAR   Garvey, M.J. and Robb, I.D., Effect of electrolytes on solution behavior of water soluble macromolecules, *J. Chem. Soc., Faraday Trans. I*, 75, 993, 1979.

1979KLE   Klein, J. and Scholz, W., Kalorimetrische Untersuchungen zum Lösungsverhalten von Polyacrylsäure und Poly(natrium acrylat) in Wasser, Formamid und Ethylenglycol, *Makromol. Chem.*, 180, 1477, 1979.

1979MOR   Morimoto, S. and Ohtani, N., Phase separation enthalpies of polymer solids in ternary polymer-solvent-nonsolvent system (Jap.), *Kenkyu Hokoku Seni Kobunshi Zairyo Kenkyushu*, 119, 35, 1979.

1979SIM   Simenido, A.V., Medved, Z.N., Denisova, L.L., Tarakanov, O.G., and Starikova, N.A., Some physicochemical properties of the mixtures of polypropylene polyols with water (Russ.), *Vysokomol. Soedin., Ser. A*, 21, 1727, 1979.

1980BAS   Basedow, A.M., Ebert, K.H., and Feigenbutz, W., Polymer-solvent interactions. Dextrans in water and DMSO, *Makromol. Chem.*, 181, 1071, 1980.

1980BIL   Bilimova, E.S., Gladkovskii, G.A., Golubev, V.M., and Medved, Z.N., Study of phase equilibriums in mixtures of poly(propylene oxide) with water (Russ.), *Vysokomol. Soedin., Ser. A*, 22, 2240, 1980.

1980DAO   Daoust, H. and Chabot, M.-A., Effect of cation size and of the presence of hydrophobic groups on heats of dilution of aqueous solutions of alkali metal and tetramethylammonium salts of the polyacrylic series, *Macromolecules*, 13, 616, 1980.

1980SKE   Skerjanc, J., Vesnaver, G., and Dolar, D., Heats of dilution of polystyrene sulfonates in dioxane-water mixtures, *Eur. Polym. J.*, 16, 179, 1980.

1980SUZ   Suzuki, H., Kamide, K., and Miyazaki, Y., Thermodynamic study on acetone solutions of cellulose diacetate by Rayleigh light scattering, *Netsusokutei*, 7, 37, 1980.

1981AHA   Aharoni, S.M., Charlet, G., and Delmas, G., Investigation of solutions and gels of poly(4-methyl-1-pentene) in cyclohexane and decalin by viscometry, calorimetry, and X-ray diffraction. A new crystalline form of poly(4-methyl-1-pentene) from gels, *Macromolecules*, 14, 1390, 1981.

1981ANS   Ansorena, F.J., Fernandez-Berridi, M.J., Elorza, J.M., Barandiaran, M.J., Guzman, G.M., and Iruin, J.J., Thermodynamics of the mixture poly(ethylene oxide)/toluene, *Polym. Bull.*, 4, 25, 1981.

1981ASP   Aspler, J.S. and Gray, D.G., Mixing of water with (hydroxypropyl)cellulose liquid crystalline mesophases, *Macromolecules*, 14, 1546, 1981.

1981DAY   Day, J.C. and Robb, I.D., Thermodynamic parameters of polyacrylamides in water, *Polymer*, 22, 1530, 1981.

1981DRU   Druz, N.I., Kreitus, A., and Chalykh, A.E., Thermodynamic parameters of mixing in the systems ethylene-vinyl acetate copolymers-hydrocarbons (Russ.), *Latv. PSR Zinat. Akad. Vestis. Kim. Ser.*, (3), 199, 1981.

1981FUJ   Fujioka, K., Nakajima, K., Baba, Y., Kagemoto, A., and Fujishiro, R., Enthalpy changes of dissociation of hemoglobin solution, *Netsusokutei* 8(3), 91, 1981.

1981HOR   Horiuchi, H. and Ohshita, T., Heat of mixing of poly(vinyl alcohol) and poly(methacrylic acid) in aqueous solution, *Kobunshi Ronbunshu*, 38, 407, 1981.

1981KOR   Koros, W.J., Smith, G.N., and Stannett, V., High-pressure sorption of carbon dioxide in solvent-cast poly(methyl methacrylate) and poly(ethyl methacrylate) films, *J. Appl. Polym. Sci.*, 26, 159, 1981.

1981MED   Medved, Z.N. and Starikova, N.A., Enthalpies of mixing of oligomers of ethylene oxide or propylene oxide with water (Russ.), *Zh. Fiz. Khim.*, 55, 2438, 1981.

1981PAO   Paoletti, S., Delben, F., and Crescenzi, V., Enthalpies of dilution of partially neutralized maleic acid copolymers in water. Correlation of experiments with theories, *J. Phys. Chem.*, 85, 1413, 1981.

1982GAR   Garcia-Fierro, J.L. and Aleman, J.V., Sorption of water by epoxide prepolymers, *Macromolecules*, 15, 1145, 1982.

1982MED   Medved, Z.N., Petrova, N.I., and Tarakanov, O.G., Comparison of thermodynamics of mixtures of oligomers of ethylene oxide or propylene oxide with water (Russ.), *Vysokomol. Soedin., Ser. B*, 24, 674, 1982.

1982MOR   Morcellet, M., Loucheux, C., and Daoust, H., Poly(methacrylic acid) derivatives 5. Microcalorimetric study of poly(*N*-methacryloyl-L-alanine) and poly(*N*-methacryloyl-L-alanine-co-*N*-phenylmethacrylamide) in aqueous solutions, *Macromolecules*, 15, 890, 1982.

1982POU   Pouchly, J., Benes, S., Masa, Z., and Biros, J., Sorption of water in hydrophilic polymers, 2. Thermodynamics of mixing of water with poly(2-hydroxyethyl methacrylate) and with poly[2-(2-hydroxyethoxy)ethyl methacrylate], *Makromol. Chem.*, 183, 1565, 1982.

1982RAE   Rätzsch, M.T. and Wohlfarth, Ch., Mischphasenthermodynamik konzentrierter Polymer-lösungen, *Wiss. Z. TH Leuna-Merseburg*, 24, 415, 1982.

1982SAX   Saxena, V. and Stern, S.A., Concentration-dependent transport of gases and vapors in glassy polymers. II. Organic vapors in ethyl cellulose, *J. Membrane Sci.*, 12, 65, 1982.

1982SH1   Sharma, S.C., Mahajan, R., Sharma, V.K., and Lakhanpal, M.L., Enthalpies of mixing of poly(tetramethylene oxides) with dioxane and cyclohexane, *Indian J. Chem.*, 21A, 682, 1982.

1982SH2   Sharma, S.C., Mahajan, R., Sharma, V.K., and Lakhanpal, M.L., Enthalpies of mixing of poly(tetramethylene oxides) with benzene and carbon tetrachloride, *Indian J. Chem.*, 21A, 685, 1982.

1982TAN   Tanaka, S., Fukagawa, O., Baba, Y., Kagemoto, A., and Fujishiro, R., Calorimetric studies on the coil-helix transition of poly(L-lysine) in water-organic solvent mixtures, *Netsusokutei*, 9(1), 2, 1982.

1982YUS   Yushkova, S. M., Guzeev, V. V., Bessonov, Yu. S., and Tager, A. A., Thermochemical study of the interaction of poly(vinyl chloride) with fillers during filling by polymerization (Russ.), *Vysokomol. Soedin., Ser. B*, 24, 483, 1982.

1983ALI   Ali, S., Thermodynamic properties of polymer solutions in compressed gases, *Z. Phys. Chem., N. F.*, 137, 13, 1983.

1983ALK   Al-Kafaji, J.K.H. and Booth, C., Enthalpy and volume changes on mixing oligo(oxyethylene)s and n-alkanes, *J. Chem. Soc., Faraday Trans I*, 79, 2695, 1983.

1983FUJ   Fujihara, I., Sakuta, T., and Kagemoto, A., Enthalpy of dilution of polystyrene/diethyl ketone, *Makromol. Chem.*, 184, 1231, 1983.

1983KIR   Kir'yanov, K.V., Determination of the enthalpy of mixing nitrocellulose with triacetin in the DAK-11 microcalorimeter (Russ.), *Termodin. Org. Soedin.*, 58, 1983.

1983ME1   Medved, Z.N. and Starikova, N.A., Enthalpies of mixing of oligomeric propylene oxides and water (Russ.), *Zh. Fiz. Khim.*, 57, 2590, 1983.

1983ME2   Medved, Z.N., Petrova, N.I., and Tarakanov, O.G., Thermodynamics of mixing of oligomeric polypropylene glycol polyols with water (Russ.), *Vysokomol. Soedin., Ser. B*, 25, 764, 1983.

1983TA1   Tager, A.A., Bessonov, Yu.S., Ikanina, T.V., Rodionova, T.A., Suvorova, A.I., and Elboim, S.A., Heats of interaction of poly(vinyl chloride) with plastificators (Russ.), *Vysokomol. Soedin., Ser. A*, 25, 1444, 1983.

1983TA2   Tager, A.A., Adamova, L.V., and Morkvina, L.I., Thermodynamics of the formation of polycomplexes of poly(acrylic acid) with poly(vinyl alcohol) and poly(ethylene oxide) (Russ.), *Vysokomol. Soedin., Ser. A*, 25, 1413, 1983.

1983WAL   Walsh, D.J., Higgins, J.S., Rostami, S., and Weeraperuma, K., Compatibility of ethylene-vinyl acetate copolymers with chlorinated polyethylenes. 2. Investigation of the thermodynamic parameters, *Macromolecules*, 16, 391, 1983.

1984DAO   Daoust, H. and St-Cyr, D., Microcalorimetric study of poly(ethylene oxide) in water and in water-ethanol mixed solvent, *Macromolecules*, 17, 596, 1984.

1984KRY   Kryukov, L.A., Sorokobatkina, M.S., and Trokhina, L.V., A method for the determination of heats of mixing between polymers or polymers and plasticizers (Russ.), *Izv. Vyssh. Uchebn. Zav., Khim. Khim. Tekhnol.*, 27, 1360, 1984.

1984PAN   Panayiotou, C.G., Statistical thermodynamics of polymer solutions, *J. Chem. Soc., Faraday Trans II*, 80, 1435, 1984.

1984VE1   Vesnaver, G., Rudez, M., Pohar, C., and Skerjanc, J., Effect of temperature on the enthalpy of dilution of strong polyelectrolyte solutions, *J. Phys. Chem.*, 88, 2411, 1984.

1984VE2   Vesnaver, G. and Skerjanc, J., Structural contributions of the counterion solvent interactions to the enthalpies of dilution of the aqueous polystyrenesulfonate solutions, *Vestn. Slov. Kem. Drus.*, 31, 325, 1984.

1985ALK   Al-Kafaji, J.K.H., Ariffin, Z., Cope, J., and Booth, C., Enthalpy and volume changes on mixing diethylene glycol di-n-alkyl ethers with diethylene glycol dimethyl ether or n-alkanes, *J. Chem. Soc., Faraday Trans I*, 81, 223, 1985.

1985EZH   Ezhov, V. S., Yushkova, S. M., Guzeev, V. V., Mozzhukhin, V. B., Malysheva, G. P., and Tager, A. A., Effect of the temperature of processing of plasticized PVC compositions on their structure, mechanical properties, and energy of interaction of the polymer with plasticizers (Russ.), *Vysokomol. Soedin., Ser. B*, 27, 385, 1985.

1985IOE   Ioelovich, M.Ya., Thermodynamics of the system amorphous regions of cellulose-water (Russ.), *Khim. Drev.*, (5), 3, 1985.

1985KIL   Killmann, E. and Bergmann, M., Microcalorimetric studies of the adsorption on N-ethylpyrrolidone, oligomeric and polymeric vinylpyrrolidone from $CHCl_3$ on silica, *Colloid Polym. Sci.*, 263, 381, 1985.

1985KIR   Kir'yanov, K.V. and Rabinovich, I.V., Enthalpies of mixing of poly(vinyl chloride) with some esters (Russ.), *Vysokomol. Soedin., Ser. B*, 27, 284, 1985.

1985RA1   Rabinovich, I.B., Khlyustova, T.B., Mochalov, A.N., Calorimetric determination of thermochemical properties and the phase diagram of cellulose nitrate-dibutyl phthalate mixtures (Russ.), *Vysokomol. Soedin., Ser. A*, 27, 525, 1985.

1985RA2   Rabinovich, I.B., Khlyustova, T.B., and Mochalov, A.N., Physico-chemical analysis of mixtures of cellulose nitrate with triacetin and thermodynamics of their mixing (Russ.), *Vysokomol. Soedin., Ser. A*, 27, 1724, 1985.

1985RA3   Rabinovich, I.B., Pet'kov, V.I., Zarudaeva, S.S., Physico-chemical analysis of mixtures of cellulose dicetate with triacetin and thermodynamic characteristics of the mixing process (Russ.), *Vysokomol. Soedin., Ser. A*, 27, 1817, 1985.

1985SAR   Sardharwalla, I. and Lawton, J.B., Calorimetric determination of the enthalpy change for the binding of methyl orange to poly(vinylpyrrolidone) in aqueous solution, *Polymer*, 26, 751, 1985.

1985SHA   Sharma, S.C. and Sharma, V.K., Enthalpies of mixing of poly(tetramethylene oxide) with benzene, methylbenzene, ethylbenzene and propylbenzene at 313.15 K, *Indian J. Chem.*, 24A, 292, 1985.

1985ZAR   Zarudaeva, S.S., Pet'kov, V.I., Rabinovich, I.B., and Kir'yanov, K.V., Thermodynamics of mixtures of cellulose diacetate with dimethyl phthalate (Russ.), *Vysokomol. Soedin., Ser. B*, 27, 1778, 1985.

1986GOL   Golova, L.K., Andreeva, O.E., Kulichikhin, V.G., Zenkov, I.D., Belousov, Yu.Ya., and Papkov, S.P., Dissolution of cellulose in mixtures of N-methylmorpholine-N-oxide with amines of various nature (Russ.), *Vysokomol. Soedin., Ser. A*, 28, 2308, 1986.

1986OVC   Ovchinnikov, E.Yu., Moseeva, E.M., Zhikharevich, L.B., and Manushin, V.I., Effect of the interaction of components of plasticizing mixtures on the plasticization of poly(vinyl chloride) (Russ.), *Vysokomol. Soedin., Ser. B*, 28, 548, 1986.

1986SH1   Sharma, S.C., Bhalla, S., and Sharma, V.K., Enthalpies of mixing of poly(tetramethylene oxide) with o-, m-, and p-xylenes and mesitylene, *Indian J. Chem.*, 25A, 131, 1986.

1986SH2   Shah, V.M., Hardy, B.J., and Stern, S.A., Solubility of carbon dioxide, methane, and propane in silicone polymers. Effect of polymer side chains, *J. Polym. Sci.: Part B: Polym. Phys.*, 24, 2033, 1986.

1986SK1   Skerjanc, J. and Fele, M., Effect of temperature on the enthalpy of dilution of weak polyelectrolyte solutions, *J. Calorim. Anal. Therm. Thermodyn. Chim.*, 17, 453, 1986.

1986SK2   Skerjanc, J., Pohar, C., and Fabjan, A., Enthalpies of dilution of poly(styrene sulfonates) in anhydrous N-methylformamide, *J. Phys. Chem.*, 90, 4364, 1986.

1987BOG   Bogdanov, B. and Mikhailov, M., Calorimetric inverstigation of water/polyoxyethylene systems, *J. Thermal Anal.*, 32, 161, 1987.

1987CAS   Castro, E.F., Gonzo, E.E., and Gottfriedi, J.C., Thermodynamics of the absorption of hydrocarbon vapors in polyethylene films, *J. Membrane Sci.*, 31, 235, 1987.

1987GOR   Goradia, U.B. and Spencer, H.G., Sorption of propane in poly(ethyl methacrylate) near T_g, *J. Appl. Polym. Sci.*, 33, 1525, 1987.

1987JAN   Janecek, H. and Turska, E., Investigation of conformational transitions in bisphenol-A polycarbonate macromolecules in mixtures of 1,1,2,2-tetrachloroethane and n-propyl alcohol, *Polymer*, 28, 847, 1987.

1987NUN   Nunes, S.P. and Wolf, B.A., On the co-occurrence of demixing and thermoreversible gelation of polymer solutions. 3. Overall view, *Macromolecules*, 20, 1952, 1987.

1987PED   Pedemonte, E. and Lanzavecchia, L., Compatible polymer mixtures. Evaluation of the interaction parameter from heat of solution measurements, *J. Calor. Anal. Therm. Thermodyn. Chim.*, 17, 519, 1987.

1987SAD   Sada, E., Kumazawa, H., Yakushij, H., Bamba, Y., Sakata, K., and Wang, S.-T., Sorption and diffusion of gases in glassy polymers, *Ind. Eng. Chem. Res.*, 26, 433, 1987.

1987SCH   Schneider, P. and Heintz, A., Enthalpies of solution of polyvinylalcohol in water and polyvinylacetate in dioxane, *Thermochim. Acta*, 119, 47, 1987.

1987SHA   Sharma, S.C., Syngal, M., and Sharma, V.K., Enthalpies and excess volumes of mixing of poly(tetramethylene oxide) fractions with tetrachloromethane, 1,2-dichloroethane and 1,1,1-trichloroethane at 313.15 K, *Indian J. Chem.*, 26A, 285, 1987.

1987TAG   Tager, A.A., Safronov, A.P., Lopyrev, V.A., Ermakova, T.G., Tatarova, L.A., Kashik, T.N., Thermodynamics of aqueous solutions of poly(1-vinylimadazole) and poly(1-vinyl-1,2,4-triazole) (Russ.), *Vysokomol. Soedin., Ser. A*, 29, 2421, 1987.

1987VES   Vesnaver, G., Kranjc, Z., Pohar, C., and Skerjanc, J., Free enthalpies, enthalpies and entropies of dilution of aqueous solutions of alkaline earth poly(styrenesulfonates) at different temperatures, *J. Phys. Chem.*, 91, 3845, 1987.

1988KAM   Kamide, K., Matsuda, S., and Saito, M., Flory enthalpy parameter at infinite dilution of polymer solutions determined by various methods, *Polym. J.*, 20, 31, 1988.

1988KAR   Karpova, A.L., Rozanova, E.A., Ostrovskii, V.E., and Timashev, S.F., Calorimetric study of the sorption of carbon dioxide and oxygen by silar block copolymer (Russ.), *Vysokomol. Soedin., Ser. B*, 30, 15, 1988.

1988SAD   Sada, E., Kumazawa, H., and Xu, P., Sorption and diffusion of carbon dioxide in polyimide films, *J. Appl. Polym. Sci.*, 35, 1497, 1988.

1989AND   Andersson, B. and Olofsson, G., Calorimetric study of binary systems of tetraethyleneglycol octylether and polyethyleneglycol with water, *J. Solution Chem.*, 18, 1019, 1989.

1989CHI   Chiou, J.S. and Paul, D.R., Gas sorption and permeation in poly(ethyl methacrylate), *J. Membrane Sci.*, 45, 167, 1989.

1989EGO   Egorochkin, G.A., Semchikov, Yu.D., Smirnova, L.A., Knyazeva, T.E., Tokhonova, Z.A., Karyakin, N.V., and Sveshnikova, T.G., Thermodynamic analysis of the copolymerization of styrene with acrylonitrile and N-vinylpyrrolidone with vinyl acetate (Russ.), *Vysokomol. Soedin., Ser. B*, 31, 46, 1989.

1989PAR   Parashar, R., Singh, H.G., and Sharma, S.C., Enthalpies of mixing of polyoxypropylene-glycols with benzene, ethanol and pyridine at 303.05 K, *Indian J. Chem.*, 28A, 317, 1989.

1989SAF   Safronov, A.P., Tager, A.A., Sharina, S.V., Lopyrev, V.A., Ermakova, T.G., Tatarova, L.A., and Kashik, T.N., Nature of hydration in aqueous solutions of poly(1-vinylazoles) (Russ.), *Vysokomol. Soedin., Ser. A*, 31, 2662, 1989.

1990COW   Cowie, J.M.G. and Swinyard, B., Location of three critical phase boundaries in poly(acrylic acid)-dioxane solutions, *Polymer*, 31, 1507, 1990.

1990DAO   Daoust, H., Darveau, R., and Laberge, F., Microcalorimetric investigation on interaction between poly(acrylic acid) and oxyethylene oligomers in water, *Polymer*, 31, 1946, 1990.

1990TAG   Tager, A.A., Safronov, A.P., Sharina, S.V., and Galaev, I.Yu., Thermodynamics of aqueous solutions of polyvinylcaprolactam (Russ.), *Vysokomol. Soedin., Ser. A*, 32, 529, 1990.

1990ZAV   Zav'yalov, N.A., Myasoedova, V.V., and Pokrovskii, S.A., Thermochemical characteristic of cellulose, its ethers and esters solutions (Russ.), *Vysokomol. Soedin., Ser. A*, 32, 2351, 1990.

1991AZI   Azizov, Sh.A., Sadykova, L.A., and Magrupov, M.A., Thermodynamic parameters of mixing of cellulose diacetate, poly(2-methyl-5-vinylpyridine) and their mixtures with dimethylformamide (Russ.), *Vysokomol. Soedin., Ser. A*, 33, 894, 1991.

1991BUR   Burmistr, M.V., Privalko, V.P., and Lipatov, Yu.S., Energetics of dissolution of alkyl-aromatic polyionenes in water (Russ.), *Dokl. Akad. Nauk Ukr. SSR*, 10, 135, 1991.

1991HEY   Hey, M.J. and Ilett, S.M., Poly(ethylene oxide) hydration studied by differential scanning calorimetry, *J. Chem. Soc., Faraday Trans.*, 87, 3671, 1991.

1991OLO   Olofsson, G., Micelle formation in non-aqueous solvents: calorimetric study of the association of poly(ethylene oxide) alkyl ethers and hexadecyltrimethylammonium bromide in formamide, *J. Chem. Soc., Faraday Trans.*, 87, 3037, 1991.

1991SAF   Safronov, A.P. and Tager, A.A., Thermodynamic criterion of the upper critical solution temperature for glassy polymers (Russ.), *Vysokomol. Soedin., Ser. A*, 33, 2198, 1991.

1991TA1   Tager, A.A. and Safronov, A.P., Complex formation in aqueous and nonaqueous solutions of poly(vinylazoles) (Russ.), *Vysokomol. Soedin., Ser. A*, 33, 67, 1991.

1991TA2   Tager, A.A., Safronov, A.P., Berezyuk, E.A., and Galaev, I.Yu., Hydrophobic interactions and lower critical solution temperatures of aqueous polymer solutions (Russ.), *Vysokomol. Soedin., Ser. B*, 33, 572, 1991.

1991TA3   Tager, A. A., Yushkova, S. M., Adamova, L. V., Kovylin, S. V., Berezov, L. V., Mozzhukhin, V. B., and Guzeev, V. V., Thermodynamics of interaction of methyl methacrylate-methacrylic acid copolymers with plasticizers and their mixtures (Russ.), *Vysokomol. Soedin., Ser. A*, 33, 357, 1991.

1991ZVE   Zverev, M.P., Zenkov, I.D., Zakharova, N.N., Zashchenkina, E.S., and Bondarenko, O.A., Properties of polyacrylonitrile and its fibers containing chemically active groups (Russ.), *Khim. Volokna*, (5), 32, 1991.

1992IV1   Ivanov, A.V., Karsakova, N.A., Shmakov, V.A., Sopin, V.F., Tsvetkov, V.G., and Marchenko, G.N., Enthal'pii vsaimodeistviya zellulosy s rastvorami azotnoi kisloty, *Zh. Prikl. Khim.*, 65, 2496, 1992.

1992IV2   Ivanov, A.V., Shmakov, V.A., Tsvetkov, V.G., Novoselova, N.V., Tsvetkova, L.Ya., and Golova, L.K., Enthal'pii vsaimodeistviya nekotorykh gidroksilsoderzhashchikh soedinenii s N-oksidom N-metilmorfolina v rastvorakh, *Izv. Vyssh. Uchebn. Zav., Khim. Khim. Tekhnol.*, 35(2), 58, 1992.

1992REI   Rein, D.H., Baddour, R.F., and Cohen, R.E., Gas solubility in glassy polymers: a correlation with excess enthalpy, *Polymer*, 33, 1696, 1992.

1992SUV   Suvorova, A.I., Safronov, A.I., Mukhina, A.Yu., and Peshekhonova, A.L., Thermodynamics of interaction of cellulose diacetate with plasticizer mixtures (Russ.), *Vysokomol. Soedin., Ser. A*, 34, 92, 1992.

1992TAG   Tager, A.A., Adamova, L.V., Safronov, A.P., Klyuzhin, E.S., and Zhigalova, E.P., Thermodynamics of poly(acrylic acid) solutions in water (Russ.), *Vysokomol. Soedin., Ser. B*, 34, 10, 1992.

1993ADA   Adamova, L.V., Klyuzhin, E.S., Safronov, A.P., Nerush, N.T., and Tager, A.A., Thermodynamics of interactions between water and copolymers of acrylate and acrylic acid (Russ.), *Vysokomol. Soedin., Ser. B*, 35, 893, 1993.

1993AUC   Aucouturier, C. and Roux, A.H., Thermodynamic study of polyethylene glycol in micellar aqueous solutions of sodium dodecylsulfate, *Calorim. Anal. Therm.*, 24, 9, 1993.

1993BRU   Brunacci, A., Yin, J., Pedemonte, E., and Turturro, A., A study on polymer-polymer interactions through mixing calorimetry, *Thermochim. Acta*, 227, 117, 1993.

1993EAG   Eagland, D., Crowther, N.J., and Butler, C.J., Interaction of poly(oxyethylene) with water as a function of molar mass, *Polymer*, 34, 2804, 1993.

1993FRE   Frezzotti, D. and Ravanetti, G.P., Evaluation of the Flory-Huggins interaction parameter for poly(styrene-co-acrylonitrile) and poly(methyl methacrylate) blend from enthalpy of mixing measurements, *J. Calorim. Anal. Therm.*, 24, 135, 1993.

1993IVA   Ivanov, A.V., Karsakova, N.A., Sopin, V.F., Tsvetkov, V.G., and Marchenko, G.N., Enthalpii vsaimodeistvija zellulosy s vodnymi rastvorami smesei azotnoi i sernoi ili azotnoi i triftoruksusnoi kislot, *Zh. Obshch. Khim.*, 63, 1519, 1993.

1993KEL   Keller, M., Heintz, A., and Lichtenthaler, R., Enthalpies of dilution of polyelectrolyte solutions and exchange enthalpies of polyelectrolyte counterions, *Thermochim. Acta*, 229, 243, 1993.

1993PER   Peregudov, Yu.S., Amelin, A.N., Perelygin, V.M., Calorimetric study of the dilution of solutions of poly(vinylbenzyltrimethylammonium) and its salts (Russ.), *Vysokomol. Soedin., Ser. B*, 35, 127, 1993.

1993SAF   Safronov, A.P., Tager, A.A., Adamova, L.V., and Klyuzhin, E.S., Thermodynamics of the interaction of polyacrylic acid with water (Russ.), *Vysokomol. Soedin., Ser. A*, 35, 702, 1993.

1993SHA   Shah, V.M., Hardy, B.J., and Stern, S.A., Solubility of carbon dioxide, methane, and propane in silicone polymers. Effect of polymer backbone chains, *J. Polym. Sci.: Part B: Polym. Phys.*, 31, 313, 1993.

1993TA1   Tager, A.A., Klyuzhin, E.S., Adamova, L.V., and Safronov, A.P., Thermodynamics of dissolution of acrylic acid-methyl acrylate copolymers in water (Russ.), *Vysokomol. Soedin., Ser. B*, 35, 1357, 1993.

1993TA2   Tager, A.A., Safronov, A.P., Sharina, S.V., and Galaev, I.Yu., Thermodynamic study of poly(N-vinyl caprolactam) hydration at temperatures close to lower critical solution temperature, *Colloid Polym. Sci.*, 271, 868, 1993.

1993ZEL   Zellner, H., Mischungsthermodynamik und Energetik der Wechselwirkungen von Polyethern in Lösungsmitteln unterschiedlicher Polarität, *Dissertation*, TU München, 1993.

1994AUC   Aucouturier, C., Rous-Desgranges, G., and Roux, A.H., Thermodynamic study of poly-ethyleneglycols in micellar solutions of water + sodium dodecylsulfate, *J. Therm. Anal.*, 41, 1295, 1994.

1994COM   Compostizo, A., Cancho, S.M., Crespo, A., and Rubio, R.G., Polymer solutions with specific interactions: Equation of state for poly(4-hydroxystyrene) + acetone, *Macromolecules*, 27, 3478, 1994.

1994FRE   Frezzotti, D. and Ravanetti, G.P., Evaluation of the Flory-Huggins interaction parameter for poly(styrene-co-acrylonitrile) and poly(methyl methacrylate) blend from enthalpy of mixing measurements, *J. Therm. Anal.*, 41, 1237, 1994.

1994HOR   Horacek, H., Extrapolation of solubility data of monomeric, oligomeric and polymeric hydrocarbons, *Macromol. Chem. Phys.*, 195, 3381, 1994.

1994KHA   Khamrakulov, G., Myagkova, N.V., and Budtov, V.P., Water sorption and diffusion in cellulose and cellulose acetates (Russ.), *Vysokomol. Soedin., Ser. B*, 36, 845, 1994.

1994KIZ   Kizhnyaev, V.N., Astakhov, M.B., and Smirnov, A.I., Hydrodynamic and thermodynamic properties of polyvinyltetrazole solutions (Russ.), *Vysokomol. Soedin., Ser. A*, 36, 104, 1994.

1994NAG   Nagahama, K., Inomata, H., and Saito, S., Measurement of osmotic pressure in aqueous solutions of poly(ethylene glycol) and poly(N-isopropylacrylamide), *Fluid Phase Equil.*, 96, 203, 1994.

1994PER   Persson, K., Wang, G., and Olofsson, G., Self-diffusion, thermal effects and viscosity of a monodisperse associative polymer: self-association and interaction with surfactants, *J. Chem. Soc., Faraday Trans.*, 90, 3555, 1994.

1994TAG   Tager, A.A., Safronov, A.P., Berezyuk, E.A., and Galaev, I.Yu., Lower critical solution temperature and hydrophobic hydration in aqueous polymer solutions, *Colloid Polym. Sci.*, 272, 1234, 1994.

1994YUS   Yushkova, S.M., Safronov, A.P., Berezyuk, E.A., Monakhova, T.G., Mozhukhin, V.B., and Guzeev, V.V., Thermodynamics of poly(vinyl chloride) interaction with low-molecular liquids (Russ.), *Vysokomol. Soedin., Ser. A*, 36, 431, 1994.

1995BAN   Banerjee, T., Chhajer, M., and Lipscomb, G.G., Direct measurement of the heat of carbon dioxide sorption in polymeric materials, *Macromolecules*, 28, 8563, 1995.

1995GRO   Grossmann, C., Tintinger, R., Zhu, J., and Maurer, G., Aqueous two-phase systems of poly(ethylene glycol) and dextran – experimental results and modeling of thermodynamic properties, *Fluid Phase Equil.*, 106, 111, 1995.

1995KIZ   Kizhnyaev, V.N., Gorkovenko, O.P., Bazhenov, D.N., and Smirnov, A.I., Interrelation between hydrodynamic and thermodynamic characteristics of solutions of poly(vinyltetrazoles) in mixed solvents (Russ.), *Vysokomol. Soedin., Ser. B*, 37, 1948, 1995.

1996ABD   Abdrakhmanova, L.A. and Khozin, V.G., Thermodynamics of binary mixtures from poly(vinyl chloride)-furan plasticizers (Russ.), *Zh. Prikl. Khim.*, 69, 486, 1996.

1996FAE   Faes, H., De Schryver, F.C., Sein, A., Bijma, K., Kevelam, J., and Engberts, J.B.F., Study of self-associating amphiphilic copolymers and their interaction with surfactants, *Macromolecules*, 29, 3875, 1996.

1996KEL   Keller, M., Lichtenthaler, R., and Heintz, A., Enthalpies of dilution of polycation solutions and exchange enthalpies of polycations, *Ber. Bunsenges. Phys. Chem.*, 100, 776, 1996.

1996LIU   Liu, Z.-H., Marechal, Ph., and Jerome, R., Intermolecular interactions in poly(vinylidene fluoride) and ε-caprolactam mixtures, *Polymer*, 37, 5317, 1996.

1996SAF   Safronov, A.P., Tager, A.A., and Koroleva, E.B., Thermodynamics of dissolution of polyacrylic acid in donor and acceptor solvents (Russ.), *Vysokomol. Soedin., Ser. B*, 38, 900, 1996.

1996VSH   Vshivkov, S.A., Rusinova, E.V., Dubchak, V.N., and Zarubin, G.B., Thermodynamics and structure of polydimethylsiloxane-methyl ethyl ketone system (Russ.), *Vysokomol. Soedin., Ser. A*, 38, 868, 1996.

1996WAN   Wang, J.-S., Naito, Y., and Kamiya, Y., Effect of the penetrant-induced isothermal glass transition on sorption, dilation, and diffusion behavior of polybenzylmethacrylate/CO_2, *J. Polym. Sci.: Part B: Polym. Phys.*, 34, 2027, 1996.

1997ARA   Aratono, M., Ohta, A., Ikeda, N., Matsubara, A., Motomura, K., and Takiue, T., Calorimetry of surfactant solutions. Measurement of the enthalpy of mixing of tetraethylene glycol monooctyl ether and water, *J. Phys. Chem. B*, 101, 3535, 1997.

1997KI1   Kizhnyaev, V.N., Gorkovenko, O.P., Safronov, A.P., and Adamova, L.V., Thermodynamics of the interaction between tetrazole-containing polyelectrolytes and water (Russ.), *Vysokomol. Soedin., Ser. A*, 39, 527, 1997.

1997KI2   Kizhnyaev, V.N., Gorkovenko, O.P., Bazhenov, D.N., and Smirnov, A.I., The solubilities and enthalpies of a solution of polyvinyltetrazoles in organic solvents (Russ.), *Vysokomol. Soedin., Ser. A*, 39, 856, 1997.

1997RAS   Rashidova, S.Sh., Voropaeva, N.L., Kalantarova, T.D., and Yushkova, S.M., Calorimetric studies of the heats of interaction between methylcellulose and poly(vinylpyrrolidone) (Russ.), *Vysokomol. Soedin., Ser. B*, 39, 556, 1997.

1997SAF   Safronov, A.P., Suvorova, A.I., Koroleva, E.V., and Maskalyunaite, O.E., Enthalpy of mixing poly(methyl methacrylate) with donor or acceptor solvents (Russ.), *Vysokomol. Soedin., Ser. A*, 39, 1998, 1997.

1998OHT   Ohta, A., Takiue, T., Ikeda, N., and Aratono, M., Calorimetric study of dilute aqueous solutions of ethylene glycol oligomers, *J. Phys. Chem. B*, 102, 4809, 1998.

1998STU   Stumbeck, M., Mischungsthermodynamik von Polyethern in Lösungsmitteln unterschied-licher Polarität. Theoretische Modelle zur Berechnung von Enthalpie und freier Enthalpie von Mischungen, *Dissertation*, TU München, 1998.

1999BON   Bondar, V.I., Freeman, B.D., and Yampolskii, Yu.P., Sorption of gases and vapors in an amorphous glassy perfluorodioxole copolymer, *Macromolecules*, 32, 6163, 1999.

1999CES   Cesaro, A., Cuppo, F., Fabri, D., and Sussich, F., Thermodynamic behavior of mixed biopolymers in solution and in gel phase, *Thermochim. Acta*, 328, 143, 1999.

1999SAF   Safronov, A.P. and Kovalev, A.A., Enthalpy of interaction of crystalline polyethyleneoxide with water, toluene, and chloroform (Russ.), *Vysokomol. Soedin., Ser. A*, 41, 1008, 1999.

1999STU   Stumbeck, M. and Killmann, E., Mixing enthalpies of polyethers in solvents of various polarity treated with a group contribution theory, *Macromol. Chem. Phys.*, 200, 348, 1999.

2000SUV   Suvorova, A.I., Safronov, A.P., and Mel'nikova, O.A., Enthalpies of interaction of cellulose derivatives with starch (Russ.), *Vysokomol. Soedin., Ser. A*, 42, 822, 2000.

2001CAS   Casarino, P., Vicini, S., and Pedemonte, E., Thermodynamics of polymer mixtures: study on the mixing process of the polystyrene/poly(vinyl methyl ether) system, *Thermochim. Acta*, 372, 59, 2001.

2001OHT   Ohta, A., Takiue, T., Ikeda, N., and Aratono, M., Calorimetric study of micelle formation in polyethylene glycol monooctyl ether solutions, *J. Solution Chem.*, 30, 335, 2001.

2001TSU   Tsuboi, A., Kolar, P., Ishikawa, T., Kamiya, Y., and Masuoka, H., Sorption and partial molar volumes of C_2 and C_3 hydrocarbons in polypropylene copolymers, *J. Polym. Sci.: Part B: Polym. Phys.*, 39, 1255, 2001.

2002ARE   Areerat, S., Hayata, Y., Katsumoto, R., Kegaswaa, T., Engami, H., and Ohshima, M., Solubility of carbon dioxide in polyethylene/titanium dioxide composite under high pressure and temperature, *J. Appl. Polym. Sci.*, 86, 282, 2002.

2002DEA   DeAngelis, M.G., Merkel, T.C., Bondar, V.I., Freeman, B.D., Doghieri, F., and Sarti, G.C., Gas sorption and dilation in poly(2,2-bistrifluoromethyl-4,5-difluoro-1,3-dioxole-*co*-tetra-fluoroethylene): comparison of experimental data with predictions of the nonequilibrium lattice fluid model, *Macromolecules*, 35, 1276, 2002.

2002SA1   Safronov, A.P. and Zubarev, A.Yu., Flory-Huggins parameter of interaction in polyelectrolyte solutions of chitosan and its alkylated derivative, *Polymer*, 43, 743, 2002.

2002SA2   Safronov, A.P. and Adamova, L.V., Thermodynamics of dissolution of glassy polymers, *Polymer*, 43, 2653, 2002.

2002SA3   Safronov, A.P., Suvorova, A.I., and Koroleva, E.V., Enthalpy of the interaction of semicrystalline polycarbonate with chloroform and dimethylformamide (Russ.), *Vysokomol. Soedin., Ser. A*, 44, 275, 2002.

2002SA4   Safronov, A.P. and Somova, T.V., Thermodynamics of poly(vinyl chloride) mixing with phthalate plasticizers (Russ.), *Vysokomol. Soedin., Ser. A*, 44, 2014, 2002.

2002SUV   Suvorova, A.I., Safronov, A.P., and Mel'nikova, O.A., Thermodynamic compatibility of starch and carboxymethyl cellulose sodium (Russ.), *Vysokomol. Soedin., Ser. A*, 44, 98, 2002.

2002THU   Thurn, T., Couderc, S., Sidhu, J., Bloor, D.M., Penfold, J., Holzwarth, J.F., and Wyn Jones, E., Study of mixed micelles and interaction parameters for triblock copolymers of the type EO_m-PO_n-EO_m and ionic surfactants: equilibrium and structure, *Langmuir*, 18, 9267, 2002.

2003CHO   Cho, E.C., Lee, J., and Cho, K., Role of bound water and hydrophobic interaction in phase transition of poly(N-isopropylacrylamide) aqueous solution, *Macromolecules*, 36, 9929, 2003.

2003JAB   Jablonski, P., Müller-Blecking, A., Borchard, W., A method to determine mixing enthalpies by DSC, *J. Therm. Anal. Calorim.*, 74, 779, 2003.

2003KSI   Ksiqzczak, A., Ksiqzczak, T., and Ostrowski, M., Intermolecular interactions and phase equilibria in nitrocellulose + s-diethyldiphenylurea system, *J. Therm. Anal. Calorim.*, 74, 575, 2003.

2004ARE   Areerat, S., Funami, E., Hayata, Y., Nakagawa, D., and Ohshima, M., Measurement and prediction of diffusion coefficients of supercritical CO_2 in molten polymers, *Polym. Eng. Sci.*, 44, 1915, 2004.

2004DEL   DeLisi, R., Lazzara, G., Milioto, S., and Muratore, N., Thermodynamics of aqueous poly(ethylene oxide)-poly(propylene oxide)-poly(ethylene oxide)/surfactant mixtures. Effect of the copolymer molecular weight and the surfactant alkyl chain length, *J. Phys. Chem. B*, 108, 18214, 2004.

2004DER   D'Errico, G., Ciccarelli, D., Ortona, O., Paduano, L., and Sartorio, R., Interaction between pentaethylene glycol n-octyl ether and low-molecular-weight poly(acrylic acid), *J. Coll. Interface Sci.*, 270, 490, 2004.

2004JAN   Jansson, J., Schillen, K., Olofsson, G., Cardoso da Silva, R., and Loh, W., The interaction between PEO-PPO-PEO triblock copolymers and ionic surfactants in aqueous solution studied using light scattering and calorimetry, *J. Phys. Chem. B*, 108, 82, 2004.

2004LOH   Loh, W., Teixeira, L.A.C., and Lee, L.-T., Isothermal calorimetric investigation of the interaction of poly(N-isopropylacrylamide) and ionic surfactants, *J. Phys. Chem. B*, 108, 3196, 2004.

2004MAL   Malik, S. and Nandi, A.K., Thermodynamic and structural investigation of thermoreversible poly(3-dodecyl thiophene) gels in the three isomers of xylene, *J. Phys. Chem. B*, 108, 597, 2004.

2004SIL   Silva, R.C. da, Loh, W., and Olofsson, G., Calorimetric investigation of temperature effect on the interaction between poly(ethylene oxide) and sodium dodecylsulfate in water, *Thermochimica Acta*, 417, 295, 2004.

2004SMI   Smith, A.L., Mulligan, R.B., and Shirazi, H.M., Determining the effects of vapor sorption in polymers with the quartz crystal microbalance/heat conduction calorimeter, *J. Polym. Sci.: Part B: Polym. Phys.*, 42, 3893, 2004.

2005BIG   Bigi, A., Comelli, F., Excess molar enthalpies of binary mixtures containing ethylene glycols of poly(ethylene glycols) + ethyl alcohol at 308.15 K and atmospheric pressure, *Thermochim. Acta*, 430, 191, 2005.

2005CAS   Castro, E., Taboada, P., and Mosquera, V., Behavior of a styrene oxide-ethylene oxide diblock copolymer/surfactant system: a thermodynamic and spectroscopy study, *J. Phys. Chem. B*, 109, 5592, 2005.

2005GOD   Godec, A. and Skerjanc, J., Enthalpy changes upon dilution and ionization of poly(L-glutamic acid) in aqueous solutions, *J. Phys. Chem. B*, 109, 13363, 2005.

2005KLO   Kloskowski, A., Chrzanowski, W., Pilarczyk, M., and Namiesnik, J., Partition coefficients of selected environmentally important volatile organic compounds determined by gas-liquid chromatography with polydimethylsiloxane stationary phase, *J. Chem. Thermodyn.*, 37, 21, 2005.

2005PRA   Prabhakar, R.S., Merkel, T.C., Freeman, B.D., Imizu, T., and Higuchi, A., Sorption and transport properties of propane and perfluoropropane in poly(dimethylsiloxane) and poly(1-trimethylsilyl-1-propyne), *Macromolecules*, 38, 1899, 2005.

2005SAN   Sannino,A., Larobina, D., Mensitieri, G., Aldi, A., and Maffezzoli, A., Simultaneous gravimetric and calorimetric analysis of chloroform sorption in nanoporous semicrystalline sPS, *J. Appl. Polym. Sci.*, 96, 1675, 2005.

2005WAN   Wang, C. and Tam, K.C., Interactions between poly(acrylic acid) and sodium dodecyl sulfate: isothermal titration calorimetric and surfactant ion-selective electrode studies, *J. Phys. Chem. B*, 109, 5156, 2005.

APPENDICES

Appendix 1   List of polymers in alphabetical order

Appendix 2   List of systems and properties in order of the polymers

Polymer(s)	Solvent(s)	Property	Page(s)
Benzylcellulose			
	benzene	$\Delta_{sol}H_B^{\infty}$	195
	cyclohexanone	$\Delta_{sol}H_B^{\infty}$	195
	trichloromethane	$\Delta_{sol}H_B^{\infty}$	195
Cellulose			
	N,N-dimethylacetamide	$\Delta_{sol}H_B^{\infty}$	196
	N,N-dimethylformamide	$\Delta_{sol}H_B^{\infty}$	196
	dimethylsulfoxide	$\Delta_{sol}H_B^{\infty}$	196
	N-methylmorpholine-N-oxide	$\Delta_{sol}H_B^{\infty}$	195
	N-methylmorpholine-N-oxide monohydrate	$\Delta_{sol}H_B^{\infty}$	196
	water	$\Delta_{sol}H_B^{\infty}$	195
Cellulose acetate			
	formic acid	$\Delta_{sol}H_B^{\infty}$	197
	methyl acetate	$\Delta_{sol}H_B^{\infty}$	197
	2-propanone	$\Delta_{sol}H_B^{\infty}$	197-198
	trichloromethane	$\Delta_{sol}H_B^{\infty}$	198
	water	$\Delta_M H_A^{\infty}$	367
Cellulose monoacetate			
	water	$\Delta_M H_A^{\infty}$	367
Cellulose triacetate			
	water	$\Delta_M H_A^{\infty}$	367
Cellulose tricarbanilate			
	cyclohexanol	$\Delta_M H$	15
	5-nonanone	$\Delta_M H$	15

Polymer(s)	Solvent(s)	Property	Page(s)
Decamethyltetrasiloxane			
	benzene	$\Delta_M H_B^\infty$	198
	1,3-dimethylbenzene	$\Delta_M H_B^\infty$	198
	2,2-dimethylbutane	$\Delta_M H$	15
	2,2-dimethylbutane	$\Delta_M H_B^\infty$	199
	n-dodecane	$\Delta_M H$	16
	n-dodecane	$\Delta_M H_B^\infty$	199
	ethylbenzene	$\Delta_M H_B^\infty$	199
	2,2,4,4,6,8,8-heptamethyl-nonane	$\Delta_M H$	16
	2,2,4,4,6,8,8-heptamethyl-nonane	$\Delta_M H_B^\infty$	199
	n-heptane	$\Delta_M H_B^\infty$	199
	n-hexadecane	$\Delta_M H$	16
	n-hexadecane	$\Delta_M H_B^\infty$	199
	n-hexane	$\Delta_M H$	16
	n-hexane	$\Delta_M H_B^\infty$	199
	n-nonane	$\Delta_M H_B^\infty$	199
	n-octane	$\Delta_M H$	16
	n-octane	$\Delta_M H_B^\infty$	200
	2,2,4,6,6-pentamethylheptane	$\Delta_M H$	17
	2,2,4,6,6-pentamethylheptane	$\Delta_M H_B^\infty$	200
	2,2,4-trimethylpentane	$\Delta_M H$	17
	2,2,4-trimethylpentane	$\Delta_M H_B^\infty$	200
Dextran			
	dimethylsulfoxide	$\Delta_{dil} H^{12}$	17
	dimethylsulfoxide	$\Delta_{sol} H_B^\infty$	200
	1,2-ethanediol	$\Delta_{sol} H_B^\infty$	200
	formamide	$\Delta_{sol} H_B^\infty$	201
	water	$\Delta_{dil} H^{12}$	17-18
	water	$\Delta_{sol} H_B^\infty$	201
Di(ethylene glycol) dibutyl ether			
	n-dodecane	$\Delta_M H$	18
	n-hexadecane	$\Delta_M H$	19
	bis(2-methoxyethyl) ether	$\Delta_M H$	19
	n-tetradecane	$\Delta_M H$	19
Di(ethylene glycol) dihexyl ether			
	bis(2-methoxyethyl) ether	$\Delta_M H$	19

Polymer(s)	Solvent(s)	Property	Page(s)
α,ω-Dihydroxy poly[di(oxyethylene) oxyadipoyl]			
	triacetin	$\Delta_M H_A^{\infty}$, $\Delta_{sol} H_{A(vap)}^{\infty}$	368
α,ω-Dihydroxy poly[di(oxyethylene) oxysuccinyl]			
	triacetin	$\Delta_M H_A^{\infty}$, $\Delta_{sol} H_{A(vap)}^{\infty}$	368
α,ω-Dihydroxy poly(hexamethylene carbonate)			
	triacetin	$\Delta_M H_A^{\infty}$, $\Delta_{sol} H_{A(vap)}^{\infty}$	368
α,ω-Dihydroxy poly(oxyethyleneo xysuccinyl)-b-poly(oxyethylene) diblock copolymer			
	triacetin	$\Delta_M H_A^{\infty}$, $\Delta_{sol} H_{A(vap)}^{\infty}$	368
α,ω-Dihydroxy poly[oxy-3-(2-methoxy ethoxy)propyleneoxysuccinyl]			
	triacetin	$\Delta_M H_A^{\infty}$, $\Delta_{sol} H_{A(vap)}^{\infty}$	369
α,ω-Dihydroxy poly(tetramethylene carbonate)			
	1,2,4-butanetriyl trinitrate	$\Delta_{sol} H_{A(vap)}^{\infty}$	369
	1,2,3-propanetriyl trinitrate	$\Delta_M H_A^{\infty}$, $\Delta_{sol} H_{A(vap)}^{\infty}$	369
	triacetin	$\Delta_M H_A^{\infty}$, $\Delta_{sol} H_{A(vap)}^{\infty}$	369
α,ω-Dihydroxy poly[tri(oxyethylene) oxysuccinyl]			
	triacetin	$\Delta_M H_A^{\infty}$, $\Delta_{sol} H_{A(vap)}^{\infty}$	369
Ethylcellulose			
	dichloromethane	$\Delta_M H_A^{\infty}$, $\Delta_{sol} H_{A(vap)}^{\infty}$	370
	1,4-dioxane	$\Delta_M H_A^{\infty}$, $\Delta_{sol} H_{A(vap)}^{\infty}$	370
	ethanol	$\Delta_M H_A^{\infty}$, $\Delta_{sol} H_{A(vap)}^{\infty}$	370
	ethyl acetate	$\Delta_M H_A^{\infty}$, $\Delta_{sol} H_{A(vap)}^{\infty}$	370
	2-propanone	$\Delta_M H_A^{\infty}$, $\Delta_{sol} H_{A(vap)}^{\infty}$	370
	tetrachloromethane	$\Delta_M H_A^{\infty}$, $\Delta_{sol} H_{A(vap)}^{\infty}$	370
	water	$\Delta_M H_A^{\infty}$	370

Polymer(s)	Solvent(s)	Property	Page(s)
Gelatine			
	water	$\Delta_{sol}H_B^\infty$	201
Guttapercha			
	trichloromethane	$\Delta_{sol}H_B^\infty$	201
Hexamethyldisiloxane			
	benzene	$\Delta_M H_B^\infty$	202
	benzene	$\Delta_M H_A$	351
	1,3-dimethylbenzene	$\Delta_M H_B^\infty$	202
	ethylbenzene	$\Delta_M H_B^\infty$	202
	n-heptane	$\Delta_M H_B^\infty$	202
	n-heptane	$\Delta_M H_A$	351
	n-hexane	$\Delta_M H$	20
	n-nonane	$\Delta_M H_A$	351
	n-octane	$\Delta_M H_A$	352
	tetradecafluorohexane	$\Delta_M H$	20
	toluene	$\Delta_M H_A$	352
	2,2,4-trimethylpentane	$\Delta_M H_A$	352
Hydroxypropylcellulose			
	1-butanol	$\Delta_M H_A^\infty$	371
	n-decane	$\Delta_M H_A^\infty$	371
	1,4-dioxane	$\Delta_M H_A^\infty$	371
	ethanol	$\Delta_M H_A^\infty$	371
	n-heptane	$\Delta_M H_A^\infty$	371
	methanol	$\Delta_M H_A^\infty$	371
	1-propanol	$\Delta_M H_A^\infty$	371
	2-propanone	$\Delta_M H_A^\infty$	371
	tetrahydrofuran	$\Delta_M H_A^\infty$	371
	toluene	$\Delta_M H_A^\infty$	371
	trichloromethane	$\Delta_M H_A^\infty$	371
	water	$\Delta_M H_A^\infty$	370-371
Lignin			
	dimethylsulfoxide	$\Delta_{dil}H^{12}$	20
Maltopentaose			
	water	$\Delta_M H_A$	352

Polymer(s)	Solvent(s)	Property	Page(s)
Maltotetraose			
	water	$\Delta_M H_A$	353
Maltotriose			
	water	$\Delta_M H$	21
	water	$\Delta_M H_A$	353
Methylcellulose			
	water	$\Delta_M H_A^\infty$	372
Natural rubber			
	benzene	$\Delta_M H$	21
	benzene	$\Delta_{sol} H_B^\infty$	202
	benzene	$\Delta_M H_A^\infty$	372
	cyclohexane	$\Delta_M H_A^\infty$	372
	n-decane	$\Delta_M H_A^\infty$	372
	n-heptane	$\Delta_M H_A^\infty$	372
	n-hexane	$\Delta_M H_A^\infty$	372
	n-nonane	$\Delta_M H_A^\infty$	372
	n-octane	$\Delta_M H_A^\infty$	372
	toluene	$\Delta_M H_A^\infty$	372
Nitrocellulose			
	2-butanone	$\Delta_{sol} H_B^\infty$	203
	butyl acetate	$\Delta_{sol} H_B^\infty$	203
	dibutyl phthalate	$\Delta_{sol} H_B^\infty$	203
	1,3-diethyl-1,3-diphenylurea	$\Delta_M H$	21
	diethyl ether	$\Delta_{sol} H_B^\infty$	204
	2,4-dinitrotoluene	$\Delta_M H$	21
	2,6-dinitrotoluene	$\Delta_M H$	21-22
	ethanol	$\Delta_{sol} H_B^\infty$	204
	ethyl acetate	$\Delta_{sol} H_B^\infty$	204
	methanol	$\Delta_{sol} H_B^\infty$	204
	2,4-pentanedione	$\Delta_{sol} H_B^\infty$	204
	2-pentanone	$\Delta_{sol} H_B^\infty$	204
	2-propanone	$\Delta_{dil} H^{12}$	22
	2-propanone	$\Delta_{sol} H_B^\infty$	205
	pyridine	$\Delta_{sol} H_B^\infty$	206
	1,2,3-triacetoxypropane	$\Delta_M H$	22
	tris(4-methylphenyl) phosphate	$\Delta_{sol} H_B^\infty$	206

Polymer(s)	Solvent(s)	Property	Page(s)
Nylon 6			
	ε-caprolactam	$\Delta_M H_A^{\infty}$, $\Delta_{sol}H_{A(vap)}^{\infty}$	372
	formic acid	$\Delta_{dil}H^{12}$	22
	formic acid	$\Delta_{sol}H_B^{\infty}$	206
	tricresol	$\Delta_{sol}H_B^{\infty}$	206
Octaacetylcellobiose			
	2-propanone	$\Delta_{sol}H_B^{\infty}$	206
	trichloromethane	$\Delta_{sol}H_B^{\infty}$	206
Octamethylcyclotetrasiloxane			
	benzene	$\Delta_M H_A$	353
	cyclopentane	$\Delta_M H$	22-23
	2,2-dimethylbutane	$\Delta_M H$	23
	2,2-dimethylbutane	$\Delta_M H_B^{\infty}$	207
	n-dodecane	$\Delta_M H$	24
	n-dodecane	$\Delta_M H_B^{\infty}$	207
	2,2,4,4,6,8,8-heptamethyl-nonane	$\Delta_M H$	24
	2,2,4,4,6,8,8-heptamethyl-nonane	$\Delta_{sol}H_B^{\infty}$	207
	n-hexadecane	$\Delta_M H$	24
	n-hexadecane	$\Delta_M H_B^{\infty}$	207
	n-hexane	$\Delta_M H$	24
	n-hexane	$\Delta_M H_B^{\infty}$	207
	n-nonane	$\Delta_M H_A$	353
	n-octane	$\Delta_M H$	24
	n-octane	$\Delta_M H_B^{\infty}$	207
	2,2,4,6,6-pentamethylheptane	$\Delta_M H$	25
	2,2,4,6,6-pentamethylheptane	$\Delta_M H_B^{\infty}$	207
	toluene	$\Delta_M H_A$	353
	2,2,4-trimethylpentane	$\Delta_M H$	25
	2,2,4-trimethylpentane	$\Delta_M H_B^{\infty}$	207
	2,2,4-trimethylpentane	$\Delta_M H_A$	354
Octamethyltrisiloxane			
	benzene	$\Delta_M H_B^{\infty}$	208
	1,3-dimethylbenzene	$\Delta_M H_B^{\infty}$	208
	ethylbenzene	$\Delta_M H_B^{\infty}$	208
	n-heptane	$\Delta_M H_B^{\infty}$	208

Polymer(s)	Solvent(s)	Property	Page(s)
Oxyethylcellulose			
	water	$\Delta_{sol}H_B^{\infty}$	208
Oxypropylcellulose			
	dichloromethane	$\Delta_M H_A^{\infty}$, $\Delta_{sol}H_{A(vap)}^{\infty}$	372
	1,4-dioxane	$\Delta_M H_A^{\infty}$, $\Delta_{sol}H_{A(vap)}^{\infty}$	372
	ethanol	$\Delta_M H_A^{\infty}$, $\Delta_{sol}H_{A(vap)}^{\infty}$	372
	2-propanone	$\Delta_M H_A^{\infty}$, $\Delta_{sol}H_{A(vap)}^{\infty}$	372
Penta(ethylene glycol)			
	water	$\Delta_M H_B^{\infty}$	208-209
Penta(ethylene glycol) dimethyl ether			
	methanol	$\Delta_M H$	25
Phenoxy (bisphenol-A polyhydroxyether)			
	acetonitrile	$\Delta_M H_A^{\infty}$	373
	benzene	$\Delta_M H_A^{\infty}$	373
	2-butanone	$\Delta_M H_A^{\infty}$	373
	chlorobenzene	$\Delta_M H_A^{\infty}$	373
	1,2-dichloroethane	$\Delta_M H_A^{\infty}$	373
	bis(2-methoxyethyl) ether	$\Delta_M H_A^{\infty}$	373
	N,N-dimethylformamide	$\Delta_M H_A^{\infty}$	373
	1,4-dioxane	$\Delta_M H_A^{\infty}$	373
	ethyl acetate	$\Delta_M H_A^{\infty}$	373
	1-propanol	$\Delta_M H_A^{\infty}$	373
	toluene	$\Delta_M H_A^{\infty}$	373
Poly(acrylic acid)			
	ethanol	$\Delta_{dil}H^{12}$	25-26
	water	$\Delta_{dil}H^{12}$	26-28
Polyacrylonitrile			
	benzene	$\Delta_{sol}H_B^{\infty}$	209
	N,N-dimethylformamide	$\Delta_{sol}H_B^{\infty}$	209-210
	dimethylsulfoxide	$\Delta_{sol}H_B^{\infty}$	210
Poly(acrylonitrile-*co*-butadiene)			
	benzene	$\Delta_{sol}H_B^{\infty}$	210-211
	benzene	$\Delta_{sol}H_{A(vap)}^{\infty}$	373

Polymer(s)	Solvent(s)	Property	Page(s)
Poly(acrylonitrile-*co*-butadiene) (*continued*)	cyclohexane	$\Delta_{sol}H_{A(vap)}{}^{\infty}$	373
	n-heptane	$\Delta_{sol}H_{A(vap)}{}^{\infty}$	373
	n-nonane	$\Delta_{sol}H_{A(vap)}{}^{\infty}$	373
	n-octane	$\Delta_{sol}H_{A(vap)}{}^{\infty}$	373
	toluene	$\Delta_{sol}H_{A(vap)}{}^{\infty}$	373
Poly(acrylonitrile-*co*-methyl acrylate)			
	1-butanol	$\Delta_{M}H_{A}{}^{\infty}$	374
	ethanol	$\Delta_{M}H_{A}{}^{\infty}$	374
	1-hexanol	$\Delta_{M}H_{A}{}^{\infty}$	374
	1-pentanol	$\Delta_{M}H_{A}{}^{\infty}$	374
	1-propanol	$\Delta_{M}H_{A}{}^{\infty}$	374
Poly(acrylonitrile-*co*-[2-(3-methyl-3-phenylcyclobutyl)-2-hydroxyethyl methacrylate]			
	benzene	$\Delta_{M}H_{A}{}^{\infty}$	374
	2-butanone	$\Delta_{M}H_{A}{}^{\infty}$	374
	n-decane	$\Delta_{M}H_{A}{}^{\infty}$	374
	1,2-dimethylbenzene	$\Delta_{M}H_{A}{}^{\infty}$	374
	n-dodecane	$\Delta_{M}H_{A}{}^{\infty}$	374
	ethanol	$\Delta_{M}H_{A}{}^{\infty}$	374
	ethyl acetate	$\Delta_{M}H_{A}{}^{\infty}$	374
	methanol	$\Delta_{M}H_{A}{}^{\infty}$	374
	methyl acetate	$\Delta_{M}H_{A}{}^{\infty}$	374
	n-nonane	$\Delta_{M}H_{A}{}^{\infty}$	374
	n-octane	$\Delta_{M}H_{A}{}^{\infty}$	374
	2-propanone	$\Delta_{M}H_{A}{}^{\infty}$	374
	toluene	$\Delta_{M}H_{A}{}^{\infty}$	374
	n-undecane	$\Delta_{M}H_{A}{}^{\infty}$	374
Poly(acrylonitrile-*co*-α-methylstyrene)			
	benzene	$\Delta_{M}H_{A}{}^{\infty}, \Delta_{sol}H_{A(vap)}{}^{\infty}$	375
	1-butanol	$\Delta_{M}H_{A}{}^{\infty}, \Delta_{sol}H_{A(vap)}{}^{\infty}$	375
	2-butanone	$\Delta_{M}H_{A}{}^{\infty}, \Delta_{sol}H_{A(vap)}{}^{\infty}$	375
	butyl acetate	$\Delta_{M}H_{A}{}^{\infty}, \Delta_{sol}H_{A(vap)}{}^{\infty}$	375
	butylbenzene	$\Delta_{M}H_{A}{}^{\infty}, \Delta_{sol}H_{A(vap)}{}^{\infty}$	375
	chlorobenzene	$\Delta_{M}H_{A}{}^{\infty}, \Delta_{sol}H_{A(vap)}{}^{\infty}$	375
	cyclohexanol	$\Delta_{M}H_{A}{}^{\infty}, \Delta_{sol}H_{A(vap)}{}^{\infty}$	375
	dichloromethane	$\Delta_{M}H_{A}{}^{\infty}, \Delta_{sol}H_{A(vap)}{}^{\infty}$	375
	1,4-dioxane	$\Delta_{M}H_{A}{}^{\infty}, \Delta_{sol}H_{A(vap)}{}^{\infty}$	375

Polymer(s)	Solvent(s)	Property	Page(s)
Poly(5,5'-bibenzimidazol-2,2'-diyl-1,3-phenylene) (*continued*)			
	2-butanone	$\Delta_{sol}H_{A(vap)}{}^{\infty}$	375
	butyraldehyde	$\Delta_{sol}H_{A(vap)}{}^{\infty}$	375
	n-decane	$\Delta_{sol}H_{A(vap)}{}^{\infty}$	375
	diethylamine	$\Delta_{sol}H_{A(vap)}{}^{\infty}$	375
	n-dodecane	$\Delta_{sol}H_{A(vap)}{}^{\infty}$	375
	ethanol	$\Delta_{sol}H_{A(vap)}{}^{\infty}$	376
	ethyl acetate	$\Delta_{sol}H_{A(vap)}{}^{\infty}$	376
	formamide	$\Delta_{sol}H_{A(vap)}{}^{\infty}$	376
	formic acid	$\Delta_{sol}H_{A(vap)}{}^{\infty}$	376
	n-heptane	$\Delta_{sol}H_{A(vap)}{}^{\infty}$	376
	nitroethane	$\Delta_{sol}H_{A(vap)}{}^{\infty}$	376
	nitromethane	$\Delta_{sol}H_{A(vap)}{}^{\infty}$	376
	1-nitropropane	$\Delta_{sol}H_{A(vap)}{}^{\infty}$	376
	n-nonane	$\Delta_{sol}H_{A(vap)}{}^{\infty}$	376
	n-octane	$\Delta_{sol}H_{A(vap)}{}^{\infty}$	376
	1-pentanol	$\Delta_{sol}H_{A(vap)}{}^{\infty}$	376
	1,2-propanediamine	$\Delta_{sol}H_{A(vap)}{}^{\infty}$	376
	1-propanol	$\Delta_{sol}H_{A(vap)}{}^{\infty}$	376
	2-propanone	$\Delta_{sol}H_{A(vap)}{}^{\infty}$	376
	propionitrile	$\Delta_{sol}H_{A(vap)}{}^{\infty}$	376
	1,1,3,3-tetramethylurea	$\Delta_{sol}H_{A(vap)}{}^{\infty}$	376
Poly[4,4'-bis(hexamethyleneoxy)-biphenyl allylmalonate]			
	n-decane	$\Delta_{M}H_{A}{}^{\infty}$, $\Delta_{sol}H_{A(vap)}{}^{\infty}$	376-377
	1,2-dimethylbenzene	$\Delta_{M}H_{A}{}^{\infty}$, $\Delta_{sol}H_{A(vap)}{}^{\infty}$	376-377
	1,3-dimethylbenzene	$\Delta_{M}H_{A}{}^{\infty}$, $\Delta_{sol}H_{A(vap)}{}^{\infty}$	376-377
	1,4-dimethylbenzene	$\Delta_{M}H_{A}{}^{\infty}$, $\Delta_{sol}H_{A(vap)}{}^{\infty}$	376-377
	toluene	$\Delta_{M}H_{A}{}^{\infty}$, $\Delta_{sol}H_{A(vap)}{}^{\infty}$	376-377
Poly(bisphenol A-isophthaloyl chloride-*co*-terephthaloyl chloride)			
	N,N-dimethylacetamide	$\Delta_{sol}H_{B}{}^{\infty}$	214
	1,1,2,2-tetrachloroethane	$\Delta_{sol}H_{B}{}^{\infty}$	214
1,2-Polybutadiene			
	benzene	$\Delta_{sol}H_{A(vap)}{}^{\infty}$	377
	cyclohexane	$\Delta_{sol}H_{A(vap)}{}^{\infty}$	377
	n-heptane	$\Delta_{sol}H_{A(vap)}{}^{\infty}$	377
	n-hexane	$\Delta_{sol}H_{A(vap)}{}^{\infty}$	377
	toluene	$\Delta_{sol}H_{A(vap)}{}^{\infty}$	377

Polymer(s)	Solvent(s)	Property	Page(s)
1,4-*cis*-Polybutadiene			
	benzene	$\Delta_{dil}H^{12}$	29-30
	benzene	$\Delta_{sol}H_{A(vap)}{}^{\infty}$	377, 378
	1,3-butadiene	$\Delta_{sol}H_{A(vap)}{}^{\infty}$	378
	n-butane	$\Delta_{sol}H_{A(vap)}{}^{\infty}$	378
	cyclohexane	$\Delta_{M}H_{B}{}^{\infty}$	215
	cyclohexane	$\Delta_{sol}H_{A(vap)}{}^{\infty}$	377-378
	cyclooctane	$\Delta_{M}H_{B}{}^{\infty}$	215
	cyclopentane	$\Delta_{M}H_{B}{}^{\infty}$	215
	cyclopentane	$\Delta_{sol}H_{A(vap)}{}^{\infty}$	378
	cis-decahydronaphthalene	$\Delta_{M}H_{B}{}^{\infty}$	215
	trans-decahydronaphthalene	$\Delta_{sol}H_{B}{}^{\infty}$	215
	n-decane	$\Delta_{sol}H_{A(vap)}{}^{\infty}$	378
	3,3-diethylpentane	$\Delta_{M}H_{B}{}^{\infty}$	215
	1,3-dimethylbenzene	$\Delta_{sol}H_{A(vap)}{}^{\infty}$	378
	2,2-dimethylbutane	$\Delta_{sol}H_{A(vap)}{}^{\infty}$	378
	2,2-dimethylpentane	$\Delta_{M}H_{B}{}^{\infty}$	215
	2,3-dimethylpentane	$\Delta_{M}H_{B}{}^{\infty}$	215
	2,4-dimethylpentane	$\Delta_{M}H_{B}{}^{\infty}$	216
	3,3-dimethylpentane	$\Delta_{M}H_{B}{}^{\infty}$	216
	n-dodecane	$\Delta_{M}H_{B}{}^{\infty}$	216
	n-dodecane	$\Delta_{sol}H_{A(vap)}{}^{\infty}$	378
	ethylbenzene	$\Delta_{sol}H_{A(vap)}{}^{\infty}$	378
	3-ethylpentane	$\Delta_{M}H_{B}{}^{\infty}$	216
	2,2,4,4,6,8,8-heptamethyl-nonanc	$\Delta_{M}H_{B}{}^{\infty}$	216
	n-heptane	$\Delta_{sol}H_{A(vap)}{}^{\infty}$	377-378
	n-hexadecane	$\Delta_{M}H_{B}{}^{\infty}$	216
	n-hexane	$\Delta_{sol}H_{A(vap)}{}^{\infty}$	377-378
	2-methylbutane	$\Delta_{sol}H_{A(vap)}{}^{\infty}$	378
	3-methylhexane	$\Delta_{M}H_{B}{}^{\infty}$	216
	2-methylpropane	$\Delta_{sol}H_{A(vap)}{}^{\infty}$	378
	n-nonane	$\Delta_{sol}H_{A(vap)}{}^{\infty}$	378
	n-octane	$\Delta_{M}H_{B}{}^{\infty}$	216
	n-octane	$\Delta_{sol}H_{A(vap)}{}^{\infty}$	378
	2,2,4,6,6-pentamethylheptane	$\Delta_{M}H_{B}{}^{\infty}$	217
	n-pentane	$\Delta_{sol}H_{A(vap)}{}^{\infty}$	378
	propane	$\Delta_{sol}H_{A(vap)}{}^{\infty}$	378
	propene	$\Delta_{sol}H_{A(vap)}{}^{\infty}$	378
	2,2,4,4-tetramethylpentane	$\Delta_{M}H_{B}{}^{\infty}$	217
	2,3,3,4-tetramethylpentane	$\Delta_{M}H_{B}{}^{\infty}$	217
	toluene	$\Delta_{dil}H^{12}$	30
	toluene	$\Delta_{sol}H_{A(vap)}{}^{\infty}$	377-378
	1,3,5-trimethylbenzene	$\Delta_{sol}H_{A(vap)}{}^{\infty}$	378

Polymer(s)	Solvent(s)	Property	Page(s)
Polybutadiene			
	acetonitrile	$\Delta_{sol}H_{A(vap)}^{\infty}$	378-379, 381
	aniline	$\Delta_{sol}H_{A(vap)}^{\infty}$	378
	benzaldehyde	$\Delta_{sol}H_{A(vap)}^{\infty}$	379
	benzene	$\Delta_{M}H$	29
	benzene	$\Delta_{sol}H_{B}^{\infty}$	214
	benzene	$\Delta_{sol}H_{A(vap)}^{\infty}$	379, 381
	benzyl alcohol	$\Delta_{sol}H_{A(vap)}^{\infty}$	379
	1-butanol	$\Delta_{sol}H_{A(vap)}^{\infty}$	379, 381
	2-butanone	$\Delta_{sol}H_{A(vap)}^{\infty}$	379, 381
	butyl acetate	$\Delta_{sol}H_{A(vap)}^{\infty}$	380-381
	butylbenzene	$\Delta_{sol}H_{A(vap)}^{\infty}$	379
	butyronitrile	$\Delta_{sol}H_{A(vap)}^{\infty}$	380-381
	1-chlorobutane	$\Delta_{sol}H_{A(vap)}^{\infty}$	380-382
	1-chloropentane	$\Delta_{sol}H_{A(vap)}^{\infty}$	380, 382
	1-chloropropane	$\Delta_{sol}H_{A(vap)}^{\infty}$	380-382
	cyclohexane	$\Delta_{sol}H_{A(vap)}^{\infty}$	379-382
	cyclohexanone	$\Delta_{sol}H_{A(vap)}^{\infty}$	379
	dibutyl ether	$\Delta_{sol}H_{A(vap)}^{\infty}$	380
	1,2-dichloroethane	$\Delta_{sol}H_{A(vap)}^{\infty}$	379
	dichloromethane	$\Delta_{sol}H_{A(vap)}^{\infty}$	379, 382
	diethyl ether	$\Delta_{sol}H_{A(vap)}^{\infty}$	380, 382
	1,4-dimethylbenzene	$\Delta_{sol}H_{A(vap)}^{\infty}$	379
	N,N-dimethylformamide	$\Delta_{sol}H_{A(vap)}^{\infty}$	379
	dipropyl ether	$\Delta_{sol}H_{A(vap)}^{\infty}$	380
	ethanol	$\Delta_{sol}H_{A(vap)}^{\infty}$	380, 382
	ethyl acetate	$\Delta_{sol}H_{A(vap)}^{\infty}$	379-382
	ethylbenzene	$\Delta_{sol}H_{A(vap)}^{\infty}$	379
	ethylcyclohexane	$\Delta_{sol}H_{A(vap)}^{\infty}$	380, 382
	n-heptane	$\Delta_{sol}H_{A(vap)}^{\infty}$	380-382
	1-heptene	$\Delta_{sol}H_{A(vap)}^{\infty}$	380-382
	n-hexane	$\Delta_{sol}H_{A(vap)}^{\infty}$	379-380, 382
	1-hexene	$\Delta_{sol}H_{A(vap)}^{\infty}$	380, 382
	methanol	$\Delta_{sol}H_{A(vap)}^{\infty}$	379-380, 382
	methyl acetate	$\Delta_{sol}H_{A(vap)}^{\infty}$	380, 382
	methylcyclohexane	$\Delta_{sol}H_{A(vap)}^{\infty}$	380, 382
	4-methyl-2-pentanone	$\Delta_{sol}H_{A(vap)}^{\infty}$	379-382
	n-octane	$\Delta_{sol}H_{A(vap)}^{\infty}$	380-382
	1-octene	$\Delta_{sol}H_{A(vap)}^{\infty}$	380-382
	n-pentane	$\Delta_{sol}H_{A(vap)}^{\infty}$	380, 382
	3-pentanone	$\Delta_{sol}H_{A(vap)}^{\infty}$	380-382

Polymer(s)	Solvent(s)	Property	Page(s)
Polybutadiene (*continued*)			
	1-propanol	$\Delta_{sol}H_{A(vap)}^{\infty}$	380-382
	2-propanol	$\Delta_{sol}H_{A(vap)}^{\infty}$	379-380, 382
	2-propanone	$\Delta_{sol}H_{A(vap)}^{\infty}$	379, 382
	propionitrile	$\Delta_{sol}H_{A(vap)}^{\infty}$	380-382
	propyl acetate	$\Delta_{sol}H_{A(vap)}^{\infty}$	380-382
	tetrachloromethane	$\Delta_{sol}H_{A(vap)}^{\infty}$	379-382
	tetrahydrofuran	$\Delta_{sol}H_{A(vap)}^{\infty}$	379-380, 382
	toluene	$\Delta_{sol}H_{A(vap)}^{\infty}$	379-382
	trichloromethane	$\Delta_{sol}H_{A(vap)}^{\infty}$	379-382
	2,2,4-trimethylpentane	$\Delta_{sol}H_{B}^{\infty}$	217
Poly(butadiene-*co*-styrene)			
	benzene	$\Delta_{M}H$	30-31
	benzene	$\Delta_{sol}H_{B}^{\infty}$	217-218
	benzene	$\Delta_{sol}H_{A(vap)}^{\infty}$	383
	cyclohexane	$\Delta_{sol}H_{A(vap)}^{\infty}$	383
	ethylbenzene	$\Delta_{sol}H_{B}^{\infty}$	219-220
	n-heptane	$\Delta_{sol}H_{A(vap)}^{\infty}$	383
	n-hexane	$\Delta_{sol}H_{A(vap)}^{\infty}$	383
	toluene	$\Delta_{sol}H_{A(vap)}^{\infty}$	383
	2,2,4-trimethylpentane	$\Delta_{sol}H_{B}^{\infty}$	219-220
Poly(1-butene)			
	cyclohexane	$\Delta_{M}H_{B}^{\infty}$	220
	cyclooctane	$\Delta_{M}H_{B}^{\infty}$	220
	cyclopentane	$\Delta_{M}H_{B}^{\infty}$	220
	cis-decahydronaphthalene	$\Delta_{M}H_{B}^{\infty}$	220
	trans-decahydronaphthalene	$\Delta_{sol}H_{B}^{\infty}$	220
	n-decane	$\Delta_{M}H_{B}^{\infty}$	221
	3,3-diethylpentane	$\Delta_{M}H_{B}^{\infty}$	221
	2,2-dimethylpentane	$\Delta_{M}H_{B}^{\infty}$	221
	2,3-dimethylpentane	$\Delta_{M}H_{B}^{\infty}$	221
	2,4-dimethylpentane	$\Delta_{M}H_{B}^{\infty}$	221
	3,3-dimethylpentane	$\Delta_{M}H_{B}^{\infty}$	221
	n-dodecane	$\Delta_{M}H_{B}^{\infty}$	221
	3-ethylpentane	$\Delta_{M}H_{B}^{\infty}$	222
	2,2,4,4,6,8,8-heptamethyl-nonane	$\Delta_{M}H_{B}^{\infty}$	222
	n-heptane	$\Delta_{M}H_{B}^{\infty}$	222
	n-hexadecane	$\Delta_{M}H_{B}^{\infty}$	222
	n-hexane	$\Delta_{M}H_{B}^{\infty}$	222

Polymer(s)	Solvent(s)	Property	Page(s)
Poly(1-butene) (*continued*)			
	3-methylhexane	$\Delta_M H_B^\infty$	222
	n-nonane	$\Delta_M H_B^\infty$	223
	n-octane	$\Delta_M H_B^\infty$	223
	n-pentane	$\Delta_M H_B^\infty$	223
	n-tetradecane	$\Delta_M H_B^\infty$	223
	n-tetradecane	$\Delta_M H_A^\infty$	384
	2,2,4,4-tetramethylpentane	$\Delta_M H_B^\infty$	223
	2,3,3,4-tetramethylpentane	$\Delta_M H_B^\infty$	224
	2,2,4-trimethylpentane	$\Delta_M H_B^\infty$	224
Poly(butyl acrylate)			
	2-propanone	$\Delta_{sol} H_B^\infty$	224
Poly(butyl methacrylate)			
	benzene	$\Delta_M H_A^\infty$, $\Delta_{sol} H_{A(vap)}^\infty$	384-385
	n-butane	$\Delta_{sol} H_{A(vap)}^\infty$	385
	1-butanol	$\Delta_{sol} H_{A(vap)}^\infty$	385
	2-butanone	$\Delta_{dil} H^{12}$	31
	2-butanone	$\Delta_M H_A^\infty$, $\Delta_{sol} H_{A(vap)}^\infty$	384-385
	butyl acetate	$\Delta_{sol} H_{A(vap)}^\infty$	385
	chlorobenzene	$\Delta_{sol} H_{A(vap)}^\infty$	385
	1-chlorohexane	$\Delta_{sol} H_{A(vap)}^\infty$	385
	1-chlorooctane	$\Delta_{sol} H_{A(vap)}^\infty$	385
	cycloheptane	$\Delta_{sol} H_{A(vap)}^\infty$	385
	cyclohexadiene	$\Delta_{sol} H_{A(vap)}^\infty$	385
	cyclohexane	$\Delta_M H_A^\infty$, $\Delta_{sol} H_{A(vap)}^\infty$	384-385
	cyclohexanone	$\Delta_M H_B^\infty$, $\Delta_{sol} H_B^\infty$	224
	cyclohexene	$\Delta_{sol} H_{A(vap)}^\infty$	385
	cyclooctane	$\Delta_{sol} H_{A(vap)}^\infty$	385
	cyclopentane	$\Delta_{sol} H_{A(vap)}^\infty$	385
	n-decane	$\Delta_M H_A^\infty$, $\Delta_{sol} H_{A(vap)}^\infty$	385
	1,1-dichloroethane	$\Delta_{sol} H_{A(vap)}^\infty$	386
	1,2-dichloroethane	$\Delta_M H_A^\infty$, $\Delta_{sol} H_{A(vap)}^\infty$	384, 386
	dichloromethane	$\Delta_M H_A^\infty$, $\Delta_{sol} H_{A(vap)}^\infty$	384-386
	1,2-dimethylbenzene	$\Delta_M H_A^\infty$, $\Delta_{sol} H_{A(vap)}^\infty$	384
	1,4-dimethylbenzene	$\Delta_M H_A^\infty$, $\Delta_{sol} H_{A(vap)}^\infty$	384-385
	1,4-dioxane	$\Delta_{sol} H_{A(vap)}^\infty$	386
	ethanol	$\Delta_{sol} H_{A(vap)}^\infty$	386
	ethyl acetate	$\Delta_{sol} H_{A(vap)}^\infty$	386
	ethylbenzene	$\Delta_M H_A^\infty$, $\Delta_{sol} H_{A(vap)}^\infty$	384, 386
	n-heptane	$\Delta_M H_A^\infty$, $\Delta_{sol} H_{A(vap)}^\infty$	385-386
	n-hexane	$\Delta_M H_A^\infty$, $\Delta_{sol} H_{A(vap)}^\infty$	385-386

Polymer(s)	Solvent(s)	Property	Page(s)
Poly(butyl methacrylate) (*continued*)			
	methyl acetate	$\Delta_{sol}H_{A(vap)}{}^{\infty}$	386
	n-nonane	$\Delta_{M}H_{A}{}^{\infty}, \Delta_{sol}H_{A(vap)}{}^{\infty}$	385-386
	n-octane	$\Delta_{M}H_{A}{}^{\infty}, \Delta_{sol}H_{A(vap)}{}^{\infty}$	385-386
	n-pentane	$\Delta_{M}H_{A}{}^{\infty}, \Delta_{sol}H_{A(vap)}{}^{\infty}$	385
	1-pentanol	$\Delta_{sol}H_{A(vap)}{}^{\infty}$	386
	2-pentanone	$\Delta_{M}H_{A}{}^{\infty}, \Delta_{sol}H_{A(vap)}{}^{\infty}$	384
	1-propanol	$\Delta_{sol}H_{A(vap)}{}^{\infty}$	386
	2-propanone	$\Delta_{sol}H_{B}{}^{\infty}$	224
	2-propanone	$\Delta_{M}H_{A}{}^{\infty}, \Delta_{sol}H_{A(vap)}{}^{\infty}$	384, 386
	propyl acetate	$\Delta_{sol}H_{A(vap)}{}^{\infty}$	386
	propylbenzene	$\Delta_{M}H_{A}{}^{\infty}, \Delta_{sol}H_{A(vap)}{}^{\infty}$	384
	tetrachloromethane	$\Delta_{M}H_{A}{}^{\infty}, \Delta_{sol}H_{A(vap)}{}^{\infty}$	384-385
	toluene	$\Delta_{M}H_{A}{}^{\infty}, \Delta_{sol}H_{A(vap)}{}^{\infty}$	384-386
	1,1,1-trichloroethane	$\Delta_{sol}H_{A(vap)}{}^{\infty}$	386
	trichloroethene	$\Delta_{M}H_{A}{}^{\infty}, \Delta_{sol}H_{A(vap)}{}^{\infty}$	385-386
	trichloromethane	$\Delta_{M}H$	31-32
	trichloromethane	$\Delta_{M}H_{B}{}^{\infty}$	224-225
	trichloromethane	$\Delta_{M}H_{A}{}^{\infty}, \Delta_{sol}H_{A(vap)}{}^{\infty}$	384-386
	2,2,4-trimethylpentane	$\Delta_{M}H_{A}{}^{\infty}, \Delta_{sol}H_{A(vap)}{}^{\infty}$	385
	n-undecane	$\Delta_{sol}H_{A(vap)}{}^{\infty}$	386
Poly(butyl methacrylate-*co*-isobutyl methacrylate)			
	cyclohexanone	$\Delta_{M}H_{B}{}^{\infty}, \Delta_{sol}H_{B}{}^{\infty}$	225
Poly(4-*tert*-butylstyrene)			
	benzene	$\Delta_{M}H_{A}{}^{\infty}$	386
	2-butanone	$\Delta_{M}H_{A}{}^{\infty}$	386
	n-decane	$\Delta_{M}H_{A}{}^{\infty}$	386
	1,2-dimethylbenzene	$\Delta_{M}H_{A}{}^{\infty}$	386
	n-dodecane	$\Delta_{M}H_{A}{}^{\infty}$	386
	ethanol	$\Delta_{M}H_{A}{}^{\infty}$	386
	ethyl acetate	$\Delta_{M}H_{A}{}^{\infty}$	386
	methanol	$\Delta_{M}H_{A}{}^{\infty}$	386
	methyl acetate	$\Delta_{M}H_{A}{}^{\infty}$	387
	n-nonane	$\Delta_{M}H_{A}{}^{\infty}$	387
	n-octane	$\Delta_{M}H_{A}{}^{\infty}$	387
	2-propanone	$\Delta_{M}H_{A}{}^{\infty}$	387
	toluene	$\Delta_{M}H_{A}{}^{\infty}$	387
	n-undecane	$\Delta_{M}H_{A}{}^{\infty}$	387

Polymer(s)	Solvent(s)	Property	Page(s)
Poly(ε-caprolactone)			
	benzene	$\Delta_M H_A^\infty$, $\Delta_{sol} H_{A(vap)}^\infty$	387, 389
	1-butanol	$\Delta_{sol} H_{A(vap)}^\infty$	387-388
	2-butanone	$\Delta_{sol} H_{A(vap)}^\infty$	387
	butyl acetate	$\Delta_M H_A^\infty$, $\Delta_{sol} H_{A(vap)}^\infty$	387-389
	tert-butyl acetate	$\Delta_M H_A^\infty$, $\Delta_{sol} H_{A(vap)}^\infty$	389
	chlorobenzene	$\Delta_M H_A^\infty$, $\Delta_{sol} H_{A(vap)}^\infty$	387, 389
	1-chlorobutane	$\Delta_{sol} H_{A(vap)}^\infty$	387
	1-chlorohexane	$\Delta_{sol} H_{A(vap)}^\infty$	387
	1-chlorooctane	$\Delta_{sol} H_{A(vap)}^\infty$	387
	1-chloropentane	$\Delta_{sol} H_{A(vap)}^\infty$	387
	cycloheptane	$\Delta_{sol} H_{A(vap)}^\infty$	387
	cyclohexadiene	$\Delta_{sol} H_{A(vap)}^\infty$	387
	cyclohexane	$\Delta_{sol} H_{A(vap)}^\infty$	387
	cyclohexene	$\Delta_{sol} H_{A(vap)}^\infty$	387
	cyclooctane	$\Delta_{sol} H_{A(vap)}^\infty$	387
	cyclopentane	$\Delta_{sol} H_{A(vap)}^\infty$	387
	n-decane	$\Delta_M H_A^\infty$, $\Delta_{sol} H_{A(vap)}^\infty$	387-389
	1,1-dichloroethane	$\Delta_{sol} H_{A(vap)}^\infty$	387
	1,2-dichloroethane	$\Delta_{sol} H_{A(vap)}^\infty$	387
	dichloromethane	$\Delta_{sol} H_{A(vap)}^\infty$	387
	1,4-dioxane	$\Delta_{sol} H_{A(vap)}^\infty$	387
	n-dodecane	$\Delta_M H_A^\infty$, $\Delta_{sol} H_{A(vap)}^\infty$	388
	ethanol	$\Delta_{sol} H_{A(vap)}^\infty$	387-388
	ethyl acetate	$\Delta_{sol} H_{A(vap)}^\infty$	387-389
	ethylbenzene	$\Delta_M H_A^\infty$, $\Delta_{sol} H_{A(vap)}^\infty$	387, 389
	n-heptane	$\Delta_M H_A^\infty$, $\Delta_{sol} H_{A(vap)}^\infty$	387-389
	n-hexane	$\Delta_M H_A^\infty$, $\Delta_{sol} H_{A(vap)}^\infty$	387-389
	isobutyl acetate	$\Delta_M H_A^\infty$, $\Delta_{sol} H_{A(vap)}^\infty$	389
	isopentyl acetate	$\Delta_M H_A^\infty$, $\Delta_{sol} H_{A(vap)}^\infty$	389
	isopropyl acetate	$\Delta_M H_A^\infty$, $\Delta_{sol} H_{A(vap)}^\infty$	389
	isopropylbenzene	$\Delta_M H_A^\infty$, $\Delta_{sol} H_{A(vap)}^\infty$	389
	methanol	$\Delta_{sol} H_{A(vap)}^\infty$	388
	methyl acetate	$\Delta_{sol} H_{A(vap)}^\infty$	387-389
	n-nonane	$\Delta_M H_A^\infty$, $\Delta_{sol} H_{A(vap)}^\infty$	387-389
	n-octane	$\Delta_M H_A^\infty$, $\Delta_{sol} H_{A(vap)}^\infty$	387-389
	n-pentane	$\Delta_M H_A^\infty$, $\Delta_{sol} H_{A(vap)}^\infty$	388-389
	1-pentanol	$\Delta_{sol} H_{A(vap)}^\infty$	387
	1-propanol	$\Delta_{sol} H_{A(vap)}^\infty$	387-388
	2-propanone	$\Delta_{sol} H_{A(vap)}^\infty$	387
	propyl acetate	$\Delta_{sol} H_{A(vap)}^\infty$	387-389
	propylbenzene	$\Delta_M H_A^\infty$, $\Delta_{sol} H_{A(vap)}^\infty$	389
	tetrachloromethane	$\Delta_{sol} H_{A(vap)}^\infty$	387
	tetrahydrofuran	$\Delta_{sol} H_{A(vap)}^\infty$	387

Polymer(s)	Solvent(s)	Property	Page(s)
Poly(ε-caprolactone) (*continued*)			
	toluene	$\Delta_M H_A^\infty$, $\Delta_{sol} H_{A(vap)}^\infty$	387, 389
	1,1,1-trichloroethane	$\Delta_{sol} H_{A(vap)}^\infty$	387
	trichloroethene	$\Delta_{sol} H_{A(vap)}^\infty$	387
	trichloromethane	$\Delta_{sol} H_{A(vap)}^\infty$	387
	n-undecane	$\Delta_M H_A^\infty$, $\Delta_{sol} H_{A(vap)}^\infty$	387-388
Poly(ε-carbobenzoxy-L-lysine)			
	1,2-dichloroethane	$\Delta_{sol} H_B^\infty$	225
Poly(ε-carbobenzoxy-L-lysine-*co*-L-phenylalanine)			
	1,2-dichloroethane	$\Delta_{sol} H_B^\infty$	225-226
Polycarbonate bisphenol-A			
	benzene	$\Delta_M H_A^\infty$	390
	butyl acetate	$\Delta_M H_A^\infty$	390
	butylbenzene	$\Delta_M H_A^\infty$	390
	tert-butylbenzene	$\Delta_M H_A^\infty$	390
	butylcyclohexane	$\Delta_M H_A^\infty$	390
	chlorobenzene	$\Delta_M H_A^\infty$	390
	cyclohexanol	$\Delta_M H_A^\infty$	390
	cyclohexanone	$\Delta_M H_A^\infty$	390
	n-decane	$\Delta_M H_A^\infty$	390
	dichloromethane	$\Delta_M H_A^\infty$	390
	1,4-dioxane	$\Delta_{dil} H^{12}$	32-33
	n-dodecane	$\Delta_M H_A^\infty$	390
	ethylbenzene	$\Delta_M H_A^\infty$	390
	methylcyclohexane	$\Delta_M H_A^\infty$	390
	2-pentanone	$\Delta_M H_A^\infty$	390
	1-propanol	$\Delta_M H_A^\infty$	390
	n-tetradecane	$\Delta_M H_A^\infty$	390
	toluene	$\Delta_M H_A^\infty$	390
	trichloromethane	$\Delta_M H_A^\infty$	390
Polycarbonate 4,4'-(1-phenylethylidene)-bisphenol			
	benzene	$\Delta_M H_A^\infty$	390
	butyl acetate	$\Delta_M H_A^\infty$	390
	butylbenzene	$\Delta_M H_A^\infty$	390
	tert-butylbenzene	$\Delta_M H_A^\infty$	390
	butylcyclohexane	$\Delta_M H_A^\infty$	390

Polymer(s)	Solvent(s)	Property	Page(s)
Polycarbonate 4,4'-(1-phenylethylidene)-bisphenol (*continued*)			
	chlorobenzene	$\Delta_M H_A^\infty$	390
	cyclohexanol	$\Delta_M H_A^\infty$	390
	cyclohexanone	$\Delta_M H_A^\infty$	390
	n-decane	$\Delta_M H_A^\infty$	390
	dichloromethane	$\Delta_M H_A^\infty$	391
	n-dodecane	$\Delta_M H_A^\infty$	391
	ethylbenzene	$\Delta_M H_A^\infty$	391
	methylcyclohexane	$\Delta_M H_A^\infty$	391
	2-pentanone	$\Delta_M H_A^\infty$	391
	1-propanol	$\Delta_M H_A^\infty$	391
	n-tetradecane	$\Delta_M H_A^\infty$	391
	toluene	$\Delta_M H_A^\infty$	391
	trichloromethane	$\Delta_M H_A^\infty$	391
Polycarbonate 4,4'-diphenylmethylidene-bisphenol			
	benzene	$\Delta_M H_A^\infty$	391
	butyl acetate	$\Delta_M H_A^\infty$	391
	butylbenzene	$\Delta_M H_A^\infty$	391
	tert-butylbenzene	$\Delta_M H_A^\infty$	391
	butylcyclohexane	$\Delta_M H_A^\infty$	391
	chlorobenzene	$\Delta_M H_A^\infty$	391
	cyclohexanol	$\Delta_M H_A^\infty$	391
	cyclohexanone	$\Delta_M H_A^\infty$	391
	n-decane	$\Delta_M H_A^\infty$	391
	dichloromethane	$\Delta_M H_A^\infty$	391
	n-dodecane	$\Delta_M H_A^\infty$	391
	ethylbenzene	$\Delta_M H_A^\infty$	391
	methylcyclohexane	$\Delta_M H_A^\infty$	391
	2-pentanone	$\Delta_M H_A^\infty$	391
	1-propanol	$\Delta_M H_A^\infty$	391
	n-tetradecane	$\Delta_M H_A^\infty$	391
	toluene	$\Delta_M H_A^\infty$	391
	trichloromethane	$\Delta_M H_A^\infty$	391
Polycarbonate 4,4'-cyclohexylidene-bisphenol			
	benzene	$\Delta_M H_A^\infty$	392
	butyl acetate	$\Delta_M H_A^\infty$	392
	butylbenzene	$\Delta_M H_A^\infty$	392
	tert-butylbenzene	$\Delta_M H_A^\infty$	392

Polymer(s)	Solvent(s)	Property	Page(s)
Polycarbonate 4,4'-cyclohexylidene-bisphenol (*continued*)			
	butylcyclohexane	$\Delta_M H_A^\infty$	392
	chlorobenzene	$\Delta_M H_A^\infty$	392
	cyclohexanol	$\Delta_M H_A^\infty$	392
	cyclohexanone	$\Delta_M H_A^\infty$	392
	n-decane	$\Delta_M H_A^\infty$	392
	dichloromethane	$\Delta_M H_A^\infty$	392
	n-dodecane	$\Delta_M H_A^\infty$	392
	ethylbenzene	$\Delta_M H_A^\infty$	392
	methylcyclohexane	$\Delta_M H_A^\infty$	392
	2-pentanone	$\Delta_M H_A^\infty$	392
	1-propanol	$\Delta_M H_A^\infty$	392
	n-tetradecane	$\Delta_M H_A^\infty$	392
	toluene	$\Delta_M H_A^\infty$	392
	trichloromethane	$\Delta_M H_A^\infty$	392
Polycarbonate 4,4'-(4-*tert*-butylcyclo-hexylidene)bisphenol			
	butyl acetate	$\Delta_M H_A^\infty$	392
	butylbenzene	$\Delta_M H_A^\infty$	392
	tert-butylbenzene	$\Delta_M H_A^\infty$	392
	butylcyclohexane	$\Delta_M H_A^\infty$	392
	chlorobenzene	$\Delta_M H_A^\infty$	392
	cyclohexanol	$\Delta_M H_A^\infty$	392
	cyclohexanone	$\Delta_M H_A^\infty$	392
	n-decane	$\Delta_M H_A^\infty$	392
	n-dodecane	$\Delta_M H_A^\infty$	392
	methylcyclohexane	$\Delta_M H_A^\infty$	392
	n-tetradecane	$\Delta_M H_A^\infty$	392
	toluene	$\Delta_M H_A^\infty$	392
Polycarbonate 4,4'-cyclohexylidene-2,2'-dimethylbisphenol			
	benzene	$\Delta_M H_A^\infty$	393
	butyl acetate	$\Delta_M H_A^\infty$	393
	butylbenzene	$\Delta_M H_A^\infty$	393
	tert-butylbenzene	$\Delta_M H_A^\infty$	393
	butylcyclohexane	$\Delta_M H_A^\infty$	393
	chlorobenzene	$\Delta_M H_A^\infty$	393
	cyclohexanol	$\Delta_M H_A^\infty$	393
	cyclohexanone	$\Delta_M H_A^\infty$	393
	n-decane	$\Delta_M H_A^\infty$	393

Polymer(s)	Solvent(s)	Property	Page(s)
Polycarbonate 4,4'-cyclohexylidene-2,2'-dimethylbisphenol (*continued*)			
	dichloromethane	$\Delta_M H_A^{\infty}$	393
	n-dodecane	$\Delta_M H_A^{\infty}$	393
	ethylbenzene	$\Delta_M H_A^{\infty}$	393
	methylcyclohexane	$\Delta_M H_A^{\infty}$	393
	2-pentanone	$\Delta_M H_A^{\infty}$	393
	1-propanol	$\Delta_M H_A^{\infty}$	393
	n-tetradecane	$\Delta_M H_A^{\infty}$	393
	toluene	$\Delta_M H_A^{\infty}$	393
	trichloromethane	$\Delta_M H_A^{\infty}$	393
Polychloroprene			
	benzene	$\Delta_{sol} H_B^{\infty}$	226
	dichloromethane	$\Delta_M H_A^{\infty}, \Delta_{sol} H_{A(vap)}^{\infty}$	393
	tetrachloromethane	$\Delta_M H_A^{\infty}, \Delta_{sol} H_{A(vap)}^{\infty}$	393
	trichloromethane	$\Delta_M H_A^{\infty}, \Delta_{sol} H_{A(vap)}^{\infty}$	393
Poly(chloroprene-*co*-methyl methacrylate)			
	dichloromethane	$\Delta_M H_A^{\infty}, \Delta_{sol} H_{A(vap)}^{\infty}$	394
	tetrachloromethane	$\Delta_M H_A^{\infty}, \Delta_{sol} H_{A(vap)}^{\infty}$	394
	trichloromethane	$\Delta_M H_A^{\infty}, \Delta_{sol} H_{A(vap)}^{\infty}$	394
Poly(p-chlorostyrene)			
	benzene	$\Delta_M H_A^{\infty}, \Delta_{sol} H_{A(vap)}^{\infty}$	395
	n-heptane	$\Delta_M H_A^{\infty}, \Delta_{sol} H_{A(vap)}^{\infty}$	395
	n-hexane	$\Delta_M H_A^{\infty}, \Delta_{sol} H_{A(vap)}^{\infty}$	395
	isopropylbenzene	$\Delta_M H_A^{\infty}, \Delta_{sol} H_{A(vap)}^{\infty}$	395
	n-pentane	$\Delta_M H_A^{\infty}, \Delta_{sol} H_{A(vap)}^{\infty}$	395
	propylbenzene	$\Delta_M H_A^{\infty}, \Delta_{sol} H_{A(vap)}^{\infty}$	395
	toluene	$\Delta_M H_A^{\infty}, \Delta_{sol} H_{A(vap)}^{\infty}$	395
Poly(1,3-cyclohexadiene)			
	benzene	$\Delta_{sol} H_B^{\infty}$	226
	toluene	$\Delta_{sol} H_B^{\infty}$	226
Poly(3,4-dichlorobenzyl methacrylate-*co*-ethyl methacrylate)			
	benzene	$\Delta_M H_A^{\infty}$	395-398
	2-butanone	$\Delta_M H_A^{\infty}$	395-398
	n-decane	$\Delta_M H_A^{\infty}$	395-398
	1,2-dimethylbenzene	$\Delta_M H_A^{\infty}$	395-398

Polymer(s)	Solvent(s)	Property	Page(s)
Poly(3,4-dichlorobenzyl methacrylate-*co*-ethyl methacrylate) (*continued*)			
	n-dodecane	$\Delta_M H_A^\infty$	395-398
	ethanol	$\Delta_M H_A^\infty$	395-398
	ethyl acetate	$\Delta_M H_A^\infty$	395-398
	methanol	$\Delta_M H_A^\infty$	395-398
	methyl acetate	$\Delta_M H_A^\infty$	395-398
	n-nonane	$\Delta_M H_A^\infty$	395-398
	n-octane	$\Delta_M H_A^\infty$	395-398
	2-propanone	$\Delta_M H_A^\infty$	395-398
	toluene	$\Delta_M H_A^\infty$	395-398
	n-undecane	$\Delta_M H_A^\infty$	395-398
Poly(diethyl maleate-*co*-vinyl acetate)			
	n-heptane	$\Delta_M H_A^\infty, \Delta_{sol} H_{A(vap)}^\infty$	398
	n-hexane	$\Delta_M H_A^\infty, \Delta_{sol} H_{A(vap)}^\infty$	398
	n-nonane	$\Delta_M H_A^\infty, \Delta_{sol} H_{A(vap)}^\infty$	398
	n-octane	$\Delta_M H_A^\infty, \Delta_{sol} H_{A(vap)}^\infty$	398
Poly(2,6-dimethyl-1,4-phenylene oxide)			
	1,2-dichlorobenzene	$\Delta_{sol} H_B^\infty$	226-227
Poly(dimethylsilmethylene)			
	n-butane	$\Delta_{sol} H_{A(vap)}^\infty$	399
	n-heptane	$\Delta_{sol} H_{A(vap)}^\infty$	399
	n-hexane	$\Delta_{sol} H_{A(vap)}^\infty$	399
	n-pentane	$\Delta_{sol} H_{A(vap)}^\infty$	399
	n-propane	$\Delta_{sol} H_{A(vap)}^\infty$	399
Poly(dimethylsiltrimethylene)			
	n-butane	$\Delta_{sol} H_{A(vap)}^\infty$	399
	n-heptane	$\Delta_{sol} H_{A(vap)}^\infty$	399
	n-hexane	$\Delta_{sol} H_{A(vap)}^\infty$	399
	n-pentane	$\Delta_{sol} H_{A(vap)}^\infty$	399
	n-propane	$\Delta_{sol} H_{A(vap)}^\infty$	399
Poly(dimethylsiloxane)			
	benzene	$\Delta_M H_B^\infty$	227
	benzene	$\Delta_M H_A$	354-356
	benzene	$\Delta_{sol} H_{A(vap)}^\infty$	399, 403
	benzene	$\Delta_M H_A^\infty, \Delta_{sol} H_{A(vap)}^\infty$	404
	bromocyclohexane	$\Delta_M H_B^\infty$	227

Polymer(s)	Solvent(s)	Property	Page(s)
Poly(dimethylsiloxane) (*continued*)			
	n-butane	$\Delta_{sol}H_{A(vap)}{}^{\infty}$	399
	1-butanol	$\Delta_{M}H_{A}{}^{\infty}, \Delta_{sol}H_{A(vap)}{}^{\infty}$	399, 404
	2-butanone	$\Delta_{M}H_{B}{}^{\infty}$	227-228
	2-butanone	$\Delta_{M}H_{A}{}^{\infty}, \Delta_{sol}H_{A(vap)}{}^{\infty}$	399, 404
	butyl acetate	$\Delta_{M}H_{B}{}^{\infty}$	228
	butyl acetate	$\Delta_{sol}H_{A(vap)}{}^{\infty}$	399
	butyl propionate	$\Delta_{M}H_{B}{}^{\infty}$	228
	chlorobenzene	$\Delta_{M}H_{B}{}^{\infty}$	228
	chlorobenzene	$\Delta_{sol}H_{A(vap)}{}^{\infty}$	400
	1-chlorobutane	$\Delta_{sol}H_{A(vap)}{}^{\infty}$	400
	1-chlorohexane	$\Delta_{sol}H_{A(vap)}{}^{\infty}$	400
	1-chlorooctane	$\Delta_{sol}H_{A(vap)}{}^{\infty}$	400
	1-chloropentane	$\Delta_{sol}H_{A(vap)}{}^{\infty}$	400
	cycloheptane	$\Delta_{sol}H_{A(vap)}{}^{\infty}$	400
	cyclohexadiene	$\Delta_{sol}H_{A(vap)}{}^{\infty}$	400
	cyclohexane	$\Delta_{M}H_{B}{}^{\infty}$	228
	cyclohexane	$\Delta_{sol}H_{A(vap)}{}^{\infty}$	400, 403
	cyclohexene	$\Delta_{sol}H_{A(vap)}{}^{\infty}$	400
	cyclooctane	$\Delta_{M}H_{B}{}^{\infty}$	229
	cyclooctane	$\Delta_{sol}H_{A(vap)}{}^{\infty}$	400
	cyclopentane	$\Delta_{M}H_{B}{}^{\infty}$	229
	cyclopentane	$\Delta_{sol}H_{A(vap)}{}^{\infty}$	400
	cis-decahydronaphthalene	$\Delta_{M}H_{B}{}^{\infty}$	229
	trans-decahydronaphthalene	$\Delta_{sol}H_{B}{}^{\infty}$	229
	decamethylcyclopenta-siloxane	$\Delta_{M}H_{A}{}^{\infty}$	405
	n-decane	$\Delta_{M}H_{B}{}^{\infty}$	229
	n-decane	$\Delta_{M}H_{A}{}^{\infty}, \Delta_{sol}H_{A(vap)}{}^{\infty}$	400, 404
	decyl acetate	$\Delta_{M}H_{B}{}^{\infty}$	230
	dibutyl ether	$\Delta_{M}H_{B}{}^{\infty}$	230
	1,1-dichloroethane	$\Delta_{sol}H_{A(vap)}{}^{\infty}$	400
	1,2-dichloroethane	$\Delta_{sol}H_{A(vap)}{}^{\infty}$	400
	dichloromethane	$\Delta_{sol}H_{A(vap)}{}^{\infty}$	400
	diethoxymethane	$\Delta_{M}H_{B}{}^{\infty}$	230
	diethyl ether	$\Delta_{M}H_{B}{}^{\infty}$	230
	3,3-diethylpentane	$\Delta_{M}H_{B}{}^{\infty}$	230
	dihexyl ether	$\Delta_{M}H_{B}{}^{\infty}$	230
	1,2-dimethoxyethane	$\Delta_{M}H_{B}{}^{\infty}$	230
	dimethoxymethane	$\Delta_{M}H_{B}{}^{\infty}$	230
	N,N-dimethylaniline	$\Delta_{M}H_{A}{}^{\infty}, \Delta_{sol}H_{A(vap)}{}^{\infty}$	404
	1,2-dimethylbenzene	$\Delta_{M}H_{B}{}^{\infty}$	231
	1,2-dimethylbenzene	$\Delta_{sol}H_{A(vap)}{}^{\infty}$	403
	1,3-dimethylbenzene	$\Delta_{M}H_{B}{}^{\infty}$	231

Polymer(s)	Solvent(s)	Property	Page(s)
Poly(dimethylsiloxane) (*continued*)			
	n-hexane	$\Delta_M H$	34
	n-hexane	$\Delta_M H_B^\infty$	237
	n-hexane	$\Delta_M H_A^\infty$, $\Delta_{sol} H_{A(vap)}^\infty$	400-404
	hexyl acetate	$\Delta_M H_B^\infty$	237
	isopropylbenzene	$\Delta_M H_B^\infty$	237
	methyl acetate	$\Delta_{sol} H_{A(vap)}^\infty$	400
	methyl butanoate	$\Delta_M H_B^\infty$	238
	methylcyclohexane	$\Delta_M H_B^\infty$	238
	methyl decanoate	$\Delta_M H_B^\infty$	238
	2-methylhexane	$\Delta_M H_B^\infty$	238
	2-methylhexane	$\Delta_{sol} H_{A(vap)}^\infty$	403
	3-methylhexane	$\Delta_{sol} H_{A(vap)}^\infty$	403
	methyl hexanoate	$\Delta_M H_B^\infty$	238
	methyl octanoate	$\Delta_M H_B^\infty$	238
	4-methyl-2-pentanone	$\Delta_M H_B^\infty$	239
	methyl propionate	$\Delta_M H_B^\infty$	239
	n-nonane	$\Delta_M H_B^\infty$	239
	n-nonane	$\Delta_M H_A$	357-358
	n-nonane	$\Delta_{sol} H_{A(vap)}^\infty$	400, 403
	n-nonane	$\Delta_M H_A^\infty$, $\Delta_{sol} H_{A(vap)}^\infty$	404
	octamethylcyclotetrasiloxane	$\Delta_M H$	34
	octamethylcyclotetrasiloxane	$\Delta_M H_B^\infty$	239
	octamethylcyclotetrasiloxane	$\Delta_M H_A$	358-359
	octamethylcyclotetrasiloxane	$\Delta_M H_A^\infty$	405
	octamethyltrisiloxane	$\Delta_M H_B^\infty$	239-240
	n-octane	$\Delta_M H$	34
	n-octane	$\Delta_M H_B^\infty$	240
	n-octane	$\Delta_M H_A$	359-360
	n-octane	$\Delta_M H_A^\infty$, $\Delta_{sol} H_{A(vap)}^\infty$	400-404
	1-octanol	$\Delta_M H_A^\infty$, $\Delta_{sol} H_{A(vap)}^\infty$	404
	2-octanone	$\Delta_M H_A^\infty$, $\Delta_{sol} H_{A(vap)}^\infty$	404
	2,2,4,6,6-pentamethylheptane	$\Delta_M H$	34
	2,2,4,6,6-pentamethylheptane	$\Delta_M H_B^\infty$	240-241
	n-pentane	$\Delta_M H_B^\infty$	241
	n-pentane	$\Delta_M H_A^\infty$, $\Delta_{sol} H_{A(vap)}^\infty$	400-404
	1-pentanol	$\Delta_M H_A^\infty$, $\Delta_{sol} H_{A(vap)}^\infty$	400, 404
	2-pentanone	$\Delta_M H_A^\infty$, $\Delta_{sol} H_{A(vap)}^\infty$	404
	pentyl acetate	$\Delta_M H_B^\infty$	241
	pentylbenzene	$\Delta_M H_A^\infty$, $\Delta_{sol} H_{A(vap)}^\infty$	404
	pentyl propionate	$\Delta_M H_B^\infty$	241
	1-propanol	$\Delta_{sol} H_{A(vap)}^\infty$	400
	2-propanone	$\Delta_M H_A^\infty$, $\Delta_{sol} H_{A(vap)}^\infty$	400, 404
	propyl acetate	$\Delta_M H_B^\infty$	241

Polymer(s)	Solvent(s)	Property	Page(s)
Poly(dimethylsiloxane) (*continued*)			
	propyl acetate	$\Delta_{sol}H_{A(vap)}^{\infty}$	400
	propylbenzene	$\Delta_M H_A^{\infty}$, $\Delta_{sol}H_{A(vap)}^{\infty}$	404
	propyl propionate	$\Delta_M H_B^{\infty}$	242
	pyridine	$\Delta_M H_A^{\infty}$, $\Delta_{sol}H_{A(vap)}^{\infty}$	404
	tetrachloromethane	$\Delta_M H_B^{\infty}$	242
	tetrachloromethane	$\Delta_{sol}H_{A(vap)}^{\infty}$	400
	n-tetradecane	$\Delta_M H_B^{\infty}$	242
	tetrahydrofuran	$\Delta_{sol}H_{A(vap)}^{\infty}$	400
	2,2,4,4-tetramethylpentane	$\Delta_M H_B^{\infty}$	242
	2,3,3,4-tetramethylpentane	$\Delta_M H_B^{\infty}$	242
	toluene	$\Delta_M H_B^{\infty}$	243
	toluene	$\Delta_M H_A$	360-362
	toluene	$\Delta_{sol}H_{A(vap)}^{\infty}$	400, 403
	toluene	$\Delta_M H_A^{\infty}$, $\Delta_{sol}H_{A(vap)}^{\infty}$	404
	1,1,1-trichloroethane	$\Delta_{sol}H_{A(vap)}^{\infty}$	400
	trichloroethene	$\Delta_{sol}H_{A(vap)}^{\infty}$	400
	trichloromethane	$\Delta_{sol}H_{A(vap)}^{\infty}$	400
	n-tridecane	$\Delta_M H_B^{\infty}$	243
	1,3,5-trimethylbenzene	$\Delta_M H_B^{\infty}$	243
	2,2,3-trimethylbutane	$\Delta_{sol}H_{A(vap)}^{\infty}$	403
	2,2,4-trimethylpentane	$\Delta_M H$	34
	2,2,4-trimethylpentane	$\Delta_M H_A$	362
	n-undecane	$\Delta_M H_B^{\infty}$	243-244
	n-undecane	$\Delta_M H_A^{\infty}$, $\Delta_{sol}H_{A(vap)}^{\infty}$	400, 404
Poly[dimethylsiloxane-*co*-methyl-(4-cyanobiphenoxy)butylsiloxane]			
	cyclohexane	$\Delta_{sol}H_{A(vap)}^{\infty}$	405
	1,2-dimethylbenzene	$\Delta_{sol}H_{A(vap)}^{\infty}$	405
	1,3-dimethylbenzene	$\Delta_{sol}H_{A(vap)}^{\infty}$	405
	1,4-dimethylbenzene	$\Delta_{sol}H_{A(vap)}^{\infty}$	405
	2,3-dimethylpentane	$\Delta_{sol}H_{A(vap)}^{\infty}$	405
	2,4-dimethylpentane	$\Delta_{sol}H_{A(vap)}^{\infty}$	405
	ethylbenzene	$\Delta_{sol}H_{A(vap)}^{\infty}$	405
	n-heptane	$\Delta_{sol}H_{A(vap)}^{\infty}$	405
	n-hexane	$\Delta_{sol}H_{A(vap)}^{\infty}$	405
	2-methylhexane	$\Delta_{sol}H_{A(vap)}^{\infty}$	405
	3-methylhexane	$\Delta_{sol}H_{A(vap)}^{\infty}$	405
	n-nonane	$\Delta_{sol}H_{A(vap)}^{\infty}$	405
	n-octane	$\Delta_{sol}H_{A(vap)}^{\infty}$	405
	n-pentane	$\Delta_{sol}H_{A(vap)}^{\infty}$	405
	toluene	$\Delta_{sol}H_{A(vap)}^{\infty}$	405
	2,2,3-trimethylbutane	$\Delta_{sol}H_{A(vap)}^{\infty}$	405

Polymer(s)	Solvent(s)	Property	Page(s)
Poly(epichlorohydrin)			
	benzene	$\Delta_{sol}H_{A(vap)}^{\infty}$	406
	1-butanol	$\Delta_{sol}H_{A(vap)}^{\infty}$	406
	2-butanone	$\Delta_{sol}H_{A(vap)}^{\infty}$	406
	butyl acetate	$\Delta_{sol}H_{A(vap)}^{\infty}$	406
	chlorobenzene	$\Delta_{sol}H_{A(vap)}^{\infty}$	406
	1-chlorobutane	$\Delta_{sol}H_{A(vap)}^{\infty}$	406
	1-chlorohexane	$\Delta_{sol}H_{A(vap)}^{\infty}$	406
	1-chlorooctane	$\Delta_{sol}H_{A(vap)}^{\infty}$	406
	1-chloropentane	$\Delta_{sol}H_{A(vap)}^{\infty}$	406
	cycloheptane	$\Delta_{sol}H_{A(vap)}^{\infty}$	406
	cyclohexadiene	$\Delta_{sol}H_{A(vap)}^{\infty}$	406
	cyclohexane	$\Delta_{sol}H_{A(vap)}^{\infty}$	406
	cyclohexene	$\Delta_{sol}H_{A(vap)}^{\infty}$	406
	cyclooctane	$\Delta_{sol}H_{A(vap)}^{\infty}$	406
	cyclopentane	$\Delta_{sol}H_{A(vap)}^{\infty}$	406
	n-decane	$\Delta_{sol}H_{A(vap)}^{\infty}$	406
	1,1-dichloroethane	$\Delta_{sol}H_{A(vap)}^{\infty}$	406
	1,2-dichloroethane	$\Delta_{sol}H_{A(vap)}^{\infty}$	406
	dichloromethane	$\Delta_{sol}H_{A(vap)}^{\infty}$	406
	1,4-dioxane	$\Delta_{sol}H_{A(vap)}^{\infty}$	406
	ethanol	$\Delta_{sol}H_{A(vap)}^{\infty}$	406
	ethyl acetate	$\Delta_{sol}H_{A(vap)}^{\infty}$	406
	ethylbenzene	$\Delta_{sol}H_{A(vap)}^{\infty}$	406
	n-heptane	$\Delta_{sol}H_{A(vap)}^{\infty}$	406
	n-hexane	$\Delta_{sol}H_{A(vap)}^{\infty}$	406
	methyl acetate	$\Delta_{sol}H_{A(vap)}^{\infty}$	406
	n-nonane	$\Delta_{sol}H_{A(vap)}^{\infty}$	406
	n-octane	$\Delta_{sol}H_{A(vap)}^{\infty}$	406
	1-pentanol	$\Delta_{sol}H_{A(vap)}^{\infty}$	406
	1-propanol	$\Delta_{sol}H_{A(vap)}^{\infty}$	406
	2-propanone	$\Delta_{sol}H_{A(vap)}^{\infty}$	406
	propyl acetate	$\Delta_{sol}H_{A(vap)}^{\infty}$	406
	tetrachloromethane	$\Delta_{sol}H_{A(vap)}^{\infty}$	406
	tetrahydrofuran	$\Delta_{sol}H_{A(vap)}^{\infty}$	406
	toluene	$\Delta_{sol}H_{A(vap)}^{\infty}$	406
	1,1,1-trichloroethane	$\Delta_{sol}H_{A(vap)}^{\infty}$	406
	trichloroethene	$\Delta_{sol}H_{A(vap)}^{\infty}$	406
	trichloromethane	$\Delta_{sol}H_{A(vap)}^{\infty}$	406
	n-undecane	$\Delta_{sol}H_{A(vap)}^{\infty}$	406
Poly(ethyl acrylate)			
	benzene	$\Delta_{sol}H_{A(vap)}^{\infty}$	407
	chlorobenzene	$\Delta_{sol}H_{A(vap)}^{\infty}$	407
	1-chlorobutane	$\Delta_{sol}H_{A(vap)}^{\infty}$	407

Polymer(s)	Solvent(s)	Property	Page(s)
Poly(ethyl acrylate) (*continued*)			
	1-chloropentane	$\Delta_{sol}H_{A(vap)}^{\infty}$	407
	cycloheptane	$\Delta_{sol}H_{A(vap)}^{\infty}$	407
	cyclohexane	$\Delta_{sol}H_{A(vap)}^{\infty}$	407
	cyclooctane	$\Delta_{sol}H_{A(vap)}^{\infty}$	407
	cyclopentane	$\Delta_{sol}H_{A(vap)}^{\infty}$	407
	n-decane	$\Delta_{sol}H_{A(vap)}^{\infty}$	407
	1,1-dichloroethane	$\Delta_{sol}H_{A(vap)}^{\infty}$	407
	1,2-dichloroethane	$\Delta_{sol}H_{A(vap)}^{\infty}$	407
	dichloromethane	$\Delta_{sol}H_{A(vap)}^{\infty}$	407
	1,4-dioxane	$\Delta_{sol}H_{A(vap)}^{\infty}$	407
	ethyl acetate	$\Delta_{sol}H_{A(vap)}^{\infty}$	407
	ethylbenzene	$\Delta_{sol}H_{A(vap)}^{\infty}$	407
	n-heptane	$\Delta_{sol}H_{A(vap)}^{\infty}$	407
	n-hexane	$\Delta_{sol}H_{A(vap)}^{\infty}$	407
	n-nonane	$\Delta_{sol}H_{A(vap)}^{\infty}$	407
	n-octane	$\Delta_{sol}H_{A(vap)}^{\infty}$	407
	tetrachloromethane	$\Delta_{sol}H_{A(vap)}^{\infty}$	407
	tetrahydrofuran	$\Delta_{sol}H_{A(vap)}^{\infty}$	407
	toluene	$\Delta_{sol}H_{A(vap)}^{\infty}$	407
	1,1,1-trichloroethane	$\Delta_{sol}H_{A(vap)}^{\infty}$	407
	trichloroethene	$\Delta_{sol}H_{A(vap)}^{\infty}$	407
	trichloromethane	$\Delta_{sol}H_{A(vap)}^{\infty}$	407
	n-undecane	$\Delta_{sol}H_{A(vap)}^{\infty}$	407
Polyethylene			
	benzene	$\Delta_{M}H_{A}^{\infty}, \Delta_{sol}H_{A(vap)}^{\infty}$	408
	1-butanol	$\Delta_{M}H_{A}^{\infty}, \Delta_{sol}H_{A(vap)}^{\infty}$	408
	2-butanone	$\Delta_{M}H_{A}^{\infty}$	408
	butyl acetate	$\Delta_{M}H_{A}^{\infty}$	408
	chlorobenzene	$\Delta_{M}H_{A}^{\infty}, \Delta_{sol}H_{A(vap)}^{\infty}$	408
	1-chlorobutane	$\Delta_{M}H_{A}^{\infty}, \Delta_{sol}H_{A(vap)}^{\infty}$	408
	chlorocyclohexane	$\Delta_{sol}H_{A(vap)}^{\infty}$	409
	1-chlorohexane	$\Delta_{M}H_{A}^{\infty}$	408
	1-chloronaphthalene	$\Delta_{sol}H_{B}^{\infty}$	244
	cyclohexane	$\Delta_{sol}H_{B}^{\infty}$	245
	cyclohexane	$\Delta_{M}H_{A}^{\infty}, \Delta_{sol}H_{A(vap)}^{\infty}$	408-409
	cyclohexanol	$\Delta_{M}H_{A}^{\infty}, \Delta_{sol}H_{A(vap)}^{\infty}$	409
	cyclohexanone	$\Delta_{M}H_{A}^{\infty}, \Delta_{sol}H_{A(vap)}^{\infty}$	408-409
	cyclopentane	$\Delta_{sol}H_{B}^{\infty}$	245
	decahydronaphthalene	$\Delta_{sol}H_{B}^{\infty}$	245
	n-decane	$\Delta_{M}H_{A}^{\infty}, \Delta_{sol}H_{A(vap)}^{\infty}$	408-409
	1,2-dichloroethane	$\Delta_{sol}H_{B}^{\infty}$	246
	1,5-dichloropentane	$\Delta_{M}H_{A}^{\infty}$	408

Polymer(s)	Solvent(s)	Property	Page(s)
Polyethylene (*continued*)			
	N,N-dimethylacetamide	$\Delta_M H_A^\infty$	408
	1,4-dimethylbenzene	$\Delta_{sol} H_B^\infty$	246
	N,N-dimethylformamide	$\Delta_M H_A^\infty$	408
	2,4-dimethylpentane	$\Delta_{sol} H_B^\infty$	246
	dimethylsulfoxide	$\Delta_M H_A^\infty$	408
	dipropyl ether	$\Delta_M H_A^\infty$	408
	dipropyl thioether	$\Delta_M H_A^\infty$	408
	n-dodecane	$\Delta_M H_A^\infty$	408
	ethyl propionate	$\Delta_M H_A^\infty$	408
	2,2,4,4,6,8,8-heptamethyl-nonane	$\Delta_{sol} H_B^\infty$	246
	n-heptane	$\Delta_M H_A^\infty$	408
	n-hexadecane	$\Delta_{sol} H_B^\infty$	247
	2-methylbutane	$\Delta_{sol} H_B^\infty$	247
	n-nonane	$\Delta_M H_A^\infty, \Delta_{sol} H_{A(vap)}^\infty$	409
	n-octane	$\Delta_M H_A^\infty$	408-409
	1-octene	$\Delta_M H_A^\infty, \Delta_{sol} H_{A(vap)}^\infty$	409
	2-pentanone	$\Delta_M H_A^\infty, \Delta_{sol} H_{A(vap)}^\infty$	408-409
	phenol	$\Delta_M H_A^\infty, \Delta_{sol} H_{A(vap)}^\infty$	409
	1,1,2,2-tetrachloroethane	$\Delta_M H_A^\infty$	408
	tetrachloromethane	$\Delta_M H_A^\infty, \Delta_{sol} H_{A(vap)}^\infty$	408
	n-tetradecane	$\Delta_M H_A^\infty$	409
	1,2,3,4-tetrahydro-naphthalene	$\Delta_{sol} H_B^\infty$	247-248
	toluene	$\Delta_{sol} H_B^\infty$	248
	toluene	$\Delta_M H_A^\infty$	408
	1,2,4-trichlorobenzene	$\Delta_{sol} H_B^\infty$	249
	1,1,2-trichloroethane	$\Delta_M H_A^\infty$	408
	trichloromethane	$\Delta_M H_A^\infty, \Delta_{sol} H_{A(vap)}^\infty$	408-409
Poly(ethylene-*co*-carbon monoxide)			
	n-octane	$\Delta_M H_A^\infty$	409
Poly(ethylene-*co*-propylene)			
	benzene	$\Delta_M H_A^\infty$	410
	tert-butylbenzene	$\Delta_M H_A^\infty$	410
	cyclohexane	$\Delta_M H_B^\infty$	249
	cyclohexane	$\Delta_M H_A^\infty$	410
	cyclooctane	$\Delta_M H_B^\infty$	249
	cyclopentane	$\Delta_M H_B^\infty$	250
	cis-decahydronaphthalene	$\Delta_M H_B^\infty$	250
	trans-decahydronaphthalene	$\Delta_{sol} H_B^\infty$	250
	n-decane	$\Delta_M H_A^\infty$	410

Polymer(s)	Solvent(s)	Property	Page(s)
Poly(ethylene-*co*-propylene) (*continued*)			
	3,3-diethylpentane	$\Delta_M H_B^{\infty}$	251
	2,2-dimethylpentane	$\Delta_M H_B^{\infty}$	251
	2,3-dimethylpentane	$\Delta_M H_B^{\infty}$	251
	2,4-dimethylpentane	$\Delta_M H_B^{\infty}$	251-252
	3,3-dimethylpentane	$\Delta_M H_B^{\infty}$	252
	n-dodecane	$\Delta_M H_B^{\infty}$	252
	ethylbenzene	$\Delta_M H_A^{\infty}$	410
	3-ethylpentane	$\Delta_M H_B^{\infty}$	252
	2,2,4,4,6,8,8-heptamethyl-nonane	$\Delta_M H_B^{\infty}$	253
	n-hexadecane	$\Delta_M H_B^{\infty}$	253
	n-hexane	$\Delta_M H_A^{\infty}$	410
	3-methylhexane	$\Delta_M H_B^{\infty}$	253-254
	n-octane	$\Delta_M H_B^{\infty}$	254
	n-octane	$\Delta_M H_A^{\infty}$	410
	2,2,4,6,6-pentamethylheptane	$\Delta_M H_B^{\infty}$	254
	2,2,4,4-tetramethylpentane	$\Delta_M H_B^{\infty}$	255
	2,2,4-trimethylpentane	$\Delta_M H_B^{\infty}$	255
	2,2,4-trimethylpentane	$\Delta_M H_A^{\infty}$	410
Poly(ethylene-*co*-propylene-*co*-diene)			
	benzene	$\Delta_{sol} H_{A(vap)}^{\infty}$	410
	n-heptane	$\Delta_{sol} H_{A(vap)}^{\infty}$	410
	n-hexane	$\Delta_{sol} H_{A(vap)}^{\infty}$	410
	n-nonane	$\Delta_{sol} H_{A(vap)}^{\infty}$	410
	n-octane	$\Delta_{sol} H_{A(vap)}^{\infty}$	410
	toluene	$\Delta_{sol} H_{A(vap)}^{\infty}$	410
Poly(ethylene-*co*-vinyl acetate)			
	acetaldehyde	$\Delta_{sol} H_{A(vap)}^{\infty}$	412
	acetic acid	$\Delta_{sol} H_{A(vap)}^{\infty}$	412
	acetonitrile	$\Delta_{sol} H_{A(vap)}^{\infty}$	412
	acrylonitrile	$\Delta_{sol} H_{A(vap)}^{\infty}$	412
	benzene	$\Delta_M H_A^{\infty}$, $\Delta_{sol} H_{A(vap)}^{\infty}$	411-412
	1-bromobutane	$\Delta_{sol} H_{A(vap)}^{\infty}$	412
	2-bromobutane	$\Delta_{sol} H_{A(vap)}^{\infty}$	412
	1-butanol	$\Delta_M H_A^{\infty}$, $\Delta_{sol} H_{A(vap)}^{\infty}$	411-413
	2-butanol	$\Delta_{sol} H_{A(vap)}^{\infty}$	412-413
	2-butanone	$\Delta_M H_A^{\infty}$, $\Delta_{sol} H_{A(vap)}^{\infty}$	411-412
	butyl acetate	$\Delta_M H_A^{\infty}$, $\Delta_{sol} H_{A(vap)}^{\infty}$	412
	chlorobenzene	$\Delta_M H_A^{\infty}$, $\Delta_{sol} H_{A(vap)}^{\infty}$	411-412
	1-chlorobutane	$\Delta_M H_A^{\infty}$, $\Delta_{sol} H_{A(vap)}^{\infty}$	411
	chlorocyclohexane	$\Delta_M H_A^{\infty}$, $\Delta_{sol} H_{A(vap)}^{\infty}$	411

Polymer(s)	Solvent(s)	Property	Page(s)
Poly(ethylene-*co*-vinyl acetate) (*continued*)	cyclohexane	$\Delta_M H_A{}^\infty$, $\Delta_{sol} H_{A(vap)}{}^\infty$	411-412
	cyclohexanol	$\Delta_M H_A{}^\infty$, $\Delta_{sol} H_{A(vap)}{}^\infty$	411
	cyclohexanone	$\Delta_M H_A{}^\infty$, $\Delta_{sol} H_{A(vap)}{}^\infty$	411
	cyclopentanone	$\Delta_{sol} H_B{}^\infty$	255-256
	n-decane	$\Delta_M H_A{}^\infty$, $\Delta_{sol} H_{A(vap)}{}^\infty$	412
	1,2-dichloroethane	$\Delta_{sol} H_{A(vap)}{}^\infty$	412
	dichloromethane	$\Delta_{sol} H_{A(vap)}{}^\infty$	412
	diethyl ether	$\Delta_{sol} H_{A(vap)}{}^\infty$	412
	1,3-dimethylbenzene	$\Delta_{sol} H_{A(vap)}{}^\infty$	412
	1,4-dimethylbenzene	$\Delta_M H_A{}^\infty$, $\Delta_{sol} H_{A(vap)}{}^\infty$	412
	1,4-dioxane	$\Delta_{sol} H_{A(vap)}{}^\infty$	412
	dipropyl ether	$\Delta_{sol} H_{A(vap)}{}^\infty$	413
	ethanol	$\Delta_M H_A{}^\infty$, $\Delta_{sol} H_{A(vap)}{}^\infty$	412-413
	ethyl acetate	$\Delta_M H_A{}^\infty$, $\Delta_{sol} H_{A(vap)}{}^\infty$	412-413
	formic acid	$\Delta_{sol} H_{A(vap)}{}^\infty$	413
	furan	$\Delta_{sol} H_{A(vap)}{}^\infty$	413
	n-heptane	$\Delta_{sol} H_{A(vap)}{}^\infty$	413
	n-hexane	$\Delta_M H_A{}^\infty$, $\Delta_{sol} H_{A(vap)}{}^\infty$	411, 413
	methanol	$\Delta_M H_A{}^\infty$, $\Delta_{sol} H_{A(vap)}{}^\infty$	412-413
	methylcyclohexane	$\Delta_{sol} H_{A(vap)}{}^\infty$	413
	2-methyl-1-propanol	$\Delta_{sol} H_{A(vap)}{}^\infty$	413
	2-methyl-1-propanol	$\Delta_{sol} H_{A(vap)}{}^\infty$	414
	nitroethane	$\Delta_{sol} H_{A(vap)}{}^\infty$	413
	nitromethane	$\Delta_{sol} H_{A(vap)}{}^\infty$	413
	1-nitropropane	$\Delta_{sol} H_{A(vap)}{}^\infty$	413
	2-nitropropane	$\Delta_{sol} H_{A(vap)}{}^\infty$	413
	n-octane	$\Delta_M H_A{}^\infty$, $\Delta_{sol} H_{A(vap)}{}^\infty$	411-413
	1-octene	$\Delta_{sol} H_{A(vap)}{}^\infty$	413
	n-pentane	$\Delta_{sol} H_{A(vap)}{}^\infty$	413
	2-pentanone	$\Delta_M H_A{}^\infty$, $\Delta_{sol} H_{A(vap)}{}^\infty$	411
	3-pentanone	$\Delta_{sol} H_{A(vap)}{}^\infty$	413
	1-propanol	$\Delta_{sol} H_{A(vap)}{}^\infty$	414
	2-propanol	$\Delta_{sol} H_{A(vap)}{}^\infty$	413-414
	2-propanone	$\Delta_M H_A{}^\infty$, $\Delta_{sol} H_{A(vap)}{}^\infty$	412
	phenol	$\Delta_M H_A{}^\infty$, $\Delta_{sol} H_{A(vap)}{}^\infty$	411
	propionitrile	$\Delta_{sol} H_{A(vap)}{}^\infty$	413
	propyl acetate	$\Delta_{sol} H_{A(vap)}{}^\infty$	413
	tetrachloromethane	$\Delta_M H_A{}^\infty$, $\Delta_{sol} H_{A(vap)}{}^\infty$	411-413
	tetrahydrofuran	$\Delta_{sol} H_B{}^\infty$	256
	tetrahydrofuran	$\Delta_{sol} H_{A(vap)}{}^\infty$	413
	toluene	$\Delta_M H_A{}^\infty$, $\Delta_{sol} H_{A(vap)}{}^\infty$	412
	trichloromethane	$\Delta_M H_A{}^\infty$, $\Delta_{sol} H_{A(vap)}{}^\infty$	411-412
	2,2,2-trifluoroethanol	$\Delta_{sol} H_{A(vap)}{}^\infty$	413
	water	$\Delta_{sol} H_{A(vap)}{}^\infty$	413

Polymer(s)	Solvent(s)	Property	Page(s)
Poly(ethylene glycol) (*continued*)			
	tetrachloromethane	$\Delta_M H, \Delta_{dil}H^{12}$	67-72
	tetrachloromethane	$\Delta_M H_A^\infty, \Delta_{sol}H_{A(vap)}^\infty$	416
	tetrahydrofuran	$\Delta_M H$	73
	tetrahydropyran	$\Delta_M H$	73-74
	toluene	$\Delta_M H_A^\infty, \Delta_{sol}H_{A(vap)}^\infty$	416
	triacetin	$\Delta_M H_A^\infty, \Delta_{sol}H_{A(vap)}^\infty$	414-415
	trichloromethane	$\Delta_M H$	74
	trichloromethane	$\Delta_M H_B^\infty$	258
	trichloromethane	$\Delta_M H_A^\infty, \Delta_{sol}H_{A(vap)}^\infty$	416
	triethylenediamine	$\Delta_M H_A^\infty, \Delta_{sol}H_{A(vap)}^\infty$	414-415
	water	$\Delta_M H, \Delta_{dil}H^{12}$	74-90
	water	$\Delta_M H_B^\infty, \Delta_{sol}H_B^\infty$	258-261
	water	$\Delta_M H_A$	362
Poly(ethylene glycol) dibutyl ether			
	1,2-dichloroethane	$\Delta_M H_A^\infty$	416
	dichloromethane	$\Delta_M H_A^\infty$	416
	ethyl acetate	$\Delta_M H_A^\infty$	416
	tetrachloroethene	$\Delta_M H_A^\infty$	416
	tetrachloromethane	$\Delta_M H_A^\infty$	416
	1,1,1-trichloroethane	$\Delta_M H_A^\infty$	416
	trichloroethene	$\Delta_M H_A^\infty$	416
	trichloromethane	$\Delta_M H_A^\infty$	416
Poly(ethylene glycol) dimethyl ether			
	1,2-dichloroethane	$\Delta_M H_A^\infty$	417
	dichloromethane	$\Delta_M H_A^\infty$	417
	ethyl acetate	$\Delta_M H_A^\infty$	417
	methanol	$\Delta_M H$	90-91
	tetrachloroethene	$\Delta_M H_A^\infty$	417
	tetrachloromethane	$\Delta_M H$	92-93
	tetrachloromethane	$\Delta_M H_B^\infty$	261
	tetrachloromethane	$\Delta_M H_A^\infty$	417
	1,1,1-trichloroethane	$\Delta_M H_A^\infty$	417
	trichloroethene	$\Delta_M H_A^\infty$	417
	trichloromethane	$\Delta_M H$	93
	trichloromethane	$\Delta_M H_B^\infty$	261
	trichloromethane	$\Delta_M H_A^\infty$	417
	2,2,2-trifluoroethanol	$\Delta_M H$	94
Poly(ethylene glycol) monododecyl ether			
	n-dodecane	$\Delta_M H_B^\infty$	262
	water	$\Delta_M H$	94

Polymer(s)	Solvent(s)	Property	Page(s)
Poly(ethylene glycol) monomethyl ether			
	1-butanol	$\Delta_M H$	94
	1-pentanol	$\Delta_M H$	94
	1-propanol	$\Delta_M H$	95
	tetrachloromethane	$\Delta_M H$	95-96
	trichloromethane	$\Delta_M H$	96
	trichloromethane	$\Delta_M H_B^\infty$	262
Poly(ethylene oxide)			
	acetonitrile	$\Delta_{sol} H_{A(vap)}^\infty$	417
	benzene	$\Delta_M H, \Delta_{dil} H^{12}$	96-97
	benzene	$\Delta_{sol} H_{A(vap)}^\infty$	417-418
	1-butanol	$\Delta_{sol} H_{A(vap)}^\infty$	418
	2-butanone	$\Delta_{sol} H_{A(vap)}^\infty$	417
	butyl acetate	$\Delta_{sol} H_{A(vap)}^\infty$	418
	chlorobenzene	$\Delta_{sol} H_{A(vap)}^\infty$	418
	cyclohexane	$\Delta_{sol} H_{A(vap)}^\infty$	417-418
	n-decane	$\Delta_M H_A^\infty, \Delta_{sol} H_A^\infty$	417-418
	dichloromethane	$\Delta_{dil} H^{12}$	98
	dichloromethane	$\Delta_M H_B^\infty, \Delta_{sol} H_B^\infty$	263
	1,2-dimethoxyethane	$\Delta_{sol} H_{A(vap)}^\infty$	417
	1,4-dioxane	$\Delta_{sol} H_{A(vap)}^\infty$	417
	n-dodecane	$\Delta_M H_A^\infty, \Delta_{sol} H_A^\infty$	417-418
	ethanol	$\Delta_{sol} H_{A(vap)}^\infty$	418
	bis(2-ethoxyethyl) ether	$\Delta_{sol} H_{A(vap)}^\infty$	418
	ethyl acetate	$\Delta_{sol} H_{A(vap)}^\infty$	417-418
	ethylbenzene	$\Delta_{sol} H_{A(vap)}^\infty$	417
	n-heptane	$\Delta_M H$	97
	n-heptane	$\Delta_M H_A^\infty, \Delta_{sol} H_A^\infty$	417-418
	n-hexane	$\Delta_M H_A^\infty, \Delta_{sol} H_A^\infty$	418
	methanol	$\Delta_{sol} H_{A(vap)}^\infty$	418
	bis(2-methoxyethyl) ether	$\Delta_{sol} H_{A(vap)}^\infty$	417
	methyl acetate	$\Delta_{sol} H_{A(vap)}^\infty$	418
	n-nonane	$\Delta_M H_A^\infty, \Delta_{sol} H_A^\infty$	418
	n-octane	$\Delta_M H_A^\infty, \Delta_{sol} H_A^\infty$	418
	n-pentane	$\Delta_M H_A^\infty, \Delta_{sol} H_A^\infty$	418
	1-propanol	$\Delta_{sol} H_{A(vap)}^\infty$	417-418
	2-propanol	$\Delta_{sol} H_{A(vap)}^\infty$	417
	2-propanone	$\Delta_{sol} H_{A(vap)}^\infty$	418
	propyl acetate	$\Delta_{sol} H_{A(vap)}^\infty$	418
	pyridine	$\Delta_{sol} H_{A(vap)}^\infty$	417
	toluene	$\Delta_{sol} H_{A(vap)}^\infty$	417-418
	trichloromethane	$\Delta_{dil} H^{12}$	98
	trichloromethane	$\Delta_M H_B^\infty, \Delta_{sol} H_B^\infty$	263

Polymer(s)	Solvent(s)	Property	Page(s)
Poly(ethylene oxide) (*continued*)			
	trichloromethane	$\Delta_{sol}H_{A(vap)}{}^{\infty}$	417
	2,2,2-trifluoroethanol	$\Delta_{sol}H_{A(vap)}{}^{\infty}$	417
	n-undecane	$\Delta_{M}H_{A}{}^{\infty}, \Delta_{sol}H_{A}{}^{\infty}$	418
	water	$\Delta_{dil}H^{12}$	98
	water	$\Delta_{M}H_{B}{}^{\infty}, \Delta_{sol}H_{B}{}^{\infty}$	263-264
Poly(ethylene oxide)-b-poly(propylene oxide) diblock copolymer			
	tetrachloromethane	$\Delta_{M}H$	98-99
Poly(ethylene oxide)-b-poly(propylene oxide)-b-poly(ethylene oxide) triblock copolymer			
	tetrachloromethane	$\Delta_{M}H$	99-100
Poly(ethylethylene)			
	benzene	$\Delta_{sol}H_{A(vap)}{}^{\infty}$	419
	n-butane	$\Delta_{sol}H_{A(vap)}{}^{\infty}$	419
	1-butanol	$\Delta_{sol}H_{A(vap)}{}^{\infty}$	419
	2-butanone	$\Delta_{sol}H_{A(vap)}{}^{\infty}$	419
	butyl acetate	$\Delta_{sol}H_{A(vap)}{}^{\infty}$	419
	chlorobenzene	$\Delta_{sol}H_{A(vap)}{}^{\infty}$	419
	1-chlorobutane	$\Delta_{sol}H_{A(vap)}{}^{\infty}$	419
	1-chlorohexane	$\Delta_{sol}H_{A(vap)}{}^{\infty}$	419
	1-chlorooctane	$\Delta_{sol}H_{A(vap)}{}^{\infty}$	419
	1-chloropentane	$\Delta_{sol}H_{A(vap)}{}^{\infty}$	419
	cycloheptane	$\Delta_{sol}H_{A(vap)}{}^{\infty}$	419
	cyclohexadiene	$\Delta_{sol}H_{A(vap)}{}^{\infty}$	419
	cyclohexane	$\Delta_{sol}H_{A(vap)}{}^{\infty}$	419
	cyclohexene	$\Delta_{sol}H_{A(vap)}{}^{\infty}$	419
	cyclooctane	$\Delta_{sol}H_{A(vap)}{}^{\infty}$	419
	cyclopentane	$\Delta_{sol}H_{A(vap)}{}^{\infty}$	419
	n-decane	$\Delta_{sol}H_{A(vap)}{}^{\infty}$	419
	1,1-dichloroethane	$\Delta_{sol}H_{A(vap)}{}^{\infty}$	419
	1,2-dichloroethane	$\Delta_{sol}H_{A(vap)}{}^{\infty}$	419
	dichloromethane	$\Delta_{sol}H_{A(vap)}{}^{\infty}$	419
	1,4-dioxane	$\Delta_{sol}H_{A(vap)}{}^{\infty}$	419
	ethanol	$\Delta_{sol}H_{A(vap)}{}^{\infty}$	419
	ethyl acetate	$\Delta_{sol}H_{A(vap)}{}^{\infty}$	419
	ethylbenzene	$\Delta_{sol}H_{A(vap)}{}^{\infty}$	419
	n-heptane	$\Delta_{sol}H_{A(vap)}{}^{\infty}$	419
	n-hexane	$\Delta_{sol}H_{A(vap)}{}^{\infty}$	419
	methyl acetate	$\Delta_{sol}H_{A(vap)}{}^{\infty}$	419

Polymer(s)	Solvent(s)	Property	Page(s)
Poly(ethylethylene) (*continued*)			
	n-nonane	$\Delta_{sol}H_{A(vap)}^{\infty}$	419
	n-octane	$\Delta_{sol}H_{A(vap)}^{\infty}$	419
	n-pentane	$\Delta_{sol}H_{A(vap)}^{\infty}$	419
	1-pentanol	$\Delta_{sol}H_{A(vap)}^{\infty}$	419
	1-propanol	$\Delta_{sol}H_{A(vap)}^{\infty}$	419
	2-propanone	$\Delta_{sol}H_{A(vap)}^{\infty}$	419
	propyl acetate	$\Delta_{sol}H_{A(vap)}^{\infty}$	419
	tetrachloromethane	$\Delta_{sol}H_{A(vap)}^{\infty}$	419
	tetrahydrofuran	$\Delta_{sol}H_{A(vap)}^{\infty}$	419
	toluene	$\Delta_{sol}H_{A(vap)}^{\infty}$	419
	1,1,1-trichloroethane	$\Delta_{sol}H_{A(vap)}^{\infty}$	419
	trichloroethene	$\Delta_{sol}H_{A(vap)}^{\infty}$	419
	trichloromethane	$\Delta_{sol}H_{A(vap)}^{\infty}$	419
Poly(ethyl methacrylate)			
	acetonitrile	$\Delta_{sol}H_{A(vap)}^{\infty}$	420
	acrylonitrile	$\Delta_{sol}H_{A(vap)}^{\infty}$	420
	benzene	$\Delta_{sol}H_{A(vap)}^{\infty}$	420
	bromobenzene	$\Delta_{sol}H_{A(vap)}^{\infty}$	420
	1-bromobutane	$\Delta_{sol}H_{A(vap)}^{\infty}$	420
	1-butanol	$\Delta_{sol}H_{A(vap)}^{\infty}$	420-421
	2-butanone	$\Delta_{sol}H_{A(vap)}^{\infty}$	420
	butyl acetate	$\Delta_{sol}H_{A(vap)}^{\infty}$	420-421
	butylamine	$\Delta_{sol}H_{A(vap)}^{\infty}$	420
	butyraldehyde	$\Delta_{sol}H_{A(vap)}^{\infty}$	420
	chlorobenzene	$\Delta_{sol}H_{A(vap)}^{\infty}$	420
	cyclohexane	$\Delta_{sol}H_{A(vap)}^{\infty}$	420
	n-decane	$\Delta_{sol}H_{A(vap)}^{\infty}$	420-421
	dipropyl ether	$\Delta_{sol}H_{A(vap)}^{\infty}$	420
	n-dodecane	$\Delta_{sol}H_{A(vap)}^{\infty}$	421
	ethanol	$\Delta_{sol}H_{A(vap)}^{\infty}$	421
	ethyl acetate	$\Delta_{sol}H_{A(vap)}^{\infty}$	421
	ethylbenzene	$\Delta_{sol}H_{A(vap)}^{\infty}$	420
	ethylcyclohexane	$\Delta_{sol}H_{A(vap)}^{\infty}$	420
	n-heptane	$\Delta_{sol}H_{A(vap)}^{\infty}$	420-421
	methanol	$\Delta_{sol}H_{A(vap)}^{\infty}$	421
	methyl acetate	$\Delta_{sol}H_{A(vap)}^{\infty}$	421
	methylcyclohexane	$\Delta_{sol}H_{A(vap)}^{\infty}$	420
	nitroethane	$\Delta_{sol}H_{A(vap)}^{\infty}$	420
	nitromethane	$\Delta_{sol}H_{A(vap)}^{\infty}$	420
	1-nitropropane	$\Delta_{sol}H_{A(vap)}^{\infty}$	420
	n-nonane	$\Delta_{sol}H_{A(vap)}^{\infty}$	420-421
	n-octane	$\Delta_{sol}H_{A(vap)}^{\infty}$	420-421
	1-octene	$\Delta_{sol}H_{A(vap)}^{\infty}$	420

Polymer(s)	Solvent(s)	Property	Page(s)
Poly(ethyl methacrylate) (*continued*)			
	1-pentanol	$\Delta_{sol}H_{A(vap)}{}^{\infty}$	420
	2-pentanol	$\Delta_{sol}H_{A(vap)}{}^{\infty}$	421
	propanoic acid	$\Delta_{sol}H_{A(vap)}{}^{\infty}$	420
	1-propanol	$\Delta_{sol}H_{A(vap)}{}^{\infty}$	420
	2-propanol	$\Delta_{sol}H_{A(vap)}{}^{\infty}$	421
	2-propanone	$\Delta_{sol}H_{A(vap)}{}^{\infty}$	420
	propionitrile	$\Delta_{sol}H_{A(vap)}{}^{\infty}$	420
	propyl acetate	$\Delta_{sol}H_{A(vap)}{}^{\infty}$	420-421
	tetrachloromethane	$\Delta_{sol}H_{A(vap)}{}^{\infty}$	420
	toluene	$\Delta_{sol}H_{A(vap)}{}^{\infty}$	420
	trichloromethane	$\Delta_{sol}H_{A(vap)}{}^{\infty}$	420
	n-undecane	$\Delta_{sol}H_{A(vap)}{}^{\infty}$	421
Poly(glycidyl methacrylate-*co*-butyl methacrylate)			
	benzene	$\Delta_{M}H_{A}{}^{\infty}, \Delta_{sol}H_{A}{}^{\infty}$	422-423
	1-chlorobutane	$\Delta_{M}H_{A}{}^{\infty}, \Delta_{sol}H_{A}{}^{\infty}$	422-423
	1-chloropropane	$\Delta_{M}H_{A}{}^{\infty}, \Delta_{sol}H_{A}{}^{\infty}$	422-423
	n-decane	$\Delta_{M}H_{A}{}^{\infty}, \Delta_{sol}H_{A}{}^{\infty}$	422-423
	1,4-dimethylbenzene	$\Delta_{M}H_{A}{}^{\infty}, \Delta_{sol}H_{A}{}^{\infty}$	422-423
	n-heptane	$\Delta_{M}H_{A}{}^{\infty}, \Delta_{sol}H_{A}{}^{\infty}$	422-423
	n-hexane	$\Delta_{M}H_{A}{}^{\infty}, \Delta_{sol}H_{A}{}^{\infty}$	422-423
	n-nonane	$\Delta_{M}H_{A}{}^{\infty}, \Delta_{sol}H_{A}{}^{\infty}$	422-423
	n-octane	$\Delta_{M}H_{A}{}^{\infty}, \Delta_{sol}H_{A}{}^{\infty}$	422-423
	n-pentane	$\Delta_{M}H_{A}{}^{\infty}, \Delta_{sol}H_{A}{}^{\infty}$	422-423
	tetrachloromethane	$\Delta_{M}H_{A}{}^{\infty}, \Delta_{sol}H_{A}{}^{\infty}$	422-423
	toluene	$\Delta_{M}H_{A}{}^{\infty}, \Delta_{sol}H_{A}{}^{\infty}$	422-423
Poly(2-hydroxyethyl acrylate)			
	benzene	$\Delta_{M}H_{A}{}^{\infty}$	424
	1-butanol	$\Delta_{sol}H_{A(vap)}{}^{\infty}$	423
	2-butanone	$\Delta_{M}H_{A}{}^{\infty}, \Delta_{sol}H_{A(vap)}{}^{\infty}$	423-424
	butyl acetate	$\Delta_{sol}H_{A(vap)}{}^{\infty}$	423
	chlorobenzene	$\Delta_{sol}H_{A(vap)}{}^{\infty}$	423
	1-chlorohexane	$\Delta_{sol}H_{A(vap)}{}^{\infty}$	423
	1-chlorooctane	$\Delta_{sol}H_{A(vap)}{}^{\infty}$	423
	n-decane	$\Delta_{M}H_{A}{}^{\infty}$	424
	1,2-dichloroethane	$\Delta_{sol}H_{A(vap)}{}^{\infty}$	423
	dichloromethane	$\Delta_{sol}H_{A(vap)}{}^{\infty}$	423
	1,2-dimethylbenzene	$\Delta_{M}H_{A}{}^{\infty}$	424
	1,4-dioxane	$\Delta_{sol}H_{A(vap)}{}^{\infty}$	423
	n-dodecane	$\Delta_{M}H_{A}{}^{\infty}$	424
	ethanol	$\Delta_{M}H_{A}{}^{\infty}, \Delta_{sol}H_{A(vap)}{}^{\infty}$	423-424

Polymer(s)	Solvent(s)	Property	Page(s)
Poly(2-hydroxyethyl acrylate) (*continued*)			
	ethyl acetate	$\Delta_M H_A^\infty$, $\Delta_{sol} H_{A(vap)}^\infty$	423-424
	ethylbenzene	$\Delta_{sol} H_{A(vap)}^\infty$	423
	methanol	$\Delta_M H_A^\infty$	424
	methyl acetate	$\Delta_M H_A^\infty$, $\Delta_{sol} H_{A(vap)}^\infty$	423-424
	n-nonane	$\Delta_M H_A^\infty$	424
	n-octane	$\Delta_M H_A^\infty$	424
	1-pentanol	$\Delta_{sol} H_{A(vap)}^\infty$	423
	1-propanol	$\Delta_{sol} H_{A(vap)}^\infty$	423
	2-propanone	$\Delta_M H_A^\infty$, $\Delta_{sol} H_{A(vap)}^\infty$	424
	propyl acetate	$\Delta_{sol} H_{A(vap)}^\infty$	424
	tetrahydrofuran	$\Delta_{sol} H_{A(vap)}^\infty$	424
	toluene	$\Delta_M H_A^\infty$, $\Delta_{sol} H_{A(vap)}^\infty$	424
	trichloroethene	$\Delta_{sol} H_{A(vap)}^\infty$	424
	trichloromethane	$\Delta_{sol} H_{A(vap)}^\infty$	424
	n-undecane	$\Delta_M H_A^\infty$	424
Poly(4-hydroxystyrene)			
	2-butanone	$\Delta_M H_A^\infty$, $\Delta_{sol} H_{A(vap)}^\infty$	424
	butyl acetate	$\Delta_M H_A^\infty$, $\Delta_{sol} H_{A(vap)}^\infty$	424
	ethyl acetate	$\Delta_M H_A^\infty$, $\Delta_{sol} H_{A(vap)}^\infty$	424
	3-pentanone	$\Delta_M H_A^\infty$, $\Delta_{sol} H_{A(vap)}^\infty$	425
	2-propanol	$\Delta_M H_A^\infty$, $\Delta_{sol} H_{A(vap)}^\infty$	425
	2-propanone	$\Delta_M H_A^\infty$, $\Delta_{sol} H_{A(vap)}^\infty$	425
	propyl acetate	$\Delta_M H_A^\infty$, $\Delta_{sol} H_{A(vap)}^\infty$	425
	tetrahydrofuran	$\Delta_M H_A^\infty$, $\Delta_{sol} H_{A(vap)}^\infty$	425
Polyindene			
	anisole	$\Delta_{sol} H_B^\infty$	264
	benzene	$\Delta_{sol} H_B^\infty$	264
	benzonitrile	$\Delta_{sol} H_B^\infty$	264
	bromobenzene	$\Delta_{sol} H_B^\infty$	265
	2-butanone	$\Delta_{sol} H_B^\infty$	265
	chlorobenzene	$\Delta_{sol} H_B^\infty$	265
	1-chlorobutane	$\Delta_{sol} H_B^\infty$	265
	1-chloroheptane	$\Delta_{sol} H_B^\infty$	265
	cyclohexane	$\Delta_{sol} H_B^\infty$	266
	N,N-dimethylaniline	$\Delta_{sol} H_B^\infty$	266
	ethyl acetate	$\Delta_{sol} H_B^\infty$	266
	ethylbenzene	$\Delta_{sol} H_B^\infty$	266
	ethyl benzoate	$\Delta_{sol} H_B^\infty$	266
	nitrobenzene	$\Delta_{sol} H_B^\infty$	267
	1-nitropropane	$\Delta_{sol} H_B^\infty$	267
	pyridine	$\Delta_{sol} H_B^\infty$	267

Polymer(s)	Solvent(s)	Property	Page(s)
Polyindene (*continued*)			
	1,1,2,2-tetrachloroethane	$\Delta_{sol}H_B^{\infty}$	267
	tetrachloromethane	$\Delta_{sol}H_B^{\infty}$	267
	1,1,1-trichloroethane	$\Delta_{sol}H_B^{\infty}$	268
	trichloromethane	$\Delta_{sol}H_B^{\infty}$	268
Poly(isobornyl methacrylate)			
	benzene	$\Delta_M H_A^{\infty}$	425
	2-butanone	$\Delta_M H_A^{\infty}$	425
	n-decane	$\Delta_M H_A^{\infty}$	425
	1,2-dimethylbenzene	$\Delta_M H_A^{\infty}$	425
	n-dodecane	$\Delta_M H_A^{\infty}$	425
	ethyl acetate	$\Delta_M H_A^{\infty}$	425
	methanol	$\Delta_M H_A^{\infty}$	425
	methyl acetate	$\Delta_M H_A^{\infty}$	425
	n-nonane	$\Delta_M H_A^{\infty}$	425
	n-octane	$\Delta_M H_A^{\infty}$	425
	2-propanone	$\Delta_M H_A^{\infty}$	425
	tetrahydrofuran	$\Delta_M H_A^{\infty}$	425
	n-undecane	$\Delta_M H_A^{\infty}$	425
Polyisobutylene			
	benzene	$\Delta_M H$, $\Delta_{dil}H^{12}$	100-101
	benzene	$\Delta_M H_B^{\infty}$	268-270
	benzene	$\Delta_M H_A^{\infty}$	425-426
	chlorobenzene	$\Delta_M H$	102
	chlorobenzene	$\Delta_M H_B^{\infty}$	270
	cyclohexane	$\Delta_M H$	102
	cyclohexane	$\Delta_M H_B^{\infty}$	271-272
	cyclooctane	$\Delta_M H_B^{\infty}$	272
	cyclopentane	$\Delta_M H_B^{\infty}$	272
	cis-decahydronaphthalene	$\Delta_M H_B^{\infty}$	272
	trans-decahydronaphthalene	$\Delta_{sol}H_B^{\infty}$	229
	n-decane	$\Delta_M H_B^{\infty}$	273
	dibutyl ether	$\Delta_M H_B^{\infty}$	273
	diethyl ether	$\Delta_M H_B^{\infty}$	273
	3,3-diethylpentane	$\Delta_M H_B^{\infty}$	273
	dihexyl ether	$\Delta_M H_B^{\infty}$	273
	2,2-dimethylpentane	$\Delta_M H_B^{\infty}$	273
	2,3-dimethylpentane	$\Delta_M H_B^{\infty}$	274
	2,4-dimethylpentane	$\Delta_M H_B^{\infty}$	274
	3,3-dimethylpentane	$\Delta_M H_B^{\infty}$	274
	dipentyl ether	$\Delta_M H_B^{\infty}$	274
	dipropyl ether	$\Delta_M H_B^{\infty}$	274

Polymer(s)	Solvent(s)	Property	Page(s)
Polyisobutylene (*continued*)			
	n-dodecane	$\Delta_M H_B^\infty$	274-275
	ethylbenzene	$\Delta_M H_B^\infty$	275
	ethyl decanoate	$\Delta_M H_B^\infty$	275
	ethyl heptanoate	$\Delta_M H_B^\infty$	276
	ethyl hexanoate	$\Delta_M H_B^\infty$	276
	ethyl nonanoate	$\Delta_M H_B^\infty$	276
	ethyl octanoate	$\Delta_M H_B^\infty$	276
	3-ethylpentane	$\Delta_M H_B^\infty$	276
	ethyl tetradecanoate	$\Delta_M H_B^\infty$	276
	2,2,4,4,6,8,8-heptamethyl-nonane	$\Delta_M H_B^\infty$	277
	n-heptane	$\Delta_M H$	103
	n-heptane	$\Delta_M H_B^\infty$	277-278
	n-hexadecane	$\Delta_M H_B^\infty$	278-279
	n-hexane	$\Delta_M H_B^\infty$	279
	2-methylbutane	$\Delta_M H_B^\infty$	280
	methylcyclohexane	$\Delta_M H_B^\infty$	280
	3-methylhexane	$\Delta_M H_B^\infty$	280
	n-nonane	$\Delta_M H_B^\infty$	280
	n-octane	$\Delta_M H_B^\infty$	281
	2,2,4,6,6-pentamethylheptane	$\Delta_M H_B^\infty$	281
	n-pentane	$\Delta_M H_B^\infty$	281-282
	tetrachloromethane	$\Delta_M H_B^\infty$	282
	n-tetradecane	$\Delta_M H_B^\infty$	283
	2,2,4,4-tetramethylpentane	$\Delta_M H_B^\infty$	283
	2,3,3,4-tetramethylpentane	$\Delta_M H_B^\infty$	283
	toluene	$\Delta_M H_B^\infty$, $\Delta_{sol} H_B^\infty$	283
	n-tridecane	$\Delta_M H_B^\infty$	284
	2,2,4-trimethylpentane	$\Delta_M H_B^\infty$	284
	2,2,4-trimethylpentane	$\Delta_M H_A^\infty$	425-426
	n-undecane	$\Delta_M H_B^\infty$	284
Poly(isobutyl methacrylate)			
	cyclohexanone	$\Delta_M H_B^\infty$, $\Delta_{sol} H_B^\infty$	285
Poly(isobutyl methacrylate-*co*-methyl methacrylate)			
	cyclohexanone	$\Delta_M H_B^\infty$, $\Delta_{sol} H_B^\infty$	285
Poly(*N*-isopropylacrylamide)			
	1-butanol	$\Delta_M H_A^\infty$	427
	water	$\Delta_{sol} H_B^\infty$	285

Polymer(s)	Solvent(s)	Property	Page(s)
1,4-*cis*-Polyisoprene			
	benzene	$\Delta_M H_A^\infty$	426
	cyclohexane	$\Delta_M H_A^\infty$	426
	n-decane	$\Delta_M H_A^\infty$	426
	n-heptane	$\Delta_M H_A^\infty$	426
	n-hexane	$\Delta_M H_A^\infty$	426
	hexafluorobenzene	$\Delta_M H_A^\infty$	426
	n-nonane	$\Delta_M H_A^\infty$	426
	n-octane	$\Delta_M H_A^\infty$	426
	toluene	$\Delta_M H_A^\infty$	426
Poly(isopropyl methacrylate)			
	benzene	$\Delta_{sol} H_{A(vap)}^\infty$	427
	1-butanol	$\Delta_{sol} H_{A(vap)}^\infty$	427
	butyl acetate	$\Delta_{sol} H_{A(vap)}^\infty$	427
	1,2-dichloroethane	$\Delta_{sol} H_{A(vap)}^\infty$	427
	ethanol	$\Delta_{sol} H_{A(vap)}^\infty$	427
	ethyl acetate	$\Delta_{sol} H_{A(vap)}^\infty$	427
	ethylbenzene	$\Delta_{sol} H_{A(vap)}^\infty$	427
	methyl acetate	$\Delta_{sol} H_{A(vap)}^\infty$	427
	2-methyl-1-propanol	$\Delta_{sol} H_{A(vap)}^\infty$	427
	2-propanol	$\Delta_{sol} H_{A(vap)}^\infty$	427
	propyl acetate	$\Delta_{sol} H_{A(vap)}^\infty$	427
	toluene	$\Delta_{sol} H_{A(vap)}^\infty$	427
	trichloromethane	$\Delta_{sol} H_{A(vap)}^\infty$	427
Poly(L-lysine)			
	methanol	$\Delta_{dil} H^{12}$	103
	water	$\Delta_{dil} H^{12}$	103
Poly[2-(3-mesityl-3-methylcyclobutyl)-2-hydroxyethyl methacrylate]			
	benzene	$\Delta_M H_A^\infty$	427
	2-butanone	$\Delta_M H_A^\infty$	427
	n-decane	$\Delta_M H_A^\infty$	427
	1,2-dimethylbenzene	$\Delta_M H_A^\infty$	428
	n-dodecane	$\Delta_M H_A^\infty$	428
	ethyl acetate	$\Delta_M H_A^\infty$	428
	methanol	$\Delta_M H_A^\infty$	428
	methyl acetate	$\Delta_M H_A^\infty$	428
	n-nonane	$\Delta_M H_A^\infty$	428
	n-octane	$\Delta_M H_A^\infty$	428
	2-propanone	$\Delta_M H_A^\infty$	428
	toluene	$\Delta_M H_A^\infty$	428
	n-undecane	$\Delta_M H_A^\infty$	428

Polymer(s)	Solvent(s)	Property	Page(s)
Poly(methacrylic acid)			
	water	$\Delta_{dil}H^{12}$	103-104
Poly(methacrylic acid) salts			
	water	$\Delta_{dil}H^{12}$	104-105
Poly(methyl acrylate)			
	benzene	$\Delta_M H_A^{\infty}$, $\Delta_{sol}H_{A(vap)}^{\infty}$	428-429
	1-butanol	$\Delta_{sol}H_{A(vap)}^{\infty}$	429
	2-butanone	$\Delta_{sol}H_{A(vap)}^{\infty}$	429
	butyl acetate	$\Delta_{sol}H_{A(vap)}^{\infty}$	429
	butylbenzene	$\Delta_M H_A^{\infty}$, $\Delta_{sol}H_{A(vap)}^{\infty}$	428
	tert-butylbenzene	$\Delta_M H_A^{\infty}$, $\Delta_{sol}H_{A(vap)}^{\infty}$	428
	butylcyclohexane	$\Delta_M H_A^{\infty}$, $\Delta_{sol}H_{A(vap)}^{\infty}$	428
	chlorobenzene	$\Delta_{sol}H_{A(vap)}^{\infty}$	429
	1-chlorobutane	$\Delta_{sol}H_{A(vap)}^{\infty}$	429
	1-chlorohexane	$\Delta_{sol}H_{A(vap)}^{\infty}$	429
	1-chlorooctane	$\Delta_{sol}H_{A(vap)}^{\infty}$	429
	1-chloropentane	$\Delta_{sol}H_{A(vap)}^{\infty}$	429
	cyclohexadiene	$\Delta_{sol}H_{A(vap)}^{\infty}$	429
	cyclohexane	$\Delta_M H_A^{\infty}$, $\Delta_{sol}H_{A(vap)}^{\infty}$	428
	cyclooctane	$\Delta_{sol}H_{A(vap)}^{\infty}$	429
	n-decane	$\Delta_M H_A^{\infty}$, $\Delta_{sol}H_{A(vap)}^{\infty}$	428-429
	cis-decahydronaphthalene	$\Delta_M H_A^{\infty}$, $\Delta_{sol}H_{A(vap)}^{\infty}$	428
	trans-decahydronaphthalene	$\Delta_M H_A^{\infty}$, $\Delta_{sol}H_{A(vap)}^{\infty}$	428
	1,1-dichloroethane	$\Delta_{sol}H_{A(vap)}^{\infty}$	429
	1,2-dichloroethane	$\Delta_{sol}H_{A(vap)}^{\infty}$	429
	dichloromethane	$\Delta_{sol}H_{A(vap)}^{\infty}$	429
	n-dodecane	$\Delta_M H_A^{\infty}$, $\Delta_{sol}H_{A(vap)}^{\infty}$	428
	1,4-dioxane	$\Delta_{sol}H_{A(vap)}^{\infty}$	429
	ethanol	$\Delta_{sol}H_{A(vap)}^{\infty}$	429
	ethyl acetate	$\Delta_{sol}H_{A(vap)}^{\infty}$	429
	ethylbenzene	$\Delta_M H_A^{\infty}$, $\Delta_{sol}H_{A(vap)}^{\infty}$	428-429
	methyl acetate	$\Delta_{sol}H_{A(vap)}^{\infty}$	429
	naphthalene	$\Delta_M H_A^{\infty}$, $\Delta_{sol}H_{A(vap)}^{\infty}$	428
	n-nonane	$\Delta_{sol}H_{A(vap)}^{\infty}$	429
	n-octane	$\Delta_M H_A^{\infty}$, $\Delta_{sol}H_{A(vap)}^{\infty}$	428
	1-pentanol	$\Delta_{sol}H_{A(vap)}^{\infty}$	429
	1-propanol	$\Delta_{sol}H_{A(vap)}^{\infty}$	429
	2-propanone	$\Delta_{sol}H_B^{\infty}$	285
	2-propanone	$\Delta_{sol}H_{A(vap)}^{\infty}$	429
	propyl acetate	$\Delta_{sol}H_{A(vap)}^{\infty}$	429
	tetrachloromethane	$\Delta_{sol}H_{A(vap)}^{\infty}$	429
	n-tetradecane	$\Delta_M H_A^{\infty}$, $\Delta_{sol}H_{A(vap)}^{\infty}$	428

Polymer(s)	Solvent(s)	Property	Page(s)
Poly(methyl acrylate) (*continued*)			
	tetrahydrofuran	$\Delta_{sol}H_{A(vap)}{}^{\infty}$	429
	1,2,3,4-tetrahydronaphthalene	$\Delta_M H_A{}^{\infty}, \Delta_{sol}H_{A(vap)}{}^{\infty}$	428
	3,3,4,4-tetramethylhexane	$\Delta_M H_A{}^{\infty}, \Delta_{sol}H_{A(vap)}{}^{\infty}$	428
	toluene	$\Delta_M H_A{}^{\infty}, \Delta_{sol}H_{A(vap)}{}^{\infty}$	428-429
	1,1,1-trichloroethane	$\Delta_{sol}H_{A(vap)}{}^{\infty}$	429
	trichloroethene	$\Delta_{sol}H_{A(vap)}{}^{\infty}$	429
	trichloromethane	$\Delta_{sol}H_{A(vap)}{}^{\infty}$	429
	2,2,5-trimethylhexane	$\Delta_M H_A{}^{\infty}, \Delta_{sol}H_{A(vap)}{}^{\infty}$	428
	3,4,5-trimethylheptane	$\Delta_M H_A{}^{\infty}, \Delta_{sol}H_{A(vap)}{}^{\infty}$	428
	2,2,4-trimethylpentane	$\Delta_M H_A{}^{\infty}, \Delta_{sol}H_{A(vap)}{}^{\infty}$	428
	n-undecane	$\Delta_{sol}H_{A(vap)}{}^{\infty}$	429
Poly(methylcyanopropylsiloxane)			
	benzene	$\Delta_M H_A{}^{\infty}, \Delta_{sol}H_{A(vap)}{}^{\infty}$	430
	1-butanol	$\Delta_M H_A{}^{\infty}, \Delta_{sol}H_{A(vap)}{}^{\infty}$	430
	n-decane	$\Delta_M H_A{}^{\infty}, \Delta_{sol}H_{A(vap)}{}^{\infty}$	430
	n-dodecane	$\Delta_M H_A{}^{\infty}, \Delta_{sol}H_{A(vap)}{}^{\infty}$	430
	n-heptane	$\Delta_M H_A{}^{\infty}, \Delta_{sol}H_{A(vap)}{}^{\infty}$	430
	n-hexane	$\Delta_M H_A{}^{\infty}, \Delta_{sol}H_{A(vap)}{}^{\infty}$	430
	n-nonane	$\Delta_M H_A{}^{\infty}, \Delta_{sol}H_{A(vap)}{}^{\infty}$	430
	n-octane	$\Delta_M H_A{}^{\infty}, \Delta_{sol}H_{A(vap)}{}^{\infty}$	430
	2-pentanone	$\Delta_M H_A{}^{\infty}, \Delta_{sol}H_{A(vap)}{}^{\infty}$	430
	pyridine	$\Delta_M H_A{}^{\infty}, \Delta_{sol}H_{A(vap)}{}^{\infty}$	430
	n-undecane	$\Delta_M H_A{}^{\infty}, \Delta_{sol}H_{A(vap)}{}^{\infty}$	430
Poly(methyl methacrylate)			
	benzene	$\Delta_{dil}H^{12}$	105-106
	benzene	$\Delta_M H_A{}^{\infty}, \Delta_{sol}H_{A(vap)}{}^{\infty}$	430-431
	1-butanol	$\Delta_M H_A{}^{\infty}, \Delta_{sol}H_{A(vap)}{}^{\infty}$	430
	2-butanone	$\Delta_M H_A{}^{\infty}, \Delta_{sol}H_{A(vap)}{}^{\infty}$	430-431
	butyl acetate	$\Delta_M H_A{}^{\infty}, \Delta_{sol}H_{A(vap)}{}^{\infty}$	430
	chlorobenzene	$\Delta_{dil}H^{12}$	106
	cyclohexanone	$\Delta_M H_B{}^{\infty}, \Delta_{sol}H_B{}^{\infty}$	285
	n-decane	$\Delta_M H_A{}^{\infty}, \Delta_{sol}H_{A(vap)}{}^{\infty}$	430-431
	1,2-dichlorobenzene	$\Delta_{dil}H^{12}$	106
	1,2-dichloroethane	$\Delta_{sol}H_B{}^{\infty}$	286
	1,2-dimethylbenzene	$\Delta_M H_A{}^{\infty}, \Delta_{sol}H_{A(vap)}{}^{\infty}$	430-431
	1,4-dioxane	$\Delta_{dil}H^{12}$	107
	1,4-dioxane	$\Delta_M H_A{}^{\infty}, \Delta_{sol}H_{A(vap)}{}^{\infty}$	431
	n-dodecane	$\Delta_M H_A{}^{\infty}, \Delta_{sol}H_{A(vap)}{}^{\infty}$	431
	ethanol	$\Delta_M H_A{}^{\infty}, \Delta_{sol}H_{A(vap)}{}^{\infty}$	430-431
	ethyl acetate	$\Delta_M H_A{}^{\infty}, \Delta_{sol}H_{A(vap)}{}^{\infty}$	430-431
	ethylbenzene	$\Delta_{sol}H_B{}^{\infty}$	286

Polymer(s)	Solvent(s)	Property	Page(s)
Poly(methyl methacrylate) (*continued*)			
	ethylbenzene	$\Delta_M H_A^\infty$, $\Delta_{sol} H_{A(vap)}^\infty$	430
	methanol	$\Delta_M H_A^\infty$, $\Delta_{sol} H_{A(vap)}^\infty$	430-431
	methyl acetate	$\Delta_M H_A^\infty$, $\Delta_{sol} H_{A(vap)}^\infty$	431
	4-methyl-2-pentanone	$\Delta_{dil} H^{12}$	108
	4-methyl-2-pentanone	$\Delta_{sol} H_B^\infty$	286-287
	4-methyl-2-pentanone	$\Delta_M H_A^\infty$, $\Delta_{sol} H_{A(vap)}^\infty$	430
	n-nonane	$\Delta_M H_A^\infty$, $\Delta_{sol} H_{A(vap)}^\infty$	431
	n-octane	$\Delta_M H_A^\infty$, $\Delta_{sol} H_{A(vap)}^\infty$	431
	2-propanone	$\Delta_{dil} H^{12}$	108-110
	2-propanone	$\Delta_{sol} H_B^\infty$	287
	2-propanone	$\Delta_M H_A^\infty$, $\Delta_{sol} H_{A(vap)}^\infty$	431
	tetrachloromethane	$\Delta_{dil} H^{12}$	110-111
	tetrachloromethane	$\Delta_M H_A^\infty$, $\Delta_{sol} H_{A(vap)}^\infty$	431
	tetrahydrofuran	$\Delta_M H_A^\infty$, $\Delta_{sol} H_{A(vap)}^\infty$	431
	toluene	$\Delta_{dil} H^{12}$	111-113
	toluene	$\Delta_{sol} H_B^\infty$	287
	toluene	$\Delta_M H_A^\infty$, $\Delta_{sol} H_{A(vap)}^\infty$	430-431
	trichloromethane	$\Delta_{dil} H^{12}$	113
	trichloromethane	$\Delta_{sol} H_B^\infty$	287-289
	trichloromethane	$\Delta_M H_A^\infty$, $\Delta_{sol} H_{A(vap)}^\infty$	430-431
	n-undecane	$\Delta_M H_A^\infty$, $\Delta_{sol} H_{A(vap)}^\infty$	431
Poly(4-methyl-1-pentene)			
	cyclohexane	$\Delta_{sol} H_B^\infty$	289
Poly[2-(3-methyl-3-phenylcyclobutyl)-2-hydroxyethyl methacrylate]			
	benzene	$\Delta_M H_A^\infty$	432
	2-butanone	$\Delta_M H_A^\infty$	432
	n-decane	$\Delta_M H_A^\infty$	432
	1,2-dimethylbenzene	$\Delta_M H_A^\infty$	432
	n-dodecane	$\Delta_M H_A^\infty$	432
	ethanol	$\Delta_M H_A^\infty$	432
	ethyl acetate	$\Delta_M H_A^\infty$	432
	methanol	$\Delta_M H_A^\infty$	432
	methyl acetate	$\Delta_M H_A^\infty$	432
	n-nonane	$\Delta_M H_A^\infty$	432
	n-octane	$\Delta_M H_A^\infty$	432
	2-propanone	$\Delta_M H_A^\infty$	432
	toluene	$\Delta_M H_A^\infty$	432
	n-undecane	$\Delta_M H_A^\infty$	432

Polymer(s)	Solvent(s)	Property	Page(s)
Poly(α-methylstyrene)			
	toluene	$\Delta_{sol}H_B^\infty$	289-291
Poly(methyl trifluoropropyl siloxane)			
	benzene	$\Delta_M H_A^\infty, \Delta_{sol}H_{A(vap)}^\infty$	432
	1-butanol	$\Delta_M H_A^\infty, \Delta_{sol}H_{A(vap)}^\infty$	432
	2-butanone	$\Delta_M H_A^\infty, \Delta_{sol}H_{A(vap)}^\infty$	432
	butyronitrile	$\Delta_M H_A^\infty, \Delta_{sol}H_{A(vap)}^\infty$	432
	N,N-dimethylaniline	$\Delta_M H_A^\infty, \Delta_{sol}H_{A(vap)}^\infty$	432
	ethyl acetate	$\Delta_M H_A^\infty, \Delta_{sol}H_{A(vap)}^\infty$	432
	n-decane	$\Delta_M H_A^\infty, \Delta_{sol}H_{A(vap)}^\infty$	432
	n-dodecane	$\Delta_M H_A^\infty, \Delta_{sol}H_{A(vap)}^\infty$	432
	n-heptane	$\Delta_M H_A^\infty, \Delta_{sol}H_{A(vap)}^\infty$	432
	n-hexane	$\Delta_M H_A^\infty, \Delta_{sol}H_{A(vap)}^\infty$	432
	n-nonane	$\Delta_M H_A^\infty, \Delta_{sol}H_{A(vap)}^\infty$	432
	n-octane	$\Delta_M H_A^\infty, \Delta_{sol}H_{A(vap)}^\infty$	432
	1-octanol	$\Delta_M H_A^\infty, \Delta_{sol}H_{A(vap)}^\infty$	432
	2-octanone	$\Delta_M H_A^\infty, \Delta_{sol}H_{A(vap)}^\infty$	432
	1-pentanol	$\Delta_M H_A^\infty, \Delta_{sol}H_{A(vap)}^\infty$	433
	2-pentanone	$\Delta_M H_A^\infty, \Delta_{sol}H_{A(vap)}^\infty$	433
	pentylbenzene	$\Delta_M H_A^\infty, \Delta_{sol}H_{A(vap)}^\infty$	433
	2-propanone	$\Delta_M H_A^\infty, \Delta_{sol}H_{A(vap)}^\infty$	433
	propylbenzene	$\Delta_M H_A^\infty, \Delta_{sol}H_{A(vap)}^\infty$	433
	pyridine	$\Delta_M H_A^\infty, \Delta_{sol}H_{A(vap)}^\infty$	433
	toluene	$\Delta_M H_A^\infty, \Delta_{sol}H_{A(vap)}^\infty$	433
	n-undecane	$\Delta_M H_A^\infty, \Delta_{sol}H_{A(vap)}^\infty$	433
	valeronitrile	$\Delta_M H_A^\infty, \Delta_{sol}H_{A(vap)}^\infty$	433
Poly[1-methyl-1-(trimethylsilyl-methyl)-1-silacyclobutane]			
	benzene	$\Delta_M H_A^\infty, \Delta_{sol}H_{A(vap)}^\infty$	433
	ethylbenzene	$\Delta_M H_A^\infty, \Delta_{sol}H_{A(vap)}^\infty$	433
	n-hexane	$\Delta_M H_A^\infty, \Delta_{sol}H_{A(vap)}^\infty$	433
	n-heptane	$\Delta_M H_A^\infty, \Delta_{sol}H_{A(vap)}^\infty$	433
	toluene	$\Delta_M H_A^\infty, \Delta_{sol}H_{A(vap)}^\infty$	433
Poly(2-methyl-5-vinyltetrazole)			
	acetic acid	$\Delta_{sol}H_B^\infty$	292
	acetonitrile	$\Delta_{sol}H_B^\infty$	292
	1,2-dichloroethane	$\Delta_{sol}H_B^\infty$	292
	N,N-diethylacetamide	$\Delta_{sol}H_B^\infty$	292
	N,N-dimethylformamide	$\Delta_{sol}H_B^\infty$	292
	dimethylsulfoxide	$\Delta_{sol}H_B^\infty$	292

Polymer(s)	Solvent(s)	Property	Page(s)
Poly(2-methyl-5-vinyltetrazole) (*continued*)			
	formamide	$\Delta_{sol}H_B^\infty$	292
	formic acid	$\Delta_{sol}H_B^\infty$	292
	nitromethane	$\Delta_{sol}H_B^\infty$	293
	pyridine	$\Delta_{sol}H_B^\infty$	293
Poly[4-(2-naphthalenesulfonyl)-styrene]			
	benzene	$\Delta_{sol}H_{A(vap)}^\infty$	433
	2-butanone	$\Delta_{sol}H_{A(vap)}^\infty$	433
	n-decane	$\Delta_{sol}H_{A(vap)}^\infty$	433
	1,2-dimethylbenzene	$\Delta_{sol}H_{A(vap)}^\infty$	433
	n-dodecane	$\Delta_{sol}H_{A(vap)}^\infty$	433
	ethanol	$\Delta_{sol}H_{A(vap)}^\infty$	434
	ethyl acetate	$\Delta_{sol}H_{A(vap)}^\infty$	434
	methanol	$\Delta_{sol}H_{A(vap)}^\infty$	434
	methyl acetate	$\Delta_{sol}H_{A(vap)}^\infty$	434
	n-nonane	$\Delta_{sol}H_{A(vap)}^\infty$	434
	n-octane	$\Delta_{sol}H_{A(vap)}^\infty$	434
	2-propanone	$\Delta_{sol}H_{A(vap)}^\infty$	434
	toluene	$\Delta_{sol}H_{A(vap)}^\infty$	434
	n-undecane	$\Delta_{sol}H_{A(vap)}^\infty$	434
Poly(octamethylene oxide)			
	benzene	$\Delta_M H_B^\infty$	293
Poly(L-ornithine)			
	methanol	$\Delta_{dil}H^{12}$	114
	water	$\Delta_{dil}H^{12}$	114
Polypentenamer			
	cyclohexane	$\Delta_M H_B^\infty$	293
	cyclooctane	$\Delta_M H_B^\infty$	293
	cyclopentane	$\Delta_M H_B^\infty$	293
	cis-decahydronaphthalene	$\Delta_M H_B^\infty$	294
	trans-decahydronaphthalene	$\Delta_{sol}H_B^\infty$	294
	3,3-diethylpentane	$\Delta_M H_B^\infty$	294
	2,2-dimethylpentane	$\Delta_M H_B^\infty$	294
	2,3-dimethylpentane	$\Delta_M H_B^\infty$	294
	2,4-dimethylpentane	$\Delta_M H_B^\infty$	294
	3,3-dimethylpentane	$\Delta_M H_B^\infty$	294
	n-dodecane	$\Delta_M H_B^\infty$	294
	3-ethylpentane	$\Delta_M H_B^\infty$	295

Polymer(s)	Solvent(s)	Property	Page(s)
Polypentenamer (*continued*)			
	2,2,4,4,6,8,8-heptamethyl-nonane	$\Delta_M H_B^\infty$	295
	n-hexadecane	$\Delta_M H_B^\infty$	295
	3-methylhexane	$\Delta_M H_B^\infty$	295
	n-octane	$\Delta_M H_B^\infty$	295
	2,2,4,6,6-pentamethylheptane	$\Delta_M H_B^\infty$	295
	2,2,4,4-tetramethylpentane	$\Delta_M H_B^\infty$	295
	2,3,3,4-tetramethylpentane	$\Delta_M H_B^\infty$	295
	2,2,4-trimethylpentane	$\Delta_M H_B^\infty$	296
Poly(2-phenyl-1,3-dioxolane-4-yl-methyl methacrylate)			
	benzene	$\Delta_M H_A^\infty$	434
	1-butanol	$\Delta_M H_A^\infty$	434
	butylamine	$\Delta_M H_A^\infty$	434
	n-decane	$\Delta_M H_A^\infty$	434
	ethanol	$\Delta_M H_A^\infty$	434
	n-heptane	$\Delta_M H_A^\infty$	434
	n-hexane	$\Delta_M H_A^\infty$	434
	n-octane	$\Delta_M H_A^\infty$	434
	1-pentanol	$\Delta_M H_A^\infty$	434
	1-propanol	$\Delta_M H_A^\infty$	434
	propylamine	$\Delta_M H_A^\infty$	434
Poly(2-phenyl-1,3-dioxolane-4-yl-methyl methacrylate-*co*-glycidyl methacrylate)			
	1-butanol	$\Delta_M H_A^\infty$	435
	n-decane	$\Delta_M H_A^\infty$	435
	ethanol	$\Delta_M H_A^\infty$	435
	n-heptane	$\Delta_M H_A^\infty$	435
	n-hexane	$\Delta_M H_A^\infty$	435
	n-octane	$\Delta_M H_A^\infty$	435
	1-pentanol	$\Delta_M H_A^\infty$	435
	1-propanol	$\Delta_M H_A^\infty$	435
Poly(m-phenylene)			
	benzene	$\Delta_{sol} H_B^\infty$	296
	N,N-dimethylacetamide	$\Delta_{sol} H_B^\infty$	296
Poly(m-phenyleneisophthalamide)			
	N,N-dimethylacetamide	$\Delta_{sol} H_B^\infty$	296
	N,N-dimethylformamide	$\Delta_{sol} H_B^\infty$	296
	1-methyl-2-pyrrolidinone	$\Delta_{sol} H_B^\infty$	297

Polymer(s)	Solvent(s)	Property	Page(s)
Polypropylene			
	benzene	$\Delta_M H_B^{\infty}$	297
	benzene	$\Delta_{sol} H_{A(vap)}^{\infty}$	437
	1-butanol	$\Delta_{sol} H_{A(vap)}^{\infty}$	437
	butyl acetate	$\Delta_{sol} H_{A(vap)}^{\infty}$	437
	chlorobenzene	$\Delta_{sol} H_{A(vap)}^{\infty}$	437
	1-chlorohexane	$\Delta_{sol} H_{A(vap)}^{\infty}$	437
	1-chloronaphthalene	$\Delta_{sol} H_B^{\infty}$	297
	1-chlorooctane	$\Delta_{sol} H_{A(vap)}^{\infty}$	437
	1-chloropentane	$\Delta_{sol} H_{A(vap)}^{\infty}$	437
	cycloheptane	$\Delta_{sol} H_{A(vap)}^{\infty}$	437
	cyclohexadiene	$\Delta_{sol} H_{A(vap)}^{\infty}$	437
	cyclohexane	$\Delta_M H_B^{\infty}$	297
	cyclohexane	$\Delta_{sol} H_{A(vap)}^{\infty}$	437
	cyclohexene	$\Delta_{sol} H_{A(vap)}^{\infty}$	437
	cyclooctane	$\Delta_M H_B^{\infty}$	298
	cyclooctane	$\Delta_{sol} H_{A(vap)}^{\infty}$	437
	cyclopentane	$\Delta_M H_B^{\infty}$	298
	cyclopentane	$\Delta_{sol} H_{A(vap)}^{\infty}$	437
	cis-decahydronaphthalene	$\Delta_M H_B^{\infty}$	298
	trans-decahydronaphthalene	$\Delta_{sol} H_B^{\infty}$	298
	n-decane	$\Delta_M H_B^{\infty}$	298
	n-decane	$\Delta_M H_A^{\infty}$	437
	1,1-dichloroethane	$\Delta_{sol} H_{A(vap)}^{\infty}$	437
	1,2-dichloroethane	$\Delta_{sol} H_{A(vap)}^{\infty}$	437
	3,3-diethylpentane	$\Delta_M H_B^{\infty}$	298
	1,2-dimethylbenzene	$\Delta_M H_B^{\infty}$	298
	1,4-dimethylbenzene	$\Delta_M H_B^{\infty}$	299
	1,4-dimethylbenzene	$\Delta_M H_B^{\infty}$	299
	2,2-dimethylpentane	$\Delta_M H_B^{\infty}$	299
	2,3-dimethylpentane	$\Delta_M H_B^{\infty}$	299
	2,4-dimethylpentane	$\Delta_M H_B^{\infty}$	299
	3,3-dimethylpentane	$\Delta_M H_B^{\infty}$	299
	1,4-dioxane	$\Delta_{sol} H_{A(vap)}^{\infty}$	437
	n-dodecane	$\Delta_M H_B^{\infty}$	299
	n-dodecane	$\Delta_M H_A^{\infty}$	437
	ethylbenzene	$\Delta_M H_B^{\infty}$	300
	ethylbenzene	$\Delta_{sol} H_{A(vap)}^{\infty}$	437
	3-ethylpentane	$\Delta_M H_B^{\infty}$	300
	2,2,4,4,6,8,8-heptamethyl-nonane	$\Delta_M H_B^{\infty}$	300
	n-heptane	$\Delta_M H_B^{\infty}$	300
	n-hexadecane	$\Delta_M H_B^{\infty}$	300
	n-hexane	$\Delta_M H_B^{\infty}$	300
	n-hexane	$\Delta_{sol} H_{A(vap)}^{\infty}$	437

Polymer(s)	Solvent(s)	Property	Page(s)
Polypropylene (*continued*)			
	3-methylhexane	$\Delta_M H_B^\infty$	301
	n-nonane	$\Delta_M H_B^\infty$	301
	n-nonane	$\Delta_{sol} H_{A(vap)}^\infty$	437
	n-octane	$\Delta_M H_B^\infty$	301
	n-octane	$\Delta_{sol} H_{A(vap)}^\infty$	437
	n-pentane	$\Delta_M H_B^\infty$	301
	n-pentane	$\Delta_{sol} H_{A(vap)}^\infty$	437
	2,2,4,6,6-pentamethylheptane	$\Delta_M H_B^\infty$	301
	1-pentanol	$\Delta_{sol} H_{A(vap)}^\infty$	437
	propyl acetate	$\Delta_{sol} H_{A(vap)}^\infty$	437
	tetrachloromethane	$\Delta_M H_B^\infty$	302
	tetrachloromethane	$\Delta_{sol} H_{A(vap)}^\infty$	437
	n-tetradecane	$\Delta_M H_A^\infty$	437
	tetrahydrofuran	$\Delta_{sol} H_{A(vap)}^\infty$	438
	1,2,3,4-tetrahydronaphthalene	$\Delta_{sol} H_B^\infty$	302
	2,2,4,4-tetramethylpentane	$\Delta_M H_B^\infty$	302
	2,3,3,4-tetramethylpentane	$\Delta_M H_B^\infty$	302
	toluene	$\Delta_M H_B^\infty$	302
	toluene	$\Delta_{sol} H_{A(vap)}^\infty$	438
	1,1,1-trichloroethane	$\Delta_{sol} H_{A(vap)}^\infty$	438
	trichloroethene	$\Delta_{sol} H_{A(vap)}^\infty$	438
	trichloromethane	$\Delta_M H_B^\infty$	302-303
	trichloromethane	$\Delta_{sol} H_{A(vap)}^\infty$	438
	2,2,4-trimethylpentane	$\Delta_M H_B^\infty$	303
	n-undecane	$\Delta_{sol} H_{A(vap)}^\infty$	438
Poly(propylene glycol)			
	anisole	$\Delta_M H$	114-115
	benzene	$\Delta_M H$	115-116
	benzene	$\Delta_M H_B^\infty$	303
	benzene	$\Delta_M H_A^\infty, \Delta_{sol} H_{A(vap)}^\infty$	438
	benzyl alcohol	$\Delta_M H$	116-117
	2-butanone	$\Delta_M H_A^\infty, \Delta_{sol} H_{A(vap)}^\infty$	438
	1,2-dichloroethane	$\Delta_M H_A^\infty, \Delta_{sol} H_{A(vap)}^\infty$	438
	dichloromethane	$\Delta_M H_A^\infty, \Delta_{sol} H_{A(vap)}^\infty$	438
	diethyl carbonate	$\Delta_M H$	117-118
	1,2-dimethylbenzene	$\Delta_M H_A^\infty, \Delta_{sol} H_{A(vap)}^\infty$	438
	1,4-dimethylbenzene	$\Delta_M H_A^\infty, \Delta_{sol} H_{A(vap)}^\infty$	438
	dimethyl carbonate	$\Delta_M H$	118-119
	ethanol	$\Delta_M H$	119-120
	ethanol	$\Delta_M H_B^\infty$	303-304
	ethylbenzene	$\Delta_M H_A^\infty, \Delta_{sol} H_{A(vap)}^\infty$	438
	methanol	$\Delta_M H$	120-121

Polymer(s)	Solvent(s)	Property	Page(s)
Poly(propylene glycol) (*continued*)			
	3-methylphenol	$\Delta_M H$	121-122
	2-pentanone	$\Delta_M H_A^{\infty}$, $\Delta_{sol} H_{A(vap)}^{\infty}$	438
	2-propanone	$\Delta_M H_A^{\infty}$, $\Delta_{sol} H_{A(vap)}^{\infty}$	438
	propylbenzene	$\Delta_M H_A^{\infty}$, $\Delta_{sol} H_{A(vap)}^{\infty}$	438
	tetrachloromethane	$\Delta_M H$	122-123
	tetrachloromethane	$\Delta_M H_B^{\infty}$	304
	tetrachloromethane	$\Delta_M H_A^{\infty}$, $\Delta_{sol} H_{A(vap)}^{\infty}$	438
	toluene	$\Delta_M H_A^{\infty}$, $\Delta_{sol} H_{A(vap)}^{\infty}$	438
	triacetin	$\Delta_M H_A^{\infty}$, $\Delta_{sol} H_{A(vap)}^{\infty}$	438
	trichloromethane	$\Delta_M H$	123
	trichloromethane	$\Delta_M H_B^{\infty}$	304
	trichloromethane	$\Delta_M H_A^{\infty}$, $\Delta_{sol} H_{A(vap)}^{\infty}$	438
	water	$\Delta_M H$	123-125
	water	$\Delta_M H_B^{\infty}$	304
	water	$\Delta_M H_A$	363
Poly(propylene glycol) dimethyl ether			
	tetrachloromethane	$\Delta_M H$	126
	trichloromethane	$\Delta_M H$	126
Poly(propylene oxide)			
	benzene	$\Delta_M H_A^{\infty}$, $\Delta_{sol} H_{A(vap)}^{\infty}$	439
	N,N-dimethylformamide	$\Delta_M H_A^{\infty}$, $\Delta_{sol} H_{A(vap)}^{\infty}$	439
	ethanol	$\Delta_M H_A^{\infty}$, $\Delta_{sol} H_{A(vap)}^{\infty}$	439
	ethyl acetate	$\Delta_M H_A^{\infty}$, $\Delta_{sol} H_{A(vap)}^{\infty}$	439
	n-heptane	$\Delta_M H_A^{\infty}$, $\Delta_{sol} H_{A(vap)}^{\infty}$	439
	n-nonane	$\Delta_M H_A^{\infty}$, $\Delta_{sol} H_{A(vap)}^{\infty}$	439
	n-octane	$\Delta_M H_A^{\infty}$, $\Delta_{sol} H_{A(vap)}^{\infty}$	439
	1-propanol	$\Delta_M H_A^{\infty}$, $\Delta_{sol} H_{A(vap)}^{\infty}$	439
	toluene	$\Delta_M H_A^{\infty}$, $\Delta_{sol} H_{A(vap)}^{\infty}$	439
Polystyrene			
	acetic acid	$\Delta_{sol} H_{A(vap)}^{\infty}$	440
	acetonitrile	$\Delta_{sol} H_{A(vap)}^{\infty}$	440
	acetonitrile	$\Delta_M H_A^{\infty}$	442
	aniline	$\Delta_{sol} H_{A(vap)}^{\infty}$	440
	anisole	$\Delta_{dil} H^{12}$	126
	anisole	$\Delta_M H_A^{\infty}$	439-445
	benzaldehyde	$\Delta_{sol} H_{A(vap)}^{\infty}$	440
	benzene	$\Delta_M H$, $\Delta_{dil} H^{12}$	126-129
	benzene	$\Delta_M H_B^{\infty}$, $\Delta_{sol} H_B^{\infty}$	305-308
	benzene	$\Delta_M H_A^{\infty}$	439-445
	benzene	$\Delta_{sol} H_{A(vap)}^{\infty}$	440, 444

Polymer(s)	Solvent(s)	Property	Page(s)
Polystyrene (*continued*)			
	dipropyl ether	$\Delta_M H_A^\infty$	442
	dipropyl thioether	$\Delta_M H_A^\infty$	442
	n-dodecane	$\Delta_M H_A^\infty$, $\Delta_{sol} H_{A(vap)}^\infty$	443
	1,2-ethanediol	$\Delta_{sol} H_{A(vap)}^\infty$	441
	ethanol	$\Delta_{sol} H_{A(vap)}^\infty$	441
	ethyl acetate	$\Delta_{dil} H^{12}$	134-136
	ethyl acetate	$\Delta_{sol} H_B^\infty$	311-312
	ethyl acetate	$\Delta_{sol} H_{A(vap)}^\infty$	441
	ethylbenzene	$\Delta_M H$, $\Delta_{dil} H^{12}$	137
	ethylbenzene	$\Delta_{sol} H_B^\infty$	312-314
	ethylbenzene	$\Delta_M H_A^\infty$	439-445
	formamide	$\Delta_{sol} H_{A(vap)}^\infty$	441
	n-heptane	$\Delta_M H_A^\infty$, $\Delta_{sol} H_{A(vap)}^\infty$	440
	n-heptane	$\Delta_M H_A^\infty$, $\Delta_{sol} H_{A(vap)}^\infty$	441-444
	n-hexadecane	$\Delta_M H_A^\infty$, $\Delta_{sol} H_{A(vap)}^\infty$	443
	n-hexane	$\Delta_M H_A^\infty$, $\Delta_{sol} H_{A(vap)}^\infty$	441-445
	methanol	$\Delta_{sol} H_{A(vap)}^\infty$	441
	2-methylheptane	$\Delta_M H_A^\infty$, $\Delta_{sol} H_{A(vap)}^\infty$	442
	2-methylpentane	$\Delta_M H_A^\infty$, $\Delta_{sol} H_{A(vap)}^\infty$	442
	3-methylpentane	$\Delta_M H_A^\infty$, $\Delta_{sol} H_{A(vap)}^\infty$	442
	2-methyl-1-propanol	$\Delta_{sol} H_{A(vap)}^\infty$	441
	naphthalene	$\Delta_M H_A^\infty$, $\Delta_{sol} H_{A(vap)}^\infty$	443
	nitrobenzene	$\Delta_{sol} H_{A(vap)}^\infty$	441
	n-nonane	$\Delta_M H_A^\infty$, $\Delta_{sol} H_{A(vap)}^\infty$	440-445
	n-octane	$\Delta_M H_A^\infty$, $\Delta_{sol} H_{A(vap)}^\infty$	440-444
	1-octene	$\Delta_M H_A^\infty$, $\Delta_{sol} H_{A(vap)}^\infty$	442
	cis-2-octene	$\Delta_M H_A^\infty$, $\Delta_{sol} H_{A(vap)}^\infty$	442
	1-octanol	$\Delta_{sol} H_{A(vap)}^\infty$	441
	n-pentane	$\Delta_{sol} H_{A(vap)}^\infty$	441
	1-pentanol	$\Delta_{sol} H_{A(vap)}^\infty$	441
	1-propanol	$\Delta_{sol} H_{A(vap)}^\infty$	441
	2-propanol	$\Delta_{sol} H_{A(vap)}^\infty$	441
	2-propanone	$\Delta_{sol} H_B^\infty$	314-315
	2-propanone	$\Delta_{sol} H_{A(vap)}^\infty$	441
	propylbenzene	$\Delta_{sol} H_B^\infty$	316
	pyridine	$\Delta_M H_A^\infty$	439-445
	pyridine	$\Delta_{sol} H_{A(vap)}^\infty$	441
	pyridine	$\Delta_M H_A^\infty$	442
	styrene	$\Delta_{sol} H_B^\infty$	316
	tetrachloromethane	$\Delta_{dil} H^{12}$	137-138
	tetrachloromethane	$\Delta_{sol} H_B^\infty$	316
	tetrachloromethane	$\Delta_{sol} H_{A(vap)}^\infty$	441
	n-tetradecane	$\Delta_M H_A^\infty$, $\Delta_{sol} H_{A(vap)}^\infty$	443
	tetrahydrofuran	$\Delta_{sol} H_{A(vap)}^\infty$	441

Polymer(s)	Solvent(s)	Property	Page(s)
Polystyrene (*continued*)			
	1,2,3,4-tetrahydronaphthalene	$\Delta_M H_A^\infty$, $\Delta_{sol} H_{A(vap)}^\infty$	443
	3,3,4,4-tetramethylhexane	$\Delta_M H_A^\infty$, $\Delta_{sol} H_{A(vap)}^\infty$	443
	toluene	$\Delta_M H$, $\Delta_{dil} H^{12}$	138-142
	toluene	$\Delta_M H_B^\infty$, $\Delta_{sol} H_B^\infty$	316-321
	toluene	$\Delta_M H_A^\infty$	439-445
	toluene	$\Delta_{sol} H_{A(vap)}^\infty$	441, 444
	toluene	$\Delta_M H_A^\infty$	442
	trichloroethene	$\Delta_{sol} H_{A(vap)}^\infty$	441
	trichloromethane	$\Delta_{dil} H^{12}$	143-144
	trichloromethane	$\Delta_M H_B^\infty$, $\Delta_{sol} H_B^\infty$	322-323
	trichloromethane	$\Delta_{sol} H_{A(vap)}^\infty$	441
	trichloromethane	$\Delta_M H_A^\infty$	442
	1,3,5-trimethylbenzene	$\Delta_{sol} H_B^\infty$	323
	1,3,5-trimethylbenzene	$\Delta_{sol} H_{A(vap)}^\infty$	444
	2,2,4-trimethylpentane	$\Delta_M H_A^\infty$, $\Delta_{sol} H_{A(vap)}^\infty$	441-442
	water	$\Delta_{sol} H_{A(vap)}^\infty$	441
Poly(styrene-*co*-acrylontrile)			
	trichloromethane	$\Delta_M H_B^\infty$, $\Delta_{sol} H_B^\infty$	324-325
Polystyrene-b-polybutadiene-b-polystyrene triblock copolymer			
	benzene	$\Delta_M H_A^\infty$, $\Delta_{sol} H_{A(vap)}^\infty$	445-446
	2-butanone	$\Delta_M H_A^\infty$, $\Delta_{sol} H_{A(vap)}^\infty$	445-446
	cyclohexane	$\Delta_M H_A^\infty$, $\Delta_{sol} H_{A(vap)}^\infty$	445-446
	1,4-dimethylbenzene	$\Delta_M H_A^\infty$, $\Delta_{sol} H_{A(vap)}^\infty$	445-446
	ethylbenzene	$\Delta_M H_A^\infty$, $\Delta_{sol} H_{A(vap)}^\infty$	445-446
	n-heptane	$\Delta_M H_A^\infty$, $\Delta_{sol} H_{A(vap)}^\infty$	445-446
	n-hexane	$\Delta_M H_A^\infty$, $\Delta_{sol} H_{A(vap)}^\infty$	445-446
	toluene	$\Delta_M H_A^\infty$, $\Delta_{sol} H_{A(vap)}^\infty$	445-446
	trichloromethane	$\Delta_M H_A^\infty$, $\Delta_{sol} H_{A(vap)}^\infty$	445-446
Poly(styrene-*co*-butyl methacrylate)			
	benzene	$\Delta_M H_A^\infty$	446
	1-butanol	$\Delta_M H_A^\infty$	446
	2-butanone	$\Delta_{dil} H^{12}$	144-145
	butyl acetate	$\Delta_M H_A^\infty$	446
	butylbenzene	$\Delta_M H_A^\infty$	446
	tert-butylbenzene	$\Delta_M H_A^\infty$	446
	butylcyclohexane	$\Delta_M H_A^\infty$	446
	chlorobenzene	$\Delta_M H_A^\infty$	446
	1-chlorobutane	$\Delta_M H_A^\infty$	446
	cyclohexane	$\Delta_M H_A^\infty$	446

Polymer(s)	Solvent(s)	Property	Page(s)
Poly(styrene-*co*-butyl methacrylate) (*continued*)			
	cyclohexanol	$\Delta_M H_A^\infty$	446
	n-decane	$\Delta_M H_A^\infty$	446
	dichloromethane	$\Delta_M H_A^\infty$	446
	n-dodecane	$\Delta_M H_A^\infty$	446
	ethylbenzene	$\Delta_M H_A^\infty$	446
	methylcyclohexane	$\Delta_M H_A^\infty$	446
	n-octane	$\Delta_M H_A^\infty$	446
	2-pentanone	$\Delta_M H_A^\infty$	446
	tetrachloromethane	$\Delta_M H_A^\infty$	446
	trichloromethane	$\Delta_M H_A^\infty$	446
	2,2,4-trimethylpentane	$\Delta_M H_A^\infty$	446
	3,4,5-trimethylheptane	$\Delta_M H_A^\infty$	446
Poly(styrene-*co*-divinylbenzene)			
	1-butanol	$\Delta_M H_A^\infty$, $\Delta_{sol} H_{A(vap)}^\infty$	447
	ethanol	$\Delta_M H_A^\infty$, $\Delta_{sol} H_{A(vap)}^\infty$	447
	methanol	$\Delta_M H_A^\infty$, $\Delta_{sol} H_{A(vap)}^\infty$	447
	1-pentanol	$\Delta_M H_A^\infty$, $\Delta_{sol} H_{A(vap)}^\infty$	447
	1-propanol	$\Delta_M H_A^\infty$, $\Delta_{sol} H_{A(vap)}^\infty$	447
Poly(styrene-*co*-ethyl acrylate)			
	n-decane	$\Delta_{sol} H_{A(vap)}^\infty$	448
	n-hexane	$\Delta_{sol} H_{A(vap)}^\infty$	448
	n-nonane	$\Delta_{sol} H_{A(vap)}^\infty$	448
Poly(styrene-*co*-maleic anhydride-*co*-methacrylic acid)			
	1-butanol	$\Delta_M H_A^\infty$, $\Delta_{sol} H_{A(vap)}^\infty$	449
	n-decane	$\Delta_M H_A^\infty$, $\Delta_{sol} H_{A(vap)}^\infty$	449
	dibutyl ether	$\Delta_M H_A^\infty$, $\Delta_{sol} H_{A(vap)}^\infty$	449
	1,4-dioxane	$\Delta_M H_A^\infty$, $\Delta_{sol} H_{A(vap)}^\infty$	449
	n-heptane	$\Delta_M H_A^\infty$, $\Delta_{sol} H_{A(vap)}^\infty$	449
	n-nonane	$\Delta_M H_A^\infty$, $\Delta_{sol} H_{A(vap)}^\infty$	449
	n-octane	$\Delta_M H_A^\infty$, $\Delta_{sol} H_{A(vap)}^\infty$	449
Poly(styrene-*co*-methyl methacrylate)			
	n-heptane	$\Delta_M H_A^\infty$, $\Delta_{sol} H_{A(vap)}^\infty$	449
	n-hexane	$\Delta_M H_A^\infty$, $\Delta_{sol} H_{A(vap)}^\infty$	449
	n-nonane	$\Delta_M H_A^\infty$, $\Delta_{sol} H_{A(vap)}^\infty$	449
	n-octane	$\Delta_M H_A^\infty$, $\Delta_{sol} H_{A(vap)}^\infty$	449

Polymer(s)	Solvent(s)	Property	Page(s)
Poly(tetramethylene oxide)			
	benzene	$\Delta_M H$	151-152
	benzene	$\Delta_M H_B^\infty$	325-326
	cyclohexane	$\Delta_M H$	152
	1,2-dichloroethane	$\Delta_M H$	153
	1,2-dichloroethane	$\Delta_M H_B^\infty$	326
	1,2-dimethylbenzene	$\Delta_M H$	153-154
	1,2-dimethylbenzene	$\Delta_M H_B^\infty$	326
	1,3-dimethylbenzene	$\Delta_M H$	154
	1,3-dimethylbenzene	$\Delta_M H_B^\infty$	326-327
	1,4-dimethylbenzene	$\Delta_M H$	154-155
	1,4-dimethylbenzene	$\Delta_M H_B^\infty$	327
	1,4-dioxane	$\Delta_M H$	155-156
	1,4-dioxane	$\Delta_M H_B^\infty$	327
	ethylbenzene	$\Delta_M H$	156-157
	ethylbenzene	$\Delta_M H_B^\infty$	327-328
	propylbenzene	$\Delta_M H$	157
	propylbenzene	$\Delta_M H_B^\infty$	328
	tetrachloromethane	$\Delta_M H$	157-158
	tetrachloromethane	$\Delta_M H_B^\infty$	328
	toluene	$\Delta_M H$	159
	toluene	$\Delta_M H_B^\infty$	329
	1,1,1-trichloroethane	$\Delta_M H$	159-160
	1,3,5-trimethylbenzene	$\Delta_M H$	160
	1,3,5-trimethylbenzene	$\Delta_M H_B^\infty$	329
Poly(tetramethyl-p-silphenylene siloxane)			
	benzene	$\Delta_M H_A^\infty, \Delta_{sol} H_{A(vap)}^\infty$	451
	1-butanol	$\Delta_M H_A^\infty, \Delta_{sol} H_{A(vap)}^\infty$	451
	1,4-dioxane	$\Delta_M H_A^\infty, \Delta_{sol} H_{A(vap)}^\infty$	451
	n-dodecane	$\Delta_M H_A^\infty, \Delta_{sol} H_{A(vap)}^\infty$	451
	n-octane	$\Delta_M H_A^\infty, \Delta_{sol} H_{A(vap)}^\infty$	451
	1-octanol	$\Delta_M H_A^\infty, \Delta_{sol} H_{A(vap)}^\infty$	451
	toluene	$\Delta_M H_A^\infty, \Delta_{sol} H_{A(vap)}^\infty$	451
	trichloromethane	$\Delta_M H_A^\infty, \Delta_{sol} H_{A(vap)}^\infty$	451
Poly(vinyl acetate)			
	acetonitrile	$\Delta_M H_A^\infty$	453
	benzene	$\Delta_{dil} H^{12}$	160-161
	benzene	$\Delta_M H_A^\infty, \Delta_{sol} H_{A(vap)}^\infty$	451-453
	1-butanol	$\Delta_M H_A^\infty, \Delta_{sol} H_{A(vap)}^\infty$	452-453
	2-butanol	$\Delta_{sol} H_{A(vap)}^\infty$	453
	2-butanone	$\Delta_{dil} H^{12}$	162-163

Polymer(s)	Solvent(s)	Property	Page(s)
Poly(vinyl acetate) (*continued*)			
	methyl acetate	$\Delta_{sol}H_B^{\infty}$	330
	2-methyl-1-propanol	$\Delta_{sol}H_{A(vap)}^{\infty}$	454
	n-nonane	$\Delta_M H_A^{\infty}$, $\Delta_{sol}H_{A(vap)}^{\infty}$	452
	n-nonane	$\Delta_{sol}H_{A(vap)}^{\infty}$	454
	n-octane	$\Delta_{sol}H_{A(vap)}^{\infty}$	454
	1-octanol	$\Delta_{sol}H_{A(vap)}^{\infty}$	454
	1-octene	$\Delta_M H_A^{\infty}$, $\Delta_{sol}H_{A(vap)}^{\infty}$	452
	1-pentanol	$\Delta_{sol}H_{A(vap)}^{\infty}$	454
	2-pentanol	$\Delta_{sol}H_{A(vap)}^{\infty}$	454
	1-propanol	$\Delta_{sol}H_{A(vap)}^{\infty}$	454
	2-propanol	$\Delta_{sol}H_{A(vap)}^{\infty}$	454
	2-pentanone	$\Delta_{sol}H_B^{\infty}$	330
	2-pentanone	$\Delta_M H_A^{\infty}$, $\Delta_{sol}H_{A(vap)}^{\infty}$	452
	2-propanone	$\Delta_{sol}H_B^{\infty}$	331
	2-propanone	$\Delta_M H_A$	363
	2-propanone	$\Delta_M H_A^{\infty}$	453
	propyl acetate	$\Delta_{sol}H_B^{\infty}$	331
	1,1,2,2-tetrachloroethane	$\Delta_M H$	166
	tetrachloromethane	$\Delta_M H$	166-168
	tetrachloromethane	$\Delta_M H_A^{\infty}$, $\Delta_{sol}H_{A(vap)}^{\infty}$	452-453
	n-tetradecane	$\Delta_{sol}H_{A(vap)}^{\infty}$	454
	tetrahydrofuran	$\Delta_{sol}H_B^{\infty}$	331
	tetrahydrofuran	$\Delta_M H_A^{\infty}$	453
	1,2,3,4-tetrahydronaphthalene	$\Delta_M H_A^{\infty}$, $\Delta_{sol}H_{A(vap)}^{\infty}$	452
	3,3,4,4-tetramethylhexane	$\Delta_M H_A^{\infty}$, $\Delta_{sol}H_{A(vap)}^{\infty}$	452
	toluene	$\Delta_{dil}H^{12}$	168
	toluene	$\Delta_M H_A^{\infty}$	453
	toluene	$\Delta_{sol}H_{A(vap)}^{\infty}$	454
	trichloromethane	$\Delta_M H_A$	363-364
	trichloromethane	$\Delta_M H_A^{\infty}$, $\Delta_{sol}H_{A(vap)}^{\infty}$	452-453
	n-undecane	$\Delta_M H_A^{\infty}$, $\Delta_{sol}H_{A(vap)}^{\infty}$	452, 454
Poly(vinyl acetate-*co*-vinyl alcohol)			
	ethanol	$\Delta_{sol}H_B^{\infty}$	331-332
	ethyl acetate	$\Delta_{sol}H_B^{\infty}$	331-332
	2-propanone	$\Delta_{sol}H_B^{\infty}$	332-333
	water	$\Delta_{dil}H^{12}$	169-171
	water	$\Delta_{sol}H_B^{\infty}$	333-335
Poly(vinyl acetate-*co*-vinyl chloride)			
	cyclohexanone	$\Delta_M H_B^{\infty}$, $\Delta_{sol}H_B^{\infty}$	335

Polymer(s)	Solvent(s)	Property	Page(s)
Poly(vinyl chloride) (*continued*)			
	dimethyl phthalate	$\Delta_M H$	176-177
	dioctyl phthalate	$\Delta_M H$	177
	1,4-dioxane	$\Delta_M H_A{}^\infty$, $\Delta_{sol} H_{A(vap)}{}^\infty$	455
	ditridecyl phthalate	$\Delta_M H$	177
	n-dodecane	$\Delta_M H_A{}^\infty$, $\Delta_{sol} H_{A(vap)}{}^\infty$	454-455
	ethanol	$\Delta_M H_A{}^\infty$, $\Delta_{sol} H_{A(vap)}{}^\infty$	455
	ethyl acetate	$\Delta_M H_A{}^\infty$, $\Delta_{sol} H_{A(vap)}{}^\infty$	455
	bis(2-ethylhexyl) phthalate	$\Delta_M H$	174-175
	n-hexane	$\Delta_M H_A{}^\infty$	454
	methanol	$\Delta_M H_A{}^\infty$, $\Delta_{sol} H_{A(vap)}{}^\infty$	455
	methyl acetate	$\Delta_M H_A{}^\infty$, $\Delta_{sol} H_{A(vap)}{}^\infty$	454-455
	naphthyltolylmethane	$\Delta_M H$	177
	n-nonane	$\Delta_M H_A{}^\infty$, $\Delta_{sol} H_{A(vap)}{}^\infty$	455
	n-octane	$\Delta_M H_A{}^\infty$, $\Delta_{sol} H_{A(vap)}{}^\infty$	455
	2-propanone	$\Delta_M H_A{}^\infty$, $\Delta_{sol} H_{A(vap)}{}^\infty$	455
	tetrachloromethane	$\Delta_M H_A{}^\infty$, $\Delta_{sol} H_{A(vap)}{}^\infty$	455
	tetrahydrofuran	$\Delta_{dil} H^{12}$	177-178
	tetrahydrofuran	$\Delta_M H_B{}^\infty$, $\Delta_{sol} H_B{}^\infty$	338-339
	tetrahydrofuran	$\Delta_M H_A{}^\infty$, $\Delta_{sol} H_{A(vap)}{}^\infty$	455
	toluene	$\Delta_M H_A{}^\infty$, $\Delta_{sol} H_{A(vap)}{}^\infty$	454-455
	trichloromethane	$\Delta_M H_A{}^\infty$, $\Delta_{sol} H_{A(vap)}{}^\infty$	455
	bis(2,5,5-trimethylhexyl) phthalate	$\Delta_M H$	175
	n-undecane	$\Delta_M H_A{}^\infty$, $\Delta_{sol} H_{A(vap)}{}^\infty$	455
	water	$\Delta_{sol} H_{A(vap)}{}^\infty$	455
Poly(vinyl chloride-*co*-vinylidene chloride)			
	trichloromethane	$\Delta_{sol} H_B{}^\infty$	339
	water	$\Delta_{sol} H_{A(vap)}{}^\infty$	455
Poly(1-vinyl-3,5-dimethyl-1,2,4-triazole)			
	N,N-dimethylformamide	$\Delta_{sol} H_B{}^\infty$	339
	water	$\Delta_{sol} H_B{}^\infty$	339
Poly(vinylidene fluoride)			
	acetophenone	$\Delta_M H_A{}^\infty$	456, 458
	2-butanone	$\Delta_M H_A{}^\infty$	456
	butyl acetate	$\Delta_M H_A{}^\infty$, $\Delta_{sol} H_{A(vap)}{}^\infty$	456-457
	butylbenzene	$\Delta_M H_A{}^\infty$	456, 458
	γ-butyrolactone	$\Delta_M H_A{}^\infty$, $\Delta_{sol} H_{A(vap)}{}^\infty$	456, 458
	chlorobenzene	$\Delta_M H_A{}^\infty$	456
	1-chlorodecane	$\Delta_M H_A{}^\infty$	456, 459

Polymer(s)	Solvent(s)	Property	Page(s)
Poly(vinyl methyl ketone) (*continued*)			
	methyl acetate	$\Delta_{sol}H_{A(vap)}{}^{\infty}$	460
	n-nonane	$\Delta_{sol}H_{A(vap)}{}^{\infty}$	460
	n-octane	$\Delta_{sol}H_{A(vap)}{}^{\infty}$	460
	1-propanol	$\Delta_{sol}H_{A(vap)}{}^{\infty}$	460
	propyl acetate	$\Delta_{sol}H_{A(vap)}{}^{\infty}$	460
	n-undecane	$\Delta_{sol}H_{A(vap)}{}^{\infty}$	460
Poly(1-vinylpyrazole)			
	acetic acid	$\Delta_{M}H_{B}{}^{\infty}$	341
	butanoic acid	$\Delta_{M}H_{B}{}^{\infty}$	341
	N,N-diethylacetamide	$\Delta_{M}H_{B}{}^{\infty}$	341
	N,N-diethylacetamide	$\Delta_{M}H_{A}{}^{\infty}$	364
	N,N-dimethylformamide	$\Delta_{M}H_{B}{}^{\infty}$	341
	N,N-dimethylformamide	$\Delta_{M}H_{A}{}^{\infty}$	364
	pentanoic acid	$\Delta_{M}H_{B}{}^{\infty}$	341
	propanoic acid	$\Delta_{M}H_{B}{}^{\infty}$	341
	water	$\Delta_{M}H_{B}{}^{\infty}$	341
Poly(1-vinyl-2-pyrrolidinone)			
	1-butanol	$\Delta_{dil}H^{12}$	178-179
	ethanol	$\Delta_{dil}H^{12}$	179
	1-propanol	$\Delta_{dil}H^{12}$	180
Poly(1-vinyl-1,2,4-triazole)			
	acetic acid	$\Delta_{M}H_{B}{}^{\infty}$	341
	butanoic acid	$\Delta_{M}H_{B}{}^{\infty}$	342
	N,N-diethylacetamide	$\Delta_{M}H_{B}{}^{\infty}$	342
	N,N-diethylacetamide	$\Delta_{M}H_{A}{}^{\infty}$	364
	N,N-dimethylformamide	$\Delta_{M}H_{B}{}^{\infty}$	342
	N,N-dimethylformamide	$\Delta_{M}H_{A}{}^{\infty}$	365
	1-methyl-2-pyrrolidinone	$\Delta_{M}H_{B}{}^{\infty}$	342
	1-methyl-2-pyrrolidinone	$\Delta_{M}H_{A}{}^{\infty}$	365
	pentanoic acid	$\Delta_{M}H_{B}{}^{\infty}$	342
	propanoic acid	$\Delta_{M}H_{B}{}^{\infty}$	342
	water	$\Delta_{M}H_{B}{}^{\infty}$	342
	water	$\Delta_{M}H_{A}{}^{\infty}$	365
Starch			
	water	$\Delta_{M}H_{A}{}^{\infty}$	365, 460

Polymer(s)	Solvent(s)	Property	Page(s)
Tetra(ethylene glycol)			
	anisole	$\Delta_M H$	180
	benzyl alcohol	$\Delta_M H$	181
	1-butanol	$\Delta_M H$	181
	dimethylsulfoxide	$\Delta_M H$	181
	ethanol	$\Delta_M H$	182
	1-pentanol	$\Delta_M H$	182
	2-phenylethanol	$\Delta_M H$	182
	3-phenyl-1-propanol	$\Delta_M H$	182-183
	1-propanol	$\Delta_M H$	183
	propylene carbonate	$\Delta_M H$	183
	tetrachloromethane	$\Delta_M H$	183-184
	tetrachloromethane	$\Delta_M H$	184
	trichloromethane	$\Delta_M H_B^\infty$	342
	water	$\Delta_M H$	184
	water	$\Delta_M H_B^\infty$	343
Tetra(ethylene glycol) diethyl ether			
	water	$\Delta_M H$	184
Tetra(ethylene glycol) dimethyl ether			
	N,N-dimethylformamide	$\Delta_M H_B^\infty$	343
	n-dodecane	$\Delta_M H$	184-185
	methanol	$\Delta_M H$	185-186
	tetrachloromethane	$\Delta_M H$	186
	tetrachloromethane	$\Delta_M H_B^\infty$	343
	trichloromethane	$\Delta_M H$	186
	trichloromethane	$\Delta_M H_B^\infty$	343
	water	$\Delta_M H$	186-187
	water	$\Delta_M H_B^\infty$	343
Tetra(ethylene glycol) monooctyl ether			
	water	$\Delta_M H_A^\infty$	365

Appendix 3   List of solvents in alphabetical order

Name	Formula	CAS-RN	Page(s)
acetaldehyde	C_2H_4O	75-07-0	375, 412
acetic acid	$C_2H_4O_2$	64-19-7	292, 340, 341, 375, 412, 440
acetonitrile	C_2H_3N	75-05-8	292, 373, 375, 378-379, 381, 412, 417, 420, 440, 442, 453
acetophenone	C_8H_8O	98-86-2	456, 458
acrylonitrile	C_3H_3N	107-13-1	375, 412, 420
allyl alcohol	C_3H_6O	107-18-6	375
2-aminoethanol	C_2H_7NO	141-43-5	414-415
aniline	C_6H_7N	62-53-3	378, 440
anisole	C_7H_8O	100-66-3	34-37, 114-115, 126, 180, 264, 439-445
aziridine	C_2H_5N	151-56-4	414-415
benzaldehyde	C_7H_6O	100-52-7	379, 440
benzene	C_6H_6	71-43-2	21, 29-31, 37-43, 96-97, 100-101, 105-106, 115-116, 126-129, 151-152, 160-161, 172-174, 195, 198, 202, 208-211, 214, 217-218, 226-227, 256-258, 264, 268-270, 293, 296-297, 303, 305-308, 325-326, 351, 353-356, 372-375, 377-379, 381, 383-387, 389-393, 395-399, 403-404, 406-408, 410-412, 415, 417-420, 422-446, 450-455, 459-460
benzonitrile	C_7H_5N	100-47-0	264
benzyl alcohol	C_7H_8O	100-51-6	43-44, 116-117, 181, 379, 441
bromobenzene	C_6H_5Br	108-86-1	265, 420
1-bromobutane	C_4H_9Br	109-65-9	375, 412, 420
2-bromobutane	C_4H_9Br	78-76-2	412
bromocyclohexane	$C_6H_{11}Br$	108-85-0	227
1,3-butadiene	C_4H_6	106-99-0	378
n-butane	C_4H_{10}	106-97-8	378, 385, 399, 450, 419
1,2,4-butanetriyl trinitrate	$C_4H_7N_3O_9$	6659-60-5	369, 415
butanoic acid	$C_4H_8O_2$	107-92-6	340-342
1-butanol	$C_4H_{10}O$	71-36-3	45-46, 94, 178-179, 181, 371, 374-375, 379, 381, 385, 387-388, 399, 404, 406, 408, 411-413, 415, 418-421, 423, 427,

Name	Formula	CAS-RN	Page(s)
decyl acetate	$C_{12}H_{24}O_2$	112-17-4	230
dibutyl adipate	$C_{14}H_{26}O_4$	105-99-7	163
dibutyl ether	$C_8H_{18}O$	142-96-1	230, 273, 380, 440, 449, 450
dibutyl malonate	$C_{11}H_{20}O_4$	1190-39-2	163
dibutyl phthalate	$C_{16}H_{22}O_4$	84-74-2	176, 203
dibutyl succinate	$C_{12}H_{22}O_4$	141-03-7	164
dichloroacetic acid	$C_2H_2Cl_2O_2$	79-43-6	28, 213
1,2-dichlorobenzene	$C_6H_4Cl_2$	95-50-1	106, 226-227, 311, 457, 459
2,2'-dichlorodiethyl ether	$C_4H_8Cl_2O$	111-44-4	457
1,1-dichloroethane	$C_2H_4Cl_2$	75-34-3	386-387, 400, 406-407, 419, 429, 437
1,2-dichloroethane	$C_2H_4Cl_2$	107-06-2	153, 164, 213, 225-226, 246, 286, 292, 326, 338, 373, 379, 384, 386-387, 400, 406-407, 412, 416-417, 419, 423, 427, 429, 437-438, 441-442, 453, 456-457
dichloromethane	CH_2Cl_2	75-09-2	98, 263, 370, 372, 375, 379, 382, 384-387, 390-394, 400, 406-407, 412, 416-417, 419, 423, 429, 438, 441, 446, 459-460
1,5-dichloropentane	$C_5H_{10}Cl_2$	628-76-2	408, 442, 453, 457
didodecyl phthalate	$C_{32}H_{54}O_4$	2432-90-8	176
diethoxymethane	$C_5H_{12}O_2$	462-95-3	230
N,N-diethylacetamide	$C_6H_{13}NO$	685-91-6	292, 340-342, 364
diethylamine	$C_4H_{11}N$	109-89-7	375, 414-415
diethyl carbonate	$C_5H_{10}O_3$	105-58-8	47, 117-118,
1,3-diethyl-1,3-diphenylurea	$C_{17}H_{20}N_2O$	85-98-3	21
diethyl ether	$C_4H_{10}O$	60-29-7	204, 230, 273, 380, 382, 412, 441
diethyl oxalate	$C_6H_{10}O_4$	95-92-1	164-165
diethyl phthalate	$C_{12}H_{14}O_4$	84-66-2	176
3,3-diethylpentane	C_9H_{20}	4032-86-4	215, 221, 230, 251, 273, 294, 298
diethyl sebacate	$C_{14}H_{26}O_4$	110-40-7	164
dihexyl ether	$C_{12}H_{26}O$	112-58-3	230, 273
diisopropyl ether	$C_6H_{14}O$	108-20-3	441
1,2-dimethoxyethane	$C_4H_{10}O_2$	110-71-4	48, 230, 417, 456
dimethoxymethane	$C_3H_3O_2$	109-87-5	48-49, 230
N,N-dimethylacetamide	C_4H_9NO	127-19-5	196, 212, 214, 296-297, 408, 453, 456-457
N,N-dimethylaniline	$C_8H_{11}N$	121-69-7	266, 404, 432
1,2-dimethylbenzene	C_8H_{10}	95-47-6	153-154, 231, 298, 311, 326, 374, 376-377, 384, 386, 395-398, 403, 405, 424-425, 428, 430-433, 436, 438, 441, 450, 455

Name	Formula	CAS-RN	Page(s)
1,3-dimethylbenzene	C_8H_{10}	108-38-3	154, 198, 202, 208, 231, 311, 326-327, 376-378, 403, 405, 412, 444, 457
1,4-dimethylbenzene	C_8H_{10}	106-42-3	154-155, 231, 246, 299, 311, 327, 376-377, 379, 384-385, 403, 405, 412, 416, 422-423, 438, 445-446
2,2-dimethylbutane	C_6H_{14}	75-83-2	15, 23, 33, 199, 207, 231, 378, 215, 221, 231, 251, 273, 294, 299
dimethyl carbonate	$C_3H_6O_3$	616-38-6	118-119
dimethylethyleneurea	$C_5H_{10}N_2O$	80-73-9	457
N,N-dimethylformamide	C_3H_7NO	68-12-2	29, 196, 209-211, 292, 296, 338-343, 364-365, 373, 379, 408, 439, 442, 453, 456-459
2,6-dimethyl-4-heptanone	$C_9H_{18}O$	108-83-8	231
2,2-dimethylpentane	C_7H_{16}	590-35-2	215, 221, 231, 251, 273, 294, 299
2,3-dimethylpentane	C_7H_{16}	565-59-3	215, 221, 232, 251, 274, 294, 299, 403, 405
2,4-dimethylpentane	C_7H_{16}	108-08-7	216, 221, 232, 246, 251-252, 274, 294, 299, 403, 405
3,3-dimethylpentane	C_7H_{16}	562-49-2	216, 221, 232, 252, 274, 294, 299
dimethyl phthalate	$C_{10}H_{10}O_4$	131-11-3	176-177
dimethylpropyleneurea	$C_6H_{12}N_2O$	89607-25-0	457
2,5-dimethylpyrazine	$C_6H_8N_2$	123-32-0	414
dimethylsulfoxide	C_2H_6OS	67-68-5	17, 20, 49-50, 181, 196, 200, 210, 292, 408, 453, 456-457
2,4-dinitrotoluene	$C_7H_6N_2O_4$	121-14-2	21
2,6-dinitrotoluene	$C_7H_6N_2O_4$	606-20-2	21-22
dioctyl phthalate	$C_{24}H_{38}O_4$	117-84-0	177
1,4-dioxane	$C_4H_8O_2$	123-91-1	32-33, 50-52, 107, 134, 155-156, 311, 327, 370-373, 375, 386-387, 400, 406-407, 412, 417, 419, 423, 429, 431, 437, 440-442, 449-451, 453, 455
1,3-dioxolane	$C_3H_6O_2$	646-06-0	52-53
dipentyl ether	$C_{10}H_{22}O$	693-65-2	232, 274
dipropyl ether	$C_6H_{14}O$	111-43-3	274, 380, 408, 413, 420, 442, 453
dipropyl thioether	$C_6H_{14}S$	111-47-7	408, 442, 453, 457
ditridecyl phthalate	$C_{32}H_{54}O_4$	119-06-2	177
dodecamethylcyclohexasiloxane	$C_{12}H_{36}O_6Si_6$	540-97-6	405
dodecamethylpentasiloxane	$C_{12}H_{36}O_4Si_5$	141-63-9	232

Name	Formula	CAS-RN	Page(s)
n-dodecane	$C_{12}H_{26}$	112-40-3	16, 18, 24, 33, 184-185, 199, 207, 216, 221, 232-233, 252, 262, 274-275, 294, 299, 374-375, 378, 386, 388, 390-393, 395-398, 404, 408, 417-418, 421, 424-425, 428, 430-433, 436-437, 443, 446, 450-458 460
1,2-ethanediamine	$C_2H_8N_2$	107-15-3	414-415
1,2-ethanediol	$C_2H_6O_2$	107-21-1	200, 441
ethanol	C_2H_6O	64-17-5	25-26, 53-59, 119-120, 179, 182, 204, 303-304, 331-332, 335, 370-372, 374, 376, 380, 382, 386-388, 395-398, 406, 412-413, 416, 418-419, 421, 423-424, 427, 429-432, 434-436, 439, 441, 447, 450, 455, 460
bis(2-ethoxyethyl) ether	$C_8H_{18}O_3$	112-36-7	418
ethyl acetate	$C_4H_8O_2$	141-78-6	134-136, 204, 233, 266, 311-312, 330-332, 370, 373-374, 376, 379-382, 386-389, 395-398, 400, 404, 406-407, 412-413, 416-419, 421, 423-425, 427-432, 434, 436, 439, 441, 450, 453, 455-458, 460
ethylbenzene	C_8H_{10}	100-41-4	137, 156-157, 199, 202, 208, 219-220, 233, 266, 275, 286, 300, 312-314, 327-328, 378-379, 384, 386-387, 389-393, 400, 403, 405-407, 410, 417, 419-420, 423, 427-430, 433, 437-446, 453, 457
ethyl benzoate	$C_9H_{10}O_2$	93-89-0	266
ethyl butanoate	$C_6H_{12}O_2$	105-54-4	233
ethylcyclohexane	C_8H_{16}	1678-91-7	380, 382, 420
ethyl decanoate	$C_{12}H_{24}O_2$	112-17-4	234, 275
ethyl dodecanoate	$C_{14}H_{28}O_2$	106-33-2	234
ethyl heptanoate	$C_9H_{18}O_2$	106-30-9	234, 276
ethyl hexanoate	$C_8H_{16}O_2$	123-66-0	234, 276
bis(2-ethylhexyl) phthalate	$C_{24}H_{38}O_4$	117-81-7	174-175
ethyl nonanoate	$C_{11}H_{22}O_2$	123-29-5	276
ethyl octanoate	$C_{10}H_{20}O_2$	106-32-1	234, 276
3-ethylpentane	C_7H_{16}	617-78-7	216, 222, 234, 252, 276, 295, 300
N-ethylpiperazine	$C_6H_{14}N_2$	5308-25-8	414-415

Name	Formula	CAS-RN	Page(s)
tetrahydrofuran	C_4H_8O	109-99-9	73-74, 177-178, 256, 331, 338-339, 371, 375, 379-380, 382, 387, 400, 406-407, 413, 419, 424-425, 429, 431, 438, 441, 453, 455
1,2,3,4-tetrahydronaphthalene	$C_{10}H_{12}$	119-64-2	247-248, 302, 428, 443, 452, 457, 459
3,3,4,4-tetramethylhexane	$C_{10}H_{22}$	5171-84-6	428, 443, 452
2,2,4,4-tetramethylpentane	C_9H_{20}	1070-87-7	217, 223, 242, 255, 283, 295, 302
2,3,3,4-tetramethylpentane	C_9H_{20}	16747-38-9	217, 224, 242, 283, 295, 302
1,1,3,3-tetramethylurea	$C_5H_{12}N_2O$	632-22-4	212-213, 376, 457
toluene	C_7H_8	108-88-3	30, 111-113, 138-142, 159, 168, 226, 243, 248, 283, 287, 289-291, 302, 316-321, 329, 352-353, 360-362, 371-387, 389-393, 395-398, 400, 403-408, 410, 412, 416-420, 422-424, 426-431, 432-434, 436, 438-446, 450-451, 453-455, 457, 459-460
m-tolyl pentafluoropropionate	$C_{10}H_7F_5O_2$	24271-51-0	436
p-tolyl pentafluoropropionate	$C_{10}H_7F_5O_2$	24271-52-1	436
m-tolyl trifluoroacetate	$C_9H_7F_3O_2$	1736-09-0	436
p-tolyl trifluoroacetate	$C_9H_7F_3O_2$	1813-29-2	436
m-tolyl trimethylsilyl ether	$C_{10}H_{16}OSi$	17902-31-7	436
p-tolyl trimethylsilyl ether	$C_{10}H_{16}OSi$	17902-32-8	436
triacetin	$C_9H_{14}O_6$	102-76-1	22, 368-369, 414-415, 438
1,2,4-trichlorobenzene	$C_6H_3Cl_3$	120-82-1	249
1,1,1-trichloroethane	$C_2H_3Cl_3$	71-55-6	159-160, 268, 386-387, 400, 406-407, 416-417, 419, 429, 438
1,1,2-trichloroethane	$C_2H_3Cl_3$	79-00-5	408, 457
trichloroethene	C_2HCl_3	79-01-6	385-387, 400, 406-407, 416-417, 419, 424, 429, 438, 441, 459-460
trichloromethane	$CHCl_3$	67-66-3	28-29, 31-32, 74, 93, 96, 98, 113, 123, 126, 143-144, 186, 195, 198, 201, 206, 224-225, 258, 261-263, 268, 287-289, 302-304, 322-325, 339, 342-343, 363-364, 371, 375, 379-382, 384-387, 390-394, 400, 406-409, 411-412, 416-417, 419-420, 424, 427, 429-431, 438, 441-442, 445-446, 451-453, 455, 459-460

Name	Formula	CAS-RN	Page(s)
tricresol	C_7H_8O	1319-77-3	206
n-tridecane	$C_{13}H_{28}$	629-50-5	243, 284, 451
triethylene diamine	$C_6H_{12}N_2$	280-57-9	414-415
2,2,2-trifluoroethanol	$C_2H_3F_3O$	75-89-8	94, 413, 417
1,3,5-trimethylbenzene	C_9H_{12}	108-67-8	160, 243, 323, 329, 378, 444
2,2,3-trimethylbutane	C_7H_{16}	464-06-2	403, 495
3,4,5-trimethylheptane	$C_{10}H_{22}$	20278-89-1	428, 446
2,2,5-trimethylhexane	C_9H_{20}	3522-94-9	428
bis(2,5,5-trimethylhexyl) phthalate	$C_{26}H_{42}O_4$	53445-26-4	175
2,2,4-trimethylpentane	C_8H_{18}	540-84-1	17, 25, 34, 200, 207, 217, 219-220, 224, 255, 284, 296, 303, 352, 354, 362, 385, 410, 425-426, 428, 441-442, 446, 459-460
tris(4-methylphenyl) phosphate	$C_{21}H_{21}O_4P$	78-32-0	206
valeronitrile	C_5H_9N	110-59-8	433
n-undecane	$C_{11}H_{24}$	1120-21-4	243-244, 284, 374, 386-398, 400, 404, 406-407, 418, 421, 424-425, 428-434, 438, 450-452, 454-455, 458, 460
water	H_2O	7732-18-5	17-18, 21, 26-28, 74-90, 94, 98, 103-105, 114, 123-125, 145-151, 169-171, 171-172, 184, 186-187, 195, 201, 208-209, 258-261, 263-264, 285, 304, 333-336, 339-343, 352-353, 362-365, 367, 370-372, 413, 441, 455, 460

Appendix 4   List of solvents in order of their molecular formulas

Formula	Name	CAS-RN	Page(s)
CCl_4	tetrachloromethane	56-23-5	67-72, 92-93, 95-96, 98-100, 110-111, 122-123, 126, 137-138, 157-158, 166-168, 183-184, 186, 242, 261, 267, 282, 302, 304, 316, 328, 343, 370, 379-382, 384-385, 387, 393-394, 400, 406-408, 411-413, 416-417, 419-423, 429, 431, 437-438, 441, 446, 452-453, 455, 459-460
$CHCl_3$	trichloromethane	67-66-3	28-29, 31-32, 74, 93, 96, 98, 113, 123, 126, 143-144, 186, 195, 198, 201, 206, 224-225, 258, 261-263, 268, 287-289, 302-304, 322-325, 339, 342-343, 363-364, 371, 375, 379-382, 384-387, 390-394, 400, 406-409, 411-412, 416-417, 419-420, 424, 427, 429-431, 438, 441-442, 445-446, 451-453, 455, 459-460
CH_2Cl_2	dichloromethane	75-09-2	98, 263, 370, 372, 375, 379, 382, 384-387, 390-394, 400, 406-407, 412, 416-417, 419, 423, 429, 438, 441, 446, 459-460
CH_2O_2	formic acid	64-18-6	22, 197, 206, 292, 376, 413
CH_3NO	formamide	75-12-7	201, 292, 376, 441
CH_3NO_2	nitromethane	75-52-5	293, 376, 413, 420
CH_4O	methanol	67-56-1	25, 59-63, 90-91, 103, 114, 120-121, 165-166, 185-186, 204, 363, 371, 374, 379-380, 382, 386, 388, 395-398, 412-413, 416, 418, 421, 424-425, 428, 430-432, 434, 436, 441, 447, 450, 455, 460
C_2Cl_4	tetrachloroethene	127-18-4	416-417

Formula	Name	CAS-RN	Page(s)
C_2HCl_3	trichloroethene	79-01-6	385-387, 400, 406-407, 416-417, 419, 424, 429, 438, 441, 459-460
C_2HCl_5	pentachloroethane	76-01-7	456-457
$C_2H_2Cl_2O_2$	dichloroacetic acid	79-43-6	28, 213
$C_2H_2Cl_4$	1,1,2,2-tetrachloroethane	79-34-5	166, 214, 267, 408, 456-457
$C_2H_3Cl_3$	1,1,1-trichloroethane	71-55-6	159-160, 268, 386-387, 400, 406-407, 416-417, 419, 429, 438
$C_2H_3Cl_3$	1,1,2-trichloroethane	79-00-5	408, 457
$C_2H_3F_3O$	2,2,2-trifluoroethanol	75-89-8	94, 413, 417
C_2H_3N	acetonitrile	75-05-8	292, 373, 375, 378-379, 381, 412, 417, 420, 440, 442, 453
$C_2H_4Cl_2$	1,1-dichloroethane	75-34-3	386-387, 400, 406-407, 419, 429, 437
$C_2H_4Cl_2$	1,2-dichloroethane	107-06-2	153, 164, 213, 225-226, 246, 286, 292, 326, 338, 373, 379, 384, 386-387, 400, 406-407, 412, 416-417, 419, 423, 427, 429, 437-438, 441-442, 453, 456-457
C_2H_4O	acetaldehyde	75-07-0	375, 412
$C_2H_4O_2$	acetic acid	64-19-7	292, 340, 341, 375, 412, 440
C_2H_5N	aziridine	151-56-4	414-415
$C_2H_5NO_2$	nitroethane	79-24-3	376, 413, 420
C_2H_6O	ethanol	64-17-5	25-26, 53-59, 119-120, 179, 182, 204, 303-304, 331-332, 335, 370-372, 374, 376, 380, 382, 386-388, 395-398, 406, 412-413, 416, 418-419, 421, 423-424, 427, 429-432, 434-436, 439, 441, 447, 450, 455, 460
C_2H_6OS	dimethylsulfoxide	67-68-5	17, 20, 49-50, 181, 196, 200, 210, 292, 408, 453, 456-457
$C_2H_6O_2$	1,2-ethanediol	107-21-1	200, 441
C_2H_7NO	2-aminoethanol	141-43-5	414-415
$C_2H_8N_2$	1,2-ethanediamine	107-15-3	414-415
C_3H_3N	acrylonitrile	107-13-1	375, 412, 420
$C_3H_3O_2$	dimethoxymethane	109-87-5	48-49, 230
C_3H_5N	propionitrile	107-12-0	376, 380-382, 413, 420
$C_3H_5N_3O_9$	1,2,3-propanetriyl trinitrate	55-63-0	369, 415
C_3H_6	propene	115-07-1	378
C_3H_6O	allyl alcohol	107-18-6	375

Formula	Name	CAS-RN	Page(s)
C_3H_6O	2-propanone	67-64-1	22, 108-110, 197-198, 205-206, 212, 224, 285, 287, 314-315, 331-333, 363, 370-372, 374, 376, 379, 382, 384, 386-387, 395-398, 400, 404, 406, 412, 416, 418-420, 424-425, 428-429, 431, 432-434, 436, 438, 441, 450, 453, 455, 458
$C_3H_6O_2$	1,3-dioxolane	646-06-0	52-53
$C_3H_6O_2$	methyl acetate	79-20-9	197, 330, 374, 380, 382, 386-389, 395-398, 400, 406, 418-419, 421, 423-425, 427-429, 431-432, 434, 436, 450, 454-455, 458, 460
$C_3H_6O_2$	propanoic acid	79-09-4	340-342, 420
$C_3H_6O_3$	dimethyl carbonate	616-38-6	118-119
C_3H_7Cl	1-chloropropane	540-54-5	380-382, 422-423
C_3H_7NO	N,N-dimethylformamide	68-12-2	29, 196, 209-211, 292, 296, 338-343, 364-365, 373, 379, 408, 439, 442, 453, 456-459
$C_3H_7NO_2$	1-nitropropane	108-03-2	267, 376, 413, 420
$C_3H_7NO_2$	2-nitropropane	79-46-9	413
C_3H_8	propane	74-98-6	378, 399
C_3H_8O	1-propanol	71-23-8	66, 95, 180, 183, 371, 373-374, 376, 380-382, 386-388, 390-393, 400, 406, 414, 417-420, 423, 429, 434-435, 439, 441, 447, 454, 460
C_3H_8O	2-propanol	67-63-0	379-380, 382, 413-414, 417, 421, 425, 427, 441, 454
C_3H_9N	propylamine	75-31-0	434
$C_3H_{10}N_2$	1,2-propanediamine	78-90-0	376
$C_4H_4N_2$	pyrazine	290-37-9	414-415
C_4H_4O	furan	110-00-9	413
C_4H_5N	pyrrole	109-97-7	414-415
C_4H_6	1,3-butadiene	106-99-0	378
$C_4H_6O_2$	γ-butyrolactone	96-48-0	456, 458
$C_4H_6O_3$	propylene carbonate	108-32-7	66-67, 183, 456-457
C_4H_7N	butyronitrile	109-74-0	380-381, 432
$C_4H_7N_3O_9$	1,2,4-butanetriyl trinitrate	6659-60-5	369, 415
$C_4H_8Cl_2O$	2,2'-dichlorodiethyl ether	111-44-4	457
C_4H_8O	2-butanone	78-93-3	31, 130-131, 144-145, 162-163, 203, 227-228, 265, 308, 329, 373-375, 379, 381, 384-387, 395-399, 404, 406, 408, 411-412, 416-417, 419-420,

Formula	Name	CAS-RN	Page(s)
C_6F_6	hexafluorobenzene	392-56-3	426, 451
C_6F_{14}	tetradecafluorohexane	355-42-0	20
$C_6H_3Cl_3$	1,2,4-trichlorobenzene	120-82-1	249
$C_6H_4Cl_2$	1,2-dichlorobenzene	95-50-1	106, 226-227, 311, 457, 459
C_6H_5Br	bromobenzene	108-86-1	265, 420
C_6H_5Cl	chlorobenzene	108-90-7	102, 106, 131, 228, 265, 270, 308-309, 330, 336, 373, 375, 385, 387, 389-393, 400, 406-408, 411-412, 418-420, 423, 429, 437, 441-442, 446, 452-453, 456
$C_6H_5NO_2$	nitrobenzene	98-95-3	267, 441
C_6H_6	benzene	71-43-2	21, 29-31, 37-43, 96-97, 100-101, 105-106, 115-116, 126-129, 151-152, 160-161, 172-174, 195, 198, 202, 208-211, 214, 217-218, 226-227, 256-258, 264, 268-270, 293, 296-297, 303, 305-308, 325-326, 351, 353-356, 372-375, 377-379, 381, 383-387, 389-393, 395-399, 403-404, 406-408, 410-412, 415, 417-420, 422-446, 450-455, 459-460
C_6H_6O	phenol	108-95-2	409, 411
C_6H_7N	aniline	62-53-3	378, 440
C_6H_8	cyclohexadiene	628-41-1	385, 387, 400, 406, 419, 429, 437
$C_6H_8N_2$	2,5-dimethylpyrazine	123-32-0	414
$C_6H_9F_3O_2$	butyl trifluoroacetate	367-64-6	436
C_6H_{10}	cyclohexene	110-83-8	385, 387, 400, 406, 419, 437
$C_6H_{10}O$	cyclohexanone	108-94-1	175, 195, 224-225, 285, 335-337, 379, 390-393, 408-409, 411, 441-442, 452, 456-459
$C_6H_{10}O_4$	diethyl oxalate	95-92-1	164-165
$C_6H_{11}Br$	bromocyclohexane	108-85-0	227
$C_6H_{11}Cl$	chlorocyclohexane	542-18-7	409, 411, 452
$C_6H_{11}NO$	ε-caprolactam	105-60-2	372
C_6H_{12}	cyclohexane	110-82-7	102, 131-134, 152, 215, 220, 228, 245, 249, 266, 271-272, 289, 293, 297, 309-310, 372-373, 377-385, 387, 400, 403, 405-412, 416-420, 426, 428, 437, 441-443, 445-446, 451-453, 459-460
C_6H_{12}	1-hexene	592-41-6	380, 382
C_6H_{12}	2-methylbutane	78-78-4	247, 280, 378

Formula	Name	CAS-RN	Page(s)
$C_6H_{12}N_2$	triethylene diamine	280-57-9	414-415
$C_6H_{12}N_2O$	dimethylpropyleneurea	89607-25-0	457
$C_6H_{12}O$	cyclohexanol	108-93-0	15, 375, 390-393, 409, 411, 446, 452-453, 459
$C_6H_{12}O$	3-hexanone	589-38-8	458
$C_6H_{12}O$	4-methyl-2-pentanone	108-10-1	108, 239, 286-287, 379-382, 430
$C_6H_{12}O_2$	butyl acetate	123-86-4	203, 228, 308, 329, 375, 380-381, 385, 387-393, 399, 406, 408, 412, 416, 418-421, 423-424, 427, 429-430, 435, 437, 441, 446, 456-457
$C_6H_{12}O_2$	*tert*-butyl acetate	540-88-5	389, 444
$C_6H_{12}O_2$	ethyl butanoate	105-54-4	233
$C_6H_{12}O_2$	isobutyl acetate	110-19-0	389
$C_6H_{12}O_2$	propyl propionate	106-36-5	242
$C_6H_{13}Cl$	1-chlorohexane	544-10-5	385, 387, 400, 406, 408, 419, 423, 429, 437, 442, 453, 456
$C_6H_{13}NO$	N,N-diethylacetamide	685-91-6	292, 340-342, 364
C_6H_{14}	2,2-dimethylbutane	75-83-2	15, 23, 33, 199, 207, 231, 378, 215, 221, 231, 251, 273, 294, 299
C_6H_{14}	n-hexane	110-54-3	16, 20, 24, 34, 199, 207, 222, 237, 279, 300, 372, 377-380, 382-383, 385-389, 395, 398-407, 410-411, 413, 418-419, 422-423, 426, 430, 432-435, 437, 441-446, 448-449, 451, 453-454, 458-460
C_6H_{14}	2-methylpentane	107-83-5	442
C_6H_{14}	3-methylpentane	96-14-0	442
$C_6H_{14}N_2$	N-ethylpiperazine	5308-25-8	414-415
$C_6H_{14}O$	diisopropyl ether	108-20-3	441
$C_6H_{14}O$	dipropyl ether	111-43-3	274, 380, 408, 413, 420, 442, 453
$C_6H_{14}O$	1-hexanol	111-27-3	374, 454
$C_6H_{14}O_3$	bis(2-methoxyethyl) ether	111-96-6	19, 373, 417, 456
$C_6H_{14}S$	dipropyl thioether	111-47-7	408, 442, 453, 457
$C_6H_{18}N_3OP$	hexamethylphosphoramide	680-31-9	457
$C_6H_{18}OSi_2$	hexamethyldisiloxane	107-46-0	236-237
C_7F_8	octafluorotoluene	434-64-0	451
C_7H_5N	benzonitrile	100-47-0	264
$C_7H_6N_2O_4$	2,4-dinitrotoluene	121-14-2	21
$C_7H_6N_2O_4$	2,6-dinitrotoluene	606-20-2	21-22
C_7H_6O	benzaldehyde	100-52-7	379, 440

Formula	Name	CAS-RN	Page(s)
$C_7H_{18}OSi$	butyl trimethylsilyl ether	1825-65-6	436
C_8H_8	styrene	100-42-5	316
C_8H_8O	acetophenone	98-86-2	456, 458
C_8H_{10}	1,2-dimethylbenzene	95-47-6	153-154, 231, 298, 311, 326, 374, 376-377, 384, 386, 395-398, 403, 405, 424-425, 428, 430-433, 436, 438, 441, 450, 455
C_8H_{10}	1,3-dimethylbenzene	108-38-3	154, 198, 202, 208, 231, 311, 326-327, 376-378, 403, 405, 412, 444, 457
C_8H_{10}	1,4-dimethylbenzene	106-42-3	154-155, 231, 246, 299, 311, 327, 376-377, 379, 384-385, 403, 405, 412, 416, 422-423, 438, 445-446
C_8H_{10}	ethylbenzene	100-41-4	137, 156-157, 199, 202, 208, 219-220, 233, 266, 275, 286, 300, 312-314, 327-328, 378-379, 384, 386-387, 389-393, 400, 403, 405-407, 410, 417, 419-420, 423, 427-430, 433, 437-446, 453, 457
$C_8H_{10}O$	2-phenylethanol	60-12-8	64-65, 182
$C_8H_{11}F_3O_2$	cyclohexyl trifluoroacetate	1549-45-7	436
$C_8H_{11}N$	N,N-dimethylaniline	121-69-7	266, 404, 432
$C_8H_{14}O_2$	cyclohexyl acetate	622-45-7	436
C_8H_{16}	cyclooctane	292-64-8	215, 220, 229, 249, 272, 293, 298, 385, 387, 400, 406, 407, 419, 429, 437
C_8H_{16}	ethylcyclohexane	1678-91-7	380, 382, 420
C_8H_{16}	1-octene	111-66-0	380-382, 409, 413, 420, 442, 452
C_8H_{16}	*cis*-2-octene	7642-04-8	442
$C_8H_{16}O$	2-octanone	111-13-7	404, 432
$C_8H_{16}O$	3-octanone	106-68-3	458
$C_8H_{16}O_2$	ethyl hexanoate	123-66-0	234, 276
$C_8H_{16}O_2$	hexyl acetate	142-92-7	237
$C_8H_{16}O_2$	pentyl propionate	624-54-4	241
$C_8H_{17}Cl$	1-chlorooctane	111-85-3	385, 387, 400, 406, 419, 423, 429, 437, 456-459
C_8H_{18}	2-methylheptane	592-27-8	442, 238, 403, 405
C_8H_{18}	n-octane	111-65-9	16, 24, 34, 200, 207, 216, 223, 240, 254, 281, 295, 301, 352, 359-360, 372-374, 376, 378, 380-382, 385-389, 395-398, 400-413, 416, 418-426, 428, 430-432, 434-437, 439-444, 446, 449-451, 454-460

Formula	Name	CAS-RN	Page(s)
$C_{12}H_{36}O_6Si_6$	dodecamethylcyclohexasiloxane	540-97-6	405
$C_{13}H_{28}$	n-tridecane	629-50-5	243, 284, 451
$C_{14}H_{26}O_4$	dibutyl adipate	105-99-7	163
$C_{14}H_{26}O_4$	diethyl sebacate	110-40-7	164
$C_{14}H_{28}O_2$	ethyl dodecanoate	106-33-2	234
$C_{14}H_{30}$	n-tetradecane	629-59-4	19, 223, 242, 283, 375, 384, 390-393, 409, 428, 437, 443, 454, 456-459
$C_{16}H_{22}O_4$	dibutyl phthalate	84-74-2	176, 203
$C_{16}H_{32}O_2$	ethyl tetradecanoate	124-06-1	276
$C_{16}H_{34}$	2,2,4,4,6,8,8-heptamethylnonane	4390-04-9	16, 24, 33, 199, 207, 216, 222, 235, 246, 253, 277, 295, 300
$C_{16}H_{34}$	n-hexadecane	544-76-3	16, 19, 24, 33, 199, 207, 216, 222, 236, 247, 253, 278-279, 295, 300, 443, 454, 456-457
$C_{17}H_{20}N_2O$	1,3-diethyl-1,3-diphenylurea	85-98-3	21
$C_{18}H_{16}$	naphthyltolylmethane	30306-53-7	177
$C_{21}H_{21}O_4P$	tris(4-methylphenyl) phosphate	78-32-0	206
$C_{24}H_{38}O_4$	dioctyl phthalate	117-84-0	177
$C_{24}H_{38}O_4$	bis(2-ethylhexyl) phthalate	117-81-7	174-175
$C_{26}H_{42}O_4$	bis(2,5,5-trimethylhexyl) phthalate	53445-26-4	175
$C_{32}H_{54}O_4$	didodecyl phthalate	2432-90-8	176
$C_{32}H_{54}O_4$	ditridecyl phthalate	119-06-2	177
H_2O	water	7732-18-5	17-18, 21, 26-28, 74-90, 94, 98, 103-105, 114, 123-125, 145-151, 169-171, 171-172, 184, 186-187, 195, 201, 208-209, 258-261, 263-264, 285, 304, 333-336, 339-343, 352-353, 362-365, 367, 370-372, 413, 441, 455, 460
H_2O_4S	sulfuric acid	7664-93-9	297

INDEX